IFAC

International Federation of Automatic Control

FUZZY INFORMATION, KNOWLEDGE REPRESENTATION AND DECISION ANALYSIS

IFAC Proceedings Series, 1984. Number 6

IFAC Proceedings Series (1984)

No. 1 ALONSO-CONCHEIRO: Real-Time Digital Control Applications

No. 2 ZWICKY: Control in Power Electronics & Electrical Drives 1983

No. 3 KLAMT & LAUBER: Control in Transportation Systems

No. 4 RIJNSDORP & PLOMP: Training for Tomorrow: Educational Aspects of Computerised Automation

No. 5 WESTERLUND: Automation in Mining, Mineral & Metal Processing 1983

No. 6 SANCHEZ: Fuzzy Information, Knowledge Representation & Decision Analysis

No. 7 PAU & BASAR: Dynamic Modelling & Control of National Economies 1983

No. 8 Van CAUWENBERGHE: Instrumentation & Automation in the Paper, Rubber, Plastics & Polymerisation Industries

No. 9 PONOMARYOV: Artificial Intelligence

No. 10 STRASZAK: Large Scale Systems: Theory & Application 1983

IFAC Related Titles

BROADBENT & MASUBUCHI: Multilingual Glossary of Automatic Control Technology

EYKHOFF: Trends and Progress in System Identification

ISERMANN: System Identification Tutorials (*Automatica Special Issue*)

FUZZY INFORMATION, KNOWLEDGE REPRESENTATION AND DECISION ANALYSIS

Proceedings of the IFAC Symposium, Marseille, France
19-21 July 1983

Edited by

E. SANCHEZ

Service Universitaire de Biomathématiques, Marseille, France

Published for the

INTERNATIONAL FEDERATION OF AUTOMATIC CONTROL

by

PERGAMON PRESS

OXFORD · NEW YORK · TORONTO · SYDNEY · PARIS · FRANKFURT

U.K.	Pergamon Press Ltd., Headington Hill Hall, Oxford OX3 0BW, England
U.S.A.	Pergamon Press Inc., Maxwell House, Fairview Park, Elmsford, New York 10523, U.S.A.
CANADA	Pergamon Press Canada Ltd., Suite 104, 150 Consumers Road, Willowdale, Ontario M2J 1P9, Canada
AUSTRALIA	Pergamon Press (Aust.) Pty. Ltd., P.O. Box 544, Potts Point, N.S.W. 2011, Australia
FRANCE	Pergamon Press SARL, 24 rue des Ecoles, 75240 Paris, Cedex 05, France
FEDERAL REPUBLIC OF GERMANY	Pergamon Press GmbH, Hammerweg 6, D-6242 Kronberg-Taunus, Federal Republic of Germany

First edition 1984

Library of Congress Cataloging in Publication Data

IFAC Symposium (1983 : Marseille, France)
Fuzzy information, knowledge representation & decision analysis.
(IFAC proceedings series)
1. Control theory – Congresses.
2. Fuzzy sets – Congresses.
I. Sanchez, Elie, 1944- II. Gupta, Madan M. III. International Federation of Automatic Control. IV. Title. V. Series.
QA402.3.1453 1983 001.53'9 83-10994

British Library Cataloguing in Publication Data

Fuzzy information, knowledge representation & decision analysis. – (IFAC proceedings series)
1. Fuzzy sets – Congresses
I. Sanchez, Elie II. Gupta, Madan M. II. Series
511.3 Q248

ISBN 0-08-030583-0

These proceedings were reproduced by means of the photo-offset process using the manuscripts supplied by the authors of the different papers. The manuscripts have been typed using different typewriters and typefaces. The lay-out, figures and tables of some papers did not agree completely with the standard requirements; consequently the reproduction does not display complete uniformity. To ensure rapid publication this discrepancy could not be changed; nor could the English be checked completely. Therefore, the readers are asked to excuse any deficiencies of this publication which may be due to the above mentioned reasons.

The Editor

Printed in Great Britain by A. Wheaton & Co. Ltd., Exeter

IFAC SYMPOSIUM ON FUZZY INFORMATION, KNOWLEDGE REPRESENTATION AND DECISION ANALYSIS

Sponsored by/Patronné par

- The International Federation of Automation Control (IFAC)
- L'Association Française pour la Cybernétique et Technique (AFCET)
- L'Association pour les Applications de l'Informatique à la Médecine (AIM)
- The International Federation for Information Processing (IFIP)
- The International Federation of Operational Research Societies (IFORS)
- The International Medical Informatics Association (IMIA)

Organized by/Organisé par

Service Universitaire de Biomathématiques
Faculté de Médecine
27 Boulevard Jean-Moulin
13385 Marseille Cédex 5, France

International Programme Committee/Comité de Programme International

L. A. Zadeh, U.S.A. (Honorary Chairman)
E. Sanchez, France (Chairman)
M. M. Gupta, Canada (Co-Chairman)
M. A. Aizerman, U.S.S.R.
K. Asai, Japan
F. Azorin, Spain
J. F. Baldwin, U.K.
C. Carlsson, Finland
L. R. Chow, China
D. Dubois, France
R. Feron, France
A. J. Fossard, France
K. S. Fu, U.S.A.
Rongjiang Guo, China
A. Kaufmann, France
H. Kwakernaak, Netherlands
P. M. Larsen, Denmark
R. Lopez De Mantaras, Spain
N. Malvache, France
E. Mamdani, U.K.
M. Mizumoto, Japan
C. Negoita, Romania
H. T. Nguyen, U.S.A.
C. Ponsard, France
H. Prade, France
V. S. Pugachev, U.S.S.R.
J. C. Simon, France
A. Straszak, Poland
Zai Fu Su, China
M. Sugeno, Japan
K. Tanaka, Japan
T. Terano, Japan
P. P. Wang, U.S.A.
Pei-zhuang Wang, China
D. Willaeys, France
R. Yager, U.S.A.
K. S. Zadeh, F.R.G.
H. J. Zimmermann, F.R.G.

National Organizing Committee/Comité Local d'Organisation

A. Kaufmann (Honorary Chairman)
M. Roux (Chairman)
E. Sanchez (Co-Chairman)
J. C. Cervera
G. Chauvet
L. Crocq
B. Fichet
M. Fieschi
A. J. Fossard
G. Giralt
J. Gouvernet
F. Gremy
M. Joubert
M. Laudet
C. Longevialle
R. Sambuc
G. Soula

PREFACE

The papers published in this volume were presented at the IFAC International Symposium on "Fuzzy Information, Knowledge Representation and Decision Analysis" held at Marseille, France, July 19-21, 1983. Preceding the Symposium a one day tutorial session was proposed to newcomers to offer them some background to better appraise the presentation of technical papers, specially in current research topics. The following communications were made at this tutorial session:

- A. Kaufmann (Grenoble, France), "Le Traitement Mathématique de l'Imprécis par la Théorie des Sous-ensembles Flous" (Mathematical Treatment of Imprecision by Fuzzy Subsets Theory).
- H. Prade (Toulouse, France), "Approximate and Plausible Reasoning: the State of the Art".
- H.-J. Zimmermann (Aachen, W.-Germany), "Decision-making in a Fuzzy Environment".
- C.A. Kulikowski (New Brunswick, U.S.A.), "A Survey on Expert Systems".
- E.H. Mamdani (London, G.B.), "Logic and PRUF. A Survey".

During the Symposium, the following four plenary lectures were given:

- L.A. Zadeh (University of California, Berkeley, U.S.A.), "Fuzzy Sets as a Basis for Knowledge Representation and Approximate Reasoning".
- C.A. Kulikowski (Rutgers University, New Brunswick, U.S.A.), "Knowledge-based Learning and Expert Systems".
- A. Kaufmann (Grenoble, France), "L'Epistémiologie de l'Incertain" ("Epistemology of Uncertainty").
- L.A. Zadeh (Berkeley, U.S.A.), "Can Expert Systems be Designed Without Using Fuzzy Logic"?

The regular papers were presented during parallel sessions. They covered a wide variety of topics ranging from hard to soft sciences. This international cooperation and exchange of information has shown to be very beneficial to all the participants. More and more theoreticians in this field are looking for applications of their ideas or for practical problems to solve. On the other hand, the applied and experienced people are sometimes in need of a conceptual framework to better their knowledge, and they found the opportunity of discussing their problems. For example, in knowledge information systems two opinions are opposed. Some researchers believe that systems will function at the (high) level of human experts only if the reasoning processes are modelled, well understood and explained by "nice" theories. Others claim that, as long as the system has good results, the methodology it uses is not really important.

In an everyday increasing range of applications and theoretical developments, it is now getting clear that fuzzy logic and derived notions are continuing to play an important role in providing natural and appropriate tools to model uncertainty.

Elie Sanchez, Editor
Service Universitaire de Biomathématiques
Faculté de Médecine de Marseille - France

ACKNOWLEDGEMENT TO REFEREES

The referees appearing in the following list are gratefully acknowledged for time and efforts they generously spent on the papers submitted for presentation at the INTERNATIONAL IFAC SYMPOSIUM on Fuzzy Information, Knowledge Representation and Decision Analysis (July 19-21, 1983, Marseille France).

ADLASSNIG K.P.
AZORIN F.
BALDWIN J.F.
BEZDEK J.
CARLSSON C.
CERRUTI U.
DE LUCA A.
DI NOLA A.
DUBOIS D.
EMPTOZ H.
FERON R.
FREKSA Ch.
FU K.S.
GOTTWALD S.
GUPTA M.M.
ILIADIS A.
JOUBERT M.
KACPRZYK J.
KAUFMANN A.
KLEMENT E.P.
KICKERT W.J.M.
KRAFT D.H.
KWAKERNAAK H.
LARSEN P.M.
LOPEZ DE MANTARAS R.
LOWEN R.
MAMDANI E.H.
MARSILI P.M.
MIZUMOTO M.
NEGOITA C.
NGUYEN H.T.
OVCHINNIKOV S.
PEI-ZHUANG WANG
PESCHEL M.
PONSARD C.
PRADE H
RUSPINI E.H.
SAMBUC R.
SANCHEZ E.
SMETS Ph.
SOULA G.
SUGENO M.
TERANO T.
TORASSO P.
TRILLAS E.
TSUKAMOTO Y.
TURKSEN I.B.
VAN NAUTA LEMKE H.R.
WILLAEYS D.
YAGER R.R.
ZADEH K.S.
ZIMMERMANN H.J.

CONTENTS

DECISION ANALYSIS

FUZZY MEASURES

SOCIO-ECONOMIC SYSTEMS

RETRIEVAL FROM FUZZY DATABASE BY FUZZY RELATIONAL ALGEBRA

M. Umano

Department of Applied Mathematics, Faculty of Science, Okayama University of Science, Ridai-cho, Okayama 700, Japan

Abstract. The real world data may have two types of ambiguity, one is in a data value itself and the other in an association between values. For representing such data, we propose a possibility-distribution fuzzy-relational model. In this model, the former ambiguity is represented by a possibility distribution and the latter by a grade of membership. The relational algebra for such fuzzy data model is defined. The traditional operations, namely, the union, intersection, difference and extended Cartesian product are similarly defined as those in fuzzy set theory. And the special relational operations, namely, the projection, join, restriction and division, are newly defined for fuzzy databases.

Keywords. Fuzzy database; fuzzy relational model of data; fuzzy relational algebra; possibility distribution; fuzzy set; fuzzy predicate.

INTRODUCTION

Database systems have been vigorously studied since Codd (1970) proposed the relational model of data. Such database systems can only deal with well-defined data. In the real world, however, there exist ambiguous data which cannot be defined in certain and well-defined forms by any means. Since in everyday life we often make decisions based on such data, the formulation and construction of a database which can represent and manipulate fuzzy data will increase the application areas of database systems and improve the interface between men and machines. We will refer to such a database as a fuzzy database (Umano, 1982).

We may have two types of ambiguity, one is in a data value itself and the other in an association between values. For such fuzzy data, we propose a possibility-distribution fuzzy-relational model, in which the former ambiguity is represented by a possibility distribution and the latter by a grade of membership.

Relational algebra for ordinary relational database has been defined by Codd (1972) which includes the traditional set operations, namely, the union, intersection, difference and extended Cartesian product, and the special relational operations, namely, the projection, join, restriction and division. We define the relational algebra for fuzzy databases. The traditional operators are similar to those in fuzzy set theory. The operations projection, join, restriction and division are newly defined for fuzzy databases.

FUZZY DATABASE

We may have two types of ambiguity in fuzzy data, one is ambiguity in a data value itself and the other in an association between values. The former ambiguity can be represented by a possibility distribution (Zadeh, 1978) and the latter by a grade of membership (Zadeh, 1965). So we propose a possibility-distribution fuzzy-relational model, in which fuzzy data are represented by fuzzy relations whose elements are possibility distributions. This model is an extension of the relational model of data by Codd (1970).

[Definition 1] A fuzzy database D_f is defined as a collection of fuzzy relations R_i, i=1,2,...,n, i.e.,

$$D_f = \{R_1, R_2, \ldots, R_n\} \quad (1)$$

in which a fuzzy relation R_i is defined as

$$\mu_{R_i}: \mathcal{P}(U_{i1})\times\ldots\times\mathcal{P}(U_{im}) \longrightarrow \mathcal{P}([0,1]), \quad (2)$$

where the symbol × denotes the Cartesian product and $\mathcal{P}(U_{ij})$, j=1,2,...,m, is a collection of all possibility distributions on a universe of discourse U_{ij}.

Note that the membership space is also extended to the collection of all possibility distributions of [0,1], so we can write a possibility distribution as a grade value. The U_{ij} is called a basic set and $\mathcal{P}(U_{ij})$ a domain. We use the attribute names to differentiate between the same domains. In the notation of a fuzzy relation in a fuzzy database, we use the table notation similar to one for relational database (Date, 1977).

R

μ	NAME	PROF	INCOME
1	Smith	Engineer	300
0.8	John	Lawyer	{0.5/490,1/500,0.6/510}p
{1/0.8,0.5/0.7}p	Richard	Lawyer	400
{1/1,0.8/0.9}p	Judy	Teacher	{0.5/350,1/360,1/370}p
{1/1,0.5/0}p	Mike	Teacher	380

S

μ	NAME	AGE
1	Smith	30
0.6	John	35
{1/1,0.7/0.9}p	Richard	{0.5/35,1/36,0.6/37}p
{1/0.8,0.6/0.7}p	Judy	{0.6/37,1/38}p
0.3	Mike	{0.6/47,1/48,0.5/49}p

Fig. 1. Fuzzy relations R and S.

[Example 1] Fig. 1 shows a fuzzy database which consists of fuzzy relations R and S whose frameworks are

R(NAME, PROF, INCOME)
S(NAME, AGE)

where NAME, PROF, INCOME in R and NAME and AGE in S are attribute names. The μ is an attribute name for membership and it is automatically added to each fuzzy realtion. Basic sets of the attributes NAME and PROF are sets of names and professions of individuals, respectively. The AGE and INCOME take a set of integers as a basic set. Note that we use the notation {...}p for a possibility distribution and {...} for a fuzzy set, but the possibility distribution which has only one element with the possibility 1 is expressed only by the element without {...}p.

FUZZY RELATIONAL ALGEBRA

Codd (1972) defined a collection of high-level operations, called relational algebra, on relations in databases.

The relational algebra consists of two groups of operations:

(1) traditional set operations --- the union, intersection, difference and extended Cartesian product,
(2) special relational operations --- the projection, join, restriction and division.

We shall define these operations for possibility-distribution fuzzy-relational model. First we define these operations for ordinary fuzzy relations whose grades and elements are not possibility distributions, and then we extend them to the case where the grades and elements contain possibility distributions.

Extension of Traditional Set Operations

For traditional set operations, the operations union and intersection have been already defined in (Zadeh, 1965) but the other operations difference and extended Cartesian product must be defined.

We shall consider two n-ary fuzzy relations R and S in a universe of discourse $U=U_1\times U_2\times\ldots\times U_n$.

(a) union $(R \cup S)$

$$R \cup S = \{\mu_R(u) \vee \mu_S(u)/u:\ u \in U\}, \qquad (3)$$

where ∨ denotes the max.

(b) intersection $(R \cap S)$

$$R \cap S = \{\mu_R(u) \wedge \mu_S(u)\ /\ u\ :\ u \in U\}, \qquad (4)$$

where ∧ denotes the min.

(c) difference $(R - S)$

$$R - S = \{(\mu_R(u)-\mu_S(u)) \vee 0\ /\ u\ :\ u \in U\}, \qquad (5)$$

where − in the right side hand denotes the ordinary subtraction. This operation is called bounded-difference in ordinary fuzzy set theory.

[Example 2] Let R and S be binary fuzzy relations as

$$R = \{0.3/\langle a,1\rangle,\ 0.9/\langle b,2\rangle,\ 0.1/\langle c,3\rangle\},$$
$$S = \{0.5/\langle b,2\rangle,\ 0.8/\langle c,3\rangle,\ 1/\langle d,4\rangle\},$$

then we have

$$R \cup S = \{0.3/\langle a,1\rangle,\ 0.9/\langle b,2\rangle,\ 0.8/\langle c,3\rangle,\ 1/\langle d,4\rangle\},$$
$$R \cap S = \{0.5/\langle b,2\rangle,\ 0.1/\langle c,3\rangle\},$$
$$R - S = \{0.3/\langle a,1\rangle,\ 0.4/\langle b,2\rangle\}.$$

(d) extended Cartesian product $(R \times S)$
By the extended Cartesian product, we get an (m+n)-ary fuzzy relation from an m-ary fuzzy relation R and an n-ary fuzzy relation S. The extended Cartesian product of fuzzy relations R in $U_1\times U_2\times\ldots\times U_m$ and S in $V_1\times V_2\times\ldots\times V_n$ is defined as

$$\begin{aligned} R \times S &= \{\mu_R(u) \wedge \mu_S(v)\ /\ \langle u,v\rangle\ :\ u \in U,\ v \in V\} \\ &= \{\mu_R(u_1,u_2,\ldots,u_m) \wedge \mu_S(v_1,v_2,\ldots,v_n)\ / \\ &\quad \langle u_1,u_2,\ldots,u_m,v_1,v_2,\ldots,v_n\rangle\ : \\ &\quad u_i \in U_i,\ v_j \in V_j\}. \end{aligned} \qquad (6)$$

[Example 3] For the same R and S as in Example 2, we have

$R \times S = \{0.3/\langle a,1,b,2\rangle,\ 0.3/\langle a,1,c,3\rangle,\ 0.3/\langle a,1,d,4\rangle,\ 0.5/\langle b,2,b,2\rangle,\ 0.8/\langle b,2,c,3\rangle,\ 0.9/\langle b,2,d,4\rangle,\ 0.1/\langle c,3,b,2\rangle,\ 0.1/\langle c,3,c,3\rangle,\ 0.1/\langle c,3,d,4\rangle\}.$

Since the grade value is only a number in [0,1] in Examples 2 and 3, we can calculate (3) - (6). But the definition of fuzzy database says that the grade values may be possibility distributions in [0,1]. In that case we use the following definition.

[Definition 2] If the binary operation $*$ is defined on any two elements u_1 and u_2 in a universe of discourse U, the binary operation $*$ on any possibility distributions Π_x and Π_y in U can be defined as

$$\Pi_x * \Pi_y = \{\pi_x(u_1)/u_1 : u_1 \in U\}p * \{\pi_y(u_2)/u_2 : u_2 \in U\}p = \{\pi_x(u_1) \wedge \pi_y(u_2) / u_1 * u_2 : u_1 \in U,\ u_2 \in U\}p. \quad (7)$$

In the case where one operand is possibility distribution and the other is not, the element u can be regarded as a possibility distribution $\{1/u\}p$ and Definition 2 can be applied.

If we have in (3) - (6) two possibility distributions Π_x and Π_y as $\mu_R(u)$ and $\mu_S(u)$, respectively, then using Definition 2 we obtain the definitions of $\vee$, $\wedge$ and $-$ for possibility distributions, i.e.,

$$\Pi_x \vee \Pi_y = \{\pi_x(u_1) \wedge \pi_y(u_2) / u_1 \vee u_2 : u_1 \in [0,1],\ u_2 \in [0,1]\}p, \quad (8)$$

$$\Pi_x \wedge \Pi_y = \{\pi_x(u_1) \wedge \pi_y(u_2) / u_1 \wedge u_2 : u_1 \in [0,1],\ u_2 \in [0,1]\}p, \quad (9)$$

$$\Pi_x - \Pi_y = \{\pi_x(u_1) \wedge \pi_y(u_2) / u_1 - u_2 : u_1 \in [0,1],\ u_2 \in [0,1]\}p, \quad (10)$$

[Example 4] For the possibility distributions in [0,1] as

$$\Pi_x = \{1/1,\ 0.8/0.9,\ 0.3/0.8\}p,$$
$$\Pi_y = \{0.3/0.6,\ 1/0.5,\ 0.4/0.4\}p,$$

we have

$$\begin{aligned}\Pi_x \vee \Pi_y &= \{1\wedge0.3 / 1\vee0.6,\ 1\wedge1 / 1\vee0.5,\ 1\wedge0.4 / 1\vee0.4,\ 0.8\wedge0.3 / 0.9\vee0.6,\ 0.8\wedge1 / 0.9\vee0.5,\ 0.8\wedge0.4 / 0.9\vee0.4,\\ &\quad 0.3\wedge0.3 / 0.8\vee0.6,\ 0.3\wedge1 / 0.8\vee0.5,\ 0.3\wedge0.4 / 0.8\vee0.4\}p\\ &= \{0.3/1,\ 1/1,\ 0.4/1,\ 0.3/0.9,\ 0.8/0.9,\ 0.4/0.9,\ 0.3/0.8,\ 0.3/0.8,\ 0.3/0.8\}p\\ &= \{1/1,\ 0.8/0.9,\ 0.3/0.8\}p,\end{aligned}$$

$$\begin{aligned}\Pi_x \wedge \Pi_y &= \{0.3/0.6,\ 1/0.5,\ 0.4/0.4,\ 0.3/0.6,\ 0.8/0.5,\ 0.4/0.4,\ 0.3/0.6,\ 0.3/0.5,\ 0.3/0.4\}p\\ &= \{0.3/0.6,\ 1/0.5,\ 0.4/0.4\}p,\end{aligned}$$

$$\begin{aligned}\Pi_x - \Pi_y &= \{0.3/0.4,\ 1/0.5,\ 0.4/0.6,\ 0.3/0.3,\ 0.8/0.4,\ 0.4/0.5,\ 0.3/0.2,\ 0.3/0.3,\ 0.3/0.4\}p\\ &= \{0.4/0.6,\ 1/0.5,\ 0.8/0.4,\ 0.3/0.3,\ 0.3/0.2\}p.\end{aligned}$$

Extensions of Special Relational Operations

For the special operations in relational algebra, Codd (1972) defined four operations, namely, the projection, join, restriction and division.

(a) projection

The projection is an operation for constructing a new relation obtained by selecting specified attributes and eliminating others of the relation. More specifically, let R be a relation whose attributes are $A_1, A_2, \ldots, An$ and r be a n-tuple in R and $r(A_{i_1}, A_{i_2}, \ldots, A_{i_k})$ be a k-tuple which contains only the value of attributes $A_{i_1}, A_{i_2}, \ldots, A_{i_k}$. Then projection of R over attributes $A_{i_1}, A_{i_2}, \ldots, A_{i_k}$ is defined as

$$R[A_{i_1}, A_{i_2}, \ldots, A_{i_k}] = \{r(A_{i_1}, A_{i_2}, \ldots, A_{i_k}) : r \in R\}. \quad (11)$$

Note that duplicated elements are eliminated.

When R is a fuzzy relation, we must compute the grade value of k-tuple. It is natural that the grade of k-tuple is a grade value of n-tuple.

[Definition 3] The projection of a fuzzy relation of R whose attributes are $A_1, A_2, \ldots, A_n$ over the attributes $A_{i_1}, A_{i_2}, \ldots, A_{i_k}$ is defined as

$$R[A_{i_1}, A_{i_2}, \ldots, A_{i_k}] = \{\mu_R(r)/r(A_{i_1}, A_{i_2}, \ldots, A_{i_k}) : \mu_R(r)/r \in R\} \quad (12)$$

where more than one same k-tuple is united to one and its grade is the maximum of them. When the grade value of n-tuple is a possibility distribution, we can compute the maximum of possibility distributions using (8).

[Example 5] The projection of a fuzzy relation R shown in Fig. 1 over the attribute PROF is

$$\begin{aligned}R[PROF] &= \{\mu_R(r)/r(PROF) : \mu_R(r)/r \in R\}\\ &= \{1/Engineer,\ 0.8/Lawyer,\ \{1/0.8, 0.5/0.7\}p/Lawyer,\ \{1/1, 0.8/0.9\}p/Teacher,\ \{1/1, 0.5/0\}p/Teacher\}\end{aligned}$$

and two Lawyers and two Teachers are united to one. We have for Lawyer:

$$\begin{aligned}0.8 &\vee \{1/0.8,\ 0.5/0.7\}p\\ &= \{1/0.8\}p \vee \{1/0.8,\ 0.5/0.7\}p\\ &= \{1/0.8,\ 0.5/0.8\}p = \{1/0.8\}p = 0.8\end{aligned}$$

and for Teacher:

$$\begin{aligned}\{1/1,\ 0.8/0.9\}p &\vee \{1/1,\ 0.5/0\}p\\ &= \{1/1,\ 0.5/1,\ 0.8/1,\ 0.5/0.9\}p\\ &= \{1/1,\ 0.5/0.9\}p.\end{aligned}$$

And we have the result shown in Fig. 2.

result	μ	PROF
	1	Engineer
	0.8	Lawyer
	{1/1,0.5/0.9}p	Teacher

Fig. 2. The result of R[PROF].

(b) join
The join operator joins two relations using common domains and generates a new, wider relation. More specifically, let θ be one of the binary comparison operators $=$, $\neq$, $<$, $>$, $\leq$ and $\geq$, then the θ-join of relations R and S over an attribute A in R and an attribute B in S is defined as

$$R[A\theta B]S = \{<r,s>: r \in R, s \in S, r(A)\, \theta\, s(B)\} \qquad (13)$$

where θ is one of the above comparison operators and the domain of attributes A and B must be the same.

When R and S are fuzzy relations, we extend an operator θ to a binary fuzzy predicate. And domains of attributes A and B do not have to be the same.

[Definition 4] The join of fuzzy relations R and S over an attribute A in R and an attribute B in S by a fuzzy predicate θ whose truth-value function is t_θ, is defiend as

$$R[\theta(A,B)]S = \{\mu/<r,s>: \mu_R(r)/r \in R,\ \mu_S(s)/s \in S,\ \mu = t_\theta(r(A),s(B)) \wedge \mu_R(r) \wedge \mu_S(s)\}. \qquad (14)$$

We can also define the join of fuzzy relations R and S over attributes A_1, A_2, ..., A_i in R and attributes B_1, B_2, ..., B_j in S by an (i+j)-ary fuzzy predicate. More details of this topic will be seen in a subsequent paper.

When attribute values are possibility distributions, we must get the truth value of fuzzy predicates for possibility distributions.

[Definition 5] Let $\theta(x,y)$ be a binary fuzzy predicate whose truth-value function is $t_\theta(u,v)$ and objects x and y have possibility distributions Π_x and Π_y in universes of discourse U and V, respectively. Then the truth value of the fuzzy predicate $\theta(x,y)$ is defined as a possibility distribution:

$$T(\theta(x,y)) = \{\pi_x(u) \wedge \pi_y(v)\ /\ t_\theta(u,v) : u \in U, v \in V\}p. \qquad (15)$$

When either x or y is a possibility distribution and the other is not, we can consider it as a possibility distribution with only one element.

Since a fuzzy predicate for possibility distributions can be computed by (15) in Definition 5 and the $\wedge$ by (9), we can compute (14) in Definition 4.

[Example 6] Let us consider a query "Get all pairs whose ages are approximately equal" for the fuzzy database of Fig. 1. This query is translated into an expression in a fuzzy relational algebra:

S[AE(AGE,AGE)]S

where AE(x,y) is a fuzzy predicate which expresses "is approximately equal" and its truth-value function may be, say,

$$t_{AE}(u,v) = \begin{cases} 1 & u = v \\ 0.8 & |u - v| = 1 \\ 0.3 & |u - v| = 2 \\ 0 & \text{otherwise} \end{cases}$$

Then we have the result as shown in Fig. 3. We shall show how to compute the garde value of the element <John, 35, Richard, {0.5/35, 1/36, 0.6/37}p> in the result. First, the truth value of the fuzzy predicate AE(x,y) is

$$\begin{aligned} & t_{AE}(35, \{0.5/35, 1/36, 0.6/37\}p) \\ &= \{0.5/t_{AE}(35,35),\ 1/t_{AE}(35,36),\ 0.6/t_{AE}(35,37)\}p \\ &= \{0.5/1,\ 1/0.8,\ 0.6/0.3\}p, \end{aligned}$$

and the minimum of this truth value and the grade values of John and Richard in S is

$$\begin{aligned} & \{0.5/1,1/0.8,0.6/0.3\}p \wedge 0.6 \wedge \{1/1,0.7/0.9\}p \\ &= \{0.5/0.6,1/0.6,0.6/0.3\}p \wedge \{1/1,0.7/0.9\}p \\ &= \{0.5/0.6,\ 0.5/0.6,\ 1/0.6,\ 0.7/0.6,\ 0.6/0.3,\ 0.6/0.3\}p \\ &= \{1/0.6,\ 0.6/0.3\}p. \end{aligned}$$

For <Richard, {0.5/35, 1/36, 0.6/37}p, Judy, {0.6/37, 1/38}p>, we have the truth value of the fuzzy predicate:

$$\begin{aligned} & t_{AE}(\{0.5/35,1/36,0.6/37\}p,\ \{0.6/37,1/38\}p) \\ &= \{0.5/0.3,\ 0.5/0,\ 0.6/0.8,\ 1/0.3,\ 0.6/1,\ 0.6/0.8\}p \\ &= \{0.6/1,\ 0.6/0.8,\ 1/0.3,\ 0.5/0\}p, \end{aligned}$$

and the result grade value:

$$\begin{aligned} & \{0.6/1,\ 0.6/0.8,\ 1/0.3,\ 0.5/0\}p \wedge \{1/1,\ 0.7/0.9\}p \wedge \{1/0.8,\ 0.6/0.7\}p \\ &= \{0.6/1,\ 0.6/0.9,\ 0.6/0.8,\ 1/0.3,\ 0.5/0\}p \wedge \{1/0.8,\ 0.6/0.7\}p \\ &= \{0.6/0.8,\ 0.6/0.7,\ 1/0.3,\ 0.5/0\}p. \end{aligned}$$

(c) restriction
The restriction constructs the subset of tuples within a relation for which a specified predicate is satisfied. More specifically, let R be a relation with attributes A_1, A_2, ..., A_n and θ be a k-ary predicate for the attribute A_{i_1}, A_{i_2}, ..., A_{i_k}, then the restriction of a relation R by a predicate θ is defined as

$$R[\theta(A_{i_1},A_{i_2},\ldots,A_{i_k})] = \{r: r \in R,\ \theta(r(A_{i_1}),r(A_{i_2}),\ldots,r(A_{i_k}))\}. \qquad (16)$$

When R is a fuzzy relation, we can extend the predicate θ to a fuzzy one. And we can have the following definition.

result

μ	NAME1	AGE1	NAME2	AGE1
1	Smith	30	Smith	30
0.6	John	35	John	35
{1/0.6, 0.6/0.3}p	John	35	Richard	{0.5/35,1/36,0.6/37}p
{0.6/0.3, 1/0}p	John	35	Judy	{0.6/37, 1/38}p
{1/0.6, 0.6/0.3}p	Richard	{0.5/35,1/36,0.6/37}p	John	35
{1/1, 0.7/0.9, 0.6/0.8, 0.5/0.3}p	Richard	{0.5/35,1/36,0.6/37}p	Richard	{0.5/35,1/36,0.6/37}p
{0.6/0.8, 0.6/0.7, 1/0.3, 0.5/0}p	Richard	{0.5/35,1/36,0.6/37}p	Judy	{0.6/37, 1/38}p
{0.6/0.3, 1/0}p	Judy	{0.6/37, 1/38}p	John	35
{0.6/0.8, 0.6/0.7, 1/0.3, 0.5/0}p	Judy	{0.6/37, 1/38}p	Richard	{0.5/35,1/36,0.6/37}p
{1/0.8, 0.6/0.7}p	Judy	{0.6/37, 1/38}p	Judy	{0.6/37, 1/38}p
0.3	Mike	{0.6/47,1/48,0.5/49}p	Mike	{0.6/47,1/48,0.5/49}p

Fig. 3. The result of S[AE(AGE,AGE)]S.

[Definition 6] The restriction of a fuzzy relation R with attributes $A_1, A_2, \ldots, A_n$ by a k-ary fuzzy predicate θ with respect to attributes $A_{i_1}, A_{i_2}, \ldots, A_{i_k}$ is defined as

$$R[\theta(A_{i_1},A_{i_2},\ldots,A_{i_k})] = \{\mu/r\colon \mu_R(r)/r \in R,\ \mu=t_\theta(r(A_{i_1}),r(A_{i_2}),\ldots,r(A_{i_k})) \wedge \mu_R(r)\}, \quad (17)$$

where t_θ is a truth-value function of the fuzzy predicate θ.

When each attribute value is a possibility distribution, the truth value of the fuzzy predicate is defined in the following.

[Definition 7] Let Π_{x_i} be possibility distributions in universes of discourse U_i, then the truth value of a k-ary fuzzy predicate $\theta(x_1,x_2,\ldots,x_k)$ whose truth-value function is $t_\theta(u_1,u_2,\ldots,u_k)$ is defined as a possibility distribution:

$$T(\theta(\Pi_{x_1},\Pi_{x_2},\ldots,\Pi_{x_k})) = \{\pi_{x_1}(u_1)\wedge\ldots\wedge\pi_{x_k}(u_k)/t_\theta(u_1),\ldots,u_k)\colon u_i \in U_i\ (i=1,2,\ldots,k)\}p. \quad (18)$$

[Example 7] Consider a query "Find all persons who are about 36 years old." to a fuzzy database shown in Fig. 1. We define a fuzzy predicate θ(x), which means "A person x is about 36 years old." and whose truth-value function is, say,

$$t_\theta(u) = \begin{cases} 1 & u = 36 \\ 0.7 & u \in \{35,37\} \\ 0.2 & u \in \{34,38\} \\ 0 & \text{otherwise} \end{cases}.$$

Then we can have an expression for the query in a fuzzy relational algebra

S[θ(AGE)].

The result is shown in Fig. 4. For John, we have

$$\mu = t_\theta(35) \wedge \mu_S(<\text{John},35>) = 0.7 \wedge 0.6 = 0.6$$

and for Richard we have

$$\mu = t_\theta(\{0.5/35,\ 1/36,\ 0.6/37\}p) \wedge \{1/1,\ 0.7/0.9\}p = \{0.5/0.7,\ 1/1,\ 0.6/0.7\}p \wedge \{1/1,\ 0.7/0.9\}p = \{1/1,\ 0.7/0.9,\ 0.6/0.7\}p.$$

On the other hand, for Mike we have

$$\mu = t_\theta(\{0.6/47,\ 1/48,\ 0.5/49\}p) \wedge 0.3 = \{0.6/0,\ 1/0,\ 0.5/0\}p \wedge 0.3 = \{1/0\}p \wedge 0.3 = 0$$

So he is not contained in the result.

result

μ	NAME	AGE
0.6	John	35
{1/1, 0.7/0.9 0.6/0.7}p	Richard	{0.5/35, 1/36, 0.6/37}p
{0.6/0.7, 1/0.2}p	Judy	{0.6/37, 1/38}p

Fig. 4. The reult of S[θ(AGE)].

(d) division

The division is used to check whether attribute values in a relation contains all attribute values in the other relation. The division was defined as an operation corresponding to the universal quantifier in the relational calculus (Codd, 1972). The division is defined using an image set, but it can be defined by a combination of the above operations. That is, the division of a relation R by a relation S over the attribute A in R and B in S can be defined as

$$R[A \div B]S = R[A'] - ((R[A'] \times S[B]) - R)[A'] \quad (19)$$

where A' denotes a complement attribute list of A.

In this point of view, we do not newly define the division of fuzzy relations but adopt the definition (19).

We have defined operations in the relational algebra for the possibility -ditribution fuzzy-relational model and illustrated them using several examples.

CONCLUSIONS

We have first defined a possibility-distribution fuzzy-relational model as a model for representing and manipulating ambiguous data and information which includes two types of ambiguity, one is in a value itself and the other in an association between values. Then we described a formulation of operations in relational algebra for such a fuzzy database.

The comparison of the fuzzy relational algebra to the fuzzy relational calculus, which has been defined in (Umano et al., 1980), is very interesting.

By the introductions of higher type possibility distributions whose values of possibility may be again a possibility distribution and higher level possibility distributions whose elements may be possibility distributions, fuzzy databases can represent a hierarchical fuzzy data and they can be used as a good tool for the representation of complex fuzzy data and knowledge.

Fuzzy database systems will find a number of applications in such fields as natural language processing, question-answering and artificial intelligence where fuzzy data play an important role in nature.

REFERENCES

Codd, E.F. (1970). A relational model of data for large shared data banks. Commu. ACM, 13, 377-387.

Codd, E.F. (1972). Relational completeness of data base sublanguages. In R. Rustin (Ed.), Data Base Systems, Prentice-Hall, Englewood Cliffs, USA, pp.65-98.

Date, C.J. (1977). An Introduction to Database Systems -- Second Edition, Addison-Wesley, Reading, USA.

Umano, M., Fukami, S., Mizumoto, M. and Tanaka, K. (1980). Retrieval processing from fuzzy databases. Tech. Rep. of IECE of Japan, 80, 204 (on Automata and Languages), pp.45-54 (in Japanese).

Umano, M. (1982). FREEDOM-0: a fuzzy database system. In M.M. Gupta and E. Sanchez (Ed.), Fuzzy Information and Decision Processes, North-Holland, Amsterdam, The Netherlands, pp.339-347.

Zadeh, L.A. (1965). Fuzzy sets. Inf. & Control, 8, 338-353.

Zadeh, L.A. (1978). Fuzzy sets as a basis for a theory of possibility. Fuzzy Sets and Systems, 1, 3-28.

INFORMATION RETRIEVAL SYSTEM USING FUZZY RELATIONAL PRODUCTS FOR THESAURUS CONSTRUCTION

L. J. Kohout*, E. Keravnou* and W. Bandler**

**Man-Computer Studies Group, Brunel University, Uxbridge, Middlesex, UK*
***Department of Mathematics, University of Essex, Colchester, Essex, UK*

Abstract. The paper describes the design of an information retrieval system processing fuzzy search requests as well as a new algorithm for the construction of a thesaurus of index terms using triangle and square products of fuzzy relations.

INTRODUCTION

The need for fuzzy methods in the automatic classification and retrieval of documents has long been evident. The appropriate theoretical framework has now been developed, along with design criteria for practical implementation. Emphasis from the beginning is on the evolutive representation of the system of descriptors or thesaurus according to the actual documents in the database; here the terms used are preordered by triangle products according to "broader than" and "more specific than", and synonymy is defined in terms of the square product.

The fundamental relation upon which these products are computed is the relevance relation between terms and documents, together with its converse; the representation of this inherently fuzzy relation is conceptually in matrix form but actually in the far more compact form of foresets and aftersets, which lend themselves to inverted file format.

Search requests, considered as weighted logical combinations of descriptors, are processed via the thesaurus through the fundamental relation, and result in appropriate documents lists, in order of decreasing relevance.

The paper is divided into two parts. Part 1 outlines some essential advantages of fuzzy information retrieval and gives a justification for a many-valued logic (MVL) extension of a Boolean information retrieval model. This extension uses fuzzy implication operators. Part 2 describes briefly two distinct fuzzy data representations and compares their efficiency.

PART 1. ON SOME ESSENTIAL ADVANTAGES OF FUZZY INFORMATION RETRIEVAL

1.1 Functions of Information Retrieval System

The function of an automatic information retrieval system is to provide the user with references to documents as the result of users' queries that specify the users' information requirements. Efficient and relevant answers to users' queries can be provided only if the automatic retrieval system utilises an adequate retrieval strategy. The purpose of an automatic retrieval strategy is to retrieve all the relevant documents, at the same time retrieving as few of the non-relevant documents as possible. Although some commercially available IR systems are now widely used, a great deal of difficult practical and theoretical problems in this field remain to be solved.

1.2 Essential Differences between Information Processing and Data Processing

Information retrieval distinguishes itself from conventional data processing by the following features:

(a) inexact matching of items
(b) necessity to specify the relevance of items
(c) polythetic classification of items as opposed to monothetic
(d) nondeterministic inductive inference

Surprisingly, all the characteristics features (a) - (d) of IR systems have also been in the forefront of the methodological discussions in the area of fuzzy systems based on multiple-valued logic [19] and it is therefore surprising that this methodology has not had, so far, any significant impact on IR. The question of relevance is the central theme of information retrieval, and it has been shown that certain types of non-boolean multiple-valued logics can treat the relevance question more adequately than probabilistic methods [2], yet IR systems are almost exclusively based on Boolean logic and/or probabilistic matching, ignoring almost totally the recent progress in the field of systems that can treat relevance more adequately.

1.3 Deficiencies of Boolean Information Retrieval Systems

1.3.1 Boolean requests. The Boolean systems ([2],[9]) use the search strategy that retrieves those documents which are "true" for the query. Each query is expressed in the form of index terms (or keywords) which are combined by the means of the usual logical operations AND, OR, NOT.

However, the Boolean Retrieval systems have some disadvantages caused by the restrictions imposed by Boolean logic. These are as follows:

(1) The Boolean cannot express the degree of relevance of the index terms to the contents of the documents.

(2) Individual users' profiles are not allowed; the system gives the same answer to different users as the response to the same request.

1.3.2 Conventional thesaurus and its deficiencies. A typical Boolean Information Retrieval System is an on-line system with a feedback provision allowing the user to change his request during one search session [2]. To use the feedback efficiently, the user has to be provided with the information that gives him guidance on how to narrow or broaden his search. This is done by giving the user access to a structured dictionary, a thesaurus, which, for any given index term, stores related terms either more general (broad terms) or more particular (specific terms). The construction of such a thesaurus is usually done either manually or automatically, based on a statistical formula [2], providing statistical associations between the terms. Because the associations are based on frequency counts, this also does not account properly for the degree of relevance of relationships between the broad and specific terms of a thesaurus.

By introducing relations based on multiple-valued logic instead of the mere two-valued logic on which conventional probability theory is based, the deficiency of a conventional thesaurus can be removed, because the degree of relevance of thesaurus terms can be dealt with. ([14],[12] and next section of this paper).

1.4 Advantages of information retrieval with MVL-extensions

All the disadvantages of the Boolean retrieval system can be removed [18] by assigning to each term and document a degree of relevance or real number in the interval [0,1], and using a multiple-valued operations AND, OR, NOT instead of Boolean ones. Where a and be are the values in [0,1] assigned respectively to two propositions, their conjunction and disjunction and the negation of the first one receive respectively the following values:

$$a \text{ AND } b = \min(a,b)$$
$$a \text{ OR } b = \max(a,b)$$
$$\text{NOT } a = 1-a$$

The thesaurus is constructed by means of multiple-valued relational products, which are calculated as follows ([7],[11]). To begin with, a number of distinct multiple-valued implication operators $\rightarrow$ are available, several of which are worth carrying along in any implementation in order to amass experimental evidence as to their differences and advantages. These include the Lukasiewicz operator $\rightarrow_5$, the Kleene-Dienes operator $\rightarrow_6$, and the Kleene-Dienes-Lukasiewicz operator $\rightarrow_{5.5}$. Where a and b are as defined above, these operators give to the expression $a \rightarrow b$ the value of the degree to which the proposition valued a materially implies the proposition valued b. These are their formulae:

$$a \rightarrow_5 b = \min(1, 1-a+b)$$
$$a \rightarrow_{5.5} b = 1-a+ab$$
$$a \rightarrow_6 b = \max(1-a, b).$$

Their magnitudes are subject to the law:

$$a \rightarrow_6 b \leq a \rightarrow_{5.5} b \leq a \rightarrow_5 b.$$

For fixed a, they are all monotonically non-decreasing in b, and for fixed b, monotonically non-increasing in a, and they all agree with the classical two-valued material implication operator at the "corners" where a and b assume the extreme values 0 or 1.

With any of these operators, and any pair of fuzzy relations R from X to Y and S from Y to Z, there are defined certain triangle products and a square product, each of which is a fuzzy relation from X to Z. Where R_{ij} is the degree to which the element x_i is R-related to the element y_j, and hence is the value a assigned to the proposition "x_i is R-related to y_j", and S_{jk} is the degree to which y_j is S-related to z_k, and hence the value b of "y_j is S-related to z_k", the triangle sub-product relation $R \triangleleft S$ tells in $(R \triangleleft S)_{ik}$ the mean degree (averaged over all the y_j) to which the R-relatedness of x_i to y_j implies the S-relatedness of y_j to z_k. The formula is

$$(R \triangleleft S)_{ik} = \frac{1}{N_j} \sum_j (R_{ij} \rightarrow S_{jk}).$$

Similarly, the triangle super-product relation $R \triangleright S$ gives the mean degree of the converse implication:

$$(R \triangleright S)_{ik} = \frac{1}{N_j} \sum_j (R_{ij} \leftarrow S_{jk})$$

(where of course $a \leftarrow b = b \rightarrow a$), and the square product relation $R \square S$ gives the mean degree of the mutual implication:

$$(R \square S)_{ik} = \frac{1}{N_j} \sum_j (R_{ij} \leftrightarrow S_{jk})$$

where $a \leftrightarrow b = \min(a \rightarrow b, a \leftarrow b)$.

Fundamental to Information Retrieval is the relation $\mathbb{D}$ (see section 2.3 below) from documents D to linguistic terms T, in which

$\mathbb{D}_{ij}$ = degree to which document d_i treats of term t_j
= degree of relevance of term t_j to document d_i
= $\mathbb{D}^T_{ji}$.

The triangle sub-product and super-product of $\mathbb{D}^T$ with $\mathbb{D}$ respectively

$(\mathbb{D}^T \triangleleft D)_{jl}$ = mean degree (averaged over documents) to which the relevance of term t_j implies the relevance of term t_l
= mean degree to which t_j is more specific than term t_l.

$(\mathbb{D}^T \triangleright D)_{jl}$ = mean degree to which term t_j is broader than term t_l.

The square product $\mathbb{D}^T \square \mathbb{D}$ has the meaning

$(\mathbb{D}^T \square D)_{jl}$ = mean degree to which terms t_j and t_l are synonymous.

In this way (see 2.3), the thesaurus of terms is constructed.

Example 1: MVL-search request:

Given an MVL-search request made of terms $\{t_{(i)}\}$, say

$$s = (t_1 \text{ AND } t_2 \text{ AND } (\text{NOT } t_3) \text{ OR } 0.5\ t_5).$$

The processor responds to S by supplying the user with the following list of documents d:

$\{d_2\}$ = 0.5
$\{d_1, d_3\}$ = 0.4
$\{d_4\}$ = 0.15

where the number at the right hand side is the degree of relevance of each document.

If the user specifies a threshold value, e.g. of 0.4, then a document with lower degree of relevance than the threshold (i.e. d_4) will not be returned. This gives the user the freedom to restrict or extend the search output according to pre-specified relevance.

Example 2: (Thesaurus requests)

After specifying the threshold of relevance of thesaurus terms, a list of broad or specific terms is provided as the response to the user request:

RELEVANCE THRESHOLD> .4

'Broad ticketisial?' - "pneumonia"

'Specific pneumonia?' - "Lobar, ticketisial, Staphylococcal, viral".

By changing the threshold of relevance the number of thesaurus terms can be extended or restricted. The examples clearly demonstrate the added flexibility gained by the inclusion of degrees of relevance.

PART 2. CONSTRUCTION OF AN INFORMATION RETRIEVAL MODEL WITH MULTIPLE-VALUED LOGIC EXTENSION

This part presents the design of a retrieval model based on multiple-valued logic, which is an extension of a Boolean retrieval model. The extension of logical and relational structures is achieved using the concepts and techniques outlined in the papers published by the authors and their co-workers [18], [14], [12], [13], [11], [7], [10].

The actual realisation of the model as an information processing system is done in terms of a sequence of virtual machines of decreasing abstraction. A brief outline of this appears in section 2.4.

The global structure of the model discussed in this section is depicted in Fig.2.1. The input of the model consists of documents, thesaurus, multiple-valued logic search requests and relational requests. The multiple-valued logic (MLV-) requests are concerned with questions related to the documents.

The relational requests (R-requests) are concerned with the questions related to the thesaurus. The ouput consists of relational output and MVL-search output. The MVL-output is the answer to the MVL-request and is a list of references to documents, whilst the R-output is the answer to an R-request and is a list of index terms.

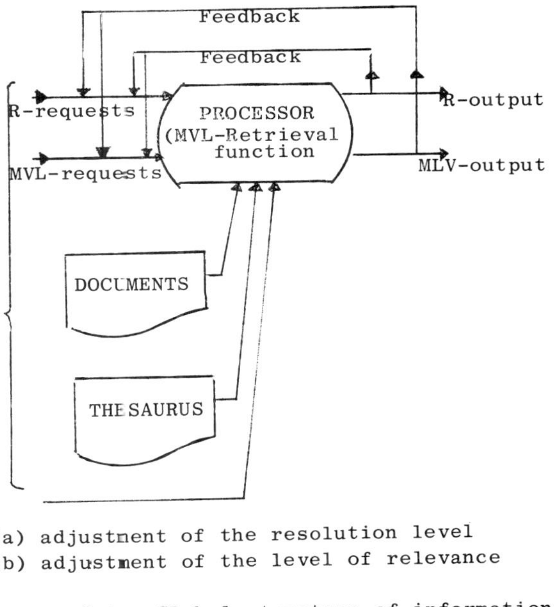

(a) adjustment of the resolution level
(b) adjustment of the level of relevance

Figure 2.1: Global structure of information retrieval system with multiple-valued logic extension.

2.2 Description of MVL- and R-requests.

A fuzzy search request of a user consists

of an element of the query set S, which is built recursively from the atomic elements - the members of the index terms set T. These primitives are combined together by the multiple-valued logic connectives AND, OR, NOT and numeric constants from the interval [0,1] (these represent the degree of relevance).

Example: $0.5(t_1 \text{ AND } t_2) \text{ OR } (t_3 \text{ AND } t_4)$

Note that each term of a compound term can be weighted by its degree of relevance. In response to the request S the processor will retrieve all documents treating either both t_1 and t_2 or both t_3 and t_4. It will then halve the degree of relevance of those treating t_3 and t_4, sort the list in decreasing order of relevance and present it to the user.

There are two types of relational requests, namely

(a) Specific t, where to is an information descriptor (index term).

(b) Broad t, where t is an index term.

In response to the request Specific t, the processor goes through the thesaurus, and provides the user with a list of terms, related to t, which are more restrictive.

Example: The following list might be provided by the processor in response to the request 'Specific pneumonia' - "Lobar, ticketisial, staphylococcal, viral".

In response to Broad t the processor goes through the thesaurus, and provides the user with a list of terms related to t which are more general.

Example: The following list might be provided by the processor to the request 'Broad ticketisial' - "pneumonia".

For a concrete example of a thesaurus using real data processed on a PDP 10 see the Appendix.

2.3 Multiple-valued representation of document-term relation

Within the processor, documents are represented in terms of the document-term Matrix $\mathbb{D}$. As discussed above in 1.4.

$\mathbb{D}_{ij}$ = the degree to which a document d_i treats the term t_j (the degree of relevance of term t_j to the content of the document d_i).

Hence, the i-th row represents an afterset of the terms that are defined by a document d_i. The j-th column represents the foreset defined by the term t_j.

2.4 MVL- and R-requests interpretation

For the interpretation of the MVL- search requests, a retrieval function is defined that maps each request to its answer which is a multiple-valued logic based subset of D:

$$R: S \rightarrow P(D)$$

The interpretation of the requests can be defined recursively (cf. Keravnou [13]). It is characterised by the requirement that it must be a homomorphism of the lattics <T, AND, OR, NOT> into the lattice $\langle P(D), \cap, \cup, \neg \rangle$.

Similarly, interpretation for R-requests can be defined. This definition is based on a set difference between the square and triangular products described in 1.4. If $aR\square S \subset aR \triangleright S$ then $aR\triangleright S - aR\square S$ gives the set for which a is considered a broad term. Similarly, if $aR\square S \subset aR\triangleleft S$ then $aR\triangleleft S - aR\square S$ gives the set of terms for which a is consididered a specific term [12], [13].

A thesaurus on which the R-requests operate is a hierarchy of index terms. This can be constructed automatically. An example of such a construction using the suite of programs described in [15] is presented in the Appendix.

2.5 Implementation details of a retrieval system with multiple-valued logic.

2.5.1 A 3-level virtual machine hierarchy

The actual implementation of the described information retrival system is constructed as a 3-level virtual machine hierarchy. The hierarchy is composed of the three distrint virtual machine levels M_3, M_2, M_1 and two interpreters $T_{3,2}$, $T_{2,1}$. The further technical details of this design can be found in Keravnou [12], [13] and Keravnou and Kohout [14].

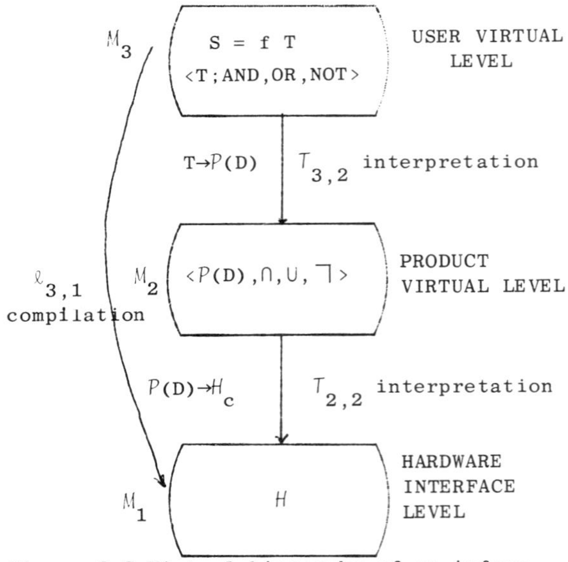

Figure 2.2 Virtual hierarchy of an information retrieval system with multiple-valued logic extension.

It should be noted that it is possible to bypass the level M_2 for certain retrival functions in order to speed up the computation. This is done by compilation transformation $\ell_{3,1}$. The compilation sets certain, often required, data structures directly, so that they need not be generated by the interpretation every time they are referred to by the system.

2.5.2 Arrays of matrices versus foreset-afterset data representation: A comparison of computation complexity. The actual computation of the relational products can be performed in two distinct ways, namely

(a) by employing the matrix representation of relations [15] or
(b) by using the foreset-afterset representation [18], [12], [13], [14].

A suite of programs for the calculation of relational products of relations represented in the matrix form has been in practical use for some time. It has been used in various medical and clinical psychological [15], [17] applications as well as in the design of some expert systems [16]. However, even a small information retrival problem employs large matrices of relational data. The scope of the problem can be illustrated by the following practical figure: taking 300 medical journals, we may get approximately 2,000 documents and 5,000 terms, i.e. we are dealing with matrices of 2,000 x 5,000 size. However, these matrices can be sparse. Hence, it is advantageous to employ the foreset-afterset representation of relations.

The estimate of the computational time for these two competing methods of relation representation is as follows:

(a) $\mathcal{O}(N.D + N^2)$ for the matrix method of product calculation,

(b) $\mathcal{O}(N.\bar{D} + N^2)$ for the foreset-afterset method.

Note that D in the (a) above represents the total number of documents, whereas $\bar{D}$ in the (b) above is the average number of documents per term. This shows that for sparse matrices, the foreset-afterset representation not only saves storage space, but is also considerably faster in relational product calculation. The problem is discussed in its full complexity in [12], [13], where the full formulas as well as further details of the implementations of fuzzy retrieval structures can be found.

REFERENCES.

1. Lancaster, F.W. (1968). Information Retrieval Systems, John Wiley, New York.

2. van Rijsbergen, C.J., (1979) Information Retrieval, 2nd edit., Butterworths London.

3. Kohout, L.J. (1975). Algebraic models in computer-aided medical diagnosis. Proc. of MEDINFO 74, North-Holland 575-9 addendum 1069-70.

4. Gaines, B.R. and Kohout, L.J. The fuzzy decade: a bibliography of fuzzy systems and closely related topics. Int.J. of Man-Machine Studies, 9, 1-68.

5. Bollmann, P and Konrad, E. Fuzzy document retrieval. Progress in Cybernetics and Systems Research, 3, Trappl, R. Klir, G.J. & Ricciardi L. (eds) Hempshere Pub.Co. and John Wiley, Washington and New York, 335-363.

6. Bandler, W. and Kohout, L.J. A Hanbook of fuzzy and crisp relations, a book to appear.

7. Kohout, L.J. and Bandler, W. (1980) Fuzzy relational products as a tool for analysis and synthesis of the behaviour of complex natural and artificial systems. In Fuzzy Sets: Theory and Application to Policy Analysis and Information Systems. Wang, S.K. and Chang, P.P. (eds) Plenum Press, New York, 341-367.

8. Bandler, W. and Kohout, L.J. (1980). Fuzzy power sets and fuzzy implication operators. Fuzzy Sets and Systems. 4, 13-30.

9. Keravnou, E., (1981). Library current awareness system (NIMROD 3). Tech. report, Dept. of Computer Science, Brunel University and National Inst. for Medical Research.

10. Willmott, R., (1980) Transitivity of inclusion and equivalence in fuzzy power-set theory. Report No.FRP-10. Dept. of Mathematics, University of Essex, Colchester. To appear in Fuzzy Sets and Systems 1982.

11. Bandler, W., and Kohout, L.J., (1980) Semantics of implication operators and fuzzy relational products. Internat.J.Man-Machine Studies, 12, 89-116. Reprinted in Fuzzy Reasoning and its Applications, Mandani, E.J. and Gaines, B.R., eds., Academic Press, London 1981.

12. Keravnou, E. Implementation of fuzzy data structures. In Systems Concepts in Medicine and Clinical Behavioural Sciences, Kohout, L.J., Bandler, W., and Stern, G. eds, to appear.

13. Keravnou E, (1982) Fuzzy Relational Products in Information Retrieval Systems, a dissertation for the degree B.Tech Dept. of Computer Science, Brunel University, Uxbrdige

14. Keravnou, E. and Kohout, L.J. (1982) System for experimental verification of deviance of fuzzy connectives in information retrieval applications. 2nd World Conf. on Mathematics at the Service of Man. Topic 7, Measuring "deviance" in non-classical logics and modelling. Las Palmas (Canary Isalnds). To appear in Proceedings thereof.

15. Bandler, W. and Kohout, L.J. (1981). The idenification of hierarchies in symptoms and patients through computation of fuzzy triangle products and closures. In Inf.Tech. for the Eighties. Parslow, R.D., ed., Heyden & Son Ltd.

16. Kohout, L.J. and Bandler, W. Approximate reasoning in intelligent relational data Bases. Ibid.

17. Kohout, L.J., Bandler, W. and Stern G. edgs., Systems Concepts in Medicine and Clinical Behavioural Sciences, a book, to appear.

18. Kohout, L.J., Keravnou, E., Bandler, W. (1982). Automatic documentary information retrieval by means of fuzzy relational products. In Fuzzy Sets and Decision Analysis. Gaines, Zadeh, Zimmermann eds, a volume in TIMS Studies in Management Science.

19. Gaines, B.R. and Kohout, L.J. (1977). The Fuzzy decade: a bibliography of fuzzy systems and closely related topics. Intl.J.Man-Machine Studies. 9, reprinted in Fuzzy Automata and Decision Processes, Gupta, M.M., Saridis, G.N. and Gaines, B.R. eds. North-Holland Elsevier, New York.

20. Bellman, R. and Zadeh, L.A. (1977). Local and fuzzy logics. In Modern Uses of Multiple-Valued Logic. G. Epstein ed., D.Riedel, Dordrecht-Boston, 103-165.

21. Sparck-Jones, K., (1972) A statistical interpretation of term specificity and its application in retrieval. J. of Documentation, 28 11-21.

APPENDIX AN EXAMPLE OF AN AUTOMATICALLY CONSTRUCTED THESAURUS

A hierarchy of index terms T, constructed automatically from the given mapping of documents D into the power set of terms $\mathcal{P}_{(T)}$. Computations were performed on PDP 10 in matrix representation.

Index terms T

1. fusion	25. glycolipid
2. competence	26. glyco-sphingo-lipids
3. phosphatidylserine	27. preparation
4. liposom#	28. sterols
5. immuno#	29. silastic
6. adjuvants	30. unilamellar
7. eliciting	31. fusion
8. antibod#	32. secretary
9. preparation	33. vesicles
10. proteo	34. surface
11. glycoprotein	35. potential
12. HJV	36. disruption
12. stearylanine	37. tetracaine
14. tyrosine	38. micelles
15. phenollyase	39. disassembly
16. character#	40. sendai
17. property	41. virus
18. haptenated	42. agglutination
19. membrane	43. glycophorin
20. Alph-tocopheral	44. lipid
21. mitochondria	45. peroxidation
22. liver	46. free
23. protein	47. abbunin-phosphol-ipid
24. H+Atpase	

T = {1,2,3,...,47} set of terms.

Documents ⟶ $\mathcal{P}$(Terms) mapping:

1. Fusion
competence
phosphatidylsenne
lipsome

2. liposome
immuno#
adjuvants
eliciting
antibody

3. preparation
proteo
liposome
glycoprotein
HJV

4. liposome
stearylanine
tyrosine
pendlyase

5. character#
immuno#
property
haptenated
liposome
membrane

6. Alpha-Tocopherol
liposom
mitochondria
liver
protein

7. Membrane
protein
liposome
mitochondrial
H+Atpase

8. anti-glycolipid
antibody
liposome

9. liposome
glyco-sphingo-lipid
preparation
character#

10. liposome
membrane
sterols
silastic

11. Unilamellar
liposome

12. membrane
fusion
secretary
vesicles
liposome

13. surface
potential
liposome

14. disruption
liposome
tetracaine
micelles

15. fusion
disassembly
sendai
virus
membrane
liposome

16. sendai
virus
liposomal
membrane
agglutination
glycophorin

17. liposomal
membrane
lipid
peroxidation

18. sterol
liposomal-free
Abbunim-phosphali-
pid

D = {1,2,...,18} set of documents.

mean = .92

sdev = .133

index of transitivity = 1

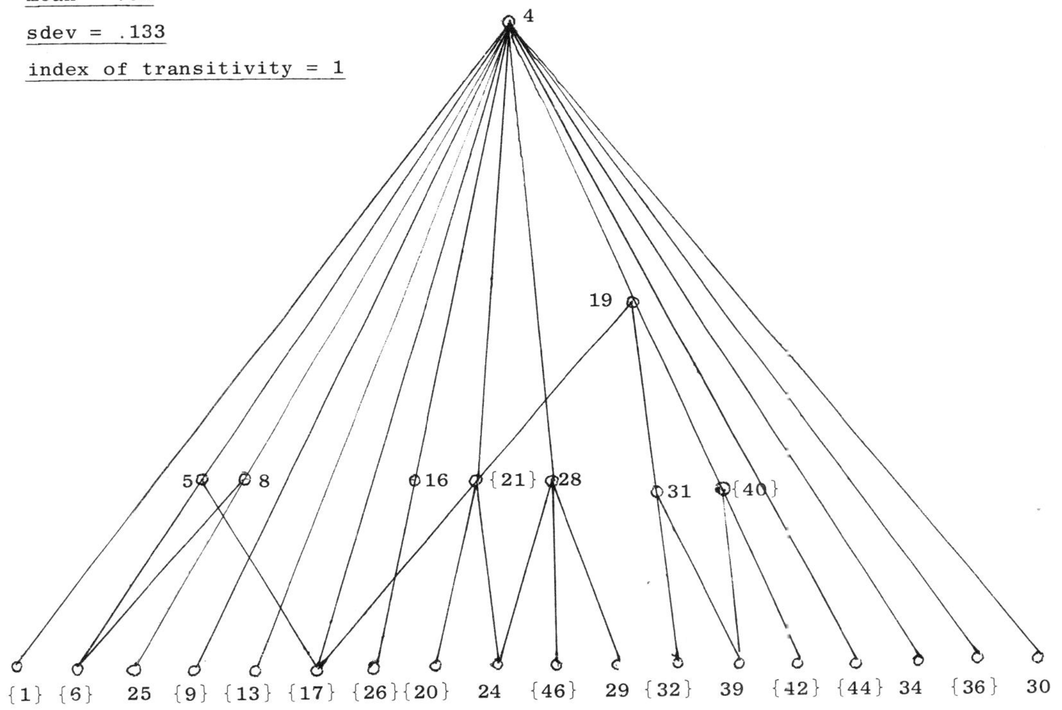

THESAURUS OF INDEX TERMS
constructed automatically from the mapping D ⟶ $\mathcal{P}$(T)

KNOWLEDGE ENGINEERING USING A FUZZY RELATIONAL INFERENCE LANGUAGE

J. F. Baldwin

Department of Engineering Mathematics, University of Bristol, Bristol, UK

Abstract. A general query language F.R.I.L. (fuzzy relational inference language) is introduced. The language F.R.I.L. can be thought of as a high level automatic inference knowledge base system, similar in applicability to PROLOG but based on the mathematics of relations instead of Predicate Calculus. The user provides a knowledge base of facts and rules from which queries can be answered automatically. An inherent parallelism is present in F.R.I.L. which will allow future computer architectures to be fully exploited. It incorporates an automated fuzzy inference mechanism and should find applications in many areas of knowledge engineering such as expert systems, linguistic controllers, intelligent data bases, modelling of complex systems and decision support systems.

Keywords. Artificial intelligence; approximation theory; computer software; decision theory; man-machine systems.

INTRODUCTION

F.R.I.L. stands for Fuzzy Relational Inference Language which can be thought of as a high level language for designing automatic inferential knowledge base systems. It is similar in applicability to PROLOG but is based upon the mathematics of relations (fuzzy and classical) instead of Predicate Calculus. The user provides a knowledge base of facts and rules from which queries can be answered automatically. Its internal automatic inference mechanism is not of a search or resolution type associated with logic inference systems but is more analagous to Gaussian Elimination for solving algebraic equations.

A complete specification of the language is not at present available and in this paper we will introduce the ideas of the language and its application via examples. An implementation of the language is available on the Honeywell computer at Bristol University. This is written in MAC-LISP and a future version for FRANZ LISP to run under UNIX will be available soon. We hope also to provide future implementations written in C and also in FORTH. The implementation to date is not complete but it does contain all commands discussed in this paper as well as more advanced commands and editing facilities which we will not discuss.

One of the motivations for designing the language was to provide a suitable means of designing expert systems in which fuzzy concepts could be included. If, for example, one considers a medical expert system, a Doctor bases his clinical decisions on information which is often imprecise, incomplete and subjective. The ability of the language F.R.I.L. to describe with varying degrees of precision the concepts and rules of medical diagnosis, some of which may be subjective, its interactive and querying feature will, for the first time, enable a dynamic dialogue between clinician and patient computer model. Work with medical staff of Frenchay Hospital, Bristol, on designing medical expert systems using F.R.I.L. has now been started in two areas:

(i) Intensive Therapy Unit - Head Injury
(ii) Management of Subarachnoid Haemorrage.

Other application work is with A.U.W.E. (Ministry of Defence) for command and control, British Aerospace on Scene Analysis, and Central Electricity Generating Board on expert systems for certain maintenance requirements.

The design of F.R.I.L. includes many ideas expressed in papers on fuzzy logic and automated inference by Baldwin (1,2,3,4,5,6), fuzzy systems by Zadeh (10,11,12,13) with special reference to the paper on PRUF (14) and Test Score Semantics (15). It has also been influenced by the work on Codd's relational Data Bases - Ullman (8) and Logic Programming - Kawalski (9). Some of the query syntax notation has been influenced by Micro Prolog developed at Imperial College.

In order to explain the concepts and constructs in F.R.I.L. examples will be used rather than attempt to give a general theoretical treatment. In this way the use of the language is made more clear. A more detailed treatment with appropriate definitions will be given in a forthcoming paper (7).

A general fuzzy knowledge base system with general deductive capability consists in the main of:-

a knowledge base of base relations, virtual relations, set theoretic relations and functions.

Base relations are tables of facts in which each tuple (rows of table) satisfies the relation to some degree χ which takes values in the interval [0 , 1]. Each column is associated with an attribute which can take values from an associated domain.

A virtual relation is a relation defined by means of a rewrite rule in terms of base relations and other virtual relations, set theoretic relations and functions or queries.

A set theoretic relation is defined by a procedure which takes as input a given tuple of values and returns a truth value in the interval [0 , 1]. For example LESS (x , y) takes two values (numbers) x and y and returns TRUE if x < y and FALSE otherwise.

An example of a function is COUNT which takes a name of a relation as argument and returns the number of tuples in the relation. The user can add new function definitions as well as relations to the knowledge base.

The process in which the user acquires information from the system is as follows:-

1. The user asks questions through the query language provided by the system.

2. The query is sent to the *base relation problem generator* which translates the query into a problem specification, containing base relations, set theoretic relations and functions only. This translation involves using the rewrite rules of the virtual relation definitions to rewrite virtual relations into base relations. The translation is done in such a way as to provide a problem reduction tree for the next stage of processing. This reduction tree breaks down the query into sub-queries if necessary and indicates how the solutions of sub-queries are to be combined.

3. The problem reduction tree is passed to the *Phase 1 process*. This determines a strategy of solution for each sub-query in terms of elementary operations on relations such as 'join', 'projection', 'cross product', 'select', etc. Phase 1 combines all solutions into a final solution which it returns.

EXAMPLE 1

Knowledge Base:-

LIKES	NAME	NAME	χ
	JIM	IRENE	1
	JOHN	HEATHER	0.7
	JOHN	MARY	0.6
	HARRY	JILL	0.4
	JILL	TOM	0.2
	IRENE	JIM	0.9
	HEATHER	JOHN	0.8

TALL	HT	χ
I-type	5 - 9	0
	5 - 10	0.6
	5 - 11	0.8
	6 - 0	1
	7 - 0	1

Domain (NAME) = {<character-string>};
Domain(HT) = [4-0 , 7 - 0].

N.B. An I-type relation allows for linear interpolation for values between any two values of a given attribute in the relation. The tuple (5 - 9 , || 0) is included so that interpolated χ values in the range 5 - 9 to 5 - 10 can be used.

PERSONS	NAME	HT	WT	χ
	JIM	6 - 1	12 - 0	1
	JOHN	5 - 9	11 - 9	1
	IRENE	5 - 5	10 - 0	1
	HEATHER	5 - 6	9 - 6	1
	MARY	5 - 3	8 - 5	1
	JILL	5 - 7	9 - 2	1
	TOM	6 - 0	13 - 5	1

HEAVY	WT	χ
I-type	10 - 9	0.1
	11 - 0	0.2
	11 - 6	0.4
	11 - 9	0.7
	12 - 0	0.8
	13 - 0	1
	14 - 0	1

N.B. In the case of PERSONS, the χ column would not normally be included since all χ values are 1.

Domain (WT) = [8 - 0 , 15 - 0]
LESS (x , y) = {(x , y) | x ε N , y ε N , x < y} where N is the set of numbers.

MOST	%	χ
	70	0
	90	1
	100	1

Domain (%) = [0 , 100]

FRIENDS(x,y) ⟵ LIKES(x,y) AND LIKES(y,x).

N.B. FRIENDS is a virtual relation and this statement can be interpreted as 'x and y are friends means that x likes y and y likes x'. Variables x , y are local to this definition.

POSSIBLE_ATHLETE(x) ← PERSONS(x,y,_)AND TALL(y)

HAS_GOOD_FRIENDS(x) ← FRIENDS(x,y) AND POSSIBLE_ATHLETE(y).

N.B. The interpretation of these statements makes sense as far as the base relations are concerned, even though their realism are open to question. The binding of variables should be obvious from these examples. The underline sign "_" that appears as a relation argument stands for an anonymous attribute variable which does not relate to the definition under consideration.

Consider the query:- Who likes a tall person?

In F.R.I.L. this query is written as

WHICH(x LIKES (x,y) AND PERSONS (y,z,_) AND TALL (x))

and returns a relation with one attribute NAME.

The problem generator for this example returns the reduction tree

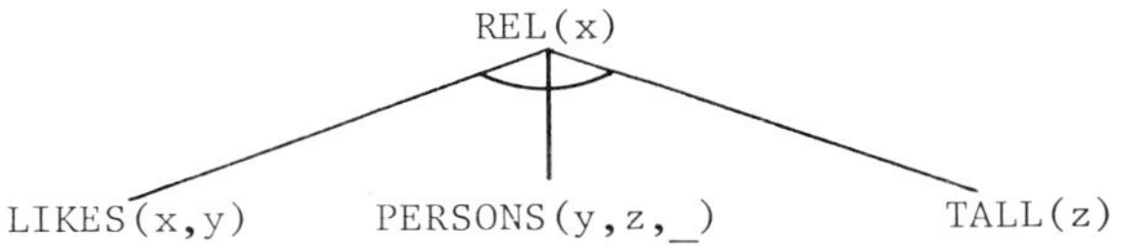

for which the leaves are base relations.

The Phase 1 process eliminates variables one at a time to produce a succession of reduction trees until the required solution is obtained. For this example

$R1(y,z) = Proj_{y \times z} PERSONS(y,z,\_)$

is formed to produce the new reduction tree

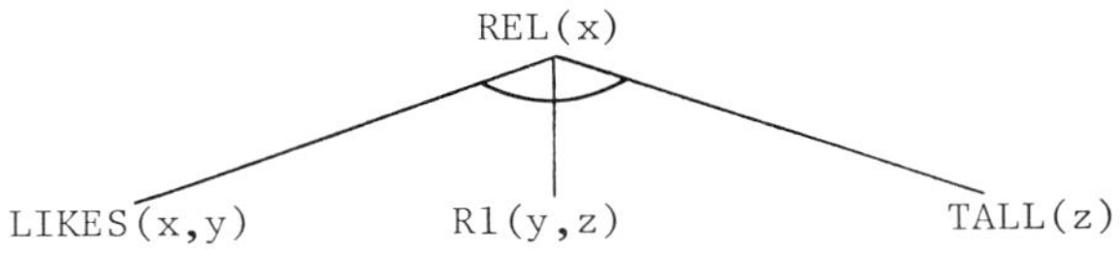

where R1 is the same relation as PERSONS with the WT column removed. Next the variable z is eliminated using

1. $R2(y,z) = R1(y,z) \sim TALL(z)$ where $\sim$ stands for 'join'

2. $R3(y) = Proj_y\ R2(y,z)$

to produce the new derivation tree

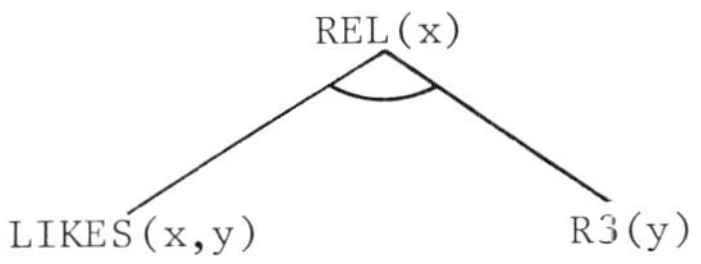

where R2 and R3 are given by

R2	NAME	HT	X
	JIM	6 - 1	1
	TOM	6 - 0	1

R3	NAME	X
	JIM	1
	TOM	1

The variable y is then eliminated using

1. $R4(x,y) = LIKES(x,y) \sim R3(y)$

2. $R5(x) = PROJ_x\ R4(x,y)$

to produce the derivation tree

where

R4	NAME	NAME	X
	JILL	TOM	0.2
	IRENE	JIM	0.9

R5	NAME	X
	JILL	0.2
	IRENE	0.9

so that R5 gives the solution to the query.

EXAMPLE 2

Knowledge base as for last example.

Query: Name those people with their weights who have a good friend,

i.e. WHICH((x,y): HAS_GOOD_FRIEND(x) AND PERSONS(x,_,y)).

In order to produce the reduction tree, the problem generator scans the query from left to right to produce the first level of the tree which is

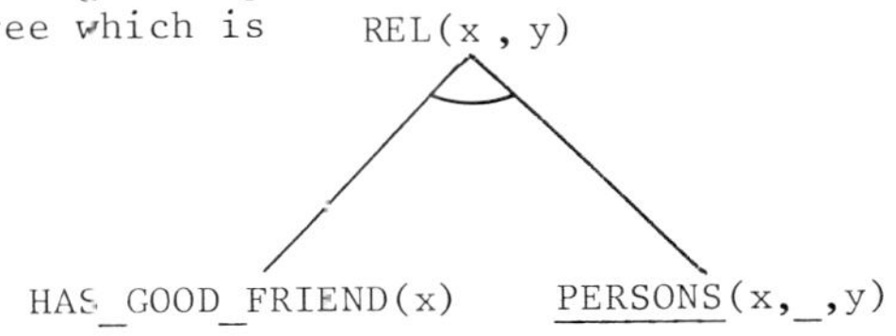

Since PERSONS is a base relations, this node is not expanded. The relation HAS_GOOD_FRIENDS is a virtual relation, so this is further expanded by forming a sub-query and deriving the derivation tree for this sub-query.

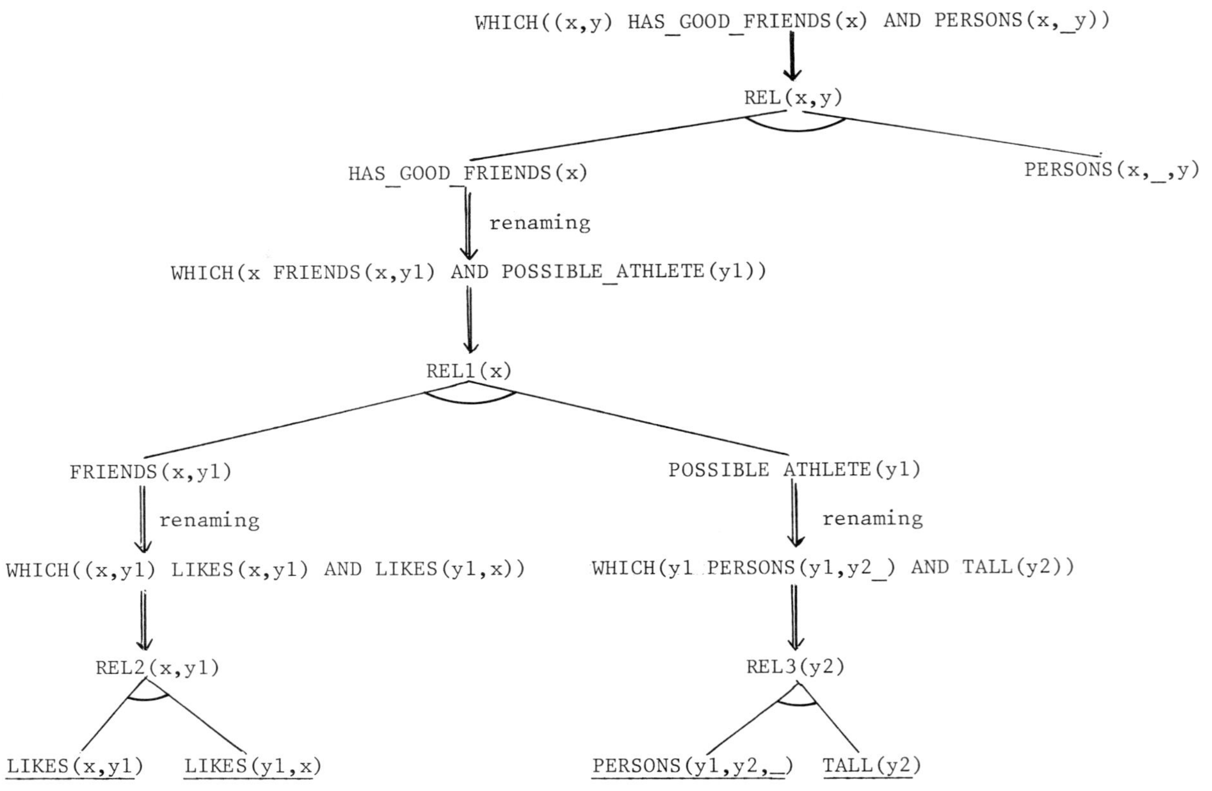

Figure 1

For a good understanding of this method, careful consideration must be given to the concepts of variable renaming and variable binding.

The sub-problems for this query can therefore be represented as

1. To solve for REL(x,y) use (REL1(x) AND PERSONS(x,_,y))
2. To solve for REL1(x) use (REL2(x,y1) AND REL3(y2))
3. To solve for REL2(x,y1) use (LIKES(x,y1) AND LIKES(y1,x))
4. To solve for REL3(y2) use (PERSONS(y1,y2,_) AND TALL(y2))

These can be compounded to obtain the final reduction tree shown in Fig. 2.

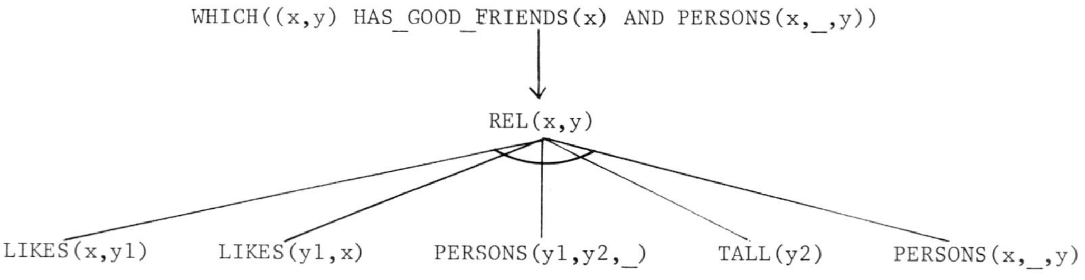

Figure 2

The Phase 1 procedure is now used in a similar way to the last example to give the solution

REL	NAME	WT	X
	IRENE	10 - 00	0.9

It should be understood that F.R.I.L. allows 'OR' connectives and more complex connectives can be used and not simply the 'AND' connective used in this paper.

Query: What is the truth that most people who are not tall are not heavy?
i.e.

```
DOES(MOST(=%DIVIDE(CARD(WHICH(x PERSONS(x,y,z)
  AND NOT TALL(y) AND NOT HEAVY(z))),
  CARD(WHICH(x PERSONS(x,y,_) AND NOT TALL(y))))))
```

Here we introduce several different notations: DOES(R) returns the max χ value for the relation R. The '=' sign signifies that for the relation MOST(x), x must be chosen as the value resulting from %DIVIDE( ). The %DIVIDE is a funtion of two arguments where each argument is a number and the result is the first number divided by the second and the results multiplied by 100. CARD(R) returns the sum of the χ values for R and thus CARD is a cardinality function. The first WHICH query returns those people who are neither tall nor heavy while the second returns those who are not tall. WHICH and DOES queries can be nested to any depth in queries or virtual relation definitions.

The relations NOT TALL and NOT HEAVY are given by

NOT TALL	HT	χ
I.TYPE	4 - 0	1
	5 - 9	1
	5 - 10	0.4
	5 - 11	0.2
	6 - 0	0

NOT HEAVY	WT	χ
I.TYPE	8 - 0	1
	10 - 8	1
	10 - 9	0.9
	11 - 0	0.8
	11 - 6	0.6
	11 - 9	0.3
	12 - 0	0.2
	13 - 0	0

These are automatically determined in F.R.I.L. Thus if

```
R1 = WHICH(x PERSONS(x,y,z) AND NOT TALL(y)
      AND NOT HEAVY(z))
R2 = WHICH(x PERSONS(x,y,_) AND NOT TALL(y))
```

then these are determined and given by

R1	NAME	χ
	JOHN	0.2
	IRENE	1
	HEATHER	1
	MARY	1
	JILL	1

R2	NAME	χ
	JOHN	1
	IRENE	1
	HEATHER	1
	MARY	1
	JILL	1

Thus CARD(R1) = 4.2;
CARD(R2) = 5

so that

DIVIDE(4.2 , 5) = 84

Let R3(x) = MOST(= 84)

then

R3	%	χ
	84	0.7

so that DOES(R3) = 0.7

giving the truth of the proposition that most people who are not tall are not heavy.

EXTENSION

There are commands in F.R.I.L. to modify the knowledge base, control flow of computation, send messages through different channels and pointers can be used for self iteration and for treating relation names as variables. The language outlined above only covers the more elementary aspects. Concepts such as truth functional modification for treating the cancatination of linguistic qualifiers with a relation, inverse truth functional modification for testing the truth of some logic proposition, editing facilities and display facilities are also included. In addition values of attributes of a relation can themselves be a relation. This allows, for example, the height of a given person to be described as VERY TALL etc.

Commands which modify the knowledge base include, for example,

UPDATE <relation_name> = WHICH(...)

which replaces tuples of <relation-name> with solution tuples of the WHICH query.

Also, commands

UPDATE_CHANGES <relation_name> = WHICH(...)

NEW <relation_name> = WHICH(...)

can be used. A relation can be given a key or keys. The value of a key of a tuple in a relation must be unique. UPDATE_CHANGES modifies <relation_name>, if any key tuple is different in the solution of the WHICH query to that of <relation_name> to that of the corresponding key tuple of the query solution. NEW simply creates a new relation in the knowledge base called <relation_name> with tuples from the WHICH query.

MESSAGE = "..." WHICH(...) ; CHANNEL

sends a message through channel CHANNEL where "..." is code for what is to be done with the solution of the WHICH query. For example, if CHANNEL is screen then, if the solution of the WHICH query is a relation of points representing triangles, the code may indicate that these triangles are to be drawn on the screen. If the solution of the WHICH query is NULL then no action is taken. The WHICH query in the above command therefore represents a TEST.

Control constructs such as SEQUENCE(...), REPEAT(...), CONTROL(...) are also included. SEQUENCE represents a sequence of commands which are obeyed one after the other. The same is true of REPEAT except that after the last command is executed control goes to the start again and exit from the loop occurs when a NULL solution to a WHICH query is found. In CONTROL the sequence is exited after completing the command of the command with the first WHICH query which returns a non NULL solution.

These extensions help in the design of expert systems and general production systems. The update represent actions in changing the short term memory of the knowledge base while MESSAGE represents actions communicated to the outside world - screens, printers, robots, machine, plants, etc.

The use of pointers is described in (7).

CONCLUSION

A very brief outline of the concepts, constructs, knowledge representation and internal mechanism for the automation of answering queries for F.R.I.L. have been given. The language should be of general use for knowledge engineering and in particular for designing expert systems, modelling complex systems and providing intelligent data bases. The ability of including fuzzy relations adds greatly to the general applicability of the language since real world problems are not always easily modelled in deterministic terms.

The language has an inherent parallelism which must be fully exploited as non Von Neumann computers become readily available.

REFERENCES

Baldwin, J.F. (1979). A new approach to approximate reasoning using a Fuzzy Logic. Fuzzy Sets and Systems, 309-325.

Baldwin, J.F. (1979). Fuzzy Logic and Fuzzy Reasoning. Int. J. Man-Mach. Stud., 11, 465-480.

Baldwin, J.F. (1979). Fuzzy Logic and its Applications to Fuzzy Reasoning. In Gupta (Ed.), Advances in Fuzzy Set Theory and its Applications, North Holland Pub. Co. pp. 93-115.

Baldwin, J.F. (1981) A Theory of Fuzzy Logic. In E. Mamdani & B. Gaines (Ed.) Fuzzy Reasoning and its Applications, Academic Press pp. 133-148

Baldwin, J.F. (1981) An Automated Fuzzy Knowledge Base. In R. Yager (Ed.) Fuzzy Systems, Pergamon Press.

Baldwin, J.F., & S.Q. Zhou (1982) A Fuzzy Relational Inference Language. (To appear)

Baldwin J.F. (1983) Fuzzy Expert Systems, International Symposium on Multiple-Valued Logic, Japan. (To appear).

Ullman, J.D. (1980) Principles of Database Systems, Pitman.

Kowalski, R. (1979) Logic for Problem Solving, North Holland Pub. Co.

Zadeh, L. (1965) Fuzzy Sets, Inf. & Control, 8, 338-365.

Zadeh, L. (1973) Outline of a New Approach to the Analysis of Complex Systems and Decision Processes, IEEE Trans. Syst., Man & Cybern., 3, 28-44.

Zadeh, L. (1975) Calculus of Fuzzy Restrictions. In Zadeh, Fu & Simura (Ed.) Fuzzy Sets and their Applications to Cognitive and Decision Processes. Academic Press.

Zadeh, L. (1978) Fuzzy Sets as a Basis for a Theory of Possibility, Fuzzy Sets and Systems, 1, 3-28.

Zadeh, L. (1981) PRUF - A Meaning Representation Language for Natural Languages. In E. Mamdani & B. Gaines (Ed.), Fuzzy Reasoning and its Applications, Academic Press.

Zadeh, L. (1981) Test-score Semantics for Natural Language and Meaning Representation via PRUF, Technical note, 247, S.R.I. International.

THE FIRST ORDER ARTIFICIAL INTELLIGENCE BIANZHENG (DIAGNOSIS) SYSTEM OF CHINESE TRADITIONAL MEDICINE (CTM)

Su Zaifu* and Wang Huaiqing**

**Department of Physics, Wuhan Medical College, Wuhan, China*
***Department of Computer Science, Huazhong Institute of Technology, Wuhan, China*

Abstract. In this paper a brief outline of distinguishing feature of Chinese traditional medicine is introduced, and a implementing fuzzy decision model of CTM, which is a induced form of pansystems medicine, is proposed. In accordance with this model, a self-adapting learning prosedure and a primary practical result, which is applied to the automatical modification of a fuzzy relation matrix, are given.

Keywords. Bianzheng; pansystems theory; fuzzy relation matrix; conditional proposition; normal-type fuzzy number; contactability; object parameter; computational method.

CTM AND ITS MATHEMECICAL REPRESENTATION

The major difference between the western medicine and CTM is that no matter what disease the patient got, in view of CTM, should be classified according to certain categories which included Bagang (the eight principal syndromes), Zangfu (Internal organs), Liuyin (the six External factors which cause disease), Yin (Nutrition), Wei (Guard), Qi (Vital of energy), Xue (State of blood), etc.

The classification process of analysis and synthesis in CTM is called Bianzheng (siminar to the analysis and differenciation of pathology in western medicine). As the Zheng is convinced, then the prescription can be given. And the whole process in clinic activity of CTM is called Bianzheng Shizhi (diagnosis and treatment based on an overall analysis of illness and the condition of the patient). And different diseases may be resulted from an identical Zheng, so that the same way of treatment can be taken, on the contrary, different Zheng may be resulted from the same disease, then the way of treatment should be varied. Zheng is the key of treatment, as considered in terms of CTM. As has demonstrated above, when a computer-aided system of BianZheng of CTM is to be established, the feature of it must be reflected, the detail information about it, consult with Su (1980, 1981) and Wang (1982).

For the convenience of explanation of the above point of view, let us take only two Zheng, Bagang and Zangfu, in consideration.

Bagang includes Yin and Yang, Exterior and Interior, Cold and Heat, Hypofunction and Hyperfunction. Take Yin and Yang as the general programme. Exterior, Cold, Hypofunction belong to Yin; and Interior, Heat, Hyperfunction belong to Yang. Any of the syndrom which embodies the position of the disease should brought into the programme of Exterior and Interior, the syndrom which embodies the character of disease should brought into the propramme of Cold and Heat, and that embodies the power to resist the disturbance should brought into the propramme of Hypofunction and Hyperfunction.

Zangfu includes heart, liver, spleen, lungs, kidney, stomach, gall, intestines and bladder. The disease of any of organs has its corresponding symptoms and signs. In order to avoid confution, it should emphasized that the meaning of Internal organs in CTM is different from that of western medicine.

The result of Bianzheng may be a composition of two or more programmes of Zheng, e.g. a Zheng may be composed of Bagang and Zangfu to form Hypofunction of hearet or Hyperfunction of spleen, etc.

A special metamedicine-like theory called pansystems medicine was developing within the framework of pansystems theory in China. From the view point of pansystems medicine, we can establish various model of the Bianzheng by using pansystems methods, including the application and treatment of fuzziness of a new type.A concrete description can be given as follows (refer to the litterature of Wu, 1982, 1983). Let B,Z,D,S be the Begang, Zangfu, Zheng and symptoms spaces respectively, then the formal pansystems (E, H) can be considered as a model of Bianzheng, where $E=B\cup Z\cup D\cup S$, $H=\{f_i\}$ ($i=1, \ldots, 6$), $f_1: B\times Z \subset D$, $f_2: S\rightarrow B$, $f_3: S\rightarrow Z$, $f_4=f_\delta: S\rightarrow D$, $f_5=\delta=(f_2\circ f_2^{-1} \vee f_3\circ f_3^{-1})$, and f_δ is the nature induction of δ, whose structure is derived from pansystems theory.

This model can be extended to various fuzziness cases according to either pansystems fashion, Zadeh's fashion, or composite fashion.e.g. let $\tilde{f}_2 \in [0.1]\uparrow(S\times B)$, $\tilde{f}_3 \in [0,1)\uparrow(S\times Z)$, $\tilde{f}_4 \in [0,1]\uparrow(S\times D)$, or let $S=[0,1]\times S'$, S' being the set of symptom names, or let $S=\prod S_i'\times L_i$, S_i' being symptom name set, L_i — corresponding poset meassure.An extension to uncertainness is $f_2\subset S\times B$, $f_3\subset S\times Z$, $f_4\subset S\times D$ and for which pansystems theory has presented more concrete investigations. The substance of Bianzheng is the formation of f_4 or $\tilde{f}_4$ or $f_6\subset Z\times B$, $\tilde{f}_6\in[0,1]\uparrow(Z\times B)$, and they usually can be described as the results of certain pansystems operations of f_2, f_3 or $\tilde{f}_2$, $\tilde{f}_3$. A useful model is

$$\tilde{f}_6 = \tilde{f}_3^{-1} * \tilde{f}_2 \qquad (1)$$

where * is the composition of algebraic product and algebraic sum. Here we take f_2 and f_3 as the fuzzy matrix.

For the sake of explaining the concept of Bianzheng decision model, let $S=\{s_1, \ldots, s_l\}$, denotes the set of symptoms, and $B=\{b_1, \ldots, b_m\}$; $Z=\{z_1, \ldots, z_n\}$ denotes the Zheng labeled with Bagangand Zangfu respectively, and $B, Z\subset D$. Further assume that the fuzzy relation matrix among S and B,Z are f_2 and f_3, and have

$$f_2 = \begin{array}{c} \\ s_1 \\ \vdots \\ s_l \end{array} \begin{array}{c} b_1 \cdots\cdots b_m \\ \left| \begin{array}{c} \\ \mu_{f_2}(i,j) \\ \\ \end{array} \right| \end{array} . \quad \begin{array}{l} i=1,\ldots,l \\ j=1,\ldots,m \end{array}$$

$$f_3 = \begin{array}{c} \\ s_1 \\ \vdots \\ s_l \end{array} \begin{array}{c} z_1 \cdots\cdots z_n \\ \left| \begin{array}{c} \\ \mu_{f_3}(i,k) \\ \\ \end{array} \right| \end{array} . \quad \begin{array}{l} i=1,\ldots,l \\ k=1,\ldots,n \end{array}$$

suppose a subset $S_i=\{s_i\}$, $s_i\in S$ expresses symptoms of a patient, then the rows will be vanished in the absence of residual symptoms. In this situation. the fuzzy relation matrix becomes

$$\tilde{f}_2 = \{\mu_{\tilde{f}_2}(u,v) \mid u\in i, v\in j\}$$
$$f = \{\mu_{\tilde{f}_3}(u,w) \mid u\in i, w\in k\}$$

by expression (1) we have

$$\tilde{f}_3^{-1} * \tilde{f}_2 = \{\mu_{\tilde{f}_3^{-1} * \tilde{f}_2}(w,v) \mid w\in k, v\in j\}$$

and the result of Zheng may be determined by the corresponding composition membership with maximum or sub-maximum that is according to the criterion of the knowledge of CTM.

The crux of the Bianzheng of CTM is to set f_2 and f_3, and it is the keylink of computer-aided Bianzheng of CTM. Of course, a high level physician may preset f_2 and f_3 with his/her experience, when the situation is very simple, but to expection, it should be much better if the relation matrix may be revised according to the patient's case which has been put into the computer in form of the fuzzy conditional proposition as "If X is F, then Y is G" e.g.:"If the symptoms of a patient (X) are 0.8/headaches, 0.6/sweat,...,(F), then the Zheng (Y) is 0.8/Exterior(G)", Where 0.8 and so on are the corresponding expression of grade of symptoms,signs or an estimation of truth value of diagnosis. So that the computer be provided with the function of learning, and the relation matrix will be completed step by step, as the input case increase, and thus the function of Bianzheng of computer will be streghthened.

ALGORITHM OF A.I. BIANZHENG SYSTEM OF CTM

Usually, the fuzzy relation matrix f_2 or f_3, which we write in simple R, where every element of it is fixed preliminarily by the medical knowledge and must be repeatedly modified through many times of testing. This will need a lot of work, as is we known, that the thinking process is very difficult to deal with and reach the best. Here we try to establish a learning system and to let the computer to modify the R by means of study in the Bianzheng process automatically. the basic thinking model is as follows.

At first the fuzzy relation matrix R could be fixed in the light of medical knowledge. As usual, in CTM, Zangfu is the main line in Bianzheng analysis. This makes us the advantage of processing the relation matrix. We take each one Zheng in Zangfu and composite with another Bianzheng programmes, in such a case, the matrix of Zangfu is decomposed into a set of vectors, so that the composition is taken between a vector and R. This makes the experts to give the R easily, but it may be very rough. If at the same time we fixed the method with which we can compute one or several object parameters (OPT), be explained in the ensuing paragraphs, then with the OPT, we can evaluate that the modified R is good or worse, the computational method of it may be described as following.

Let an element in R be changed to a direction, if the OPT is better, then let the element change to the direction continually, if it is worse, then let the element be changed to another direction, if the OPT is kept in constant, choose another element to go on. Passing several times modifing, every element in R all will be approached to the best values. The fuzzy conditional propsitions which are put into the computer are the basis of learning. As all knows, the system has some intelligence, but due to the learning method of the system is fixed preliminarily, it can'tbe changed, so that the system is called a first order A.I. system.

The learning system is detailed as following.

Suppose the input and the output of the idea model are known, i.e. the Biabzheng of every patient can be fixed perfectly. The schematic diagram of learning is shown in Fig.1, and explains as

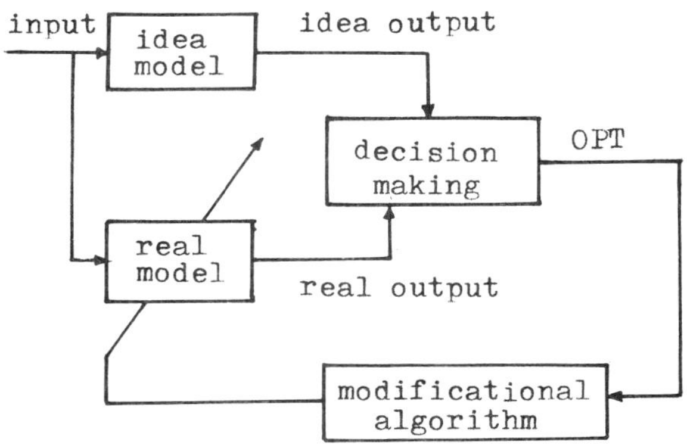

Fig.1. The schematic diagram of learning process

<u>The flow of computer algorithm.</u>
1. Store in computer the collected fuzzy proposition as initial conditions, and each of it is the result of an accurate Bianzheng of a patient.
2. Set initial values for fuzzy relation matrix, such as f_2 or f_3, etc.
3. Compute the OPT under composition rule of the fuzzy reference, which is selected suitably.
4. If the OPT satisfy the preset fittingness criterion, then goto 8, else
5. Compare the OPT after learning with that before, and see wether it is better or worse or keep in constant, if worse, then goto 7, if keep in constant, then goto8,else
6. Learn once again, i.e. to modify the relation matrix according to the defined operation, and after this, goto 3
7. The current stage of study failed, change the initial value or lower the level of fittingness criterion, then goto 3
8. Stop self-adaptive modification.

<u>The selection of OPT</u>. Assume there are k conditional propositions and one fuzzy relation matrix R. Suppose some one of the k conditional proposition is in form as "If the fuzzy subset of symptom is S_1, then the fuzzy subset of Zheng is D_1", the fuzzy relation matrix R which is passed l times of learning, is $R^{(l)}$; then we can obtained the l time result $D^{(l)}$ under composition operation between S_1 and $R^{(l)}$. Let

$$D_1=\mu(d_1)/d_1+\dots+\mu(d_m)/d_m$$
$$D^{(l)}=\mu^{(l)}(d_1)/d_1+\dots+\mu^{(l)}(d_m)/d_m$$

We can calcultate the contactability e_i, and the contactability of two normaltype fuzzy muber $n_1(c_1)$ and $n_2(c_2)$ is defined as

$$\Delta(n_1(c_1),n_2(c_2))=\exp\{-(n_1-n_2)^2/(c_1-c_2)^2\}$$

in this paper, the membership of rela-matrix are dealt as normal-type numbers.

Let

$$e_1=\Delta(\mu(d_1),\mu^{(l)}(d_1))$$
$$\dots\dots\dots\dots\dots\dots$$
$$e_m=\Delta(\mu(d_m),\mu^{(l)}(d_m))$$

the OPT$=\bigwedge_{i=1}^{k}\bigwedge_{j=1}^{m} e_{ij}$ is selected, and the maxmum of OPT is taken as criterion of decision-making, where k is the number of fuzzy conditional propositions which has been put into the computer. The effect of the criterion of such decision making will make the relation matrix in the operation of composition to have the maximum contactability for every cases of Zheng, that is the least contactability attain the largest.

<u>Way of learning</u>. In the Bianzheng of CTM, the initial value r_{ij} of any element in R need not be very exact. For redusing the calculation, the search-algorithm with the constant step is adapted.

Let the constant step, i.e. the modifying value, be ξ, and the fuzzy relation matrix, which is passed l times learning, is $R^{(l)}$, take any element r_{ij} in $R^{(l)}$ and to modify it. At first we add the increment ξ to r_{ij}, other element is fixed for the same time, calculate the OPT and to meassure wether the new OPT is better or not, if it is better, the step is right, let $r_{ij}^{(l+1)} = r_{ij}^{(l)} + \xi$; if not, we give the $-\xi$ to r_{ij}, and calculate OPT again, if the new one is better then the old, let $r_{ij}^{(l+1)} = r_{ij}^{(l)} - \xi$. If both of the direction are worse, let $r_{ij}^{(l+1)} = r_{ij}^{(l)}$. Thus every element in $R^{(l)}$ all are midified. After this the l+1 times of learning is finished. It is easy to show that the algorithm is convergency.

As is well known, the amount of calculation will be very large, if the algorithm is unoptimized, fortunately, here may decrease about 3 powers of amount of calculation when some skillful coding is adapted. Thus the optimized algorithm may be used in the practise.

TEST WITH COMPUTER

According to the preceding context, the experimental program CTMoo with PASCAL is written for the Bianzheng of Hypofunction of Xin(heart). There are Two text files being established, the one is to store the fuzzy relation matrix, the other is to store the data of clinical cases, fuzzy conditional propositions, which can be put into the computer continuously and the fuzzy relation is modified along with them.

As has been stated in the above part, the selection of OPT is a very important thing to be taken into account, what we selected is the result of Bianzheng of the computer as compared with that of the physician. And let the minimum value of contactability of them to be raised step by step in the procedure of learning and reach the maximum as it is in convergency.

For example, there are five cases of patient as following

NO.1 Symptoms(0.7/palpitation, 0.4/chest stuffy, 0.7/isomsia, 0.7/weakness, 1.0/slow and inregular pulses). Diagnosis of physician(0.6/Xin Qixu).

NO.2 Symptoms(0.7/palpitation, 0.6/light headed, 0.4/short breath, 0.7/weakness, 1.0/slow and thread-like pulse, 0.7/pale).
Diagnosis of physician(0.8/Xin Yangxu, 0.8/Xin Qixu).

NO.3 Symptoms (0.4/palpitation, 1.0/heart pain, 1.0/chest stuffy, 0.4/light headed, 1.0/short breath, 0.7/weakness, 0.4/pale).
Diagnosis of physician(0.8/Xin Qixu).

NO.4 Symptoms(1.0/palpitation 0.7/chest stuffy, 0.4 short breath, 1.0/red tip of tongue, 1.0/slow and irregular pulse).
Diagnosis of physician(0.7/Xin Qixu, 0.7/Xin Yangxu).

NO.5 Symptoms(0.7/palpitation, 0.7/chest stuffy, 0.7/heart pain, 0.7 short breath, 0.7/sweat, 1.0/night sweat, 0.7/weakness, 1.0/slow and irregular pulse).
Diagnosis of physician(0.8/Xin Qixu).

The diagnosis of above 5 cases which are given by the physician may be written into a matrix as following:

z_1	z_2	z_3	z_4
0.6	0.0	0.0	0.0
0.8	0.8	0.0	0.0
0.8	0.0	0.0	0.0
0.7	0.7	0.0	0.0
0.8	0.0	0.0	0.0

Where z_1, z_2, z_3, z_4, represent Xin Qixu, Xin Yanxu, Xin Yinxu and Xin Xurxu respectively, the diagnosis of the same cases given by computer with initial relation matrix are

z_1	z_2	z_3	z_4
.80	.72	.63	.63
.86	.81	.47	.76
.85	.86	.76	.58
.86	.82	.80	.54
.91	.90	.80	.51

The contactability of the two diagnosis are

Δe_1	Δe_2	Δe_3	Δe_4
.62	.47	.69	.59
.69	.99	.98	.70
.97	.55	.70	.91
.77	.86	.49	.85
.87	.50	.63	.96

obviously, while before learning the OPT is equal to .47 (with underline in above list).

After 11 times of learning the modified matrix is shown in the right of Table 1, and the variation of OPT in each learning is plotting in Fig.2, with the new matrix we have the diagnosis of computer

z_1	z_2	z_3	z_4
.68	.43	.38	.43
.80	.72	.21	.60
.80	.56	.37	.40
.77	.59	.40	.39
.86	.63	.56	.27

After learning, the contactability of diagnosis between the computer and the physician are

Δe_1	Δe_2	Δe_3	Δe_4
.93	.84	.92	.86
1.00	.93	.90	.89
.99	.92	.99	1.00
.95	.87	.97	.99
.96	.86	.93	.96

as indicated by the result of contactability, the OPT is raised from .47 to

.84 and the diagnosis of the computer after learning is much better than that before.

The premary result of testing has been demonstrated this method provided with rapid calculation, well convergency and satisfy the demand of the application of Bianzheng system.

The character of this method only demands a rough filled matrix and put the formal conditional proposition into the computer, then the computer can calculate automatically all element of these fuzzy matrix more procisely, and R will be improved step by step. Of course, this method except to supply a useful tool for the study of computer-aided Bianzheng of CTM, but also can be used in another type of diagnosis or in general, to establish a fuzzy relation matrix, as the situation is complex.

As a prospect, we will study continuously the high order A.I. system with algorithm that can be varied, and improve this system to use for the CTM practise.

I wish to thank Prof. Wu Xuenou for his suggestions regarding pansystems analysis, Dr. Sanchez for his valuable idea regarding the manuscript and the encouragment of my friend Mr. Cao Defun.

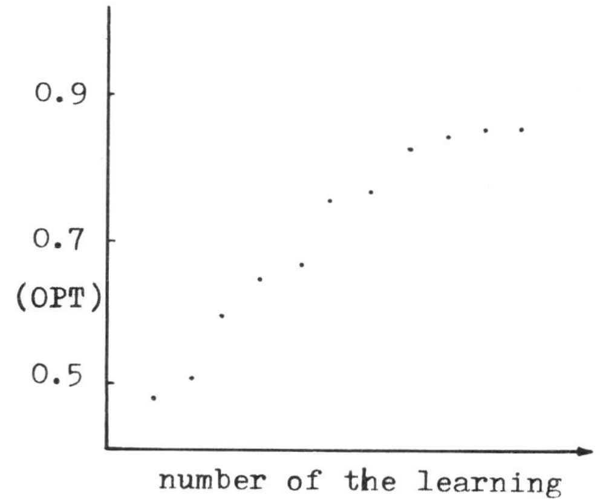

Fig. 2. The graph expresses the increase of OPT after each of learning

TABLE 1 The Fuzzy relation Matrix before and after learning

Before learning				After learning			
.6	.5	.6	.5	.5	.3	.3	.3
.2	.4	.2	.0	.2	.1	.1	.0
.0	.0	.4	.4	.0	.0	.4	.4
.4	.4	.5	.0	.3	.0	.2	.0
.2	.0	.3	.5	.1	.0	.2	.4
.0	.0	.2	.2	.0	.0	.2	.2
.2	.0	.0	.0	.2	.0	.0	.0
.4	.4	.2	.2	.4	.4	.0	.1
.4	.3	.0	.0	.4	.0	.0	.0
.0	.0	.3	.0	.0	.0	.3	.0
.0	.0	.2	.0	.0	.0	.2	.0
.0	.2	.0	.2	.0	.2	.0	.2
.3	.2	.0	.2	.2	.0	.0	.0
.2	.3	.0	.0	.2	.3	.0	.0
.2	.0	.2	.0	.2	.0	.2	.0
.2	.0	.0	.0	.2	.0	.0	.0
.4	.3	.0	.1	.4	.3	.0	.1
.0	.0	.2	.0	.0	.0	.0	.0
.2	.2	.0	.3	.2	.2	.0	.3
.0	.0	.0	.0	.0	.0	.0	.0
.3	.0	.0	.0	.3	.0	.0	.0
.4	.4	.0	.0	.2	.3	.0	.0
.2	.2	.0	.2	.2	.3	.0	.1
.0	.0	.0	.0	.0	.0	.0	.0
.0	.0	.3	.0	.0	.0	.3	.0
.2	.2	.0	.6	.2	.2	.0	.6
.4	.0	.0	.2	.4	.0	.0	.2
.0	.4	.0	.0	.0	.4	.0	.0
.0	.0	.0	.0	.0	.0	.0	.0
.0	.0	.0	.0	.0	.0	.0	.0
.0	.0	.0	.0	.0	.0	.0	.0

continued......

REFERENCE

Kaufmann, A. (1980). Applications des Sous-Ensembles Flous a la Recherche Opertionelle el a La Gestion. Materials of Seminar on fuzzy sets and Applications in Huazhong Inst. of Techn.

Sanchez, E. (1977). Solution in Composite Fuzzy Relation Equations : Application to Medical Diagnosis in Brouwerian Logic. Fuzzy Automata and Decision Process (M.M. Gupta and G.N. Sandis Eds.). North-Holland, New York.

Su Zaifu (1980). The Fuzzy Modal in Disease-Diagnosis (Bianzheng) of Chinese Traditional Medicine. J. of Huazhong Institute of Technology , Special issue on Fuzzy Mathemetics.

Su Zaifu (1981). Application of Fuzzy Relation Equation in Disease-Diagnosis and on Treatment of Chinese traditional Medicine. Science Exploration, 1, pp. 121.

Wang Huaiqing, Su Zaifu (1982). The Application of Fuzzy Mathemetics in Diagnosis by Computer in Traditional Chinese Medicine. Fuzzy Mathematics, 2,1, pp.91.

Wu Xuemou (1982,1983). Pansystem Methodology: Concepts, Theorems and applications (I),(II),(III),(IV). Science Exploration, 1,2,4;1.

THE APPLICATION OF POSSIBILITY THEORY TO THE EVALUATION OF ANAESTHETIC AVAILABILITY OF OPERATION UNDER ACUPUNCTURE ANAESTHESIA

Mian Ouyang* and Su Zaifu*

Department of Mathematics, Wuhan University, Wuhan, Hubei, People's Republic of China

***Department of Physics, Wunan Medical College, Wuhan, Hubei, People's Republic of China*

Keywords. Possibility theory; operation under acupuncture anaesthesia; evaluation of naethetic availability; Pain-Index.

Summary

It is difficult to evaluate the anaethetic availability of operation under acupuncture anaesthesia, because none of the essay method (except lingiustic method) is ready accessible to usual practise. This paper put forward a method which is based on the concept of Zadeh for the purpose of evaluation.

The possibility evaluation are taken as the following:

1. Select a numerical value which expresses the anaethetic availability, we call it Pain-Index and is defined as

$$P = \sum_i 10^{r_i} t_i$$

where the subscript i is a mark to express the step in various operation, and t_i, r_i are records of the corresponding time and pain response respectively. The pain response is devided into 4 grades, which are marked with "0", "+", "++", "+++" to express "no pain", "slight pain", "mid pain" and "strict pain" respectively. And as we know, that the grades are closely related to the pain response, so that we need a value of P, be used as a standard, to evaluate the availability of acupuncture anaethesia.

2. Classify the cases of operation under acupuncture anaesthesia into 4 groups as traditional method, and plot a histogram of PI with respect to the number of cases.

3. We take as S-type function, by the histogram, and draw up a suitable possibility distribution.

4. We assign a qualification to the possibilty distribution and the linguistic representation is expressed by the valus of α-possibility.

For example, we take 26 cases of appendectomy under acupuncture anaethesia as a random sample, and cases is evaluated by the traditional method and ours, the results are the same, but the later is better applied to some special cases.

References

1. Sanchez, E. (1980). Linguistic approach in fuzzy logic of the WHO classification of dyslipoproteinemias. A Paper in International Congress on Applied Systems Research and Cybernetics, Acapulco, Mexico.

2. Zadeh, L.A. (1978). Fuzzy sets as a basic for a theory of possibility. Fuzzy sets and Systems, 10, 3-28.

FUZZY SET THEORY IN MEDICAL SCIENCES

M. M. Gupta, R. R. Martin-Clouaire and P. N. Nikiforuk

Cybernetics Research Laboratory, College of Engineering, University of Saskatchewan, Saskatoon, Saskatchewan, Canada S7N 0W0

Abstract. Fuzziness is an inherent characteristic in all decision making problems especially in medical sciences. The medical information base refers to the physician's knowledge of clues and diseases. The notion of diseases is subject to uncertainty. This uncertainty is due to both the fuzziness arising from biological variations in a clinical class and the ambiguity associated with the practice of summarizing two different collections of clues under a same disease name.

In this paper, we present how the clinical based information can be processed using the fuzzy set theory.

Keywords. Fuzzy set theory; medical diagnosis; medical decision making.

EVOLUTION OF MEDICAL DIAGNOSIS

One of the most important and crucial tasks in the medical sciences consists of finding the patient's disease(s) and is (are) causing discomfort or dysfunction. A patient's history taking and physical examination are diagnostic procedures dating back to antiquity. In the 19th century supplementary procedures were added from the newer fields of roentgenology, microbiology and biochemistry. The use of necropsy and microscopy created new forms of medical evidence. In the late 19th and 20th centuries there were further technological advances in different domains like electron microscopy, biophysics, immunology, physiology. These discoveries changed the nosology and as a consequence the diagnosis techniques involved in the determination of the patient's disease(s) went through changes too.

As the volume of information used by the physician has increased in recent decades, it has become evident that the individual practitioner can no longer hope to acquire enough expertise to manage adequately the full range of clinical problems he will encounter. In other words, modern medicine has to deal with the organization and analysis of large amounts of complex data. The digital computer can be one of the doctor's most useful assistants in helping solve these problems, and its potential capabilities are more and more attractive since its price, power, size and reliability have improved dramatically.

Besides, the advent of the digital computer has been a great stimulus for trying to systematize the thought process which the clinician uses every day but cannot define. One of the first attempts utilizing the computer as an aid in medical diagnosis for a partial simulation of its process was made by Ledley and Lusted in 1959 [1]. They proposed that this problem might be considered in terms of symbolic logic and conditional probability (Bayes' theorem). A few well defined problem areas of medicine can be managed effectively by this technique as in [2], but it has its limitations too because statistics and incidences of clues and diseases often differ. This is due to the fact that in most cases diseases are described qualitatively but are not clearly defined so as to eliminate all other possibilities. Many prior discussions of the shortcomings of Bayesian inference have concentrated on the lack of independence of many of the conditional probabilities.

FUZZY SETS IN MEDICAL SCIENCES

Thus, research on medical diagnosis has served to emphasize the need for better methods of collecting, coding and processing medical information and to demonstrate the inadequacy of conventional mathematical methods for dealing with such problems. In a 1962 paper, Professor Zadeh [3] summed up the situation as follows: "In fact, there is a fairly wide gap between what might be regarded as 'animate' system theorists and 'inanimate' systems theorists at the present time, and it is not at all certain that this gap will be narrowed, much less closed, in the near future. There are some who feel that this gap reflects the fundamental inadequacy of the conventional mathematics, the mathematics of precisely defined points, functions, sets, probability measures, etc..., for coping with the analysis of biological systems and that to deal effectively with such systems, which are generally orders of magnitude more complex than man made systems, we need a radically different kind of mathematics, the mathematics of fuzzy or cloudy quantities which are not describable in terms of probability distributions". A couple of years later, Zadeh [4,5] provided the mathematical formalism of the fuzzy set theory which has the ultimate aim, according to him [6], to represent how the human mind perceives and manipulates imprecise and qualitative information.

The use of fuzzy sets for medical diagnosis was first suggested by Zadeh himself [7]. Sanchez [8,9] studied the representation of medical knowledge by means of fuzzy relations. Sambuc [10] defined and applied the concept of ϕ-fuzzy functions to thyroid pathology. Fuzzy clustering algorithms [11] have been tested on different classifications as in [12] of ECG patterns, in [13] of chromosomes and in [14] for the diagnosis of hypertension. Smets [15] used Shafer's belief functions [16] in order to model the symptom-disease relationship. Artificial intelligence approaches have been outlined by Wechsler [17] and Shortliffe [18]. The latter used production rules as a representation for fuzzy knowledge and applied his method to infectious diseases. Lastly, Sanchez et al. [19] described an application to the classification of dyslipoproteinemias based on the Zadeh's possibility measure [20].

ACKNOWLEDGEMENTS

This work is supported by the NSERC of Canada under grants A-5625 and A-1080.

REFERENCES

[1] Ledley, R.S. and Lusted, L.B. (1959). "Reasoning Foundations of Medical Diagnosis", Science, 130:9-21.
[2] Bishop, C.R. and H.R. Warner (1969). "A Mathematical Approach of Medical Diagnosis: Application of Polycythemic States Utilizing Clinical Findings with Values Continuously Distributed", Comp. Biomed. Res., 2:486-493.
[3] Zadeh, L. (1962). "From Circuit Theory to System Theory", Proc. IRE, 50:856.
[4] Zadeh, L.A. (1965). "Fuzzy Sets", Memo. ERL, No. 64-44, Univ. of California, Berkeley.
[5] Zadeh, L.A. (1965). "Fuzzy Sets", Information and Control, 8:338-353.
[6] Zadeh, L.A. (1973). "Outline of a New Approach to the Analysis of Complex Systems and Decision Processes", IEEE Trans. Syst. Man, Cybern., 3:28-44.
[7] Zadeh, L.A. (1969). "Biological Application of the Theory of Fuzzy Sets and Systems", In Biocybernetics of the Central Nervous System (PROCTOR, L.D. ed.), Little Brown, Boston, Mass., 199-212.
[8] Sanchez, E. (1977). "Solutions in Composite Fuzzy Relation Equations: Application to Medical Diagnosis in Browerian Logic", In Fuzzy Automata and Decision Processes (GUPTA, M.M. et al. eds), North Holland: 221-234.
[9] Sanchez, E. (1976) "Resolution of Composite Fuzzy Relation Equations", Information and Control, 30:38-48.
[10] Sambuc, R. (1975). "Fonction ϕ-Floues Applications à l'Aide au Diagnostic en Pathologie Thyroidienne", Thèse de Médecine, Univ. de Marseille.
[11] Bezdek, J. (1977). "Prototype Classification and Feature Selection with Fuzzy Sets", IEEE Trans. Syst. Man and Cybern., 7:87-92.
[12] Albin, M. (1975). "Fuzzy Sets and Their Application to Medical Diagnosis and Pattern Recognition", Ph.D. Thesis, Univ. of California, Berkeley.
[13] Lee, E.T. (1975). "Shape Oriented Chromosome Classification", IEEE Trans. Syst. Man, Cybern. 5: 629-632.
[14] Fordon, W.A. (1976). "Computer-Aided Differential Diagnosis of Hypertension", Ph.D. Thesis, Purdue University, West Lafayette, In.
[15] Smets, Ph. (1981). "Medical Diagnosis: Fuzzy Sets and Degree of Belief", Fuzzy Sets and Systems, 5:259-266.
[16] Shafer, G. (1976). "A Mathematical Theory of Evidence", Princeton University Press, Princeton, N.J.
[17] Wechsler, H. (1976). "A Fuzzy Approach to Medical Diagnosis", Int. J. Biomed. Comput., 7:191-203.
[18] Shortliffe, E.H. (1976). "Computer-Based Medical Consultations: MYCIN", American Elsevier, North Holland, New York.
[19] Sanchez, E., J. Gouvernet, R. Bartolin, and L. Vovan (1981). "Linguistic Approach in Fuzzy Logic of the WHO Classification of Dyslipoproteinemias". In Recent Developments in Fuzzy Set and Possibility Theory (YAGER, R.R. ed), Pergamon Press, Elmsford, New York.
[20] Zadeh, L.A. (1978). "Fuzzy Sets as a Basis for a Theory of Possibility", Fuzzy Sets and Systems 1:3-28.
[21] Martin-Clouaire, R.R. (1982). "A Computer-Aided Medical Diagnosis Method Based on a Fuzzy Set Theoretical Approach". M.Sc. Thesis, Univ. of Saskatchewan.

EMERGENCE AND POTENTIALITY IN INTERCONNECTED SYSTEMS

I. Redon*, D. Morand* and C. Gaudeau**

**Ecole Supérieure d'Informatique Electronique Automatique, 9 rue Vésale, 75005 Paris, France*
***Laboratoire de Bio-Informatique (Laboratorie de Physiologie), U.E.R. Médecine, Université François Rabelais, 2 bis Bd Tonnellé, 37032 Tours, France*

Abstract. This paper deals with the notion of potentiality and the "emergence" of new properties when systems are interconnected. Several methods are considered to show emergence. First, we describe an aspect of the mereologie : fragments. Then we shall show " emergence" by transfert function with two methods. First, we introduce potential variables in the input and output, complex numbers in the state representations matrix and after fuzzy systems.

Keywords. System theory, observability, fuzzy set theory, modelling, biomedical.

INTRODUCTION

The approach of the concept of potentiality in life systems or physical systems has been most often done by using the system theory. This approach allows to represent biological and physiological functions and to describe the complexity of self-organized systems.

Nevertheless, the method used in systemics take into account only the parts of the system which can be controlled at the input level and the parts that can be observed on the output : hence these methods are not complete.

They are not complete because there exist potentialities or latent properties in the isolated systems. These properties, that can't be at first observed, may appear when several systems are associated. We shall say then that there is " emergence" of new potentialities. (QUASTHER G., 1970)

The " mereologie" (LAFORGE) is a logical approach which can be applied to complex modelling systems with variables that can't be observed. We shall deal more accurately later with one of the aspects of this theory : the breakdown of systems into fragments (LAFORGE 1973-74-77; LAFORGE & GAUDEAU, 1977).
We shall then show different methods which allow demonstration of emergence of new potentialities when several systems are associated We shall end with the fuzzy systems theory (ZADEH, SANCHEZ) that is to be compared with the other methods.

I. FRAGMENTS

We shall break down the systems into fragments which allow to account for the non-observable variables. (GAUDEAU, CRENDAL & Coll. 1979)

I.1 System representation

As it is shown in figure 1, four systems appear.
S1 can be observed and is controllable
S2 can't be observed but is controllable
S3 can't be controllable but is observed

S4 is neither observable neither controllable.

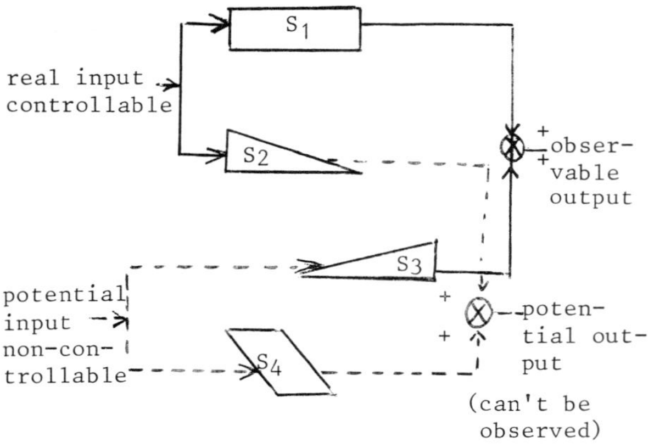

Fig. 1 Break down of a system into fragments.

The diagram demonstrates the fact that the observable output may partially depend on non-observable but potential input and in the same way the control of an observable input may influence a non-observable output. This model is then much more complete.

I.2 Association of systems

The association of several systems will show clearly the fact that some observable variables depend on non-observable variables.

Consider two compatibles systems, S and S'. Let us associate the non-observable part of S to the non-controlable part of S' as shown in fig. 2.(COBELLI, 1976 ; CHI-TSONG CHEN, 1967, 68)

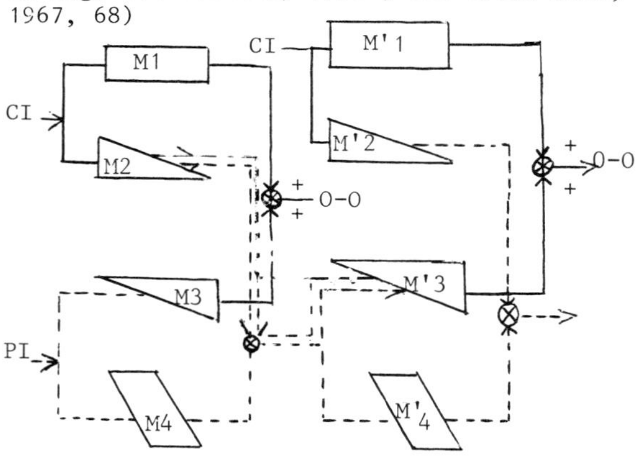

Fig. 2. Association of 2 systems S and S' broken down into fragments.

O-O : observable output P-I : potential input
P-O : potential output
C-I : commandable input

The double dotted path shows that the association of two fragments M2 and M'3 allows demonstration of a phenomenon which first was only potential, i.e. the inobservable output of system S.
The break down into fragments allows to model mathematically the possible emergence of new properties. The method can be applied to several domains, particularly to biomedical domain throughout the compartments theory.

II. TRANSFER MATRIX

II.1 Creating potential variables at the input and output levels

We may demonstrate emergence (of new potentialities) when several systems are associated, by breaking down the input and output variables (for every system) into 2 parts :

- the observable one (indexed 0)
- the non-observable one (indexed I)

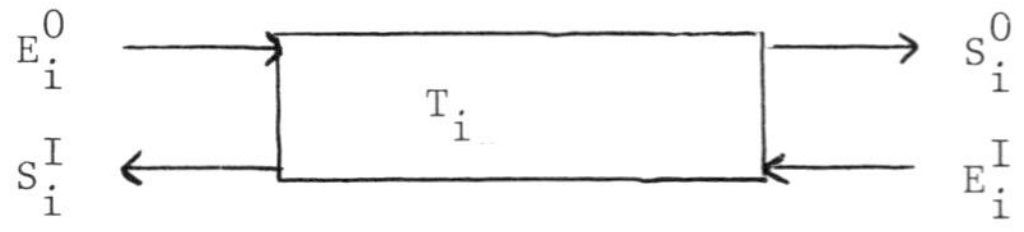

Fig. 3 system representation

For every system i, let T_i be the transfer matrix :

$$T_i = \begin{pmatrix} K & L \\ M & N \end{pmatrix}; \text{ K, L, M, N are four matrical blocks}$$

Remark If $E_i^I = 0$; the matricial equation becomes : $S_i^0 = K_i \ E_i^0$, equation used in the classical system theory.

Consider the inter-connection of 2 defined systems as shown in fig. 4.

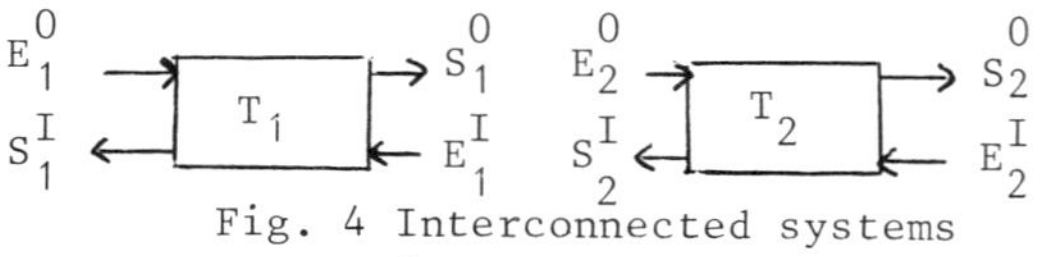

Fig. 4 Interconnected systems

The input $E = \begin{pmatrix} E_1^0 \\ E_2^I \end{pmatrix}$ of the bulksystem gives birth to an output $S : \begin{pmatrix} S_2^0 \\ S_1^I \end{pmatrix}$.

If we take into account the matricial equations $\begin{pmatrix} S_1^0 \\ S_1^I \end{pmatrix} = \begin{pmatrix} K_1 & L_1 \\ M_1 & N_1 \end{pmatrix} \begin{pmatrix} E_1^0 \\ E_1^I \end{pmatrix}$

and $\begin{pmatrix} S_2^0 \\ S_2^I \end{pmatrix} = \begin{pmatrix} K_2 & L_2 \\ M_2 & N_2 \end{pmatrix} \begin{pmatrix} E_2^0 \\ E_2^I \end{pmatrix}$

With $S_1^0 = E_2^0$ and $S_2^I = E_1^I$; and if we assume that $E_2^I = 0$, we obtain :

$$\boxed{S_2^0 = K_2 \, G \, K_1 \, E_1^0}$$

This can be represented by :

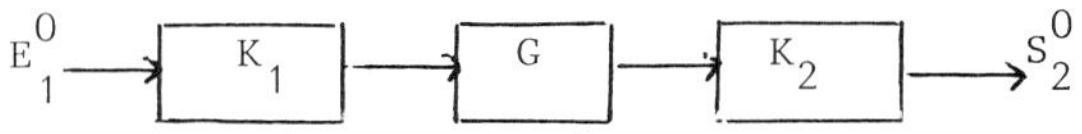

Where $G = (\, Id - L_1 M_2)^{-1}$

Remark. We used this representation to model rumour phenomenon and we obtained quite satisfying results.

In the classical theory we should have had :

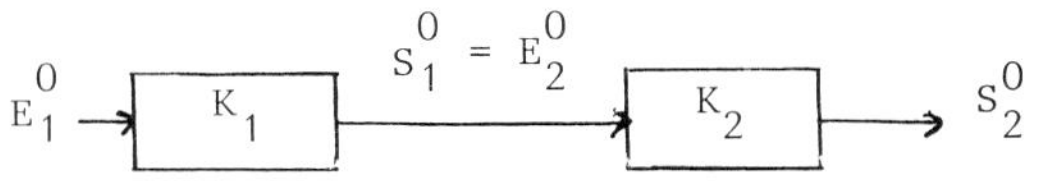

Where : $S_2^0 = K_2 \, K_1 \, E_1^0$

The G matrix demonstrates the part of S_1 and S_2 which was first inobservable but potential It characterizes then the emergence of a potentiality shown by the interconnection.

II.2 Introducing complex numbers in the state representation matrices (CORBELLI, 1976)

The representation of a system by its state equations (JORDAN) is given by :

$$\dot{X} = AX + BE$$
$$Y = CX + DE$$

E set of the inputs X set of the states Y set of the outputs

The transfer function depends on A, B, C, D.

$$G\,(\,S) = C\,(\,pI - A\,)^{-1}\, B + D$$

We shall break down every matrix into a real part and an imaginary part.

e.g : $A = A^0 + iA^i$

A^0 real part (—> observable part)

A^i imaginary part (—> non-observable part)

All the matrices indexed ^{0}, are observable

All the matrices indexed ^{i}, can't be observed.

Let us consider the classical association of two systems as shown in fig. 5.

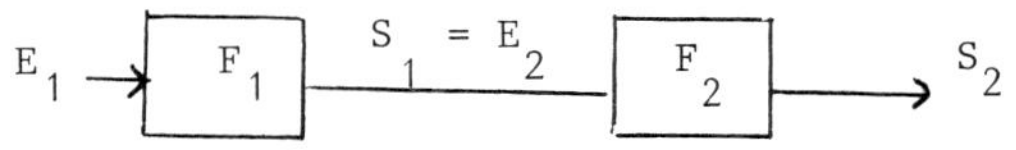

Fig. 5. System association

For both system 1 and 2, we break down the matrix of state representation. This leads to the following equations :

$$\text{I} \begin{cases} \dot{X}_1^0 = (A_1^0 + i\,A_1^I)\,X_1 + (B_1^0 + i\,B_1^i)\,E_1 \\ S_1 = (C_1^0 + i\,C_1^I)\,X_1 + (D_1^0 + i\,D_1^i)\,E_1 \end{cases}$$

And if $S_1 = E_2$

$$\text{II} \begin{cases} \dot{X}_2^0 = (A_2^0 + i\,A_2^i)\,X_2 + (B_2^0 + iB_2^i) \\ \qquad \left[(C_1^0 + iC_1^i)X_1 + (D_1^0 + iD_1^i)\,E_1\right] \\ S_2 = (C_2^0 + iC_2^i)\,X_2 + (D_2^0 + iD_2^i) \\ \qquad \left[(C_1^0 + iC_1^i)\,X_1 + (D_1^0 + iD_1^i)\,E_1\right] \end{cases}$$

The system II leads to the following equation

$$S_2 = \left[\,(D_2^0 C_1^C - D_2^i C_1^i) + i\,(D_2^0 C_1^i + D_2^i C_1^0)\right] X_1$$
$$+ \,(\,C_2^0 + i\,C_2^i\,)\,X_2$$
$$+ \left[(D_2^0 D_1^0 - D_2^i D_1^i) + i\,(D_2^i D_1^0 + D_2^0 D_1^i)\right] E_1$$

And we have :

$$\text{Re}\,(S_2) = (\,D_2^0 C_1^0 - D_2^i C_1^i)X_1$$
$$+ C_2^0 X_2$$
$$+ (D_2^0 D_1^0 - D_2^i D_1^i)\,E_1$$

We can see that the observable part of S_2 (that is to say the real part: $\text{Re}(S_2)$) depends on the observable parts of D_2, C_1 and D_1. If we associate the system S_1 with several other systems, with particular conditions, we can find D_1^i and C_1^i.

Let H be the transfer function of the bulk system. With similar computations we can show that :

$$Re(H) \neq Re(F_1) \times Re(F_2)$$

Where Re (H) is the observable part of H Re (F_1) and Re (F_2) are respectively the transfer functions of the isolated systems 1 and 2.

The observable part of H also depends on some inobservable parts of both systems.

III. FUZZY SYSTEMS

In the system theory every property is true or false. There is no intermediate possibility between these two limits. We can think, that a theory which would take into account the intermediate possibilities (i.e. some non-observable or potential properties) would allow to demonstrate the " emergence" of new properties when several systems are associated. That's why the fuzzy systems theory based on concept of fuzzy sets seems to be interesting. (ZADEH, 1965 ; SANCHEZ 1976)

III.1 State representation

The basic idea of fuzzy sets theory is the concept of "appurtenance function". It's a kind of extension of the concept of " characteristic function" in the classical sets theory.

Let A be a non fuzzy set and μ_A its characteristic function μ_A is defined by :

$$\begin{cases} \mu_A(x) = 1 \text{ if } x \in A \\ \mu_A(x) = 0 \text{ if } x \notin A \end{cases}$$

Let $\underset{\sim}{A}$ be the fuzzy set on A and $\mu_{\underset{\sim}{A}}$ it's " appartenance function".

$$\mu_{\underset{\sim}{A}}(x) = i \quad \text{with } 0 \leqslant i \leqslant 1$$

means that x belong to A with a degree : i.

The discrete time state representation of a deterministic system is governed by the equations :

$$\begin{cases} x(t+1) = f(x(t),u(t)) \\ y(t) = g(x(t),u(t)) \end{cases}$$

Where f is the transition function and g is the output function.

x : state variable. $x \in X$: set of the states

u : input variable. $u \in U$: set of the inputs

y : output variable. $y \in V$: set of the outputs

If the system is non deterministic, it will be described as :

$$\begin{cases} A(t+1) = F(x(t), u(t)) \\ B(t) = G(x(t), u(t)) \end{cases}$$

with $A(t+1) \in P(X)$ and $B(t) \in P(V)$ P(X) and P(V) are respectively the sets of the subsets of X and V.

A fuzzy system will be described as follows:

$$\text{I} \begin{cases} \mu_{A(t+1)}(x(t+1)) = \mu_{\underset{\sim}{A}(t)\, 0\, \underset{\sim}{R}(t)}(x(t);x(t+1)/x(t),u(t)) \\ \mu_{B(t)}(y(t)) = \mu_{\underset{\sim}{A}(t)\, 0\, \underset{\sim}{T}(t)}(x(t);y(t)/x(t),u(t)) \end{cases}$$

We can remark that A (t+1) becomes $\underset{\sim}{A}$ (t+1), fuzzy set on X, and B(t) becomes $\underset{\sim}{B}(t)$: fuzzy set on V. f and g are replaced by the fuzzy relations R and T. The " appurtenance functions " associated with R and T are μ_R and μ_T.

$\mu_R(x(t+1) / x(t),u(t))$ transition from x at time t to x at time t+1 parameterized by u.

$\mu_T(y(t) / x(t), u(t))$ relation between x and y parameterized by u.

I is equivalent to II (because of the definition of the composition of fuzzy relations).

$$\text{II} \begin{cases} \mu_{A(t+1)}(x(t+1)) = \sup_{x(t)} (\min(\mu_{\underset{\sim}{A}(t)}(x(t)); \mu_{\underset{\sim}{R}(t)}(x(t+1);x(t),u(t)) \\ \mu_{B(t)}(y(t)) = \sup_{x(t)} (\min(\mu_{\underset{\sim}{A}}(x(t)); \mu_{\underset{\sim}{T}}(y(t)/x(t),u(t)) \end{cases}$$

We can express the fact that inputs can be observed by considering the set of the inputs as a fuzzy set. We assume then that there exists at each time a dependance between the states and the inputs. This dependance is expressed by the fuzzy relation : $\underset{\sim}{Q}(t)$.

The fuzzy system is then governed by :

$$\text{II}\begin{cases}\mu_{\underset{\sim}{A}(t+1)}(x(t+1))= \sup_{u(t)}(\sup_{x(t)} (\min(\mu_{\underset{\sim}{Q}(t)}(x(t),u(t)) ; \\ \quad \mu_R(x(t+1)/x(t),u(t))) \\ \mu_{\underset{\sim}{B}(t)}(y(t))= \sup_{u(t)} (\sup_{x(t)} (\min(\mu_{\underset{\sim}{Q}(t)}(x(t),u(t)) ; \\ \quad \mu_T(y(t)/x(t),u(t)))\end{cases}$$

Remark : $\mu_{\underset{\sim}{A}(t)}$ is replaced there by $\mu_{\underset{\sim}{Q}(t)}$ and another sup appears.

If the inputs and the states are not interactive then :

$$\mu_{\underset{\sim}{Q}(t)}(x(t),u(t))= \min (\mu (x(t), \mu (u(t)))$$

Association of fuzzy systems

We associate two fuzzy sets W and W' to reach a shared goal. We assume that W and W' have fuzzy inputs parameterized by states. This is represented by the fuzzy relations $\underset{\sim}{Q}(t)$ and $\underset{\sim}{Q}'(t)$ respectively for W and W'.

We also assume that the state of W influence W' and reciprocally. We represent this influence by the fuzzy transition relations $\underset{\sim}{R}(x(t))$ for W and $\underset{\sim}{R}'(x(t))$ for W'.

The association of W ans W' is represented as follows :

$$\mu_{A(t+1)}(x(t+1)) = \sup_{u(t)} (\sup_{x(t)} (\min/(\mu_{\underset{\sim}{Q}(t)}(x(t),u(t)); \mu_{\underset{\sim}{R}}(x(t+1)/x(t),u(t)))$$

And we have the same equation with A', W',Q', and R'.

The two systems will try to evolve on shared states to reach their goal. Then, they must be compatible. In the best case, the fuzzy sets of W and W' are the same. Compatibility can be expressed by the Hamming's distance which is the distance currently used in the fuzzy sets theory.

The concept of potentiality is then introduced at 3 levels :

- first level : the states.
  X is the set of all the possible states of W and W' as well as those that may be reached during the experiment.
- second level : the inputs.
  There is a dependance between the inputs and the states.
- third level :
  W influence W' and reciprocally. It is interesting to introduce the fuzzy systems to demonstrate potentiality but the implementation is difficult and complex.

CONCLUSION

Different methods, interesting in a mathematical point of view, allow the modelling of emergence (either at the level of the variables, or at the level of the systems). On the other hand, these methods fast become complex and difficult to implement when the number of systems associated is too high.

Similar approaches to potentiality have been realized by Leguizamon in using the "category " theory and the concept of extrinsic energy considered as a form of non observable potentiality. (LEGUIZAMON , 1980 - 1982)

According to Lupasco, the concept of potentiality seems to be useful in (the domain of) quantum mechanics (potentiality of the momentum and energy). Lupasco thinks that the concept of potentiality would allow to define an equilibrium position of the physical as well as biological systems.(LUPASCO, 1941, 1951, 1970 ; NICOLESCU, 1982)

We thanks very much Claudia REBOCHO for collaboration. And Société de Bio-Informatique for help.

REFERENCES

QUASTHER G.(1970)
Emergence in Biological organisations.
Yale University Press.

NICHOLSON S.(1972)
Structure of interconnected systems.
IEEE Control Engineering

SANTOS E.S. (1973)
Fuzzy sequential functions.
J.CYBERN 3 N° 3. 15-31

NEGOITA C.V and RALESCU, D.A (1974) Fuzzy systems and artificial intelligence. Kybernotes 3 173-178

ZADEH L.A (1965) Fuzzy sets and systems Proc. Symp. Syst. Theory Polytech. Inst. BROOKLYN N.Y. pp. 29-39.

BIBLIOGRAPHIE

CHI-TSONG CHEN., and CHARLES A., DESOER., (1967) Contracbility and Observability of Composite Systems IEEE Transactions on Automatic Control.Vol. AC-12 N° 14 August

CHI-TSONG CHEN., (1968) Representation of Linear Time Invariant Composite IEEE Transactions on Automatic Control. Vol. 13 N°3 June.

COBELLI C., GIORGIO ROMANIN., JACUR (1976) Controllability, observability and Structural Identifiability of Multy-input and Multy-output Biological Compartimental Systems IEEE Transactions on Biomedical Engineering.Vol. BME 23 N°2 March.

GAUDEAU C., (1975) Sur la Reconnaissances des états d'un système physiologique et de ces processus aléatoires. Identification par des modèles stochastiques.(Application aux systèmes cardio-vasculaires et digestif. TH. Biol. Hum. Université de TOURS

GAUDEAU C., CRENDAL W., VINCENT S.,GALEA J.M

LAFORGE J.M., (1979)Notion de fragment et de territoire dans l'identification des Bio-systèmes complexes R.B.M Vol. 1, N° 5 Sept/Oct. p. 377-379.

LAFORGE J.M, (1973) Classe méréolotique et Construction logique PH. D. Université de GAND.

LAFORGE J.M, (1974)Fondements pour une méréologie ensembliste Rev. Logique et Analyse N° 17 p. 65-66.

LAFORGE J.M, (1977) Méréologie logique des blocs pseudo-topologiques sur des repères Comm. ARK'ALL . N°2 p. 89-110.

LAFORGE J.M, (1977) Méréologie logique des blocs pseudo-topologiques sur des repères Comm. 6ème Congrés International de Biologie Mathématique N° 61-78 p. 21-42.

LAFORGE J.M, et GAUDEAU C., (1977) Notions de fragment et de territoire dans l'analyse des systèmes complexes C.R Congrés A.F.C.E.T Maitrise des systèmes,Versailles

LEGUIZAMON A.C, (1980) Concept of Energy in Biological Systems and the Effects of Irradiations of Low Energies on Enzyme-Substrate Systems Bulletin of Mathematical Biology. Vol. 42 p. 161-172 Pergamon Press Ltd.

LEGUIZAMON A.C, (1982)Teoria de Categorias en Biologia Relacional. Anal. Acad. Nac. Cs. Ex; Fis. Nat. Buenos Aires, Tomo 34.

LUPASCO S., l'expérience micro-physique et la pensée Humaine (1941)P.U.F PARIS.

LUPASCO S.(1951) Le principe d'antagonisme et la logique de l'energie (prolegonismes à une science de la Contradiction) Hermann,Collection " Actualités Scientifiques et Industrielles " N° 1133.

LUPASCO S. (1970) Les Trois Matières 10/18 Julliard p. 150.

NICOLESCU B.(1982) LUPASCO et la Genèse de la Réalité, 3 Mill. N° 3.

SANCHEZ E.,SAMBUC R.(1976)"Relations Floues Fonctions O-floues.Application à l'aide au diagnostic en pathologie thyrodienne". Journée d'informatique Médicale de TOULOUSE. Proc: TAYLOR & FRANCIS Ltd. LONDON.

DECISION ANALYSIS OF A NEW KIND OF VISUAL ILLUSION

Ma Mou-chao* and Lin Lai-fu**

**Institute of Psychology, Academia Sinica, Beijing, People's Republic of China*
***Department of Mathematics, Beijing Normal University, Beijing, People's Republic of China*

ABSTRACT. 1. The Multistage Evaluation Method, used in the present investigation, was based on binary fuzzy sets theory. It is suitable not only to quantify fuzziness of visual illusion, but also to make decision analysis on it.
2. This new geometrical illusion is composed of two components, i.e. orientation and contrast. The compositive operation of fuzzy matrix was used to describe their interaction.
3. Operator of Max-Min was proved to be better for description of this interaction.

Keywords: Decision analysis; visual illusion; empirical operator; compositing experiment; human information processing.

INTRODUCTION

Former investigations have indicated that any illusion is always related with some factor or factors. To find and to establish this relationship is an important aspect of illusion study. Kummapas, T. M (1955); Weintraub, D.J. and Krantz, D. H(1971) have carried out the empirical formulas that are used to predict the magnitude of this illusion. But it is obvious that the prediction dealt with only the relationship between the magnitude of the geometrical illusion and the individul factor. In respect to decision analysis of complex geometrical illusion, a significant report rarely appeared. One of the reasons seems to be lacking a more suitable technique that is able to reveal the implication of the illusion.

Recently, a new kind of Multistage Evaluation Method based on fuzzy sets model of category judgment has been proved to be useful to measure psychological fuzziness (Ma Mou-chao 1982; Ma Mou-chao & Cao Zhi-qiang 1982 a.b; Ma Mou-chao & Cao Zhi-qiang 1983). Fortunitely, it was successful that application of this technique to measure visual illusion (Ma Mou-chao & Liu Lai-fu 1983). In this way it is natural to explore further how to make decision analysis on it and how to establish an appropriate mathematical model, if it is possible.

PROCEDURE OF THE MULTISTAGE EVALUATION OF VISUAL ILLUSION

When the Multistage Evaluation Method was used to measure an illusion, it is necessary to establish a scale of illusion, where the scale was divided in to several (e.g. five) categories, labelling as $\ll$ $<$ $=$ $>$ $\gg$. Because the categories in terms of natural language were fuzzy conceptions, it might be imagined that there was parellel physical continuum (i.e., fuzzy subsets) to correspond with them. For the sake of simplifying computation, we replaced those fuzzy subsets with these number -2, -1, 0, +1, +2, respectively. In case of the measurement observer is asked to establish a correspondence between perceptions and catogories, in which the degree of correspondence was decribed by number between 0 and 10. Concrete procedure: first, 78 observers were required to choose one from the assigned categoies in which there was the highest correspondence of all. Then, the paired comparision was put down at two categories of abjacent peak until all categories are evaluated. The evaluation could be assignment cf score directly according to own judgment, and could also be oral decription in terms of adjective. After that, it was convenient to transform those adjective into scores. There was a relstionship of the transformation between adjectives and scores the following:

complete	noncorrespondence	0
moderate	noncorrespondence	1-2
little	noncorrespondence	3-4
medium		5
little	correspondence	6-7
moderate	correspondence	8-9
complete	correspondence	10

The mean of the observers' the scores was obtained and then it was divided by 10, and the result was transformed into number in close interval [0, 1] which was membership grade on each category.

THE MATHEMATICAL MODEL OF VISUAL ILLUSION

Assuming that U is all of n categories on visual scale. Fuzzy values (here, representative values) of category boundaries on the scale are denoted as x_i (i=1,2,...n). A geometrical illusion configuration G can induce a fuzzy illusion $\underset{\sim}{G}$ which is a fuzzy set in universe of discourse U, and its membership function is $\mu_{\underset{\sim}{G}}(x)$, $x \in U$. It can also be renoted

$$\underset{\sim}{G}=\sum_{i=1}^{n}\mu_i \big| x_i$$

by Zadeh's notation, where x_i denote some categories in U; μ_i are grades of membership of $\underset{\sim}{G}$ on x_i

We call

$$\Delta x=\left\{\frac{1}{A_0}\sum_{i=1}^{n}x_i\mu_i-x_0\right\},\quad A_0=\sum_{i=1}^{n}\mu_i$$

the strength of induced illusion where x_0 is the ratio of comparing quantity to standard one in the geometrical configuration G; Δx gives an average index of degree of illusion, which can be either positive or negative. When $\Delta x>o$ the comparing part in the configuration is overestimation, and if $\Delta x<o$ it is underestimation. Under the condition of the effect of illusive factor a geometrical configuration in which the illusive cue is involved induces corresponding illusion. The former can be regarded as a fuzzy set on U, and the later can also be thought as anotherone on U. Therefore this effect of the illusive factor can be regarded as a fuzzy transformation from U into itself. For any $\underset{\sim}{A}\in\mathcal{F}(U)$ there exists a corresponding fuzzy set $B=TA\in\mathcal{F}(U)$ If we assume $\underset{\sim}{R}$ as a binary fuzzy relations on U determined by this illusive factor, then there is

$$\underset{\sim}{B}=\underset{\sim}{R}\circ\underset{\sim}{A}$$

where

Under the condition of some illusive factor, the binary fuzzy relation on U can be demonstrated by membership grades on every category in the scale which are obtained by the Multistage Evaluation Method.

The operators V^* Λ^* in fuzzy transformation are 'and' and 'or' of fuzzy set. Basing on data obtained, the degree of fit can be compared for different operators in the field of fuzzy sets.

1. Probabilistic Sum product $\hat{+}$ $\cdot$
2. Bezdek Harris $\boxplus$ $\boxdot$
3. Hamacher $\overset{+}{\cup}_p$ $\dot{\cup}_p$
4. Dinstain $\overset{+}{\xi}$ $\dot{\xi}$
5. Yager $\overset{Y}{\smile}_p$ $\overset{\lambda}{}_p$
6. Zadch $\vee$ $\wedge$

(these formulas see following appendix*)

Therefore, the effect of an illusive cue is demonstrated by a binary fuzzy relation R on U. Different effects of illusive cue can be described by different fuzzy relations.

If a geometrical illusive configuration is compounded by two illusive cues, we denote fuzzy relations corresponding cues as R, S, then the fuzzy transformation producted by this goometrical illusive configuration can be regarded as a composition of these two fuzzy relations, Ro or SoR.

For the purpose of comparision, theoritical curve with observational one degree of nearness is given according to the following formula

$$\rho_{\underset{\sim}{A}\cdot\underset{\sim}{B}}=1-\sum_{i=1}^{n}\frac{\mu_{\underset{\sim}{A}\cup\underset{\sim}{B}}(x_i)}{\sum_{i=1}^{n}\mu_{\underset{\sim}{A}\cup\underset{\sim}{B}}(x_i)}\left|\mu_{\underset{\sim}{A}}(x_i)-\mu_{\underset{\sim}{B}}(x_i)\right|$$

$$\mu_{\underset{\sim}{A}}=\sum_{i=1}^{n}\mu_{\underset{\sim}{A}}(x_i)\Big|\,x_i\,,\quad \mu_{\underset{\sim}{B}}=\sum_{i=1}^{n}\mu_{\underset{\sim}{B}}(x_i)\Big|\,x_i$$

$$\mu_{\underset{\sim}{A}\cup\underset{\sim}{B}}(x_i)=\sum_{i=1}^{n}\max\left[\mu_{\underset{\sim}{A}}(x_i),\ \mu_{\underset{\sim}{B}}(x_i)\right]\Big|\,x_i$$

EXPERIMENT AND ITS RESULT

a. Comparision of strength of the illusion for several configurations. These configurations used by the present invesitigation were shown in fig. I.A. B. C. D. E.

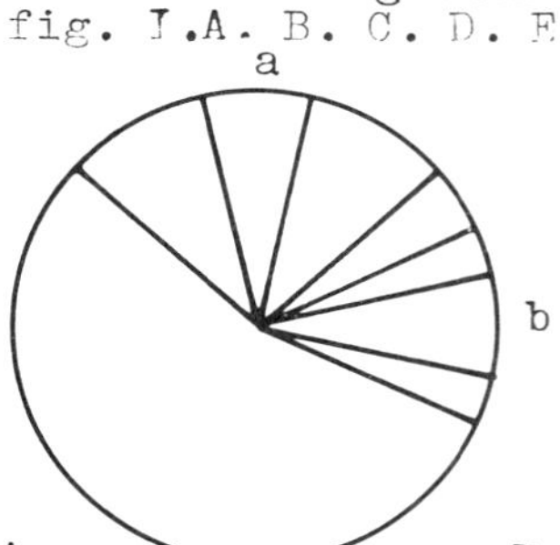

A

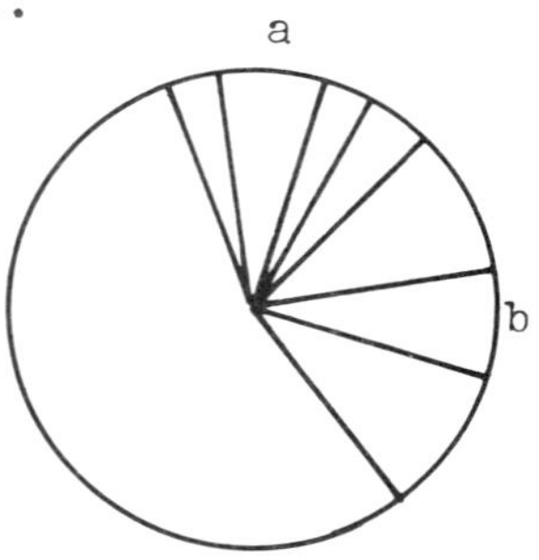

B

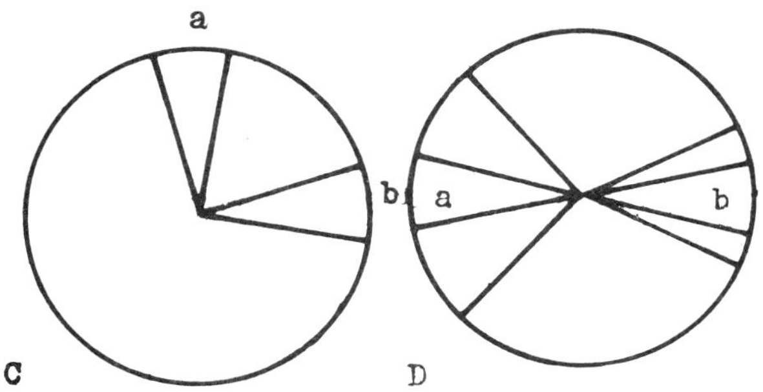

C D

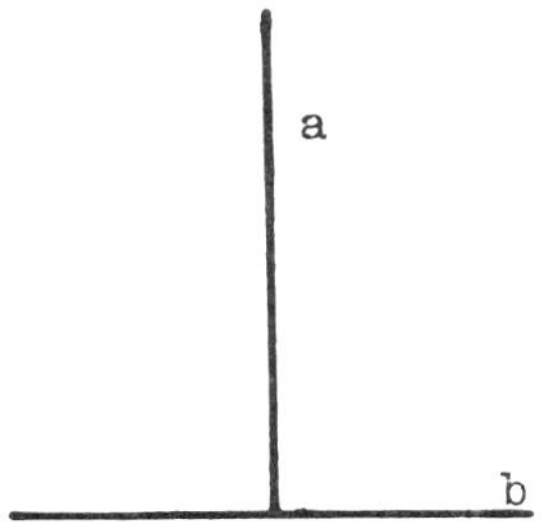

E

let's call configuration A "positive composition"; B "negative composition"; C "orientation"; D "contrast". The configuration E is the well known vertical-horizontal illusion. In any given configuration "a" denotes the comparing stimulus, and "b" denotes standard one.

78 observers participated the trials. 45 of them were pupils from secondary school. The others were students, The former and the latter conducted the trials separatly. The configurations were orthogonal to observer's line of vision. There was no limit of time of seeing the configuations for observer.

Table 1. The comparison of the strenght of for the give configurations

Ss \ config.	A	B	C	D	E
Students	.52	.47	.25	.29	*.84
Pupils	.30	.27	.17	.04	.48

*datum is from another article c.f. Ma Mou-chao & Liu Lai-fu (1983)

It was seen clearly from table I that the order of strength of any given illusion from the horizontal illusion configuration, "positive composition", "negative composition", and individual cues either "orientation" after "contrast" or vise versa.

b. "composition" model's experiment. In those configurations (except vertical-horizontal). A and B seem to be composited of two components C and D in different ways. Now the problem to be solved was what rule this composition obeys. Assuming that the composition was the interaction between both components. In this way, the experimental design of a matrix 5×5 order was adopted; 5 experimental variables were devoted to each components. In addtion to these, there were also two compositive effects to be considered. The total included 12 trials in which each observer has to accomplish.

In order to make the above assumption more acceptable the configurations A and B were modified as. A' and B' in Fig2.

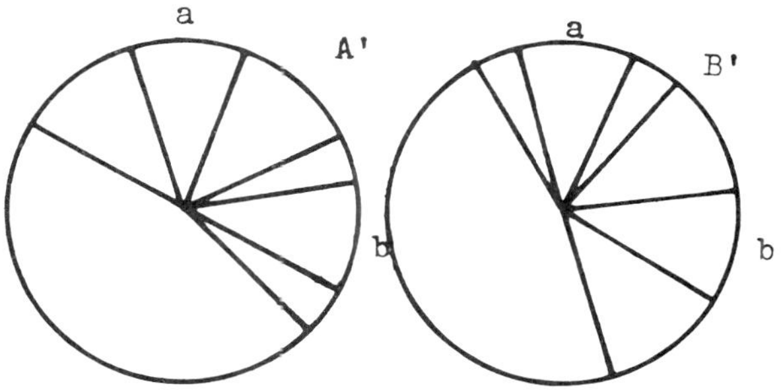

Basing on data obtained the different operators listed above were examined and their degree of fit of the theoritical value to the observational one was assessed. It was noted that the degree of fit was thought to be identical with the degree of nearness.

Table2 The comparison of the degree of nearness obtained by different kinds of 'and', 'or'

composit. deg. \ deg. nearness \ operators	$\boxplus$ $\boxdot$	$\overset{+}{\cup}_3$ $\overset{\cdot}{\cup}_3$	$\overset{+}{\cup}_4$ $\overset{\cdot}{\cup}_4$	$\overset{+}{\xi}$ $\overset{\cdot}{\xi}$	$\hat{+}$ $\cdot$
positive	.802	.739	.742	.741	.754
negative	.837	.710	.713	.742	.727
composit. deg. \ deg. nearness \ operators	$\curlyvee_2$ $\curlywedge_2$	$\curlyvee_3$ $\curlywedge_3$	$\curlyvee_4$ $\curlywedge_4$	$\curlyvee_5$ $\curlywedge_5$	$\vee$ $\wedge$
positive	.767	.845	.881	.906	.947
negative	.750	.813	.850	.872	.921

Degree of nearness of observational value itself is denoted as 1

It was clearly shown from Table2 that operator of composition.

There are two ways of the effect in the composition, 'orientation' V. 'contrast' (PoQ) and 'contrast' v. 'orientation' (QoP).

Table3 The comparision of both ways of the effect

composit. deg. / nearness / composit. ways	positive	negative
PoQ	.947	.921
QoP	.874	.891

It was obvious that PoQ is better than QoP. Under the way of PoQ, the theoritical curve was given according to operator of Max-Min, and theoritical and the observational curves were shown in Fig.2

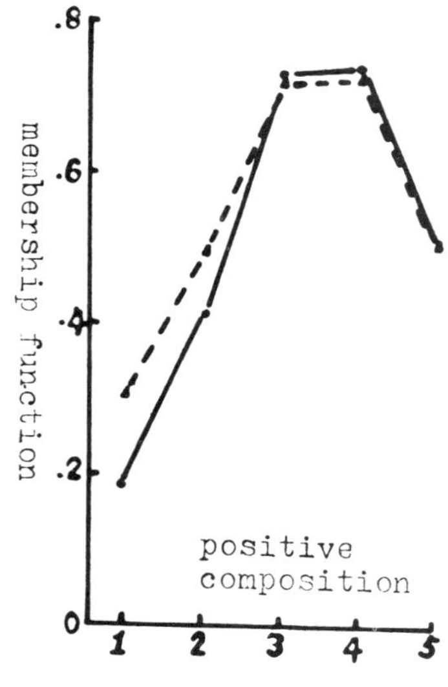

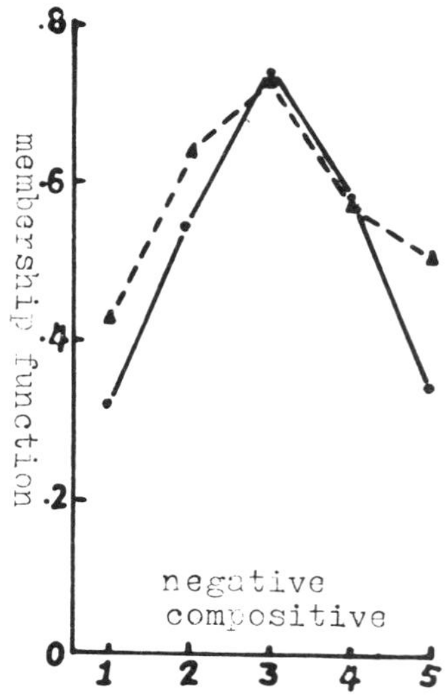

Fig.2, the comparision of the theoritical curve with observational one. ('...' the theoritical curve. '__' the observational curve)

It was shown that the shapes of both curves were in general similar, particularly in the medium parts but there was some departure of the theoritical data from observational ones in the extremes.

DISCUSSION

What implication was the finding that PoQ is better than QoP? The most favourable meanings that could be indicated were as follows:
One of the meanings might implicate that more attention was paid to the part expressing effect of the orientation than that part expressing contrast. In general, the PoQ meant that the former is a strong component in complex configuration. Another meaning might implicate that when seeing the complex configuration, first, the orientation effect came into being then the contrast one was in order.

Another problem is that why there is a departure in the extreme on the scale. The so-called end effect may possibly be a reason where it makes psychological quantity in the extreme suppress, so that real judgment could not be correctly reflected. Another reason is that there might exist a reaction with less confidence at least in the extreme, consequently, in which it led error to appear, if using the Multistage Evaluation Method.

CONCLUSION

A new kind of Multistage Evaluation Method was employed by the present investigation, which, in fact, was based on binary fuzzy sets. It was very suitable to make decision analysis on compositing effect of the illusion.

A complex geometrical configuration if illusion was given. Using the operation of composition was described very well.

The empirical result indicated that operator of Max-Min is the best among those available operators of 'and', 'or', for the effect of composition in illusive configuration.

REFERENCES

Kunnapas, T.M. (1955) "The vertical-

horizontal illusion and the visual field" J. Exper. psychol. V, 53, n,6
Mou-chao, Ma (1982) "Psychological Fuzziness and one Fuzzy Set Method" in press.
Mou-chao, Ma & Lai-fu, Liu (1983) " A New Approach to Investigating Visual Illusions -- An Attempt to Measure Visual Illusions by the Multistage Evaluation Method" Acta Psychologica Sinica N.3
Mou-chao, Ma & Zhi-qiang, Cao (1982a) "A Kind of Multi-Dimensions of Decision Making and Its Mathematical Model" in "A General Survey of Sytems Methodology" proceedings of the twenty-sixth annual meeting of the society for general systems research with the American association for the advancement of science. Washington, D.C, Jan, 5-9
Mou-chao, Ma & Zhi-qiang, Cao (1982b) "The Multistage Evaluation Method in Psychological Measurement: An Application of Fuzzy Sets Theory to Psychology" in "Approximate Reasoning in Decision Analysis" editors Dr. M.M. Gupta and Dr. E. Sanchez
Mou-chao, Ma & Zhi-qiang, Cao (1983) "A Fuzzy Set Model For Categorical Judgment and One Fuzzy Set Method" Acta Psychologica Sinica N.2
Weintraub, D.J., & Krantz, D.H. (1971) "The Poggendoff illusion: Amputations, rotations, and other perturbations" perception and psychophysics, Vol 10, 257-264
Yong-yi, Cheng (1982) "Exploration of Fuzzy Operator (I)" Fuzzy Mathematics. V.2, N.2p1-9

Prof.. of Exprimental Psychology Siegen K. Chou, Department of Psychology, Beijing University, has read and polished the original English of this report.

*

$$a \hat{+} b = a+b-ab, \quad a \cdot b = ab$$

$$a \boxplus b = \min(a+b,1), \quad a \boxdot b \quad \max(0,a+b-1)$$

$$a \overset{+}{\underset{p}{\vee}} b = \frac{a \hat{+} b-(1-p)ab}{p+[(1-p)(a \hat{+} b)]}, \quad a \overset{\cdot}{\underset{p}{\vee}} b = \frac{ab}{p+(1-p)(a \hat{+} b)}$$

$$a \overset{+}{\varepsilon} b = \frac{a+b}{1-ab}, \quad a \overset{\cdot}{\varepsilon} b = \frac{ab}{1+(1-a)(1-b)}$$

$$a \underset{p}{Y} b = \max[1,(a^p+b^p)^{\frac{1}{p}}]$$

$$a \underset{p}{\curlywedge} b = 1-\min\{1,[(1-a)^p+(1-b)^p]\}^{\frac{1}{p}}$$

$$a \vee b = \max(a,b), \quad a \wedge b = \min(a,b)$$

APPLICATION OF FUZZY CONTROLLER TO AUTOMOBILE SPEED CONTROL SYSTEM

S. Murakami

Department of Computer Science, Kyushu Institute of Technology 1-1 Sensuicho, Tobata, Kitakyushu 804, Japan

Abstract. This paper deals with the synthesis of fuzzy logic controller by using the fuzzy reasoning based on Lukasiewicz logic proposed by Y.Tsukamoto and its application to automobile speed control system. The controller is composed of a set of linguistic control rules which are conditional linguistic statements expressing the relationship between inputs and outputs. The inputs are the 'error' which is the difference between the set point and the present speed, and the 'change in error' which is the difference between error values at the present and last sampling instants. The output, that is, manipulated variable is inferred by using truth qualification, converse of truth qualification and fuzzy modus ponens which are based on the fuzzy logic introducing the linguistic truth value.
An algorithm to compute the values of manipulated variable on the microprocessor (Intel 8085) is developed in order to make a real time speed control of autombile. We discuss the results of road test of automobile (TOYOPET CROWN '72) equipped with the microprocessor and we also discuss about usefulness of fuzzy logic controller applied to the automobile speed control system.

Keywords. Fuzzy controller; Linguistic control rules; Fuzzy reasoning; Automobile speed control; Digital control.

INTRODUCTION

Since a fuzzy logic control was proposed by E. H. Mamdani[1974], many studies applied to industrial process such as heat exchanger process, warm water plant, sinter plant and so on have been done. To synthesize a fuzzy controller, so called linguistic control rules are used. These rules are composed by implications such as "if x (output) is A then y (control) is B", which are described by fuzzy relations, $R = A \times B$. Given implications and a lingustic proposition, that is, "x is A'", then "y is B'" is inferred by $B' = A' \circ R$. This fuzzy reasoning is called as the compositional rule of inference.

In this paper, we synthesize a fuzzy logic controller by using the fuzzy reasoning based on linguistic truth value (LTV) proposed by Y. Tsukamoto[1979]. And we apply it to the automobile speed control system. The fuzzy controller is composed of a set of linguistic control rules which are conditional linguistic statements of relationship between inputs and outputs. The outputs are the 'error' which is the difference between the set point and the present speed of automobile and the 'change in error' which is the difference between error values at the present and the last sampling instants. The outputs, that is, manipulated variable is inferred by using truth qualification, converse of truth qualification and fuzzy modus ponens.

An algorithm to compute the value of manipulated variable on the microprocessor is developed. This algorithm is implemnted by the machine language in order to make a real time speed control of automobile. We carried out the road tests of automobile (TOYOPET CROWN '72, automatic) equipped with the micoprocessor and recorders. The characteristics of this fuzzy controller are investigated through these experiments and computer simulations. Finally, we discuss about usefulness of fuzzy logic controller applied to the automobile speed control system.

FUZZY REASONING

Linguistic Truth Value

Let a fuzzy proposition, P, which includes fuzzy predicates, be described as follows,

$$P = \text{"X is A"}$$

where X is a name of object and A represents an attributional property of X and A is given as a fuzzy subset of U. Now, the truth value of fuzzy proposition P is denoted by $\underline{P} = \tau$, which is defined as a fuzzy subset of universe of truth values, $V \in [0,1]$. If τ is represented in a linguistic way, it is called as the linguistic truth value. The examples of linguistic truth values are true, false, very true, more or less true, rather false, etc.

Truth Qualification (TQ) and Converse of Truth Qualification (CTQ)

Given the linguistic proposition of the form as

$$P = \text{"X is A" is } \tau ,$$

TQ implied to obtain the fuzzy predicate, B, equivalent to the following proposition,

"X is B",

where A and B are fuzzy subset of U. Let μ_A and μ_B be membership functions mapping from U to V. Then, using the diagram of Truth Qualification as shown in Fig.1, $\mu_B(u)$ is given as follows,

$$\mu_B(u) = \mu_\tau(\mu_A(u)) \; , \; u \in U$$

Conversely, it is called as Converse of Truth Qualification to find a linguistic truth value provided that A and B are specified. In this case, using the diagram of Converse of Truth Qualification shown in Fig.2, $\mu_\tau(v)$ is given as follows,

$$\mu_\tau(v) = \mu_B(\overset{-1}{\mu_A}(v))$$

where μ_A is one-to-one mapping from U to V.

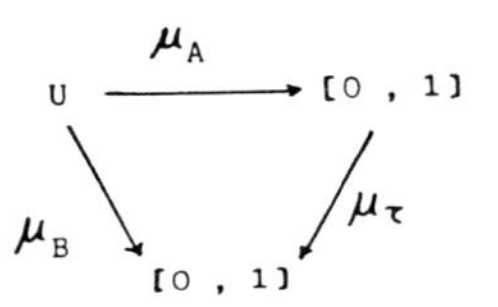

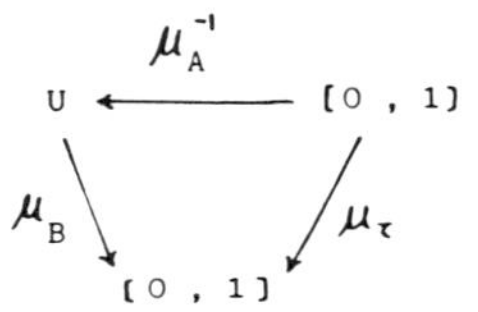

Fig. 1. Diagram of TQ. Fig. 2. Diagram of CTQ.

Fuzzy Reasoning (Fuzzy Modus Ponens)

Let the numerical truth values of propositions, P, Q and R ($\triangleq$ P → Q) be denoted as |P|, |Q| and |R|, respectively. By using the Lukasiewicz infinite valued logic, the numerical truth value of implication, R, is given as

$$|P \to Q| = (1 - |P| + |Q|) \wedge 1$$

When the truth status of propositions, P and Q, are given by linguistic truth values, the following representation is obtained,

$$\mu_{\underset{\sim}{P} \to \underset{\sim}{Q}}(r) = \sup_{(p,q) \in f^{-1}(r)} [\mu_{\underset{\sim}{P}}(p) \wedge \mu_{\underset{\sim}{Q}}(q)]$$

where $f(p,q) \triangleq (1 - p + q) \wedge 1$ and p, q, $r \in [0,1]$.

Fuzzy modus ponens is to infer the truth status of Q, $\underset{\sim}{Q}$, provided that the truth status of implication R,$\underset{\sim}{R}$, and proposition P, $\underset{\sim}{P}$. Let the α -level sets of $\underset{\sim}{P}$ and $\underset{\sim}{R}$ be as follows,

$$\underset{\sim}{P}^\alpha = [p_1(\alpha), p_2(\alpha)] \; , \; p_1(\alpha), p_2(\alpha), \alpha \in [0,1]$$
$$\underset{\sim}{R} = [r_1(\alpha), 1] \; , \; r_1(\alpha), \alpha \in [0,1]$$

then the α -level set of $\underset{\sim}{Q}$ is obtained as

$$\underset{\sim}{Q}^\alpha = [(p_1(\alpha) - (1 - r_1(\alpha))) \vee 0, 1] \qquad (1)$$

where $\underset{\sim}{R}$ is assumed to be possible true and convex, and also $\underset{\sim}{P}$ is assumed to be normal and convex. $\mu_{\underset{\sim}{Q}}(q)$ is given from $\underset{\sim}{Q}^\alpha$ as follows,

$$\mu_{\underset{\sim}{Q}}(q) = \sup_{\alpha \in [0,1]} [\alpha \wedge \chi_{\underset{\sim}{Q}^\alpha}(q)] \qquad (2)$$

where $\chi_{\underset{\sim}{Q}^\alpha}(q)$ is the characteristic function defined as

$$\chi_{\underset{\sim}{Q}^\alpha}(q) = \begin{cases} 1 \; , & \text{if } \mu_{\underset{\sim}{Q}}(q) \geq \alpha \\ 0 \; , & \text{if } \mu_{\underset{\sim}{Q}}(q) < \alpha \end{cases}$$

PROCEDURE TO OBTAIN FUZZY CONTROL

Assumed that the linguistic control rules are given by the following forms,

$$R_i = \text{If x is } P_i \text{ then y is } Q_i$$
$$R_i \text{ is } \tau_i \;\Leftrightarrow\; \underset{\sim}{R_i} = \tau_i \qquad (i = 1,2,\ldots,n)$$

where x is the input into the fuzzy controller and y is output from the fuzzy controller. P_i and Q_i are fuzzy subsets on the supporting sets, X and Y, respectively. Now, when the input into the fuzzy controller is given as follows,

"x is D"

let's describe the procedure to obtain the output of fuzzy controller,

1) Let "x is D" be equivalent to "x is P_i" is $\underset{\sim}{P_i}$. Now that the membership functions of D and P_i are specified, by using CTQ, the linguistic truth value of P_i, that is $\underset{\sim}{P_i}$, is given as follows,

$$\underset{\sim}{P_i} = \int_V \mu_D\,(\overset{-1}{\mu_{P_i}}(v))/v$$

2) By using fuzzy modus ponens (Eq.(1) and (2)), $\underset{\sim}{Q_i}$ is determined, since $\underset{\sim}{P_i}$ and $\underset{\sim}{R_i}$ have been obtained.

3) The fuzzy set, C_i, on the output (control) supporting set, Y, is obtained by using TQ. Now that we have

"y is C_i" = "y is Q_i" is $\underset{\sim}{Q_i}$,

we get

$$C_i = \int_Y \mu_{\underset{\sim}{Q_i}}(\mu_{Q_i}(y))/y$$

4) Operating D to all R_i (i = 1,2,...,n), output (control) is determined by the intersection of the fuzzy sets C_i (i = 1,2,...,n) as

$$C = \bigcap_i C_i$$

5) If a non-fuzzy value of output is required as in process control, then we may choose y^* for which

$$\mu_C(y^*) = \underset{y \in Y}{\mathrm{Max}}\ \mu_C(y)$$

This is the basic procedure to determine an actual control by using fuzzy informations.

ALGORITHM TO DETERMINE MANIPULATED VARIABLE OF AUTOMOBILE SPEED CONTROL SYSTEM

First, we mention about the outline of automobile speed control system which is shown in Fig.3. The speed of vehicle is detected by the tachogenerator and it is converted into digital signals through A/D converter. The digital data of vehicle speed is taken into the fuzzy controller. Hardwares of fuzzy controller consist of CPU (Intel 8085), 4K ROM and 16K RAM. This micorporcessor calculates the manipulated variable according to the procedure mentioned in the prvious section. The calculated value is then converted into analogue signal through D/A converter. By this analogue signal, the actuator is manipulated and the throttle valve of vehicle is opened or closed. Hence, the speed of vehicle is controlled.

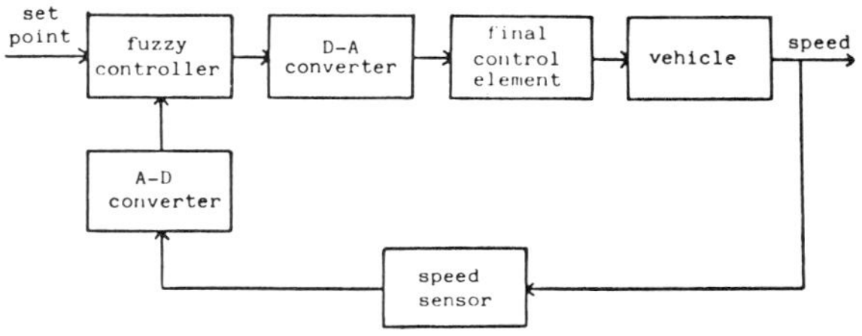

Fig. 3. Automobile speed control system using fuzzy controller.

In the next place, we present an algorithm to determine the manipulated variable of automobile speed control system. The inputs to fuzzy controller are the 'error' and 'change in error', where the sampling period is set on 1 second. According to the procedure to obtain fuzzy control mentioned in previous section, a change in manipulated variable is inferred. Hence, the manipulated variable at the present instant is obtained by adding the one at the last instant to the change in manipulated variable.

The following linguistic control rules of fuzzy controller are used in this paper,

LCR-1 : If e_k is P_e , then Δu_k is PU_e
LCR-2 : If e_k is N_e , then Δu_k is NU_e
LCR-3 : If Δe_k is $P_{\Delta e}$, then Δu_k is $PU_{\Delta e}$
LCR-4 : If Δe_k is $N_{\Delta e}$, then Δu_k is $NU_{\Delta e}$

where

$$e_k = r - x_k,$$
$$\Delta e_k = e_k - e_{k-1},$$
$$\Delta u_k = u_k - u_{k-1},$$

and r is the desired speed (set point), x_k is the speed at time k , and u_k is the manipulated vaiable at time k. And P., N., PU., NU. are fuzzy subsets on the corresponding supproting sets. Letters, P and PU, imply 'positive", and N and NU imply "negative". Each linguistic control rule is considered to be derived from respective goal which can be competitive to each other. An actual control will be determined through a compromise among different criteria. Let us suppose that the truthness of each rule is true.

Now, the membership function of P., N., PU., and NU. are to be characterized as follows,

$$\mu_{P_e}(e) = \frac{1}{\pi}\tan^{-1}(a\cdot e) + 0.5,$$
$$\mu_{N_e}(e) = \frac{1}{\pi}\tan^{-1}(-a\cdot e) + 0.5,$$
$$\mu_{P_{\Delta e}}(\Delta e) = \frac{1}{\pi}\tan^{-1}(a'\cdot\Delta e) + 0.5,$$
$$\mu_{N_{\Delta e}}(\Delta e) = \frac{1}{\pi}\tan^{-1}(-a'\cdot\Delta e) + 0.5,$$
$$\mu_{PU_e}(\Delta u) = \frac{1}{\pi}\tan^{-1}(b\cdot\Delta u) + 0.5,$$
$$\mu_{NU_e}(\Delta u) = \frac{1}{\pi}\tan^{-1}(-b\cdot\Delta u) + 0.5,$$
$$\mu_{PU_{\Delta e}}(\Delta u) = \frac{1}{\pi}\tan^{-1}(b'\cdot\Delta u) + 0.5,$$
$$\mu_{NU_{\Delta e}}(\Delta u) = \frac{1}{\pi}\tan^{-1}(-b'\cdot\Delta u) + 0.5,$$

where

$$a = \tan(0.45\cdot\pi)/r,$$
$$a' = \tan(0,45\cdot\pi)/c_{max}\ (\doteq a/c),\ c = c_{max}/r,$$
$$b = \tan(0.45\cdot\pi)/u_{max},$$
$$b' = \tan(0.45\cdot\pi)/d_{max}\ (\triangleq b/d),\ d = d_{max}/u_{max},$$

and, c_{max}, u_{max}, and d_{max} are scaling factors which are parameters to be determine the size of respective supporting set.

Given inner data of e_k and Δe_k, we can determine the non-fuzzy control Δu_k^* according to the procedure mentioned in pevious section. To simplify the computation, we assume that the inputs, e_k and Δe_k, are non-fuzzy values. The algorithm to determine the actual control (Δu_k^*) at time k is summarized as follows,

1) If $e_k > 0$ and $\Delta e_k > 0$, and
   i) $a\cdot e_k = c/(a\cdot\Delta e_k)$, then $\Delta u_k^* = \sqrt{d}/b$
   ii) $g > 0$, then $\Delta u_k^* = (-(1 + d) + h)/2gb$
   iii) $g < 0$, then $\Delta u_k^* = (-(1 + d) - h)/2gb$
   iv) $(a/b)\cdot e_k = (a\cdot d/b\cdot c)\cdot\Delta e_k$,
   then $\Delta u_k^* = (a/b)\cdot e_k$
2) If $e_k < 0$ and $\Delta e_k < 0$, and
   i) $a\cdot e_k = c/(a\cdot\Delta e_k)$, then $\Delta u_k^* = -\sqrt{d}/b$
   ii) $g > 0$, then $\Delta u_k^* = (-(1 + d) - h)/2gb$
   iii) $g < 0$, then $\Delta u_k^* = (-(1 + d) + h)/2gb$
   iv) $(a/b)\cdot e_k = (a\cdot d/b\cdot c)\cdot\Delta e_k$,
   then $\Delta u_k^* = (a/b)\cdot e_k$
3) If $e_k \geq 0$ and $\Delta e_k \leq 0$, or $e_k \leq 0$ and $\Delta e_k \geq 0$, and
   i) $g = 0$, then $\Delta u_k^* = 0$
   ii) $g \neq 0$, then $\Delta u_k^* = (-(1 + d) + h)/2gb$

where

$$g = (a \cdot e_k + (a/c) \cdot \Delta e_k)/(1 - a \cdot e_k \cdot (a/c) \cdot \Delta e_k)$$

$$h = \sqrt{(1 + d)^2 + 4g^2 d}$$

Using the above algorithm, the manipulated variable at time k, u_k, whitch is the input into the final control element, is determined as

$$u_k = u_{k-1} + \Delta u_k^*$$

SIMULATION

To investigate the characteristics of fuzzy logic controller developed in this paper, some computer simulations of automobile speed control system were carried out. As a controlled system, we used an automobile, TOYOPE CROWN ('72, Automatic). The transfer function of the controlled system including the final control element was approximately identified from the step response as follows,

$$G(s) = \frac{35.78}{(1 + 1.728\ s)(1 + 16.848\ s)} \cdot e^{-0.4s}$$

Fig.4 shows the result of computer simulation when the set point is set at 60 Km/h for the first 80 seconds and then is changed to 40 Km/h. In this case, as the controlled variable (speed) gets very near the set point, the speed once slows down slightly and then settles up at the set point. PI controller could not realize such a response. The changing pattern of manipulated variable shows that it is close to the human operation of speed control. The larger c_{max} and u_{max} are, the faster the rise time become as shown in Fig.5. But, in this case, an overshoot is occurred.

Fig.6 show the result of simulation in the case where the control is suspended between at 81 second and at 95 second and then the control is again started from 95 second. In this case, very slight overshoot is observed, since the manipulated variable from 95 second gets larger.

Fig.7 shows the result in the case where the gain of controlled system is forced to change after 60 second by using a sine curve of which amplitude is an half of system gain. This corresponds to the case where the vehicle is supposed to run on an uphill and downhill road. In spite of rather big change of gain, the automobile is well controlled by using this fuzzy logic controller.

From these simulation results, we could obtain such suggestions that the fuzzy controller is very usefull for the automobile speed control system and that the speed control system which is robust to the disturbances and the change of system characteristics can be constructed.

EXPERIMENT

To make real time road test, the experiment

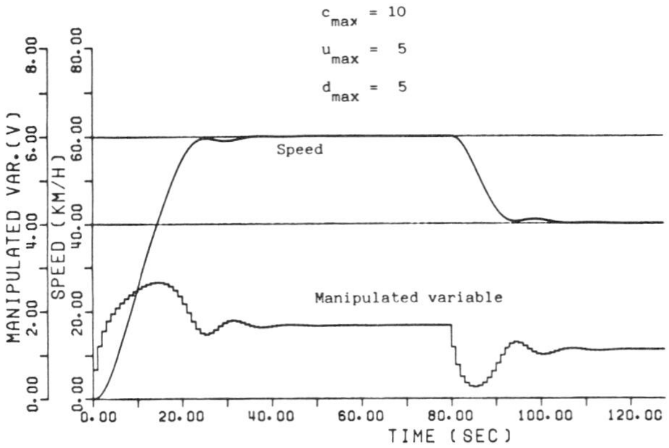

Fig. 4. Result of simulation when set point is set at 60 Km/h and thwen changed to 40 Km/h.

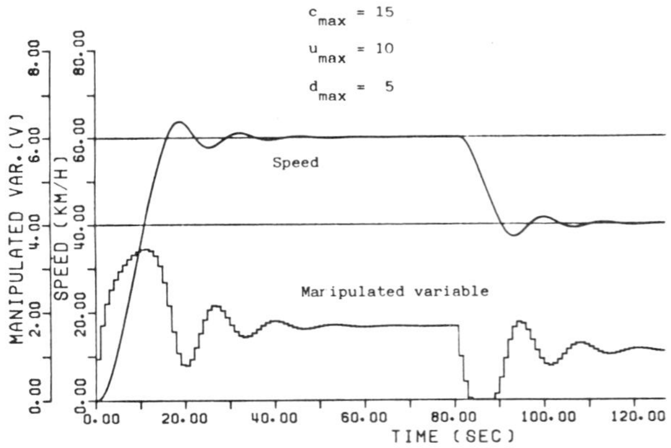

Fig. 5. Result of simulation when scaling factors are changed.

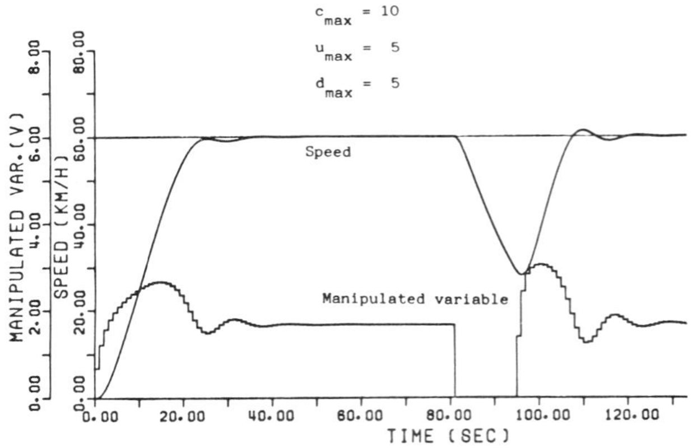

Fig. 6. Result of simulation when control is suspended.

vehicle was equipped with a microprocessor which consisted of CPU (Intel 8085), 4K ROM and 16K RAM. It was also equipped with devices to record the speed and the manipulated variable. The algorithm to calculate the maniplated variable mentioned in previous section was programmed using the machine language in ROM. Computing time to obtain a manipulated variable was about 50 msec in the average.

The first road test was done on the flat road. Fig.8 shows its result. The set point was first set at 40 Km/h and then changed up to 60 Km/h, The response of speed seems to have the same characteristic as the results of computer simulations. That is, as the speed of vehicle gets very near the set point, it slows down and then settles up to the steady state. And also an overshoot is not observed in both cases. The changes of manipulated variable in the steady state was not too large for the driver to feel on his body.

Next, we carried out a road test on the interval from Murasakigawa Interchange to Kurosaki Interchange of Kitakyushu Highway. The distance was about 12 Km. The interval has the uphill and downhill road with the ascent of maximum 4.7 % and the descent of maximum 3.7 %. In Fig.9, the result of road test is shown. The set point was set at 60 Km/h, since the speed limit of the highway is 60 Km/h. The speed of automobile is well controlled ever on the uphill and downhill road. The fluctuation of speed from the set point stay within ±2 Km/h.

We can find that the value of manipulated variable in the steady state in case of the first road test (set point 60 Km/h) is less than that in case of the second road test. This is because we used the high octane gasoline in the first road test and the regular gasoline in the second road test by mistake. In spite of the kind of gasoline used, the control performance has not great difference. From these road tests of automobile speed control system with the fuzzy logic controller, we can conclude that the fuzzy logic controller developed in this paper is robust to the change of gain of controlled vehicle and the disturbances. And, by using this fuzzy controller, a stable and comfortable driving at a desired speed could be realized.

CONCLUDING REMARKS

We could synthesize the automobile speed control system using fuzzy logic controller which has robustness. The results of road tests show usedfulnesss of the fuzzy logic controller. The fuzzy controller developed here is said to be suitable to control such systems (processes) that the system gain frequently changes and that the characteristics of controlled system are not precisely known.

We are going to develop the fuzzy control system which includes the mechanism to learn the driver's driving manner. And also we are to investigate qualitative properties of scaling factors (parameters to determine sizes of supporting sets) which will affect control performances.

REFERENCES

Mamdani, E. H. (1974). Application of Fuzzy Algorithms for Control of a Simple Dynamic Plant, Proc. IEE, 121, 12, 1585-1588.

Tsukamoto, Y. (1979). Fuzzy Logic Based on Lukasiewicz Logic and Its Application to Diagnosis and Control, Doctoral dissertation (T.I.T).

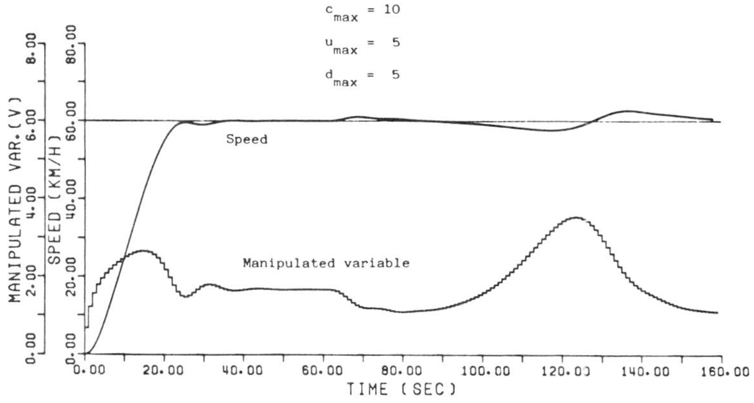

Fig. 7. Result of simulation when system gain is forced to change by sine curve.

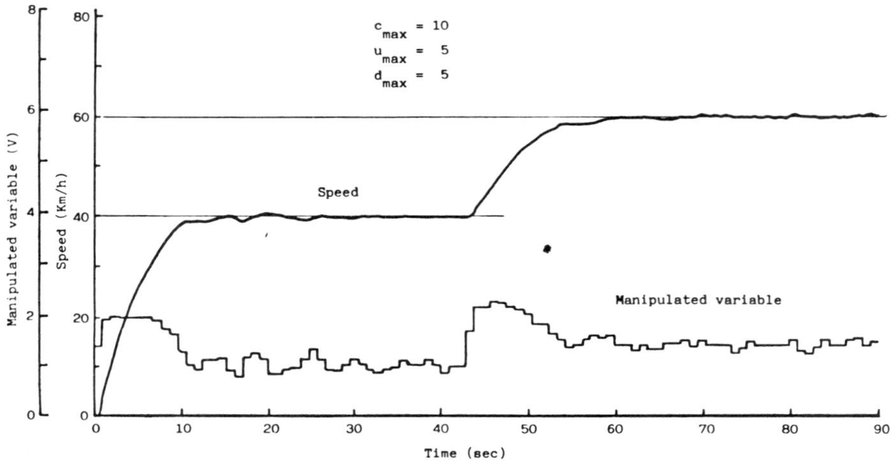

Fig. 8. Result of road test on flat road.

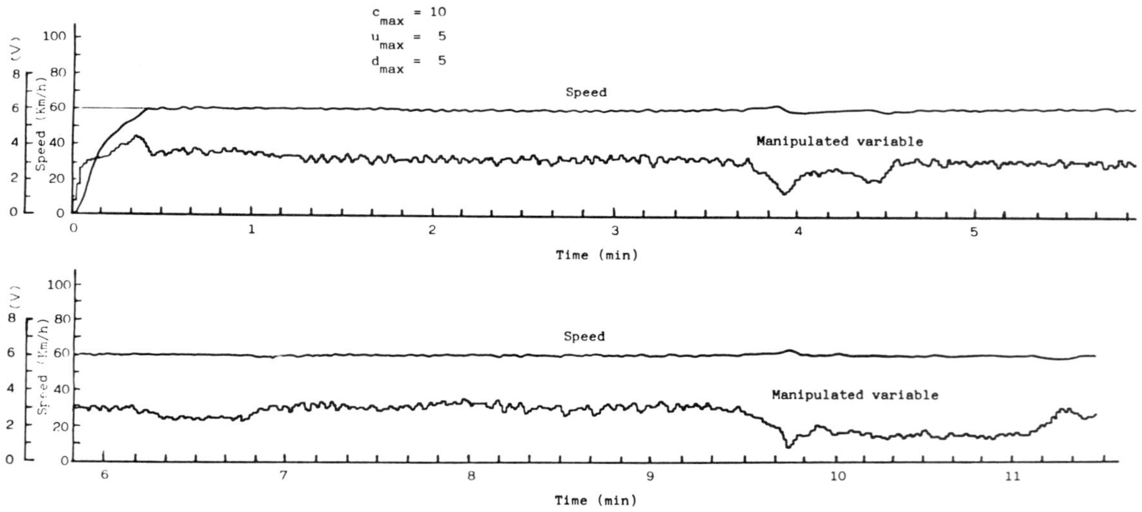

Fig. 9. Result of road test on Kitakyushu Highway.

CONTROL OF COMPLEX PROCESSES BY USING FUZZY PROBABILISTIC CONTROLLER

E. Czogala*, W. Pedryck and L. Walichiewicz

Department of Automatic Control and Computer Sci. Silesian Technical University, Gliwice, Poland

Abstract. The paper deals with a class of heuristic algorithms of control of complex or ill-defined processes, where a concept of probabilistic set is proposed. A so-called fuzzy probabilistic controller is constructed. The controller discussed consists of a set of control rules which describes the control policy and preserve approximate nature of human reasoning.

Keywords. Probabilistic set, controller, decision-making.

INTRODUCTION

Zadeh's (1965) linguistic approach to fuzzy systems has motivated works dealing with synthesis and several important aspects of analysis of decision-making algorithms called fuzzy controllers and investigated by many researchers (cf. Mamdani 1976, Kickert and Mamdani 1978, Tong 1977. Hirota (1981) has introduced an idea of probabilistic sets regarding the value of membership function of a fuzzy set as a random variable depending on parameter. This concept seemed to be introduced for the reason of ambiguity and subjectivity of human observers (decision-makers) cannot be uniquely expressed in [0,1] - interval. Taking into account this notion, it is possible to consider algorithms of decision-making where two forms of uncertainty could be covered, viz. fuzziness and randomness. The paper is organized as follows. The basic notions such as: probabilistic set ,its distribution function representation, some functions on probabilistic sets are presented in Section 2. Section 3 provides eessential considerations which are useful in construction of the controller. Concluding remarks are presented in Section 4.

*The final version of this paper was written after one of the authors (E. Czogała) had been awarded by Alexander von Humboldt Foundation receiving the research fellowship.

BASIC NOTIONS AND DEFINITIONS

Following Hirota (1981) the background idea of probabilistic sets lies in regarding the value of membership functions of fuzzy sets as a random variables depending on parameters.

A probabilistic set A on the total space $\mathcal{X}$ is defined by $(\Omega_c, \mathcal{B}_c)$ - valued random variable on $(\Omega, \mathcal{B}, P)$ for each $x \in \mathcal{X}$ and such correspondence is called a defining function of probabilistic set denoted also by μ_A i.e.

$$\mu_A : \mathcal{X} \times \Omega \to \Omega_c \qquad (1)$$
$$(x, \omega) \to A(x, \omega)$$

where $\mu_A(x, \cdot)$ is the $(\mathcal{B}, \mathcal{B}_c)$ - messurable function for each fixed $x \in \mathcal{X}$. $(\mathcal{B}, \mathcal{B}_c)$ - measurability of this function means that

$$\forall_{x \in X} \forall_{z \in \Omega_c} \{\omega : \mu_A(x, \omega) < z\} \in \mathcal{B} \qquad (2)$$

Treating the defining function as a random with parameter is possible to introduce the distribution function description (representation) of probabilistic set.

Considering n-dimensional distribution for any set of numbers $x_1, x_2, \ldots, x_n \in \mathcal{X}$ (n is chosen arbitrarily) and respective set of $z_1, z_2, \ldots, z_n \in \Omega_c$ it can be defined (for probabilistic set A) by the following equality:

$$F_{A(x_1)\ A(x_2)\ \ldots\ A(x_n)}(z_1, z_2, \ldots, z_n) =$$

$$= P(\{\omega:\ A(x_1,\omega)<z_1,\ A(x_2,\omega)<z_2, \ldots,$$

$$\ldots, A(x_n,\omega)<z_n\}) \quad (3)$$

Of course, the symmetry and compatibility conditions are satisfied for the above given distribution functions. Such conditions hold for the respective n-dimensional density functions as well.

Now taking into account the notion of probabilistic set and its distribution function we will find the distribution functions of the following important functions:

$$W = \min(X_1, X_2, \ldots, X_n) \quad (4)$$

$$M = \max(X_1, X_2, \ldots, X_n) \quad (5)$$

where $X_1, X_2, \ldots, X_n$ are independent probabilistic sets defined on the following universes of discourse

$$X_1;\ \mathscr{X}^1 = \{x_1^1, x_2^1, \ldots, x_K^1\}$$

$$X_2;\ \mathscr{X}^2 = \{x_1^2, x_2^2, \ldots, x_L^2\} \quad (6)$$

$$\vdots$$

$$X_n;\ \mathscr{X}^n = \{x_1^n, x_2^n, \ldots, x_M^n\}$$

Assuming that $P(X_j(x_{j_k}^k, \omega) < w) = F_{X_j(x_{j_k}^k)}(w)$ we have

$$F_W(w) = P(W<w) = 1 - P(W \geq w) =$$

$$= 1 - P(X_1 \geq w)\ P(X_2 \geq w) \ldots P(X_n \geq w) =$$

$$= 1 - \prod_{j=1}^{n} \left[1 - F_{X_j(x_{j_k}^k)}(w)\right] \quad (7)$$

and

$$F_M(w) = P(M<w) =$$

$$= P(X_1<w,\ X_2<w, \ldots, X_n<w) =$$

$$= P(X_1<w)\ P(X_2<w) \ldots P(X_n<w) =$$

$$= \prod_{j=1}^{n} F_{X_j(x_{j_k}^k)}(w) \quad (8)$$

The above obtained distribution functions of min and max functions of the collection of probabilistic sets come into prominence in further considerations.

A CONCEPT OF A FUZZY PROBABILISTIC CONTROLLER

Let us briefly remind the foundations of a fuzzy controller as given in Mamdani,(1976).

This controller is in fact a decision-making algorithm using linguistic rules to describe the control policy and its important feature is that it tries to preserve the approximate nature of human reasoning than to avoid it. The concept of the controller bases on the theory of fuzzy sets proposed by Zadeh which makes it possible to treat linguistic, imprecise rules of control mathematically.

The fuzzy controller consists of a collection of linguistic rules which describe individual control situations. Each rule connects the process outputs (or state variables) with changes of concrete values of control variables A control rule, i, may be regarded as an implication:

$$X_{1(i)} \cap \ldots \cap X_{n(i)} \Rightarrow U_{1(i)} \cap \ldots \cap U_{p(i)} \quad (9)$$

where

$X_{k(i)}$ - is the fuzzy value of the process output variable, k, defined on the fixed universe of discourse $\mathscr{X}^k$; $k = 1, 2, \ldots, n$

$U_{l(i)}$ - is the fuzzy value of the control variable, l, defined on the fixed universe of discourse $\mathbb{U}^l$; $l = 1, 2, \ldots, p$

These fuzzy values are expressed as fuzzy sets of the respective universe of discourse by corresponding memebership functions.

Each rule produces a fuzzy relation in(n + p)dimensional space $\mathscr{X}^1 \times \ldots \times \mathscr{X}^n \times \mathbb{U}^1 \times \ldots \times \mathbb{U}^p$ given by the product:

$$R_i = X_{1(i)} \times \ldots \times X_{n(i)} \times$$

$$\times U_{1(i)} \times \ldots \times U_{p(i)} \quad (10)$$

The overall relation R is calculated as the union of individual rules, i.e.

$$R = \bigcup_{i=1}^{N} R_i \qquad (11)$$

Thus, the entire fuzzy control algorithm is represented by the single fuzzy relation R. The function of the controller is to infer from the actual values of the process outputs Xa_k (k = 1,2,...,n) the actual values of control to apply Ua_l (l=1,2,...,p).

The outputs from the controller can be obtained by using the compositional rule of inference

$$\bar{U}a = Xa_1 \circ \ldots \circ Xa_n \circ R \qquad (12)$$

The result is a composed fuzzy set $\bar{U}a$ in p - dimensional space $U^1 \times \ldots \times U^p$. The fuzzy sets describing the fuzzy values of individual control variables can be obtained by projecting the composed fuzzy set on the respective universes of discourse:

$$Ua_l = \mathrm{proj}_{U^l} \bar{U}a \qquad (13)$$

For practical reasons a decision is required concerning the deterministic control values to be applied. One possible way is to choose the element of the respective universe, corresponding to the maximum membership value of the calculated set Ua_l. If this value is not unique, the mean value of these elements is chosen.

Very often in such a situation we deal with a group of experts (e.g. human operators of the process) where each of them provides a different opinion on the intrepretation of linguistic terms in which the control policy is described. In this situation the designer of the control system faces the difficulty of how to fix the proper value of the degree of membership for elements of respective universes according to the lingustic terms used in the rules of control.

The problem arises because the membership function must be accurately given, and on the other hand, the features of this function, as shape, values, have an important effect on the performance of the controller. To make matters worse the subjectivity of human observers may be followed by the varying character of the process, the particular situation and so on.

The concept of probabilistic sets (Hirota, 1981) gives the possibility of a formal approach to the mentioned problems. A probability space $(\Omega, \mathcal{B}, P)$ called parameter space, represents the standards of judgment of imprecise, linquistic terms.
A set of all degrees of membership is called characteristic space $(\Omega_c, \mathcal{B}_c)$.

It may be assumed that if the standard $\omega \in \Omega$ is fixed, the degree of membership can be determined, thus the concept of probabilistic sets includes the concept of fuzzy sets (Zadeh, 1965). The degree of membership is given by a parametric random variable instead of an accurately given membership function in case of a fuzzy set. This parameter is an element of the universe. It seems important that it is not necessary to know the exact probabilistic structure of parameter space, but it may estimated statistically.

The idea of a fuzzy probabilistic controller arises from the formalization of linguistic rules of control by means of probabilistic sets. More particularly the synthesis of the controller is based on the distribution function of probabilistic sets. Hence the decision making may be supported by moment analysis, using a probabilistic measure and operating with data, thus being possible in a flexible, analytical manner.

The basis of synthesis of a fuzzy probabilistic controller is a collection of linguistic rules. The difference between a fuzzy and fuzzy probabilistic approach is that now we use probabilistic sets instead of fuzzy ones to describe the linguistic variables. In the same way as for a fuzzy controller we can achieve the fuzzy probabilistic relation R, which creates the "memory" of the controller

$$R = \bigcup_{i=1}^{N} R_i = \bigcup_{i=1}^{N} \{X_{1(i)} \cap \ldots \cap X_{n(i)} \Rightarrow$$

$$\Rightarrow U_{1(i)} \cap \ldots \cap U_{p(i)}\} \qquad (14)$$

This can be denoted by means of the defining functions:

$$\mu_R = \mu_R(x^1_{j_1}, \ldots, x^n_{j_n}, u^1_{s_1}, \ldots, u^p_{s_p}, \omega) =$$

$$= \max_{1 \leq i \leq N} \min(\mu_{X_{1(i)}}, \ldots, \mu_{X_{n(i)}}, \mu_{U_{1(i)}}, \ldots, \mu_{U_{p(i)}}) \qquad (15)$$

The fuzzy probabilistic relation R makes it possible to calculate the output of the controller $\bar{U}a$ (index "a" means actual) in the same way as in the case of fuzzy controllers:

$$\bar{U}a = Xa_1 \circ \dots \circ Xa_n \circ R \tag{16}$$

and:

$$Ua_l = \operatorname{proj}_{U^l} \bar{U}a \tag{17}$$

Using defining functions we can express this as follows:

$$\mu_{\bar{U}a}(u^1_{s_1},\dots,u^p_{s_p},\omega) = \max_{X^1}\dots\max_{X^n} \min(\mu_{Xa_1},\dots,\mu_{Xa_n},\mu_R) \tag{18}$$

$$\mu_{Ua_l}(u^l_{s_l},\omega) = \max_{U^1}\dots\max_{U^{l-1}} \max_{U^{l+1}}\dots\max_{U^p} \mu_{\bar{U}a}(u^1_{s_1},\dots,u^p_{s_p},\omega) \tag{19}$$

Defining functions are random variables so we may use distribution functions. To make it clear, let us accept the following designations:

$$x^k_{j_k} \in X^k \equiv y^k_{j_k} \in Y^k \qquad k = 1,2,\dots,n$$

$$u^m_{s_m} \in U^m \equiv y^{n+m}_{j_{n+m}} \in Y^{n+m} \qquad m = 1,2,\dots,p$$

$$x_{j(i)} \equiv Y_{j(i)} \qquad j = 1,2,\dots,n \tag{20}$$

$$U_{l(i)} \equiv Y_{n+l(i)} \qquad l = 1,2,\dots,p$$

$$F_{Y_{q(i)}(y^q_{j_q})}(w) = F_{Y_{q(i)}}(w)$$

Assuming that Y_i; $i = 1,2,\dots,n+p$ are independent, the distribution function of R is equal to:

$$F_R(w) = F_{R(y^1_{j_1},\dots,y^{n+p}_{j_{n+p}})}(w) = \prod_{i=1}^{N}\Big(1 - \prod_{q=1}^{n+p}(1 - F_{Y_{q(i)}}(\omega))\Big) \tag{21}$$

The application of the compositional tule of inference yields:

$$F_{\bar{U}a}(w) = F_{\bar{U}a(y^{n+1}_{j_{n+1}},\dots,y^{n+p}_{j_{n+p}})}(w) = \prod_{Y^1}\dots\prod_{Y^n}\Big(1-\prod_{q=1}^{n}(1-F_{Xa_q}(w))(1-F_R(w))\Big) \tag{22}$$

The distribution function describing the individual control variables is equal to:

$$F_{Ya_{n+l}}(w) = \prod_{Y^{n+1}}\dots\prod_{Y^{n+l-1}}\ \prod_{Y^{n+l+1}}\dots\prod_{Y^{n+p}} F_{\bar{U}a}(w) \tag{23}$$

In our opinion it should be stressed that the distribution function of Ua_l provides more information than the membership function. It should be also pointed out that the concept of a fuzzy controller is embeded in the idea of the fuzzy probabilistic controller presented above. It is obvious because of the fact that if the degree of membership is given by a fixed value, the distribution function is a "unit step" function, and the results of operating on the distribution function are the same as max, min operations on the membership function. Let us illustrate it by an example. Let R and X be a fuzzy relation and fuzzy set, respectively so, that:

$$\mu_R(x_j,u_k) = w_{jk}$$
$$\mu_X(x_j) = w_j \qquad j=1,2,\dots,n;\quad k = 1,2,\dots,p \tag{24}$$

Thus their distribution functions are equal to:

$$F_{R(x_j,u_k)}(w) = \begin{cases} 0 & w < w_{jk} \\ 1 & \text{otherwise} \end{cases}$$
$$F_{X(x_j)}(w) = \begin{cases} 0 & w < w_j \\ 1 & \text{otherwise} \end{cases} \tag{25}$$

Then the composition of F_R and F_X is equal to,

$$F_{U(u_k)}(w) = \prod_{j=1}^{p}\big(F_{X(x_j)}(w) + F_{R(x_j,u_k)}(w) - F_{X(x_j)}(w)F_{R(x_j,u_k)}(w)\big) = \begin{cases} 0 < w\ \max(\min(w_1,w_{1k}),\dots,\min(w_p,w_{pk})) \\ 1 \text{ otherwise} \end{cases} \tag{26}$$

The same result will be achieved if we find the distribution function of the membership function of the set U given by the compositional rule of inference:

$$\mu_U(u_k) = \max_{1 \leq j \leq p} \min(\mu_R(x_j, u_k), \mu_X(x_j)) \qquad (27)$$

CONCLUDING REMARKS

The concept of the probabilistic set discussed here forms an extension of the notion of the fuzzy set, facilitating the handling fuzzy, nonfuzzy and probabilistic (stochastic) form of information in a unique way, so it can be more flexible tool for constructing decision-making algorithms than discussed until now.

The idea of the probabilistic set can be helpful and convenient when designing the controller in situations we deal with a group of experts of a concrete system and each of them presents different control strategy.

The formalisation of control rules by means of probabilistic sets provides much more information about the performance of the controller than it was possible to obtain discussing fuzzy controller. Of course, the price we have to pay for it, is concerned with more complicated methods of operating with the distribution functions representing the probabilistic sets. The calculations of the moments of defining function of the respective probabilistic sets point out the influence of the factor of randomness in control strategy; moreover sensitivity analysis can be performed.

Summerizing, the advantage of this approach is twofold,

- it is possible to combine both forms of uncertainty (randomness and fuzziness) in the synthesis of the controller,
- distribution function representation of probabilistic sets of state and control makes possible to proceed further considerations e.g. the calculation of deterministic control values in analytical manner.

REFERENCES

Hirota K. (1981). Concepts of probabilistic sets. Fuzzy Sets and Systems, 5, 31-46.

Kickert W.M., and E.M. Mamdani. Analysis of fuzzy logic controller. Fuzzy Sets and Systems, 1, 29-44.

Mamdani E.H. (1976). Advances in the linguistic synthesis of fuzzy controllers. Int. J. Man - Mach. Stud., 8, 669-678.

Tong R.M. (1977). A control engineering review of fuzzy systems. Automatica, 13, 559-569.

Zadeh L.A. (1965). Fuzzy sets. Inf. Control, 8, 338-353.

DERIVATION OF FUZZY CONTROL RULES FROM HUMAN OPERATOR'S CONTROL ACTIONS

T. Takagi and M. Sugeno

Department of Systems Science, Tokyo Institute of Technology 4259 Nagatsuta, Midori-ku, Yokohama 227, Japan

Abstract.

So called fuzzy control has been developed for the purpose of realization of a man-like controller with the aid of computer. However, most of fuzzy controllers have two very important problems. First is a defect of the reasoning algorithm. Second is a way to acquire control rules, particulaly the precise parameters in the rules. In this paper considering the above problems, we propose a realistic fuzzy reasoning algorithm and a method to identify control rules from human operator's actual control actions. Further we examine the performance of the proposed algorithm by applying it to water cleaning process control.

Keywords. Identification; Industrial control; Fuzzy Reasoning; Fuzzy Control; Modelling

1. INTRODUCTION

Among many fields related to fuzzy theory, fuzzy control is one of the most interesting fields since we have many real situations where fuzzy control is effectively applicable. As far as fuzzy control is concerned, we have had many theoretical and experimental studies so far. We can say that we are now prepared to deal with practical control problems. For instance a fuzzy controller for cement kiln control has been developed by F. L. Smidth Co.[2].

However there still remain some important problems, practical rather than theoretical.

1) fuzzy reasoning as a tool to describe processes
2) derivation of fuzzy control rules
3) design of conversational fuzzy controller
4) experience of real apllications

This paper discusses the first two problems from a practical point of view. It presents a very simple method of fuzzy reasoning for both fuzzy control and fuzzy modelling, and shows how to derive fuzzy control rules from human operator's control actions by using real data.

2. A SIMPLIFIED METHOD OF FUZZY REASONING

2-1 Some Problems of Past Reasoning Methods

Ordinary fuzzy reasoning is characterized by two parts: choice of unimodal fuzzy variables as are shown in Fig. 1 and fuzzy relational composition. There is a composition-like method used in practice but it is not so different from an ordinary one anyway. This reasoning method has some difficulties when we apply it especially to control problems.

In fuzzy control we use fuzzy implications called control rules such as

$$x \text{ is } A_1 \text{ and } y \text{ is } B_1 \rightarrow z \text{ is } C_1 \quad (1)$$

$$x \text{ is } A_2 \text{ and } y \text{ is } B_2 \rightarrow z \text{ is } C_2\text{, etc.,} \quad (2)$$

where the problem is to infer "z is z_0" from a given premise "x is x_0" and "y is y_0".

We shall denote the membership function of a fuzzy variable A by A(x) in the sequel. The difficulties arise in the following parts.

First, if we choose fuzzy variables shown in Fig. 1, it is very difficult to distinguish two points x_0 and x_1 where $A_1(x_0) = A_1(x_1)$. To overcome this difficulty, we have to choose many variables such as "very small", "small", "medium" and so on.

Secondly, perhaps due to the above reason, if we use fuzzy relational composition, in general many control rules are necessary with respect to each dimension of the premise as has been seen in the past studies of fuzzy control. Suppose five fuzzy variables at one dimension, then the number of control rules in case of n-dimensions becomes 5^n. It causes a big problem in practical application, of course.

Finally, if we look at ordinary fuzzy reasoning as a mathematical tool to describe processes, it is not so powerful partly because of its nonlinearity. Suppose that we are given an underlying relation among three variables such as

$$z = ax + by. \tag{3}$$

Choose fuzzy variables A_1 and A_2 for x, B_1 and B_2 for y, then we can obtain by simple calculation of fuzzy numbers

$$C_{ij} = aA_i + bB_j, \tag{4}$$

$$i = 1, 2 \text{ and } j = 1, 2.$$

Then we have four implications

$$x \text{ is } A_i \text{ and } y \text{ is } B_j \rightarrow z \text{ is } C_{ij} \tag{5}$$

and from these follow fuzzy relations R_{ij}. The whole relation is given by

$$R = \bigcup_{i,j} R_{ij}. \tag{6}$$

Now given a premise "x is x_0 and y is y_0" where x_0 is located between A_1 and A_2, y_0 between B_1 and B_2, the value of z is inferred as a fuzzy number with the membership function $R(x_0, y_0, z)$ which is not equal to the expected value, i.e., $z_0 = ax_0 + by_0$ even if we take the value of z maximizing $R(x_0, y_0, z)$.

2-2 Format of Implications

The implication R we suggest in this paper is of the format

$$R : \text{If } f(\ x_1 = A_1, \ \cdots\cdots, \ x_k = A_k\ )$$

$$\text{then } y = g(\ x_1, \ \cdots\cdots, x_k\ ) \tag{7}$$

where each variable have following meaning.

- y : variable of the consequence whose value is inferred.
- $x_1 - x_k$: variables of the premise that appear also in the part of the consequence
- $A_1 - A_k$: fuzzy sets representing the area in which the rule R should be applied for a reasoning corresponding to $x_1 - x_k$, respectively.
- f : logical function defining the proposition of the premise.
- g : function which implies the value of y when $x_1 - x_k$ satisfy the premise condition.

Example

$$R : \text{If } x_1 = \text{small and } x_2 = \text{big}$$

$$\text{then } y = x1 + x2 \tag{8}$$

This implication states that if x1 is small and x2 is big, then the value of y would be equal to the sum of x1 and x2.

2-3 Algorithm of Presented Fuzzy Reasoning

The authors have suggested a method of multi-dimensional fuzzy reasoning[4] based on Lukasiewicz logic where fuzzy variables are assumed to have monotone membership functions which do not cause the case as $A_1(x_0) = A_1(x_1)$ for $x_0 \neq x_1$. The method enables one to represent any linear relation by using, for example, four implications in two dimensional case.

Now we suggest a simplified method of reasoning by developing the above idea. For the sake of simplicity, especially from a practical point of view, all the fuzzy variables are assumed to have linear membership functions.

Suppose that we have implications R^i (i = 1, $\cdots$, n) of the above format. When we are given

$$(\ x_1 = x_1^o, \ \cdots\cdots, \ x_1 = x_k^o\ )$$

where $x_1^o - x_k^o$ are singletons, the predicted value of y is inferred in the following steps.

1) For each implication R^i, the value of y^i is calculated by the function g^i concerned with the consequence

$$y^o = g^i(\ x_1, \ \cdots\cdots, \ x_k\ ) \tag{9}$$

2) The truth value of the proposition " $y = y^i$ " is calculated by the equation

$$/\ y = y^i\ / = /\ f^i(\ x_1^o = A^i_1, \ \cdots\cdots,$$

$$x_k^o = A^i_k\ )\ / \wedge /\ R^i\ / \tag{10}$$

where /*/ means the truth value of proposition *. The truth value

$$/\ f^i(\ x_1^o = A^i_1, \ \cdots\cdots, \ x_k^o = A^i_k\ )\ /$$

is found according to the logical function f^i as a function of $/\ x_1^o = A^i_k\ /$, $\cdots\cdots$, $/\ x_k^o = A^i_k\ /$, where

$/\ x^o = A\ / = A(x^o)$,

i.e., the grade of the meambership of x^o.

It has been reported by Tsukamoto [5] that the value suggested in equation (10) is the least estimation of the truth value of the consequence in Goguen's multivalued logic.

For simplicity in this paper we assume

$$/\ R^i\ / = 1, \tag{11}$$

so the truth value of consequence is obtained as

$$/\ y = y^i\ / = /\ f^i(\ x_1 = A^i_1, \ \cdots\cdots,$$

$$x_k = A^i_k\ )\ / \tag{12}$$

3) The final output y infered from n implications is calculated as the weighted average of all y^i's with the weights $/ y = y^i /$:

$$y = \frac{\sum / y = y^i / \times y^i}{\sum / y = y^i /}$$

$$= \frac{\sum /f^i(x_1 = A^i_1, \cdots, x_k = A^i_k)/ \times g^i(x_1, \cdots, x_k)}{\sum /f^i(x^i = A^i 1, \cdots, x_k = A^i_k)/} \quad (13)$$

Example

Suppose that we have the following three implications,

R^1 : IF x_1 = $small_1$ and x_2 = $small_2$

then $y^1 = x_1$ (14)

R^2 : If $x_1 = big_1$

then $y^2 = 2 \times x_1$ (15)

R^3 : If $x_2 = big_2$

then $y^3 = 3 \times x_2$ (16)

Table 1 shows that reasoing process by each implication when we are given $x_1 = 12$, $x_2 = 5$.

The column "Premise" in Table 1 shows the membership functions of the fuzzy sets small and big in the premises. The column "Consequence" shows the value of y^i calculated by the function g^i of each implication and "Tv" shows the truth value of $/y = y_i/$ which is calculated, according to the logical connective in the premise. For example,

$$/ y = y^i / = / x_1{}^o = small_1 / \wedge / x_2{}^o = small_2 /$$

$$= small_1(x_1{}^o) \wedge small_2(x_2{}^o)$$

$$= 0.25 \quad (17)$$

The value inferred by the implications is obtained

$$y = \frac{0.25 \times 17 + 0.2 \times 24 + 0.375 \times 15}{0.25 + 0.2 + 0.375}$$

$$\simeq 17.8 \quad (18)$$

3. DERIVATION OF FUZZY CONTROL RULES

3-1 Three Types of Derivation

There are, in general, four methods to derive fuzzy control rules. We shall briefly discuss them. In this paper we deal with the third one.

1) based on operator's experiance and/or control engineer's knowledge

Most of the reported fuzzy controllers have been designed by this method[3]. One of those aspects are hueristics, trial and error, or something like that. So the method is not always generalized nor formalized up to so called a design procedure of a fuzzy controller.

Here we just mention the following. Sometimes we face the fact that an operator cannot tell linguistically in details why he takes such and such a control action in some particular situation.

2) based on the fuzzy model of a process

Some studies[1, 4] have been reported on this basis. The idea itself is nothing but what is adopted in classic or modern control theory. So it seems to be very desirable and could be said logical in a sense. It also provides a way to theoretically analyse fuzzy control systems. A problem here is concerned with a method of fuzzy identification of a process by which we mean parameter identification in a fuzzy model. We need to this purpose a general mathemataical tool to describe a process just like a linear equation. The method of fuzzy reasoning discussed in this paper has been developed also under this respect.

3) based on operator's control actions

First let us notice that deriving fuzzy control rules is the same as making a fuzzy model of operator's control.

If we have a way to express a human operator's control actions by a set of fuzzy control rules, this method becomes very useful when operation data are available. Fortunately in cases where computer control is planned in place of human control it is easy to obtained a lot of data concerned with operator's control actions. To the aim of identification we have to make a format of a control rule, and reduce the problem to parameter identification.

4) based on leraning

Theoretically the self-organizing design of a fuzzy controller[6] is vary interesting. There are, however, is many difficulties in applying this method to industrial process except the case where experiments using real plants are possible.

3-2 Algorithm to Derive Control Rules

Now let us consider an algorithm of the identification. The format of a control rule is the same to that previously discussed.

When a set of data $(x1j, x2j, \cdots, xkj)$ $\rightarrow yj$ ($j = 1, \cdots, n$) is given, the outline of the algorithm is the following.

1) Divide fuzzily k-dimentional space concerned with the variables $x_1 \sim x_k$ into some number of areas. It is performd by human operator's or engineer's knowledge, e.g., what type of propositions to be set for instance, $A_1(x_1)$ incresing, $A_2(x_2)$ decreasing so and so. These areas mean the premises of rules. In each area the following three steps are performed.

2) Form the logical function f and calculate the truth value

$$/ f(x_1 = A_1, \cdots\cdots, x_k = A_k)/$$

corresponding to each data $x_{1j} - x_{kj}$ (j = 1, ••• , m) available in this area, which is called a weighting factor w_j to j-th datum.

3) Calculate the parameters $p_0, \cdots\cdots, p_k$ in equation (19) by the weighted linear regression analysis using the data with weighting factor w_j.

$$y = p_0 + p_1 \times x_1 + \cdots\cdots + p_1 \times x_k \quad (19)$$

4) Rule R in this area is finally expressed as

$$R : \text{If } f(x_1 = A_1, \cdots\cdots, x_k = A_k)$$

$$\text{then } y = p_0 + p_1 \times x_1 + \cdots\cdots + p_k \times x_k \quad (20)$$

The weighted regression is derived from natural extension dividing the input space into fuzzy subspaces instead of crisp ones.

Other methods deriving control rules roughly discussed above are desirable to be combined with each other since for instance the obtained data may not always cover the whole control area: in some area there may be no data or few for identifidation.

4. FUZZY MODELLING OF HUMAN OPERATOR'S CONTROL ACTIONS

4-1 Water Cleaning Process

We shall now show a real example where operator's control actions are fuzzily modelled.

The control process is a water cleaning process for civil water supply as is illustrated in Fig. 2. In the process, turbed river water first comes into a mixing tank where chemical products called PAC and also chlorine are put and mixed in the water. Then the mixed water flows into a sedimentation tank where the turbed part of water is cohered with the aid of PAC and settled to the bottom. After sedimentation which takes about 3 - 5 hours depending on the capacity of the tank, the treated water finally flows into a filtration tank producing clean water. Chlorine is put only for sterilization of water.

The main control problem of a human operator in this process is to decide the amount of PAC to be put so that the turbidity of the treated water is kept below a certain level. There is the optimal amount not too little nor much depending on the properties of the turbed water. The amount of PAC must be controlled also from an economical point of view.

The process is characterized by a lack of physical model, thus the great importance of operator's experience, big change of the turbidity of river water and the fact that turbidity itself is not clearly defined nor accurately measured.

However, a number of variables influencing sedimentaion process have been found so far which can be measured. Let us first list up all the variables concerned.

TB1 : turbidity of the original water(ppm)
TB2 : turbidity of the treated water(ppm)
PAC : amount of PAC(ppm)
TE : temparture of water(°C)
PH : pH
AL : alkalinity
CL : amount of chlorine(ppm)

For example if TE is lower, then PAC is more necesarry. Both PH and AL affect nonlinearly the necessary amount of PAC. The optimal PAC depends on these variables: the relation among them is not clear. There are some other variables influencing the process, e.g., plankton in the river water which increases in Spring time but cannot be measured at present.

In most of water cleaning plants a statistical model has been built. However the models are are not accurate. These cover only steady state, i.e., a small range of TB1. TB1 increases for example 100 times more when it rains. So an operator controls PAC taking into account of TB1, TE, PH, AL, TB2. Now our process can be illustrated as in Fig. 3.

4-2 Decision of Conrol Rules

The authors have got a lot of operation data where all the variables are measured every hour for four months. That is, the number of data is 24 hours x 30 days x 4 months = 2880. Table 2 shows a part of them.

Among the data we have used about 600 for the identification which are taken in June and July. June in Japan is a rainy season and July is summer.

According to the identification algorithm discussed previously, 64 control rules are derived which can be called a fuzzy model of operator's control as is stated, where a control rule is of the form

(TB1, TB2, PH, TE) → PAC (21)

where AL is omitted for simplicity. Some of them are shown in Fig. 4.

4-3 Results of Fuzzy Control

The results obtained from a fuzzy model are shown in Table 3 as well as operator's control and the results of a statistical model. Table 2 shows that an operator's control actions are well modelled in the form of fuzzy control rules.

We have divided the original data set into two sets; identifying data and testing one. The testing data is used as we can test the performance of the identified rules in various cases.

The statistical model is represented in equation (22) that is often used in water cleaning process.

$$D = 9.11\sqrt{TB1} - 79.8pH + 12.7Cl + 1255.6 \quad (22)$$

The average of absolute differences between operator's action and the results of fuzzy model, and statistical model for 38 testing data are

fuzzy model	:	72.6
satatistical model	:	128.0

These results show the excellence of the fuzzy model.

5. CONCLUSIONS

The authors started this work to implement a fuzzy controller in a water cleaning plant. To this aim it is necessary to derive control rules from operator's control actions since we faced the fact that operators could not tell well what they ware doing when they ware interviewed.

In this paper we have presented a fuzzy reasoning algorithm and a rule identification algorithm based on the operator's control action's. Further applying these algorithms to the control of water cleaning process we have observed the efficiency of the proposed methods.

The algorithm for deriving control rules still depends partly on try and error. Both theoretical study and experience are necessary to systemize it.

The suggested method concerned with reasoning has been developed as a general tool for fuzzy modelling of a sytem. To this respect fuzzy modelling of a dynamical system is now also under consideration.

REFERENCES

[1] Braae, M., and D. A. Rutherford (1979). Theoretical and Linguistic Aspects of the Fuzzy Logic Controller, Automatica, 15, 553-557

[2] Holmblad, L. P., and J. J. Ostergaard (1982). Control of a Cement Kiln by Fuzzy Logic, in Fuzzy Information and Decision Process (M. M. Gupta and E. Sanchez eds.) North-Holland

[3] Mamdani, E. H., and S. Assilian (1975). An Experiment in Linguistic Synthesis with a Fuzzy Logic Controller, Int. J. Man-Mach. Stud., 7, 1-13

[4] Sugeno, M. and T. Takagi (1982). A New Approach to Designe of Fuzzy Controller, in Advances in Fuzzy Set Theory and Applications, (P.P Wang ed.) Plenum

[5] Tsukamoto, Y., (1975). An approach to fuzzy reasoning method, in Advances in Fuzzy Set Theory and Applications (M.M. Gupta ed.), pp 407-428

[6] Yamazaki, T., and E. H. Mandani (1982). On the Performance of a Rule-based Self-organising Controller, Proc. of IEEE Conf. of Applications of Adaptive and Mutivariable Control, Hull

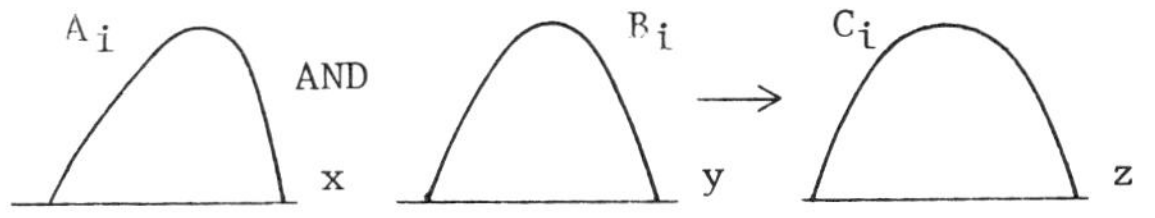

Fig. 1. Ordinary Fuzzy Implication

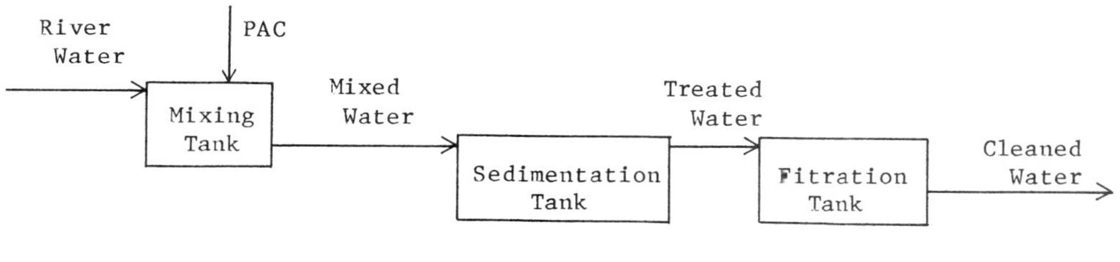

Fig. 2. Water Cleaning Process

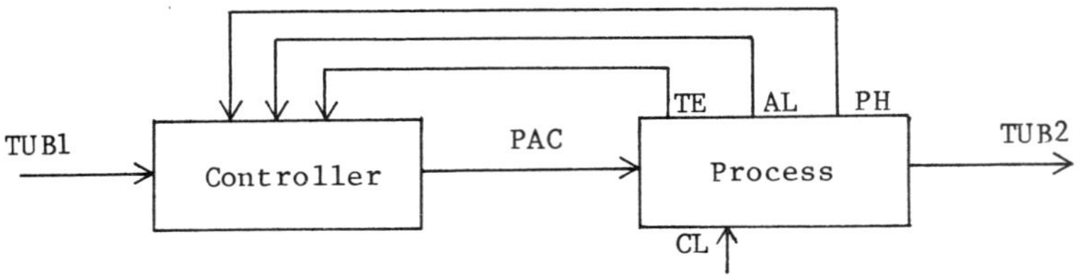

Fig. 3. Diagram of Control Process

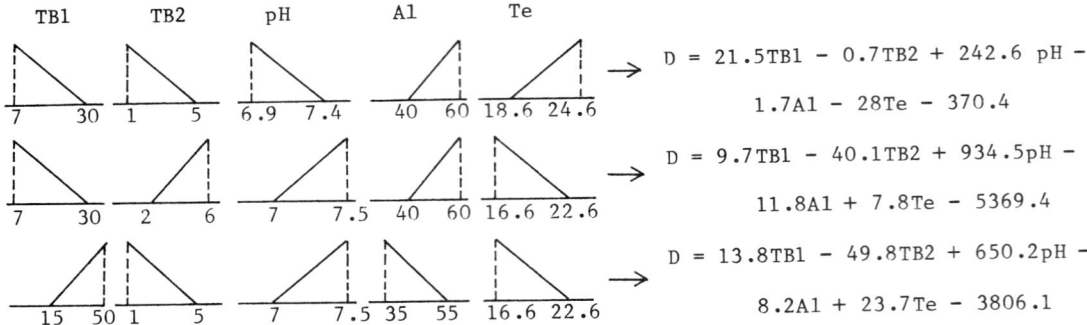

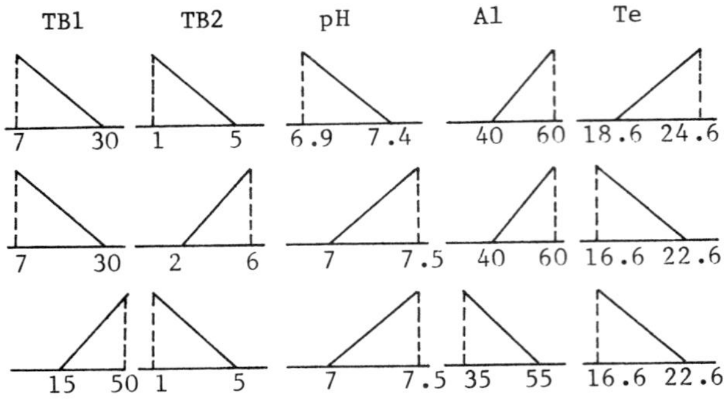

→ D = 21.5TB1 − 0.7TB2 + 242.6 pH − 1.7A1 − 28Te − 370.4

→ D = 9.7TB1 − 40.1TB2 + 934.5pH − 11.8A1 + 7.8Te − 5369.4

→ D = 13.8TB1 − 49.8TB2 + 650.2pH − 8.2A1 + 23.7Te − 3806.1

Fig. 4. Example of Control Rules

TABLE 1 Reasoning Algorithm

Rule	Premise		Consequence	Tv
R1	small (0, 16) → .25	small (0, 8) → .375	y = 12 + 5 = 17	.25 ∧ .375 = .25
R2	big (10, 20) → .2		y = 2 × 12 = 24	.2
R3		big (2, 10) → .375	y = 3 × 5 = 15	.375
	x_1 = 12	x_2 = 5		

TABLE 2 Operation Data

TUB1	PH	TE	PAC	TUB2
10.0	7.1	18.8	1300	1.0
17.0	7.0	18.6	1300	1.0
22.0	7.3	19.4	1400	2.0
50.0	7.1	19.5	1400	1.0
9.0	7.3	23.3	900	4.0
11.0	7.1	20.7	900	1.0
12.0	7.2	21.3	900	3.0
14.0	7.2	23.6	900	4.0
35.0	7.0	17.8	1200	1.0
20.0	7.0	16.6	1100	1.0
20.0	6.9	17.8	1100	1.0
18.0	7.1	17.3	1100	1.0
12.0	7.2	18.8	900	3.0
8.0	7.2	18.0	1000	1.5
11.0	7.1	19.2	1000	2.0
50.0	7.0	18.0	1200	1.5
35.0	7.0	17.7	1200	1.5
30.0	7.0	17.3	1100	1.5
16.0	7.1	19.3	1100	3.0

TABLE 3 Result of Fuzzy Control

Operator	Statistical Model	Fuzzy Model
1300	994.7	1105.7
1300	995.9	973.5
1300	1119.6	1102.0
1400	1151.1	1242.0
1400	1409.4	1472.4
900	1066.4	920.1
900	1068.9	952.5
900	1012.3	865.9
1200	1286.8	1171.7
1200	1246.8	1154.3
1100	1151.4	1131.4
1100	1199.5	1145.6
1100	1159.4	1114.2
1000	985.7	930.2
1000	1009.3	941.6
1000	1038.2	940.7
1200	1398.3	1219.6
1200	1290.6	1164.3

SOME OF THE PROPERTIES OF FUZZY DISCRETISATION

D. Willaeys

Laboratoire d'Automatique et Systèmes Homme-Machine, Laboratoire Associé au C.N.R.S. n°370, Université de Valenciennes et du Hainaut Cambrésis, 59326 Valenciennes, France

Abstract. The representation of subjective informations with fuzzy sets or distributions of possibility brings up the problem of numerizing these data for processing it by a computer. Two types of representation may be used for this : the first type consists in assigning to the value of the membership function of a fuzzy set, an explicit equation ; the second type requires the discretisation of the referential into a certain number of segments, and labeled by a point. A fuzzy set may be defined by assigning a value to the membership function to each segment. This latter technique has the disavantage of requiring, in order to get acceptable accuracy, the memorization of a large number of values for each fuzzy set or fuzzy relation. In order to partially overcome these disavantages, we proposed a method called "Fuzzy discretisation" which consists in cutting up a real space, not into disconnected segments, but rather into a series of non-disconnected fuzzy sets. This article's object is to extend the definition of this method to n-ary fuzzy relations and to demonstrate that using it alters only slightly the data treated.

Keywords. Fuzzy discretisation ; fuzzy chopping.

INTRODUCTION

The use of a computer for processing subjective information formalized by fuzzy relations brings up the problem of numerizing these data.

Two types of representation may be used for this. The first type consists in assigning to the value of the membership function of a fuzzy set, A of X = R, an explicit equation for instance, the function S proposed by ZADEH, or perhaps triangular or bell-shaped functions, which rather easily allow arithmetic operations on fuzzy numbers (DUBOIS, PRADE, 1978).

The second type requires the discretisation of the X referential into a certain number of segments, generally equal, and labeled by a point x_i ; X is then represented by a discrete space $\{x_i\}$. A sub-set A of X may be defined in $\{x_i\}$, by assigning a value of the membership function $\mu_A(x_i)$ to each segment x_i.

This latter technique is mainly used for implanting fuzzy algorithms in an industrial computer (KING, 1975 ; TONG, 1976) ; it has the disavantage of requiring, in order to get acceptable accuracy, the memorization of a large number of values for each fuzzy sub-set.

In the case of multidimensional fuzzy relations, the number of values to be stored in the computer memory grows as the relation's n-dimensional power. Thus, if a discrete set of N points constitutes each dimension of the relation, N^n values must be stored - which can interfere with the use of mini -or micro- computers, both from the viewpoint of memory loading and from that of processing calculation speed.

In order to partially overcome these disavantages, we proposed (WILLAEYS, 1978, 1981) a method called "Fuzzy discretisation" which consists in cutting up a real space, not into disconnected segments, but rather into a series of non-disconnected fuzzy sets.

This article's object is to extend the definition of 'Fuzzy Discretisation " to n-ary fuzzy relations, and to demonstrate that using it alters only slightly the data treated.

FUZZY DISCRETISATION OF A REAL SPACE

Definition

Let us take a set X = R and series $\{\chi_j\}$ of fuzzy subsets of X the membership function $\mu_{\chi_j}(x)$ of which is such that

$$\exists_j : \mu_{\chi_j}(x) \neq 0 \quad \forall x \in X \qquad (1)$$

that is, that each point of X is "known" in at least one fuzzy subset χ_j.
Now let us take a fuzzy subset $A \subset X$ having the membership function $\mu_A(x)$. A may be defined in relation to the series $\{\chi_j\}$, used as the "fuzzy referential χ" by the expression :

$$\mu_A(\chi_j) = \bigvee_x(\mu_A(x) \wedge \mu_{\chi_j}(x)) \qquad (2)$$

By this definition, a point x of X is defined in χ by :

$$\mu_x(\chi_j) = \mu_{\chi_j}(x)$$

and will be known at most by :

$$\bigvee_j \mu_{\chi_j}(x)$$

If A is an ordinary set, then (2) is written :

$$\mu_A(\chi_j) = \bigvee_{x \in A} \mu_{\chi_j}(x)$$

the fuzzy set A is totally known in χ if $A \subseteq \bigcup_j \chi_j$.

Note that definition (2) has the same equation as the measure of possibility of a fuzzy subset proposed by ZADEH (1978) : if $\mu_\chi(x)$ is considered as the possibility distribution of a variable χ, $\mu_A(\chi)$ is the measure of possibility that χ is A.

Reconstitution of a Fuzzy Subset

Reciprocally, it is also possible to define the opposite passage, from χ to $X_R \subseteq X$, X_R being the best knowledge of X in χ, by :

$$\mu_{A_R}(x) = \bigvee_j (\mu_A(\chi_j) \wedge \mu_x(\chi_j)) \qquad (3)$$

$$= \bigvee_j (\mu_A(\chi_j) \wedge \mu_{\chi_j}(x))$$

X can be totally reconstituted if it is completely known in χ, that is :

$$X_R = X \quad \text{if } \bigcup_j \chi_j = X$$

$$\text{or yet if } \bigvee_j \mu_{\chi_j}(x) = 1 \quad \forall x$$

Discretisation and Reconstitution of a Fuzzy Relation

By extension to a multidimensional space, definitions (2) and (3) apply to fuzzy relations in the following manner :
Let R be a fuzzy relation in X x Z = R^2, with the membership function $\mu_R(x,z)$ and a series $\{\gamma_\ell\}$ on X x Z, the discretisation of R in $\{\gamma_\ell\}$ used as a "fuzzy referential γ", is then defined by :

$$\mu_R(\gamma_\ell) = \bigvee_x \bigvee_z (\mu_R(x,Z) \wedge \mu_{\gamma_\ell}(x,z)) \qquad (4)$$

and the reconstitution, by :

$$\mu_{R_R}(x,z) = \bigvee_{\gamma_\ell} (\mu_R(\gamma_\ell) \wedge \mu_{\gamma_\ell}(x,z)) \qquad (5)$$

If the series $\{\chi_j\}$ on X and $\{\zeta_k\}$ on Z exist, the series $\{\gamma_\ell\}$ can be defined by the direct product $\{\gamma_\ell\} = \{\chi_j\} \times \{\zeta_k\}$.

Definitions (4) and (5) then become :

$$\mu_R(\chi_j,\zeta_k) = \bigvee_x \bigvee_z (\mu_R(x,z) \wedge \mu_{\chi_j}(x) \wedge \mu_{\zeta_k}(z)) \qquad (6)$$

and

$$\mu_{R_R}(x,y) = \bigvee_{\chi_j} \bigvee_{\zeta_k} (\mu_R(\chi_j,\zeta_k) \wedge \mu_{\chi_j}(x) \wedge \mu_{\zeta_k}(z)) \qquad (7)$$

Equation of the Discretisation and Reconstitution Fuzzy Relations

In order to streamline the notation, it is useful to formulate these definitions by means of fuzzy relation equations. Definition (2) corresponds not only to the definition of the measure of possibility, but also to the composition of fuzzy relations ; for this definition matches a fuzzy subset of χ to a fuzzy subset of X = R.

If we note as f the fuzzy function of X in χ having the membership function :

$$\mu_f(x,\chi_j) = \mu_{\chi_j}(x) = \mu_x(\chi_j)$$

and if we note as α the fuzzy subset of χ with the membership function $\mu_A(\chi_j)$, then we can write :

$$\alpha = f \circ A$$

In like manner, definition (3) can be written :

$$A_R = f^T \circ \alpha$$

noting f^T transposed from the relation f such that :

$$\mu_{f^T}(\chi_j,x) = \mu_f(x,\chi_j)$$

Finally, for fuzzy relations, by noting as ρ the fuzzy subset of γ arising from $R \subset X \times Z$ by the relation q of X x Z in γ, having the membership function $\mu_q(x,z,\gamma_\ell)$, then relation (4) is written :

$$\rho = q \circ R$$

and (5) becomes :

$$R_R = q^T \circ \rho \text{ with } \mu_{q^T}(\gamma\ell \,;\, x,z) = \mu_q(x,z,\gamma_\ell)$$

In the case that $\{\gamma_\ell\} = \{\chi_j\} \times \{\zeta_k\}$, relation q has as its membership function :

$$\mu_q(x,z\ ;\ \chi_j,\zeta_k) = \mu_{\chi_j}(x) \wedge \mu_{\zeta_k}(z)$$

From now on, only this case will be considered for fuzzy relations.

COMPARATIVE PROPERTIES OF FUZZY SETS'UNION AND INTERSECTION IN REAL SPACE AND DISCRETE SPACE

Comparison of the Discretisation of Union and Intersection with These Operations in the Initial Space

The properties of the composition allow us to write :

Proposition P1. If $A \subseteq B$ then $\alpha \subseteq \beta$, and if $A = B$ then $\alpha = \beta$
This proposition proceeds from :

$$\alpha = f \circ A \quad \text{and} \quad \beta = f \circ B$$

Likewise,
Proposition P2. If $\alpha \subseteq \beta$ then $A_R \subseteq B_R$, and if $\alpha = \beta$ then $A_R = B_R$ due to the fact $A_R = f^T \circ \alpha$ and $B_R = f^T \circ \beta$

If we consider union
Proposition P3. $\alpha \cup \beta = f \circ (A \cup B)$ and $(A \cup B)_R = A_R \cup B_R$ for

$$f \circ (A \cup B) = (f \circ A) \cup (f \circ B) \text{ and}$$

$$(A \cup B)_R = f^T \circ (\alpha \cup \beta) = f^T \circ \alpha) \cup (f^T \circ \beta)$$

Finally for intersection
Proposition P4. $\alpha \cap \beta \supseteq f \circ (A \cap B)$ and $(A \cap B)_R \subseteq A_R \cap B_R$ for

$$f \circ (A \cap B) \subseteq (f \circ A) \cap (f \circ B) = \alpha \cap \beta$$

and

$$f^T \circ (\alpha \cap \beta) \subseteq (f^T \circ \alpha) \cap (f^T \circ \beta) = A_R \cap B_R$$

consequently, passing from X to χ does not modify either inclusion or union, and overestimates intersection.

Properties of the Reconstitution

Now let us consider the problem of reconstitution (3)

$$A_R = f^T \circ \alpha = f^T \circ (f \circ A) = (f^T \circ f) \circ A$$

$f^T \circ f$ having the membership function

$$\mu_{f^T \circ f}(x,x_R) = \underset{\chi_j}{V}(\mu_{\chi_j}(x) \wedge \mu_{\chi_j}(x_R)) \qquad (8)$$

This fuzzy relation can be split up as follows :

$$f^T \circ f = D \cup (f_T \circ f) \qquad (9)$$

with $\mu_D(x,x_R) = \underset{\chi_j}{V}\ \mu_{\chi_j}(x)$ if $x = x_R$
$= 0$ elsewhere ;

for (8) can in fact be written :

$$\mu_{f^T \circ f}(x,x_R) = \underset{\chi_j}{V}(\mu_{\chi_j}(x_R) \wedge \mu_{\chi_j}(x_R))\ V\ \underset{\chi_j}{V}(\mu_{\chi_j}(x) \wedge \mu_{\chi_j}(x_R))$$

$$= \left[\underset{\chi_j}{V}(\mu_{\chi_j}(x_R))\right]\ V\ \left[\underset{\chi_j}{V}(\mu_{\chi_j}(x) \wedge \mu_{\chi_j}(x_R))\right]$$

with relation (9), it is possible to write :

$$A_R = (D \cup (f^T \circ f)) \circ A = (D \circ A) \cup (f^T \circ f \circ A) \qquad (10)$$

and thus to obtain :

$$A_R \supseteq D \circ A \qquad (11)$$

which leads to the following proposition :
Proposition P5. If $A \subseteq \underset{j}{\cup}\chi_j$ then $A \subseteq A_R$

for if $A \subseteq \underset{j}{\cup}\chi_j$ then $D \circ A = A$

In fact, $\mu_{D\circ A}(x_R) = \underset{x}{V}\ (\mu_A(x) \wedge \mu_D(x,x_R))$

$$= \mu_A(x_R) \wedge \mu_D(x,x_R)$$

$$= \mu_A(x_R) \wedge (\underset{\chi_j}{V}\ \mu_{\chi_j}(x_R))$$

and as

$$\mu_A(x) \leqslant \underset{\chi_j}{V}\ \mu_{\chi_j}(x),\ D \circ A = A \qquad (12)$$

Starting with proposition P5, we can again find the property previously demonstrated in (WILLAEYS, 1978).

Proposition P6 : If $X_R = X$, then $A \subseteq A_R$.
If $\underset{\chi_j}{V}\ \mu_{\chi_j}(x) = 1$, the diagonal fuzzy relation D is unitary, i.e., D = I. Relation (11) ,therefore, is written : $A_R \supseteq I \circ A \Rightarrow A_R \supseteq A$

To sum up, if fuzzy subset A is completely known in χ, that is, if A is enclosed in X_R (that being the best knowledge of X in χ), then A is augmented by A_R.

It is still possible to refine our "a priori" knowledge of A_R by remarking that in the general case of the fuzzy composition R on X x Y with a fuzzy subset $A \subseteq X$, the result $B \subseteq Y$ is such that :

$$B = R \circ A \subseteq \underset{X}{Proj}\ R \cap \underset{X}{Proj}\ A \subseteq \underset{X}{Proj}\ A \qquad (13)$$

for $\mu_B(y) = \underset{x}{V}(\mu_A(x) \wedge \mu_R(x,y)) \leqslant \left[\underset{x}{V}(\mu_A(x))\right] \wedge \left[\underset{x}{V}\mu_R(x,y)\right]$

therefore $\mu_B(y) \leqslant \underset{x}{V}\ \mu_A(x) \quad \forall y$

If we go back to relation (10)

$$A_R = (D \circ A) \cup (f^T \circ f \circ A)$$

and consider set $\{x_m\}$ of X such that :

$$\mu_A(x_m) = \underset{x}{V}\ \mu_A(x)$$

i.e., the set of points where the membership function is maximum, always respecting $A \subseteq X_R$, for these points x_m, $D \circ A = A$ and therefore :
$A_R = (D \circ A) \cup (f^T \circ f \circ A) = A$ for $x = x_m$

In fact, on one hand relation (12) yields $D \circ A = A$ if A is known in X_R and on the other hand, relation (13) applied to $f^T \circ f \circ A$ allows the deduction that

$$f^T \circ f \circ A \subseteq \underset{x}{\mathrm{Proj}}\, A$$

Thus, $A_R = (D \circ A) \cup (f^T \circ f \circ A)$ will be written, in the particular case of points x_m :

$$\text{for } x \in \{x_m\} \quad A_R = A \cup (f^T \circ f \circ A) = (\underset{x}{\mathrm{Proj}}\, A) \cup (f^T \circ f \circ A)$$

and as $f^T \circ f \circ A \subseteq \underset{x}{\mathrm{Proj}}\, A \quad \forall x$

for $x \in \{x_m\}$ then $A_R = A$

or yet $\mu_{A_R}(x_m) = \mu_A(x_m)$

This allows us to write the following proposition :

Proposition P5'. If $A \subseteq \underset{j}{\cup} \chi_j$ then $A \subseteq A_R$, and particularly, $A = A_R$ for all those points where $\mu_A(x)$ is maximal.

If we now examine the set $\{x_0\} \subset X$ such that $\mu_A(x_0) \geq \underset{\chi_j}{\vee} \mu_{\chi_j}(x_0)$,

that is, the points where A is not entirely known in X_R, then $D \circ A = \underset{j}{\cup} \chi_j$ for $x \in \{x_0\}$

since D is the diagonal relation, and relation (13) applied to $(f^T \circ f) \circ A$ can also be written :

$$(f^T \circ f) \circ A \subseteq \underset{x}{\mathrm{Proj}}(f^T \circ f) \cap \underset{x}{\mathrm{Proj}} A \subseteq \underset{x}{\mathrm{Proj}}(f^T \circ f)$$

therefore :

$$\mu_{(f^T \circ f) \circ A}(x_R) \leq \underset{x}{\vee} \left(\underset{\chi_j}{\vee} (\mu_{\chi_j}(x) \wedge \mu_{\chi_j}(x_R)) \right) = \underset{\chi_j}{\vee} (\mu_{\chi_j}(x_R) \wedge (\vee \mu_{\chi_j}(x)))$$

$$\mu_{(f^T \circ f) \circ A}(x_R) \leq \underset{\chi_j}{\vee} (\mu_{\chi_j}(x_R))$$

and $A_R = (D \circ A) \cup (f^T \circ f \circ A)$ will be written, in the particular case of the points of $\{x_0\}$:

$$\text{for } x \in \{x_0\} \quad A_R = (\underset{j}{\cup} \chi_j) \cup (f^T \circ f \circ A)$$

$$A_R = \underset{j}{\cup} \chi_j \text{ since } f^T \circ f \circ A \subseteq \underset{j}{\cup} \chi_j$$

This allows us to state the following proposition :

Proposition P5''. For every x where $\mu_A(x) \leq \underset{\chi_j}{\vee} \mu_{\chi_j}(x)$ then $\mu_A(x) \leq \mu_{A_R}(x)$ and particularly $\mu_A(x) = \mu_{A_R}(x)$ if $\mu_A(x) = \underset{x}{\vee} \mu_A(x)$

For every x where $\mu_A(x) \geq \underset{\chi_j}{\vee} \mu_{\chi_j}(x)$ then $\mu_{A_R}(x) = \vee \mu_{\chi_j}(x)$

Using these propositions, it is now possible to compare the reconstitutions of union and intersection with these same operations carried out in the initial referential.

Comparison of the Reconstitution of Union and Intersection to These Operations in the Initial Space

With the previously defined notations, we can write :

Proposition P7. $f^T \circ (\alpha \cup \beta) \supseteq A \cup B$ if A and $B \subseteq \underset{j}{\cup} \chi_j$

Propositions P2 and P3 let us write :

$$f^T \circ (\alpha \cup \beta) = f^T \circ f \circ (A \cup B) = (A \cup B)_R$$

and P5 then gives $(A \cup B)_R \supseteq A \cup B$

Likewise, comparison of intersection lets us write :

Proposition P8. $f^T \circ (\alpha \cap \beta) \supseteq A \cap B$ if A and $B \subseteq \underset{j}{\cup} \chi_j$

This follows in fact from propositions P2, P4 and P5 :

$$f^T \circ (\alpha \cap \beta) \supseteq f^T \circ f \circ (A \cap B) = (A \cap B)_R \supseteq A \cap B$$

To sum up, if we effect unions or intersections in χ, the reconstitution of the results includes those obtained by these same operations effected directly in X.

In order to be able to use this fuzzy discretisation within the framework of fuzzy relations, it remains to be demonstrated, this same property of increase, for the composition of fuzzy relations.

COMPARATIVE PROPERTIES OF THE COMPOSITION OF FUZZY RELATIONS IN THE INITIAL SPACE AND THE DISCRETE SPACE

Comparison Between the Discrete and Non Discrete Compositions

Let us take two fuzzy relations R_1 of $X \times Z$ and R_2 of $Z \times T$, and their associates ρ_1 of $\chi \times \zeta$ and ρ_2 of $\zeta \times \tau$ by transformation (6) using the following passage functions :

f of X in χ with $\mu_f(x, \chi_j) = \mu_{\chi_j}(x)$

g of Z in ζ with $\mu_g(z, \zeta_k) = \mu_{\zeta_k}(z)$

h of T in τ with $\mu_h(t, \tau_\ell) = \mu_{\tau_\ell}(t)$

χ, ζ, τ being the series of fuzzy subsets defined respectively on X, Z, T the transform ρ_3 of $R_2 \circ R_1 \subset X \times T$ in $\chi \times \tau$ has the membership function :

$$\mu_{\rho_3}(\chi_j, \tau_\ell) = \underset{x}{\vee} \underset{t}{\vee} \left[\underset{z}{\vee} (\mu_{R_1}(x,z) \wedge \mu_{R_2}(z,t)) \wedge (\mu_{\tau_\ell}(t) \wedge \mu_{\chi_j}(x)) \right] \quad (14)$$

the membership function of $\rho_2 \circ \rho_1$ can be modified as follows :

$$\mu_{\rho_2 \circ \rho_1}(\chi_j, \tau_\ell) = \underset{\zeta_k}{\vee} \left[\mu_{\rho_1}(\chi_j, \zeta_k) \wedge \mu_{\rho_2}(\zeta_k, \tau_\ell) \right]$$

$$= \underset{\zeta_k}{\vee}\left[\left[\underset{x}{\vee}\ \underset{z}{\vee}\left[\mu_{R_1}(x,z)\wedge\mu_{\chi_j}(x)\wedge\mu_{\zeta_k}(z)\right]\right]\wedge\left[\underset{z}{\vee}\ \underset{t}{\vee}\left[\mu_{R_2}(z,t)\wedge\mu_{\zeta_k}(z)\wedge\mu_{\tau_\ell}(t)\right]\right]\right]$$

$$= \underset{\zeta_k}{\vee}\ \underset{x}{\vee}\ \underset{t}{\vee}\left[\left[\underset{z}{\vee}(\mu_{R_1}(x,z)\wedge\mu_{\chi_j}(x)\wedge\mu_{\zeta_k}(z))\right]\wedge\left[\underset{z}{\vee}(\mu_{R_2}(z,t)\wedge\mu_{\zeta_k}(z)\wedge\mu_{\tau_\ell}(t))\right]\right]$$

$$\geqslant \underset{\zeta_k}{\vee}\ \underset{x}{\vee}\ \underset{t}{\vee}\left[\underset{z}{\vee}\left[\mu_{R_1}(x,z)\wedge\mu_{\chi_j}(x)\wedge\mu_{R_2}(z,t)\wedge\mu_{\zeta_k}(z)\wedge\mu_{\tau_\ell}(t)\right]\right]$$

$$= \underset{\zeta_k}{\vee}\ \underset{x}{\vee}\ \underset{t}{\vee}\left[\mu_{\chi_j}(x)\wedge\mu_{\tau_\ell}(t)\wedge\left[\underset{z}{\vee}(\mu_{R_1}(x,z)\wedge\mu_{R_2}(z,t)\wedge\mu_{\zeta_k}(z))\right]\right]$$

$$= \underset{x}{\vee}\ \underset{t}{\vee}\left[\mu_{\chi_j}(x)\wedge\mu_{\tau_\ell}(t)\wedge\left[\underset{z}{\vee}(\mu_{R_1}(x,z)\wedge\mu_{R_2}(z,t)\wedge(\underset{\zeta_k}{\vee}\ \mu_{\zeta_k}(z)))\right]\right]$$

If $\mu_{R_1}(x,z)$ or $\mu_{R_2}(z,t)$ are less than $\underset{\zeta_k}{\vee}\ \mu_{\zeta_k}(z)$, whatever z et t, then we find $\mu_{\rho_3}(\chi_j,\tau_\ell)$; this allows us to write :

Proposition P9. If $\underset{x}{\text{Proj}}R_1$ or $\underset{t}{\text{Proj}}R_2 \subseteq \underset{k}{\cup}\zeta_k$ then $\rho_3 \subseteq \rho_2 \circ \rho_1$

If we examine the particular case where one of the fuzzy relations is a fuzzy subset $B = R \circ A$, with $B \subset Z$, $R \subset X \times Z$, $A \subset X$ by noting respectively $\beta \subset \zeta$, $\rho \subset \chi \times \zeta$, $\alpha \subset \chi$ their transform by f et g with

$$\mu_f(x,\chi_j) = \mu_{\chi_j}(x) \text{ et } \mu_g(z,\zeta_k) = \mu_{\zeta_k}(z)$$

we obtain the following proposition :

Proposition P10. If $A \subseteq \underset{j}{\cup}\chi_j$ then $\beta \subseteq \rho \circ \alpha$

It now remains to examine the reconstitution of a fuzzy relation.

Comparison of the Reconstitution of a Fuzzy Relation With the Original Relation

Let us then take $R \subset X \times Z$ and its transform $\rho \subset \chi \times \zeta$ by the function $q = f \times g$ having the membership function :

$$\mu_q(x,z,\chi_j,\zeta_k) = \mu_{\chi_j}(x)\wedge\mu_{\zeta_k}(z)$$

the reconstitution R_R of R can be written :

$$R_R = q^T \circ \rho = (q^T \circ q) \circ R$$

By following the same reasoning as for propositions P5, P5', P5", R_R can be written :

$$R_R = (q^T \circ q) \circ R = (D_2\ \cup(q^T \circ q)) \circ R$$

Noting for D_2, the diagonal relation :

$$\mu_{D_2}(x,z,z_R,z_R) = \underset{\chi_j}{\vee}\ \underset{\zeta_k}{\vee}\ (\mu_{\chi_j}(x)\wedge\mu_{\zeta_k}(z))$$

if $x = x_R$ and $z = z_R$
$=0$ elsewhere which allows us to get this inequation :

$$R_R \supseteq D_2 \circ R \qquad (15)$$

Noting $\overline{\chi_j}$ the cylindrical extension of χ_j to ζ and $\overline{\zeta_k}$ the cylindrical extension of ζ_k to χ with the membership functions :

$$\mu_{\overline{\chi_j}}(x,z) = \mu_{\chi_j}(x) \qquad \forall z \text{ and}$$

$$\mu_{\overline{\zeta_k}}(x,z) = \mu_{\zeta_k}(z) \qquad \forall x$$

equation (15) then allows us to extend P5 to fuzzy relations by :

Proposition P11. If $R \subseteq (\underset{j}{\cup}\overline{\chi_j}) \cap (\underset{k}{\cup}\overline{\zeta_k})$ then $R \subseteq R_R$

Likewise, following the pathway that led to proposition P5' if $R \subseteq (\underset{j}{\cup}\chi_j) \cap(\underset{k}{\cup}\zeta_k)$ then $D_2 \circ R = R$ and relation (13) extended to the composition of two relations leads to :

$$(q^T \circ q) \circ R \subseteq \underset{x,z}{\text{Proj}}\ R$$

Subsequently, considering couples of points $(x,z)_m$ such that $\mu_R(x,z)$ is maximal and therefore $\mu_R(x,z) = \underset{x,z}{\text{Proj}}\ R$

Then $R_R = (D_2 \circ R) \cup (q^T \circ q) \circ R$
$= R \cup (q^T \circ q) \circ R$

When considering points $(x,z)_m$ we can write :

$$\forall (x,z)_m \quad R_R = \underset{x,z}{\text{Proj}}\ R \quad \text{since}$$

$$(q^T \circ q) \circ R \subseteq \underset{x,z}{\text{Proj}}\ R$$

which leads to the extension of P5'

Proposition P11'. If $R \subseteq (\underset{j}{\cup}\overline{\chi_j}) \cap (\underset{k}{\cup}\overline{\zeta_k})$

then $R \subseteq R_R$, and $R = R_R$ for each doublet (x,z) for which $\mu_R(x,z)$ is maximal.

Finally, for the set of points $\{x_0,z_0\} \subseteq X \times Z$ such that :

$$\mu_R((x,z)_0) \geqslant (\underset{\chi_j}{\vee}\ \mu_{\chi_j}(x_0)) \wedge (\underset{\zeta_k}{\vee}\ \mu_{\zeta_k}(z_0))$$

relation (13) lets us write that :

$$(q^T \circ q) \circ R \subseteq \text{Proj}\ (q^T \circ q) \subseteq (\underset{j}{\cup}\overline{\chi_j}) \cap (\underset{k}{\cup}\overline{\zeta_k})$$

and consequently :

$$R_R = D_2 \circ R \ \cup (q^T \circ q) \circ R$$

becomes for all (x_0,z_0) :
$\forall (x,z) \in \{x_0,z_0\}$

$$R_R = (\underset{j}{\cup}\overline{\chi_j}) \cap (\underset{k}{\cup}\overline{\zeta_k}) \cup (q^T \circ q) \circ R$$

$$= (\underset{j}{\cup}\overline{\chi_j}) \cap (\underset{k}{\cup}\overline{\zeta_k})$$

Proposition P5" extended to fuzzy relations then gives rise to the following proposition P11" :

Proposition P11". For all couples $(x,z) \in X \times Z$ where $R \subseteq (\underset{j}{\cup}\overline{\chi_j}) \cap (\underset{k}{\cup}\overline{\zeta_k})$ then $R_R \supseteq R$;

particularly $R_R = R$ if $\mu_R(x,z)$ is maximal. For all other points, $R_R = (\underset{j}{\cup}\overline{\chi_j}) \cap (\underset{k}{\cup}\overline{\zeta_k})$

It is now possible, using propositions P10 and P11 to state the comparative property of composition of two fuzzy relations in the real referential and the discrete referential.

Comparison of the Reconstitution of the Composition With the Original Composition

Using R_1 of X x Z, R_2 of Z x T and $R_3 = R_2 \circ R_1$ of X x T and their respectives transforms ρ_1 of $\chi \times \zeta$, ρ_2 of $\zeta \times \tau$ by functions $q = f \times g$, $r = q \times h$, $s = h \times f$, this property is stated :

Proposition P12. If $\underset{x}{\text{Proj}}\ R_1$ or $\underset{t}{\text{Proj}}\ R_2$ are totally known in ζ and if R_3 is totally known in $\chi \times \tau$, then :

$$s^T \circ (\rho_2 \circ \rho_1) \supseteq R_2 \circ R_1$$

In fact, the proposition P9 allows us to write $\rho_3 \subseteq \rho_2 \circ \rho_1$, therefore :

$$s^T \circ (\rho_2 \circ \rho_1) \supseteq s^T \circ \rho_3$$

as $s^T \circ \rho_3 = (s^T \circ s) \circ R_3 = (R_2 \circ R_1)_R$

and $(R_2 \circ R_1)_R \supseteq R_2 \circ R_1$ by proposition P11

then $s^T \circ (\rho_2 \circ \rho_1) \supseteq R_2 \circ R_1$

that is, the reconstitution of the composition of two fuzzy relations in the discrete referential incloses this composition in the original référential.

CONCLUSION

This set of properties thus allows us to effect operations on fuzzy subsets ou fuzzy relations indiscriminately in the real referential or an equivalently fuzzy discrete one.

The working principle of this technique, which will be illustrated by examples in the oral presentation, allows us either to effect our calculations in the discrete referential and to make decisions on the basic of the discrete results or to proceed with a reconstitution of the results in the original referential, for example, in order to proceed with a visual presentation of the data.

REFERENCES

Dubois D., Prade F. (1978). Operations on fuzzy numbers. International journal on system sciences , Vol. 9.

King P.J., Mandani E.H. (1975). The application of fuzzy control systems to industrial processes. 6ème Congrès IFAC, Boston.

Tong R.M. (1976). Analysis of fuzzy control algoritms using the relation matrix. International journal of Man-Machine studies, Vol. 8.

Willaeys D., Malvache N. (1978). Utilisation de la discrétisation floue pour le traitement d'informations floues par calculateur. Colloque international sur la théorie des sous-ensembles flous. Marseille.

Willaeys D. (1981). The use of fuzzy sets for the treatment of fuzzy information. International journal of fuzzy sets and system, Vol. 5, n°3.

Zadeh L.A. (1978). Fuzzy sets as a basis for a theory of possibility. International journal of fuzzy sets and system, Vol. 1.

LINGUISTIC PHASE PLANE — FUZZY CONTROL FOR NONLINEAR SYSTEMS

Deng Ju-Long* and Li Cong-xi**

**Dept. of Automation, Huazhong University of Science and Technology, Wuhan, China*
***Dept. of Mech. Engn. Huazhong University of Science and Technology Wuhan, China*

Abstract. In this paper, the linguistic phase plane of fuzzy control is developed, and some definition and theorem are proposed. According to this new concept and approaches, the fuzzy controller of hydraulic position servo realized by microcomputer is built, it gives quite satisfactory results in experiments on a practical system containing multistage relays , higher order dynamic equation and some hydraulic soft values.

Keywords. Fuzzy control, nonlinear system, lingustic phase plane, phase plane, control theory.

LINGUISTIC PHASE PLANE

Fuzzy controller have been initiated by E.H. Mamdani and others since 1974 (1-3). The control information of the fuzzy controller is presented by some linguistic variables such as: positive big (pb), negativ big (nb) etc.

The normal linguistic variable can be described by fuzzy number $\underset{\sim}{N}$ (4-5), see Fig 1., that

$$\underset{\sim}{N} = (m,\ \alpha,\ \beta)_{LR}$$

$$\mu(x) \leq 1,\quad \mu(x) = const,\quad x \in [c,\ d]$$

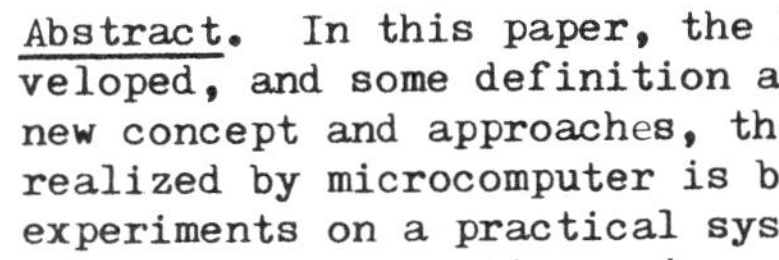

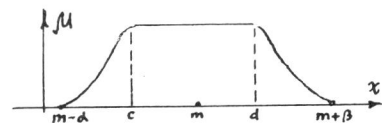

Fig. 1.

Now, let us consider the set of $\underset{\sim}{N}$ having same L.R. Denote the univers of discourse of $\underset{\sim}{N}$ by X and the universe of all fuzzy subset of $\underset{\sim}{N}$ by Y that

$$Y \triangleq \left\{ f \mid f : X \rightarrow [\lambda, 1],\ 0 < \lambda \leq 1 \right\}$$

In abbreviation, we do not distinguish the difference between a fuzzy subset and it's membership function, so that a fuzzy subset f of X can be defined by

$$f : X \rightarrow [\lambda, 1],\quad 0 < \lambda \leq 1 .$$

According to the equivalence relation R (denote by symbol $\sim$ or π), we have

$$[x] = \left\{ z \mid f(z) = f(x),\ z \in X \right\}$$

thus we call $[x]$ the equivalence class. When $[x]$ corresponds to fuzzy number $\underset{\sim}{N}$, therefore, we may denote

$$[x] \triangleq [f^{-1}(\underset{\sim}{N})]$$

where f^{-1} is only symbol. Now, we denote the quotient set of X by X/π

$$X/\pi = \left\{ [x], [y], [z], \cdots \right\}$$

Definition: When $[\alpha] \cap [\beta] = \phi$
$\forall [\alpha], [\beta] \in X/\pi$
and

$$x_i \leq x_{i+1},\quad \forall x_i, x_{i+1} \in [x],\ i = 0, 1, \cdots n$$

$$y_i \leq y_{i+1},\quad \forall y_i, y_{i+1} \in [y],\ i = 0, 1, \cdots m$$

$$z_i \leq z_{i+1},\quad \forall z_i, z_{i+1} \in [z],\ i = 0, 1, \cdots \ell$$

$$x_n \leq y_0,\quad y_m \leq z_0,\quad \cdots$$

Thus, we extract the elements from equivalence class $[\alpha]$ which belongs to x/π and set it as the elements of ordered set $\{x/\pi\}$ whose elements are arranged according to their magnitude, and denote it as follows

$$\{x/\pi\} = \left\{ \langle x \rangle, \langle y \rangle, \cdots \right\}$$

then $\{x/\pi\}$ becomes a full ordered set, Therefore, we call $\{x/\pi\}$ the discrete linguistic coordinate axis in terms of the geometrical meaning.

For a fuzzy controller, we usually denote linguistic variables for the error and the derivatives of error by E and $\dot{E}$ respectively. Then we call $\{E/\pi\}$ and $\{\dot{E}/\pi\}$ the discrete horizontal and vertical axis. We also call plane corresponding to these linguistic axices the linguistic phase plane.

Remark: We usually take the place of equivalence class $[x]$ with chopping set N_λ that is

$$[x] \Rightarrow N_\lambda \triangleq \left\{ x \mid \mu_{\underset{\sim}{N}}(x) \geq \lambda \right\}$$

where $\mu_{\underset{\sim}{N}}$ is the membership function of $\underset{\sim}{N}$ and $\Rightarrow$ means correspondence. Therefore, we have the linguistic coordinate axis as follows

$$\{x/\pi\}_\lambda = \{\langle N_\lambda\rangle, \langle M_\lambda\rangle \cdots\}$$

$$\{x/\pi\}_\lambda \Rightarrow \{x/\pi\}$$

where

$$\langle N_\lambda\rangle \Rightarrow \langle x\rangle, \ \langle M_\lambda\rangle \Rightarrow \langle y\rangle, \cdots$$

Definition If

1° $N_\lambda \cap M_\lambda \neq \phi$

2° $\exists \mu_i(x_\nu) = \sup_j \mu_j(x_\nu), \ x_\nu \in N_\lambda \cap M_\lambda$

3° $x_\nu \in \{x/\pi\}_\lambda$

and arranging it in coordinate axis in terms of magnitude, then we name the linguistic phase plane as subcontinuous plane.

Remark: All of linguistic phase plane and axis are almost discrete or subcontinuous. In abbreviation , we do not distinguish the difference between discrete and subcontinuous so as to $\{x/\pi\}$ and $\{x/\pi\}_\lambda$, $\langle x\rangle$ and $\langle N_\lambda\rangle$, $[x]$ and $\langle x\rangle$ hereafter.

Definition Let region S_{pq} in linguistic phase plane be

$$S_{pq} = \{(x,y) \mid x \in \bigcup_{i=p_1}^{p_\alpha} \langle x\rangle_i, \ y \in \bigcup_{j=q_1}^{q_\beta} \langle y\rangle_j, \ \langle x\rangle_i \in \{E/\pi\}, \ \langle y\rangle_j \in \{\dot{E}/\pi\}\}$$

and exists equivalence class $[c]_{pq}$ like follows

$$[c]_{pq} = \{C_{pq}(S_i) \mid C_{pq}(S_j) = C_{pq}(S_i), \ C_{pq}: R^2 \to R^1, \ S_i, S_j \in S_{pq}\}$$

$$[c]_{pq} = [f^{-1}(\underset{\sim}{C}_{pq})]$$

then we call $\underset{\sim}{C}_{pq}$ the pq- th linguistic control variable.

Definition For a fuzzy controller, let X and Y represent error E and derivatives of error Ė respectively. Given the function (continuous or step) y_1 , y_2 , we have

$$S^*_{pq} = \{(x,y) \mid x \in \bigcup_k [x]_k, \ y \in \bigcup_\ell [y]_\ell, \ y(x) \, \mathbf{R}_1 (y_1(x), y_2(x))\}$$

$$[C^*]_{pq} = \{C^*_{pq}(S_i) \mid C^*_{pq}(S_j) \, \mathbf{R}_2 \, C^*_{pq}(S_i), \ C^*_{pq}: R^2 \to R^1, \ S_i, S_j \in S^*_{pq}\}$$

$$[C^*]_{pq} = [f^{-1}(\underset{\sim}{C}^*_{pq})]$$

where R_1 is a certain relation between $y(x)$ and $y_1(x)$, $y_2(x)$. $\mathbf{R}_2$ is a certain relation between C^*_{pq} , when $\mathbf{R}_i, i=1,2$ imply that:

1° $R_1: \ y_1(x) \geq y(x) \geq y_2(x)$

$R_2: \ C^*_{pq}(S_j) = C^*_{pq}(S_i)$

we call $\underset{\sim}{C}^*_{pq}$ the linguistic phase tape.

2° $R_1: \ y_1(x) \geq y(x) \geq y_2(x)$

$R_2: \ C^*_{pq}(S_j) = C^*_{pq}(S_i) = 0$

we call $\underset{\sim}{C}^*_{pq}$ the empty linguistic phase tape (or empty tape) and denote it by $\underset{\sim}{C}^\circ_{pq}$.

3° $R_1: \ y(x) = \frac{1}{2}(y_1(x) + y_2(x))$

we call $\underset{\sim}{C}^*_{pq}$ the mean phase tape.

Definition The bound between two linguistic phase tape is called the switching line.

Definition Since every point in coordinate axis represents the certain information, thus we call point (x,y) in linguistic phase plane the unit information.

Proposition 1 The minimum length of trajectory corresponds to minimum information.
Proof: omitted.

Proposition 2 Let $a=(x_i, y_i)$, $b=(x_j, y_j)$ be points on neighborhood of empty tape and let C_a and C_b be control variable in point a and b respectively. Therefore, the optimal control which satisfies the criterion of minimum dynamic energy is

IF $|x_j| \geq |x_i|$, $|y_i| > |y_j|$

THEN $C_a < C_b$ (when $C_b > 0$ then $C_a < 0$)

IF $|x_i| > |x_j|$, $|y_i| > |y_j|$

THEN $C_b \leq C_a$

Proof: Since y_i , y_j is the dynamic energy possessed by system. x_i , x_j is the deviation of this system, in order to assure the satisfied response, thus for more (less) deviation, we must pay more (less) control energy.

For example A part of fuzzy control table which is used by hydraulic position servo is shown in Table 1.
From Table1, we have
1". The quotient set of empty tape is

$$[x]_0 = \{[(x_1, y_1)], [(x_2, y_2)], [(x_3, y_3)]\}$$

$$x_1 \in [f_e^{-1}(PS)] \ , \ y_1 \in [f_r^{-1}(NB)] \cup [f_r^{-1}(NM)]$$

$$x_2 \in [f_e^{-1}(0)] \ , \ y_2 \in [f_r^{-1}(NS)] \cup [f_r^{-1}(0)] \cup [f_r^{-1}(PS)]$$

$$x_3 \in [f_e^{-1}(NS)] \ , \ y_3 \in [f_r^{-1}(PB)] \cup [f_r^{-1}(PM)]$$

where f_e and f_r are membership function of error and derivatives of error.

TABLE 1

Ė \ E (C)	NB	NM	NS	0	PS	PM	PB
PB	NM	NS	0	PS	PM	PB	PB
PM	NM	NS	0	PS	PM	PB	PB
PS	NB	NM	NS	0	PS	PM	PB
0	NB	NM	NS	0	PS	PM	PB
NS	NB	NM	NS	0	PS	PM	PB
NM	NB	NB	NM	NS	0	PS	PM
NB	NB	NB	NM	NS	0	PS	PM

2" The quotient set of linguistic phase tape for linguistic control variable PM, is

$$[f_c^{-1}(PM)] = \{[(x_4, y_4)], [(x_5, y_5)], [(x_6, y_6)]\}$$

$$x_4 \in [f_e^{-1}(PB)], \quad y_4 \in [f_r^{-1}(NM)] \cup [f_r^{-1}(NB)]$$

$$x_5 \in [f_e^{-1}(PM)], \quad y_5 \in [f_r^{-1}(NS)] \cup [f_r^{-1}(PS)] \cup [f_r^{-1}(0)]$$

$$x_6 \in [f_e^{-1}(0)], \quad y_6 \in [f_r^{-1}(PB)] \cup [f_r^{-1}(PM)]$$

where f_c is membership function of control.

3" The possible quotient set of control from initial point (x_0 , y_0) to origin is as follows

$$[c] = \{f_c^{-1}(NB), f_c^{-1}(NM), f_c^{-1}(NS), f_c^{-1}(0)\}$$

$$x_0 \in [f_e^{-1}(NB)], \quad y_0 \in [f_r^{-1}(0)]$$

the linguistic phase tape corresponding to $[c]$ is

$$[x]_c = \{[(x_7, y_7)], [(x_8, y_8)], [(x_9, y_9)], [(x_{10}, y_{10})]\}$$

$$x_7 \in [f_e^{-1}(NB)], \quad y_7 \in [f_r^{-1}(0)] \cup [f_r^{-1}(PS)]$$

$$x_8 \in [f_e^{-1}(NM)], \quad y_8 \in [f_r^{-1}(PM)] \cup [f_r^{-1}(PB)]$$

$$x_9 \in [f_e^{-1}(NS)], \quad y_9 \in [f_r^{-1}(PS)]$$

$$x_{10} \in [f_e^{-1}(0)], \quad y_{10} \in [f_r^{-1}(0)]$$

Control of linguistic trajectory

Consider the following charecteristics of relay, it is shown in (Fig2).

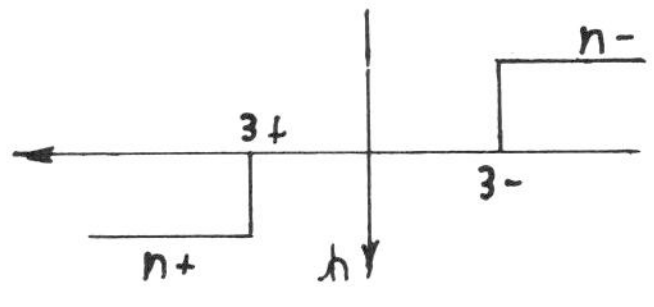

Fig. 2.

$$\psi(\cdot) = \begin{cases} +u, & x \in [+\varepsilon, +\infty) \\ 0, & x \in [-\varepsilon, +\varepsilon] \\ -u, & x \in (-\infty, -\varepsilon] \end{cases}$$

Obviously, it is three equivalence classes of x that

$$[x_1] \Rightarrow +u \Rightarrow PB, \quad x_1 \in [+\varepsilon, +\infty)$$

$$[x_2] \Rightarrow 0 \Rightarrow 0, \quad x_2 \in [-\varepsilon, +\varepsilon]$$

$$[x_3] \Rightarrow -u \Rightarrow NB, \quad x_3 \in (-\infty, -\varepsilon]$$

Let us consider the following dynamic system

$$\dot{y} + ay + bx = \psi(y, x, u), \quad y = \frac{dx}{dt}, \qquad (1)$$

denote the eignvalue of this system by

$$\lambda = R_e(\lambda) + I_m(\lambda)$$

where $R_e(\lambda)$ and $I_m(\lambda)$ are real and imaginary part of λ respectively. The mean phase tape of system (1) in the neighborhood of equilibrium point can be expressed as follows

$$\sigma_i(x, y) = \{(x, y) \mid \dot{y} + ay + bx = \psi, \ \lambda_1 \mathbf{R}_i \lambda_2\}$$
$$i = 1, 2 \cdots 6$$

where R_i is the i-th relation between λ_1 and λ_2.

The difference of trajectory is due to the various relation R_i between λ_1 and λ_2.

1. centered point

$$R_1: \lambda_1 R_1 \lambda_2 \Rightarrow R_e(\lambda_1) = R_e(\lambda_2) = 0, \ I_m(\lambda_1) = -I_m(\lambda_2) \neq 0$$

2. stable focus

$$R_2: \lambda_1 R_2 \lambda_2 \Rightarrow R_e(\lambda_1) = R_e(\lambda_2) < 0, \ I_m(\lambda_i) \neq 0, \ i = 1, 2$$

3. unstable focus

$$R_3: \lambda_1 R_3 \lambda_2 \Rightarrow R_e(\lambda_1) > 0, \ R_e(\lambda_2) > 0, \ I_m(\lambda_i) \neq 0, \ i = 1, 2$$

4. stable node

$$R_4: \lambda_1 R_4 \lambda_2 \Rightarrow R_e(\lambda_1) < 0, \ R_e(\lambda_2) < 0, \ I_m(\lambda_i) = 0, \ i = 1, 2$$

5. unstable node

$$R_5: \lambda_1 R_5 \lambda_2 \Rightarrow R_e(\lambda_1) > 0, \ R_e(\lambda_2) > 0, \ I_m(\lambda_i) = 0, \ i = 1, 2$$

6. saddle point

$$R_6: \lambda_1 R_6 \lambda_2 \Rightarrow R_e(\lambda_1) > 0, \ R_e(\lambda_2) < 0, \ I_m(\lambda_i) = 0, \ i = 1, 2$$

Theorem For the systems having multistage relay, the optimal switching line which satisfies the criterion of minimum information and satisfied response is broken line $\{\rho\}$

$$\{\rho\} = \{\rho_1(m_1, b_1), \rho_2(m_2, b_2), \cdots \rho_n(m_n, b_n)\}$$

$$\rho_i(m_i, b_i) = \{(x, y) \mid y = m_i x + b_i\}, \quad i = 1, 2, \cdots n$$

where m_i, b_i are real numbers,

Proof: As well known, the emptyswitching line plays a decisive role for control precision. Let $l(y(x))$ be a part of switching line, according to proposition 1, the optimal switching lines should satisfy:

$$l(y(x)) = \min$$

we know that

$$l(y(x)) = \int_{x_1}^{x_2} (1 + \dot{y}^2)^{\frac{1}{2}} dx$$

By using the Euler condition, that is

$$F_y - F_{x\dot{y}} - F_{y\dot{y}}\dot{y} - F_{\dot{y}\dot{y}}\ddot{y} = 0$$

$$F(x, y, \dot{y}, \ddot{y}) = (1 + \dot{y}^2)^{\frac{1}{2}}$$

the solution in terms of the above condition is

$$y = c_1 x + c_2$$

where c_1 and c_2 are real numbers.

The empty switching line in second quadrant is the bound between control "PS" and "NS". According to proposition 2, the optimal empty switching line on second quadrant is strainght line having negative slope. Scie it is multistage relay, thus the empty switching line is broken line, that difference slope is due to various limit control of a certain relay. Since the final point may be arbitrary point on horizontal axis, thus all of switching line located on whole phase plane should be parallel.

Remark: How to dertermine the slope of switching line? Since the slope depends on parameters of system, we determine it by means of a special program (Fig 3), it is an indefinityly cyclic program it constantly samples, searches, and sends out the search results until the valuable switching line is obtained. After that the system comes to normal work accoring to the obtained phase plane and portrait.

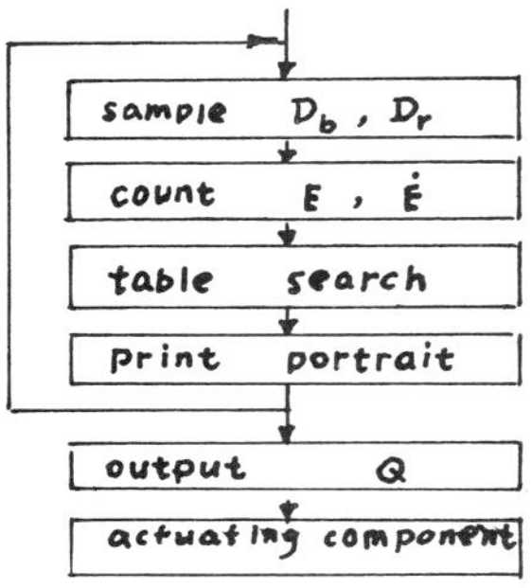

Fig. 3.

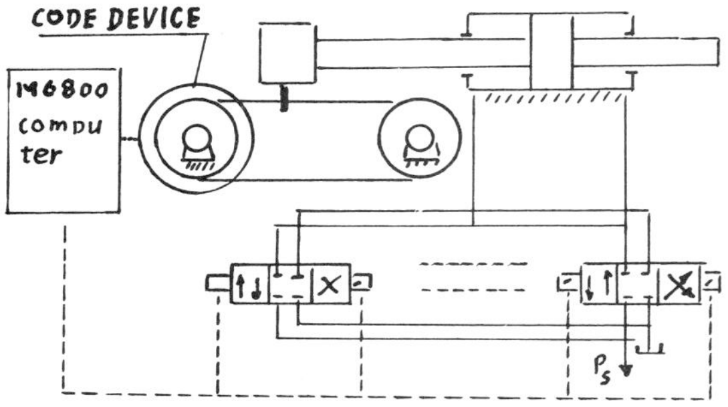

Fig. 4.

An instance

Fig. 4. illustrates a hydraulic servo system composed of four on-off valves with different flowrates, whose block diagram is shown in Fig5, in which x_p represents the displacement of the postion, A_p is the net area of this piston, K_x is the feedback coefficient, E_i and Q_i (i=1,...8) are the control valves and the positive and negative saturation flowrates of valves, and $G_o(s)$ is the transfer function of the linear link in hydraulic system

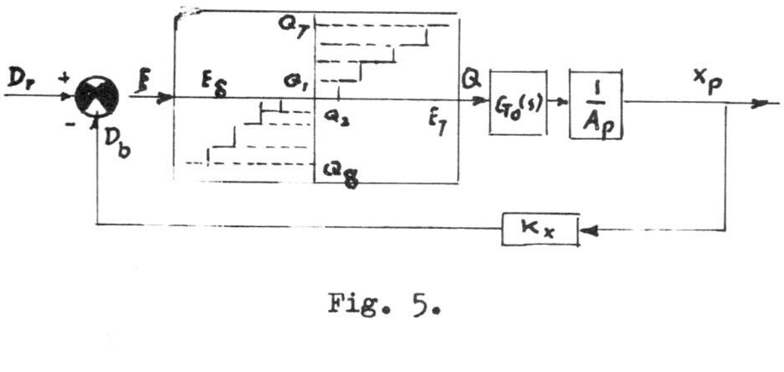

Fig. 5.

Fig 6. shows the block diagram of the fuzzy system.

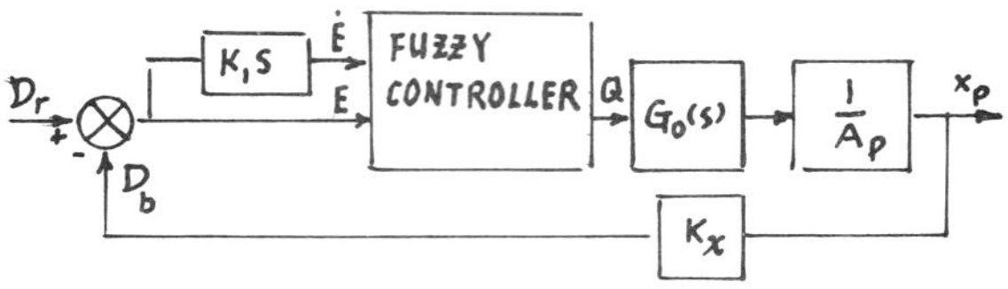

Fig. 6.

Fig 7. is the lingustic phase plane of Fig 6.

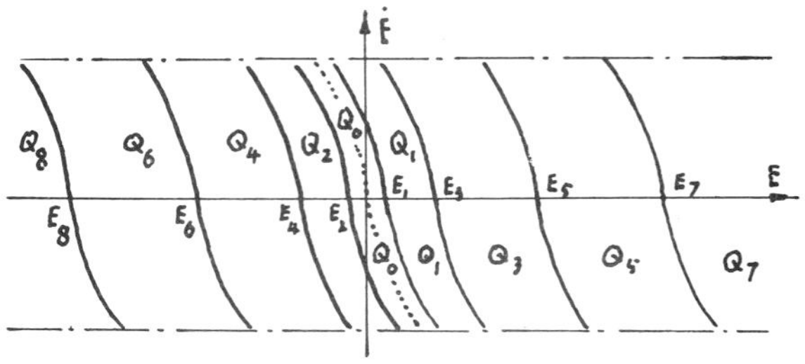

Fig. 7.

Although it is a higher order (more than 6) nonlinear system having four relays, it gives a satisfactory results in experiments by means of fuzzy control with linguistic phase plane. Fig 8. shows it's response of piston displacement after adopting fuzzy control. Fig 9 is the response without fuzzy control.

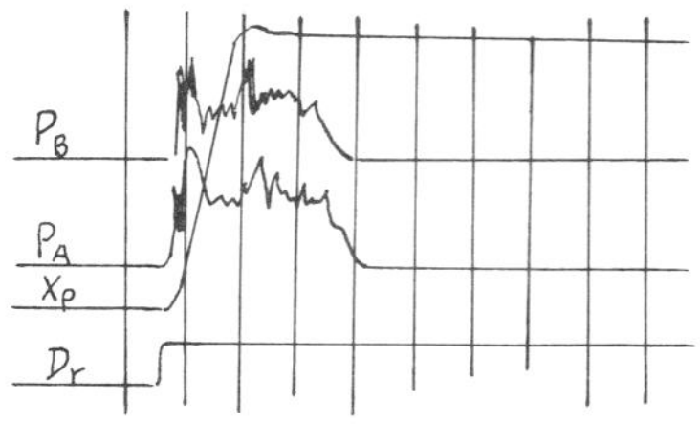

Fig. 8.

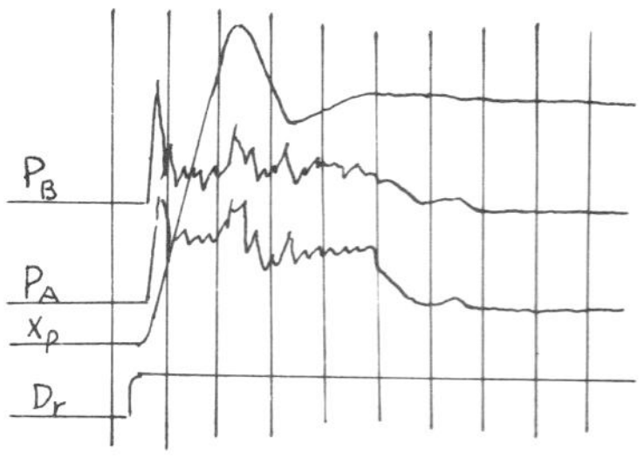

Fig. 9.

Conclution

1. By comparing the control of linguistic phase plane with conventional phase plane, we know that

*since the linguistic coordinate axis and phase tape are piecewise equivalence in length and in area respectively, the

control variable of a certain trajectory is piecewise constant. However, the trajectory of conventional control takes on only one value except o. For a multistage control, the higher precision can be obtained by means of that when the piston is far from the required positioning point (namely error is great), the valve of large flowrate is turned on to give the piston a high speed. When the piston is close to the point (namely error is small), the valve of large flowrate is turned off and the one of the small flowrate or the negative control valve should be turned on, so as to enable the piston to reach the positioning point quicly, smoothly and without overregulation. Thus we say: with regared to precision , the fuzzy control excels the conventional control especially the varying range of displacement is wide.

For the nonlinear systems containing soft (grey) values, relay, higher order dynamic equation, the conventional phase plane is inadequate, since it only suits the second order systems. However, the linguistic phase plane is suitable for this case by virtue of it's wide linguistic phase tape. In fact, the exact typical trajectory is not necessary in terms of wide phase tape. Thus we say: the linguistic phase plane excels the conventional control in adaptive ability.

2. Experimental results show that the system is suitable for use to in industry by virtue of it's satisfied response that the piston to reach the positioning point quicly, smoothly, and without overregulation, flexible control, little effect of load variation, high stability and precision.

References

1. Mamdani E.H. Application of fuzzy algorithm for control of simple dynamic plant. Proc. IEE 121 1585-1588 (1974)

2. Pappis C.P. Mamdani E.H. A fuzzy logic controller for a traffic junction. IEEE Trans. Syst. Man. & Cyber. SMC-7 No 10 (1976)

3. Tong. R.M. A control engineering review of fuzzy systems. Automatic Vol 13 pp559-569

4. Dubois D. Prade H. Operation on fuzzy numbers. Int. J. System Sci. 2(1978), pp213-227

5. Dubois D. Prade H. Fuzzy real algobra some results. Fuzzy Sets and systems 2(1979) pp327-348

6. Gupta M.M. Feedback control applications of fuzzy set theory: A survey. Proc. 8th IFAC, Kyoto, Japan, 1981

7. Wang P.Z. & Lou S.P. The responsibility of a fuzzy controller. Proc. 8th IFAC Kyoto. Japan

8. Li Cong-zi, Huang Shu-huai, Deng Ju-long, Time optimal fuzzy control of on-off hydraulic position servo by microcomputer. Report 4th IFAC/IFIP Sym. on Int. cont. problems in manufacturing technology U.S.A. 1982

9. Deng Ju-long, Control problems of grey systems. Systems & control letters, North-holland publishing company, Vol1, No 5 1982

A GENERALIZED FORMULATION OF MULTISTAGE DECISION-MAKING AND CONTROL UNDER FUZZINESS

J. Kacprzyk*

*Machine Intelligence Institute, Hagan School of Business, Iona College
New Rochelle, N.Y. 10801, USA*

Abstract. Multistage decision-making(control) under fuzziness consists basically in finding an optimal sequence of controls which maximizes the membership function of the fuzzy decision. The fuzzy decision is mainly the intersection of fuzzy constraints imposed on the successive controls applied and fuzzy goals imposed on the successive states attained. This is equivalent to determining an optimal sequence of controls which "best satisfies the fuzzy constraints and fuzzy goals at all the control stages". The purpose of this paper is to propose a model in which the strict "all" may be replaced by a milder requirement, e.g., "most", "much more than 50%", etc. A calculus of linguistically quantified statements is used.

Keywords. Multistage decision-making under fuzziness; fuzzy control; linguistic quantifiers.

INTRODUCTION

The growing use of control models, not only for "hard", e.g., technological, but also "soft", e.g., socioeconomic, problems makes the ability to account for softness extremely relevant. Traditional hard control models based upon the "precise" conventional mathematics do not provide adequate tools. The advent of fuzzy sets provided means for handling imprecision. Bellman & Zadeh's(1970) decision-making in a fuzzy environment provided a control related framework.

The fuzzy environment consists of fuzzy constraints imposed on the successive controls (inputs) applied, and of fuzzy goals imposed on the successive states(outputs) attained. The fuzzy decision is the intersection of the fuzzy constraints and fuzzy goals. The problem is to find an optimal sequence of controls to maximize the membership function of the fuzzy decision.

Semantically, this problem is equivalent to finding an optimal sequence of controls which "best satisfies the fuzzy constraints and fuzzy goals at all the control stages". The "all" may obviously seem to be an unnecessarily strict requirement. Thus, we propose here a model in which "all" may be replaced by some milder requirement given by a linguistic quantifier, e.g., "most", "much more than 50%", etc. We will therefore seek an optimal sequence of controls which "best satisfies the fuzzy constraints and fuzzy goals at most, much more than 50%, etc. of the control stages".

*on leave from: Systems Research Institute
Polish Academy of Sciences
ul. Newelska 6
01-447 Warsaw, Poland

Elements of a calculus of linguistically quantified statements proposed by Yager(1981,1982) are used. The solution boils down to finding a path in some decision tree.

MULTISTAGE DECISION-MAKING (CONTROL) UNDER FUZZINESS - THE CONVENTIONAL APPROACH

To provide a point of departure, let us briefly recall the classic Bellman & Zadeh's(1970) model. In its basic form, the system under control is deterministic and described by

$$x_{t+1}=f(x_t,u_t) \qquad t=0,1,... \qquad (1)$$

where $x_t, x_{t+1} \in X=\{s_1,...,s_n\}$ are the states at the control stages t and t+1, respectively, and $u_t \in U=\{c_1,...,c_m\}$ is the control at t.

At each t, u_t is subjected to a fuzzy constraint C^t and a fuzzy goal G^{t+1} is imposed on x_{t+1}. The fuzzy decision is

$$\mu_D(u_0,...,u_{N-1} \mid x_0)=\mu_{C^0}(u_0) \wedge \mu_{G^1}(x_1) \wedge ... \wedge \mu_{C^{N-1}}(u_{N-1}) \wedge \mu_{G^N}(x_N)= = \bigwedge_{t=0}^{N-1} (\mu_{C^t}(u_t) \wedge \mu_{G^{t+1}}(x_{t+1})) \qquad (2)$$

where N is a fixed and specified termination time, and x_{t+1}'s are given by Eq.(1).

An optimal sequence of controls $u_0^*,...,u_{N-1}^*$ is sought, such that

$$\mu_D(u_0^*,...,u_{N-1}^* \mid x_0)= = \max_{u_0,...,u_{N-1}} \mu_D(u_0,...,u_{N-1} \mid x_0) \qquad (3)$$

i.e. that best satisfies the fuzzy constraints and fuzzy goals at all the control stages. To emphasize this "all", let us rewrite Eq.(3) as

$$\mu_D(u_0^*,\ldots,u_{N-1}^* \mid x_0,\text{"all"}) = \max_{u_0,\ldots,u_{N-1}} \mu_D(u_0,\ldots,u_{N-1} \mid x_0,\text{"all"}) \quad (4)$$

This problem may be solved by dynamic programming (Bellman & Zadeh, 1970) or branch-and-bound (Chang & Pavlidis, 1977; Kacprzyk,1978b).

The above basic formulation was then considerably extended, mostly by: (1) the type of termination time(fixed and specified, implicitly given, fuzzy, and infinite), and (2) the type of system under control (deterministic, stochastic, and fuzzy). For a survey, see Kacprzyk (1982a, 1983). Here we will consider only the basic case, Eqs.(1)-(3), i.e. with the deterministic system under control and fixed and specified termination time, because the purpose is to introduce some more basic generalization.

A GENERALIZED FORMULATION OF MULTI-STAGE DECISION-MAKING (CONTROL) UNDER FUZZINESS

First, let us rewrite the conventional formulation Eqs.(1)-(3). At each t, the fuzzy constraint C^t and fuzzy goal G^{t+1} must be satisfied. This may be written as the following statements

$$\begin{cases} P_1: \text{"}C^0 \text{ and } G^1 \text{ are satisfied"} \\ \cdots\cdots\cdots\cdots \\ P_N: \text{"}C^{N-1} \text{ and } G^N \text{ are satisfied"} \end{cases} \quad (5)$$

Evidently, C^t and G^{t+1} are satisfied to some extent depending on the particular u_t applied and x_{t+1} attained. The (degree of) truth of P_{t+1} is now introduced as

$$\text{truth } P_{t+1} = \text{truth("}C^t \text{ and } G^{t+1} \text{ are satisfied")} = \mu_{C^t}(u_t) \wedge \mu_{G^{t+1}}(x_{t+1}) \quad (6)$$

Example 1. If $C^t = 0.5/c_1 + 1/c_2$ and $G^{t+1} = 0.3/s_1 + +0.7/s_2 + 1/s_3$, then for $u_t = c_1$ and $x_{t+1} = s_2$

$$\text{truth } P_{t+1} = 0.5 \wedge 0.7 = 0.5$$

Therefore

$$\mu_D(u_0,\ldots,u_{N-1} \mid x_0,\text{"all"}) = \text{truth}(P_1 \text{ and } \ldots \ldots P_N) = \text{truth } P_1 \wedge \ldots \wedge \text{truth } P_N = (\bigwedge_{t=0}^{N-1} \mid \text{"all"}) \, (\mu_{C^t}(u_t) \wedge \mu_{G^{t+1}}(x_{t+1})) \quad (7)$$

and the problem given by Eq.(3), or equivalently Eq.(4), becomes: find $u_0^*,\ldots,u_{N-1}^*$, such that

$$\mu_D(u_0^*,\ldots,u_{N-1}^* \mid x_0,\text{"all"}) = \max_{u_0,\ldots,u_{N-1}} \mu_D(u_0,\ldots,u_{N-1} \mid x_0,\text{"all"}) = \max_{u_0,\ldots,u_{N-1}} (\bigwedge_{t=0}^{N-1} \mid \text{"all"}) (\mu_{C^t}(u_t) \wedge \mu_{G^{t+1}}(x_{t+1})) \quad (8)$$

The essence of the generalization proposed is to replace the above strict "all" by some milder requirement given by a linguistic quantifier, e.g., "most", "much more than 50%", etc. Let us generally denote such a linguistic quantifier by Q. In this case, the fuzzy decision becomes

$$\mu_D(u_0,\ldots,u_{N-1} \mid x_0, Q) = (\bigwedge_{t=0}^{N-1} \mid Q)(\mu_{C^t}(u_t) \wedge \mu_{G^{t+1}}(x_{t+1})) \quad (9)$$

and the problem is to find $u_0^*,\ldots,u_{N-1}^*$, such that

$$\mu_D(u_0^*,\ldots,u_{N-1}^* \mid x_0, Q) = \max_{u_0,\ldots,u_{N-1}} \mu_D(u_0,\ldots,u_{N-1} \mid x_0, Q) \quad (10)$$

Thus, if Q="most", we seek an optimal sequence of controls to "best satisfy the fuzzy constraints and fuzzy goals at most of the control stages"

For the solution of the above problem, we will employ the so called substitution method based on a calculus of linguistically quantified statements proposed by Yager(1981, 1982).

Solution by the substitution method

To emphasize the essence of the method, let us be more specific and assume that the termination time is N=3, i.e. we have the statements P_1, P_2 and P_3 of the type of Eq.(5). We introduce the set

$$V = \{v\} = \{P_1, P_2, P_3, P_1 \text{ and } P_2, P_1 \text{ and } P_3, P_2 \text{ and } P_3, P_1 \text{ and } P_2 \text{ and } P_3\} \quad (11)$$

which contains all the possible combinations for fulfillment of the subsequent fuzzy constraints and fuzzy goals.

The quantifier Q is now defined as a fuzzy set in V, $Q \subseteq V$; $\mu_Q(v)$ indicates the degree to which v satisfies the meaning of Q.

Example 2. For N=3 and V as in Eq.(11), "all" is evidently given by

$$\mu_{\text{"all"}}(v) = \begin{cases} 1 & \text{for } v = P_1 \text{ and } P_2 \text{ and } P_3 \\ 0 & \text{for other } v \in V \end{cases} \quad (12)$$

and

$$\mu_{\text{"most"}}(v) = \begin{cases} 1 & \text{for } v = P_1 \text{ and } P_2 \text{ and } P_3 \\ 0.7 & \text{for } v \in \{P_1 \text{ and } P_2, P_1 \text{ and } P_3, P_2 \text{ and } P_3\} \\ 0.2 & \text{for } v \in \{P_1, P_2, P_3\} \end{cases} \quad (13)$$

Now, for each $v \in V$ we calculate $\mu_D(\cdot \mid \cdot)$ as follows: if $v = P_{k1}$ and ... and P_{kp}, then

$$\mu_D(\cdot \mid \cdot) = u_T(v) = \text{truth}(P_{k1} \text{ and } \ldots \text{ and } P_{kp}) = \text{truth } P_{k1} \wedge \ldots \wedge \text{truth } P_{kp} = \bigwedge_{t=k1}^{kp} (\mu_{C^{t-1}}(u_{t-1}) \wedge \mu_{G^t}(x_t)) \quad (14)$$

The(degree of) truth of the statement "the fuzzy constraints and fuzzy goals are satisfied at Q control stages" is now expressed as the possibility (see Zadeh, 1978) that $v \in V$ both satisfies the meaning of Q and is true, that is

$$\mu_D(u_0,\ldots,u_{N-1} \mid x_0,Q)=\mathrm{Poss}(Q \cap T)=$$

$$=\max_{v \in V} (\mu_Q(v) \wedge \mu_T(v)) \qquad (15)$$

Example 3. Let N=3, V be as in Eq.(11), and

$C^0=0.5/c_1+1/c_2 \quad G^1=0.1/s_1+0.6/s_2+1/s_3$

$C^1=1/c_1+0.7/c_2 \quad G^2=0.6/s_1+1/s_2+0.5/s_3$

$C^2=1/c_1+0.6/c_2 \quad G^3=1/s_1+0.8/s_2+0.3/s_3$

If now, e.g., $u_0=c_1$, $u_1=c_2$, $u_2=c_1$, and $x_1=s_2$, $x_2=s_2$, $x_3=s_2$, then we obtain for the particular v's:

$\mu_T(P_1 \text{ and } P_2 \text{ and } P_3)=\mu_T(P_1 \text{ and } P_2)=$
$=\mu_T(P_1 \text{ and } P_3)=\mu_T(P_1)=0.6$

$\mu_T(P_2 \text{ and } P_3)=\mu_T(P_2)=0.6$

$\mu_T(P_3)=0.8$

Thus, for Q="most", Eq.(15) yields

$$\mu_D(u_0,u_1,u_2 \mid x_0,\text{"most"})=\max_{v \in V} (\mu_{\text{"most"}}(v) \wedge \mu_T(v))=(1 \wedge 0.5) \vee (0.7 \wedge 0.5) \vee (0.7 \wedge 0.5) \vee (0.7 \wedge 0.6) \vee (0.2 \wedge 0.5) \vee (0.2 \wedge 0.6) \vee (0.2 \wedge 0.8)=0.6$$

Let us note that $\mu_D(u_0,u_1,u_2 \mid x_0,\text{"all"})=0.5$.

In the control problem considered, we seek therefore $u_0^*,\ldots,u_{N-1}^*$, such that

$$\mu_D(u_0^*,\ldots,u_{N-1}^* \mid x_0,Q)= \max_{u_0,\ldots,u_{N-1}} \max_{v \in V} (\mu_Q(v) \wedge \mu_T(v)) \qquad (16)$$

Given below are some properties which will ultimately lead to a solution algorithm.

Some properties and the solution algorithm

It can readily be seen that the solution of Eq.(16) virtually involves the following two issues: (1) the determination of $\mu_D(\cdot \mid \cdot,Q)$ for a fixed sequence of controls due to Eq.(15), and (2) the determination of an optimal sequence of controls due to Eq.(16).

Determination of $\mu_D(\cdot \mid \cdot,Q)$. First, let us limit the class of possible quantifiers Q to the monotonic quantifiers for which there holds

$$\mu_Q(v_3) \geqslant \mu_Q(v_1) \vee \mu_Q(v_2) \qquad (17)$$

for any $v_1,v_2,v_3 \in V$, such that $v_3=v_1$ and v_2. Thus, Eq.(17) indicates the fact that the more the fuzzy constraints and fuzzy goals are satisfied, the better. Such quantifiers are the only ones that are practically relevant in control.

We introduce now the α-level set of Q

$$Q_\alpha=\{v \in V: \mu_Q(v) \geqslant \alpha\} \quad \text{for all } \alpha \in (0,1] \qquad (18)$$

For any $v \in Q_\alpha$, $v=P_{k1}$ and ... and P_{kp}, we define its length as

$$g_\alpha=g_\alpha(P_{k1} \text{ and } \ldots \text{ and } P_{kp})=p \qquad (19)$$

and the minimum length of any $v \in Q_\alpha$ is

$$\bar{g}_\alpha=\min_{v \in Q_\alpha} g_\alpha \qquad (20)$$

Example 4. For N=3, V as in Eq.(11), and "most" as in Eq.(13), we have

$(\text{"most"})_{0.7}=\{P_1 \text{ and } P_2, P_1 \text{ and } P_3, P_2 \text{ and } P_3, P_1 \text{ and } P_2 \text{ and } P_3\}$

$g_{0.7}(P_1 \text{ and } P_2)=g_{0.7}(P_1 \text{ and } P_3)=$
$=g_{0.7}(P_2 \text{ and } P_3)=2$

$g_{0.7}(P_1 \text{ and } P_2 \text{ and } P_3)=3$

and therefore $\bar{g}_{0.7}=2$.

If now $M=\{(\mu_{C^0}(u_0) \ \mu_{G^1}(x_1)),\ldots,(\mu_{C^{N-1}}(u_{N-1}) \ \mu_{G^N}(x_N))\}$, which is evidently a set of real numbers for fixed $u_0,\ldots,u_{N-1}$ and $x_0,x_1,\ldots,x_N$, and $a\bar{g}_\alpha$ is the $\bar{g}_\alpha$-th largest element of M, then we have an important property.

Proposition 1. For any monotonic Q, there holds

$$\mu_D(\cdot \mid \cdot,Q)=\max_{v \in V} (\mu_Q(v) \wedge \mu_T(v))= \max_{\alpha \in (0,1]} (\alpha \wedge \max_{v \in Q_\alpha} \mu_T(v))= \max_{\alpha \in (0,1]} (\alpha \wedge a\bar{g}_\alpha)= \max_{i=1,\ldots,N} (a_i \wedge b_i) \qquad (21)$$

where a_i is the i-th largest element of M, and b_i is the grade of membership of the elements of length i in Q.

Proof. First, let us notice that

$$\max_{v \in V} (\mu_Q(v) \wedge \mu_T(v))= \max_{\alpha \in (0,1]} (\alpha \wedge \max_{v \in Q_\alpha} \mu_T(v))$$

is a result known from, e.g., fuzzy mathematical programming (Tanaka, Okuda & Asai,1974). And next,we must prove that

$$a\bar{g}_\alpha=\max_{v \in Q_\alpha} \mu_T(v)$$

Let us denote a generic element of Q_α by P_{k1} and ... and P_{kp}, $p=\bar{g}_\alpha, \bar{g}_\alpha+1,\ldots,N$. Now, Eq.(14) implies that we need only consider $p=\bar{g}_\alpha$, because $\mu_T(v)$ cannot be higher for any $p > \bar{g}_\alpha$. The maximization in Eq.(21) is therefore to proceed over v's of length $\bar{g}_\alpha$. Evidently, "$\wedge$" used in Eq.(21) implies that the highest value of $\mu_T(v)$ is obtained for such $v \in V$ which contains the first $\bar{g}_\alpha$ highest elements of M. This value is obviously equal to $a\bar{g}_\alpha$, the $\bar{g}_\alpha$-th largest element of M.Q.E.D.

Example 5. For the data as in Example 3, we have

$$\mu_D(u_0,u_1,u_2 \mid x_0,\text{"most"})= \max_{i=1,2,3} (a_i \wedge b_i)=$$

$=(0.8 \wedge 0.2) \vee (0.6 \wedge 0.7) \vee (0.5 \wedge 0.1) = 0.6$

i.e. the same result as that obtained in Example 3 in a straightforward way.

<u>Determination of $u_0^*, \ldots, u_{N-1}^*$.</u> First, let us represent the control process as a decision tree of the type shown in Fig.1. Its nodes represent the consecutive states attained (x_0 in the root), and its arcs are associated with the controls applied and the values of $\mu_{C^t}(u_t) \wedge \mu_{G^{t+1}}(x_{t+1})$. To each path (sequence of the consecutive $u_0, \ldots, u_{N-1}$ and the corresponding $x_1, \ldots, x_N$) there corresponds some value of

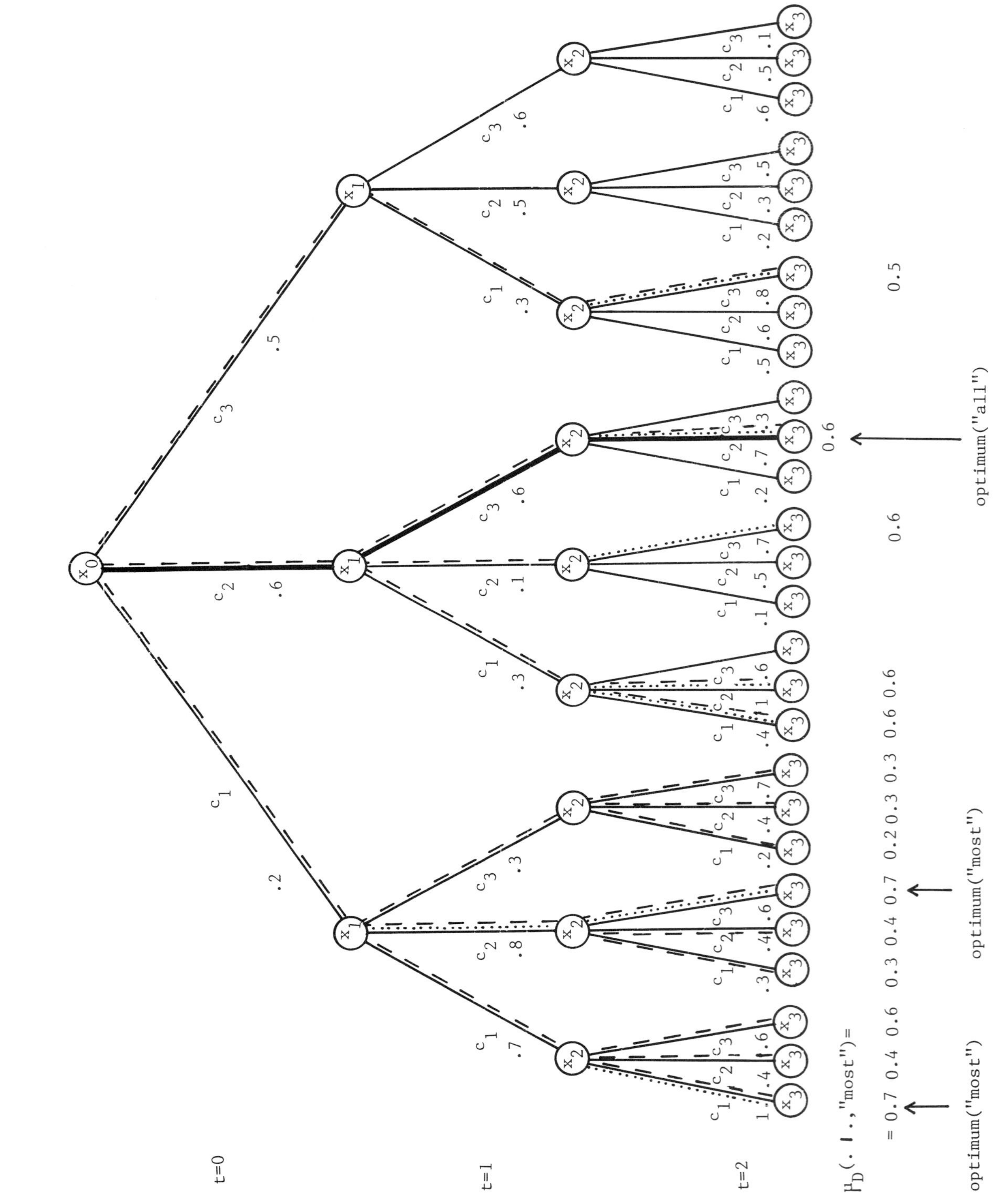

Fig.1. Decision tree with the problem specifications and solutions

$\mu_D(\cdot \mid \cdot)$. The determination of $u_0^*,\dots,u_{N-1}^*$ is therefore equivalent to finding a path in the decision tree, such that Eq.(16) holds. The full enumeration (27 paths in Fig.1) is evidently possible, but inefficient. We will therefore device an implicit enumeration algorithm neglecting nonpromising paths. It is based on the following properties.

Proposition 2. For any monotonic Q and any $u_0,\dots,u_{N-1}$ we have

$$\mu_D(u_0,\dots,u_{N-1} \mid x_0,Q) \geqslant$$
$$\geqslant \mu_D(u_0,\dots,u_{N-1} \mid x_0,\text{"all"}) \qquad (22)$$

and in particular

$$\mu_D(u_0^*,\dots,u_{N-1}^* \mid x_0,Q) \geqslant$$
$$\geqslant \mu_D(u_0^*,\dots,u_{N-1}^* \mid x_0,\text{"all"}) \qquad (23)$$

Proof. Let us note that P_1 and ... and $P_N \in V$, and we may write $V=\{\,.,\ P_1 \text{ and } \dots \text{ and } P_N\}$. Therefore

$$\mu_D(u_0,\dots,u_{N-1} \mid x_0,Q)=\max_{v\in V}\,(\mu_Q(v)\wedge\mu_T(v))=$$
$$=(\mu_Q(P_1 \text{ and } \dots \text{ and } P_N)\wedge\mu_T(P_1 \text{ and } \dots$$
$$\dots P_N))\vee \max_{v\in V-\{P_1 \text{ and } \dots \text{ and } P_N\}}(\mu_Q(v)\wedge$$
$$\wedge\mu_T(v))=\mu_D(u_0,\dots,u_{N-1} \mid x_0,\text{"all"})\vee$$
$$\vee \max_{v\in V-\{P_1 \text{ and } \dots \text{ and } P_N\}}(\mu_Q(v)\wedge\mu_T(v))$$

which obviously implies Eq.(22). And since it holds for any sequence of controls, then for an optimal sequence of controls in particular, i.e. Eq.(23).Q.E.D.

We have now the necessary condition for a path (its corresponding sequence of controls) to correspond to an optimal solution of the generalized problem given by Eq.(16).

Proposition 3. If $u_0^*,\dots,u_{N-1}^*$ is an optimal solution of Eq.(16) with some monotonic quantifier Q, then in its corresponding path there exists at least one arc, such that for u_t^*, $t\in\{0,1,\dots,N\}$, associated with this arc there holds

$$\mu_{C^t}(u_t^*)\wedge\mu_{G^{t+1}}(f(x_t,u_t^*)) \geqslant$$
$$\geqslant \mu_D(u_0^*,\dots,u_{N-1}^* \mid x_0,\text{"all"}) \qquad (24)$$

Proof. For each u_t^*, we evidently have

$$\mu_D(u_0^*,\dots,u_{N-1}^* \mid x_0,Q) \leqslant$$
$$\leqslant \mu_{C^t}(u_t^*)\wedge\mu_{G^{t+1}}(f(x_t,u_t^*))$$

and due to Proposition 2

$$\mu_D(u_0^*,\dots,u_{N-1}^* \mid x_0,Q) \geqslant$$
$$\geqslant \mu_D(u_0^*,\dots,u_{N-1}^* \mid x_0,\text{"all"})$$

and they both imply Eq.(24). Q.E.D.

Thus, if Eq.(24) holds, such a path is promising; it may correspond to an optimal solution.

The properties stated lead to the following algorithm for determining the solution of the problem considered given by Eq.(16):

1. Construct the decision tree of the type as in Fig.1.
2. Solve the conventional problem given by Eq.(4): find $u_0^*,\dots,u_{N-1}^*$, such that

$$\mu_D(u_0^*,\dots,u_{N-1}^* \mid x_0,\text{"all"})=$$
$$=\max_{u_0,\dots,u_{N-1}} \mu_D(u_0,\dots,u_{N-1} \mid x_0,\text{"all"}) \qquad (25)$$

3. Find such arcs in the decision tree that

$$\mu_{C^t}(u_t)\wedge\mu_{G^{t+1}}(x_{t+1}) >$$
$$> \mu_D(u_0^*,\dots,u_{N-1}^* \mid x_0,\text{"all"}) \qquad (26)$$

   Let us notice that we consider in Eq.(26) only " > ", because we already have some arcs for which "=", i.e. among those corresponding to the solution of the conventional problem.
4. Determine all the paths from x_0 to x_N containing the arcs found in Step 3.
5. For each path found in Step 4, find the value of $\mu_D(u_0,\dots,u_{N-1} \mid x_0,Q)$ due to Eq.(21), and take as the optimal one(s) (its corresponding $u_0^*,\dots,u_{N-1}^*$) that (those) for which $\mu_D(\cdot \mid \cdot,Q)$ attains its maximum value.

Example 6. Let us solve the problem: find u_0^*,u_1^*,u_2^*, such that

$$\mu_D(u_0^*,u_1^*,u_2^* \mid x_0,\text{"most"})=$$
$$=\max_{u_0,u_1,u_2} \mu_D(u_0,u_1,u_2 \mid x_0,\text{"most"}) \qquad (27)$$

with "most" given by Eq.(13), and the problem specifications as in Fig.1 (the particular values of the states are obviously irrelevant).

The consecutive steps of the algorithm are:

1. The decision tree is constructed as shown in Fig.1.
2. The solution of the conventional problem given by Eq.(4) is: $u_0^*=c_2$, $u_1^*=c_3$, $u_2^*=c_2$ with $\mu_D(\cdot \mid \cdot,\text{"all"})=0.6$. The corresponding path is shown as "▬▬".
3. The arcs satisfying Eq.(26) are shown as "• • •".
4. The paths containing the arcs found in Step 3 are shown as "- - -".
5. For each path found in Step 4, the values of $\mu_D(\cdot \mid \cdot,\text{"most"})$ are given below the leaves of the decision tree.
   There are therefore two optimal solutions:
   - $u_0^*=c_1$, $u_1^*=c_1$, $u_2^*=c_1$; and
   - $u_0^*=c_1$, $u_1^*=c_2$, $u_2^*=c_3$,

   and for both $\mu_D(u_0^*,u_1^*,u_2^* \mid x_0,\text{"most"})=0.7$.

CONCLUDING REMARKS

The generalization proposed makes it possible to neglect an incidentally low fulfillment of the fuzzy constraints and/or fuzzy goals at some (not many) control stages. This may help make control models more flexible and closer to the real, practical perception of the nature and aim of control

Finally, let us note that the generalization has obviously nothing to do with the notion of fuzzy termination time as introduced by Fung & Fu(1977) and Kacprzyk(1977, 1978a).

REFERENCES

Bellman, R.E., and L.A. Zadeh(1970). Decision-making in a fuzzy environment. Manag. Sci. 17, 151-169.

Chang, R.L.P., and Pavlidis T.(1977). Fuzzy decision tree algorithms. IEEE Trans. Syst. Man & Cybern. SMC-7, 28-39.

Fung, L.W., and K.S. Fu(1977). In M.M. Gupta, G.N. Saridis, and B.R. Gaines(Eds.). Fuzzy Automata and Decision Processes. North-Holland, Amsterdam.

Kacprzyk, J.(1977). Control of a nonfuzzy system in a fuzzy environment with fuzzy termination time. Syst. Sci. 3, 320-331.

Kacprzyk, J.(1978a). Decision-making in a fuzzy environment with fuzzy termination time. Fuzzy Sets & Syst. 1, 169-179.

Kacprzyk, J.(1978b). A branch-and-bound algorithm for the multistage control of a nonfuzzy system in a fuzzy environment. Contr.& Cybern. 7, 51-64.

Kacprzyk, J.(1982a). Multistage decision processes in a fuzzy environment: a survey. In M.M. Gupta, and E.Sanchez (Eds.). Fuzzy Information and Decision Processes. North-Holland. Amsterdam.

Kacprzyk, J.(1982b). A generalization of multistage decision-making and control in a fuzzy environment. Busefal (forthcoming).

Kacprzyk, J.(1983). Multistage Decision-Making under Fuzziness: Theory and Applications. Verlag TÜV Rheinland, Cologne.

Tanaka, H., T. Okuda, and K. Asai(1974). On fuzzy mathematical programming. J. Cybern. 3, 37-46.

Yager, R.R.(1980). Quantified propositions in a linguistic logic. Techn. Rep. #RRY 80-07, Iona College, New Rochelle,N.Y.

Yager, R.R.(1982). Quantifiers in the formulation of multiple objective decision functions. Techn. Rep. #MII-108, Machine Intelligence Institute, Iona College, New Rochelle, N.Y.

Zadeh, L.A.(1978). Fuzzy sets as a basis for a theory of possibility. Fuzzy Sets & Syst. 1, 3-28.

ANALYSIS OF FUZZINESS IN IMAGE DISCRIMINATION PROCESSES

O. G. Chorayan

Department of Physiology and Research Institute of Neurocybernetics, Rostov State University, Rostov on Don, USSR

Abstract. The paper prssents the results of study of fuzzy logic in image discrimination processes using the account of fuzzy probability. A method of calculation of choice fuzzy probability is described. To determine the fuzzy probability α-leveled fuzzy sets is suggested.

Keywords. fuzzy probability; decision-making; image discrimination process; membership function; α-leveled fuzzy sets.

One of the main reasons why man is included into modern complex biotechnical systems of control and communication is that he possesses a unique ability to make rational decisions in conditions of information shortage. There are some grounds to consider this ability to be connected with the skill to operate with the fuzzy concepts to apply the fuzzy logic to practical production activity. Uncertainty of situations in which man as a rule lives and works makes researchers to widely use probabilistic methods to analysis of neuro- and psychophysiological mechanisms of purposeful behaviour. However, utilization of probabilistic methods alone is not sufficient enough, for usually we have uncertain incomplete knowledge of distribution of probabilities in concrete tasks. Besides, in most problematic situations being practically solved by man there is uncertainty primarily related to vagueness uncertainty of the concepts formed and fixed in the information thesaurus of a person that makes decision (Zadeh, I965, I973; Jumarie, I978, I980; Chorayan, I979, I982; Negoita, I980, I98I).

Decision-making processes while pattern recognizing in the condition of various information uncertainty are convenient objects for neuro- and psychophysiological analysis of the complexhierarchically process of accumulation of information about the image to be recognized and comparison of information at hand with the idea about this image kept in mind. To apply the theory of fuzzy sets to studies in image discrimination psychophysiological mechanisms it is necessary to develop and test method of fuzzy phenomena probabilistic characteristics determination in the course of fuzzy logic of man's reasoning in a problematic situation.

The aim of the present work was to develop a method of calculation of choice fuzzy probability and alternative actions value, and to analyze the dynamics of the index in the process of vague discrimination in human.

The students were suggested to identify an image on the basis of the successive characteristics representation. Information was given in the form in the 5 slides with gradually increasing semantics (additional properties of the object). For example a student is to recognize " a tram". In the instructions 5 possible objects are mentioned: a ship / x_I /, a carriage / x_2 /, a tram / x_3 /, a trolleybus / x_4 /, a bus / x_5 /. The student is proposed to evaluate the degree of identity of the represented image with one of the abovesaid objects by the following linguistic terms: "may be", "yes, most probably", "no, most probably", "no", "yes" and also by numerical values of the membership function corresponded to the following linguistic terms: "yes"- I,0, "no"- 0, "may be"- 0,5, "yes, most probably"-0,8, "no, most probably"-0,2. Thus, for instance, the first stage (the first slide) of "tram" discrimination can be described by the following fuzzy set:

$$A=\left\{\frac{0}{x_I}, \frac{0,8}{x_2}, \frac{I,0}{x_3}, \frac{0,5}{x_4}, \frac{0,2}{x_5}\right\}$$

The subjective probabilities have the following meaning :

$p(x_I) = 0,3$; $p(x_2) = 0,3$; $p(x_3) = 0,4$;

$p(x_4) = 0$; $p(x_5) = 0$.

Then according to Zadeh (I968) the probability of the fuzzy image representation and identification of the pattern is determined as:

$$p(A) = 0 \cdot 0,3 + 0,8 \cdot 0,3 + I,0 \cdot 0,4 + 0,5 \cdot 0 + 0,2 \cdot 0 = 0,64 .$$

Yager (I977) rightly supposes that the probability of a fuzzy phenomenon should be determined not by a single number, but by the fuzzy probability, a fuzzy subset which membership function numerical values correspond to the share of the phenomena in question in the compound fuzzy set. To determine the fuzzy probabilities of Yager we should introduce α-leveled fuzzy sets:

$$A_\alpha = \{x \in X \mid \mu_A(x)\}$$

where A_α - the subset of elements with the membership function more than α . In the considered case α- leveled fuzzy sets are as follows :

$A_{\alpha_I} = \{x_I, x_2, x_3, x_4, x_5\}$

when $\alpha \geqslant 0$;

$A_{\alpha_2} = \{x_2, x_3, x_4, x_5\}$

when $\alpha > 0$;

$A_{\alpha_3} = \{x_2, x_3, x_4\}$

when $\alpha > 0,2$;

$A_{\alpha_4} = \{x_2, x_3\}$

when $\alpha > 0,5$;

$A_{\alpha_5} = \{x_3\}$

when $\alpha > 0,8$.

Accordingly $p(A_\alpha)$ assumes the following values:

$p(A_{\alpha_I}) = I,0$; $p(A_{\alpha_2}) = 0,7$;

$p(A_{\alpha_3}) = 0,7$; $p(A_{\alpha_4}) = 0,7$;

$p(A_{\alpha_5}) = 0,4$.

As a result we have the fuzzy probability:

$$p(A_\alpha) = \left\{\frac{0}{I,0}, \frac{0,2}{0,7}, \frac{0,5}{0,7}, \frac{0,8}{0,7}, \frac{I,0}{0,4}\right\}$$

The final results obtained for all the examinees in the course of pattern recognizing point at gradual increase in the amount of values of the fuzzy set calculated according Zadeh (I968) . In the case of fuzzy probability of Yager(I977) there is observed the increase in the denominator numerical values. There is an exceptions when an examinee fails to identify the image. This enables us to conclude that the abovementioned algorithm of fuzzy probability account may be used for predicting results of the recognizing process being one of the widely spread types of decision-making in man-machine systems functioning in the condition of data shortage.

References

Chorayan, O.G. (I979) . Fuzzy Algorithm of Thought Processes. State Univ. Publ. House, Rostov on Don.

Chorayan, O.G. (I982) Stochasticity, probability and fuzziness in the brain activity. BUSEFAL ,N IO, 74-82.

Jumarie, G. (I978) Relativistic fuzzy sets as a mean to introduce human factors in pattern recognition systems. Int. Conf. on Cybernet. and Soc., Tokyo, pp.I-I7

Jumarie, G. (I980) Relativistic fuzzy sets. Toward a new approach to subjectivity in human systems. Math. Sci. hum. ,pp 39-75 .

Negoita ,C.V. (I980) Pullback versus feedback. Human Systems Management , I , 7I-76.

Negoita ,C.V. (I98I) The current interest in fuzzy optimization. Fuzzy Sets a. Systems, 6 , 26I-269.

Yager ,R.R. I977 Multiple objective decision-making using fuzzay sets. J. Man-Machine Studies, 9, 375-382.

Zadeh ,L.A. (I965) Fuzzy sets. Inform. a. Control,8, 338-353.

Zadeh, L.A. (I968) Probability measures of fuzzy events. J.Math. Anal. Applicat. ,23 , 42I-427.

Zadeh ,L.A. (I973) Outline of a new approach to the analysis of complex systems and decision processes. IEEE Trans. Syst. Man a. Cybern. SMC-3, 28-44

A FUZZY MODEL APPLICATION TO AN ADAPTIVE ARC WELDING ROBOT

D. V. Lakov and G. N. Nachev

BAS, Institute of Industrial Cybernetics and Robotics Akad. G. Bonchev st B1 12, Sofia 1113, Bulgaria

Abstract. Basic problem in the arc welding is a good technology performance in the varying vorking conditions. The necessary adaptation is performed by noncontact sensing of the welding edges. The shape and geometry information of the sensed edges is intrinsic indefinite by nature and therefore ill-defined. This presumes the use of fuzzy models for control. In the article a fuzzy estimate of the robot's states by a noncontact sensor is described. The sensor gives direct measured membership function of robot's position. Interpretation of this estimate is performed using fuzzy set powers and solution rule forming for robot's trajectory adjustment. By this way the human interpretation which reflects the internal structure of the model through man's insight, is changed by fuzzy estimate by robot itself using phisical properties of the object. The proposed fuzzy model is algorithmized, programmed and under test in real working conditions with an arc welding adaptive robot.

Keywords. Fuzzy models; adaptive control; arc welding robots; teaching; noncontact sensors; microprocessors.

PROBLEM DEFINITION

The robot arc welding process requires precise definition of torch motion, position and orientation of the parts to be welded, arc tention and current, a good developed teaching stratecy for making quality welds. At present achievement of these conditions are accomplished by fixing of exactly preprocessed parts and continuous-path contouring. The teaching is made through arc welding trajectory recordings by means of specified points, certain technology parameters between them and welding start-stop. The teaching is made only once for a given type of parts with presumption of repeatability of shape and dimentions.

In the real working conditions due to slight deviations in the shape and dimentions of the parts the technological process is disturbed. This imposes additional tuning of the motion for every next part. Another restriction is a large time expenditure and leaves some possibility to produce waste. A necessity of seam tracking arrises. There are tracking system in which a continuous sensing of the joint before the torch is performed. Based on the information for geometry deviations compensative motions are performed, superimposed on the basic motion. The problem solution encounters difficulties in the sense that the estimating procedure is sophisticated and leads to large time deficiency in the on-line tracking.

As a shape and size are varying and the sensor signal is noisy it is natural to use estimating procedures based on fuzzy models.

MODEL IMPLEMENTATION

On the Fig. 1. a noncontact laser sensor is shown mounted on the welding torch in such maner that the sensed area is all the time ahed of it. The sensible element of the sensor is CCD-line installed in a plain perpendicular to the moving direction with its sensible side looking to the welding parts.(Fig. 2) If the laser beam scans welding parts crosswise to the joint the CCD-line registered reflection will image the joint profile in the certain point.

Let us accept that the image produced by the laser sensor gives via terms of fuzzy sets one point of the joint. If the trajectory is sensed with sufficient rate it may be defined as a fuzzy curve in the sense of [1].

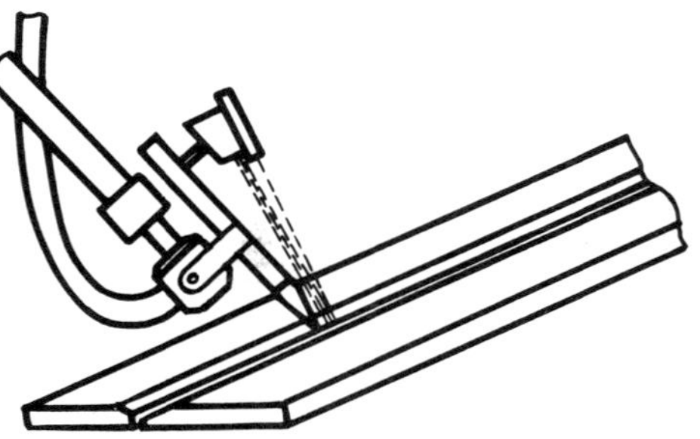

Fig. 1. Welding torch with noncontact laser sensor.

For any noncontact sensor's position the powers of the two sets: $P_{s_l} = P_{s_r}$ are calculated and the straight line L —— L limiting the two equipower sets is found. In the common case it does not coincide with the center point of the CCD-line. The center point of the CCD-line corresponds to the position of the torch in the intersection plain - the upper diagrams in Fig. 2. Therefore a torch motion to the left or right for nullification of the difference is necessary - lower left or right diagram in Fig. 2. This corrective action is calculated stepwise during the arc welding process, depending on the welding rate, the distsnce between the noncontact sensor and the torch, maximal admissible side deviation of the torch and inertia of the robot's hands.

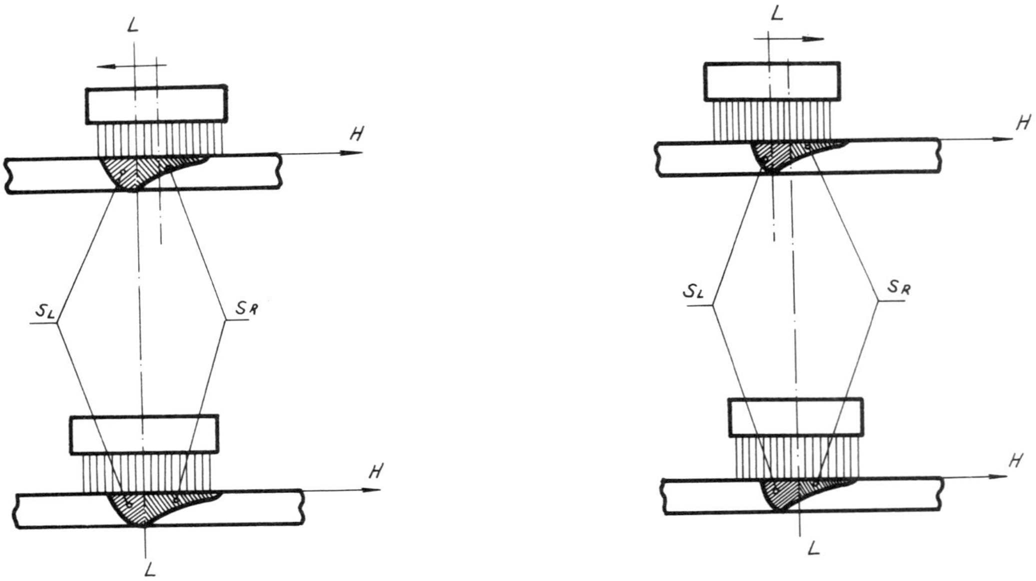

Fig. 2. Fuzzy estimates via CCD-line registrations

The reflected and registered light is representative for the joint intersection, which must be filled with the welding material. This image may be interpreted as a fuzzy set, which will define the torch position. The control may be realized on the bases of CCD-line registrations set the welding torch in the intersection image's center of gravity. If we try to let the registrations of the noncontact sensor to define two equipower fuzzy sets: S_l - left and S_r - right defined on the common continuum H - perpendicular direction of the joint line.

$$S_l : H \rightarrow (0,1) \qquad P_{s_l} = P_{s_r}$$
$$S_r : H \rightarrow (0,1)$$

SYSTEM REALIZATION

The control system working in accordance with thus proposed model is built by two microprocessors communicating via dual-port RAM. The first processor is reserved for calculation of the actual torch position - the offset respectively, and the other for interpolation, control and teaching. The program for offset's calculation via equipower's fuzzy sets is executed for 160 microseconds, thus allowing the noncontact sensor to give information every one millisecond.

CONCLUSIONS

Seam tracking with the noncontact sensor functioning according to the method described is effective by admissible side deviations of the actual contour up to 20 mm. Trajectory correction may be done directly taking into account the calculated offset or through appropriate autoregressive model.

One further development of the proposed method is connected with consecutive scanning procedure in longitudinal direction of the joint line for actual trajectory prediction.

The model may be used in various welding tipes such as for instance: T-joint, V-joint, etc., when above defined conditions are met.

REFERENCES

Zadeh, L.A. (1975). The concept of a linguistic variable and its application to approximate reasoning -iii, Information sciences, 9, 43 - 80.

Negoita, C.V., Ralescu, D.A. (1975). Applications of fuzzy sets to system analysis, Birkhauser verlag, Basel und Stuttgart

STABILIZING AND IDENTIFICATION OF THE OPPONENT IN NONLINEARIZABLE DIFFERENTIAL GAME WITH INCOMPLETE INFORMATION

J. M. Skowronski

Department of Mathematics, University of Queensland, St Lucia, Qld 4067, Australia

Abstract. Applying the Liapunov design technique and the theory of nonlinear adaptive identifiers, models are constructed to provide state and opponent's control information for feedback strategies in a two players competitive nonlinearizable differential game. The information is of prescribed accuracy and obtained after a prescribed time interval. The identifiers allow to attain the standard objectives: playability for collision (capture) with a target, avoidance of a target, and possible escape from a target. The linear game observers of the Luenberger type follow as a particular case. The method is illustrated by an example.

Keywords. Competitive qualitative differential game, nonlinear adaptive identifiers, Liapunov function.

INTRODUCTION

Most of the presently studied dynamic models are required to be robust with respect to a bounded uncertainty. In particular they are governed by a state dependent (feedback) control program of an active agent, who must override a time dependent unknown but bounded input attributed to a passive opponent in the game-against-nature. In the genuine competitive game with two active agents ≡ players the problem is doubled: each player chooses a feedback strategy against all options of his opponent. Either way, there is a penalty for lack of information on inputs which we would like to avoid. On the other hand, the state information needed for the strategies is often not completely observable, the observation different for each player. Moreover, even if it was observable, the game equations still might not have unique solutions due to the imput uncertainty, thus in turn generating uncertain states. Consequently we face the problem of simultaneous identification of input and state, at the same time securing the objectives of the game.

Traditionally, stochastic methods are applied to such a task, but on many occasions they may prove unfeasible. For instance when we can make no probabilistic assumptions about the random variables involved, or when the system is observed and controlled during a single run, etc. The alternative is to employ to our game the adaptive identifiers, cf. Luenberger-Rhodes (1969) for the state and Krassovski (1974) for the input identification, Galperin-Skowronski (1983) for nonlinear observability. The theory of identifiers considers the state and system parameters given, and constructs an auxiliary model called identifier whose states and parameters converge ultimately to these of the identified system. At present this theory is mainly linear and identifies constant parameters, cf. Narendra (1976). It uses the technique of subtracting the equations of the identified system and its identifier thus producing an "error equation". Then the asymptotic stability of its trivial solution (zero-error) secures the required convergence. Unfortunately even if our system was linear or linearizable, the insertion of strategies usually produces strong nonlinearity, which makes the subtraction of equations impractical. Moreover, the inputs to be identified are never constant. Thus we propose to achieve the required convergence by securing some type of Liapunov stability of the diagonal set in the Cartesian product of state-input spaces of our game and of some nonlinear identifier, cf. Skowronski (1981).

STATEMENT OF THE PROBLEM

Consider the differential game

$$\dot{x} = f(x,t,u^1,u^2) , \qquad (1)$$

where $x(t)$ is an *uncertain* state-information vector in the known bounded plying region $\Delta \subset \mathbb{R}^N$, for all $t \in J_o^f = [t_o,t_f]$, $t_f < \infty$, and $u^i(t) \in U_i \subset \mathbb{R}^{r_i}$, $t \in J_o^f$, are control vectors to be selected by two players $i=1,2$ from a known compact U_i. The functions $f_1(\cdot),\ldots,f_N(\cdot)$, $f = (f_1,\ldots,f_N)^T$, are assumed known.

The state observation available to the player i

is given by the read-out n-vector $y^i(t)$, $t \in J_o^f$, in the known band $Y^i \subset \mathbb{R}^n$, $n \le N$. In particular y^i may be represented by a set of observable components of $x(t)$. Moreover player i measures the noisy estimate of the change in his opponent control

$$\dot{u}^j = g^i(t,w) \tag{2}$$

with $w(t)$, $t \in J_o^f$, s-vector of uncertainty parameters bounded in known $W \subset \mathbb{R}^S$. To accomodate the noises and possible discontinuities of $u^i(t)$, the feedback program for choosing control is made set valued, $P^i(\cdot): \Delta \times \mathbb{R} \times U_j \to$ subsets of U_j, defined by $u^i(t) \in P^i(x(t),t,u^j(t))$, $t \in J_o^f$.

We want to concentrate on identification, thus the game-objectives are common and simple. *Player 1*, wants to force the solutions of (1) into a given target $\theta \subset \Delta$ (*collision*), then he may want them to stay in θ indefinitely (*capture*). *Player 2*, opposes the first task of 1 i.e. wants the solutions to *avoid* θ, and, if this is impossible, at least to make them *escape* from θ in specified time.

Following the theory of identifiers, cf. §1, we design the auxiliary model

$$\dot{z}^i = f^i(z^i,t,u^i,v^j,y^i) \tag{3}$$

$$\dot{v}^j = g_a^i(z^i,t,v^j,y^i)\ ,\ i=1,2 \tag{4}$$

with $z^i(t) \in \Delta \subset \mathbb{R}^N$, $v^j(t) \in U_j \subset \mathbb{R}^{r_j}$, $t \in J_o^f$, known vectors and $(f_1^i,\dots,f_N^i)^T = f^i$, $(g_{a1}^i,\dots,g_{aN}^i)^T = g_a$ known functions such as to satisfy the definitions below, securing the objectives. To this end, we introduce the vectors $(x(t),u^j(t))^T = \xi^i(t)$, $(z^i(t),v^j(t))^T = \zeta^i(t)$ and $(\xi^i(t),\zeta^i(t))^T \in z^i(t) \in \Delta^2 \times U_j^2$ in the product space $\mathbb{R}^{2(N+r_j)}$, where $\Delta^2 = \Delta \times \Delta$, $U_j^2 = U_j \times U_j$, $i,j=1,2$, $i \ne j$. Then combining the right hand sides of (1), (2), (3), (4) into one vector $F^i(z^i,u^i,y^i,w,t)$ we obtain for each player i the contingent product system

$$\dot{z}^i \in \{F^i(z^i,u^i,y^i,t) \,|\, u^i \in P^i(x,t,u^j),\ w \in W,\ y^i \in Y^i,\ j \ne i\}. \tag{5}$$

For suitable f, f^i, g^i, g_a^i, P^i, given $z^{io} = z^i(t_o) \in \Delta^2 \times U_j^2$, (5) has absolutely continuous solutions $k(z^{io},t_o,\cdot): J_o^f \to \Delta^2 \times U_j^2$, forming the family $K_i(z^{io},t_o)$, cf. Filippov (1971).

Define the diagonal $M^i = \{z^i \in \Delta^2 \times U_j^2 \mid \|\zeta^i - \xi^i\| = 0\}$ and its neighbourhood, $M_\eta^i = \{z^i \in \Delta^2 \times U_j^2 \mid \|\zeta^i - \xi^i\| < \eta^i\}$, $\eta^i > 0$, in $\mathbb{R}^{2(N+r_j)}$, with $\|\cdot\|$ some norm in this space.

Then let Δ_i be a subset of Δ required by the player i for his objective, $Z_i = \Delta_i \times \Delta_i \times U_j^2$ and $M^i = M^i \cap Z_i$, $M_\eta^i = M_\eta^i \cap Z_i$.

Definition 1. The model (3), (4) is a *stabilizing η^i-identifier on Δ_i* of (1), (2), iff given η^i, Δ_i there are f^i, g_a^i, P^i such for some $T_i > 0$ and for each $k(\cdot) \in K_i(z^{io},t_o)$ we have $k(z^{io},t_o,t) \in M_\eta^i$ for all $t \ge t_o + T_i$, $(z^{io},t_o) \in Z_i \times \mathbb{R}$. We adjust the origin of $\mathbb{R}^N$ such that it is in the given target $\theta \subset \Delta$. Let $\theta^4 = \theta \times \theta \times \mathbb{R} \times \mathbb{R}$, $\theta_\eta^4 = M_\eta^1 \cap \theta^4$.

Definition 2. A stabilizing η^1-identifier is *playable on $\Delta_1 \subset \Delta$ for collision* with $\theta \subset \Delta_1$ iff the strategy $P^{i=1}$ of Def. 1 is such that for some $T_c > T_1$ each $k(\cdot) \in K_1(z^{10},t_o)$ implies $k(z^{10},t_o,t) \in \theta_\eta^4$ for all $t \ge t_o + T_c$, $(z^{10},t_o) \in M_\eta^1 \times \mathbb{R}$. When T_c is given a-priori (desired) we have the *collision after* T_c.

Definition 3. A stabilizing η^2-identifier is playable on $\Delta_2 \subset \Delta$, $\Delta_2 \cap \theta = \emptyset$, for *avoidance* of θ, iff the strategy $P^{i=2}$ of Def. 1 is such that $k(\cdot) \in K_2(z^{20},t_o)$ implies $k(z^{20},t_o,t) \cap \theta^4 = \emptyset$ for all $t \in J_o^f$, $(z^{20},t_o) \in Z_2 \times \mathbb{R}$.

Definition 4. A stabilizing η^2-identifier is playable on $\Delta_2 \subset \Delta$ for *ultimate avoidance* of θ, $\theta \cap \Delta_2 = \emptyset$, iff the strategy $P^{1=2}$ of Def. 1 implies that there is a constant $T_A > 0$ such that $k(\cdot) \in K_2(z^{20},t_o)$ yields $k(z^{20},t_o,t) \cap \theta^4 = \emptyset$ for all $t \ge t_o + T_A$, $(z^{20},t_o) \in Z_s \times \mathbb{R}$. When T_A is given a-priori we have *avoidance after* T_A.

SUFFICIENT CONDITIONS

Consider a candidate set $\Delta_i \subset \Delta$ and let $N(\partial Z_i)$ denote a neighbourhood of its boundary ∂Z_i. Let also $N_\varepsilon^i = N(\partial Z_i) \cap \overline{Z}_i$, $CM_\eta^i = Z_i - M_\eta^i$, $S_\eta^i(\text{open}) \supset \overline{CM_\eta^i}$. Furthermore introduce two C^1-functions $V_s(\cdot): N_\varepsilon^i \times \mathbb{R} \to \mathbb{R}$, and $V_\eta(\cdot): S_\eta^i \times \mathbb{R} \to \mathbb{R}$ and define

$$v_s = \inf V_s(z^i,t) \,|\, (z^i,t) \in \partial Z_i \times \mathbb{R}\,,$$

$$v_\eta^- = \inf_\eta V_\eta(z^i,t) \,|\, (z^i,t) \in (\partial M_\eta^i \cap CM_\eta^i) \times \mathbb{R}\,;$$

$$v^+ = \sup V_\eta(z^i,t) \,|\, (z^i,t) \in (Z_i \cap \overline{CM}^i) \times \mathbb{R}\,.$$

Theorem 1. Given Δ_i, η^i, the model (3), (4) satisfies Def. 1 if there are V_s, V_η such that

(i) $V_s(z^i,t) < v_s$;

(ii) $v_\eta^- \le V_\eta(z^i,t) \le v^+$; and

for every $u^i \in P^i(z^i,t,v^j)$, there is $T_i > 0$ such that

$$\text{(iii)} \quad \frac{\partial V_s}{\partial t} + \nabla_{z^i} V_s(z^i,t)\cdot F^i(z^i,u^i,y^i,w,t) < 0; \qquad (6)$$

$$\text{(iv)} \quad \frac{\partial V_\eta}{\partial t} + \nabla_{z^i} V_\eta(z^i,t)\cdot F^i(z^i,u^i,-^i,w,t) \le -\frac{v_\eta^+ - v_\eta^-}{T_i}, \qquad (7)$$

for all $y^i \in Y^i$, $w \in W$.

(v) $V_\eta(z^i,t) \le v_\eta^-$, $(z^i,t) \in M_\eta^i \times \mathbb{R}$.

<u>Proof</u>. We use here similar arguments to those in Galperin-Skowronski (1983). Suppose some $k(z^{io},t_o,t)$, $(z^{io},t_o) \in Z_i$, crosses ∂Z_i. Then there is $t_1 > t_o$ such that $k(t_1) \in \partial Z_i$ and by (i), $V_s(k(t_1),t_1) \ge v_s \ge V_s(k(t_o),t_o)$ contradicting (iii). Hence all solutions $k(z^{io},t_o,\cdot) \in K_i(z^{io},t_o)$ from Z^i stay in this set.

Integrating (7) along any $k(z^{io},t_o,\cdot)$, $(z^{io},t_o) \in CM_\eta^i$ we obtain $\dot V_\eta(k(t),t) \le -(v_\eta^+-v_\eta^-)/T_i$ or

$$t \le t_o + T_i[v_\eta(z^{io},t_o)-v_\eta(z^i,t)/v_\eta^+-v_\eta^-].$$

By (ii), we have $V_\eta(z^{10},t)-V(z^i,t) \le v_\eta^+-v_\eta^-$ implying that the fractional multiplier of T_i is ≤ 1, whence the solution must leave $\overline{CM}_\eta^i$ after some T_i. Since the solutions may not leave Z_i, they must enter M_η^i at $t_i > t_o + T_i$. There is no return to $\overline{CM}_i$. Indeed, if there were $t_2 > t_i$ such that $k(t_2) \subset \overline{CM}_\eta^i$, then by (ii), (v) $V_\eta(k(t_2),t_2) \ge v_\eta^- \ge V_\eta(k(t_i),t_i)$ contradicting (iv). The above applies to any solution from Z_i and hence all of them stay in M_η^i after $t \ge t_o + T_i$, QED.

Let $C\theta^4 = Z_i - \theta^4$ and $S_\theta^1(\text{open}) \supset \overline{C\theta}^4$. Consider $V_\theta : S_\theta^1 \times \mathbb{R} \to \mathbb{R}$ with

$$v_\theta^- = \sup V_\theta(z^1,t) \mid (z^1,t) \in \partial\theta^4 \times \mathbb{R};$$
$$v_\theta^+ = \inf V_\theta(\mu^1,t) \mid (z^1,t) \in \partial Z_1 \times \mathbb{R}.$$

<u>Theorem 2</u>. Given Δ_1, η^1, $\Theta \subset \Delta_1$ the model (3), (4) satisfies Def. 2 if there are C^1-functions V_θ, V_η(as in Theorem 1) such that the conditions (ii), (iv), (v) of Theorem 1 hold and moreover

(i)' $v_\theta^- \le V_\theta(z^1,t) \le v_\theta^+$;

for every $u^1 \in P^1(z^1,t,v^2)$, there is $T_\theta > 0$ such that

$$\text{(iii)'} \quad \frac{\partial V_\theta}{\partial t} + \nabla_{z^1} V_\theta(z^1,t)\cdot F^1(z^1,u^1,y^1,w,t) \le -\frac{v_\theta^+ - v_\theta^-}{T_\theta} \qquad (8)$$

for all $y^1 \in Y^1$, $w \in W$.

<u>Proof</u>. By the same argument as in the proof of Theorem 1 the solutions do not leave Z_1. Integration of (8) along any solution in $C\theta^4$ gives

$$t \le t_o + T_\theta \frac{V_\theta(z^{10},t_o) - V_\theta(z^1,t)}{v_\theta^+ - v_\theta^-}$$

and in view of (i)' the fractional coefficient ≤ 1 implementing entrance to $\mathring{\theta}^4$ and no return by the same argument as for M_η^i. Then after t_o+T_c, $T_c = \max(T_1,T_\theta)$ all solutions are in θ_η^4, QED.

Similar theorems can be built for Definitions 3,4. However the implementation of Theorem 1 may be made easier by using the following approach. A stabilizing η^i-identifier satisfies Definitions 2,3,4, if there are additional C^1-functions $V_2 : Z_1 \times \mathbb{R} \to \mathbb{R}$, $V_3 : Z_2 \times \mathbb{R} \to \mathbb{R}$, $V_4 : Z_2 \times \mathbb{R} \to \mathbb{R}$ such that conditions for collision, capture, avoidance and ultimate avoidance proved in Skowronski (1981a), Skowronski-Vincent (1982), Leitmann-Skowronski (1977), (1983), hold, respectively. Limited space prevents us from quoting these conditions. Note however, that letting $V_s(z^1,t) = V_2(\xi^1,t) + V_2(\zeta^1,t)$ and $V(z^1,t) = |V_2(\xi^1,t) - V_2(\zeta^1,t)|$, one satisfies Theorem 1 thus obtaining a stabilizing η^1-identifier playable on Δ_1 for capture. Identical subsituation of V_3 and V_4 yields the combined Definitions 1,3 and 1,4, respectively. The quoted papers give methods for finding V_2, V_3, V_4, hence also for implementation of Theorem 1. Obviously, in similar fashion we may obtain V_2, V_η and thus stabilizing η^i-identifier playable for many other properties not defined in this paper, both qualitative and optimal, provided there are Liapunov type sufficient conditions proved for the particular playability and the corresponding Liapunov function is found.

COMPARISON WITH LUENBERGER OBSERVER

The existing linear methods do not allow the similtaneous identification of state and parameters. Let us choose state identification $\equiv$ observability and follow Galperin-Skowronski, (1983). Suppose (1) takes the form

$$\dot x = Ax + Bu^1 + Cu^2, \quad y^i = G^i x, \; i=1,2 \qquad (9)$$

with A, B, C, G^i constant matrices of order

$N\times N$, $N\times r_1$, $N\times r_2$, $n\times N$, respectively. Choose (3) in terms of the Luenberger observer

$$\dot{z}^i = (A-K^iG^i)z^i + K^iy^i + Bv^1 + Cv^2, \quad i=1,2 \qquad (10)$$

with K^i some constant $N\times N$ matrix. We assume u^1, u^2 given: $u^i(t) \equiv v^i(t)$, $i=1,2$. Subtracting (9) from (10)

$$\dot{z}^i - \dot{x} = (A-K^iG^i)(z^i-x), \quad i=1,2. \qquad (11)$$

For suitable K^i Luenberger obtains $z^i-x \to 0$, $t\to\infty$. The stability is obviously assured and we check the identification by letting $V_\eta = (z^i-x)^T(z^i-x)$ which satisfies (ii). Then also $\dot{V}_\eta = (\dot{z}^i-\dot{x})^T(z^i-x) + (z^i-x)^T(\dot{z}^i-\dot{x})$ and substituting (9), (10), $\dot{V}_\eta = -(z^i-x)^TQ^i(z^i-x)$, where $-Q^i = (A-K^iG^i)^T + (A-K^iG^i)$. With Luenberger's K^i's the matrices Q^i are positive definite allowing (iv).

EXAMPLE

Condider the game

$$\left.\begin{aligned}\dot{x}_1 &= x_2\\ \dot{x}_2 &= -u^1 - x_1 - u^2x_1^3\end{aligned}\right\} \qquad (12)$$

where $(x_1(t),x_2(t)) \in \Delta_1 : |x_1(t)|, |x_2(t)| < 1$, $\forall t$, with the read-out of player 1: $y(t) = x_2(t)$, $\forall t\in[t_0,t_f]$. It is a nonlinear damped oscillator controlled by a damping function recognized as a strategy $u^1 = p^1(x_1,x_2,u^2)$, where $p^1(\cdot)$ is continuous function and x_1, u^2 are to be identified. The latter is a "nature" subject to limitation $U_2 : u^2 = \text{const} \in [0,1]$. We specify U_1 by assuming the damping viscous and with bounded power:

(i) $p^1(x_1,0,u^2) = 0$, $\forall x_1 \in \Delta_1$, $u^2 \in U_2$; (13)

(ii) there is $M > 0$ such that

$$0 < p^1(x_1,x_2,u^2)x_2 < M \qquad (14)$$

for $x_2 \neq 0$, $x_2 \in \Delta_1$ and any $x_1 \in \Delta_1$, $u^2 \in U_2$. As $u^2 = \text{const}$ we have

$$\dot{u}^2 = g(t,w) \equiv 0 , \qquad (15)$$

the noise w influencing the value of the constant only. We let the target $\theta = \{(x_1,x_2) \mid x_1=0,\ x_2=0\}$, and attempt to produce an η^1-identifier which will be playable on Δ_1 for collision with (0,0). To this aim we design the model (3) in the form

$$\left.\begin{aligned}\dot{z}_1 &= y\\ \dot{z}_2 &= -\ell - z_1 - v^2z_1^3\end{aligned}\right\} \qquad (16)$$

where $\ell > 0$, and we let the adaptive law (4) be

$$\dot{v}^2 = -mv^2, \quad v^2(t_o) > 0, \qquad (17)$$

where $m > \ell$. Hence the system (5) is jointly represented by (12), (15), (16) and (17) in the space $\mathbb{R}^6$ of vectors $z = (x_1,x_2,u^2,z_1,z_2,u^2)$.

Consider now the total energies, of the oscillator

$$H(x_1,x_2,u^2) = \tfrac{1}{2}x_2^2 + \tfrac{1}{2}x_1^2 + \tfrac{1}{4}u^2x_1^4 , \qquad (18)$$

and the model

$$H(z_1,z_2,v^2,y) = \tfrac{1}{2}y^2 + \tfrac{1}{2}z_1^2 + \tfrac{1}{4}v^2z_1^4 . \qquad (19)$$

We choose

$$V_\theta(z) = H(x_1,x_2,u^2) + H(z_1,z_2,v^2,y) + \tfrac{1}{2}(u^2)^2 + \tfrac{1}{2}(v^2)^2.$$

The condition (i)' of Thoorem 2 is obviously satisfied by such function. In order to check on (iii)', we calculate

$$\dot{V}_\theta(k(t)) = \dot{H}(x_1,x_2,u^2) + \dot{H}(z_1,z_2,v^2,y) + v^2\dot{v}^2, \qquad (20)$$

where

$$\dot{H}(x_1,x_2,u^2) = -u^1x_2,\quad \dot{H}(z_1,x_2,v^2,y) = -\ell y. \qquad (21)$$

Substituting $u^1 = p^1(x_1,x_2,u^2)$ and (17),

$$\dot{V}_\theta = -p^1(x_1,x_2,u^2)x_2 - \ell y - m(v^2)^2.$$

The condition (iii)' of Theorem 1 holds if

$$\dot{\tilde{V}}_\theta = \max_y \dot{V}_\theta(z,y) = -p^1x_2 - \ell - m(v^2)^2 \le -C_1 \qquad (22)$$

for $C_1 = \dfrac{v_\theta^+ - v_\theta^-}{T_\theta}$. This is obviously satisfied, in view of (14), with a suitable choice of M. Now, designate $\delta H(z) = H(x_1,x_2,u^2) - H(z_1,x_2,v^2,y)$, and choose

$$V_\eta(z) = |\delta H(z)| + |v^2-u^2|, \qquad (23)$$

satisfying (ii), (v) immediately. As the norm $\|\cdot\|$ is defined by (23), $z \in CM^1$ implies $\delta H(z) \neq 0$, $v^2 - u^2 \neq 0$. Hence, if we start from $z_1(t_o) = 1$, $z_2(t_o) = 1$, $v^2(t_o) = 1$ with $x_1(t_o)$, $x_2(t_o) \in \Delta_1$, $u^2 \in U_2 = [0,1]$ we must have $\delta H(z(t_o)) > 0$, $v^2(t_o) - u^2 > 0$ which in view of $\delta H(z) \neq 0$, ($\Rightarrow \delta H(z(t)) \neq 0$, $\forall t$) and (17) yields $\delta H(z(t)) > 0$, and $v^2(t) - u^2 > 0$, for as long as we are in CM^i_η. Hence

$$\dot{V}_\eta(k(t)) = \dot{H}(x_1,x_2,u^2) - \dot{H}(z_1,z_2,v^2,y) + \dot{v}^2.$$

Substituting (21) and (17),

$$\dot{V}_\eta(k(t),t) = -u^1x_2 + \ell y - mv^2$$

The condition (iv) holds if

$$\dot{\tilde{V}}_\eta(k(t),t) = \max_y \dot{V}_\eta(k(t),t) = -u^1x_2 + \ell - mv^2 \le -\frac{v_\eta^+ - v_\eta^-}{T_1}$$

for some $T_1 > 0$. This is obviously satisfied in view of (17). Then by Theorem 2 our task is achieved.

REFERENCES

Filippov, A.F. (1971). The existence of solutions of generalized differential equations, Math. Notes, Vol 10, pp.608-611.

Galperin, E.A., J.M. Skowronski (1983). Playable asymptotic observers for differential games with incomplete information. Proc. Optimization Days. Montreal, 1983 - to appear.

Krassovskii, N.N. (1974). Game theoretic control and problems of stability, Prob. Control and Information Thoory, Vol. 3 (3), pp. 171-182.

Leitmann, G. and J. Skowronski. (1977). Avoidance Control, J.Opt.Th. and Appl., Vol. 23, pp. 581-591.

Leitmann, G. and J. Skowronski. (1983). A note on avoidance control. J.Dyn.Syst. Meas. Control (ASME) - to appear.

Luenberger, D.G. and I.B. Rhodes, (1969). Differential games with imperfect state information, IEEE Trans.Autom.Control, Vol. AC-14, pp. 29-38.

Narendra, K.S. (1976). Stable identification schemes, in System Identification. Acad. Press, pp. 165-209.

Skowronski, J.M., (1981). Adaptive identification of models stabilizing under uncertainty. Lecture Notes in Biomathematics, Vol. 40, pp. 64-78. Springer.

Skowronski, J.M. (1981a). Collision with capture and escape, Israel J. Technology, Vol. 18, pp. 70-75.

Skowronski, J.M. and T.L. Vincent. (1982). Playability with and without capture, J.Opt.Th. & Appl., Vol. 36, pp. 111 128.

SEPARATION FUNCTIONS AND MEASURES OF FUZZINESS

C. Dujet

Centre de Mathématiques INSA et Université Lyon I 69622 Villeurbanne Cedex, France

Abstract. A more general axiomatic approach for the concept of separation in the setting of fuzzy sets is given :
for this purpose, participation measures are used, and extended to a class of fuzzy sets. A new notion of "nearness" (proximity) between fuzzy sets gets away from this study, taking in account the "average behaviour" of a fuzzy set.
In the final section, entropies and separation functions are compared.

Keywords. Fuzzy set - Lattice - valuation - separating power - fuzzy measure-participation measure.

INTRODUCTION

In some previous papers, I elaborated a notion of separation in the setting of fuzzy sets ; the philosophical starting point could be summarized by the two following sentences :

"Although this may seem a paradox, all science is dominated by the idea of approximation" (B. Russel)

"Nature may appear to us a s if everything was linked or as if everything was separated" (Simmel).

The notion I introduced was to take in account the "average behaviour" of a fuzzy set, and more specifically, was to try to evaluate, how much a fuzzy set f of un universe E was "separating" the elements of E , compared to the ideal case, when f should take only the values zero or one, in which case E is "naturally" separated into two complementary sets.

This concept of separation has led to new notions of complementation and partition in the class of fuzzy sets (See Dujet [1] [2], as well as new orderings, which are not depending on the usuel ones and may be useful in classification problems.

Besides, I have already shown, how the separting power of a fuzzy set turns out to be a generalized inforamation of KAMPE de FERIET upon the class of fuzzy sets, under the new ordering I defined. Dujet [3] [4].

However, the notion of separation was based on a valuation upon the lattice of fuzzy sets, and restricted to the case of a finite universe. To escape from the limit imposed by a valuation in problems where a hierarchical order is no need, I try first to use the participation measures of Tsichritzis [5], which proceed from the same point of view : to express an "average behaviour".

I - SEPARATING POWER AND PARTICIPATION MEASURE

We first recall sorme definitions which are needed for the following.

Definition 1. - Participation measure of Tsichritzis

Given a countable set U , a participation measure m on U is defined as as mapping from subsets of U to the real numbers satisfying the following axioms :

m1. $m(\emptyset) = 0$
m2. $\forall A \in \mathscr{P}(u), \forall B \in \mathscr{P}(u)$, if $A \subset B$, then $m(A) \leqslant m(B)$.
m3. $\forall A \in \mathscr{P}(u)$, $m(A) + m(\bar{A}) = 1$, where $\bar{A}$ is the complement of A with respect to U.

Remark : a participation measure is a fuzzy measure which moreover satisfies m3.

PREMIMINARIES NOTATIONS

Given a distributive lattice H, a H-fuzzy set is any couple (A,α) where A is an ordinary set, α is a mapping from A to H. (PONASSE [6]).

Given a referential E , we shall denote by $\mathscr{C}(E)$ the class of the fuzzy subsets of E : that is to say the H-fuzzy sets (A,α) where A runs over $\mathscr{P}(E)$.

With the union, intersection defined in [6], $\mathscr{C}(E)$ is a lattice, to be distinguished from the lattice $\mathscr{L}(E)$:

$$\mathscr{L}(E) = \{(E,f),\ f : E \to H\}$$

whe shall denote by f, g... the elements of $\mathscr{L}(E)$.

Given f belonging to $\mathscr{L}(E)$, we can consider

a class of fuzzy sets, denoted by $\mathscr{C}_f(E)$, defined as follows :

$$\mathscr{C}_f(E) = \{(A,f/A) \; ; \; A \in \mathscr{P}(E)\} ,$$

where f/A is the restriction of the mapping f to A .

Whe shall use the notation f_A instead of (A, f/A).

According to the definitions of union and intersection in $\mathscr{C}(E)$, it is easy to see that the class $\mathscr{C}_f(E)$ is stable for these operations.

Definition 1.2 - Separating power

Given a finite set E , given a valuation v upon (E), a v-separating power in E is the mapping ⋆ from $\mathscr{L}(E) \times \mathscr{P}(E)$ into the real line, defined as follows :

$$f \star A = \frac{1}{|A|} v(f_A) - \frac{1}{|\bar{A}|} v(f_{\bar{A}}) \quad \text{if} \quad A \notin \{\emptyset, E\}$$

$$f \star A = 0 \quad \text{if} \quad A = \emptyset \quad \text{or} \quad A = E .$$

Notation - $|A|$ denotes the cardinal of A

Definition 1.3 - Separating index of a fuzzy set .

Given a v-separating power in E , the v-separating index of a fuzzy set of E , denoted by $S_v(f)$, is defined as follows :

$$S_v(f) = \text{Sup}\,\{|f \star A|,\; A \in \mathscr{P}(E)\}.$$

Proposition 1.1

Let f be a fuzzy set of E .
Let v be a valuation upon the lattice $\mathscr{C}(E)$.
The set function m on E defined by :

$$m(A) = \frac{1}{v(f_E)} v(f_A), \; A \in \mathscr{P}(E) \quad \text{is a parti-}$$

cipation measure on E , providing the valuation v satisfies the two following conditions : i) $v(\emptyset) = 0$
ii) $v(A,\alpha) \in R^+$, $\forall (A,\alpha) \in \mathscr{C}(E)$.

It is easy to see that conditions i) and ii) imply the monotony of v with respect to inclusion order in $\mathscr{C}(E)$, hence m satisfies axiom. m2.

Axiom m3 follows from the property of a valuation :

$$v(f_A) + v(f_{\bar{A}}) = v(f_A \vee f_{\bar{A}}) \quad \text{(because of ii))}$$

$$\text{"} \qquad \text{"} \qquad = v(f_E)$$

From this considerations, the separating index of a fuzzy set of E may be rewritten as follows :

$$\text{(I)} \quad S_V(f) = v(f_E) \underset{A \in \mathscr{P}(E)}{\text{Sup}} \{ | \frac{1}{|A|} m(A) - \frac{1}{|\bar{A}|} m(\bar{A}) | \}$$

when E is finite.

This leads us to try to derive the separating power a fuzzy set from a participation measure on E , but the point is, that the set-function m which appears above in (1) must be generate in an uniform way for every fuzzy set of E.

On the other hand, it would be pleasant, to deduce a separating power from somewhat a "participation" measure, because it would be then intuitively acceptable with the concept of separation and what is expected from it : the more "participation" we have on the bi-partition (A, $\bar{A}$) of E , the more "separation" we shall find on E , by mean of a fuzzy set f , according to relation (1).

Keeping in mind the necessity to generate uniformly a set-function on E , we are led to adapt the notion of participation defined on an ordinary subset of E to one defined on a fuzzy subset of E.

II - EXTENSION OF PARTICIPATION MEASURES

The aim is to elaborate a function defined on the class of the fuzzy subsets of a referential E (finite or not), which would express some idea of participation, and should generalize the participation measures on E of Tsichritzis.

So, we would like some similar axioms, and we have therefore to cope with the notion of complementation in lattice of fuzzy sets (as well as in the lattice of fuzzy subsets).The lattice of fuzzy sets is not a complemented one, with respect to the union and intersection operations.

But we are concerned now with the concept of "participation" of a fuzzy set of E in E , and in this way, it seems natural enough to think, that there should be a relation between the "participation" conveyed by a fuzzy set and the "participation" conveyed by its negation.

I stress, this approach is quite different from my previous one : in my first papers on the concept of separation, my purpose was, by the mean of this concept, to find non-pointwise negation and partition in the lattice of fuzzy sets.

Here I shall use the negation to define the participation measure and then the separating power and index.

So from now, we assume the lattice of fuzzy sets of E has a negation (most usually : Zadeh's negation : $\neg f(x) = 1 - f(x)$, see [7] when the set of membershipgrades is [0,1].)

Remark : if the lattice of fuzzy sets of E has a negation (noted $\neg$) the negation in the lattice $\mathscr{C}(E)$ of fuzzy subsets of E is defined by : $\neg(A, \alpha) = (A, \neg\alpha)$, because it is obvious, a H-fuzzy set and its negation have the same support set.
Therefore the following definition :

Definition 2.1 - Fuzzy participation measure on E

Given a referential set E, given a negation $\neg$ defined in the lattice $\mathscr{C}(E)$ (class of fuzzy subsets of E), a fuzzy participation measure on E is defined as a mapping from $\mathscr{C}(E)$ into the real numbers, satisfying the following axioms :

m'1. $m(\emptyset) = 0$ where $\emptyset$ is the empty fuzzy set
m'2. $m(A,\alpha) + m(\neg(A,\alpha)) = 1 \quad \forall A \in \mathscr{P}(E)$, $\forall \alpha \in \mathscr{L}(A)$.
m'3. if $(A,\alpha) \subseteq (A,\beta)$, then $m(A,\alpha) \leqslant m(A,\beta)$.

Remark : if α is the mapping from A on zero, then the support set of (A,α) is empty and (A,α) will be identified to the empty fuzzy set.

Notation : we use also the term F-participation measure on E.

Proposition 2.1

The restriction of a fuzzy participation measure on E to the class of ordinary subsets of E is a participation measure on E.

Proof :

Given A an ordinary set of E, A may be identified to a fuzzy subset of E by the mean of its characteristic function 1_A.

So we put $A \equiv (E, 1_A)$, for every A belonging to (E). Let m be a F-participation measure on E, we have :

$$m(E,1_A) + m(\neg(E,1_A)) = 1$$

but $\neg(E,1_A) = (E,\neg 1_A) = (E,1_{\bar{A}}) = \forall A \in \mathscr{P}(E)$.

(We stress, that for every negation it is possible to define in $\mathscr{C}(E)$, the property : $\neg 1_A = 1_{\bar{A}}$ always holds. See Yager [8].

hence follows $m(A) + m(\bar{A}) = 1 \quad \forall A \in \mathscr{P}(E)$ (axiome m2 satisfied). Let A and B belonging to $\mathscr{P}(E)$ such that A is included in B. $A \subset B$ implies $1_A \leqslant 1_B$;

by axiom m'3, it follows : $m(E,1_A) \leqslant m(E,1_B)$

hence : if $A \subset B$, then $m(A) \leqslant m(B)$ (axiom m3 satisfied).
Axiom m1 is obviously satisfied, and the proposition 2.1 is proved.

Now, we will see that it is possible to generate from a fuzzy participation measure on E a set-function on E in an uniform way for every fuzzy set of E.
Given a fuzzy set f of E, and given a Fuzzy participation measure m on E we can consider the restriction m' of m to the class $\mathscr{C}_f(E)$ of fuzzy subsets of E.

$$\mathscr{C}_f(E) = \{(A,f/A),\ A \in \mathscr{P}(E)\}$$

Let φ_f be the mapping from $\mathscr{P}(E)$ on to $\mathscr{C}_f(E)$; the mapping $m'_o\,\varphi_f$ is the set-function looked for

$$\mathscr{P}(E) \xrightarrow{\ \varphi_f\ } \mathscr{C}_f(E) \xrightarrow{\ m'\ } R$$

$$A \longrightarrow (A,f/A) \longrightarrow m(A,f/A) = m_f(A)$$

$$m_f$$

Thus we have defined a mapping from the lattice of fuzzy sets of E $\mathscr{L}(E)$ in the class of set-functions on E ; this mapping will be called a "participation generator" and denoted by g_m

$$g_m : \mathscr{L}(E) \longrightarrow \{\text{set-functions on } E\}$$
$$f \longrightarrow m_f$$

We still need to introduce the following function :
for every set function θ on E, let us denote by θ^c the set-function on E defined as follows :

$$\theta^c(A) = \theta(\bar{A}) \ , \ \forall A \in \mathscr{P}(E).$$

Now we are ready to define the separating index of a fuzzy set, derived from a Fuzzy participation measure, in the same way that we have defined the separating index derived from a valuation.

Definition 2.2 - Separation norm :

Given a fuzzy participation measure m on a referential E, a separation norm N_S on E is a mapping from $\mathscr{L}(E)$ into the set of real numbers defined as follows :

$N_S(f) = \|m_f - m_f^c\|$ for every fuzzy set f of E, where m_f is the image of f by the participation generator, and $\| \ \|$ is the L_1-norm, i.e.

$$\|m_f - m_f^c\| = \text{Sup}\ \{|m_f(A) - m_f^c(A)|,\ A \in \mathscr{P}(E)\}$$

Remark : $N_S(f)$ may be called the m-separating index of f.

Example : in the case of a finite referential E, a fuzzy participation measure on E may be defined as the following mapping :

$$(A,\alpha) \to \frac{1}{|A|} \sum_{x \in A} \alpha(x) \qquad \forall A \in \mathscr{P}(E)$$
$$\forall \alpha : A \to [0,1] \ ,$$

where the negation in $\mathscr{C}(E)$ is
$\neg(A,\alpha) = (A, 1-\alpha)$.

Proposition 2.2.

A separation norm on E is symmetric with respect to the negation in $\mathscr{L}(E)$ that is to say :

$$N_S(f) = N_S(\neg f) \ , \ \forall f \ , \ f \in \mathscr{L}(E).$$

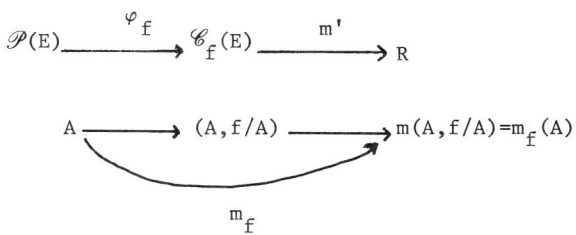

Proof : Let m be the fuzzy participation measure, from which N_S is derived. Then we have $m(A, f/A) + m(A, \neg f/A) = 1 \quad \forall A \in \mathcal{P}(E)$

$$N_S(f)=\mathrm{Sup}\ \{|m(f_A)-m(f_{\bar{A}})|,\ A \in \mathcal{P}(E)\}$$
$$=\mathrm{Sup}\ \{|1-m(\neg f_A)-(1-m(\neg f_{\bar{A}}))|,\ A \in \mathcal{P}(E)\}$$
$$=\mathrm{Sup}\ \{|m(A, \neg f/A)-m(\bar{A}, \neg f/\bar{A})|,\ A \in \mathcal{P}(E)\}$$
$$= N_S(\neg f).$$

Proposition 2.3

A separation norm on E is maximal for the fuzzy sets of E whose membership function is a boolean charateristic function.

Proof

Let m be the fuzzy participation measure on E, N_S the separation norm derived from it ; from the axioms of a F-participation measure on E the image of a fuzzy subset (A,α) of E is a real number belonging to the interval [0,1] ; this implies, the image of a fuzzy set f of E by the separation norm N_S is also a real number belonging to the unit interval. Let $(E,1_A)$ be a fuzzy set of E, whose membership function is the characteristic function of an ordinary subset A of E.
Let us calculate $N_S(E,1_A)$.

$$N_X(E,1_A) = \mathrm{Sup}\{|m(X,1_A/X)-m(\bar{X},1_{A/\bar{X}})|, X \in \mathcal{P}(E)\}$$

When X=A, $m(A,1_{A/A})-m(\bar{A},1_{A/\bar{A}})=m(A,1)-m(\bar{A},0)$ or $m(\bar{A},0)=m(\emptyset)$ (=0) and $m(A,1)=1$ (by axiom m'2) thus we obtain :

$$N_S(E,\ 1_A) = 1$$

For our concern : to evaluate to which extent a fuzzy set of a referential set E is "separating" E, we dispose now of two classes of functions which assign to a fuzzy set a positive real number measuring this separation :

- the separating index, derived from a valuation
- the separation norm, derived from a fuzzy participation measure.

From the common properties of these functions, it is possible to get clear the axiomatic framework of the concept of separation in the setting of fuzzy sets.

Definition 2.3 - Separation functions

Given a set E (finite or not), given a negation $\neg$ in the lattice $\mathcal{L}(E)$ of fuzzy sets of E, a separation function is a mapping s from $\mathcal{L}(E)$ into the unit interval of real numbers, satisfying the following axioms :

S1. $s(\emptyset) = 0$
S2. $s(f) = s(\neg f) \ \forall f \in \mathcal{L}(E)$.
S3. $s(1_A) = 1 \quad \forall A,\ A \neq \emptyset,\ A \neq E,\ A \in \mathcal{P}(E)$.

Remark : The intersection of the two classes of separation functions we have described above is not empty : it contains the function of the example §2.

III - SEPARATION FUNCTIONS, MEASURE OF FUZZINESS AND ENTROPIES

Our basic idea is that, the more a fuzzy set is "near" an ordinary set, the less it is fuzzy ; another way to express this idea is to say : the more a fuzzy set is "separating" its support set, the less it is fuzzy. Therefore, a separating index or a separation norm looks like a good measure of fuzziness.

Moreover, by using a fuzzy participation measure on a referential set E, the lattice of fuzzy sets of E will be provided with a quasi-distance, from which it is possible to characterize fuzziness in another way, as proposed by Yager [8] : "the measure of fuziness of a concept is related to the distinction between the concept and its negation".

Proposition 3.1

Given a fuzzy participation measure on E, the mapping d_m which assigns to every couple (f,g) of fuzzy sets of E the L_1-norm of the set function $g_m(f)-g_m(g)$ on E, where g_m is the participation generator associated to m, is a quasi-distance on E.
The mapping d_m is defined as follows :

$$d_m(f,g) = \|m_f - m_g\|$$
$$=\mathrm{Sup}\{|m(A,f/A)-m(A,g/A)|, A \in \mathcal{P}(E)\}$$

This leads to define a "proximity" or "nearness" relation in the lattice of fuzzy sets :

Definition 3.1 - Nearness in $\mathcal{L}(E)$

Given any fuzzy sets f, g, h of E, f is said "nearer g than h" if the distance d_m between f and g is less than the distance between f and h.

Proposition 3.2

Given a fuzzy participation measure m on E, we have the following result :

$\forall f, f \in \mathcal{L}(E)$, $d_m(f, \neg f) = 0$ if and only if $m_f(A) = \frac{1}{2}$, $\forall A \in \mathcal{P}(E)$.

Proof :

$$d_m(f, \neg f) = \|m_f - m_{\neg f}\|$$

$$d_m(f, \neg f) = 0 \Leftrightarrow \mathrm{Sup}\{|m(A,f/A)-m(A, \neg f/A|,\ A \in \mathcal{P}(E)\} = 0$$
$$m(A,f/A)-m(A,\neg f/A)=0 \ \forall A \in \mathcal{P}(E)$$
$$m(A,f/A)=m(A,\neg f/A) \ \forall A \in \mathcal{P}(E).$$

Besides, m(A, f/A) + m(A, ⌐ f/A) = 1 according to Axiom m'2, thus we obtain $d_m(f, \neg f) = 0 \Leftrightarrow m(A,f/A) = \frac{1}{2} \quad \forall A \in \mathscr{P}(E)$.

Corollary

If the set of membershipgrades is [0,1], and if the fuzzy participation measure m is the one of Example §2, then $d_m(f, \neg f) = 0$ is equivalent to $f(x) = \frac{1}{2}$, $\forall x \in E$.

This shows that the positive real number $d_m(f, \neg f)$ associated to f is a good measure of the fuzziness of f, from the Yager's point of view : if the negation of f is among the closiest fuzzy sets to f, then the participation of f on any subset of the referential is constant and eqnal to $\frac{1}{2}$, which do express that f is as fuzzy as possible.

Thus, two semantically different points of view (Dujet-Yager) may complete themselves by a combination of the two norms :

$$\|m_f - m_{\neg f}\| \text{ and } \|m_f - m_f^c\| .$$

Now we have to compare entropies and separation functions. First we recall the definitions of a non-probabilistic entropy, as stated by de Luca and Termini [9].

Definition 3.2 - Entropy or H-function on $\mathscr{L}(E)$

An entropy is any map $h : \mathscr{L}(E) \to R$ such that
i) h is isotone with the ordrer $\leqslant_s$
ii) h(f) = 0 if and only if f is a boolean characteristic function.

N.B. The order $\leqslant_s$ is the "sharpened" relation introduced by Trillas and Riera [10]
h is said symmetric if $h(f) = h(\neg f)$ for every $f \in \mathscr{L}(E)$.

Proposition 3.3

If h is a symmetric entropy on $\mathscr{L}(E)$, then the function 1-h is a separation function on $\mathscr{L}(E) - \{\emptyset, E\}$.

This follows obviously from def 3.2 and def 2.3, and points out the links and differences between an entropy and a separation function.

To conclude this study, I would like to observe that a fuzzy participation measure on a referentiel set E may infer a notion of "contiguity" in the lattice of fuzzy sets of E, in a natural enough way, and analoguous to the concept of contiguity introduced on probability measures by G. Roussas [11], which provides a powerful mathematical tool with applications in statistics. So it may be expected, that a notion of "contiguity" on fuzzy participation measures would be potentially a main concept for theory and applications.

NOTIONS OF CONTIGUITY IN FUZZY SETS

Let $\{f_n\}$ and $\{g_n\}$ be sequences of fuzzy sets of a referential E.

Definition 3.3

Given a fuzzy participation measure m on E, the sequences $\{f_n\}$ and $\{g_n\}$ of fuzzy sets of E are said to be contiguous if the sequence of real numbers $d_m(f_n, g_n)$ tends to zero.

Concluding Remarks

My purpose of giving a more precise axiomatic approach of the concept of separation, I introduced in the fuzzy sets, is achieved. By doing it, I face unexpected outcomes, which open the way to many other inverstigations.

Particularly, it is likely, there would be major interests to investigate connections between a fuzzy participation measure and a possibility measure of Zadeh [12], and to develop and extend the notion of contiguity (slightly touched above) to the possibility measures. Already, Dubois and Prade [13] have remarked some link between a possibility measure and a participation measure. Separation functions and more specifically separation norms should be very useful in Pattern Recognition and Clustering Problems or problems which need a systemic approach.

This notions are beginning to be applied to modelize a Pattern Recognition Problem in Radiologie and good results are hoped.

BIBLIOGRAPHY

Dujet, Ch. Separation dans les ensembles flous et complémentation induite. Actes de la Table Ronde CNRS sur le flou LYON 1980

Dujet, Ch. Valuation et Séparation dans les ensembles flous. Journées Mancelles : Informations et Questionnaires 15-17 Sept. 80 . Publications CNRS

Dujet, Ch. Ensembles co-flous et information, basés sur le pouvoir séparateur d'un ensemble flou, thèse de 3e cycle - Sept. 81

Dujet, Ch. Separation and Information in the setting of fuzzy sets IFAC Symposium on Theory and Applications of Digital Control, New-Delhi (India) Jan 1982.

Tsichritzis, D. Participation measures, Journal of Math. Analysis and Applications 36, 60-72 (1971)

Ponasse, D. Séminaire Mathématique Floue (1979-1980) Publications Dept. Math. Université Lyon 1.

Zadeh, L. A. Fuzzy sets, Inf. and Control 8, 338-353 (1965).

Yager, R. On the measure of Fuzziness and Negations. II lattices Inf and Control 44, 236-260 (1980)

De Luca and Termini , Entropy and Energy measures of a fuzzy set. Advances in fuzzy set theory and Applications (Yager-Gupta Editors) North-Holland (1979).

Trillas E., and Riera, T. , Entropies in finite fuzzy sets. Inform. Sciences, 15, 159-168 (1978)

Roussas, G., Contiguity of probability measures Cambridge University Press (1972).

Zadeh, L.A., Fuzzy sets as a basis for a theory of possibility Fuzzy sets and systems 1, N°1, 3-28 (1978).

Dubois, D., and Prade, H., A class of fuzzy measures based on triangular norms. International Journal of general Systems Vol 8 N°1 1982.

MEASUREMENT OF FUZZINESS: AN INTERPRETATION OF THE AXIOMS OF MEASUREMENT

I. B. Turksen

Department of Industrial Engineering, University of Toronto, Toronto, Ontario, Canada M5S 1A4

Abstract. Deterministic axioms form the foundations of the measurement theory. In every measurement experiment of fuzziness, however, human responses almost always display inconsistencies with a random behavior pattern. In this paper, it is suggested that measurement axioms be re-stated in probabilistic terms and be reinterpreted fuzzily. A type II fuzzy set interpretation of the response frequencies is discussed for the measurement of "highness" of the sound of a note on a piano in order to display the behavior pattern of a human subject in such a measurement experiment.

Keywords. Measurement, fuzzy, axioms, deterministic, probabilistic, possibilistic, interpretation, experimental, plausibility.

INTRODUCTION

The Fundamental Measurement of fuzziness may be regarded as the construction of homomorphisms from empirical relational structures that express membership in fuzzy sets. In particular, if $< \Theta, \gtrsim_A >$ is an empirical relational structure of a fuzzy concept (attribute) A over a set of objects Θ and $< Re, \geq >$ is a numerical relational structure, then a real valued function ϕ, called a membership function, on Θ is a homomorphism if it takes the empirical relation $\gtrsim_A$ into the numerical relation $\geq$ (Turksen, 1979; Turksen and Norwich, 1981; Norwich and Turksen, 1982, 1982, 1982; Turksen, 1982).

This view of measurement of fuzziness is a special case of the general theory of measurement that is expressed abstractly as follows: If $< \Theta, R_1, \ldots, R_m >$ is an empirical relational structure and $< Re, S_1, \ldots, S_m >$ is a numerical relational structure, then a real-valued function ϕ on Θ is a homomorphism if it takes each R_i into S_i, $i = 1,2,\ldots,m$ (Krantz, Luce, Suppes, and Tversky, 1971).

In this general and abstract approach, foundational analysis of measurement consists of clarifying assumptions and stating representation and uniqueness theorems for the construction of such homomorphisms. Furthermore, this relational structure approach unifies many historically, interesting examples of measurement structures. The key examples are axiomatization of utility measurement by Von Neuman and Morgenstern (1953), Savage (1954), Suppes and Winet (1955), Davidson, Suppes and Siegel (1957).

Almost all of these measurement theoretic studies include a set of deterministic axioms that attempt to clarify: (i) behavioural aspects of subject's responses and (ii) procedures for constructing numerical assignments. These behavioural axioms are generally stated as weak order, weak transitivity and weak monotonicity, whereas the structure axioms are known as solvability and Archimedian conditions. For instance in a decision theory framework, theories of "subjective choice" are formulated with the assumption of a "rational behavioral" (Suppes, 1955; Savage, 1954). For example, P. Suppes (1955) presents eleven axioms for "rational subjective choice" and separates them into two categories: (i) rationality axioms, and (ii) structure axioms. It is assumed that the "pure" rationality axioms, deterministic as they are, should be satisfied by any rational, reflective person in a decision making situation. On the otherhand, the "pure" structure axioms impose limitations on the kind of situations to which an analysis may be applied.

Since there is no really "satisfactory" definition of rationality, these theories have a somewhat tenious quality for empirisists. The experimental literature concerned with the validity of these theories are indeed very few and only test certain limited aspects of these theories, simply because these theories are expressed in a deterministic manner and do not lend themselves to an "exact" analysis in their relation to experimental data. For example, in measurement experiments of fuzziness based on an axiomatic formalism, subjects almost always exhibit inconsistencies among and ambiguities within their responses to given stimuli (Turksen, 1979; Turksen and Norwich, 1981; Norwich and Turksen, 1981, 1982; Turksen, 1982). These inconsistent and ambiguous response behaviour of subjects display anything but "rational" or deterministic behaviour. Thus historical debate on "rationality" or determinism and scarce analysis of experimental data strongly suggest that every axiomatic theory of measurement whether it be for the measurement of fuzziness or preference needs to be modified. Such modifications has to account for the inexactness and randomness on the one hand and the insufficiency of data, on the other, that are inescapable components of every empirical study of human subjects.

In an attempt to remedy these deficiencies, it is suggested in this paper, that the deterministic "rationality" axioms of measurement theory firstly be restated in probabilistic terms and then secondly be reinterpreted in possibilistic terms.

A measurement theory can be validated either logically or experimentally. In this paper an empirical validation is suggested with a definition of a "possibilistic measure of satisfaction" for axioms of measurement theory in order to capture properly the imprecise and inconsistent nature of human responses in measurement experiments of fuzziness.

AXIOMATIC FOUNDATIONS

Every foundational analysis of the measurement theory includes definitions of "weak order" and "algebraic-difference structure" and theorems of "ordinal" and "interval" scales. These essentials are briefly reviewed here in the language of Krantz et. al. (1971).

Definition 1. Let Θ be set and $\gtrsim_A$ be a binary relation on Θ, i.e., $\gtrsim_A$ is a subset of $\Theta' = \Theta \times \Theta$. The relational structure $< \Theta, \gtrsim_A >$ is a weak-order if and only if, iff, for all $\theta_i, \theta_j, \theta_k \in \Theta$, the following two axioms are satisfied:

1. Connectedness: Either $\theta_i \gtrsim_A \theta_j$ or $\theta_j \gtrsim_A \theta_i$.
2. Transitivity: If $\theta_i \gtrsim_A \theta_j$ and $\theta_j \gtrsim_A \theta_k$, then $\theta_i \gtrsim_A \theta_k$.

It is noted that this special weak-order is reflexive and antisymmetric.

Theorem 1. Suppose that Θ is a finite nonempty set. If $< \Theta, \gtrsim_A >$ is a weak-order, then there exists a real-valued function ϕ on Θ such that for all $\theta_i, \theta_j \in \Theta$, $\theta_i \gtrsim_A \theta_j$ iff $\phi(\theta_i) \geq \phi(\theta_j)$. Moreover, ϕ' is another real-valued function on Θ with the same property iff there is a strictly increasing function f, with domain and range equal to Re, such that for all $\theta \in \Theta$, $\phi'(\theta) = f[\phi(\theta)]$, i.e., ϕ is an ordinal scale.

Definition 2. Suppose Θ is a nonempty set and $\gtrsim_{A'}$ is a quaternary relation on Θ'. The relational structure $< \Theta', \gtrsim_{A'} >$ is an algebraic-difference structure iff, $\theta_i, \theta_j, \theta_k, \theta_l, \theta'_i, \theta'_j$, and $\theta'_k \in \Theta$ and all sequences $\theta_1, \theta_2, \ldots, \theta_i, \ldots \in \Theta$ the following five axioms are satisfied:

1. $< \Theta', \gtrsim_{A'} >$ is a weak order.
2. Sign Reversal: If $\theta_j\theta_i \gtrsim_{A'} \theta_l\theta_k$, then $\theta_k\theta_l \gtrsim_{A'} \theta_i\theta_j$.
3. Weak-Monotonicity: If $\theta_j\theta_i \gtrsim_{A'} \theta'_j\theta'_i$, and $\theta_k\theta_j \gtrsim_{A'} \theta'_k\theta'_j$, then $\theta_k\theta_i \gtrsim_{A'} \theta'_k\theta'_i$.
4. Solvability: If $\theta_j\theta_i \gtrsim_{A'} \theta_l\theta_k \gtrsim_{A'} \theta_i\theta_i$, then there exists $\theta'_l, \theta''_l \in \Theta$ such that $\theta'_l\theta_i \sim_{A'} \theta_l\theta_k \sim_{A'} \theta_j\theta''_l$.
5. Archimedian Condition: If $\theta_1, \theta_2, \ldots, \theta_i, \ldots$, is a strictly bounded standard sequence ($\theta_{i+1}\theta_i \sim \theta_2\theta_1$ for every θ_i, θ_{i+1} in the sequence; not $\theta_2\theta_1 \sim \theta_1\theta_i$, and there exists $\theta'\theta'' \in \Theta$ such that $\theta'\theta'' > \theta_i\theta_1 > \theta''\theta'$ for all θ_i in the sequence), then it is finite.

Theorem 2. If $< \Theta', \gtrsim_{A'} >$ is an algebraic-difference structure, then there exists a real-valued function ϕ on Θ such that, for all $\theta_i, \theta_j, \theta_k, \theta_l \in \Theta$, $\theta_j\theta_i \gtrsim_{A'} \theta_l\theta_k$ iff $\phi(\theta_j) - \phi(\theta_i) \geq \phi(\theta_l) - \phi(\theta_k)$.
Moreover, ϕ is unique up to a positive linear transformation, i.e., if ϕ' has the same properties as ϕ, then there are real constants α, β, with $\alpha > 0$, such that $\phi' = \alpha\phi + \beta$, i.e., ϕ is an interval scale.

A Probabilistic Interpretation

The relational structure $< \Theta, \gtrsim_A >$ is a weak-order and $< \Theta' \gtrsim_{A'} >$ is an algebraic-difference structure if our subjects behave "rationally" and deterministically. We know from our experiments that human subjects behave rather irrationally and nondeterministically (Turksen and Norwich, 1981; Norwich and Turksen, 1981, 1982; Turksen, 1982). Therefore, the "rationality" and deterministic behaviour assumptions must be put aside and all the axioms of the measurement theory must be restated in probabilistic terms.

There are hints in Krantz et. al. (1971, p. 3), Coombs et. al. (1970, p. 156) and Debrue (1958) for probabilistic treatments. However, our formulation is more general. This is done with the following definitions and conjectures.

Definition 1'. Let Θ be a set and $\gtrsim_A$ be a binary relation on Θ, i.e., $\gtrsim_A$ is a subset of Θ'. The relational structure $< \Theta, \gtrsim_A >_p$ is a weak probabilistic order if for all θ_i, θ_j, θ_k, $\varepsilon\ \Theta$, the following two axioms are satisfied probabilistically above a probability threshold p:

1. Probabilistic Connectedness:
Either $P\{\theta_i \gtrsim_A \theta_j\} \geq p$ or $P\{\theta_j \gtrsim_A \theta_j\} \geq p$ where $P\{\theta_i \gtrsim_A \theta_j\}$ is the probability that θ_i will be in relation $\gtrsim_A$ to θ_j. It is defined as:

$$P\{\theta_i \gtrsim_A \theta_j\} = \lim_{n\to\infty} \frac{n_1(\theta_i \gtrsim_A \theta_j)}{n}, \quad 0 \leq n_1 \leq n, \tag{1}$$

where n_1 is the number of trials in which a subject states that $\theta_i \gtrsim_A \theta_j$ and n is the total number of trials, i.e.,
$n = n_1(\theta_i \gtrsim_A \theta_j) + n_2(\theta_j \gtrsim_A \theta_i)$.

2. Probabilistic Transitivity:
If $P\{\theta_i \gtrsim_A \theta_j\} \geq p$ and $P\{\theta_j \gtrsim_A \theta_k\} \geq p$ then $P\{\theta_i \gtrsim_A \theta_k\} \geq p$.

Conjecture 1'. Suppose that Θ is a finite nonempty set. If $< \Theta, \gtrsim_A >_p$ is a probabilistic-weak-order then there exists a random valued function ϕ_p on Θ such that for all $\theta_i, \theta_j\ \varepsilon\ \Theta$: $P\{\theta_i \gtrsim_A \theta_j\} \geq p$ iff $P\{\phi_p(\theta_i) \geq \phi_p(\theta_j)\} \geq p$. Moreover, ϕ'_p is another random valued function on Θ with the same property iff there exists a strictly increasing function f with domain and range equal to Re, such that for all $\theta\ \varepsilon\ \Theta$: $P\{\phi'_p(\theta) = f[\phi_p(\theta)]\} \geq p$, i.e., ϕ_p is a probabilistic-ordinal scale.

Definition 2'. Suppose Θ is a nonempty set and $\gtrsim_{A'}$ is quaternary relation on Θ, i.e., a binary relation on Θ'. The relational structure $< \Theta', \gtrsim_{A'} >_p$ is a probabilistic-algebraic-difference structure iff, θ_i, θ_j, θ_k, θ_1, θ'_i, θ'_j and $\theta'_k\ \varepsilon\ \Theta$ and all sequences $\theta_1, \theta_2, \ldots, \theta_i, \ldots\ \varepsilon\ \Theta$ the following axioms are satisfied probabilistically above a threshold probability p:

1. $< \Theta', \gtrsim_{A'} >_p$ is a probabilistic-weak-order.

2. Probabilistic-Sign Reversal:
If $P\{\theta_j\theta_i \gtrsim_{A'}, \theta_1\theta_k\} \geq p$ then $P\{\theta_k\theta_1 \gtrsim_{A'}, \theta_i\theta_j\} \geq p$.

3. Probabilistic-Weak-Monotonicity:
If $P\{\theta_j\theta_i \gtrsim_{A'}, \theta'_j\theta'_i\} \geq p$ and $P\{\theta_k\theta_j \gtrsim_{A'}, \theta'_k\theta'_j\} \geq p$, then $P\{\theta_k\theta_i \gtrsim_{A'}, \theta'_k\theta'_i\} \geq p$.

4. Probabilistic-Solvability:
If $P\{\theta_j\theta_i \gtrsim_{A'}, \theta_1\theta_k \gtrsim_{A'}, \theta_i\theta_i\} \geq p$ then there exists θ'_1, $\theta''_1\ \varepsilon\ \Theta$ such that $P\{\theta'_1\theta_i \sim_{A'}, \theta_1\theta_k \sim_{A'}, \theta_j\theta''_1\} \geq p$.

5. Probabilistic Archmedian Condition:
This means the probability that "no interval is infinitely large with respect to any smaller one whose jnd, just noticable difference, is not zero" is greater than or equal to a probability measure p. In other words, the probability of the gain in A-ness in going from θ_i to θ_j can only be exceeded by a finite number of successive jnd gains starting at θ_i is larger than or equal to a probability measure p.

Conjecture 2'. If $< \Theta', \gtrsim_{A'} >_p$ is a probabilistic-algebraic-difference structure, then there exists a random valued function ϕ_p on Θ such that, for all $\theta_i, \theta_j, \theta_k, \theta_1\ \varepsilon\ \Theta$, $P\{\theta_j\theta_i \gtrsim_{A'}, \theta_1\theta_k\} \geq p$ iff $P\{\phi_p(\theta_j)-\phi_p(\theta_i) \geq \phi_p(\theta_1)-\phi_p(\theta_k)\} \geq p$. Moreover, ϕ_p is unique up to a probabilistic-positive linear transformation, i.e., if ϕ'_p has the same properties as ϕ_p, then there are real constants α, β, with $\alpha>0$, such that $P\{\phi'_p = \alpha\phi_p + \beta\} \geq p$, i.e., ϕ_p is a probabilistic-interval scale. It is to be noted that the computations of the left hand sides, of all the inequalities, in Definitions 1', 2' and Conjectures 1', 2', are to be interpreted in accordance with the probability calculus and the classical rules of inference.

A Possibilistic Interpretation

In classical (objectivistic) view, the empirical probability of an event is the limit of the ratio between the number m of occurrences of a specified event E over n trials in an experimental universe U as n becomes "large", i.e.,

$$P(E) = \lim_{n(U)\to\infty} \frac{m(E)}{n(U)} \tag{2}$$

The implication of this view is that response data must be observed by an analyst an infinite number, or at least a large number of times for every experimental setting in a measurement study. This is generally possible in measurement experiments of physical attributes. However, it is almost an impossibility when the response data are to be provided by human subjects as in the case of subjective evaluations in general and measurement of fuzziness in particular. It is a fact that human subjects do get bored or tired easily or long run experiments are too costly. In these cases, only a few points of response data may be observed under ordinary circumstances. Thus probabilities of fuzzy events may not in general be computed due to lack of sufficiently "large" trials. Hence the deterministic axioms of measurement theory need to be reinterpreted for the cases of insufficient data.

In such situations, let the event E be the validation of an axiom by a human response and m(E) the number of occurrences of this event over all the trials n in an experimental universe U. Then let the possibilistic "degree of satisfaction" Π of an axiom be defined as:

$$\Pi(\text{axiom}) = \frac{m(E)}{n(U)}. \tag{3}$$

Definition 1". Let Θ be a set and $\gtrsim_A$ be a binary relation on Θ, i.e., $\gtrsim_A$ is a subset of Θ'. The relational structure $<\Theta, \gtrsim_A>_\Pi$ is a possibilistic-weak-order iff for all θ_i, θ_j, $\theta_k \in \Theta$, the following two axioms are satisfied possibilistically above a possibility threshold π:

1. Possibilistic Connectedness:
Either $\Pi\{\theta_i \gtrsim_A \theta_j\} \geq \pi$ or $\Pi\{\theta_j \gtrsim_A \theta_i\} \geq \pi$, where $\Pi\{\theta_i \gtrsim_A \theta_j\}$ is the possibility that θ_i will be in relation $\gtrsim_A$ to θ_j. It is defined as:

$$\Pi\{\theta_i \gtrsim_A \theta_j\} = \frac{n_1(\theta_i \gtrsim_A \theta_j)}{n}, \quad 0 \leq n_1 \leq n, \tag{4}$$

where n_1 is the number of trials in which a subject states that $\theta_i \gtrsim_A \theta_j$ and n is the total number of trials, i.e., $n = n_1(\theta_i \gtrsim_A \theta_j) + n_2(\theta_j \gtrsim_A \theta_i)$. It is noted that Π is a proportion and not a probability.
2. Possibilistic Transitivity:
If $\Pi\{\theta_i \gtrsim_A \theta_j\} \geq \pi$ and $\Pi\{\theta_j \gtrsim_A \theta_k\} \geq \pi$ then $\Pi\{\theta_i \gtrsim_A \theta_k\} \geq \pi$.

Conjecture 1". Suppose that Θ is a finite nonempty set. If $<\Theta, \gtrsim_A>_\Pi$ is a possibilistic-weak-order, then there exists a possibilistic-real-valued function ϕ_Π on Θ such that for all θ_i, $\theta_j \in \Theta$, $\Pi\{\theta_i \gtrsim_A \theta_j\} \geq \pi$ if $\Pi\{\phi_\Pi(\theta_i) \geq \phi_\Pi(\theta_j)\} \geq \pi$. Moreover, ϕ'_Π is another possibilistic-real-valued function on Θ with the same property iff there exists a strictly increasing function f with domain and range equal to Re, such that for all $\theta \in \Theta$, $\Pi\{\phi_\Pi(\theta) = f[\phi_\Pi(\theta)]\} \geq \pi$, i.e., ϕ_Π is a possibilistic-ordinal scale.

Definition 2". Suppose Θ is a nonempty set and $\gtrsim_{A'}$ is a quaternary relation on Θ, i.e., a binary relation on Θ'. The relational structure $<\Theta', \gtrsim_{A'}>_\Pi$ is a possibilistic-algebraic-difference structure iff, θ_i, θ_j, θ_k, θ'_i, θ'_j and $\theta'_k \in \Theta$ and all sequences $\theta_1, \theta_2, \ldots, \theta_i, \ldots, \in \Theta$ the following axioms are satisfied possibilistically above a possibility threshold π:

1. $<\Theta', \gtrsim_{A'}>_\Pi$ is a possibilistic-weak-order.
2. Possibilistic Sign Reversal:
If $\Pi\{\theta_j\theta_i \gtrsim_{A'} \theta_1\theta_k\} \geq \pi$ then $\Pi\{\theta_k\theta_1 \gtrsim_{A'} \theta_i\theta_j\} \geq \pi$.
3. Possibilistic-Monotonicity:
If $\Pi\{\theta_j\theta_i \gtrsim_{A'} \theta'_j\theta'_i\} \geq \pi$ and $\Pi\{\theta_k\theta_j \gtrsim_{A'} \theta'_k\theta'_j\} \geq \pi$ then $\Pi\{\theta_k\theta_i \gtrsim_{A'} \theta'_k\theta'_i\} \geq \pi$.
4. Possibilistic-Solvability:
If $\Pi\{\theta_j\theta_i \gtrsim_{A'} \theta_1\theta_k \gtrsim_{A'} \theta_i\theta_i\} \geq \pi$ then there exists θ'_1, $\theta''_1 \in \Theta$ such that $\Pi\{\theta'_1\theta_i \sim_{A'} \theta_1\theta_k \sim_{A'} \theta_j\theta''_1\} \geq \pi$.
5. Possibilistic-Archimedian Condition:
This means the possibility that "no interval is infinitely large with respect to any smaller one whose jnd is not zero" is greater than or equal to possibility measure π. That is, possibility of the gain in A-ness in going from θ_i to θ_j can be exceeded by a finite number of successive jnd gains starting from θ_i is greater than or equal to a possibility measure π.

Conjecture 2". If $<\Theta', \gtrsim_{A'}>_\Pi$ is a possibilistic-algebraic-difference structure, then there exists a possibilistic real-valued function ϕ_Π on Θ such that, for all θ_i, θ_j, θ_k, $\theta_1 \in \Theta$, $\Pi\{\theta_j\theta_i \gtrsim_{A'} \theta_1\theta_k\} \geq \pi$ iff $\Pi\{\phi_\Pi(\theta_j)-\phi_\Pi(\theta_i) \geq \phi_\Pi(\theta_1)-\theta_\Pi(\theta_k)\} \geq \pi$. Moreover, ϕ_Π is unique up to a possibilistic-positive linear transformation, i.e., if ϕ'_Π has the same properties as ϕ_Π, then there are real constants α, β, with $\alpha>0$, such that $\Pi\{\phi'_\Pi = \alpha\phi_\Pi+\beta\} \geq \pi$, i.e., ϕ_Π is a possibilistic-interval scale. It is further to be noted that the computations on the left hand sides, of all the inequalities in Definitions 1", 2" and Conjectures 1", 2", are to be interpreted in accordance with the possibility calculus and the inference rules of approximate reasoning (see for example Turksen and Yao, 1982).

EXPERIMENTAL RESULTS

Various experiments were conducted to determine membership functions empirically. Some of these experiments were reported in previous papers (Turksen, 1979; Turksen and Norwich, 1981; Norwich and Turksen, 1981, 1982; Turksen, 1982). Other experiments will be reported from time to time in the future. These experiments have included visual, aural, and other sense stimuli. In this paper we will review an experiment with aural stimuli.

A fundamental measurement experiment was preformed using the stimulus as the sound of a note on a piano. The player was instructed to strike the note with the same strength in order to provide an identical intensity. Thus the isolated attribute A was the "highness" of the pitch of a sound. The subject was asked to respond to the question: "How true is it that this note is high?" and was instructed to provide a response moving a pointer on a scale. The scale was a straight line with two end points marked B_L and B_U. There were no numbers visible to the subject. The otherside of the line scale was marked $B_L = 0,.1,.2,\ldots,.8,.9,1. = B_U$ and was visible only to the analyst. This method of measurement is known as the method of direct rating. In the experiment, the notes were limited to one octave on the piano, i.e., from the note "DO" up to the following note "DO'". That is the set of notes was {DO, RE, MI, FA, SO, LA, SI, DO'} $= \Theta = \{\theta_i | i = 1,2,\ldots,8\}$. For these eight notes, ten replications were made of each in random order. The responses of a subject are shown in Table 1. At this point, if one were to prefer statistical interpretation of the response data, one could compute the means and s^2 of the sample responses for each $\theta \varepsilon \Theta$ as shown in Table 1.

One could also interpret the responses in terms of probability distributions by observing empirical histograms. For example, for θ_1 = DO and θ_2 = RE, one would have the histograms as in Fig. 1.

However, since n = 10 is small compared to $n \to \infty$, it would be improper to interpret the results in terms of probability distributions. Otherwise, one would have to ask how good are these estimates $\hat{P}$ of P? biased or unbiased? sufficient or not? consistent or not? etc. None of these questions can properly be answered with a small sample.

On the otherhand, we could interpret the results as fuzzy membership values based on limited responses i.e. subjective evaluations of an individual in this experimental setting. In this sense, one needs to reinterpret the histograms as type II fuzziness. For example, for θ_1 = "DO" and θ_2 = "RE" one would look at the same histograms with fuzzy set perspective as in Fig. 2.

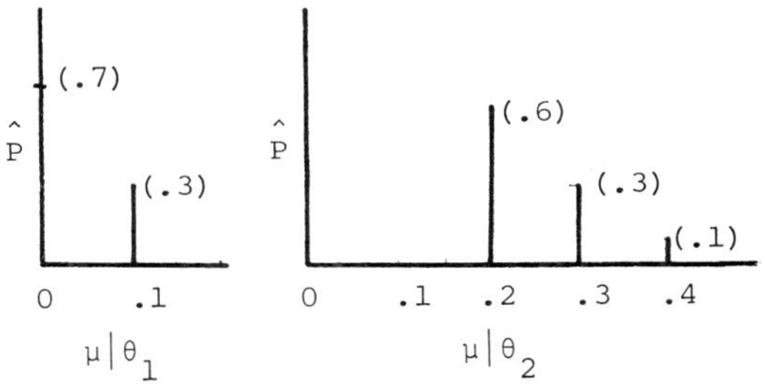

Fig. 1. Probability Histograms

where (·) shows the fraction of occurrences $\hat{P}$ an estimate of P as the probability that a certain membership value $\mu|\theta$ has occurred.

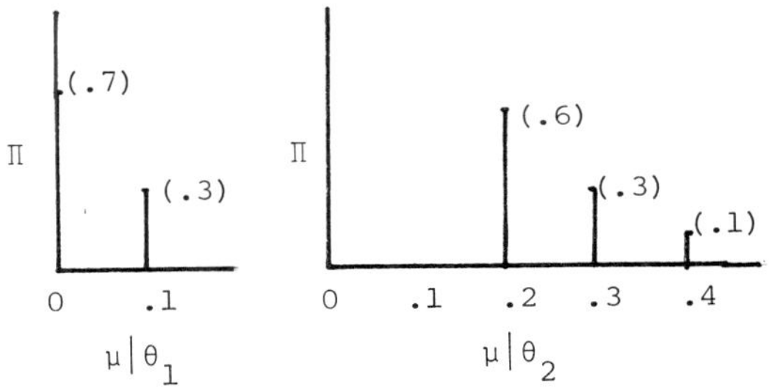

Fig. 2. Possibility Histograms

where (·) shows again the fraction of occurrences Π as the possibility that a certain membership value $\mu|\theta$ has occurred.

The validation of the axioms of measurement theory considered earlier may be investigated within the framework stated in the last section, i.e., probabilistically and/or possibilistically depending on how appropriate one finds each interpretation with respect to a proper set of subjective responses. For each axiom, one has to set up an essentially new experiment.

TABLE 1 Responses of a Subject: Mean and s^2 for $\theta \varepsilon \Theta$

N / T	DO	RE	MI	FA	SO	LA	SI	DO'
1	.0	.2	.3	.5	.6	.8	.8	1.0
2	.1	.2	.2	.5	.6	.7	.9	.8
3	.0	.4	.3	.4	.6	.6	.7	.9
4	.0	.2	.3	.4	.4	.8	.9	1.0
5	.0	.3	.5	.4	.5	.7	.9	1.0
6	.1	.3	.3	.5	.6	.6	.9	1.0
7	.0	.3	.3	.5	.6	.6	.8	1.0
8	.0	.2	.2	.6	.6	.8	.9	1.0
9	.0	.2	.4	.5	.6	.6	.8	.9
10	.1	.2	.3	.5	.7	.6	.8	1.0
Mean	.03	.25	.31	.48	.58	.68	.84	.96
$s^2 X10^4$	21	42	77	29	62	84	49	49

CONCLUSIONS

Some basic axioms of measurement theory were restated and reinterpreted probabilistically and possibilistically in order to overcome the inconsistency and randomness displayed by the responses of human subjects in measurement experiments of fuzziness. It appears more and more plausible, with the accumulation of various response data, that fuzzy set interpretation of the axioms of measurement is better suited for reasonable explanation of human responses. A detail account of an experimental validation of these axioms within the stated framework is left for a future paper.

ACKNOWLEDGEMENT

The research is supported by the Natural Science and Engineering Council of Canada, Grant No. A-7698.

REFERENCES

Coombs, C.H., R.M. Dawes and A. Tversky (1970). Mathematical Psychology. Prentice-Hall, Englewood Cliffs, New Jersey.

Davidson, D., P. Suppes and S. Seigel (1957). Decision Making: An Experimental Approach. Stanford: Stanford University Press.

Debrue, G. (1958). Stochastic Choice and Cardinal Utility. Econometrica, 26, 440-444.

Guilford, J.P. (1954). Psychometric Methods, 2nd ed. McGraw-Hill, New York.

Krantz, D.H., R.D. Luce, P. Suppes and A. Tversky (1971). Foundations of Measurement, Vol. 1, Academic Press, New York.

Neumann, J. Von and O. Morgenstern (1953). The Theory of Games and Economic Behaviour. Princeton, New Jersey Princeton University Press, 3rd ed.

Norwich, A.M. and I.B. Turksen (1982). The Fundamental Measurement of Fuzziness, in R.R. Yager (Ed.), Fuzzy Sets and Possibility Theory. Pergamon Press, New York. 49-60.

Norwich, A.M. and I.B. Turksen (1982). The Construction of Membership Functions, in R.R. Yager (Ed.), Fuzzy Sets and Possibility Theory. Pergamon Press, New York. 61-67.

Norwich, A.M. and I.B. Turksen (1982). Meaningfulness in Fuzzy Set Theory, in R.R. Yager (Ed.), Fuzzy Sets and Possibility Theory. Pergamon Press, New York. 68-74.

Norwich, A.M. and I.B. Turksen (1982). Stochastic Fuzziness, in M.M. Gupta and E.E. Sanchez (Eds.), Fuzzy Information and Decision Processes. North-Holland, Amsterdam. 13-22.

Savage, L.J. (1954). The Foundations of Statistics. New York, John Wiley.

Suppes, P. (1955). The Role of Subjective Probability and Utility in Decision Making. Proceedings Third Berkeley Symposium on Mathematical Statistics and Probability. 61-73.

Suppes, P. and M. Winet (1955). An Axiomatization of Utility Based on the Notion of Utility Differences. Management Science, 1. 259-270.

Suppes, P. and J.L. Zinnes (1963). Basic Measurement Theory, in R.D. Luce, R.R. Bush and E. Galanter (Eds.), Handbook of Mathematical Psychology, Vol. 1. John Wiley and Sons, New York. 1-76.

Thole, U., H.J. Zimmerman and P. Zysno (1969). On the Suitability of Minimum and Product Operators for the Intersection of Fuzzy Sets. Fuzzy Sets and Systems 2. 167-180.

Torgerson, W.S. (1958). Theory and Methods of Scaling. John Wiley and Sons, New York.

Turksen, I.B. and A.M. Norwich (1981). Measurement of Fuzziness. Proceedings of the International Conference on Policy Analysis and Information Systems. August 19-22. Taipei, Taiwan, Meadea Enterprises Co., Ltd., Taipei, Taiwan. 745-754.

Turksen, I.B. (1979). Measurement of Linguistic Variables ... Proceedings of the 23rd Annual North American Meeting. Society for General Systems Research. January 3-6. Houston, Texas. 278-284.

Turksen, I.B. (1982). Stochastic and Fuzzy Fuzzy Sets. Proceedings of the 2nd World Conference on Mathematics at the Service of Man. June 28-July 3. Las Palmas, Canary Islands, Spain. 649-654.

Turksen, I.B. and D.D. Yao (1982). Bounds on Fuzzy Inference. Proceedings of the Sixth European Meeting on Cybernetics and Systems Research. Vienna. April 13-16. 729-734.

Zadeh, L.A. (1965). Fuzzy Sets. Information and Control, 8. 338-353.

Zadeh, L.A. (1975). The Concept of a Linguistic Variable and its Application to Approximate Reasoning, I, II, III. Information Sciences 8. 301-357.

THE EXTENDED FUZZY OPERATORS AND THE FUZZY TRUTH-VALUED POSSIBILITY DEGREE

Zhang Wenxiu and Le Huiling

Department of Mathematics, Xi'an Jiaotong University, Xi'an, Shaanxi Province, The People's Republic of China

Abstract. In this paper, the extended fuzzy operators are introduced by the extension principle. We have proved that the extended fuzzy operators are also t(s)-norms for a new ordering. The fuzzy truth-valued possibility degree for a fuzzy σ-algebra is introduced. The fuzzy integral with regard to the fuzzy truth-valued possibility degree is defined. Lastly, the fuzzy truth-valued logic, which is defined by the fuzzy truth-valued possibility degree, will be considered.

Keywords. Extended fuzzy operator; t-norm; s-norm; fuzzy truth-valued possibility degree; fuzzy integral; fuzzy-valued logic.

THE EXTENDED FUZZY OPERATORS

Let L be a partial ordering set with the minimal element 0 and the maximal element 1.

Definition 1 A t-norm is a mapping $T: L^2 \to L$ which satisfies the following properties:

(1) $T(0,0)=0$, $T(a,1)=a$;

(2) $a\le c$, $b\le d \Rightarrow T(a,b)\ T(c,d)$;

(3) $T(a,b)=T(b,a)$;

(4) $T(T(a,b),c)=T(a,T(b,c))$.

If a mapping $S: L^2 \to L$ satisfies (2).(3).(4) and (1') $S(1,1)=1$, $S(a,0)=a$ $(a\in L)$,

then S is called a s-norm.

Suppose $L=[0,1]$, then

$T_1(a,b)=a\wedge b$

$T_2(a,b)=a\cdot b$

$T_3(a,b)=\max(0,\ a+b-1)$

are all t-norms, and

$S_1(a,b)=a\vee b$

$S_2(a,b)=a+b-a\cdot b$

$S_3(a,b)=\min(1,\ a+b)$

are all s-norms.

Definition 2 Suppose that T is a t-norm on [0,1] and S is a s-norm on [0,1]. The extended fuzzy operators T and S are defined by

$$\mu_{A\tilde{T}B}(z) = \bigvee_{z=T(x,y)} \mu_A(x)\wedge\mu_B(y)$$

$$\mu_{A\tilde{S}B}(z) = \bigvee_{z=S(x,y)} \mu_A(x)\wedge\mu_B(y)$$

where $A,B\in F([0,1]) = \{A;\ \mu_A: [0,1]\to[0,1]\}$.

Obviously, we can see that

$A\tilde{T}B,\ A\tilde{S}B \in F([0,1])$.

We denote

$A \cap B = A\tilde{T}_1B$ $\quad A\cup B = A\tilde{S}_1B$

$A \oslash B = A\tilde{T}_2B$ $\quad A\bullet B = A\tilde{S}_2B$

$A \cap B = A\tilde{T}_3B$ $\quad A\cup B = A\tilde{S}_3B$

Definition 3 Suppose that $A,B\in F([0,1])$, A is called to be implicated in B if and only if the following properties are satisfied:

(1) If $X_0\in A_\alpha$, there exist a Y_0 such that $Y_0\in B_\alpha$ and $X_0\le Y_0$, we write $\widetilde{\sup}\ A_\alpha \le \widetilde{\sup}\ B_\alpha$;

(2) If $Y'\in B_\alpha$, there exist a X' such that $X'\in A_\alpha$ and $X'\le Y'$, we write $\widetilde{\inf}\ A_\alpha \le \widetilde{\inf}\ B_\alpha$.

where A_α and B_α are the open α-cuts of A and B respectively, we denote $A \precsim B$.

If $A,B\in F([0,1])$, we have

$A\cap B \precsim A\oslash B \precsim A\cap B \precsim A\cup B \precsim A\bullet B \precsim A\cup B$.

Let

$$\mu_1(X)=\begin{cases}1, & X=1\\ 0, & X\ne 1\end{cases} \qquad \mu_0(X)=\begin{cases}1, & X=0\\ 0, & X\ne 0\end{cases}$$

1 and 0 are the maximal and minimal element of F([0,1]) for partical ordering $\precsim$ respectively. F([0,1]) is a partial ordering set for the new ordering $\precsim$.

Theorem 1 If T and S are a t-norm and a s-norm on [0,1] respectively, then $\tilde{T}$ and $\tilde{S}$ are a t-norm and a s-norm on F([0,1]) respectively.

Proof: Suppose that $\tilde{T}(A,B)=A\tilde{T}B$

(1) Let $\tilde{T}(0,0)=A$. Because $T(0,0)=0$,

We have $1 \geq \mu_{\underset{\sim}{A}}(0) \geq \mu_{\underset{\sim}{O}}(0) \wedge \mu_{\underset{\sim}{O}}(0)=1$, i.e. $\mu_{\underset{\sim}{A}}(0)=1$. On the other hand, because $T(x,y) \leq x \wedge y$, we have $T(x,y)>0$ if and only if $x>0$ and $y>0$. Hence $z>0 \Rightarrow \mu_{\underset{\sim}{A}}(z)=0$. The proof of $\tilde{T}(\underset{\sim}{A},\underset{\sim}{1})=\underset{\sim}{A}$ is analogical.

(2) Suppose that $\underset{\sim}{A} \lessapprox \underset{\sim}{C}$ and $\underset{\sim}{B} \lessapprox \underset{\sim}{D}$. Let

$$\underset{\sim}{E}=\tilde{T}(\underset{\sim}{A},\underset{\sim}{B}) \qquad \underset{\sim}{F}=\tilde{T}(\underset{\sim}{C},\underset{\sim}{D})$$

If $z_0 \in E_\alpha$, there are a x_0 and a y_0 such that $z_0=T(x_0,y_0)$ and $\mu_{\underset{\sim}{A}}(x_0) \wedge \mu_{\underset{\sim}{B}}(y_0) > \alpha$, i.e. $x_0 \in A_\alpha$ and $y_0 \in B_\alpha$. Because $\underset{\sim}{A} \lessapprox \underset{\sim}{C}$ and $\underset{\sim}{B} \lessapprox \underset{\sim}{D}$, there are a x' and a y' such that $x' \geq x_0$, $x' \in C_\alpha$ and $y' \geq y_0$, $y' \in D_\alpha$. So that $z'=T(x',y') \geq T(x_0,y_0)=z_0$ and $z' \in F_\alpha$. Finally $\widetilde{\sup}\, E_\alpha \leq \widetilde{\sup}\, F_\alpha$. The proof of of $\widetilde{\inf}\, E_\alpha \leq \widetilde{\inf}\, F_\alpha$ is analogical. Thus, we have $\underset{\sim}{A}\tilde{T}\underset{\sim}{B} \lessapprox \underset{\sim}{C}\tilde{T}\underset{\sim}{D}$.

(3) and (4) are apparent.

Analogically, we can prove that $\tilde{T}$ is a s-norm on $F([0,1])$. It follows from theorem 1 that $\cap, \odot, \circledcirc$ are all t-norms on $F([0,1])$ and $\cup, \oplus, \uplus$ are all s-norms on $F([0,1])$.

THE FUZZY TRUTH-VALUED POSSIBILITY DEGREE

Definition 4 A fuzzy truth number is a $\underset{\sim}{A}$ with the following properties:

(1) $\underset{\sim}{A} \in F([0,1])$;

(2) $\underset{\sim}{A}$ is normal, i.e. there is a x_0 such that $\mu_{\underset{\sim}{A}}(x_0)=1$;

(3) $A_\alpha=\{x; \mu_{\underset{\sim}{A}}(x) \geq \alpha\}$ $(\alpha \leq 1)$ is a closed interval.

If $\underset{\sim}{A}$ is a fuzzy truth number, then $\underset{\sim}{A}$ is convex. We denote

$$F^*([0,1])=\{\underset{\sim}{A};\ \underset{\sim}{A} \text{ is a fuzzy truth number}\}$$

Theorem 2 If t-norm T is continuous on $[0,1]\times[0,1]$, then

$$\underset{\sim}{A},\ \underset{\sim}{B} \in F^*([0,1]) \Rightarrow \underset{\sim}{A}\tilde{T}\underset{\sim}{B} \in F^*([0,1])$$

The s-norm S has the analogical property.

Proof: Suppose that $\mu_{\underset{\sim}{A}}(x_0)=\mu_{\underset{\sim}{B}}(y_0)=1$ and $\underset{\sim}{D}=\underset{\sim}{A}\tilde{T}\underset{\sim}{B}$, then we have $\mu_{\underset{\sim}{A}}(z_0)=1$, where $z_0=T(x_0,y_0)$. i.e. $\underset{\sim}{D}$ is normal. Let

$$a_\alpha=T(\inf A_\alpha,\ \inf B_\alpha)$$
$$b_\alpha=T(\sup A_\alpha,\ \sup B_\alpha)$$

we have $D_\alpha \subset [a_\alpha,\ b_\alpha] \subset D$

In fact, if $z \in D_\alpha$, there exists a x_0 and a y_0 such that $z=T(x_0,y_0)$ and $x_0 \in A_\alpha$, $y_0 \in B_\alpha$. So $a_\alpha \leq z \leq b_\alpha$, i.e. $D_\alpha \subset [a_\alpha,\ b_\alpha]$. Let $A_\alpha=[x_1,x_2]$ and $B_\alpha=[y_1,y_2]$ such that $a_\alpha=T(x_1,y_1)$ and $b_\alpha=T(x_2,y_2)$. We have $a_\alpha \in D_\alpha$ and $b_\alpha \in D_\alpha$. If $z_0 \in [a_\alpha,\ b_\alpha]$, because T is continuous, there exists a x_0 and a y_0 such that $z_0=T(x_0,y_0)$ $(x_1 \leq x_0 \leq x_2,\ y_1 \leq y_0 \leq y_2)$. Then

$$\mu_{\underset{\sim}{D}}(z_0) \geq \mu_{\underset{\sim}{A}}(x_0) \wedge \mu_{\underset{\sim}{B}}(y_0)$$

$$\geq \mu_{\underset{\sim}{A}}(x_1) \wedge \mu_{\underset{\sim}{A}}(x_2) \wedge \mu_{\underset{\sim}{B}}(y_1) \wedge \mu_{\underset{\sim}{B}}(y_2) \geq \alpha$$

Therefore $[a_\alpha,\ b_\alpha] \subset D_\alpha$ and

$$D_\alpha = \bigcap_{\lambda < \alpha} [a_\lambda,\ b_\lambda]$$

i.e. D_α is a closed interval. Above fact implies $\underset{\sim}{A}\tilde{T}\underset{\sim}{B}$ is a fuzzy truth number. Similarilly, we can prove that $\underset{\sim}{A}\tilde{S}\underset{\sim}{B} \in F^*([0,1])$, when S is continuous s-norm and $\underset{\sim}{A}$, $\underset{\sim}{B} \in F^*([0,1])$. In particular, if $\underset{\sim}{A}$, $\underset{\sim}{B} \in F^*([0,1])$, we have $\underset{\sim}{A}\cap\underset{\sim}{B}$, $\underset{\sim}{A}\cup\underset{\sim}{B} \in F^*([0,1])$. We further define

$$\mu_{\underset{\sim}{A}^c}(x)=\mu_{\underset{\sim}{A}}(1-x) \quad (x \in [0,1])$$

If $\underset{\sim}{A} \in F^*([0,1])$, then $\underset{\sim}{A}^c \in F^*([0,1])$.

Theorem 3 $(F^*([0,1]), \cup, \cap, c)$ has properties:

(1) It is a distributive lattice;

(2) $\underset{\sim}{A} \cup \underset{\sim}{A}=\underset{\sim}{A} \cap \underset{\sim}{A}=\underset{\sim}{A}$;

(3) $\underset{\sim}{A} \cap \underset{\sim}{O}=\underset{\sim}{O}$, $\underset{\sim}{A} \cup \underset{\sim}{1}=\underset{\sim}{1}$, $\underset{\sim}{A} \cup \underset{\sim}{O}=\underset{\sim}{A} \cap \underset{\sim}{1}=\underset{\sim}{A}$;

(4) $(\underset{\sim}{A}^c)^c=\underset{\sim}{A}$;

(5) $(\underset{\sim}{A} \cup \underset{\sim}{B})^c=\underset{\sim}{A}^c \cap \underset{\sim}{B}^c$, $(\underset{\sim}{A} \cap \underset{\sim}{B})^c=\underset{\sim}{A}^c \cup \underset{\sim}{B}^c$.

The proof of this theorem follows directly from verifying.

Theorem 4 $\underset{\sim}{A} \lessapprox \underset{\sim}{B} \Leftrightarrow \underset{\sim}{A} \cap \underset{\sim}{B}=\underset{\sim}{A} \Leftrightarrow \underset{\sim}{A} \cup \underset{\sim}{B}=\underset{\sim}{B}$ $(\underset{\sim}{A},\underset{\sim}{B} \in F^*([0,1]))$

Proof: If $\underset{\sim}{A} \cap \underset{\sim}{B}=\underset{\sim}{A}$, then $(\underset{\sim}{A} \cap \underset{\sim}{B})_\alpha=A_\alpha$ $(\alpha < 1)$. Suppose $z \in A_\alpha$, there exists a x and a y such that $z=x \wedge y$ and $x \in A_\alpha$, $y \in B_\alpha$. Because $z \leq y$, we have $\widetilde{\sup}\, A_\alpha \leq \widetilde{\sup}\, B_\alpha$. If $y \in B_\alpha$ and $x \in A_\alpha$, $z=x \wedge y$ and $z \in (\underset{\sim}{A} \cap \underset{\sim}{B})_\alpha = A_\alpha$, So $\widetilde{\inf}\, A_\alpha \leq \widetilde{\inf}\, B_\alpha$. We can see that $\underset{\sim}{A} \lessapprox \underset{\sim}{B}$. Conversely, suppose that $\underset{\sim}{A} \lessapprox \underset{\sim}{B}$. If $z \in A_\alpha$, there exists a y such that $y \geq z$ and $y \in B_\alpha$. This fact implies $z=z \wedge y \in (\underset{\sim}{A} \cap \underset{\sim}{B})_\alpha$. i.e. $A_\alpha \subset (\underset{\sim}{A} \cap \underset{\sim}{B})_\alpha$.

If $z \in (\underset{\sim}{A} \cap \underset{\sim}{B})_\alpha$, there exists a x and a y such that $z=x \wedge y$ and $x \in A_\alpha$, $y \in B_\alpha$. If $x=z$, $z \in A_\alpha$. If $x > z$, $y=z$. In second case, because $\widetilde{\inf}\, A_\alpha \leq \widetilde{\inf}\, B_\alpha$, there is a x' such that $x' \in A_\alpha$ and $x' \leq z$. In terms of covexity of $\underset{\sim}{A}$, $[x',x] \subset A_\alpha$. Thus, $z \in A_\alpha$, i.e. $(\underset{\sim}{A} \cap \underset{\sim}{B}) \subset A_\alpha$.

Above fact implies $\underset{\sim}{A}\cap\underset{\sim}{B}=\underset{\sim}{A}$. The proof of $\underset{\sim}{A}\cup\underset{\sim}{B}=\underset{\sim}{B}\Leftrightarrow\underset{\sim}{A}\lesssim\underset{\sim}{B}$ is analogical.

Suppose that $A_n\in F^*([0,1])$ $(n\geq 1)$, Let

$$a_\alpha = \sup_n \inf (\underset{\sim}{A}_n)_\alpha$$
$$b_\alpha = \sup_n \sup (\underset{\sim}{A}_n)_\alpha$$
$$c_\alpha = \inf_n \inf (\underset{\sim}{A}_n)_\alpha$$
$$d_\alpha = \inf_n \sup (\underset{\sim}{A}_n)_\alpha$$

then $a_\alpha\leq b_\alpha$ and $c_\alpha\leq d_\alpha$. When α increases, we have a_α and c_α increase, b_α and d_α decrease. Let

$$\mu_{\underset{\sim}{A}}(x) = \bigvee_{\alpha\in[0,1]} \alpha\cdot\chi_{[a_\alpha,\, b_\alpha]}(x)$$
$$\mu_{\underset{\sim}{B}}(x) = \bigvee_{\alpha\in[0,1]} \alpha\cdot\chi_{[c_\alpha,\, d_\alpha]}(x)$$

We denote

$$\underset{\sim}{A} = \bigcup_{n=1}^{\infty}\!\!\cdot\ \underset{\sim}{A}_n \qquad \underset{\sim}{B} = \bigcap_{n=1}^{\infty}\!\!\cdot\ \underset{\sim}{A}_n$$

then $A_\lambda = \bigcap_{\alpha<\lambda}[a_\alpha,b_\alpha]$ and $B_\lambda=\bigcap_{\alpha<\lambda}[c_\alpha,d_\alpha]$ are the closed intervals. We can prove that

$$(\cup\underset{\sim}{A}_n)^c = \cap\, \underset{\sim}{A}_n^c$$
$$(\cap\underset{\sim}{A}_n)^c = \cup\, \underset{\sim}{A}_n^c$$
$$\cap\, A_n \lesssim \underset{\sim}{A}_n \lesssim \cup\, A_n$$

Definition 5 A fuzzy truth-valued σ-algebra is $F^*\subset F^*([0,1])$ with the following properties:

(1) $\underset{\sim}{1},\underset{\sim}{0}\in F^*$;
(2) $\underset{\sim}{A}\in F^*\Rightarrow\underset{\sim}{A}^c\in F^*$;
(3) $\underset{\sim}{A}_n\in F^*(n\geq 1)\Rightarrow\cup\underset{\sim}{A}_n\in F^*$.

Definition 6 Suppose that $F\subset F(X)$ is a fuzzy σ-algebra. A fuzzy truth-valued possibility degree is a mapping from F to F^* with the following properties:

(1) $\sigma(\phi)=\underset{\sim}{0}$, $\sigma(X)=\underset{\sim}{1}$;
(2) $\sigma(\underset{\sim}{A}^c)=\sigma(\underset{\sim}{A})^c$;
(3) $\sigma(\cup\underset{\sim}{A}_n)=\cup\sigma(\underset{\sim}{A}_n)$.

Let $\Phi=\{[a,b];\ a,b,\in[0,1]\}$. We define

$$\cup[a_n,\ b_n] = [\sup a_n,\ \sup b_n]$$
$$\cap[a_n,\ b_n] = [\inf a_n,\ \inf b_n]$$
$$[a,b]^c = [1-b,\ 1-a]$$

If $a_1\leq a_2$, $b_1\leq b_2$, we denote $[a_1,\ b_1]\lesssim[a_2,\ b_2]$. Then Φ is a soft algebra. $[0,0]$ and $[1,1]$ are the minimal and the maximal element of respectively.

Definition 7 F is a fuzzy σ-algebra. A interval measure is a mapping m from F to Φ with the following properties:

(1) $m(\phi)=[0,0]$, $m(X)=[1,1]$;
(2) $m(\underset{\sim}{A}^c) = m(\underset{\sim}{A})^c$;
(3) $m(\cup\underset{\sim}{A}_n) = \cup m(\underset{\sim}{A}_n)$.

Theorem 5 If $\sigma: F\to F^*$ is a fuzzy truth-valued possibility degree. Let

$$m_\alpha(\underset{\sim}{A}) = \sigma(\underset{\sim}{A})_\alpha \quad (\underset{\sim}{A}\in F)$$

$(\alpha>0)$, then $m_\alpha: F\to\Phi$ is a interval measure. If $\{m_\alpha;\ \alpha>0\}$ has the following properties:

(1) $m_\alpha: F\to\Phi\ (\alpha>0)$ is a interval measure;
(2) $\alpha<\beta\Rightarrow m_\alpha(\underset{\sim}{A})\supset m_\beta(\underset{\sim}{A})\quad(\underset{\sim}{A}\in F)$.

Then $\sigma(\underset{\sim}{A})=\bigcup_{\lambda\in[0,1]} m_\lambda(\underset{\sim}{A})$ is a fuzzy truth-valued possibility degree.
The proof of this theorem follows directly from verifying.

Suppose that $\sigma: F\to F^*$ is a fuzzy truth-valued possiblity degree and $F_\alpha=\{x;\ h(x)\geq\alpha\}\in F$ $(\alpha\leq 1)$. The integral of h(x) with respact to the fuzzy truth-valued possibility degree is defined as

$$\int_{\underset{\sim}{A}} h(x)\circ\sigma(\cdot) = \bigcup_{\alpha\in[0,1]} [\alpha\cap\sigma(\underset{\sim}{A}\cap F_\alpha)]$$

We can see that, the integral with respect to a fuzzy truth-valued possibility degree $\sigma(\cdot)$ has analogical properties with Sugeno's fuzzy integral.

THE FUZZY TURTH-VALUED LOGIC

Let L be a complete distributive lattic (Propositional set) and for all $x\in X$, $\sigma(x,\ p)$ $(p\in L)$ be a fuzzy truth-valued possibility degree. $\sigma(x, p)$ is called the fuzzy truth value of propositional "x is p". The fuzzy truth value of "p implicates q" is $\sigma(x, p \to q) = \sigma(x,p)^c \dot{\cup} \sigma(x,q)$. The fuzzy truth value of "p equals q" is $\sigma(x, p \Leftrightarrow q) = \sigma(x, p \to q) \dot{\cap} \sigma(x, q \to p)$. The fuzzy truth value of negative proposition $\bar{p}$ of p is $\sigma(x,\bar{p}) = \sigma(x,p)^c$. $(X, L, \sigma(x,p))$ is called the fuzzy truth-valued logic or ultrafuzzy logic. We can see that the fuzzy truth-valued logic has the following properties:

(1) $\sigma(x, p \to p) = \sigma(x, p \vee \bar{p})$;

(2) $\sigma(x, p \Leftrightarrow p) = \sigma(x, p \vee \bar{p})$;

(3) $\sigma(x, p \Leftrightarrow \bar{p}) = \sigma(x, p \wedge \bar{p})$;

(4) $\sigma(x, p \to (p \vee \bar{p})) = \sigma(x, p \vee \bar{p})$;

(5) $\sigma(x, (p \vee \bar{p}) \to p) = \sigma(x, p)$;

(6) $\sigma(x, (p \wedge \bar{p}) \to p) = \sigma(x, p \vee \bar{p})$;

(7) $\sigma(x, p \to (p \wedge \bar{p})) = \sigma(x, p)$;

(8) $\sigma(x, p \to (q \to r)) = \sigma(x, (p \wedge q) \to r)$.

Analogically, we can define the fuzzy truth-valued logic by $\odot$ and $\oplus$ or $\mathring{\cap}$ and $\mathring{\cup}$.

REFERENCES

Zadeh. L. A. (1978) Fuzzy sets as a basis for a theory of possibility. Fuzzy Sets and Systems 1(1).

Sugeno. M and Terano. T (1975) Analysical representative of fuzzy system. Fuzzy automata.

Dubois. D and Prade. H (1980) Fuzzy Sets and Systems, theory and application. New York.

Klement. K. P. (1980) Fuzzy σ-algebra and fuzzy measurable function. Fuzzy Sets and Systems 4.

Weng Peizhuang (1982) Fuzzy sets and categories of fuzzy sets. Advances in Mathematics, China. Vol 11, No 1.

Zhang Wenxiu and Zhao Ruhuai (1982) Theory of the possibility degree. Advance in Mathematics. China. Vol 11, No 3.

A NOTION OF FUZZY CARDINAL

O. Botta and M. Delorme

Université Claude Bernard — Lyon 1, 43 boulevard du 11 novembre 1918, 69622 Villeurbanne cedex, France

Keywords. Fuzzy sets - modal connectives - successors - ordinals - cardinals.

INTRODUCTION

We propose the cardinal notion which came out during the research of a fuzzy sets axiomatic theory [1] (we call them J-sets, according to the lattice J of both membership degrees and truth values).

Our method is similar to the classical set theory in Logic, e.i. cardinal notions came after ordinal ones. When J is a chain, particulary [0,1], we find fuzzy cardinals of litterature [3], from the first articles of de Luca and Termini [2] to the recent publications of Zadeh [6].

PRELIMINARIES

Let $(J, \leq, \wedge, \vee)$ be a complete lattice with a zero-element 0, a unit element 1, three unary operations - negation $\neg$, necessity $\Box$ and possibility $\Diamond$ - and a binary operation, denoted by $\to$, such that for all α, β, γ in J : $\neg 0 = 1$, $\neg 1 = 0$, $\Box\alpha = 1$ if $\alpha = 1$ and $\Box\alpha = 0$ if not, $\Diamond\alpha = 0$ if $\alpha = 0$ and $\Diamond\alpha = 1$ if not, $\alpha \to \beta = 1$ iff $\alpha \leq \beta$ and $1 \to 0 = 0$, $\alpha \to (\beta \to \alpha) = 1$ and $(\alpha \to (\beta \to \gamma)) \to ((\alpha \to \beta) \to (\alpha \to \gamma)) = 1$.
So, J is an implicative lattice, and the following equations are satisfied :
$(\alpha \to \beta) \to ((\beta \to \gamma) \to (\alpha \to \gamma)) = 1$
and $(\beta \to \gamma) \to ((\alpha \to \beta) \to (\alpha \to \gamma)) = 1$.
With the complementation defined by α in J and $\bar{\alpha} = \alpha \to 0$, J is a Heyting lattice [5]. But this complementation is not used in this paper. We write $\alpha \leftrightarrow \beta$ for $\alpha \to \beta \wedge \beta \to \alpha$.

We are making use of a Kelley-Morse-like-impredicative-class-theory as meta-theory ; thus a formula, with possible quantified classes, defines a class. The following definition is then justified :

the class $V^{(J)}$ is the smallest class X such that :

- F1 : $\emptyset$ is an object of X
- F2 : If a is a non-empty object of X, a is a mapping from a set (denoted dom a) in $J_\star = J - \{0\}$.
- F3 : If a is a non-empty object of X, every element of dom a is an object of X.
- F4 : If a is a function with values in $J_\star$ and if any element of dom a is an object of X, a is an object of X.

The objects of $V^{(J)}$ are called *J-sets*.
(If J is a Boolean Algebra, in spite of the analogy, we find only a part of classical Boolean model V^{J}).

Let a and b be $V^{(J)}$-objects. We will say that *a is a J-element of b* iff $a \in$ dom b, and we will write it *aEb*.
$\emptyset$ has no J-elements.

Our aim was to make $(V^{(J)},E)$ a model of a modal theory.

We are using a first-order language with equality, including a binary predicate, the logical symbols of which have been enlarged with both modal connectors $\Box$ and $\Diamond$.
The closed formulae, with constants in $V^{(J)}$, usually defined (taking account of the canonical adjunctions relative to $\Box$ and $\Diamond$) form a class ; constructions and proofs by induction about the closed-formulae complexity are possible thanks to the impredicative class theory which we are working with.

With any closed-formula A we associate an element $\|A\|$ of J (truth value of A in $V^{(J)}$), defined by induction, in the following way :
$\|a = b\| = 1$ iff $a = b$, $\|a = b\| = 0$ if not.
$\|aEb\| = b(a)$.
$\|\neg A\| = \neg\|A\|$.
$\|AkB\| = \|A\| k \|B\|$, avec $k = \wedge, \vee, \to, \leftrightarrow$.
$\|\Box A\| = \Box\|A\|$, $\|\Diamond A\| = \Diamond\|A\|$.

$$\|\forall x A(x)\| = \bigwedge_{a \text{ in } V^{(J)}} \|A(a)\| ,$$

$$\|\exists x A(x)\| = \bigvee_{a \text{ in } V^{(J)}} \|A(a)\| .$$

A closed-formula A is said to be valid (in $V^{(J)}$) iff $\|A\| = 1$; and to be verified if $\|A\| \neq 0$ (the assertion "a is a J-element of b" does not mean that aEb is valid in $V^{(J)}$, but only that aEb is verified).

We will say two J-sets are J-isomorphic if and only if there exists a bijection φ from dom a into dom b such that, for any x in dom a, $a(x) = b(\varphi(x))$. A J-set with finite domain is said to be finite.

If a and b are two J-sets, the class of $V^{(J)}$-objects such that $\| dEa \vee dEb \| \neq 0$ is a set denoted dom(a ∪ b), and a ∪ b is the J-set $\{(x,a(x) \vee b(x))/x \in \text{dom}(a \cup b)\}$. If we adopt the symbol $< \alpha a, \beta b > = \{(\{(a,\alpha),(b,\beta)\},1),(\{(a,\alpha)\},1)\}$, where a and b are J-sets, and α, β elements of $J_\star$, we introduce
$a \times b = \{(< a(x)x,b(y)y >,a(x) \wedge b(y))/ (x,y) \in \text{dom } a \times \text{dom } b, a(x) \wedge b(y) \neq 0\}$.

Let us situate in this frame the J-integers set and the J-ordinals class, which are useful for defining both finite and infinite cardinals of $V^{(J)}$.

THE TREE OF J-INTEGERS. THE J-ORDINALS.

The notion of J-ordinal is based on the notion of successor, especially multiple-successor, to avoid a necessarily arbitrary or reducing choice : we call J-successor of a J-set a every J-set like $a \cup \{(a,\alpha)\}$, with $\alpha \in J_\star$.

In order to obtain a collection of J-integers, we use the same proceeding as the classic one : we begin with $\emptyset$ and build the different J-successors of $\emptyset$.

We put $\emptyset = 0$, $\emptyset \cup \{(\emptyset,\alpha)\} = \{(\emptyset,\alpha)\} = 1_\alpha$, $1_\alpha \cup \{(1_\alpha,\beta)\} = \{(\emptyset,\alpha),(1_\alpha,\beta)\} = 1_{\alpha,\beta}$, which leads to the following construction.
Let $\alpha_1,\ldots,\alpha_p$ be a finite sequence of $J_\star$-elements. We define $n_{\alpha_1\ldots\alpha_p}$ by recurrence on p by setting : if p = 1, $n_{\alpha_1} = \{(\emptyset, \alpha_1)\} = 1_{\alpha_1}$;
if p > 1,
$n_{\alpha_1\ldots\alpha_p} = n_{\alpha_1\ldots\alpha_{p-1}} \cup \{(n_{\alpha_1\ldots\alpha_{p-1}}, \alpha_p)\}$.

Then, for any integer $p \geq 1$ and finite sequence of $J_\star$-elements, $n_{\alpha_1\ldots\alpha_p}$ is a J-set, the domain's cardinal of which is p ; moreover, for integers p, q, $J_\star$-elements α_i, β_j, $n_{\alpha_1\ldots\alpha_p} = n_{\beta_1\ldots\beta_q}$ if and only if p = q and, for any i, $1 \leq i \leq p$, $\alpha_i = \beta_i$.

The set of the finite sequences of $J_\star$-elements is denoted by $\mathscr{S}_f(J_\star)$. The domain of the J-set $\{(\emptyset,1)\} \cup \bigcup_{(\alpha_1\ldots\alpha_p) \in \mathscr{S}_f(J_\star)} \{(n_{\alpha_1\ldots\alpha_p},1)\}$ is denoted $\mathbb{N}^{(J)}$, classical set of J-integers. (A finite J-set is isomorphic to a J-integer, and conversely.

More generally, if λ is an ordinal, $(\alpha_1 \ldots \alpha_\lambda)$ a λ-$J_\star$-elements sequence [3], we define a J-ordinal $n_{\alpha_1\ldots\alpha_\lambda}$ by induction on λ, as follows :
If $\lambda = 0$, any λ-sequence is empty, we take $n_{\alpha_1\ldots\alpha_\lambda} = \emptyset$; if λ is an isolated ordinal, $\lambda = \mu + 1$,
$n_{\alpha_1\ldots\alpha_{\mu+1}} = n_{\alpha_1\ldots\alpha_\mu} \cup \{(n_{\alpha_1\ldots\alpha_\mu},\alpha_{\mu+1})\}$;
if λ is a limit-ordinal,
$n_{\alpha_1\ldots\alpha_\lambda} = \bigcup_{\mu < \lambda} \{(n_{\alpha_1\ldots\alpha_\mu}, \alpha_{\mu+1})\}$.
So we obtain J-sets, and the class $O^{(J)}$ of J-ordinals (isolated or limit according to λ). The J-integers are the finite J-ordinals.

Every J-successor of a J-ordinal is a J-ordinal, and for $\alpha \in J_\star$, we define the successor function s_α from $O^{(J)}$ into $O^{(J)}$ such that $s_\alpha(n) = n \cup \{(n,\alpha)\}$.

Every J-ordinal n validates
$\forall x \forall y(xEn \wedge yEn \rightarrow xEy \vee yEx \vee x = y)$.
E induces on $O^{(J)}$ and on every J-ordinal a strict ordering (always denoted E). E is a well-ordering on every J-ordinal, but not total ordering on $O^{(J)}$, nor on $\mathbb{N}^{(J)}$ (for example for α, β of $J_\star$, $\| 1_\alpha E 1_\beta \| = 0$).
Therefore the J-set of J-integers is not a J-ordinal, althougt it is the J-set of the finite J-integers.
We can resume about J-integers :
$\mathbb{N}^{(J)}$ is a tree of height ω ; its branches, of length ω, are the J-ordinals defines by ω ; its p-levels ($p \in \mathbb{N}$) are the maximal anti-chains.
When J is denumerable, $\mathbb{N}^{(J)}$ is an ω-normal-tree.
We remark that if, for a J-set a, $a^\diamond$ is $\{x^\diamond/x \in \text{dom } a\}$, defined by induction on the rank, and if n is a J-ordinal defined by the λ-sequence $\alpha_1,\ldots,\alpha_\lambda$, $n^\diamond = \lambda$.

OPERATIONS ON $O^{(J)}$.

We define on $O^{(J)}$ an addition and a multiplication as following :

J-addition.

If a is a J-ordinal, we define a + 0 = a, $a + s_\alpha(n) = s_\alpha(a+n)$, and if n is limit, $a + n = \bigcup_{m \in \text{dom } n} (a+m)$. The operation on $O^{(J)}$ which associates the J-ordinals x, y with x + y is called J-addition. We prove, by induction on $n^\diamond$, that a + n is a J-ordinal defined by $a^\diamond + n^\diamond$, such that $a + n = n_{\alpha_1\ldots\alpha_\lambda\beta_1\ldots\beta_\mu}$, where $a = n_{\alpha_1\ldots\alpha_\lambda}$, and $n = n_{\beta_1\ldots\beta_\mu}$. Lastly $s_\alpha(n) = n + 1_\alpha$, for every n of $O^{(J)}$.

This addition is associative, and non commutative. As in the classical case, it allows the definition of a compatible but non total ordering.

J-multiplication.

We define the right-J-multiplication by 1_{α_o} in the following way :
$0.1_{\alpha_o} = 0$, $s_\alpha(n).1_{\alpha_o} = (n+1_\alpha)1_{\alpha_o} = n.1_{\alpha_o} + 1_{\alpha\wedge\alpha_o}$, and if n is limit, $n.1_{\alpha_o} = \bigcup_{m \in \text{dom } n} m.1_{\alpha_o}$.
We can now define the left-J-multiplication by the J-ordinal b :
$b.0 = 0$, $b.s_\alpha(n) = b(n+1_\alpha) = b.n + b.1_\alpha$ and if n is limit, $b.n = \bigcup_{m \in \text{dom } n} b.m$.

We prove by induction on $n^{\diamond}$ that b.n is a J-ordinal such that
$b.n = n_{\beta_1 \wedge \alpha_1 \dots \beta_\mu \wedge \alpha_1 \dots \beta_\mu \wedge \alpha_\lambda}$, where
$b = n_{\beta_1 \dots \beta_\mu}$ and $n = n_{\alpha_1 \dots \alpha_\lambda}$, writing where we cancell $\beta_j \wedge \alpha_i$ when $\beta_j \wedge \alpha_i = 0$. We remark so that the rank of the product of two J-ordinals can be less than the product of the ranks.

The J-multiplication so defined is associative and left distributive. It is not commutative, nor self-cancelling nor right-distributive. There are zero-divisors.

If a is a set, we define a J-set $\hat{a}$ by induction: if $a = \emptyset$, $\hat{a} = \emptyset$; if $a \neq \emptyset$, $\hat{a} = \{(\hat{b},1)/b \in a\}$; $\hat{a}$ is said to be standard. We remark the standard J-ordinals are the J-ordinals index by sequences of 1, and they form an isomorphic chain to that of the ordinals, on which the J-addition and the J-multiplication recovered their classical properties. Then $O^{(J)}$ fans out, enlarging according to the lattice J.

DEFINITION OF THE J-CARDINALS.
OPERATIONS.

Finite case.

We remark that every J-isomorphism defines an equivalence relation on $\mathbb{N}^{(J)}$. Thus we will call J-cardinal of $n_{\alpha_1 \dots \alpha_p}$, denoted $\overline{\overline{n_{\alpha_1 \dots \alpha_p}}}$, the mapping of the set of the J-integers isomorphic to $n_{\alpha_1 \dots \alpha_p}$ on 1. It is independant of the order of the sequence $\alpha_1 \dots \alpha_p$. Finally, we call J-cardinal of a finite J-set a, denoted card(a), the J-cardinal of one of the isomorphic J-integers.

General case.

We prove, using choice axiom of ZFC, that every J-set is isomorphic to a J-ordinal. As in the finite case above, every J-isomorphism induces on $O^{(J)}$ an equivalence relation. However, in the finite case, two J-isomorphic J-integers have domain with the same number of elements, whereas, in the non finite case, two J-isomorphic J-ordinals are defined by two isomorphic classical ordinals possibly distinct.

Let $n = n_{\alpha_1 \dots \alpha_{n^{\diamond}}}$ be a J-ordinal and $\overline{\overline{n^{\diamond}}}$ the cardinal of the ordinal $n^{\diamond}$. We shall denote C_n the class of the J-ordinals J-isomorphic to n, and by D_n the class of the C_n-objects x such that $x^{\diamond} = \overline{\overline{n^{\diamond}}}$. D_n is the set of C_n-objects which are minimal for the ordering induced by E.

Let n be a J-ordinal. The constant mapping which maps every element of D_n on 1 is a J-set, called J-cardinal of n and denoted $\overline{\overline{n}}$. Let a be an J-set. We call J-cardinal of a the J-cardinal of a J-ordinal which is J-isomorphic to a, denoted card(a).

Operations on the J-cardinals.

The J-addition and J-multiplication induce, on the class $\mathbb{C}^{(J)}$ of the J-cardinals two commutative operations defined by
$\overline{\overline{n}} + \overline{\overline{m}} = \overline{\overline{n+m}}$ and $\overline{\overline{n}}.\overline{\overline{m}} = \overline{\overline{n.m}}$,

Proposition : If a and b are two J-sets such that dom a $\cap$ dom b $= \emptyset$,
card(a $\cup$ b) = card(a) + card(b)
card(a $\times$ b) = card(a) . card(b)
This desired result generalises the classical result of the classical set theory. Besides, all the previously defined notions give the classical results when $J = \{0,1\}$, or when we consider only the standard J-sets.

REPRESENTATION THEOREM.

Proposition : There exists a bijection between $\mathbb{C}^{(J)}$ and the class of the mappings from $J_\star$ onto the class of cardinals.

For a J-cardinal $n = n_{\alpha_1 \dots \alpha_\lambda}$, we put down $\psi(n) = f$, where $f(\alpha)$ is the cardinal of the set of index i such that $\alpha_i = \alpha$. Obviously, if n is finite, $\psi(n)$ is a mapping from $J_\star$ in $\mathbb{N}$. For example, $\psi(4_{\alpha\alpha\beta\gamma}) = f$, with $f(\alpha) = 2$, $f(\beta) = f(\gamma) = 1$ et $f(\delta) = 0$ everywhere else.

ψ is a bijection which is an isomorphism for the addition, since
$$\psi(\overline{\overline{n_{\alpha_1 \dots \alpha_\lambda}}} + \overline{\overline{n_{\beta_1 \dots \beta_\mu}}}) = \psi(\overline{\overline{n_{\alpha_1 \dots \alpha_\lambda \beta_1 \dots \beta_\mu}}})$$
$$= \psi(\overline{\overline{n_{\alpha_1 \dots \alpha_\lambda}}}) + \psi(\overline{\overline{n_{\beta_1 \dots \beta_\mu}}}).$$

In the finite case, the J-cardinals form a set, and we have : "there exists an isomorphism for addition between the set of finite J-cardinals and the set of the mappings from $J_\star$ into $\mathbb{N}$ "almost everywhere null" (nul except for a finite numbers of elements).

SPECIAL CASE : J IS A CHAIN

For the J-cardinals we can then choose, by a canonical method in the class of the J-isomorphic J-ordinals the element, whose index is in mounting order, and call it J-cardinal. In this case, the J-cardinals are J-ordinals $n_{\alpha_1 \alpha_2 \dots \alpha_p}$ with $\alpha_1 \leq \alpha_2 \leq \dots \alpha_p$. We obtain a class of J-ordinals in which the J-addition and the J-multiplication are commutative.

The tree of the J-ordinals is as follows :

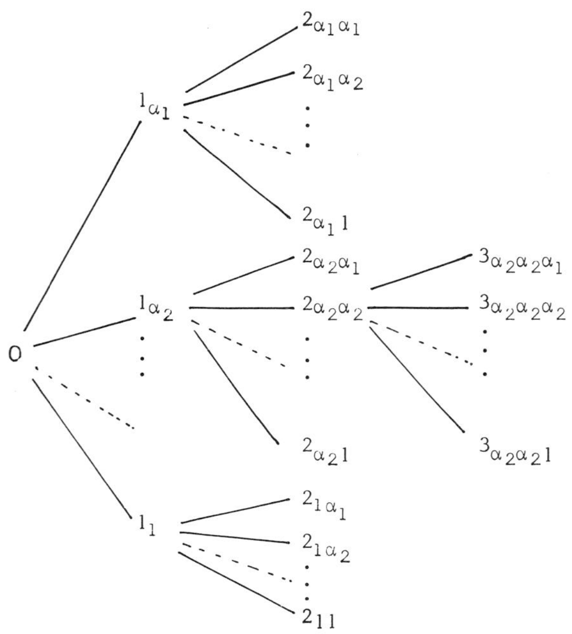

These J-cardinals, in the finite case, can represent, theoretically, some finite fuzzy cardinals defined in the litterature [2] .

REFERENCES

Botta, O. Delorme, M. "A f-set universe" in "Fuzzy information and decision processes". North Holland.

Deluca, E. and S. Termini (1972). A definition of a non-probabilistic entropy in the sitting of fuzzy sets theory. Information and control, Vol. 20.

Dubois, D. "A new definition of the fuzzy cardinality of finite fuzzy sets preserving the classical additivity property". Busefal (Automn 1981).

Jech, T. Set theory. Academic Press.

Rasiowa, H. An algebraic approach to non classical logics. North Holland.

Zadeh, L.A. (1982). A computional approach to fuzzy quantifiers in natural languages. Memorandum No. UCB/ERL M 83/36. Berkeley.

FUZZY FILTERS — A GENERALIZATION OF CREDIBILITY MEASURES

U. Höhle

Fachbereich Mathematik, Universität-Gesamthochschule Wuppertal, Gaußstr. 20, D-56 Wuppertal 1, Federal Republic of Germany

Abstract. The purpose of this paper is to emphasize the importance of credibility measures for a theory of fuzzy filters, to which every concept of fuzzy convergence is closely related. Moreover we use fuzzy filters as a basis for the definition of fuzzy points and fuzzy subsets.

Keywords. Fuzzy filters, 1-filters, credibility measures, necessity measures, fuzzy points, fuzzy subsets.

FUZZY FILTERS AND CREDIBILITY MEASURES

Let X be an arbitrary nonvoid set and $P(X)$ be the power set of X. A mapping $\mu : P(X) \to [0,1]$ (i.e. a fuzzy subset μ of $P(X)$, cf. [7]) is called a *fuzzy filter on* X iff μ satisfies the following conditions

(F1) $\mu(\phi) = 0$, $\mu(X) = 1$

(F2) $\mu(A) + \mu(B) \leq \mu(A \cup B) + \mu(A \cap B)$.

A fuzzy filter μ is a fuzzy *ultrafilter* iff μ fulfills the additional property

(F3) $1 - \mu(A) \leq \mu(\complement A) \quad \forall A \in P(X)$.

From (F1) - (F3) we infer that the set of all fuzzy ultrafilters on X coincides with the set of all *finitely* additive probability measures on $P(X)$; this situation is not very surprising : Even in the crisp case ordinary ultrafilters on X and 2-valued, finitely additive probability measures on $P(X)$ are equivalent notions.

Theorem 1 For every fuzzy filter μ on X and for every crisp subset A of X there exists a fuzzy ultrafilter ν on X provided with the properties

$$\mu(A) = \nu(A) \quad ,$$

$$\mu(B) \leq \nu(B) \qquad \forall B \in P(X) \quad .$$

Proof. On the set of all fuzzy filters we consider a partial ordering $\preceq$ defined as follows

$$\eta_1 \preceq \eta_2 \iff \eta_1(B) \leq \eta_2(B) \quad \forall B .$$

(a) We fix a crisp subset A of X. If μ is a fuzzy filter on X, then $\hat{\mu}$ defined by

$$\hat{\mu}(B) = \mu(A \cup B) + \mu(A \cap B) - \mu(A)$$

is also a fuzzy filter on X. In particular $\hat{\mu}$ has the following properties

$$\hat{\mu}(A) = \mu(A) \quad , \quad \hat{\mu}(\complement A) = 1 - \mu(A) \quad ,$$

$$\mu \preceq \hat{\mu} \quad .$$

(b) Setting $Z = \{\eta \ , \ \mu \preceq \eta\}$ we obtain that Z is nonvoid, and that $(Z, \preceq)$ is inductively ordered. By virtue of *Zorn's lemma* Z has a maximal element η_0. If η_0 is not a fuzzy ultrafilter, then there exists a crisp subset C and a fuzzy filter $\hat{\eta}_0$ with the properties (cf. step (a))

$$\eta_0 \preceq \hat{\eta}_0 \quad , \quad \eta_0(\complement C) \lneq \hat{\eta}_0(\complement C) \quad ;$$

but this is a contradiction to the maximality of η_0. Hence η_0 is a fuzzy ultrafilter satisfying the desired properties . Q.E.D.

Corollary 2 (Structure theorem for fuzzy filters) Let μ be a fuzzy filter on X. Then the relation

$$\mu(A) = \inf \{\nu(A) \mid \mu \preceq \nu \ , \ \nu \text{ fuzzy ultrafilter}\}$$

holds true .

Definition 3 Let $\mu : P(X) \to [0,1]$ be a mapping equipped with (F1) .

(a) μ is called a credibility measure on $P(X)$ iff μ satisfies the following condition (cf. [5] , [6])

$$\text{(C)} \quad \sum_{i=1}^{n} (-1)^{i-1} \sum_{1 \leq j_1 < \ldots < j_i \leq n} \mu\Big(\bigcap_{k=1}^{i} A_{j_k}\Big) \leq \mu\Big(\bigcup_{i=1} A_i\Big) \quad .$$

(b) μ is said to be a necessity measure on $P(X)$ iff μ fulfills the additional property (cf. [5])

(N) $\mu(A \cap B) = \mathrm{Min}(\mu(A), \mu(B))$.

Hierarchy. Necessity measures $\Longrightarrow$ $\Longrightarrow$ credibility measures $\Longrightarrow$ fuzzy filters .

In the following we give a characterization of necessity measures by 1-filters .

Definition 4 A nonvoid subset $\mathbb{F}$ of $[0,1]^X$ is a 1-*filter on* X iff F satisfies the following conditions (cf. [2] , [3])

$$(F1) \begin{cases} (h \in [0,1]^X \text{ with } \forall \varepsilon > 0 \quad \exists f_\varepsilon \in \mathbb{F} \\ \text{s.t. } f_\varepsilon(r) - \varepsilon \leq h(r) \quad \forall r \in X) \\ \Longrightarrow \quad h \in \mathbb{F} \end{cases}$$

$$(F2) \quad f_1, f_2 \in \mathbb{F} \quad \Longrightarrow \quad f_1 \wedge f_2 \in \mathbb{F}$$

$$(F3) \quad \sup\{f(r), r \in X\} = 1 \quad \forall f \in \mathbb{F}.$$

Let $N(X)$ be the set of all necessity measures on $P(X)$ and $F_1(X)$ be the set of all 1-filters on X . We introduce a mapping $\Theta : F_1(X) \longrightarrow N(X)$ by

$$\Theta(\mathbb{F})(A) = \sup_{f \in \mathbb{F}}(\inf_{r \notin A}(1 - f(r))) \qquad A .$$

Theorem 5 The mapping Θ is bijective.

Proof (a) (Injectivity) *Step 1* . For every pair $(\alpha, A) \in [0,1] \times P(X)$ we define a fuzzy subset $h_{(\alpha,A)}$ by

$$h_{(\alpha,A)} = \begin{cases} 1 & , \; r \in A \\ 1-\alpha & , \; r \notin A \end{cases} .$$

The axiom (*F*1) implies

$$\Theta(F)(A) \geq \alpha \iff h_{(\alpha,A)} \in \mathbb{F} \qquad (*)$$

Step 2 . Let $\mathbb{F}_1$ and $\mathbb{F}_2$ are 1-filters with $\Theta(\mathbb{F}_1) = \Theta(\mathbb{F}_2)$. By virtue of (*) the relation

$$\{f \in \mathbb{F}_1, \text{ f 2-valued}\} = \{f \in \mathbb{F}_2, \text{ f 2-val.}\}$$

holds . Since every fuzzy subset of X can be uniformly approximated by simple fuzzy subsets , we infer from (*F*1) and (*F*2) that $\mathbb{F}_1 = \mathbb{F}_2$.

(b) (Surjectivity) Let μ be a necessity measure. Since $[0,1]$ is a completely distributive lattice, it is easy to show that

$$\mathbb{F}_\mu = \{f \in [0,1]^X, \inf_{r \notin A}(1 - f(r)) \leq \mu(A)\}$$

is a 1-filter satisfying the equation

$$\Theta(\mathbb{F}_\mu) = \mu .$$

Q.E.D.

From theorem 5 we conclude that for every necessity measure μ there exists an ordinary ultrafilter *U* , which is finer than μ - i.e. $\mu(A) \leq 1_U(A) \quad \forall A \in P(X)$. This statement is not longer true for arbitrary credibility measures; but in view of corollary 2 any credibility measure can be written as an infimum of a family of finitely additive probability measures (= fuzzy ultrafilters) .

FUZZY SUBSETS AND FUZZY POINTS

Since every ordinary filter *F* on X being stable for arbitrary intersections - i.e.

$$\{A_i, i \in I\} \subseteq F \quad \Longrightarrow \quad \bigcap_{i \in I} A_i \in F \quad -$$

can be identified with an ordinary, nonvoid subset of X , we introduce the notion of a fuzzy nonvoid subset of X as follows : A *fuzzy nonvoid subset of* X is a fuzzy filter μ on X satisfying the additional property

$$(F4) \quad \begin{array}{l} \{A_i, i \in I\} \downarrow \text{ directed downward:} \\ \inf_{i \in I} \mu(A_i) = \mu(\bigcap_{i \in I} A_i) \end{array} .$$

Obviously (F4) is a τ-smoothness condition, to which a fuzzy filter can be subjected .

For every τ-smooth necessity measure μ on $P(X)$ there exists an unique normal fuzzy subset g of X (i.e. $\sup\{g(r), r \in X\} = 1$) in the sense of Zadeh ([7]) satisfying the condition

$$\mu(A) = \inf_{r \notin A}(1 - g(r)) \quad \forall A \in P(X).$$

Hence every normal fuzzy subset in the sense of Zadeh can be identified with a fuzzy nonvoid subset equipped with axiom (N) . If g is not a normal fuzzy subset of X , but satisfies the inequality $1 \leq \sum_{r \in X} g(r)$, then g can be considered as a fuzzy nonvoid subset μ_g as follows

$$\mu_g(A) = \mathrm{Max}(1 - \sum_{r \notin A} g(r), 0) \qquad A.$$

Obviously μ_g is in general not a credibility measure on $P(X)$.

In the following we study the interrelations between fuzzy nonvoid subsets and random sets. First we recall some properties of random sets (cf. [4]) : On $P(X)$ we consider the interval topology $\tau_{\subseteq}$ - i.e. the coarsest topology such that each order interval $[A,B] = \{C, A \subseteq C \subseteq B\}$ is closed. It is well known that $(P(X), \tau_{\subseteq})$ is a totally disconnected, compact space. Each regular Borel probability measure η on $P(X)$ (or space law, cf. [4]) is a *random set in* X .

Theorem 6 (a) Every random set η in X with $\eta(\{\phi\}) = 0$ induces a τ-smooth credibility measure μ_η by

$$\mu_\eta(A) = \eta([\phi,A]) \quad \forall A \in P(X) \quad .$$

(b) Let μ be a credibility measure on $P(X)$. Then the following assertions are equivalent

(i) μ is τ-smooth - i.e. is a fuzzy nonvoid subset of X

(ii) There exists an unique random set η in X equipped with the property $\mu(A) = \eta([\phi,A]) \quad \forall A$.

Proof. The assertion (a) and the implication (ii) $\Rightarrow$ (i) in assertion (b) are obvious. Therefore we only verify the implication (i)$\Rightarrow$(ii). Since μ is a credibility measure on $P(X)$, we infer from the dual version of theorem 3.3 in [1] that there exists a regular Borel probability measure m_μ on the set $F(X)$ of all ordinary filters on X satisfying the following condition

$$m_\mu(\{F \in F(X),\ A \in F\}) = \mu(A) \quad \forall A \quad .$$

Further the mapping $\Psi : F(X) \rightarrow P(X)$ determined by

$$\Psi(F) = \bigcap \{A,\ A \in F\}$$

is Baire measurable. In particular we have

$$\Psi^{-1}([\phi,A]) = \bigcap_{p \notin A} \{F,\ \complement\{p\} \in F\} \qquad (**)$$

We denote by η the image measure of m_μ under Ψ. Obviously η is a random set in X. By virtue of (F4) and (**) the relation

$$\eta([\phi,A]) = \mu(A) \quad \forall A \in P(X)$$

holds. Q.E.D.

In view of the preceding theorem 6 τ-smooth credibility measures and random sets η in X with $\eta(\{\phi\}) = 0$ are equivalent notions. Therewith every "nonvoid" random set is a fuzzy nonvoid subset, but not vice-versa.

Interpretation. Let μ be a fuzzy nonvoid subset of X. According to theorem 6 we interpret $\mu(A)$ as the degree that the fuzzy nonvoid subset μ is contained in the crisp subset A.

Since stable ordinary ultrafilters can be identified with ordinary points, we introduce the notion of fuzzy points as follows : A *fuzzy point in* X is a fuzzy ultrafilter on X equipped with property (F4). Evidently discrete Radon probability measures on X and fuzzy points in X are equivalent notions. A fuzzy point η in X is crisp iff η fulfills the axiom (N). If η is a fuzzy point, then we use an interpretation for η, which differs essentially from the usual interpretation of probability measures : $\eta(A)$ is the *degree*, in which a fuzzy point η "hits the ordinary subset A of X"; therefore the axioms (F1) - (F3) can be understood as follows : A fuzzy point never hits the empty set, a fuzzy point in X hits always the basic set X, or the fuzzy point *lives* in X; the degree that the fuzzy point η hits A is high iff the degree that η hits the complement of A is small.

Let μ be a fuzzy nonvoid subset of X. A fuzzy point η is *"an element of"* μ iff $\mu(A) \leq \eta(A) \quad \forall A \in P(X)$. In particular we define a fuzzy relation between fuzzy points and fuzzy nonvoid subsets of X as follows :

$$E(\eta,\mu) = \inf_{A \in P(X)} (\mathrm{Min}(1-\mu(A) + \eta(A)\ ,\ 1)$$

$E(\eta,\mu)$ is the *degree* that η is "an element of" μ. In particular $E(\eta,\mu) = 1$ iff η is "an element of" μ.

Theorem 7 Let X be an arbitrary, nonvoid set, g be a normal fuzzy subset of X in the sense of Zadeh. Then there *exists* a family $\{\nu_i\ ,\ i \in I\}$ of fuzzy points in X such that

$$\inf_{r \notin A} (1 - g(r)) = \inf_{i \in I} \nu_i(A) \quad \forall A\ ;$$

i.e. every *normal* fuzzy subset g in the sense of Zadeh can be considered as a collection of fuzzy points ν, which are "an element of" g.

The proof of theorem 7 relies on the following

Lemma 8 Let $(\alpha_k)_{k \in \mathbb{N}}$ be an increasing sequence in $[0,1]$ with the properties

$$\alpha_k \lneq \alpha_{k+1} \quad , \sup\{\alpha_k,\ k \in \mathbb{N}\} = 1 \quad .$$

Further let g be a fuzzy subset of $\mathbb{N}$ with $g(\mathbb{N}) = \{\alpha_k\ ,\ k \in \mathbb{N}\}$. Then for all $n_o \in \mathbb{N}$ there exists a fuzzy point ν_{n_o} in $\mathbb{N}$ satisfying the conditions

1^0 $\inf_{r \notin A} (1 - g(r)) \leq \nu_{n_o}(A) \quad \forall A$,

2^0 $\inf_{r \notin \complement\{n_o\}} (1 - g(r)) = \nu_{n_o}(\complement\{n_o\})$.

Proof. $A_k = \{n \in \mathbb{N},\ g(n) = \alpha_k\}$. Let k_o be the positive integer with $n_o \in A_{k_o}$; and for every $k \geq k_o+1$ we choose an element $n_k \in A_k$. Setting $\beta_{k_o} = \alpha_{k_o}$, $\beta_k = \alpha_k - \alpha_{k-1}$ for all $k \geq k_o+1$, we obtain that $\nu_{n_o} : P(\mathbb{N}) \rightarrow [0,1]$ given by

$$\nu_{n_o}(A) = \sum_{n_k \in A} \beta_k$$

is a σ-additive probability measure. Obviously ν_{n_o} fulfills the condition 2^0. In order to prove 1^0 we proceed as follows :

Case 1 . $\complement A \cap A_k = \phi$ for all $k \geq k_o$; this implies : $\nu_{n_o}(A) = 1$.

Case 2 . $\complement A \cap A_k \neq \phi$ for infinitely many $k \geq k_o$; this implies : $\inf_{n \notin A} (1 - g(n)) = 0$.

Case 3 . $\complement A \cap A_k \neq \phi$ for finitely many $k \geq k_o$; $k_m = \text{Max}\{k, \complement A \cap A_k \neq \phi\}$; then we obtain the following relation

$$\inf_{n \notin A} (1 - g(n)) = 1 - \alpha_{k_m} = \sum_{k=k_m+1}^{\infty} \beta_k =$$

$$= \nu_{n_o}\Big(\dot{\bigcup_{k=k_m+1}} A_k\Big) \leq \nu_{n_o}(A) .$$

Proof of theorem 7 : For every normal fuzzy subset h of X (in the sense of Zadeh) and for every point $x_o \in X$ with $h(x_o) \neq 0$ there exists a fuzzy subset g of X satisfying the following conditions

$\{x, g(x) \neq 0\} = \{\alpha_k , k \in \mathbb{N}\}$ with

$\alpha_k \lneq \alpha_{k+1}$, $g(x_o) = h(x_o)$, $g \lesssim h$, $\sup\{g(x) , x \in X\} = 1$;

hence the assertion follows from lemma 8 . Q.E.D.

CONCLUSION

From a filter theoretical point of view fuzzy filters, which are not credibility measures, do not permit any probabilistic interpretation, even though any fuzzy filter is an infimum of a family of finitely additive probability measures. Moreover it seems to be very interesting to consider fuzzy filters as a framework for the study of the correlations between fuzzy subsets and fuzzy points.

REFERENCES

[1] Höhle, U., Klement, E.P. (1982). Plausibility measures - a general framework for possibility measures and fuzzy probability measures, Institut für Mathematik der Universität Linz (Austria) , Institutsbericht Nr. 221 .

[2] Höhle, U. (1982). Probabilistic topologies induced by L-fuzzy uniformities, manuscripta math., 38, 289 - 323 .

[3] Lowen, R. (1982). Fuzzy neighborhood spaces, Fuzzy Sets and Systems, 7, 165-189 .

[4] Matheron, G. (1975). Random Sets and Integral Geometry. Wiley, New York, London .

[5] Prade, H. (1979). Nomenclature of fuzzy measures, Proceedings of the first International Seminar on Fuzzy Set Theory at Linz (Ed. E.P. Klement), 8 - 25 , Linz (Austria) .

[6] Shafer, G. (1976). A mathematical theory of evidence, Princeton University Press .

[7] Zadeh, L.A. (1978). Fuzzy sets as a basis for a theory of possibility, Fuzzy Sets and Systems, 1 , 3-28

[8] Zadeh, L.A. (1965). Fuzzy sets, Information and Control, 8 , 338-353.

AN ANALYSIS ON FUZZY MEMBERSHIP RELATION IN FUZZY SET THEORY

Liu Ying-ming

Institute of Mathematics, Sichuan University, Chengdu, China

Abstract. In the previous paper (Liu, 1982 b) we gave some mutually equivalent systems of axioms, which seem intuitively to be evident, for determining the neighborhood structure in fuzzy topological spaces from the standpoint of topology and fuzzy set theory. We also proved the theorem that the unique neighborhood structure satisfying one of these systems of axioms is exactly the Q-neighborhood structure (Pu and Liu, 1980a; Liu,1983) which has already been playing and will continually play a significant role in the research of fuzzy topology. But the neighborhood structure is determined by a fuzzy membership relation between the fuzzy points and the fuzzy sets. Therefore we presently pay attention to this relation and analyse it from the viewpoint of fuzzy set theory. The four principles for determining the fuzzy membership relation are established. We also show that the unique fuzzy membership relation is exactly the Q-relation which determines the Q-neighborhood structure mentioned above.

Keywords. Set theory; L-fuzzy set ; fuzzy membership relation; quasi-coincidence relation; fuzzy topology; Q-neighborhood; Multiple-choice principle.

INTRODUCTION

Since the fuzzy set theory was founded (Zadeh, 1965), the significant advances have been gained both in its theoretical research and in its applications. The concept of fuzzy set is more general than the concept of ordinary set and the ordinary set theory is the basis of modern mathematics, so the establishment of fuzzy set theory has naturally some influences on the varied area of pure mathematics. In particular, we may develope the topological space theory in such a more general framework, i.e. the framework of fuzzy sets. In fact, there have been a lot of important works on fuzzy topology. Now both deepth and range of the research of the fuzzy topology have attained to the same extent as those of ordinary topological space theory. These works have already deepened our knowledge about some basic concepts and related results in ordinary topological spaces and also have brougt to the attention of algebraistes and logicians in the multiple-valued logic. In the early stage of the research of fuzzy topology, there were a lot of defficulties when we wanted to introduce a reasonable concept of fuzzy point and to establish a satisfactory neighborhood structure because of the serious limitation of notion of traditional neighborhood in fuzzy topological spaces. In face of such a situation, so called "pointless approach" appeared. This pointless approach was specially applied to investigate those topological properties not relating to the notion of "point". Though some profound results are yielded via the pointless approach (Goguen,1967;Hutton, 1977), the limitation of this approach is evident. For instance, it is unavoidable to deal with the notion of point in the important imbedding theory (Liu, 1983). We introduced the concept of fuzzy point and defined a fuzzy membership relation (i.e. quasi-coincidence relation) between the fuzzy points and the fuzzy sets (Pu and Liu, 1980a). Thus we sucessfully established a satisfactory fuzzy Moore-Smith convergence theory. Afterwards the quasi-coincidence relation (for short, Q-relation) and the corresponding neighborhood structure (called Q-neighborhood) play an important role in the research of many problems, such as fuzzy product and quotient space (Pu and Liu,1980b), fuzzy compactness (Liu, 1981; Wang,1983), fuzzy uniformity and imbedding theory (Liu,1983). From the standpoint of topology and fuzzy set theory, we have already given several systems of axioms to characterize the Q-neighborhood structure (Liu,1982b). But the neighborhood structure is determined by a more basic relation, namely fuzzy membership relation between fuzzy points and fuzzy sets. By its very nature, the fuzzy membership relation does not connect with the topology. Consequently in this paper we shall further analyse the fuzzy membership relation only from the angle of fuzzy set theory. The new characterizations of Q-relation, which are obtained via this analysis, will be helpful to deepen the knowledge about both the intension of Q-neighborhood and the limitation of traditional neighborhood structure.

PRELIMINARIES

In this paper, X denotes a non-empty ordinary set and L denotes a fuzzy lattice, i.e. a completely distributive lattice with an order inversing involution. For each $a \in L$, a' is called its complement. The greatest element and the least element of L are denoted by 1 and 0 respectively. A map from X to L is said to be an L-fuzzy set (simply, fuzzy set or set) in X. The collection of all the fuzzy sets in X, denoted by L^X, can be naturally seen as a fuzzy lattice and its lattice operations and complementary operation are pointwise induced by the corresponding operations in the lattice L. A fuzzy set in X is called a fuzzy point (simply, point) iff it takes the value 0 for all $y \in X$ except one, say $x \in X$. If its value at x is $\lambda \in L$ ($\lambda \neq 0$), we denote it by x_λ. The symbol $\lhd$ denotes a fuzzy membership relation between the fuzzy points and the fuzzy sets. The notation "$x_\lambda \lhd A$" means that there exists relation $\lhd$ between the point x_λ and the set A. The negation of the relation $\lhd$ is denoted by $\not\lhd$. N denotes the collection of all the positive integer. K_λ denotes the fuzzy set taking the constant value λ on X. We also write K_1 and K_0 as X and $\emptyset$ respectively.

Definition 1. The fuzzy point x_λ is said to be contained in a fuzzy set A, or to belong to A, denoted by $x_\lambda \in A$, iff $\lambda \leq A(x)$. x_λ is said to be quasi-coincident with A iff $\lambda \not\leq A'(x)$.

Evidently both the fuzzy belonging relation and the quasi-coincidence relation are generalizations of ordinary belonging relation in set theory. The quasi-coincidence relation has already been given (Liu,1983). When the fuzzy lattice L=[0,1] , we have $\lambda' = 1-\lambda$, hence the formula $\lambda \nleq A'(x)$ is equivalent the formula $\lambda + A(x) > 1$, i.e. in this special case, the above definition of quasi-coincidence reduces to Definition 2.2' of (Pu and Liu, 1980a).

In the early stage of development of fuzzy set theory, the value set L (i.e. range domain) was usually the interval [0,1] . Later people discovered that the membership grades of some fuzzy sets (at the same point) may be non-comparable, so the more general notion of L-fuzzy set was introduced, namely the value set [0,1] was replaced by the fuzzy lattice L. Comparing with the richness of structure in the interval [0,1] , there exist only order relation and involution (i.e. complementary operation) in fuzzy lattice L. Hence we can concentrate on our target and more easily determine the characterization of the fuzzy membership relation.

SOME PRINCIPLES DETERMINING FUZZY MEMBERSHIP RELATION

In this paragraph we shall establish some principles to determine the fuzzy membership relation. Some discussions on these principles will also be given.

I. Extension principle Restricting the relation $\lhd$ to the ordinary set theory, $\lhd$ will become the usual belonging relation $\in$. Precisely, for any fuzzy lattice L, if p and A are an ordinary point and an ordinary subset in X respectively, then $p \lhd A$ iff $p \in A$.

Since the fuzzy sets (points) take the ordinary sets (points, resp.) as special case, it is natural to put forward the Extension principle.

For a fuzzy point $x_\lambda = p$ and a fuzzy set A, the relation $p \lhd A$ or not must be determined by the membership grades of p and A at each point of X. But p takes zero value except at the point x, so the relation $p \lhd A$ or not must be determined by a relation between $p(x) = \lambda$ and $A(x) = \mu$. In view of there is only order relation and involution in L, the relation between λ and μ will be described by a system of formulae about λ and μ expressed in terms of the order relation and involution. (e.g. $\lambda \nleq \mu'$ and so on). In addition, this system of formulae will be valid not only for some pair of λ and $\mu = A(x)$, but for any fuzzy lattice L and any $\lambda \in L$ $(\lambda \neq 0)$ and any A(x) as well. According to the consideration above, we raise the following:

II. Value set determination principle. The fact that $x_\lambda \lhd A$ or not is completely determined by a system of formulae about λ and A(x) expressed in terms of the order relation and involution. Moreover, the system of formulae is valid not only for some pair λ and A(x), but for any fuzzy lattice L and any $\lambda \in L(\lambda \neq 0)$ and any A(x) as well.

III. Maximum and minimum principle. For any fuzzy point p, $p \lhd X$ and $p \not\lhd \emptyset$, where X and $\emptyset$ denote the greatest L-fuzzy set and the least L-fuzzy set on X respectively.

IV Multiple-choice principle. Suppose that A_i is a family of L-fuzzy sets. If $x_\lambda \lhd \bigvee\{A_i\}$ (the union of these A_i), then there exists a A_i such that $x_\lambda \lhd A_i$.

Two remarks on the above four principles:

(1) In ordinary set theory, the Multiple-choice principle is obvious. But in the fuzzy set theory, it is not still valid for the fuzzy belonging relation (see Definition 1). In fact, we have the following counterexample. Take fuzzy lattice L=[0,1] , $\lambda = 1$ and $x \in X$.

Consider the union A of fuzzy sets $A_n = x_{1-1/n}$ ($n \in N$). Obviously, $x_\lambda \in A$, but $x_\lambda \notin$ any A_n. The invalidity of the principle IV for relation $\in$ may be a important cause to lead to the serious limitation of corresponding neighborhood structure --- the traditional neighborhood system --- in the research of fuzzy topological spaces. This limitation was clearly realized in the early stage of research of fuzzy topological spaces. In order to preserve the principle IV, Wong (1974) abandoned the priciple I and gave pecular definitions of "point" and "belong to" relation which do not take the ordinary notion of point and "belong to" as special cases. Thus Wong developed his theory of fuzzy topological spaces. But there are two many things to take care of at the same time, so the results obtained have a great deal of divorce from the intuition of the ordinary general topology. Our primitive consideration to introduce the concept of Q-neighborhood just stems from an exploration of certain kind of fuzzy neighborhood structure satisfying both the principle I and the principle IV. We always emphasize the fundamental importance of the principl IV. (cf. the note before Theorem 1 in (Liu, 1982b))

(2) The principle II has been indicated (Liu,1982b). But as shown in the next paragraph, at present the content of the expression "formulae about λ and μ expressed in terms of the order relation and involution" is more plentiful. Besides, noting that the order relation in a lattice has more properties, e.g. there exist the intersection $\wedge$ and the union $\vee$, we can understand the above principle II as follows: "$x_\lambda \lhd A$ or not is completely determined by a system of formulae about λ and $A(x)$ expressed in terms of $\leqslant$, $\wedge$, $\vee$ and involution." For this comprehension on the Value-set determination principle, the corresponding investigation will be given in another paper.

DETEMINATION ON FUZZY MEMBERSHIP RELATION

The partial ordering in the fuzzy lattice L is denoted by $\leqslant$. The meaning of the following nine symbols:

$$\nleqslant, <, \nless, =, \neq, \geqslant, \ngeqslant, >, \ngtr$$

is obvious. These nine symbols with the symbol $\leqslant$ are called as relation symbol. According to the principle II, the fuzzy membership relation $\lhd$ between a point x_λ and a set A is determined by a system of formulae about λ, $A(x)=\mu$, λ' and μ' expressed in terms of the above ten relation symbols. We call such a system satisfying the aforementioned four principles as a "rational system" respect to x_λ and A (for short, R-system).

Theorem 1 The rational system respect to x_λ and A must consist of the single formula: $\lambda \nleqslant A'(x)$. In other words, the quasi-coincidence relation is a unique fuzzy membership relation satisfying the above four principles.

Proof First we make some general discussion on R-systems. Without loss of generality, we assume that each formula appears at most once in a R-systems.

(1) We may assume that the formula of R-system can not be formed by single element of λ, λ', μ and μ'. In fact, these formula is either false (e.g. $\lambda' \neq \lambda'$) or identical (e.g. $\mu = \mu$). In the latter case, noting the single idendity can not form a R-system (otherwise we have $x_1 \lhd \phi$ contradicting with the principle I and III), we may delete this idendical formula and do not cause the problem of "empty system".

(2) Any formula of R-system can not be

formed by the pair of λ and λ'. In fact, take $L=[0,1]$. For such a formula (e.g. $\lambda \nleq \lambda'$), it is easy to see that there exists $\rho \in L$ such that $\rho \neq 0$ and ρ does not satisfy this formula. Hence for the point x_ρ and any fuzzy set A, $x_\rho \not\triangleleft A$. Especially $x_\rho \not\triangleleft X$ contradicting with the principle III. Now we may assume that λ and λ' do not appear in each formula of R-system at the same time. Noting the involution is order-reversing, we may further assume that the element λ' never appears in any formula of R-system. (If necessary, we may take complement such that the element λ' is replaced by λ.)

(3) Since for any relation symbol, its "reversing relation" symbol is still a relation symbol, we may assume that the element μ', if exists, alway appears on the right of a formula of R-system.

Now we connect λ and μ with the ten relation symbol to get the following ten formulae:

(1) $\lambda \nleq \mu$,(2) $\lambda \geqslant \mu$,(3) $\lambda \neq \mu$,
(4) $\lambda \not< \mu$,(5) $\lambda > \mu$,(6) $\lambda \leq \mu$,
(7) $\lambda \ngeq \mu$,(8) $\lambda = \mu$,(9) $\lambda < \mu$,
(10) $\lambda \not> \mu$.

We claim that none of the above ten formulae can appear in R-system. First none of the formulae (1),(3),(5),(7) and (9) can appear in R-system. Otherwise take $L=[0,1]$. For $\lambda=1=\mu$ none of these five formulae are valid, so $x_1 \not\triangleleft X$ contradicting to the principle I. Secondly, none of the formulae (2), (4) and (8) can appear in R-system. Otherwise take $L=[0,1]$. For $\lambda=\frac{1}{2}$ and $\mu=1$ none of these three formulae are valid, so $x_{\frac{1}{2}} \not\triangleleft X$ contradicting to the principle III. Finally we discuss the formulae (6) and (10). Take $L=[0,1]$, $\lambda=1$ and $\mu_n = 1-\frac{1}{n+2}$ ($n \in N$). For λ and any μ_n these two formulae are not valid, so $x_1 \not\triangleleft$ any K_{μ_n}. (K_{μ_n} is a fuzzy set taking constant value μ_n). On other hand, we have $\bigvee \{K_{\mu_n}\} = X$. According to the principle IV, we have $x_1 \not\triangleleft X$ cortradicting to the principle I. Hence neither of the formulae (6) and (10) can appear in R-system also.

Now we connect λ and μ' with the ten relation symbols to get the other ten formulae as follows:

(1) $\lambda \leq \mu'$,(2) $\lambda \geqslant \mu'$,(3) $\lambda \neq \mu'$,
(4) $\lambda \not< \mu'$,(5) $\lambda > \mu'$,(6) $\lambda \leq \mu'$,
(7) $\lambda \ngeq \mu'$,(8) $\lambda = \mu'$,(9) $\lambda < \mu'$,
(10) $\lambda \not> \mu'$.

We claim that none of the formulae (6), (7),(8),(9) and (10) can appear in R-system. On the contrary, take $L=[0,1]$. For $\lambda = \mu = 1$ none of these five formulae are valid. Hence $x_1 \not\triangleleft X$ contradicting to the principle. Also the formula (5) does not appear in R-system. Otherwise, take $L=\{0,1,\rho,\rho'\}$, where ρ is not comparable with ρ'. Obviously this L, called the diamond of type I, is a fuzzy lattice. For $\lambda = \rho$ and $\mu = \rho$ or ρ', the formula (5) is not valid. Hence we have $x_\rho \not\triangleleft K_\rho$ and $x_\rho \not\triangleleft K_{\rho'}$. In view of $K_{\rho'} \vee K_\rho = X$, according to the principle IV we have $x_\rho \not\triangleleft X$. This is a contradiction to the principle III. In a word in the avove ten formulae about λ and μ', only the formulae (1),(2),(3) and (4) may appear in R-system.

Now we consider all the formulae about μ and μ' as follows:

(1') $\mu \leq \mu'$,(2') $\mu \geqslant \mu'$,(3') $\mu \neq \mu'$,
(4') $\mu \not< \mu'$,(5') $\mu > \mu'$,(6') $\mu \leq \mu'$,
(7') $\mu \ngeq \mu'$,(8') $\mu = \mu'$,(9') $\mu < \mu'$,
(10') $\mu \not> \mu'$.

None of the formulae (6'),(7'),(8'),(9') and (10') can appear in R-system. Otherwise, for any fuzzy lattice L and $\mu = 1 \in L$ none of these five formulae are valid. Hence $x_1 \not\triangleleft X$ contradicting to the principle I.

Neither of the formulae (2') and (5') can appear in R-system. Otherwise, take L as the above diamond of type I. For

$\mu = \rho$ or ρ' neither of these two formulae are valid. Hence $x_1 \not\triangleleft K_\rho$ and $x_1 \not\triangleleft K_{\rho'}$. According to the princeple IV, we have $x_1 \not\triangleleft (K_\rho \vee K_{\rho'})=X$ contradicting to the principle I.

Neither of the formulae (1') and (3') can appear in R-system. Otherwise take $L=\{0,1,\rho_1,\rho_2\}$, where ρ_1 is not comparable with ρ_2 and $\rho_1' = \rho_1$ and $\rho_2' = \rho_2$. Obviously L, called the diamond of type II, is a fuzzy lattice. For $\lambda=1$ and $\mu = \rho_1$ neither of these two formulae are valid. Hence we have $x_1 \not\triangleleft K_{\rho_1}$ and $x_1 \not\triangleleft K_{\rho_2}$. According to the principle IV, $x_1 \not\triangleleft (K_{\rho_1} \vee K_{\rho_2})=X$ contradicting with the principle I.

To sum up, there remains the formula (4') $\mu \not< \mu'$ with the aforementioned four formulae: (1) $\lambda \nleq \mu'$, (2) $\lambda \geq \mu'$, (3) $\lambda \neq \mu'$ and (4) $\lambda \not< \mu'$, which may appear in R-system.

Now consider those systems which consist of a single formula of the aforementioned five formula: (1) --- (4) and (4').

The formulae (2) and (4) are not permitted. Otherwise take $L=[0,1]$. For $\lambda=1$ and $\mu=0$ these two formulae are valid. Hence $x_1 \triangleleft \emptyset$ contradicting to the principle I.

The formula (3) is not permitted. Otherwise take $L=[0,1]$. For $\lambda=1/2$ and $\mu=0$ the formula (3) holds. Hence $x_{\frac{1}{2}} \triangleleft \emptyset$ contradicting to the principle III.

The formula (4') is not permitted. Otherwise take $L=[0,1]$ and $\mu=1/2$ and $\mu_n=1/2 - 1/n+2$ ($n \in N$). For each μ_n the formula (4') is not true, namely any fuzzy point $p \not\triangleleft K_{\mu_n}$. On other hand, for $\mu = \frac{1}{2}$ the formula (4') holds, hence $p \triangleleft K_\mu$. But $K_\mu = \vee K_{\mu_n}$, this is a contradiction to the principle IV.

Hier only the formula (4) remains. In fact, it is permitted to form a R-system and it exactly determines the quasi-coinciden relation. Now the principle I, II and III are clearly satisfied. Suppose that $x_\lambda \triangleleft \vee A_j$. Put $A_j(x)=\mu_j$. If $x_\lambda \not\triangleleft$ each A_j, i.e. $\lambda \leq \mu_j'$, then $\lambda \leq \wedge \mu_j'$. In virtue of Lemma 1 in (Liu, 1982a) $\wedge \mu_j' = (\vee \mu_j)'$, so $\lambda \leq (\vee \mu_j)'$. The latter contradicts to the assumption: $x_\lambda \triangleleft \vee A_j$. Hence there exists a j such that $\lambda \nleq \mu_j'$, i.e. $x_\lambda \triangleleft A_j$. That is, the principle IV holds.

Now consider these systems which consist of two formulae of the aforementioned five formulae.

It is easy to see that the system of formulae (1) and (2) and the system of formulae (2) and (3) is equivalent to the formula (5). Similarly, the systems of formulae (1) and (3), of the formulae (1) and (4) and of the formulae (3) and (4) are all equivalent to the formula (1). The system of the formulae (2) and (4) is equivalent to the formula (2). Therefore each system of two formulae of the formulae (1) --- (4) is either equivalent to the formula (1) or is not R-system.

As for the systems consisting of the formula (4') and one of the formulae (1) --- (4), none of them are R-system. In fact, take $L=[0,1]$, $\lambda=1$, $\mu=1/2$ and $\mu_n=1/2 - 1/n+2$ ($n \in N$). Following the argument on the system consisting of the single formula (4'), we easily prove the just above claim.

Now consider those systems which consist of three or more formulae of the aforementioned five formulae. In this case, at least some two formula of the formulae (1) --- (4) must appear and they are equivalent to a single formula (see the just above discussion). Therefore via mathematical induction those systems are either equivalet to the system consisting of the single formula (1) or not rational system. Thus we complete the proof.

INDEPENDENCE ON PRINCIPLES DETERMINING FUZZY MEMBERSHIP RELATION

Proposition 1 If a fuzzy membership relation satisfies the principle II and III, then it also satisfies the principle I.

Proof According to the principle II, we assume that the fuzzy membership relation $\triangleleft$ is determined by a system S about λ, λ', μ and μ' in terms of the ten relation symbols. Let $x=x_1$ and A be an ordinary point and an ordinary subset of X respectively. Naturally $A(x)=1$ or 0. According to the principle III, we have $x_1 \triangleleft X$ and $x_1 \ntriangleleft \emptyset$, hence if and only if $A(x)=1$, i.e. $x \in A$ we have $x_1 \triangleleft A$. In other words, this fuzzy membership relation $\triangleleft$ satisfies the principle I.

Proposition 1 says that the principle I may be deleted, but in view of the intuition of the principle I and its convenience in some discussion (cf. the remark in § 2 of (Wong,1974)) we still retain the principle I.

Theorem 2 The principles II, III and IV are independent each other.

Proof (1) Independence of the principle II. Take a fuzzy membership relation $\triangleleft$ as follows:

For a fuzzy point p and a fuzzy set A, $p \triangleleft A$ iff $A \neq \emptyset$.

Obviously this relation satisfies the principles III and IV. But it does not satisfy the principle II. Otherwise by Proposition 1, it also satisfies the principle I. This contradictes to the fact that the unique fuzzy membership relation satisfying all the four principles is quasi-coincident relation.

(2) Independence of the principle III. Consider a fuzzy membership relation $\triangleleft$ determined by the following system:

$$\begin{cases} \lambda \nless \mu, \\ \lambda \nleq \mu', \end{cases}$$

This relation $\triangleleft$ naturally satisfies the principle II. Now assume that $x_\lambda \triangleleft \bigvee A_j$. If $x_\lambda \ntriangleleft$ each A_j, then $\lambda \leq A_j(x)=\mu_j$ or $\lambda \leq \mu'_j$. If for some j we have $\lambda < \mu_j$, then $\lambda < \bigvee \mu_j$. Hence $x_\lambda \ntriangleleft \bigvee A_j$. If for each j we have $\lambda \leq \mu'_j$ then $\lambda \leq \bigwedge \mu'_j = (\bigvee \mu_j)'$ [Lemma 1 (Liu, 1981)]. Hence also $x_\lambda \ntriangleleft \bigvee A_j$. Thus the principle IV hold for this relation $\triangleleft$. As for the principle III, take $L=[0,1]$ and $\lambda=1/2$. For any μ, if $1/2 \nleq \mu'$, then $1/2 < \mu$. Thus $x_{\frac{1}{2}} \ntriangleleft$ any fuzzy set, i.e. the principle III is not true for $\triangleleft$.

(3) Independence of the principle IV. The fuzzy membership relation $\triangleleft$ determined by the single formula $\lambda \leqslant \mu$ (i.e. the fuzzy belonging relation) obviously satisfies the principle II and III, but the principle IV.

Proposition 2 The principles I, II and IV do not imply the principle III. This fact can be seen in the argument on the independence of the principle III of Theorem 2.

REFERENCES

Goguen, J.A.(1967). L-fuzzy sets. J. Math. Anal. Appl., 18, 145-174.

Hutton, B.(1977). Uniformity on fuzzy topological spaces. J.Math. Anal. Appl., 58, 559-571.

Liu, Ying-ming,(1981). Compactness and Tychonoff theorem in fuzzy topological spaces. Acta Math. Sinica, 24, 260-269. (in Chinese).

Liu, Ying-ming (1982a). On fuzzy Stone-Cech compactification, Kexue Tongbao, 27, 799.

Liu, Ying-ming (1982b). Neighborhood structures in fuzzy topological spaces. Kexue Tongbao, 27, 1243-1244.

Liu, Ying-ming (1983). A pointwise Characterization of complete regularity and imbedding theorem in fuzzy topological spaces. Scientia Sinica, Series A, 243-253.

Pu Paoming and Liu Ying-ming (1980a). Fuzzy topology I, Neighborhood Structures of a fuzzy points and Moore-Smith convergence. J.Math. Anal. Appl., 76, 571-599.

Pu Paoming and Liu Ying-ming (1980b). Fuzzy topology II, product and quotient spaces. J.Math.Anal.Appl., 77, 20-37.

Wang, Guo-jun (1979). Topological molecular lattices. Sanshi Sida Xuebao, 1, 1-15. (in Chinese).

Wang, Guo-jun (1983). A new fuzzy compactness defined by fuzzy nets. J.Math. Anal. Appl., (to appear).

Wong, C.K., (1974). Fuzzy points and local properties of fuzzy topology. J.Math. Anal. Appl., 46, 316-382.

Zadeh, L.A. (1965). Fuzzy sets. Inform. Control, 8, 338-353.

EMBEDDING A FUZZY ORDERING INTO A FUZZY-LINEAR FUZZY ORDERING (SZPILRAJN-MARCZEWSKI-LIKE THEOREMS)

N. Blanchard

Département de Mathématiques, Université Claude-Bernard — Lyon I, F 69622 Villeurbanne Cedex, France

Summary : The aim of the paper is a critical study of various systems of axioms defining a fuzzy ordering on a fuzzy set. This study is done first by the way of an assessment of the theorems concerning the embedding of any fuzzy ordering into a (weakly, strongly) fuzzy-linear fuzzy ordering, the extension being a good one or not, and then by the way of an assessment of the systems of axioms in which any fuzzy ordering may be identified with the intersection of all the (weakly, strongly) fuzzy-linear fuzzy orderings which extend it.

Une version française de l'article est disponible auprès de l'auteur.

Keywords : Fuzzy relations ; fuzzy orderings ; fuzzy chains ; Szpilrajn-Marczewski.

AMS classification : 03E72 ; secondary classification : 06A05.

INTRODUCTION

As above mentioned, the matter of this paper is giving some fuzzy equivalents of Szpilrajn-Marczewski's theorem : "any ordering may be embedded into a linear ordering". Let us precise the intentions of the author. Szpilrajn's theorem is a very powerful tool when studying ordered sets and the relations theory. It would be agreeable and useful to have such a tool when studying fuzzy orderings on fuzzy sets. Nevertheless, having such a tool was not the initial motivation of this work. The author wished to be able to judge the well-foundedness of various systems of axioms defining a fuzzy ordering on a fuzzy set and some related notions - Such a judgement is basically supported by three criteria:

a) The first one is the appropriateness of the axioms to the needs of the users. Such an appropriateness is essential, but judging it is not the author's business.
b) There exist some (mathematical) natural examples of a fuzzy ordering. It is desirable that these natural models satisfy the chosen axioms. This will be the second criterion.
c) The above study being done and leaving one free to choose various systems, we shall test the skills of the accepted systems for being good mathematical tools, in other words, their ability of giving basic theorems. That will be the matter of this paper.

Let us say at once that the author failed in this project : several systems of axioms give rise to Szpilrajn-Marczewski - like theorems. Nevertheless, there is no matter of being disappointed. Just as different (non equivalent) definitions of a fuzzy topology must be chosen in various occasions, it is clear that different notions of fuzzy orderings must be used in order to modelize concrete situations of different nature. In the multiplicity of possible systems of axioms, let us see only an additional proof of the richness of the theory of fuzzy sets.

I - THE POSSIBLE AXIOMS FOR A FUZZY ORDERING

Some part of the below mentioned axioms were given elsewhere (Blanchard, 1982) and their reasons for being were explained. Nevertheless, for the reader convenience, we reproduce here the initiated discussion, so as some new notions. Obviously a fuzzy ordering on a fuzzy set is a *fuzzy relation* which satisfies some axioms of *fuzzy reflexivity*, *fuzzy antisymetry fuzzy transitivity*. Now we must give some sense to the four words we used.

In all what follows, $\tilde{A} = (A, \alpha)$ will denote some fuzzy set, i.e the datum of an (ordinary) set A and of an (ordinary) function α from A to some complete distributive lattice J . We denote $N_u(\tilde{A})$ the α-cut $\{x \in A/\alpha(x) \leq u\}$.

I-1 Definition

We call fuzzy relation on the support A of $\tilde{A}$ the datum of an (ordinary) function R from A×A to J.

I-2 Definition

We call fuzzy relation on the fuzzy set $\tilde{A}$ the datum, for any $(x,y) \in A \times A$, of a monotonic function :
$\varphi(x,y) : t \to \varphi(x,y)(t) = R_t(x,y)$ from $[0, \alpha(x) \wedge \alpha(y)]$ to J.

I-3 Commentary :

I-1 is a very manageable notion, but the relation in question is a fuzzy relation on A and not a fuzzy relation on $\tilde{A}$. Its interpretation is evident : $R(x,y) = u$ means that the degree of plausibility of x being related to y is equal to u. This definition is convenient for providing the set A simultaneously with a fuzzy structure $\tilde{A}$ and with a fuzzy ordering R, $\tilde{A}$ and R being linked by the further mentioned axioms.

It is more difficult to define the notion of a fuzzy relation on a *fuzzy* set $\tilde{A}$ = the notion which seems extremely natural and which is used notably by Cerruti (1982) : R is a function from $A \times A$ to J satisfying $R(x,y) \leqslant \alpha(x) \wedge \alpha(y)$ (i.e. the graph of R is a fuzzy subset of $\tilde{A} \times \tilde{A}$), this notion generally cannot be used in order to define a fuzzy ordering for reasons which will appear in § I-5. The interpretation of definition I-2 is the following : given x and y in $\tilde{A}$, they belong to any α-cut $N_t(\tilde{A})$, for any choice of $t \leqslant \alpha(x) \wedge \alpha(y)$. Then writing $\varphi(x,y)(t) = u$ (or $R_t(x,y) = u$) means : when you consider that x and y are in the α-cut $N_t(\tilde{A})$, the degree of plausibility for x being related to y is u. With regard to the monotonicity imposed to $\varphi(x,y)$, it means : • either we consider that, when $t > t'$ (and thus $N_t(\tilde{A}) \subseteq N_{t'}(\tilde{A})$), we are more demanding in N_t than in $N_{t'}$ before saying that xRy, and thus $\varphi(x,y)(t) \leqslant \varphi(x,y)(t')$, •or we consider that when $t > t'$ we discriminate in N_t some relations that we did not discriminate in $N_{t'}$, and then $\varphi(x,y)(t) \geqslant \varphi(x,y)(t')$.

Definition I-2 is interesting principally every where one introduces first a fuzzy structure $\tilde{A}$ and then a fuzzy ordering. Let us remark that I-1 is a particular case of I-2, where $\varphi(x,y)$ is a constant function from $[0, \alpha(x) \wedge \alpha(y)]$ to J that we identify to some number $R(x,y)$.

I-4 Definition

A fuzzy relation R on A is said to be i-fuzzy-reflexive if and only if it satisfies axiom R_i where :

R1 = R2 = R3 $\quad R(x,x) \geqslant \alpha(x)$

R4 $\quad R(x,x) = 1$

I-5 Definition

The fuzzy relation R on A is said to be i-fuzzy-antisymetric if and only if it satisfies the axiom A_i, where :

A1 $\quad R(x,y) \wedge R(y,x) > 0 \Rightarrow x=y$;

A2 $\quad R(x,y) \wedge R(y,x) \geqslant \alpha(x) \wedge \alpha(y) \Rightarrow x=y$;

A3 $\quad [R(x,y) \wedge R(y,x) > \{t / t < \alpha(x) \wedge \alpha(y)\}] \Rightarrow x=y$;

A3* $\quad [R(x,y) \wedge R(y,x) > \{t / t < \alpha(x) \vee \alpha(y)\}] \Rightarrow x=y$;

A4 $\quad R(x,y) \wedge R(y,x) = 1 \Rightarrow x = y$.

Axiom A3 may be more pleasantly written :
when J is a complete dense chain :
A'3 $\quad R(x,y) \wedge R(y,x) > \alpha(x) \wedge \alpha(y) \Rightarrow x=y$;
When J is a finite chain :
A"3 $\quad R(x,y) \wedge R(y,x) \geqslant \alpha(x) \wedge \alpha(y) \Rightarrow x = y$.

I-6 Definition

A fuzzy relation R on A is said to be i-transitive if and only if it satisfies axiome T_i where :

T1 $\quad R(x,z) \geqslant \bigvee_{y \in A} R(x,y) \wedge R(y,z) \wedge \alpha(x) \wedge \alpha(y) \wedge \alpha(z)$;

T2 $\quad \begin{matrix} R(x,y) \geqslant \alpha(x) \wedge \alpha(y) \\ R(x,y) \geqslant \alpha(y) \wedge \alpha(z) \end{matrix} \Rightarrow R(x,z) \geqslant \alpha(x) \wedge \alpha(z)$;

T3=T4 $\quad R(x,y) \geqslant \bigvee_{\alpha(y) \geqslant \alpha(x) \wedge \alpha(z)} R(x,y) \wedge R(y,z)$;

T3$_\star$ $\quad R(x,y) \geqslant \bigvee_{\alpha(y) \geqslant \alpha(x) \vee \alpha(z)} R(x,y) \wedge R(y,z)$

T3* $\quad R(x,z) \geqslant \bigvee_{y \in A} R(x,y) \wedge R(y,z)$.

I-7 Definition

A fuzzy relation R on A is said to be an i-fuzzy ordering on $\tilde{A}$, or, as well, R is said to satisfy system i, if and only if R satisfies R_i, A_i, T_i.

I-8 Commentary :

In actual fact, systems 1.2.3.4 are fuzzy orderings defined on the support A of $\tilde{A}$ which are strongly connected to the fuzzy structure of $\tilde{A}$. The axioms A_i express the degree up to which one agrees to go against a perfect antisymetry (A1) considering the feeble degrees of membership of x and y. The axioms T_i express in what cases one agrees to go against a perfect transitivity (T3*), either considering the feeble degrees of membership of the elements, or demanding the "passing in transit" only for elements of great level of fuzziness (T3) ; cf also §I-9. System 1 is constructed in order that the relations $\leqslant_u$ defined on the α-cuts $N_u(\tilde{A})$ by $x \leqslant_u y$ iff $R(x,y) \geqslant u$ be an inductive system of ordinary orderings ; system 2 is constructed in order to the relations $x \leqslant_u y$ iff $R(x,y) \geqslant \alpha(x) \wedge \alpha(y)$ be an inductive system. System 3 corresponds to my intuitive idea of constructing a fuzzy ordering *after* setting a fuzzy structure $\tilde{A}$ on the set A. On the one hand Zadeh (1971), on the other hand Ovchinnikov (1981) considered some axioms of reflexivity, antisymetry, transitivity for a fuzzy relation which (when considering the particular case of a crisp fuzzy structure on A) coincide with R1,R2,R3,R4,A1,A2,A3,A3*, A4,T1,T2,T3$_\star$,T3*,T4 (but not with T2).

I-9 Fuzzy orderings derived from I-2

The axioms of fuzzy reflexivity, fuzzy antisymetry, fuzzy transitivity which are naturally linked to I-2 are :

$\varphi(x,x) \geqslant id_{[0,\alpha(x)]}$

$\varphi(x,y) \wedge \varphi(y,x) \geqslant id_{[0,\ \alpha(x) \wedge \alpha(y)]} \Rightarrow x = y$

$\varphi(x,z) \geqslant \bigvee_{\alpha(y) \geqslant \alpha(x) \wedge \alpha(z)} \varphi(x,y) \wedge \varphi(y,z).$

When one inerprets $\varphi(x,y)(t) = u$ by saying: "In $N_t(\hat{A})$ x is discriminated from y up to degree u", the axiom of antisymetry means that whenever we cannot discriminate x from y in any level t better than with degree t, x and y are equal (indiscriminable). The condition "$\alpha(y) \geqslant \alpha(x) \wedge \alpha(z)$" which occurs in transitivity may be interpreted by saying that, when two functions are equal or comparable, their supports $[0,\ \alpha(x) \wedge \alpha(z)]$ and $[0,\ \alpha(x) \wedge \alpha(y)] \cap [0,\ \alpha(y) \wedge \alpha(y)]$ must be equal.

I-10 Definition

An i-fuzzy ordering S on a fuzzy set $\tilde{A}$ is said to be an i-reinforcement (or i-extension) of the j-fuzzy ordering R iff $S(x,y) \geqslant R(x,y)$ for any x and y in A. We also say that the fuzzy ordering R is embedded in the fuzzy ordering S.

I-11 Definition

An i-reinforcement S of a j-fuzzy ordering R is said to be a good reinforcement (or a good extension) of R iff :

$R(x,y) \neq 0 \Rightarrow [S(x,y) = R(x,y) \text{ and } S(x,y) = R(y,x)]$

I-12 Definition

An i-fuzzy ordering R is said to be strongly fuzzy-linear iff : $R(x,y) \vee R(y,x) \geqslant \alpha(x) \wedge \alpha(y)$ for any x and y in A.

I-13 Definition

An i-fuzzy ordering R is said to be weakly f.-linear iff $R(x,y) \vee R(y,x) > 0$ for any x and y in A.

II - THE LINK BETWEEN THE VARIOUS SYSTEMS OF AXIOMS

II-1 The obvious facts :

R4 ⇒ R2 = R3　　　$T3^{\star}$　　　↗↗ T1

A_1 ⇒ A2 ⇒ A3 ⇒ A4　　　↘↘ T3 ⇒ $T3_{\star}$

↓↓

$A3^{\star}$

II-2 Theorem

When J is a complete chain, any 3-fuzzy ordering (R3, A3, T3) may be embedded into a reinforced 3-fuzzy ordering (R3, A3, $T3^{\star}$).

Proof : We introduce the transitive envelope S of R :

$S(x,y) = \bigvee_{n \in N^{\star},\ t_0 = x,\ t_n = y} R(t_0,t_1) \wedge \ldots \wedge R(t_{n-1},t_n)$

The antisymetry of S may be proved by verifying first that :

$S(x,y) = [\bigvee_{\alpha(t) < \alpha(x) \wedge \alpha(y)} R(x,t) \wedge R(t,y)] \vee R(x,y).$

For details, see Blanchard (1982).

II-3 Theorem :

When J is a complete chain, any 4-fuzzy ordering may be embedded in a reinforced 4-fuzzy ordering (R4, A4, $T4^{\star} = T3^{\star}$).

III- THE THEOREMS OF EMBEDDING ANY FUZZY ORDERING INTO A FUZZY-LINEAR FUZZY ORDERING.

III-1 Theorem :

When the lattice J is entire (the infimum of two non-zero elements is a non-zero element), any 1-fuzzy ordering may be reinforced into a strongly fuzzy-linar fuzzy ordering.

Proof : The binary relation $\leqslant$ defined on A by $x \leqslant y$ iff $R(x,y) > 0$ is an ordering. This ordering may be reinforced into a linear ordering $\precsim$ (thanks to Szpilrajn-Marczewski's theorem). Then the fuzzy binary relation S which is defined on A by:

$S(x,y) = R(x,y) \vee (\alpha(x) \wedge \alpha(y))$ si $x \precsim y$,

$S(x,y) = R(x,y)$ si $x \not\precsim y$.

actualizes an extension of R which is a strongly fuzzy linear 1-fuzzy ordering.

III-2 Remark : The above proof uses Szpilrajn Marczewski's theorem, and so the axiom of choice. For an actual construction of $\precsim$, then of S, when A is finite, take pattern by III-5.

III-3 Remark

$R(x,y) \geqslant \alpha(x) \wedge \alpha(y) \Rightarrow$

$[S(x,y) = R(x,y)$ et $S(y,x) = R(y,x)]$

III-4 Theorem :

Any 2-fuzzy ordering may be reinforced into a strongly linear 2-fuzzy ordering.

Proof : The same as in III-1 (with different calculations !), and the same remarks.

III-5 Theorem :

When J is a complete chain, any 3-fuzzy ordering satisfying a reinforced axiom of antisymetry ($A3^{\star}$) may be reinforced into a strongly fuzzy-linear fuzzy ordering (the reinforcement also satisfies $A3^{\star}$).

Proof of the theorem and construction of the reinforcement when A is finite

A detailed proof was given in Blanchard (1982) A resume of its may be the following : Step 1: the ordinary binary relation defined on A by $x \leqslant y$ iff $(R(x,y) > \alpha(x) \wedge \alpha(y)$ or $x = y)$ is reflexive and acyclic ; step 2 : this relation may be reinforced into a linear ordering on A ; step 3 : if a and b are note strongly comparable, and if $a \precsim b$, then the relation S defined by

$$S(x,y)=R(x,y)\ [R(x,a)\wedge\alpha(a)\wedge\alpha(b)\wedge R(b,y)]$$

is a reinforcement of R which makes a and b strongly comparable (the existence of $\precsim$ is used to proove the fuzzy antisymetry of S) ; step 4 : the theorem is valid when A is finite (by iteration) ; step 5 : the theorem is valid for any set A (by logic compacity). Such a proof uses in step 2 Szpilrajn-Marczewski's theorem. In other words, it gives no construction. When one needs an actual construction, I suggest, when A is finite to replace steps 1 and 2 above mentioned by the detailed following construction, wich takes pattern by a classical one which is used by Zadeh (1971). Unfortunately, so as the proof propounded in Blanchard (1982), this proof is valid only when J is a complete chain.

Construction of S : We assume that A is finite. Let us regard the binary relation (which generally is not an ordering) $\leqslant$ defined on A by $x \leqslant y$ iff $(x = y$ or $R(x,y) > \alpha(x) \wedge \alpha(y))$. Let C_0 be the set of minimal elements for this relation. Then C_0 is not void : let $x_1 \in A$ such that $\alpha(x_1)$ be minimal among the $\alpha(x)$. If x_1 is minimal for $\leqslant$, then C_0 is not void ; if x_1 is not minimal, then there exists $y \neq x_1$ such that $R(y,x_1) > \alpha(y)\ \ \alpha(x_1)$. Let x_2 be of smallest degree of membership among the degrees of membership of such y. Then :

$R(x_2,x_1) > \alpha(x_1) \wedge \alpha(x_2)$ and

$R(y,x_1) > \alpha(y) \wedge \alpha(x_1) \Rightarrow \alpha(y) \geqslant \alpha(x_2)$.

by iteration, when knowing x_k , we construct, if x_k is not in C_0 , x_{k+1} such that

$R(x_{k+1},x_k) > \alpha(x_k) \wedge \alpha(x_{k+1})$ and

$R(y,x_k) > \alpha(y) \wedge \alpha(x_k) \Rightarrow \alpha(y) \geqslant \alpha(x_{k+1})$.

We proove that the so constructed x_k are all different, which proves (A being finite) that the process must stop, in other words that x_k belongs to C_0 for some k . We entirely construct C_0 by the same process, with every element of A playing the role of x_1 , one after another. In an analogous manner, we call C_1 the set of the minimal elements of $D_1 = A\ \ C_0$, C_2 the set of the minimal element of $D_2 = D_1\ \ C_1$ and so on. The so-constructed C_k are a disjoint covering. By construction, if $x \in C_k$ and $y \in C_\ell$ and $\ell \geqslant k$, $y \neq x$, then $R(y,x) < \alpha(x)\ \ \alpha(y)$. Let us rank the elements of A, taking first the elements of C_0 with an arbitrary ordering, then the elements of C_1 , and so on. So we have an actual construction of an ordering $\precsim$ on A reinforcing $\leqslant$. So let $A = \{x_1, x_2, \ldots, x_n\}$ (where $i \leqslant j \Rightarrow x_i \precsim x_j$) ; the proof may be continued as in Blanchard (1982), the steps 3 and 4 being possibly made more systematical (and computable) by making first x_2 comparable with x_1, then x_3 comparable with x_2 , and so on . The (very little) following example will show the process :

Let $J = [0,1]$ and let $\tilde{A} = (A, \alpha)$ and R be defined by :

$$: \alpha(x) = 33 \ ; \ \alpha(y) = 1 \ ; \ \alpha(z) = 5 \ ;$$

$$R(x,x)=1 \ ; \ R(y,y)=1 \ ; \ R(z,z)=.25 \ ; R(x,y)=0;$$

$$R(y,x)=.3 \ ; \ R(x,z)=R(z,x)=.25 \ ; \ R(y,z)=.25 \ ;$$

$$R(z,y)=0.$$

$$\begin{array}{c} \\ x \\ y \\ z \end{array} \begin{array}{c} x \quad\quad z \\ \begin{pmatrix} 1 & 0 & 1 \\ .3 & 1 & .25 \\ .25 & 0 & .5 \end{pmatrix} \end{array}$$

matrix of R

$$\begin{array}{c} \\ x \\ y \\ z \end{array} \begin{array}{c} x \quad\quad y \quad\quad z \\ \begin{pmatrix} 1 & 0 & 1 \\ 1 & 1 & .3 \\ .25 & 0 & .5 \end{pmatrix} \end{array}$$

Strong transitivised matrix

$$\begin{array}{c} \\ x_1=y \\ x_2=x \\ x_3=z \end{array} \begin{array}{c} x_1=y \quad x_2=x \quad x_3=z \\ \begin{pmatrix} 1 & 1 & .3 \\ 0 & 1 & 1 \\ .25 & 0 & .5 \end{pmatrix} \end{array}$$

matrix of S after a new ranking of A.

$$\begin{array}{c} \\ x_1=y \\ x_2=x \\ x_3=z \end{array} \begin{array}{c} x_1=y \quad x_2=x \quad x_3=z \\ \begin{pmatrix} 1 & 1 & 1 \\ 0 & 1 & 1 \\ .25 & .25 & .5 \end{pmatrix} \end{array}$$

matrix of the strong fuzzy-linear reinforcement T of R.

III-6 Theorem :

When J is a complete chain, any 4-fuzzy ordering may be reinforced into a strongly fuzzy-linear 4-fuzzy ordering.

Proof : The same plan as in III-5, reduced to setps 3,4,5.

IV- THE THEOREMS OF WELL -EXTENDING A FUZZY ORDERING INTO A WEAKLY FUZZY-LINEAR FUZZY ORDERING

IV-1 Remark : In this paragraph, we shall come up against some difficulties when trying to extend to the case of fuzzy sets with infinite support some results which are valid for fuzzy sets with finite support. Generally, it will be possible extending them provided that : $\bigwedge\{R(x,y)/x \in A,\ y \in A,\ R(x,y) \neq 0\} \neq 0$. Let us remark that without this condition, there is no theoretical objection against the existence of a good reinforcement of R into a weakly fuzzy-linear fuzzy ordering nevertheless there is some theoretical objection against the existence of a reinforcement where the "reinforced" values remain smaller than the non-null initial values.

IV-2 Theorem :

If $m = \{R(x,y)/x \in A, y \in A, R(x,y) \neq 0\}$ is not null, then any i-fuzzy ordering on $\tilde{A}$ may be extended into a weakly fuzzy-linear i-fuzzy ordering.

Proof : We consider again the ordinary linear ordering introduced in III.1 , III.4, or III.5 according to the studied fuzzy ordering. In any case we proof that S defined by :

$$S(x,y) = R(x,y) \vee m \quad \text{if} \quad x \gtrless y$$
$$= R(x,y) \quad \text{if} \quad x \not\lessgtr y$$

is convenient.

IV-3 Remark : The above condition is satisfied especially when J is an entire distributive lattice (the infimum of two non-zero elements is a non-zero element)and $\alpha(A)$ is finite (which is automatically actualized whenever J is finite or A is finite).

V - THE THEOREMS OF EMBEDDING ANY FUZZY ORDERING INTO A PRODUCT OF FUZZY CHAINS.

V-1 Theorem

Any 2-fuzzy ordering is equal to the intersection of all the strongly fuzzy-linear 2-fuzzy orderings which extend it.

Proof : Let x and y belong to A. Let us show that there exists some reinforcement S of A into a strongly fuzzy-linear fuzzy ordering S satisfying $S(x,y) = R(x,y)$.

a) If $R(x,y) \vee R(y,x) \geqslant \alpha(x) \wedge \alpha(y)$, then any reinforcement constructed as in III-4 is convenient.
b) If $R(x,y) \vee R(y,x) \not\geqslant \alpha(x) \wedge \alpha(y)$ (and then x and y non comparable for $\leqslant$ defined on A in § III-4, then we reinforce ordering $\leqslant$ on A into a linear ordering $\leqslant_1$ satisfying $y <_1 x$, and then we continue as in III-4.

A fortiori :

V-2 Theorem :

Any 2-fuzzy ordering is the intersection of all the weakly fuzzy linear 2-fuzzy orderings which extend it.

V-3 Theorem

Any 3-fuzzy ordering is the intersection of all the weakly fuzzy-linear fuzzy orderings which extend it.

Proof : See Blanchard (1982)

VI - OPEN QUESTIONS :

Does there exist some theorem of extending any 3-fuzzy ordering into a strongly linear 3-fuzzy ordering without assuming a reinforced antisymetry ($A3^{\star}$). Especially, may the (weakly fuzzy-linear) fuzzy ordering on the fuzzy integers (Blanchard, 1981) be embedded into a strongly fuzzy-linear 3-fuzzy ordering ?

BIBLIOGRAPHY :

Blanchard, N.(1981). Theories cardinale et ordinale des ensembles flous. *Ph. D. Thesis* - *Lyon*

Blanchard, N.(1982). A few fuzzy theorems analougous to some classical theorems concerning ordered sets. In E.P. Klement (Ed.), *Proceedings of the 4th international seminar on fuzzy set theory*. Johannes Kepler Universität, Linz, Austria.

Bonnet, R., and M. Pouzet (1982). Linear extensions of orderd sets. In I. Rival (Ed.), *Ordered sets Reidel Publishing* company, pp. 125-170

Cerruti, U. (1982). Graphs and Fuzzy Graphs. In M. Gupta and E. Sanchez (Ed.), *Fuzzy information and decision processes*. North Holland. pp. 123-131.

Ovchinnikov, S.V.(1981). Structure of fuzzy binary relations. *Fuzzy sets and systems* vol 6, N°2.

Zadeh, L.A. (1971). Similarity relations and fuzzy orderings. *Inf. Sci.* Vol. 3. pp. 117-200

FUZZY STATEMENT FORMATION BY MEANS OF LINGUISTIC MEASURES

Y. Tsukamoto and Y. Hatano

Department of Control Engineering, Tokyo Institute of Technology, Meguro-ku, Tokyo, Japan

Abstract. The new concept of linguistic measure is introduced by fuzzifying the parameter in the isomorphic mapping transforming probability to λ-fuzzy measure. Linguistic measures are defined as fuzzy subsets of [0,1] and their labels are related to several natural languages used under uncertainty. The mathematical developments are described around the triad consisting of probability, linguistic measure and linguistic truth value. The applicability of the concept is suggested by considering fuzzy representation problems.

Keywords. Probability; Fuzzy measures; Isomorphism; Linguistic Measure; Linguistic Truth Value, Fuzzy Representation problem.

INTRODUCTION

In everyday life we use many types of expressions under uncertainty, e.g., It is possible that ..., It is doubtful that ..., It is not very probable that ..., etc. This paper is concerned with the representation problem of uncertainty including both probability and subjective standpoint. Particularly, we consider the problem of finding an appropriate expression, provided that the probability of some fuzzy or nonfuzzy event is given. There are not a few background mathematics related to the present study. In addition to the ordinary probability theory and possibility theory, the recent study by Banon has enabled us to deal with many sorts of fuzzy measures in a more systematic way. Further there have appeared some of applicational studies of the triangular norms. Schwyla and, more recently, Kruse have investigated the isomorphism between additive measures and fuzzy measures.

The concept of linguistic measure is introdeced in between probabilistic uncertainty and subjective attitudes. According to Zadeh 's context of linguistic variable, it means a linguistic value taken by the linguistic variable the name of which is subjective belief or doubt. The mathematical formalization of linguistic measures is closely related to the existence of the isomorphism between probability and fuzzy measures. In particular the multicative generator of T_S-norm investigated by Frank is the isomorphic mapping of probability measure into λ-fuzzy measure proposed by Sugeno, which plays a very important role in this paper.

By experiment, the definitions of the primary terms of linguistic measures are given as fuzzy subsets of [0,1]. Finally, as an applicational study, a method of composing fuzzy sentences automatically is presented.

MATHEMATICAL PRELIMINARIES

Let R be the set of real numbers and $\mathcal{B}$ Borel field of R. In this study set function means every mapping of $\mathcal{B}$ to the real interval [0,1] unless otherwise specified. Let ∅ denote empty set. First we define two kinds of measures that are characterized by monotonicity.

[Def.1] C-measure is defined as a set function having the properties;

$$C(\emptyset) = 0 \quad (2.1.a)$$
$$C(R) = 1 \quad (2.1.b)$$
$$A \supset B \Rightarrow C(A) \geqq C(B). \quad (2.1.c)$$

[Def.2] D-measure is a set function having the properties;

$$D(\emptyset) = 1 \quad (2.2.a)$$
$$D(R) = 0 \quad (2.2.b)$$
$$A \supset B \Rightarrow D(A) \leqq D(B). \quad (2.2.c)$$

In the above, C(A) and D(A) may be regarded as the degrees of subjective belief and doubt ,respectively, in the proposition, "$W \in A$", where W is an indicated variable. Then, probabilty measure is a C-measure having the properties;

$$\forall A,B \in \mathcal{B},\ A \cap B = \emptyset : P(A \cup B) = P(A) + P(B) \quad (2.3.a)$$
$$(A_n \uparrow A) \Rightarrow (P(A_n) \uparrow P(A)). \quad (2.3.b)$$

λ-fuzzy measure denoted by G_s is also a C-measure characterized by ,for $\forall s \in]0,\infty[$,

$$\forall A,B \in \mathcal{B},\ A \cap B = \emptyset : G_s(A \cup B) = G_s(A) + G_s(B) + (s-1)G_s(A)G_s(B) \quad (2.4.a)$$

$$(A_n \uparrow A) \Rightarrow (G_s(A_n) \uparrow G_s(A)). \quad (2.4.b)$$

Note that the parameter is denoted by s, but not λ that is usually used.

[Def.3] Let F and G be two set functions. Then we say that G is quasi-isomorphic to F whenever

$$\exists_{\phi} \in G(\mathcal{B})^{F(\mathcal{B})} : G = \phi \circ F \qquad (2.5)$$

[Def.4] We say that there exists isomorphism between G and F if the function, ϕ, in the definition 3 is continuous and bijective.

[Proposition 1] For each $s \in]0,\infty[$, there exists a probability distribution which is isomorphic to a given distribution of λ-fuzzy measure.

(As for the proof of this proposition, see /Schwyla/, /Kruse/ or /Tsukamoto/)

In effect, it is easy to see that the following function satisfies the conditions as stated in the definition 4,

$$\phi_s(u) = \begin{cases} \dfrac{s^u - 1}{s - 1} & \text{if } s \neq 1 \\ u & \text{if } s = 1 \end{cases} \qquad (2.6)$$

Thus a probability system for a given λ-fuzzy measure can be derived from the following equation as;

$$\forall A \in \mathcal{B} : P(A) = \phi_s^{-1}(G_s(A)). \qquad (2.7)$$

Conversely, if a probability distribution is given we can also find the corresponding λ-fuzzy measure by the following,

$$\forall A \in \mathcal{B} : G_s(A) = \phi_s(P(A)). \qquad (2.8)$$

Let us now define two special C-measures.

[Def.5] 0-1 possibility measure focused on $B \in \mathcal{B}$ is denoted by Π_B^* and is defined as a set function from $\mathcal{B}$ to {0,1} having the properties; for $\forall A \in \mathcal{B}$,

$$\Pi_B^*(A) = \begin{cases} 1 & \text{if } A \cap B \neq 0 \\ 0 & \text{otherwise.} \end{cases} \qquad (2.9)$$

[Def.5] 0-1 necessity measure focused on $B \in \mathcal{B}$ is a C-measure defined by, for $\forall A \in \mathcal{B}$,

$$N_B^*(A) = \begin{cases} 1 & \text{if } A \supset B \\ 0 & \text{otherwise.} \end{cases} \qquad (2.10)$$

Let a probability distribution be given by a probability density function denoted by ρ, and let T be the support set as;

$$T = \{x \in R : \ (x) > 0\}. \qquad (2.11)$$

Then, we have

[Proposition 2] Π_T^* is quasi-isomorphic to P through the mapping given by

$$\phi_0(u) = \lim_{s \to 0} \phi_s . \qquad (2.12)$$

(Proof) By Eq.(2.6) we obtain

$$\phi_0(u) = \begin{cases} 1 & \text{if } u > 0 \\ 0 & \text{if } u = 0. \end{cases} \qquad (2.13)$$

On the other hand, by the definition 5,

$$\Pi_T^*(A) = \begin{cases} 1 & \text{if } A \cap T \neq \emptyset \\ 0 & \text{if not} \end{cases}$$

,so that we can write as

$$\forall A \in \mathcal{B} : \Pi_T^*(A) = \phi_0(P(A)) \text{ a.e.} \qquad (2.14)$$

□

[Proposition 3] N_T^* is quasi-isomorphic to P by

$$\phi_\infty \triangleq \lim_{s \to \infty} \phi_s . \qquad (2.15)$$

(Proof) Since

$$\phi_\infty(u) = \begin{cases} 1 & \text{if } u = 1 \\ 0 & \text{if } u < 1, \end{cases} \qquad (2.16)$$

we have by the definition 6

$$\forall A \in \mathcal{B} : N_T^*(A) = \phi_\infty(P(A)). \qquad \square \ (2.17)$$

[Lemma 1] ϕ_s defined by (2.6) is strictry increasing with respect to u for $\forall s \in]0,\infty[$.

Now let

$$\psi_u(s) \triangleq \phi_s(u) \qquad \text{for } \forall s \in [0,\infty], \ \forall u \in [0,1], \qquad (2.18)$$

where ϕ_0 and ϕ_∞ are defined by Eqs. (2.13) and (2.16), respectively.

[Lemma 2] ψ_u is a decreasing function of s for all $u \in]0,1[$.

To discuss such negative words as impossible, incredible, unbelievable, dubious, or doubtful etc., we have to prepare some of d-measures as given in the definition 2. Let η_s denote Sugeno's negation as ;

$$\eta_s : [0,1] \to [0,1]$$

$$\eta_s(y) = (1-y) / (1+(s-1)y) , \ \forall s \in]0,\infty[\qquad (2.19)$$

[Def.7] The set function given by

$$\forall A \in \mathcal{B} : D_s(A) = \eta_s(G_s(A)) , \ \text{for each } s \in]0,\infty[\qquad (2.20)$$

are D-measure and we say that D_s is dual to G_s.

[Proposition 4] There exists a λ_d-measure distribution that is isomorphic to a given λ-fuzzy measure by η_s and also to a probability distribution by $\eta_s \circ \phi_s$ for all $s \in]0,\infty[$.

(proof) The proof is quite easy and is ommited.

Futhermore, considering the polar cases, we define two sorts of D-measures as ;

$$\exists B \in \mathcal{B}, \ \forall A \in \mathcal{B} : I_B^*(A) = \begin{cases} 1 & \text{if } A \cap B = \emptyset \\ 0 & \text{if otherwise.} \end{cases} \qquad (2.21)$$

$$\forall A \in \mathcal{B} : \Xi_B^*(A) = \begin{cases} 1 & \text{if } A \not\supset B , \\ 0 & \text{if otherwise.} \end{cases} \qquad (2.22)$$

Let T be the nonfuzzy set given by (2.11). Then , in the similar way as stated in the proposition 2 and 3, these two D-measures are quasi-isomorphic to P and we can write as ;

$$\forall A \in \mathcal{B} : I_T^*(A) = \phi_\infty((1-P(A))) \ (\triangleq D_\infty(A)), \qquad (2.23)$$

$$\forall A \in \mathcal{B} : \Xi_T^*(A) = \phi_0((1-P(A))) \ (\triangleq D_0(A)), \qquad (2.24)$$

D_∞ may be called 0-1 impossibility measure, which is dual to 0-1 necessity measure. By the proposition 4 and Eq.(2.18), we summarize more simply as ;

$$\forall A \in \mathcal{B} : D_s(A) = \phi_s(P(\bar{A})) = \psi_{1-P(A)}(s) \quad \text{for } s \in [0,\infty] \qquad (2.25)$$

Let us define θ_u as,

$$\theta_u : [0,\infty] \to [0,1]$$
$$\theta_u(s) = (\eta_s \circ \phi_s)(u) \quad (= \psi_{1-u}(s)). \tag{2.26}$$

[Lemma 3] θ_u is an increasing function of s for all $u \in]0,1[$.

(proof) This lemma follows directly from lemma 2, Eqs.(2.19) and (2.26). □

In the sequel, we use the conventions $1/0=\infty$ $1/\infty=0$, and

$$[x] \triangleq \begin{cases} 1 & \text{if } x=1, \\ 0 & \text{if } 0 \leq x < 1. \end{cases} \tag{2.27}$$

DEFINITION OF LINGUISTIC MEASURES

The values taken by G_s and D_s could be regarded as numerical truth values attached to a probabilistic uncertainty from a subjective view-point to be characterized by the value of the parameter s. In addition to the existence of the isomorphisms as shown in the propositions 1 to 4, what is of key importance for the justification of introducing linguistic measures are the property of the monotonicity as shown in the lemma 2 and 3, and furthermore, the capability of very clear interpretations about the polar cases as shown in propositions 2,3 and (2.23).

The background idea behind linguistic measures is to fuzzify the parameter s appeared in the isomorphic mapping ϕ_s of probability space into fuzzy measure space. However, it seems to be rather difficult to directly handle it since s variates from 0 to ∞. So let us first introduce a continuous and bijective mapping of [0,1] onto $[0,\infty]$, which is denoted by ξ. Let γ_u be the following composite function as

$$\forall u \in [0,1] \; : \; \gamma_u \triangleq \psi_u \circ \xi \tag{3.1}$$

where ψ_u is given by Eq.(2.18). Futhermore, let us pose the following constraint to ξ as

$$\forall h \in [0,1] \; : \; \gamma_u(h) \Big|_{u=0.5} = h, \tag{3.2}$$

then can be defined as follows,

$$\xi : [0,1] \to [0,\infty]$$
$$\forall h \in [0,1] \; : \; \xi(h) = \left(\frac{1-h}{h}\right)^2 \tag{3.3}$$

and the inverse function of ξ is written as

$$\forall s \in [0,\infty] \; : \; \xi^{-1}(s) = \frac{1}{1+\sqrt{s}} \tag{3.4}$$

Here we readily obtain

$$\forall s \in [0,\infty] \; : \; \xi^{-1}(s)+\xi^{-1}(1/s) = 1 \tag{3.5}$$

Similary we define $_u : [0,1] \to [0,1]$ as

$$\delta_u \triangleq \theta_u \circ \xi \qquad \text{for each } u \in [0,1] \tag{3.6}$$

where θ_u is given by Eq.(2.24).

Let us now give the difinition of linguistic measure.

[Def.8] A linguistic measure is a fuzzy set of the real interval, [0,1], and it is denoted by H.

This unit interval is the domain of ξ. In particular, the primary term set of linguistic measures consists of several special fuzzy subsets which are normal convex.

[Def.9] We call linguistic measures LCM and LDM when they are combined with γ_u and δ_u, respectively.

In the above definition, LCM and LDM are the abbreviations from linguistic C-measure and linguistic D-measure, respectively.

[Lemma 4] γ_u(resp. δ_u) is a continuous and strictly increasing function of h for all $u \in]0,1[$. Depending on whether u=1 or 0, it takes the constant values 1 or 0,(resp. 0 or 1) for all $h \in [0,1]$.

The figure of $\gamma_u(\delta_u)$ is shown in Fig. 1.

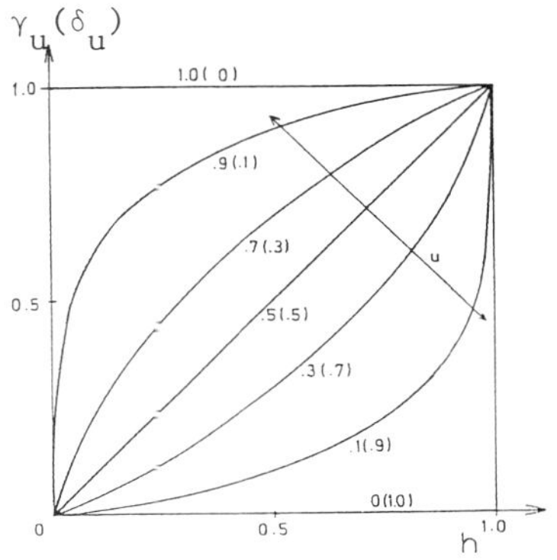

Fig.1 Functions of γ_u and δ_u

[Lemma 5] $\forall u \in]0,1[\; : \; \psi_u^{-1}((1-v)) = 1/(\psi_u^{-1}(v))$ (3.7)

$\forall u \in]0,1[\; : \; \theta_u^{-1}((1-v)) = 1/(\theta_u^{-1}(v))$ (3.8)

(proof) Let

$\psi_u(s) = v$, for $u \in]0,1[$,

then we have by Eq.(2.18) for all $s \in [0,\infty]$

$$\psi_{1-u}(1/s) = \frac{(1/s)^{1-u}-1}{(1/s)-1}$$
$$= 1 - \psi_u(s)$$
$$= 1 - v$$

Considering the inverse function of the above two representations, we readily obtain (3.7). The similar procedure is also valid for the derivation of (3.8). □

Let us now consider the transformation of a linguistic measure H by γ_u. The fuzzy set H will induce another fuzzy set of [0,1] which we denote simply by V. By the application of the extension principle to γ_u, V can be expressed as ;

for each $v \in [0,1]$,

$$\mu_V(v) = \begin{cases} \sup\limits_{h \in \gamma_u^{-1}(v)} \mu_H(h) & \text{if } \gamma_u^{-1}(v) \neq \emptyset, \\ 0 & \text{if } \gamma_u^{-1}(v) = \emptyset \end{cases} \tag{3.9}$$

By the lemma 4, (3.9) can be written as follows ;

for each $v \in [0,1]$,

$$\mu_V(v) = \begin{cases} \mu_H(\xi^{-1}(\psi_u^{-1}(v))) & \text{if } u \in]0,1[\quad (3.10.a) \\ [v] & \text{if } u=1 \quad (3.10.b) \\ [1-v] & \text{if } u=0 \quad (3.10.c) \end{cases}$$

It should be noticed that the fuzzy set V can be regarded as a linguistic truth value proposed by Zadeh. In effect, let us consider the following sentence as ;

"It is L that X is F." (3.11)

where L is a generic label assigned to a

linguistic measure H such as credible, possible, or probable etc., X is the name of some object and F stands for a fuzzy or nonfuzzy predicate. Suppose that the probability that "x is F" is correct takes a rather low value. If we substitute "credible" into L in the above sentence, the truth status of the whole statement, (3.11), will be very low. But, instead, if "possible" is used as L, then the truth status should become high. Thus, the fuzzy set V may be taken as the linguistic truth value representing the truth status of the whole statement, (3.11).

We can also deal with the case where L in (3.11) is given by a negative form. Using the relation as ;

$$\delta_u = \gamma_{1-u} \quad , \qquad (3.12)$$

we have instead of (3.10)
for each $v \in [0,1]$,

$$\mu_V(v) = \begin{cases} \mu_H(\xi^{-1}(\psi_{1-u}^{-1}(v))) & \text{if } u \in]0,1[\quad (3.13.a) \\ [1-v] & \text{if } u=1 \quad (3.13.b) \\ [v] & \text{if } u=0 \quad (3.13.c) \end{cases}$$

Now we show several labels attached to the linguistic measures as depicted in Fig. 2 where $L_i (i=1,2,\ldots,7)$ represent respectively 1-necessary, certain, credible, probable, plausible, possible, and 1-possible.

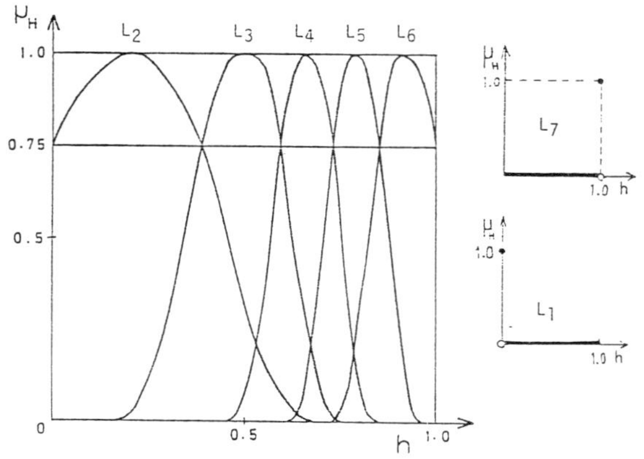

Fig.2 Primary Linguistic Measures

In Fig. 2, L_1 and L_7 stem from the extension of 0-1 necessity measure and 0-1 possibility measure, respectively, and the membership functions are the ones obtained from a simple experiment. In this experiment, fifteen subjects were asked to assign a probability by use of interval to each of the labels from L_2 to L_6. The value of u was set simply by the mean value, and the shape of each membership function was determined by use of S-function so that it satisfies

$$\tau_0 = \gamma_u(L_i) \quad i=2,3,\ldots,6 \quad , \qquad (3.14)$$

and λ-completeness for $\lambda=0.75$, where τ_0 is a special linguistic truth value that represents true.

When considering LDM's in Fig. 2, we may assign to L_i's$(i=1,\ldots,6)$ impossible, unbelievable, dubious, uncertain, questionable, and fallible in ascending order, where L_7 is difficult to be named since such a case very rarely occurs. However, these membership functions still need to be improved by the further experiments. More strictly, the influences of the situation or semantics will have to be taken into consideration for such definitions as done above.

MATHEMATICAL PROPERTIES OF LINGUISTIC MEASURE

Now we can develope some of mathematical properties along the triad composed of probability, linguistic measures, and linguistic truth values. Let $\mathcal{H}$ and $\mathcal{V}$ be the sets of all the linguistic measures and linguistic truth values, respectively, and let ζ denote a mirror image mapping, $\zeta:[0,1]\to[0,1]$, defined by

$$\zeta(\mu_H(h)) = \mu_H(1-h) \quad (\triangleq \mu_{H^*}(h)) \qquad (4.1)$$

[Lemma 6] $\quad u \in [0,1]$

$$\zeta \circ \gamma_u = \gamma_{1-u} \circ \zeta \qquad (4.2)$$

(proof) Let, in the sence of the extension principle,

$V_1 = (\zeta \circ \gamma_u)(H)$ and $V = (\gamma_{1-u} \circ \zeta)(H)$

then we have by Eq.(3.10.a), for each $v \in]0,1[$,

$$\mu_{V_1}(v) = \mu_H(\xi^{-1}(\psi_u^{-1}(1-v)))$$
$$\mu_{V_2}(v) = \mu_H(1-\xi^{-1}(\psi_{1-u}^{-1}(v)))$$

So, it suffices to show that, for each $v \in [0,1]$,

$$\xi^{-1}(\psi_u^{-1}(1-v)) = 1 - \xi^{-1}(\psi_{1-u}^{-1}(v)),$$

but, this is obvious from (3.5) and the lemma 5. It is easy to see by Eqs.(3.10.b) (3.10.c) that the lemma holds when u=1 or 0. □

Using the relation as ;

$$\zeta^{-1} \circ \zeta = \text{Identity Mapping},$$

Eq.(4.2) can be rewritten as ;

$$u \in [0,1] \; : \; \gamma_u = \zeta \circ \gamma_{1-u} \circ \zeta \; . \qquad (4.3)$$

Let

$V(F,H) \triangleq$ L.T.V. of "It is H that X is F." (4.4)

$(F,H,V) \triangleq$ "It is H that X is F" is V. (4.5)

$\overline{F} \triangleq$ the complement of F,

then, the lemma 6 can be expressed simply as ;

$$V^*(F,H) = V(\overline{F},H^*) \qquad (4.6)$$

or, equivalently,

$$(F,H,V) = (\overline{F},H^*,V^*) \qquad (4.7)$$

In order to make clear the meaning of the above equalities, let us consider the case in which F is a nonfuzzy event. We define the two special linguistic truth values as ;

$$\text{c.t.} \triangleq \text{completely true} = \int_{[0,1[} 0/v \cup 1/1 \quad , \qquad (4.8)$$

$$\text{c.f.} \triangleq \text{completely false} = 1/0 \cup \int_{]0,1]} 0/v \quad , \qquad (4.9)$$

where the integral stands for union. For instance, let

H = 1-necessaly

then, (4.7) means that

"F is 1-necessary" is c.t.
= "$\overline{F}$ is 1-possible" is c.f.

Futhermore, in the case where u=1, that is, "X is F" is self-evident, we have

$$\gamma_u |_{u=1} = 1 \quad \text{for all } h \in [0,1] , \qquad (4.10)$$

so that, by Eq.(3.10.b), we obtain

$$V(F,H) = \text{c.t.} \quad \text{for all normal } H \in \mathcal{H} \quad (4.11)$$

What (4.11) means is that if F is asserted with probability one then its truth status must take the highest one from any subjective point of view. The similar discussion could be done for the case when u=0.

Let us now consider the case of linguistic D-measures, where H is replaced by the notation Q.

[Lemma 7] $(F,Q,V) = (\overline{F},Q^*,V^*)$ (4.12)

(proof) It suffices to show that

$$\forall u \in [0,1] : \delta_u = \zeta \circ \delta_{1-u} \circ \zeta , \qquad (4.13)$$

and this is obvious from Eqs.(4.3) and (3.12). □

Now let Q be a dual LDM to H that is LCM. Then we have

[Lemma 8] If Q is dual to H, then

$$(F,H,V) = (F,Q^*,V^*) \qquad (4.14)$$

(proof) By Eqs.(4.3) and (3.12) we have

$$\gamma_u = \zeta \circ \delta_u \circ \zeta ,$$

which shows that the lemma is valid. □

We now obtain an interesting proposition from the above lemmas.

[Proposition 5] If Q is dual to H,

$$F \text{ is } H = \overline{F} \text{ is } Q. \qquad (4.15)$$

(proof) By applying the lemma 7 to the right side of (4.14), we readily have

$$(F,H,V) = (\overline{F},Q,V)$$

, that is,

$$V(F,H) = V(\overline{F},Q) ,$$

which directly represents (4.15). □

For example, since impossible is dual to necessary, the above proposition asserts

F is necessary = $\overline{F}$ is impossible ,

which is one accepted in modal logic as a common sense.

Now suppose that the probability of "X is F" and the linguistic truth value, V(F,H), are given beforehand. Then, the problem od finding the linguistic measure can be solved by use of (3.10), that is,

$$\text{for each } h \in [0,1], \quad \mu_H(h) = \mu_V(\gamma_u(h)) \quad \text{for } u \in]0,1[\qquad (4.16)$$

When the probability is one and zero, we can only have the following special linguistic measures for c.t. and c.f. as ; respectively

(F,1-nec,c.t.) and (F,poss,c.f.) ,

where the latter can be rewritten by lemma 8 as ;

(F,imposs,c.t.)

One can easily extend the above discussions to the case where the probability is given by a fuzzy probability.

COMPOSING FUZZY SENTENCES

The analysis developed so far may be applied to the synthesis problem of composing an appropriate vague sentence under the situation in which only the probabilistic knowledge is given. Here we confine ourselves with a special form of the sentence as given by (3.11). That is, the problem is to find L under some criterion in the following,

It is L that X is F. (5.1)

It would be rather difficult to talk about what is a good sentence or proper information in the case where it includes more or less of uncertainty. In this study two different kinds of criteria are introduced to evaluate the quality of such fuzzy information. First it should be composed so that the truth status of the sentence becomes as high as possible. Second, the sentence to be composed should not have too much redundancy. In this respect one can refer to the paper by Yager where two indexes, truth and specificity, are considered. In the present case we consider the degrees of truth and non-redundancy which are defined as follows;

$$t = \frac{1}{2}(\mathrm{Sup}(V \cap V^*) + \mathrm{Inf}(V \cup \overline{V}^*)) \qquad (5.2)$$

$$n = \frac{1}{2}(\mathrm{Sup}(L \cap L^*) + \mathrm{Inf}(L \cup \overline{L}^*)) \qquad (5.3)$$

where

$$V^* = \int_{[0,1]} v/v \qquad (5.4)$$

$$L^* = \int_{[0,1]} (1-h)/h. \qquad (5.5)$$

In other words, t and n represent the consistency degrees of V and L with, respectively, V* and L*. It is easy to see that these two criteria are in the competitive relation. So let us define the degree of the good quality of a fuzzt statement by the following scaler index as;

$$q = \sqrt[3]{t^2 n} . \qquad (5.6)$$

Then the design problem is reduced to the problem of finding such L that maximizes the index q. Fig.3 shows the relationship between the primary linguistic measures and q with probability as a parameter, where the integers denote in ascending order *necessary*, *very certain*, *certain*, *very credible*, *credible*, *very probable*, *probable*, *very plausible*, *plausible much possible*, *possible* and *1-possible*.
Another alternate method is to find L that maximizes n defined by (5-3) with the constraint as t=t*.
Such algorithms as stated above enable us to compose automatically a fuzzy sentence which seems to be suitable with the vagueness possessed by "X is F". In fact, the methods have

been applied to the fuzzy representation problems in the fault diagnosis with fuzzy information, and to modelling of human decision making process under probabilistic uncertainty /Hatano et al./.
More generally, it is expected that the linguistic measure theory developed in this study can apply to giving appropriate descriptions to numerical outcomes stemming from the applications of fuzzy mathematics to real issues,

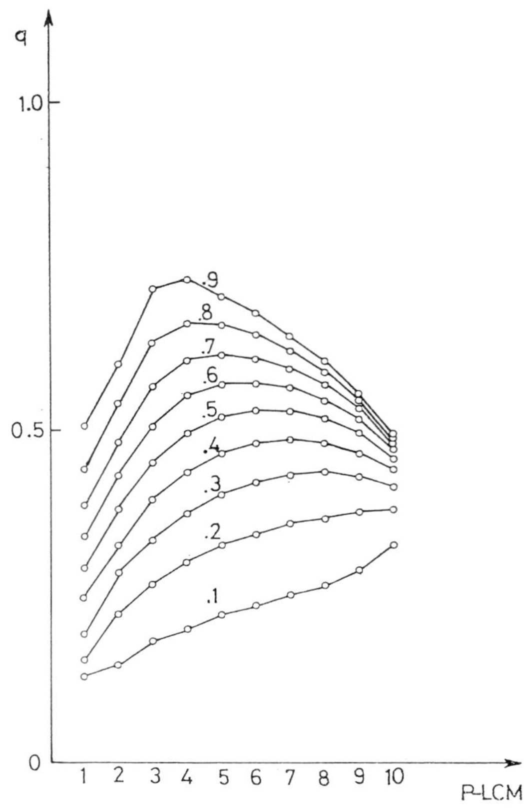

Fig.3 Relation Between Index, q, and Linguistic C-measres

REFERENCES

Banon, G. (1981). Distinction between several subsets of fuzzy measures. Fuzzy Sets and Systems, 5, 291-305.

Dubois, D., and H. Prade. A class of fuzzy measure based on triangular norms. J. General Systems, 8, 1, 43-61, (1982).

Frank, M.J. (1976). On the simultaneous associativity of F(x,y) and x+y-F(x,y). Aequ. Math., 19, 194-226.

Kruse, R. (1982). A note on λ-additive fuzzy measures. Fuzzy Sets and Systems, 8, 219-222.

Hatano, Y., Y. Tsukamoto and M. Sugeno. (1983). Linguistic measure and fuzzy representation problems. Proc. 1-st Symposiam on Knowledge Eng., (in Japanese).

Klement, E.P. (1980). Characterization of fuzzy measures constructed by means of triangular norms. Institutsberichit Nr.180, Johannes Kepler Univ., Austria.

Sugeno, M. (1974). Theory of fuzzy integral and its applications. Thesis, Tokyo Institute of Technology, Tokyo, Japan.

Tsukamoto, Y. (1982). A measure theoretic approach to evaluation of fuzzy set defined on probability space. Fuzzy Mathematics, 3, 89-98.

Schwyla, W. (1980). About the isomorphism between some Sugeno measures and classical measures. Proc. 2-nd Int. Seminar on Fuzzy Set Theory, Linz.

Yager, R.R. (1982). Measuring the quality of linguistic forecasts. Iona College Press, New York.

Zadeh, L.A. (1980). The concept of a linguistic variable and its application to approximate reasoning. Information Sciences, 8, 199-249.

FORMALISATION OF AN APPROXIMATE REASONING: THE ANALOGICAL REASONING

L. Bourrelly, E. Chouraqui and M. Ricard

Laboratoire d'Informatique pour les Sciences de l'Homme, Centre National de la Recherche Scientifique, 31, Chemin Joseph Aiguier — 13402 Marseille Cedex 9, France

Abstract. This paper presents the description and the formalisation of a model for reasoning by analogy. This model is based upon the similarity of proportions between specific elements of the object language. Such a similarity is evaluated by a filtering fonction established from a specific operation of matching so called the $\Longrightarrow$-unification.

Keywords. Knowledge representation, reasoning by analogy, inference, matching, unification, similarity of structures, control structure.

I. INTRODUCTION

In order to increase the cognitive potential of problem solving systems applied to the domains of empirical knowledge, several research projects in artificial intelligence (knowledge representation, expert systems, natural language understanding, etc.) have been and are still concerned with the design of approximate reasoning models {2}, {8}, {9}, {13}, {14}, {17}. By this we understand reasoning in which there are no assertions on the validity of the results, because the associated theoretical models do not depend on classical deductive logic. Among all these modes of reasoning, analogical reasoning keeps a privileged position.

Analogy has always been considered an *essential factor of invention*, as is underlined for example by DOROLLE in {7} : "La faculté inventive par excellence, c'est la faculté d'identification, la faculté de percevoir des ressemblances et des différences. Et le jeu de cette faculté suppose une singulière aptitude à penser par analogie". But logicians do not trust analogy as a proof instrument and the logical description of this mode of reasoning is complicated by the semantical looseness of the word. Consequently, analogical reasoning does not allow one to conclude with certainty as is the case with deductive reasoning. Moreover, it is very difficult to evaluate the probability of corrections of the obtained results. Also, apart from a few attempts mainly at an epistemological level, reasoning by analogy has been very little studied on its own, although it is frequently used in the domain of empirical knowledge, for example in law, medecine, social sciences, economics or archaeology (see on this subject {6}, {11}, {12}). This is why the argumentative value of this type of reasoning has been more systematically studied by logicians interesting themselves in natural language as mentionned in {10} and {15}. In the first paper there is an attempt to bring out some elements for a natural logic and hence to study the logical and discursive processes producing analogies. The second paper deals with the elaboration of schemas of inference based upon the similarities of proportions. It includes an evaluation of the role played, in the scientific discourse, by the argumentation as a validation of the obtained results when these discourses relate to uncompletely described universe. It is another weakness of analogical reasoning to try to produce new knowledge from a universe whose description is neither exhaustive nor universal.

These different projects never reached a level of formalisation allowing an efficient automisation of this type of reasoning.

Nevertheless, research in artificial intelligence, in order to understand and automatise the operation of certain modes of "natural reasoning", has been of increasing interest not only in deductive reasoning but also in *approximate reasoning*. In this framework, reasoning by analogy plays a very important role, altough most publications concerning this sector of activities do not refer explicitly to analogy, as in {8} for example. It is KLING {13} who defines for the first time, to our knowledge, a paradigm for reasoning by analogy. From there, other recent studies - particularly in knowledge representation - have explicitly used this mode of reasoning, founded essentially upon the similarity of structures which by the concept of "Matching", well known in artificical intelligence, allows, evaluation in diverse ways ; for example {2}, {9}, {14}.

This communication presents a model for reasoning by analogy which was created in specific symbolic system for representing real world knowledge, the ARCHES system {4} ; used

essentially for applications concerning human sciences.
Like every formal system, ARCHES is composed of two interdependent components. The first component relates to the modalities of knowledge representation which are determined by the object language of ARCHES, and the algebraical organization of its elements. Thus this system allows representation of each set of real facts (objects of the material culture, factual assertions, events, etc.) whose description and organization are adequate to its architecture. The second component relates to the inferential activity of ARCHES whose objective is to obtain new knowledge - by problem solving - from the facts which have been recorded in the corresponding knowledge base. In order to do this, we have first defined a mode of deductive reasoning which is founded on the logical and semantic properties of the representatical elements of the object language. Then, we determine a specific resolution principle which allows the construction of the corresponding deductive solver. Then, the inferential activities of the system are extended in defining - in addition to deductive reasoning - a particular type of reasoning by analogy. The description of this mode of reasoning naturally uses certain elements defining the modalities of representation and the treatment of the ARCHES system ; elements which we exposed in summary in the second paragraph. Next we present the hypotheses on which analogical reasoning is based (see § III.) ; hypotheses which are at the origin of the construction of the analogical model, permitting the definition of a particular mode of analogical reasoning (see § IV). Finally, we examine the effective modalities of the execution of this reasoning in the ARCHES system (see § V.).

II. KNOWLEDGE REPRESENTATION IN ARCHES SYSTEM.

The representation of facts is determined by the composition of entities constructed from the general notion of *concepts*. Concepts are sets in which the elements, called *individuals*, denote unambiguous and distinct observable terms which produce facts expressing *states* and/or *changes of state*. For example statements like "Peter sleeps peacefully", "Mary's dress is made of bright red wool", "The amphora N1 is localized in the oven F1",and "Marseilles has a temperate climate" represent elementary facts which are susceptible to be conveyed by ARCHES. In these statements PETER, N1, F1, and MARSEILLES are individuals which belong respectively to the concepts PERSON, AMPHORA, OVEN and CITY. Concepts and individuals are data for the ARCHES system, and therefore depend upon the applications considered.

Each fact is represented by a formula giving a description of an individual. More precisely : if C denotes a concept, x a variable or a constant of individual and y a description then C(x,y) is a formula or *structure* of ARCHES' symbolic system. The set of structures determines the object language of ARCHES. Each individual b belonging to concept C corresponds to one *these* and one only, - i.e. one structure which is always true whatever the interpretation in ARCHES -, noted $C(b,\mathfrak{D}(b))$ in which $\mathfrak{D}(b)$ is the description of b representing the facts which are connected with b. The theses are recorded in the knowledge base associated with the ARCHES system : they describe the set of individuals which are the object of study. We call the set of theses which are constructed from a given concept a *field*. A *description* consists of descriptive elements, or *descriptive terms*, connected by connectors which are defined in ARCHES by rules comparable to those used in natural deduction methods : these are the and of addition (*), the non exclusive or (+), negation ($\neg$), and the immediate future (F), the mediate future (G), and the and of succession ($\circ$) which allow expression of the evolution of the studied facts (see on this subject {5}). The descriptive terms permit the representation of properties and in a more general manner the state relations which characterise the individuals. They are composed of four basic entities : feature, class, operator and the functional symbol $. The *features* permit the representation of the individuals' distinctive characters : for example, from the preceding statements, quote SLEEP, PEACEFUL, RED, CLEAR, TEMPERATE, etc. Those of a given semantic nature are regrouped in sets called *classes*. The class symbols express the semantic scope of the state relations which are attested to in the elementary descriptions. Thus the features RED, TEMPERATE, and F1 refer respectively to the classes COLOUR, CLIMATE and LOCALIZATION. The relationships which exist between the classes and the features are expressed by particular functional symbols, in general n-ary, called *operators*. The operators allow specification of the semantic nature of the state relations which characterise individuals. For example, the statement "Automobile whose seating capacity is inferior or equal to 5" shows the kind of relationship which can exist between classes and features : the feature "number of seats" (i.e. numerical value 5) is related to its class CAPACITY by the arithmetical operator .LE. ($\leqslant$). Thus this statement leads to the descriptive term .LE.(CAPACITE,5). Likewise the unary operators ISA and IN permit the construction of descriptive terms ISA(CLIMATE,TEMPERATE) and IN(LOCALIZATION,INS(OVEN ,F1)), conveying the descriptive elements associated with the statements "Marseilles has a temperate climate" and "the amphora is in the oven F1" (INS is a *link* indicating that F1 is an *instance* of the OVEN concept). Moreover properties and state relations can be *described locally*, as in the following two statements : "Amphoras stamped T_1 of type P_1" and "Automobile of a dark blue colour". The characterization of features by descriptive terms permits the precise representation of this descriptive situation (properties of properties, state relations made precise by these properties ; state relations of state relations, etc.). The relation which expresses the attribution of a descriptive element to a feature is represented by the *functional symbol \$*. The functional symbol \$ thus allows the representation of local descriptions of state relations. (Example : ISA(COLOUR,\$(BLUE,ISA(HUE,BRIGHT))) ; for more detail see {3}).

The algebraic structure of sets of descriptions determines the modalities of derivation of the descriptions carried out by a *deduction relation* noted $\Longrightarrow$. This relation is defined in part by the semantic and logical properties of the descriptive terms (decomposition rules, inheritance rules, etc.), and in part by the formation rules of descriptions by way of connectors and descriptive terms. It allows the determination of particular modalities of unification of all descriptive pairs (D,D') ; modalities which are expressed by the predicate $\breve{U}(D,D')$ and defined as follows : the predicate $\breve{U}(D,D')$ is true if and only if there exists a description D_1 and a substitution δ_1 such that $D \Longrightarrow D_1$ and $D_1\delta_1 = D'\delta_1$. The resolution principle of the ARCHES symbolic system and the corresponding deductive solver are based upon this particular unification called $\Longrightarrow$-unification. We will show that the notion of similarity introduced in our analogical model is equally evaluated with the help of the $\Longrightarrow$-unification (see § V.2.). To do this we will use certain definitions which are summarised in the following. We call *schema of description* any description in which all the features are represented by variables. It is bound if and only if at least one of these variables is bound. We thus define the predication SQTLIE(D) which is true if and only if D is a bound schema of description. Finally, we call *sub-structure* of a structure $S(x,\mathcal{D}(x))$ all formulas of the form $S(x,d_x)$ such that $\mathcal{D}(x) \Longrightarrow d_x$. Subsequently, we will designate two such structures by P and T (thus recorded in ARCHES), by A and B two sub-structures of P, and by C and D two sub-structures of T.

III. HYPOTHESES OF THE FOUNDATIONS OF ANALOGICAL REASONING.

The analogical paradigm formalized in the ARCHES system is founded on the similarity of sub-structures whose most general formulation is :

"A is to B what C is to D"

It expresses a resemblance of relationships for which the perfect type is the geometrical proportion (the symbol == expressing the resemblance between the relationships $\frac{A}{B}$ and $\frac{C}{D}$) :

$$\frac{A}{B} == \frac{C}{D}$$

This paradigm corresponds to our objectives, because it effectively permits one to deduce, from "certain similarities", new knowledge (D) from known knowledge (A,B, and C), eventually incomplete (in effect in the general case A and C are respectively different from structures P and T, which are incompletely described). The analogical inference corresponding to this paradigm is expressed as follows (fig. 1.) :

A of P resembles C of T
BUT B derives from P

THEREFORE D derives from T

This analogical inference provides conditional assertions whose *logical validity is not postulated* because it operates on incompletely described universes. Obviously the non-control of validity can produce new knowledge which renders the ARCHES system contradictory, as in the following example.

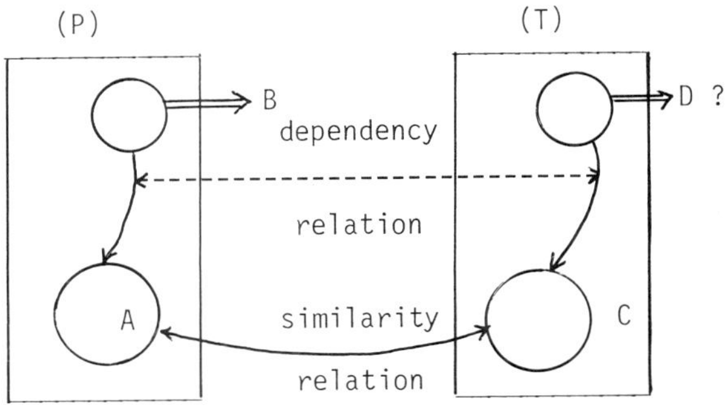

Fig. 1. Representation of the analogical paradigm in the ARCHES system.

Example 1.

Let P and T be the following two structures :

STAR(EARTH,$\mathcal{D}$(EARTH)) (P)
STAR(MOON,$\mathcal{D}$(MOON)) (T)

in which $\mathcal{D}$(EARTH) and $\mathcal{D}$(MOON) are respectively the descriptions of the EARTH and the MOON inside the STAR field.
We ask :

A=STAR(EARTH,ISA(SHAPE,ROUND)) with : $\mathcal{D}$(EARTH) $\Longrightarrow$ ISA(SHAPE,ROUND)
B=STAR(EARTH,ISA(PEOPLING,MAN)) with : $\mathcal{D}$(EARTH) $\Longrightarrow$ ISA(PEOPLING,MAN)
C=STAR(MOON,ISA(SHAPE,ROUND)) with : $\mathcal{D}$(MOON) $\Longrightarrow$ ISA(SHAPE,ROUND)

We hope to demonstrate the statement "D=STAR(MOON,ISA(PEOPLING,MAN))?", knowing that it could not be proved by means of deductive reasoning. The analogical inference rule defined below specifies this with the help of the four sub-structures A,B,C and D, in the following manner :

The EARTH and the MOON have the same shape
BUT The EARTH is inhabited by men

THEREFORE The MOON is inhabited by men

Although there is a total similarity between the descriptions carried by the sub-structures A and C (i.e. the *same* shape), this inference is manifestly non-valid. This invalidity is due first of all to the fact that the inference is based on the resemblance of two very incomplete descriptions of the EARTH and the MOON : in effect, only one property is taken into account to apply the inference rule. But equally it is due to the fact that the properties "the EARTH is round" and "the EARTH is inhabited by men" hold absolutely no semantic relation to each other (evident and/or known), or more generally no dependency relation.

This example (and the analysis of its invalidity) is highly representative of the problems which arise in the carrying out of analogical reasoning. It permits us to state the hypotheses which are at the base of the analogical paradigm of the ARCHES system.

Hypothesis 1. *Analogical reasoning is carried out each time that a statement D, sub-structure of a structure T, could not be proved by deductive reasoning.*

This hypothesis clearly indicates that one of our essential preoccupations is to extend the inferential activity of the ARCHES system. If deductive reasoning is of interest for a large class of problems, it is obvious that this type of reasoning is insufficient in order to derive new knowledge from incompletely described domains ; such is the objective of analogical reasoning.

Hypothesis 2. *If B and D, sub-structures of structures P and T respectively, had been proved in ARCHES and resemble one another, then there exist at least two statements A and C, sub-structures of P and T respectively and distinct from B and D, which resemble one another.*

Similarity is a vague and very general notion, the mesure of which depends not only on the organization of the information to be compared, but also on the problems to be resolved. The pertinence of this similarity increases the *degree of probability* in an anaological inference of conclusion D when one knows the other three terms A, B and C. This degree of probability depends on two factors : on the one hand, it is as great as the similarity between the sub-structures A and C is strong, that is to say that similarity leads to *identity* ; on the other hand it is as great as the quantity of information carried by the sub-structures A and C is important, that is to say the quantity leads to *exhaustivity* of information carried by P and T. Thus the limit of analogical reasoning in ARCHES is determined by two particular values : the identity representing maximum similarity, and the exhaustivity expressed by the taking into account of a completely described universe ; it is therefore deductive reasoning.

Hypothesis 3. *If A' (respectively C'), sub-structure of P (respectively T), does not resemble any sub-structure of T (respectively P), then A' (respectively C') has no influence in the starting of an analogical inference.*

The objective of this hypothesis is to take into account,in the analogical reasoning of ARCHES, only the *information which tends to facilitate analogical inference* (which could be qualified as positive criteria, see hypothesis 2) to the detriment of *those which tend to oppose the starting of analogical inference* (which could be qualified as negative criteria). A priori there is no reason to give preference to positive criteria over negative criteria. The latter might even play an essential role in the control of validity in analogical inference. This is the case in particular in example 1 ; in effect the taking into account of the negative criterion "Presence/Absence of an atmosphere surrounding a planet" would have forbidden the starting of analogical inference. However these choices find justification in the definition of hypothesis 4.

Hypothesis 4. *An analogical inference can only be started if an effective dependency relation exists, noted DEPENDENCY, on the one hand between sub-structures A and B, and on the other hand between sub-structures C and D.*

This dependency relation indicates the realisation that B (respectively D) is conditioned by the existence of A (respectively C) (fig. 1.). The degree of probability of conclusion D is as high as the relation of dependency is strong. *Dependency relations* can be fixed by experts using scholarship and savoir-faire in the domains in which they operate. They *define the hypotheses* which contribute to the resolution of given problems, their validation constituting an essential and indispensable stage in all coherent experimental proceedings. Nonetheless, they can just as well be determined automatically by the ARCHES system (see § V.I.). In example 1 there is no dependency between the state relations (SHAPE,ISA) and (PEOPLING,ISA). And yet if one had indicated to the system that the realization of the relation (PEOPLING,ISA) was conditioned by the existence of an atmosphere surrounding the planets (state relation (ELEMENT,ISA) for example) then the analogical inference could not have been started. Note that this dependency relation serves as a vehicule for negative criteria (see hypothesis 3).

Hypothesis 5. *Sub-structures B and D carry descriptions which have common elements of description.*

In other words the essential objective of inferial activity of analogical reasoning as defined by ARCHES is *to transfer a partial description* (that carried by B) of structure P to structure T.

These five hypotheses, which determine the methodological framework of a particular mode of analogical reasoning, are at the origin of the construction of the analogical model set out in ARCHES (see § IV.2.), in which the functioning is previously analysed using one particular example (see § IV.1.).

IV. THE ANALOGICAL MODEL IN THE ARCHES SYSTEM.

IV.1. Analysis of an example.

Example 2.

Let us consider the fiels CITY as describing the geography of European cities (human, physical, economic and climatic geography,...). Let P and T be the two structures associated with the cities Marseilles and Rome :

CITY(MARSEILLES,𝔇(MARSEILLES)) (P)
CITY(ROME,𝔇(ROME)) (T)

Let us further designate the following sub-structures A, B and C :

A=CITY(MARSEILLES,ISA(SITUATION,COASTAL)* ISA(LATITUDE,MEDIUM))
B=CITY(MARSEILLES,ISA(CLIMATE,TEMPERATE))
C=CITY(ROME,ISA(SITUATION,COASTAL)* ISA(LATITUDE,MEDIUM))

We hope to demonstrate the statement "D=CITY(ROME,ISA(CLIMATE,x))?" knowing that this could not be proved with the help of deductive reasoning (hypothesis 1). To do this, *we suppose that the climate of European cities depends uniquely on their geographical location and their latitude*, (hypothesis 4) :

DEPENDENCY(ISA(CLIMATE,x),ISA(LOCATION,y))
DEPENDENCY(ISA(CLIMATE,x),ISA(LATITUDE,z))

Let us designate by U_E and R_E sets whose elements are respectively features and state relations represented by pairs (class, operator) which are components of description E. Let $R_A \cup U_A$, $R_B \cup U_B$, $R_C \cup U_C$ and $R_D \cup U_D$ be the sets associated with sub-structures A, B, C and D respectively ; and let f be the function which transforms $R_A \cup U_A$ into $R_C \cup U_C$ and $R_B \cup U_B$ into $R_D \cup U_D$.

This transformation necessarily supposes that $R_A \cap R_C \neq \phi$ and $R_B \cap R_D \neq \phi$, the result of which is that f cannot exist unless certain state relations are common respectively to sub-structures A and C on the one hand and to sub-structures B and D on the other hand. Furthermore it takes into account the semantics of the features : two features can be associated in this transformation only if they belong to the same class, and the elements which are identical in the target and source set correspond to one another in a one to one manner.

To demonstrate statement D, one has to be able to determine the value of the function f at the TEMPERATE point : x=f(TEMPERATE). Note to this effect that f represents the *identify function* for all points other than the TEMPERATE point. This identity function simply indicates a maximum similarity between the descriptions carried by A and C (hypothesis 2). Thus we are in an extremely favourable case : on the one hand the dependency criteria are verified ; on the other hand the sub-structures A and C carry identical descriptions. *Thus analogical reasoning consists of inferring that f equally represents the identity function to the TEMPERATE point* (which conforms with hypothesis 5) ; the result being :x=f(TEMPERATE)=TEMPERATE. Statement D is thus demonstrated.

It is possible to find counter-examples of the same nature where analogical inference leads to a contradiction. For example NAPLES and NEW-YORK have the same geographical location and comparable latitudes ; nonetheless they have very different climates. But in this case NEW-YORK coes not belong to the set of European cities whose climate supposedly depends only or their geographical location and their latitude. If one were to extend the geographical area of investigation, one whould obviously have to complete and/or modify the criteria of dependency. For this example in particular it would be necessary to precise that the climate depends also on LONGITUDE and the nature of the OCEAN CURRENTS. This examplifies the hypothetical character of the dependency criteria which only experts can evaluate and justify according to their erudition.

IV.2. Definition of the analogical model.

The analogical model of ARCHES is defined as a particular mapping between two structures P and T, whose evaluation which is founded on relatonships of similarity between sub-structures pernits the derivation of new knowledge from T by transferring a partial description from P to T.

More precisely the analogical model is composed of the following elements :

α) A structure T from which sub-structure D must be derived (which cannot be proved by way of deductive reasoning). Let R_D and U_D respectively be the set of pairs (class, operator) (i.e. state relations) and the set of features which compose sub-structure D.

β) A relation of dependency between $R'_D \subset R_D$ and a set R_e of state relations given a priori.

γ)B, a sub-structure of P, which creates the sets R_B and U_E, and which resembles D. B and D are said to resemble one another if and only if $R_B = R_D$; the result is that R_B depends equally upon R_e.

δ) Two sub-structures of P and T, A and C respectively, which create sets R_A, R_C, U_A and U_C such that $R_e \subset R_A$ and $R_e \subset R_C$ (thus $R_A \cap R_C \neq \phi$).

ε) A mapping f wich allows the transformation of $R_A \cup U_A$ into $R_C \cup U_C$ and $R_B \cup U_B$ into $R_D \cup U_D$. The function f has no existence unless the two following conditions are verified : /1/ $R_B = R_D$; /2/ $R_A \cap R_C \neq \phi$ (conditions imposed by points (γ) and (δ) of the model).

(ζ) The method of evaluation of the function f is defined as follows : /1/ the elements which are identical in the two groups of sets correspond in a one to one manner, thus in particular $f(R_B)=R_D=R_B$ and $f(R_e)=R_e \subset R_A \cap R_C$; /2/ the mapping respects the semantics of features: two features correspond to one another if and only if they belong to the same class ; /3/ the evaluation f is not defined for the other elements.

η) The analogical evaluation rule of the proof of D is expressed as follows : if f represents locally the identity function in the transformation $U_A \rightarrow U_C$, that is to say that there exists $U'_A \subset U_A$ such that $f(U'_A)=U'_A \subset U_C$ and if U'_D designates the set of all the variables contained in U_D, eventually empty, such that $U'_D=f(U'_B)$ with $U'_B \subset U_B$, then f represents the identity function for the points of U'_B, from with $U'_D=U'_B$. If $U_B \cap U_D \neq \phi$ after evaluation, then D is *analogically inferred*. In this case, B and D carry descriptions which have common elements of description (precisely those defined by $U_B \cap U_D$), which conforms entirely with hypothesis 5 underlying the definition of the analogical model.

We have noted that it is the identify function which allows the evaluation of the similarity between sub-structures A and C. If f never represents the identity function, then A and C are different and sub-structure D cannot be proved by analogy.

V. MODALITES OF THE EXECUTION OF ANALOGICAL REASONING IN THE ARCHES SYSTEM.

V.1. The dependency graphs.

In this paragraph we define the way in which ARCHES represents the dependency which exists between state relations, conforming with the analogical model (see § IV.2.) ; point (β)). These dependency relations are expressed with the help of the two-place predicate DEPENDENCY (see hypothesis 4, § III.) : the first argument indicates the descriptive term whose existence is subordinated to the realization of the second argument, which also represents a descriptive term. Naturally a given descriptive term can depend on several other descriptive terms. Generally several dependency relations are given for each application, and each particular set of dependency relations is called a dependency graph.

The given dependency graph determines the particular ARCHES interpretation which establishes the nature of the relationships of subordination between the descriptive terms. To change the elements of this graph would amount to changing the interpretation. Thus *these dependency relations are hypotheses* which are determined a priori by those who employ the ARCHES system. Nonetheless, for each given application, ARCHES can automatically determine all the dependency relations whose first arguments have been established by its users. This mode of automatic evaluation of dependency graphs is interesting since it can play a heuristic role in its users' discovery of new dependency relations.

V.2. Evaluation of similarity by the ⟹-unification operation.

If E_1 and E_2, two schemas of bound descriptions, are equal then they possess common descriptive elements. In fact, if the sets of state relations (pair(class, operator)) and the features associated with E_1 and E_2 respectively are designated by R_1, R_2, U_1 and U_2, then : $R_1=R_2$ and $U_1 \cap U_2 \neq \phi$.

The result of this is that the *two schemas of bound description E_1 and E_2 resemble one another in the manner of the analogical model* (see § IV.2., point (η)). It is therefore possible to evaluate the similarity of any two sub-structures with the help of the matching concept defined in ARCHES, i.e. the operation of ⟹-unification. More precisely, two sub-structures $P(x,\mathcal{A}_x)$ and $T(y,\mathcal{L}_y)$ resemble each other in the ARCHES system if and only if the three following relations are simultaneously verified : $\vec{U}(\mathcal{A}_x,E)$, $\vec{U}(\mathcal{L}_y,E)$ and SQTLIE(E). In other words two sub-structures resemble one another if and only if the same schema of bound description can be derived from their descriptions.

V.3. The rule of analogical inference.

The definition of the analogical model and the method of evaluation of the similarity by way of the ⟹-unification operation allows the definition of the rule of analogical inference in the ARCHES system.

This definition presupposes the following conditions :
We hope to demonstrate the sub-structure $T(y,\mathcal{E}_y)$ in the structure $T(y,\mathcal{D}(y))$ which could not be proved by way of deductive reasoning (see § IV.2. ; point (α)) to do this, a structure $P(x,\mathcal{D}(x))$ is given from which $P(x,\mathcal{E}_x)$ could be derived such that $\mathcal{E}_x$ and $\mathcal{E}_y$ resemble one another in the manner of the analogical model. Moreover, if there exist two descriptions $\mathcal{A}_x$ and $\mathcal{L}_y$ which resemble each other equally in the manner of the analogical model, then it is inferred that the statement $T(y,\mathcal{E}_y)$ is analogically demonstrated ; which leads to following rule of inference :

$$\frac{\begin{array}{ll} P(x,\mathcal{D}(x)) \vdash P(x,\mathcal{E}_x) & \\ \vdash \vec{U}(\mathcal{E}_x,E1)\ ;\ \vdash \vec{U}(\mathcal{E}_y,E1)\ ; & \left.\right\} \text{existence of } \mathcal{E}_x \text{ which resembles } \mathcal{E}_y \\ \vdash \text{SQTLIE}(E1) & \\ P(x,\mathcal{D}(x)) \vdash P(x,\mathcal{A}_x) & \\ T(y,\mathcal{D}(y)) \vdash T(y,\mathcal{L}_y) & \\ \vdash \vec{U}(\mathcal{A}_x,E2)\ ;\ \ \vec{U}(\mathcal{L}_y,E2)\ ; & \left.\right\} \text{existence of } \mathcal{A}_x \text{ and } \mathcal{L}_y \text{ which resemble one another} \\ \vdash \text{SQTLIE}(E2) & \\ \vdash T(y,\neg\mathcal{E}y) & \text{consistency} \end{array}}{P(x,\mathcal{D}(x)),T(y,\mathcal{D}(y)) \vdash T(y,\mathcal{E}_y)}$$

V.4. Modalities of the use of the rule of analogical inference.

The application of the rule of analogical inference depends on the dependency graph, that is to say on the interpretation established a priori by the users of the ARCHES system (see § V.1.). In fact, the conclusion of this rule cannot be admitted unless the elements which compose $\mathcal{E}_x$ and $\mathcal{E}_y$ depend on those composing $\mathcal{A}_x$ and $\mathcal{L}_y$. Consequently the descriptive terms composing these elements must be able to be ⟹-unified at the vertices of this graph in such a manner that the dependency relations are respected (i.e. from $\mathcal{E}_x$ and from $\mathcal{E}_y$ towards $\mathcal{A}_x$ and $\mathcal{L}_y$).

This mode of using the rule of analogical inference - *filtered by the dependency graphs* - contributes to the production from a set of logically true (i.e. valid) structures, sets of structures whose truth depends on interpretations defined by these dependency graphs. Our system thus permits us to build for each interpretation, *the set of ARCHES theorems formed in two parts* :one part *valid* corresponding to the ARCHES theses and to the statements proved with the help of deductive reasoning ; the other *satisfiable* produced by the application of the rule of analogical inference.

VI. CONCLUSION.

The definition of the rule of analogical inference, the method of evaluating similarity and finally the exploration modalities of dependency graphs permit us to formally construct the solver which executes analogical reasoning in the ARCHES system. This solver is naturally articulated on the deductive demonstrator in the sense that its operations depend equally on the mode of deductive reasoning. The two principal elements which organize these two solvers are in the process of being implemented, for which we use the PROLOG language.

REFERENCES.

{1} BLANCHE R. - Le Raisonnement. P.U.F., Paris, 1973.

{2} CHEN D.T-W. , FINDLER N. - Toward analogical reasoning in problem solving by the computers.*Journal of cybernetics*, vol.9, 369-397, 1979.

{3} CHOURAQUI E. - Construction of data structures for representing real-world knowledge.*Congrès EURO-IFIP-79*, London, 25-28 Septembre 1979.

{4} CHOURAQUI E. - Contribution à l'étude théorique de la représentation des connaissances, le système symbolique ARCHES. *Thèse d'Etat*, Nancy, 16 octobre 1981.

{5} CHOURAQUI E. - ARCHES, un système symbolique de représentation et de traitement de connaissances. *Congrès AFCET Informatique 1981*, Gif-sur-Yvette, 18-20 Novembre 1981.

{6} DE COSTER M. - L'analogie en sciences humaines. P.U.F., Paris, 1978.

{7} DOROLLE M. - Le raisonnement par analogie. P.U.F., Paris, 1949.

{8} GELERNTER H. - Realization of a geometry - theorem proving machine. *Computer and thought*, McGraw-Hill book company, Inc. N.Y., 134-152, 1963.

{9} GICK N.L., HOLYOAK K.J. - Analogical problem solving. *Cognitive psychology*, vol. 12, n° 3, 306-355, july 1980.

{10} GRIZE J.B. - Le discours analogique. *Colloque international sur "La représentation des connaissances et raisonnement dans les sciences de l'homme"*, St Maximin, IRIA, 428-435,1979.

{11} Groupe de Travail sur l'Analogie. *Recueil de textes présentés aux journées sur l'analogie*, M.S.H., n° 1, Mai 1980.

{12} KALINCWSKI G. - Le raisonnement en sciences juridiques. *Colloque international sur "La représentation des connaissances et raisonnement dans les sciences de l'homme"*, St Maximin, IRIA 476-485, 1979.

{13} KLING R.E. - A paradigm for reasoning by analogy. *Artificial Intelligence*, vol. 2, 147-178, 1971.

{14} KLIX F., Van der MEER E. - Analogical reasoning - An approach to mechanisms underlying human intelligence performances.*Human and artificial intelligence*, KLIX F. (ed.), North-Holland Publishing Company, Amsterdam, 193-241, 1979.

{15} PERELMAN Ch. - Le champ de l'argumentation. Presses Universitaires de Bruxelles, 1976.

{16} POLYA G. - Mathematics and plausible reasoning. Princeton University Press, Princeton, 1962.

{17} WINSTON P.H. - Learning and reasoning by analogy. *Communications of the A.C.M.*, vol. 23, n° 12, 689-703, Décembre 1980.

DEGREE OF ABSTRACTION OF CONCEPTS AND LONGITUDINAL REASONING

Li Tai-Hang

Shanghai Institute of Computer Technology, 30 Hu Nan Road Shanghai, China

Abstract. This paper gives a precise definition of degree of abstraction of concepts, and thence derives a series of criteria and algorithms of longitudinal (deepening) reasoning. Also, the iterative self-generation formula of membership degree is obtained as a subsidiery result which makes the self-adjusting and self-learning of fuzzy reasoning possible. Some of the results in this paper have been applied in a practical intelligence diagnosis system and a fuzzy control system with complicated processes.

Keywords. N-th grade concept space; degree of abstraction of concepts; longitudinal reasoning; traverse reasoning; backward reasoning; iterative generation formula of membership degree

A. DEGREE OF ABSTRACTION OF CONCEPTS

Assume x to be objectively observable space, and denote its map in a certain thinking space as $p^{(0)}(x)$. In $p^{(0)}(X)$, the so-called concept will be a certain subset of $p^{(0)}(X)$. Its totality will be denoted as $p^{(1)}(x)=\{A^{(1)} \mid A^{(1)} \subseteq P^{(0)}(X)\}$.
$A^{(1)}$ may be either a clear subset or a fuzzy subset of $p^{(0)}(x)$. However, a map $\mu_{A^{(1)}}: p^{(0)}(x) \rightarrow [0,1]$ does always exist, if there is $A^{(0)} \in p^{(0)}(x)$, then $\mu_{A^{(1)}}(A^{(0)})$ may be called the membership degree of $A^{(0)}$ to concept $A^{(1)}$. If $A^{(1)}$ is a clear subset of $p^{(0)}(x)$, then $\mu_{A^{(1)}}(A^{(0)})$ only takes a value 0 or 1.
Similarly, if denote $p^{(2)}(x)=\{A^{(2)} \mid A^{(2)} \subseteq P^{(1)}(X)\}$, then a map $\mu_{A^{(2)}}: p^{(1)}(x) \rightarrow [0,1]$, $\mu_{A^{(2)}}(A^{(1)})$ will be called the membership degree of concept $A^{(1)}$ to concept $A^{(2)}$, in which $A^{(2)} \in p^{(2)}(x)$, $A^{(1)} \in p^{(1)}(x)$.
Thus, in this way, there may be $P^{(n)}(X) = \{A^{(n)} \mid A^{(n)} \subseteq P^{(n-1)}(X)\}$, map $\mu_{A^{(n)}}: P^{(n-1)}(X) \rightarrow [0,1]$ may exist. $\mu_{A^{(n)}}(A^{(n-1)})$ is called the membership degree of concept $A^{(n-1)}$ to concept $A^{(n)}$, in which $A^{(n)} \in P^{(n)}(X)$, $A^{(n-1)} \in P^{(n-1)}(X)$.
We may call $p^{(0)}(x)$ as impression space, or concept space of 0 level, $p^{(1)}(x)$ as concept space of 1 level, $p^{(n)}(x)$ as concept space of n level.
Obviously, there will be $p^{(0)}(x) \subseteq p^{(1)}(x) \subseteq \ldots \subseteq p^{(n)}(x)$.
Definition 1: If concept $A \in p^{(i)}(x) - p^{(i-1)}(x)$, the i will be called as degree of Abstraction of concept, and denoted as $\alpha(A) = i$.
This definition states that, if i is the degree of abstraction of A, then A will only belong to concept space of $p^{(i)}(x)$ or to concept space of higher levels, than $p^{(i)}(x)$.
If a concept space of n level, varicus kinds of concepts will a degree of abstraction from 0 to n may spread throughout.

B. LONGITUDINAL REASONING, TRAVERSE REASONING AND BACKWARD REASONING

Denote "$\longmapsto$" as a symbol of concept reasoning in the n-th grade concept space.
Definition 2.1: In the n-th grade concept space, if $A_1 \longmapsto A_2$, $\alpha(A_1) < \alpha(A_2)$, then $A_1 \longmapsto A_2$ will be called longitudinal reasoning and denoted as $A_1 \underset{a}{\longmapsto} A_2$.
Definition 2.2: In the n-th grade concept space, if $A_1 \longmapsto A_2$, $\alpha(A_1) = \alpha(A_2)$, then $A_1 \longmapsto A_2$ will be called traverse reasoning and denoted as $A_1 \underset{b}{\longmapsto} A_2$.
Definition 2.3. In the n-th grade concept space, if $A_1 \longmapsto A_2$, $\alpha(A_1) > \alpha(A_2)$ then $A_1 \longmapsto A_2$ will be called backward reasoning and denoted as $A_1 \underset{c}{\longmapsto} A_2$.
A concept reasoning proceeding in the n-th grade concept space is nothing but a synthesis of the three foregoing reasoning types, so it may briefly be denoted as $\overset{n}{\underset{i=1}{\longmapsto_k}} A_i \quad k \in \{a,b,c\}$.

C. TRUSTABILITY OF LONGITUDINAL REASONING

Definition 3.1: If $A_1 \underset{a}{\longmapsto} A_k$, then membership degree $\mu_{A_k}(A_1)$ is reasoning trustability too, and denoted as $Q_{A_k}(A_1)$.

Definition 3.2: If $A_1 \underset{a}{\longmapsto} A_k$, then $1-Q_{A_k}(A_1)$ will be called reasoning suspectability of $A_1 \underset{a}{\longmapsto} A_k$ and denoted as $\eta_{A_k}(A_1)$

But in actual thinking process, the longitudinal reasoning from A_1 to A_k is usually promoted by a combination of many conditions. Assume $A_2, A_3 \ldots A_i$ to be these conditions and their degree of abstraction are all less than $\alpha(A_k)$, then this reasoning may be denoted as $A_1 \overset{A_2 \cdots A_i}{\underset{a}{\longmapsto}} A_k$. Its reasoning trustability will be denoted as $Q_{A_k}(A_1 \wedge A_2 \wedge \cdots \wedge A_i)$ and its suspectability denoted as $\eta_{A_k}(A_1 \wedge A_2 \wedge \cdots \wedge A_i)$.

As the more factors to promote the implementation of reasoning, the less should be the reasoning suspectibility therefore it follow:

$$\eta_{A_k}(A_1 \wedge A_2 \wedge \cdots \wedge A_i) \leq \min\{\eta_{A_k}(A_1), \eta_{A_k}(A_2), \cdots \eta_{A_k}(A_i)\}$$

Now let H be a set of operators Hs, and Hs satisfying:

If there are real numbers $a_1, a_2, \ldots a_i$ all $\in [0,1]$, then

(1). $Hs(a_1, a_2, \ldots, a_i) \in [0,1]$;

(2). $Hs(a_1, a_2, \ldots a_i) \leq \min\{a_1, a_2, \ldots, a_i\}$

Then, in different thinking systems, there may be different operators $Hs \in H$, to make

$$\eta_{A_k}(A_1 \wedge A_2 \wedge \cdots \wedge A_i) = Hs(\eta_{A_k}(A_1), \eta_{A_k}(A_2), \cdots \eta_{A_k}(A_i))$$

Obviously, conventional multiplication operator "$\cdot$" $\in$ H.

Theorem 3.1: If $A_1 \overset{A_2, \cdots, A_i}{\underset{a}{\longmapsto}} A_k$, then reasoning trustability

$$Q_{A_k}(A_1 \wedge A_2 \wedge \cdots \wedge A_i) = 1 - Hs(1-\mu_{A_k}(A_1), \cdots, 1-\mu_{A_k}(A_i))$$

Proof: This will be proved by iterative application of Definition 3.1 and 3.2.

Definition 3.3: If $A_1 \underset{a}{\longmapsto} A_2$, and more over, $A_2 \underset{a}{\longmapsto} A_3$, then the trustability from $A_1 \underset{a}{\longmapsto} A_2 \underset{a}{\longmapsto} A_3$ will be called process trustability and denoted as $Q_{A_3}(A_1, A_2)$.

If the trustability of $A_2 \underset{a}{\longmapsto} A_3$ is 1, obviously the trustability of $A_1 \underset{a}{\longmapsto} A_2 \underset{a}{\longmapsto} A_3$ should be equal to the trustability of $A_1 \underset{a}{\longmapsto} A_2$. Otherwise, the former should be less than the latter, *i.e.* $Q_{A_3}(A_1, A_2) \leq Q_{A_2}(A_1)$. For the same reason, there should be $Q_{A_3}(A_1, A_2) \leq Q_{A_3}(A_2)$

Then

$$Q_{A_3}(A_1, A_2) \leq \min(Q_{A_3}(A_2), Q_{A_2}(A_1))$$

i.e $Q_{A_3}(A_1, A_2) = Hs'(Q_{A_3}(A_2), Q_{A_2}(A_1))$ $Hs' \in H$

Generally, if $A_1 \underset{a}{\longmapsto} A_2 \underset{a}{\longmapsto} \cdots \underset{a}{\longmapsto} A_k$, process trustability will be denoted as $Q_{A_k}(A_1, A_2, \ldots, A_{k-1})$, and also

$$Q_{A_k}(A_1, A_2, \cdots, A_{k-1}) = Hs'(Q_{A_k}(A_{k-1}), \cdots, Q_{A_2}(A_1))$$

If we denote suspectability(called process suspectability) at that time as $\eta_{A_k}(A_1, A_2, \cdots, A_{k-1})$, by applying Definition, it is easy to prove

Theorem 3.2: $\eta_{A_k}(A_1, \cdots, A_{k-1}) = 1 - Hs'(1-\eta_{A_k}(A_{k-1}), \cdots, 1-\eta_{A_2}(A_1))$

If the process of conceptual reasoning be $A_1 \underset{a}{\longmapsto} A_2 \overset{A_3}{\underset{a}{\longmapsto}} A_4$, then process trustability will be denoted as $Q_{A_4}((A_1, A_2) \wedge A_3)$, now there exists

Theorem 3.3: $Q_{A_4}((A_1, A_2) \wedge A_3) = 1 - Hs(1 - Hs'(Q_{A_4}(A_2), Q_{A_2}(A_1)), 1 - Q_{A_4}(A_3))$

Proof: As $Q_{A_4}(A_1, A_2) = Hs'(Q_{A_4}(A_2), Q_{A_2}(A_1))$ and by applying Theorem 1, we obtain:

$$Q_{A_4}((A_1, A_2) \wedge A_3) = 1 - Hs(1 - Q_{A_4}(A_1, A_2), 1 - Q_{A_4}(A_3))$$
$$= 1 - Hs(1 - Hs'(Q_{A_4}(A_2), Q_{A_2}(A_1)), 1 - Q_{A_4}(A_3))$$

Q.E.D

If the process of conceptual reasoning be $A_1 \overset{A_2}{\underset{a}{\longmapsto}} A_3 \underset{a}{\longmapsto} A_4$, then the process trustability may be denoted as $Q_{A_4}(A_1 \wedge A_2, A_3)$, and we have

Theorem 3.4: $Q_{A_4}(A_1 \wedge A_2, A_3) = Hs'(Q_{A_4}(A_3), 1 - Hs(1-Q_{A_3}(A_2), 1-Q_{A_3}(A_1)))$

Same proof as Theorem 3.3

The computation of trustability in the process of longitudinal reasoning, will play an important role in the simplification of reasoning process and the selection and conclusion.

In a practical thinking simulation system, for convenience, we may take all $Hs = Hs' =$ operator "$\cdot$", then the previous formula will be derived to be separately:

(1) $\eta_{A_k}(A_1 \wedge A_2 \wedge \cdots \wedge A_i) = \prod_{j=1}^{i} \eta_{A_k}(A_j)$

(2) $Q_{A_k}(A_1 \wedge A_2 \wedge \cdots \wedge A_i) = 1 - \prod_{j=1}^{i}(1 - Q_{A_k}(A_j))$

(3) $Q_{A_k}(A_1, A_2, \cdots, A_{k-1}) = \prod_{j=1}^{k-1} Q_{A_{j+1}}(A_j)$

(4) $\eta_{A_k}(A_1, A_2, \cdots, A_{k-1}) = 1 - \prod_{j=1}^{k-1}(1 - \eta_{A_{j+1}}(A_j))$

(5) $Q_{A_4}((A_1, A_2) \wedge A_3) = 1 - (1 - Q_{A_4}(A_2) Q_{A_2}(A_1))(1 - Q_{A_3}(A_1))$

(6) $Q_{A_4}(A_1 \wedge A_2, A_3) = Q_{A_4}(A_3)(1 - (1 - Q_{A_3}(A_2))(1 - Q_3(A_1)))$

Of course, formula 5 and 6 can further be unified and popularized into a more generalized case.

Examples of application

Example 1:

Assume the membership degree of feeling hot (α(feeling hot)=1) to catching cold to be 0.3, the membership degree of dreading coldness (α(dreading coldness)=1) to catching cold to be 0.5, then, in accordance with formula(2), the trustability of catching cold deduced from both feeling hot and dreading coldness concurrently is:

$$Q_{A_3}(A_1 \wedge A_2) = 1-(1-0.3)(1-0.5) = 0.65$$

Example 2:

If the membership degree of bodily tempreture 37.8°c (α(37.8°c)=0) to feeling hot is 0.8, the membership degree of feeling hot to catching cold is 0.3, then, according to formula(3), the trustability of catching cold deduced from bodily tempreture 37.8°C is:

$$Q_{A_3}(A_1, A_2) = 0.8 \cdot 0.3 = 0.24$$

In fact, 0.24 may be seen to be membership degree of bodily tempreture 37.8°c to catching cold.

Example 3:

In Example 2, if bodily tempreture 37.8°c with concurrent dreading cold, then, according to formula(5), the trustability of catching cold deduced from the above conditions is:

$$Q_{A_4}((A_1, A_2) \wedge A_3) = 1-(1-0.8 \cdot 0.3)(1-0.5) = 0.62$$

(Because at this time, the reasoning process is 37.8°c $\underset{a}{\longmapsto}$ feeling hot $\overset{\text{dreading cold}}{\underset{a}{\longmapsto}}$ catching cold)

Obviously, these consequences of foregoing examples are all comparatively reasonable.

D. INTERATIVE GENERATION OF MEMBERSHIP DEGREE

Inorder to show the random adjustment in practice and flexibility of reasoning, membership degree should not be empirical or fixed.

Let λ^+ denote the positive effect of action upon the external observable space taken by the system as a result of reasoning (may also be the correct evaluation of reasoning conclusion from outside world), and let λ^- denote the negative one (may also be the erroneous evaluation of reasoning conclusion from outside world) and its feed back evaluation in thinking system be functions $f(\lambda^+)>0$ and $g(\lambda^-)>0$ respectively.

After N times thinking reasoning, there always exists corresponding positive and negative accumulation of evaluation between any pair of A_{i-1} and A_i (of course $\alpha(A_{i-1})<\alpha(A_i)$), and be denoted respectively.

$$F(N)\Big|_{A_{i-1}}^{A_i}=\sum_{j=1}^{N} f(\lambda_j^+) \quad \text{and} \quad G(N)\Big|_{A_{i-1}}^{A_i}=\sum_{j=1}^{N} g(\lambda_j^-)$$

(Here, if the jth thinking does not pass through $A_{i-1}\overset{}{\underset{a}{\mapsto}} A_i$, then $f(\lambda_j^+)=g(\lambda_j^-)=0$)

In non-ambiguous condition, positive and negative accumulation of evaluation may be simply denoted as F(N) and G(N) respectively.

Obviously, the trustability of reasoning $A_{i-1}\underset{a}{\mapsto} A_i$ will increase in value with the increase of F(N) and decrease in value with the increase of G(N).

Furthermore, if $N\to\infty$, then:

When $F(N)=\omicron(G(N))$, i.e. $F(N)/G(N)\to\infty$, there should be $Q_{A_i}^{(N)}(A_{i-1})\to 1$

When $F(N)=\omicron(G(N))$, i.e. $F(N)/G(N)\to 0$, there should be $Q_{A_i}^{(N)}(A_{i-1})\to 0$

Or, $Q_{A_i}^{(N)}(A_{i-1})=L(F(N)/G(N)$, where L is a transformation, i.e.

$L: [0,\infty)\to[0,1)$.

There are many such transformation functions L(x). Because there is a function D(x) Satisfying:

When $x\to 0$, then $D(x)\to 0$; $X\to\infty$, then $D(x)\to\infty$, then $L(x)=D(x)/D(x)+c$(constant $c\neq 0$) will be the result.

In practical systems, we may take the simplest case where $D(x)=x$, $c=1$, then $L(x)=x/x+1$ and the

$$Q_{A_i}^{(N)}(A_{i-1})=\frac{F(N)/G(N)}{F(N)/G(N)+1}=\frac{F(N)}{F(N)+G(N)}$$

If when $N\to\infty$, there does exist a stable membership degree $\mu_{A_i}^{(N)}(A_{i-1})$, according to Definition 3.1, when $A_{i-1}\underset{a}{\mapsto} A_i$, then:

$$Q_{A_i}^{(N)}(A_{i-1})=\mu_{A_i}^{(N)}(A_{i-1}) ,$$

$$\mu_{A_i}(A_{i-1})=\lim_{N\to\infty}\mu_{A_i}^{(N)}(A_{i-1})=\lim_{N\to\infty}Q_{A_i}^{(N)}(A_{i-1})$$

$$=\lim_{N\to\infty}\frac{F(N)}{F(N)+G(N)}$$

this is a self-generation function of membership degree.

Unfortunately, how ever, this is not very practical formula, because single reasoning $A_{i-1}\underset{a}{\mapsto} A_i$ will oppear very scarcely during thinking process.

Most of them will be reasoning $A_{i-1}\overset{B_1\cdots B_L}{\underset{a}{\longmapsto}} A_i$ promoted by multiple factors, therefore the above limit will be difficult to approach.

Thus, we may repeat the above proof, where

$$Q_{A_i}^{(N)}(A_{i-1}\wedge B_1\wedge\cdots\wedge B_L)=\frac{F(N)}{F(N)+G(N)}$$

For easiness in writing symbols, let $A_{i-1}=B_0$, then

$$Q_{A_i}^{(N)}(B_0\wedge B_1\wedge\cdots\wedge B_L)=\frac{F(N)}{F(N)+G(N)}$$

When the operator $H_S=$"•", form formula(1)

$$\eta_{A_i}^{(N)}(B_0\wedge B_1\wedge\cdots\wedge B_L)=\prod_{r=0}^{L}\eta_{A_i}^{(N)}(B_r)$$

Now let

$$\eta_{A_i}^{(N)}(B_j)=\eta_{A_i}^{(N-1)}(B_j)+\varepsilon\,\eta_{A_i}^{(N-1)}(B_j) \quad (j=0,1,\cdots,L)$$

then

$$\frac{F(N)}{F(N)+G(N)}=Q_{A_i}^{(N)}(B_0\wedge B_1\wedge\cdots\wedge B_L)=1-\eta_{A_i}^{(N)}(B_0\wedge B_1\cdots\wedge B_L)$$

$$=1-\prod_{r=0}^{L}(1+\varepsilon)\eta_{A_i}^{(N-1)}(B_r)$$

thence

$$\varepsilon=\left(\frac{G(N)}{F(N)+G(N)}\right)^{\frac{1}{L+1}}\left(\prod_{r=0}^{L}\eta_{A_i}^{(N-1)}(B_r)\right)^{-\frac{1}{L+1}}-1$$

therefore

$$\eta_{A_i}^{(N)}(B_j)=\left(\frac{G(N)}{F(N)+G(N)}\right)^{\frac{1}{L+1}}\left(\prod_{r=0}^{L}\eta_{A_i}^{(N-1)}(B_r)\right)^{-\frac{1}{L+1}}\eta_{A_i}^{(N-1)}(B_j)$$

Again according to Definition 3.2, we obtain

$$\mu_{A_i}^{(N)}(B_j)=1-\eta_{A_i}^{(N)}(B_j)$$

$$=1-\left(\frac{G(N)}{F(N)+G(N)}\right)^{\frac{1}{L+1}}\left(\prod_{r=0}^{L}(1-\mu_{A_i}^{(N-1)}(B_r))\right)^{-\frac{1}{L+1}}(1-\mu_{A_i}^{(N-1)}(B_j)) \qquad (j=0,1,\cdots,L)$$

This is a generalized and practical interative generation formula of membership degree.

Let $L=0$, this formula will be degraded to be

$$\mu_{A_i}^{(N)}(B_0)=1-\frac{G(N)}{F(N)+G(N)}=\frac{F(N)}{F(N)+G(N)}$$

it just coinsides with the above conclusion.

But in fact, it is impossible that B_1, B_2, ..., B_L is a factor to help to realize of every reasoning from B_0 to A_i , that is, it is only possible that a certain combination of them is a factor to help to realize of reasoning.

at that time, a more generalized formular that can be proved is:

$$\mu_{A_i}^{(N)}(B_j)=1-\left(\frac{\sum_{r=1}^{N}g(\lambda^-_{r.i.Sm})}{\sum_{r=1}^{N}f(\lambda^+_{r.i.Sm})+\sum_{r=1}^{N}g(\lambda^-_{r.i.Sm})}\right)^{\frac{1}{m+1}}\left(\prod_{r=0}^{m}(1-\mu_{A_i}^{(N-1)}(B_r))\right)^{-\frac{1}{m+1}}(1-\mu_{A_i}^{(N-1)}(B_j))$$

$$\left(\begin{array}{l} j=0,1,\cdots,l \\ 0 \le m \le l \\ 1 \le s \le \frac{l!}{m!(l-m)!} \end{array}\right)$$

Where S_m is a combined serial member that m factors to help to realize will occur amongst l possible factors to help to realize with m floating randomly between 0 and l.

Example of application

In a practical computer diagnosis system, according to experts' personal experiences, the initial membership degree of condition of illness such as cough, chest pain, running a fever, yellow phleagm ...... to "hot lung" (acute pneumonia) are 0.15,0.20,0.25, 0.15, ... respectively the foreging formulas are used in this system we take:

$$\lambda^+ = \begin{cases} 0 & \text{if condition of illness is getting neither better nor worse,} \\ 1 & \text{if condition of illness is getting better} \\ 2 & \text{if condition of illness is vanishes} \end{cases}$$

$$\lambda^- = \begin{cases} 0 & \text{if condition of illness vanishes or is getting better} \\ 1 & \text{if condition of illness is getting neither better nor worse} \\ 2 & \text{if condition of illness is getting worse} \end{cases}$$

In addition,

Through review test of more than 280 different examples of illness, the fougoing membership degree becomes 0.12,0.23,0.19,0.13, ... respectively then, the percentage of hits used in diagnosis becomes higher and higher.

How to create membership degree has long been an obstacle of Fuzzy mathematics. We have seen that this obstacle can be easily over taken if we observe this kind of problems during thinking process.

Obviously, because the choice of feedback evaluation functions f and g is different from operator H_s, different generating formulas of membership degree will be obtained. In fact, so is the case of objective nature. Even in identical envivonment different thinking systems are not in synchronization in respect of self-regulation and self-completion.

As change of membership degree will give direct effect on the computation of reasoning trustability, the process of reasoning and the choice of final decision will make random adjustment according to the actual result of feedback. Even confronting identical problem two consecutive reasoning processes of a thinking system will never be unchanged.

Note: The difference between the symbol "$\longmapsto$" and the traditional symbol "$\Longrightarrow$" is that the former may not necessarily be definite reasoning. Besides longitudinal reasoning discussed in this paper, there is also traverse reasoning and backward reasoning which will be discussed seperately.

UPPER AND LOWER POSSIBILITIES INDUCED BY A MULTIVALUED MAPPING

D. Dubois* and H. Prade**

*Cert/Dera BP 4025 31055 Toulouse Cédex (France)
**LSI Université P. Sabatier 31062 Toulouse Cédex (France)

Abstract. In a seminal paper, fifteen years ago, Dempster introduced the concepts of upper and lower probabilities by considering a probability space, together with a multivalued mapping. In this paper, we study the analogous situation when a possibility space is given and the same multivalued mapping. Then, we get upper and lower possibilities. While an upper possibility is still a possibility, a lower possibility is no longer a possibility usually. These concepts are applied to an approximate reasoning problem which turns out to be a "possibilistic" version of a classical problem dealt with in decision theory.

Keywords. Upper and lower probability, possibility, approximate reasoning, decision theory.

INTRODUCTION

In a seminal paper /1/, about fifteen years ago, Dempster introduced the concepts of upper and lower probabilities by considering a probability space $(X,\mathcal{A},P)$ together with a multivalued mapping Γ from X to a set S. Then, Γ induces upper and lower probabilities on S, or if we prefer Shafer's terminology /9/, plausibility and belief functions on S.

In this paper, after a brief recalling of Dempster's approach we examine the situation of two multivalued series mappings Γ_1 from X to S_1 and Γ_2 from S_1 to S_2. Then, we focus our attention on the particular case where Γ induces possibility and necessity measures on S (in Shafer's terminology :"consonant plausibility and belief functions"), since Zadeh's possibility measures /12/ are an important particular case of plausibility functions while necessity measures, which have been introduced by duality (see Dubois, Prade /2/), are belief (or, if we prefer, credibility) functions. Next, the more general case, where Γ is a fuzzy mapping rather than a multivalued mapping, is considered. Then, a rule of combination, analogous to Dempster's rule /1/, /9/, is proposed for combining possibility measures. Lastly, applications to approximate reasoning and decision theory are briefly presented. The concluding remark sketches another possible application.

DEMPSTER'S APPROACH

Let $(X,\mathcal{P}(X),P)$ be a probability space, where X is finite and $\mathcal{P}(X)$ denotes the set of subsets of X, and Γ be a multivalued mapping from X to S, i.e. $\forall x \in X, \Gamma(x) \subseteq S$. This model corresponds to a random experiment where the outcome cannot be precisely observed but can only be located in a subset of possible eventualities. Given a subset A of S, Γ induces two noticeable subsets on X :

$$\Gamma^{-1*}(A) = \{x \in X, \Gamma(x) \cap A \neq \emptyset\} \quad (1)$$

and

$$\Gamma_*^{-1}(A) = \{x \in X, \Gamma(x) \neq \emptyset, \Gamma(x) \subseteq A\} \quad (2)$$

Viewing $\Gamma(x)$ as the set of the possible images of x by an ill-known ordinary mapping f, $\Gamma^{-1*}(A)$ is the set of elements of X whose images by f possibly belong to A while $\Gamma_*^{-1}(A)$ is the set of elements of X whose images by f necessarily belong to A. When Γ is single-valued, $\Gamma_*^{-1}(A) = \Gamma^{-1*}(A) = \Gamma^{-1}(A)$.

Then, the plausibility and the credibility of A respectively defined by (from now we use this terminology which seems to be more adequate from a semantic point of view) :

$$\forall A \subseteq S,\ Pl(A) = \frac{P(\Gamma^{-1*}(A))}{P(\Gamma^{-1*}(S))} \quad (3)$$

$$\forall A \subseteq S,\ Cr(A) \quad \frac{P(\Gamma_*^{-1}(A))}{P(\Gamma_*^{-1}(S))} \quad (4)$$

where $\Gamma^{-1*}(S) = \Gamma_*^{-1}(S) = \{x \in X, \Gamma(x) \neq \emptyset\}$ with $P(\Gamma^{-1}(S)) \neq 0$. Cr(A) and Pl(A) are lower and upper bounds of the unknown probability $P(\{x \in X, f(x) \in A\})$. Pl and Cr are completely determined by the 2^m quantities (where $|S| = m$) $P(\{x \in X, \Gamma(x)=F\})$; since the subsets $\{x \in X, \Gamma(x)=F\}$, with $F \subseteq S$, form a partition of X, we have :

$$\sum_{F \subseteq S} P(\{x \in X, \Gamma(x) = F\}) = 1 \quad (5)$$

Then the expressions of Pl et Cr in terms of the so-called basic probability assignment m (see Shafer /9/) from S to [0,1] defined by:

$$m(\emptyset) = 0$$

$$\forall\, F \subseteq S, F \neq \emptyset, m(F) = \frac{P(\{x \in X, \Gamma(x) = F\})}{1 - P(\{x \in X, \Gamma(x) = \emptyset\})} \qquad (6)$$

are given by

$$\forall\, A \subseteq S,\ Pl(A) = \sum_{F \cap A \neq \emptyset} m(F) \qquad (7)$$

$$\forall\, A \subseteq S,\ Cr(A) = \sum_{F \subseteq A} m(F) \qquad (8)$$

Note that $\sum_{F \subseteq S} m(F) = 1$; thus, $Cr(S) = Pl(S) = 1$.

we have

$$\forall\, A \subseteq S,\ Pl(A) = 1 - Cr(\bar{A}) \qquad (9)$$

Let $\mathcal{C}$ be the set of probability measures P such that :

$$\forall\, A \subseteq S,\ Cr(A) \leq P(A) \leq Pl(A) \qquad (10)$$

where Cr and Pl are credibility and plausibility functions defined from a given basic probability assignment m. Let V be a real-valued function defined over S. Dempster /1/ proved that, if E(V) denotes the usual expectation of V calculed on the basis of a probability P, the lower and upper bounds of E(V) when P ranges on $\mathcal{C}$,

$$E_*(V) = \min_{P \in \mathcal{C}} E(V) \qquad (11)$$

and

$$E^*(V) = \max_{P \in \mathcal{C}} E(V) \qquad (12)$$

are given by :

$$E_*(V) = \int_{-\infty}^{+\infty} v\, d\, F^*(v) \qquad (13)$$

and

$$E^*(V) = \int_{-\infty}^{+\infty} v\, d\, F_*(V) \qquad (14)$$

where

$$F^*(v) = Pl(\{s \in S,\ V(s) \leq v\}) \qquad (15)$$

and

$$F_*(v) = Cr(\{s \in S,\ V(s) \leq v\}) \qquad (16)$$

provided that the measurability requirements hold. Then, it can be checked that (see for instance Smets /10/)

$$E_*(V) = \sum_{A \subseteq S} m(A).\min_{s \in A} V(s) \qquad (17)$$

and

$$E^*(V) = \sum_{A \subseteq S} m(A).\max_{s \in A} V(s) \qquad (18)$$

Note that

$$Pl(A) = E^*(\chi_A) \qquad (19)$$

and

$$Cr(A) = E_*(\chi_A) \qquad (20)$$

where χ_A denotes the characteristic function of A.

REITERATING THE PROCESS OF GENERATION OF UPPER AND LOWER PROBABILITIES

Let Γ_1 be a multivalued mapping from the probability space $(X, \mathcal{P}(X), P)$ to a set S_1 and Γ_2 be a multivalued mapping from S_1 to S_2 such that $\forall s \in S_1, \Gamma_2(s) \neq \emptyset$. Then, if $\Gamma_i^{-1*}(A)$ and $\Gamma_{i*}^{-1}(A)$ are the maximal and minimal inverse images of A by Γ_i, defined by (1) and (2) we have :

$$\forall A \subseteq S_2, \Gamma_1^{-1*}(\Gamma_2^{-1*}(A)) = (\Gamma_2 \circ \Gamma_1)^{-1*}(A) \qquad (21)$$

$$\forall A \subseteq S_2, \Gamma_{1*}^{-1}(\Gamma_{2*}^{-1}(A)) = (\Gamma_2 \circ \Gamma_1)_*^{-1}(A) \qquad (22)$$

with $(\Gamma_2 \circ \Gamma_1)(x) = \bigcup_{s \in \Gamma_1(x)} \Gamma_2(s)$

Proof

$$\Gamma_2^{-1*}(A) = \{s \in S_1, \Gamma_2(s) \cap A \neq \emptyset\}$$

$$\Gamma_1^{-1*}(\Gamma_2^{-1*}(A)) = \{x \in X, \Gamma_1(x) \cap \{s \in S_1, \Gamma_2(s) \cap A \neq \emptyset\} \neq \emptyset\}$$

$$= \{x \in X, (\bigcup_{s \in \Gamma_1(x)} \Gamma_2(s)) \cap A \neq \emptyset\}$$

$$\Gamma_{2*}^{-1}(A) = \{s \in S_1, \Gamma_2(s) \subseteq A\} \quad \text{since } \forall s, \Gamma_2(s) \neq \emptyset$$

$$\Gamma_{1*}^{-1}(\Gamma_{2*}^{-1}(A)) = \{x \in X, \Gamma_1(x) \neq \emptyset, \Gamma_1(x) \subseteq \{s \in S_1, \Gamma_2(s) \subseteq A\}\}$$

$$= \{x \in X, \Gamma_2 \circ \Gamma_1(x) \neq \emptyset, (\bigcup_{s \in \Gamma_1(x)} \Gamma_2(s)) \subseteq A\}$$

Q.E.D

An upper and a lower plausibility function can be defined on S_2 by :

$$\forall\, A \subseteq S_2, Pl^*(A) = Pl(\Gamma_2^{-1*}(A)) \qquad (23)$$

$$\forall\, A \subseteq S_2, Pl_*(A) = Pl(\Gamma_{2*}^{-1}(A)) \qquad (24)$$

since $\Gamma_2^{-1*}(S_2) = \Gamma_{2*}^{-1}(S_2) = S_1$ and $Pl(S_1) = 1$

From (3) and (21) we get :

$$Pl^*(A) = \frac{P((\Gamma_2 \circ \Gamma_1)^{-1*}(A))}{P((\Gamma_2 \circ \Gamma_1)^{-1*}(S_2))} \qquad (25)$$

which shows that an upper plausibility function is still a plausibility function.

Similarly, an upper and a lower credibility function can be defined on S_2 by :

$$\forall\, A \subseteq S_2, Cr^*(A) = Cr(\Gamma_2^{-1*}(A)) \qquad (26)$$

$$\forall\, A \subseteq S_2, Cr_*(A) = Cr(\Gamma_{2*}^{-1}(A)) \qquad (27)$$

and (4) and (22) yields

$$Cr_*(A) = \frac{P((\Gamma_2 \circ \Gamma_1)_*^{-1}(A))}{P((\Gamma_2 \circ \Gamma_1)_*^{-1}(S_2))} \qquad (28)$$

which shows that a lower credibility function is still a credibility function.
It can be checked that :

$$\forall A \subseteq S_2, Pl^*(A) = 1 - Cr_*(\bar{A}) \geq Pl_*(A) \quad (29)$$

$$\forall A \subseteq S_2, Pl_*(A) = 1 - Cr^*(\bar{A}) \quad (30)$$

since

$\Gamma_{2*}^{-1}(A) = \overline{\Gamma_2^{-1*}(\bar{A})}$ when $\forall s \in S_1, \Gamma_2(s) \neq \emptyset$

However, Pl_* and Cr^* are not credibility or plausibility functions generally, since it may happen that $\exists$ A and B such that $Pl_*(A) > Cr^*(A)$ and $Cr^*(B) > Pl_*(B)$. Nevertheless, Pl_* and Cr^* are non-decreasing functions with respect to set-inclusion.

UPPER AND LOWER POSSIBILITIES

The formulae (7) and (8) enable to express a plausibility and a credibility function in terms of a basic probability assignment m. A subset F such that $m(F)>0$ is called a focal element.

Possibility (resp.necessity) measures are plausibility (resp. credibility) functions whose focal elements can be ordered in a nested sequence (Shafer /9/ ; see also Dubois, Prade /5/).

Noticing that if a collection of subsets of S_1, $\{\Gamma_1(x)\}_{x \in I}$ can be ordered in a nested sequence, then the collection of subsets of S_2, $\{(\Gamma_2 o \Gamma_1)(x)\}_{x \in I}$ can also be ordered in a nested sequence; from (6), (25), and (28) it is clear that an upper possibility (resp. a lower necessity) measure is still a possibility (resp. necessity) measure.

Let Π be a possibility measure defined on S_1. The multivalued mapping Γ_2 (it is no longer supposed that $\forall s \in S_1, \Gamma_2(s) \neq \emptyset$) induces an (upper) possibility measure Π^* and a lower possibility measure Π_* (which is not a possibility measure in general). We have :

$$\forall A \subseteq S_2, \Pi^*(A) = \frac{\Pi(\Gamma_2^{-1*}(A))}{\Pi(\Gamma_2^{-1*}(S_2))} \quad (31)$$

$$\forall A \subseteq S_2, \Pi_*(A) = \frac{\Pi(\Gamma_{2*}^{-1}(A))}{\Pi(\Gamma_{2*}^{-1}(S_2))} \quad (32)$$

Π_* and Π^* can be expressed in terms of "a basic possibility assignment" n from $\mathcal{P}(S_2)$ to $[0,1]$, defined by

$$n(\emptyset) = 0$$

$$\forall F \subseteq S_2, F \neq \emptyset, \; n(F) = \frac{\Pi(\{s \in S_1, \Gamma_2(s) = F\})}{\Pi(\{s \in S_1, \Gamma_2(s) \neq \emptyset\})} \quad (33)$$

then, we have

$$\forall A \subseteq S_2, \Pi^*(A) = \max_{A \cap F \neq \emptyset} n(F) \quad (34)$$

and

$$\forall A \subseteq S_2, \Pi_*(A) = \max_{F \subseteq A} n(F) \quad (35)$$

since for a possibility measure Π, we always have $\Pi(A \cup B) = \max(\Pi(A), \Pi(B))$. Note that $\max_{F \subseteq S_2} n(F) = 1$

Lastly, we have

$$\forall A \subseteq S_2, \; \max(\Pi^*(A), \Pi_*(\bar{A})) = 1 \quad (36)$$

since

$\Gamma_2^{-1*}(A) \cup \Gamma_{2*}^{-1}(\bar{A}) = \{s \in S_1, \Gamma_2(s) \neq \emptyset\}$

<u>N.B.</u>: Sanchez /8/ has defined the upper inverse R^{-1*} and the lower inverse R_*^{-1} of a fuzzy relation on $X \times 2^S$, respectively by :

$$\forall x \in X, \forall A \subseteq S, \mu_{R^{-1*}}(A,x) = \sup_{A \cap F \neq \emptyset} \mu_R(x,F) \quad (37)$$

$$\forall x \in X, \forall A \subseteq S, \mu_{R_*^{-1}}(A,x) = \sup_{\substack{F \neq \emptyset \\ F \subseteq A}} \mu_R(x,F) \quad (38)$$

we recognize (34) and (35) in (37) and (38)

Γ IS A FUZZY MAPPING

A fuzzy relation is a fuzzy set on a cartesian product. A fuzzy relation R on X x S induces a fuzzy mapping Γ defined by

$$\forall x \in X, \forall s \in S, \mu_{\Gamma(x)}(s) = \mu_R(x,s) \quad (39)$$

$\mu_{\Gamma(x)}(s)$ estimates to what extent s belongs to the image of x by Γ. Clearly a multivalued mapping is a particular case of fuzzy mapping. When Γ is fuzzy, the focal element defined via (6), becomes fuzzy too.

Yager /11/ has proposed an extension of the formulae (7) and (8) which define a plausibility and a credibility function in terms of their basic probability assignments, to the case where the focal elements are fuzzy sets rather than crisp sets and where the events may also be fuzzy sets of S.

Then, we have

$$Pl(A) = \sum_F m(F) \cdot \sup_{s \in S} \min(\mu_A(s), \mu_F(s)) \quad (40)$$

$$Cr(A) = \sum_F m(F) \cdot \inf_{s \in S} \max(\mu_A(s), 1-\mu_F(s)) \quad (41)$$

provided that the number of focal elements is finite. When the focal elements are non fuzzy (40) and (41) reduces to (19) and (18) with $V = \mu_A$, i.e. the plausibility (resp. the credibility) of a fuzzy event is the upper (resp. lower) expectation of its membership function (see Smets /10/).

Notes that $\sup_s \min(\mu_A(s), \mu_F(s))$ is a degree of intersection which estimates to what extent $A \cap F$ is non empty while $\inf_s \max(\mu_A(s), 1-\mu_F(s))$ estimates to what extent F is included in A (see Dubois, Prade /4/) ; thus (40) and (41) clearly generalize (7) and (8). However, other scalar indices of intersection or of inclusion may be used.

In case of upper and lower possibilities, when Γ is non fuzzy, n can be viewed as the membership function of a fuzzy set of crisp sets (the focal elements) ; when Γ is fuzzy, one is led to the calculation of the upper and the lower possibility of an event out of the knowledge of a level 2 fuzzy set, i.e. a fuzzy set of fuzzy sets. Then, (34) and (35) may be generalized by :

$$\Pi^*(A) = \max_F \min(n(F), \mathrm{Inter}(A,F)) \quad (42)$$

$$\Pi_*(A) = \max_F \min(n(F), \mathrm{Incl}(A,F)) \quad (43)$$

where Inter(A,F) and Incl(A,F) respectively denote an index of intersection and an index of inclusion (of F in A).

N.B.: Note that in (42) and (43) "min" has been chosen, rather than the product for aggretating n(F) with the index because in possibility theory "min" plays a role which is analogous to the role played in probability theory by the product used in (40) and (41).

A RULE OF COMBINATION ANALOGOUS TO DEMPSTER'S RULE

Dempster's rule of combination of evidence (see Dempster /1/, Shafer /9/, enables to combine two bodies of evidence respectively represented by two basic probability assignments m^1 and m^2. The resulting basic probability assignment m is calculated as :

$$\text{i)}\ m(\emptyset) = 0$$

$$\text{ii)}\ \forall F \subseteq S,\ F \neq \emptyset,\ m(F) = \frac{\sum_{ij,\ G_i \cap H_j = F} m^1(G_i) \cdot m^2(H_j)}{1 - \sum_{ij,\ G_i \cap H_j = \emptyset} m^1(G_i) \cdot m^2(H_j)} \quad (44)$$

The normalization of m (F) in formulae (44) has been criticized by Zadeh /13/. See also Dubois, Prade /5/, Prade /7/.

By analogy, the following rule may be proposed for combining two basic possibility n^1 and n^2 into the basic possibility assignment n :

$$\text{i)}\ n(\emptyset) = 0$$

$$\text{ii)}\ \forall F \neq \emptyset,\ n(F) = \max_{i,j,\ G_i \cap H_j = F} \min(n^1(G_i), n^2(H_j)) \quad (45)$$

Generally, a possibility measure Π on S can be defined via a so-called possibility distribution π from S to [0,1] such that : (S finite)

$$\forall A \subseteq S,\ \Pi(A) = \sup_{s \in A} \pi(s) \quad (46)$$

Then, $\pi(s) = \Pi(\{s\})$. Let π^1 and π^2 be the possibility distributions respectively associated to the possibility measures generated by n^1 and n^2 (by means of formula (34)) ; let π the possibility distribution associated to the possibility measure generated by n (calculated from n^1 and n^2 by (45)) ; then, it can be checked that

$$\pi = \min(\pi^1, \pi^2) \quad (47)$$

while Dempster rule applied to the same possibility measures yields a plausibility measure such that : (Dubois-Prade /5/)

$$Pl(\{s\}) = \pi^1(s) \cdot \pi^2(s) .$$

APPLICATION TO APPROXIMATE REASONING

We consider a collection of propositions of the form :

"if we are in situation i, then X is Ai"

where $i \in I = [\![1, n]\!]$, and the Ai's are fuzzy sets restricting the possible values of some attribute of X. Some possibility distribution on I is known. Let $\pi(i)$ be the possibility of being in situation i. Then, the question to answer is "How possible is 'X is E' ?" where E is some fuzzy attribute value. A similar problem has been solved by Zadeh /14/ in the case where a probability distribution (rather than a possibility distribution) is available on the situations.

Then, using (42) and (43) we can compute the upper an the lower possibility of E :

$$\Pi^*(E) = \max_i \min(\pi(i), \sup_s \min(\mu_E(s), \mu_{A_i}(s))) \quad (48)$$

$$\Pi_*(E) = \max_i \min(\pi(i), \inf_s \max(\mu_E(s), 1-\mu_{A_i}(s))) \quad (49)$$

If the degrees of possibility $\pi(i)$ are not precisely known, but their possible values are represented by fuzzy numbers $Q(i)$, we are led to the computation of expressions such as :

$$\sup \min(\mu_{Q(1)}(q_1), \ldots, \mu_{Q(n)}(q_n)) \quad (50)$$
$$z = \max_i (\min(k(i), q_i))$$
$$1 = \max_i q_i$$

where the $k(i)$'s are scalars and the extra constraint $1 = \max_i q_i$ guarantees that the q_i's form a possibility distribution.

N.B.: Note that when $E=\{s\}$, (48) reduces to

$\Pi^*(\{s\}) = \pi^*(s) = \max_i \min(\pi(i), \mu_R(i,s))$

with $\mu_R(i,s) = \mu_{A_i}(s)$, i.e. the usual scheme of approximate reasoning in possibility theory (R being a fuzzy relation whose membership function μ_R can be viewed as a conditional possibility distribution).

For $E=\{s\}$, (49) gives $\pi_*(s) = 0$ except if $\exists\, i, \forall\, s' \neq s, \mu_{A_i}(s') < 1$ and $\pi(i) \neq 0$.

APPLICATION TO DECISION THEORY

The above application can be restated in terms of decision theory. Now, the situations i are the possible consequences of some decision D, and A_i is the reward of D if situation i follows D. In the usual setting of utility theory the assumptions are by far idealistic since a precise knowledge of the probabilities (it is assumed that it makes sense to speak in terms of probabilities) of consequences and precise reward assessments are required.

If precise probabilities $p(i)$ are available, formulae (40) and (41) enable to compute the plausibility and the credibility of a reward E. When, the probabilities are fuzzy numbers P_i, we are led to the computation of expressions such as :

$$\sup_{p_i} \min(\mu_{P_1}(p_1) \ldots \mu_{P_n}(p_n)) \quad (51)$$
$$z = \sum_i k(i) \cdot p_i$$
$$1 = \sum_i p_i$$

where the $k(i)$'s are scalars. Using results of Dubois, Prade /3/, the computation of expressions such as (51) is easy to perform without approximation.

If possibilities, rather than probabilities, are available the upper and lower possibilities of getting a reward E can be computed using formulae (48), (49) and (50). Moreover, expectations can be computed using recents results of Dubois, Prade /6/.

CONCLUDING REMARK

This paper has reconsidered Dempster's approach from a possibility point of view and sketched some applications. We may think of other applications, for instance, in pattern recognition problems, the situations i may be interpreted as the possible configurations of a set of objects and the A_i's as sets of interpretations.

REFERENCES

/1/ Dempster,A.P.(1967) Upper and lower probabilities induced by a multivalued mapping.Ann.Math.Statist.vol.38,pp 325-339

/2/ Dubois D., Prade H.,(1980) Fuzzy sets and systems : theory and applications. Vol.114,in Mathematics in Sciences and Eng. Series,Academic Press,New York

/3/ Dubois D.,Prade H.,(1981) Additions of interactive fuzzy numbers. IEEE Automatic Control, vol AC-26, n°4, pp 926-936

/4/ Dubois D., Prade H.,(1982) A unifying view of comparison indices in a fuzzy set-theoretic framework ; in:Fuzzy Set and Possibility Theory : Recents Developments (R.R.Yager ed) Pergamon Press, pp 3-13

/5/ Dubois D., Prade H.,(1982) On several representations of an uncertain body of evidence; in:Fuzzy Information and Decision Processes (M.M.Gupta, E.Sanchez, eds) North Holland, pp 167-181

/6/ Dubois D., Prade H.(1982) Upper and lower possibilistic expectations and some applications. Proc.4 Int.Seminar on Fuzzy Set Theory (E.P.Klement ed)Linz, Austria, Sept.13-17, 1982

/7/ Prade H.(1982) Modèles mathématiques de l'imprécis et de l'incertain en vue d'applications au raisonnement naturel. Thèse d'Etat Université Paul Sabatier, Juin 1982, 358 p.

/8/ Sanchez E.(1979) Inverses of fuzzy relations. Application to possibility distributions and medical diagnosis. Fuzzy Sets and Systems vol.2, pp 75-86

/9/ Shafer G.(1976) A mathematical theory of evidence. Princeton University Press, Princeton, 298 p.

/10/ Smets P. (1981) The degree of belief in a fuzzy event. Information Sci. vol.25 pp. 1-19

/11/ Yager R.R. (1982) Generalized probabilities of fuzzy events from fuzzy belief structures. Tech. Rep. MII 203, Iona College, New Rochelle, NY (19p.)

/12/ Zadeh L.A. (1978) Fuzzy sets as a basis for a theory of possibility. Fuzzy sets and Systems, vol.1 pp 3-28

/13/ Zadeh L.A. (1979) On the validity of Dempster's rule of combination of evidence. Memo UCB/ERL M79/24, University of California, Berkeley (12p)

/14/ Zadeh L.A. (1979) Fuzzy sets and information granularity. in Advances in Fuzzy Set Theory and Applications (M.M. Gupta, R.K. Ragade, R.R. Yager eds) North Holland, Amsterdam pp 3-18

FUZZY REASONING WITH VARIOUS FUZZY INPUTS

M. Mizumoto

Department of Management Engineering, Osaka Electro-Communication University, Neyagawa, Osaka 572, Japan

Abstract. This paper compares inference results of a fuzzy conditional inference with a fuzzy input A' and a fuzzy conditional A ⟹ B under several translating rules for the conditional A ⟹ B by Zadeh, Mamdani and Mizumoto when a fuzzy input A' is a fuzzy set obtained by attaching to the fuzzy set A a linguistic hedge such as *slightly*, *sort of*, *highly* and so on. It is shown that the translating rule Rs proposed before by the author can get reasonable inference results which fit our intuition. Moreover, a new composition called "max-$\triangle$ composition" is introduced and it is shown that the inference results for various fuzzy inputs A' are better than those under the ordinal compositional rule of inference which uses "max-min composition."

Keywords. Fuzzy reasoning; fuzzy conditional inference; linguistic hedge

INTRODUCTION

In our daily life we often make such an inference of the form:

Prem 1:	If x is A then y is B	
Prem 2:	x is A'	(1)
Cons:	y is B'	

where A, A', B, B' are fuzzy concepts. In order to make such an inference, Zadeh (1975) suggested an inference rule called "compositional rule of inference" which infers B' of Cons from Prem 1 and 2 by taking the max-min composition of fuzzy set A' and the fuzzy relation which is translated from the fuzzy conditional proposition "If x is A then y is B." In this connection, he (1975), Mamdani (1977) and Mizumoto et al. (1979, 1982) suggested several translating rules for translating the fuzzy proposition "If x is A then y is B" into a fuzzy relation.

In Mizumoto (1979, 1982) we compared inference results by their translating rules only when A' of Prem 2 is A, *very* A (= A^2), *more or less* A (= $A^{0.5}$) and *not* A (= ¬A), most of which are special case of A^{α}.

It will be of interest to obtain and discuss inference results under other kinds of A'. In this paper we obtain inference results when A' is a fuzzy set obtained by attaching to the fuzzy set A a linguistic hedge such as *slightly*, *sort of*, *highly*, INT, WEAK, MIDI and so on, and discuss which translating rule can get reasonable inference results.

FUZZY CONDITIONAL INFERENCE

We shall consider the form of inference of (1) in which a fuzzy conditional proposition "If x is A then y is B" is contained. The inference may be viewed as fuzzy modus ponens which reduces to the classical modus ponens when A' = A and B' = B.

For simplicity, we shall rewrite (1) as

$$\frac{\begin{array}{l} A \Rightarrow B \\ A' \end{array}}{B'} \qquad (2)$$

where A, A', B, B' are fuzzy concepts which are represented by fuzzy sets in universes of discourse U, U, V and V, respectively.

The fuzzy conditional A ⟹ B may represent a certain relationship between A and B. From this point of view, Zadeh (1975), Mamdani (1977) and Mizumoto et al. (1979, 1982) proposed several translating rules for translating A ⟹ B into a fuzzy relation in U x V.

Let A and B be fuzzy sets in U and V, respectively, and let x and ⊕ be cartesian product and bounded-sum for fuzzy sets, respectively. Then the following fuzzy relations in U x V can be translated from A ⟹ B. The fuzzy relations Rm and Ra were proposed by Zadeh, Rc by Mamdani, and the others by Mizumoto by introducing the implications of many-valued logic systems. For example, Ra (arithmetic rule) is given as

$$Ra = (\neg A \times V) \oplus (U \times B) \qquad (3)$$

$$= \int_{U \times V} 1 \wedge (1 - \mu_A(u) + \mu_B(v)) \,/\, (u,v).$$

It is noted that this rule is based on the implication rule of Lukasiewicz's logic, i.e.,

$$a \to b = 1 \wedge (1 - a + b), \quad a,b \in [0,1] \qquad (4)$$

Therefore, as other translating rules, it is possible to introduce other implication rules of many-valued logic systems to a translating rule for A ⟹ B (cf. Mizumoto (1982)).

Now, let $\mu_A(u) = a$ and $\mu_B(v) = b$, then we have such translating rules as

$$\text{Rm:}\quad (a \wedge b) \vee (1 - a). \tag{5}$$

$$\text{Ra:}\quad 1 \wedge (1 - a + b). \tag{6}$$

$$\text{Rc:}\quad a \wedge b. \tag{7}$$

$$\text{Rs:}\quad \begin{cases} 1 & \ldots\; a \leqq b, \\ 0 & \ldots\; a > b. \end{cases} \tag{8}$$

$$\text{Rg:}\quad \begin{cases} 1 & \ldots\; a \leqq b, \\ b & \ldots\; a > b. \end{cases} \tag{9}$$

$$\text{Rb:}\quad (1 - a) \vee b. \tag{10}$$

$$R_\Delta:\quad \begin{cases} 1 & \ldots\; a \leqq b, \\ \frac{b}{a} & \ldots\; a > b. \end{cases} \tag{11}$$

In the fuzzy modus ponens of (2), the consequence B' can be deduced from Prem 1 and 2 by taking the max-min composition "o" of the fuzzy set A' and the fuzzy relation obtained in (5)-(11) (the compositional rule of inference). For example, the consequence Ba' by the rule Ra is given as

$$Ba' = A' \circ Ra. \tag{12}$$

$$\mu_{Ba'}(v) = \bigvee_u \{\mu_{A'}(u) \wedge \mu_{Ra}(u,v)\} \tag{13}$$

$$= \bigvee_u \{\mu_{A'}(u) \wedge [1 \wedge (1 - \mu_A(u) + \mu_B(v))]\}.$$

In the same way, we have

$$Bm' = A' \circ Rm. \tag{14}$$

$$Bc' = A' \circ Rc.$$

$$\vdots$$

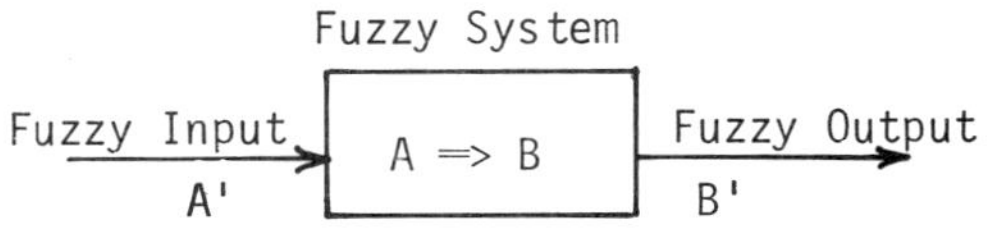

Fig. 1. Fuzzy system (A => B) with fuzzy input A' and fuzzy output B'.

The fuzzy modus ponens of (2) represents that the consequence B' is deduced when the premise A' is given under the condition A => B. If we regard the fuzzy conditional A => B (that is, fuzzy relation) as a fuzzy system, then A' and B' correspond to "fuzzy input" and "fuzzy output," respectively (See Fig.1). It will be interesting to discuss what kinds of fuzzy outputs B' are obtained when various kinds of fuzzy inputs A' are input to the fuzzy system.

LINGUISTIC HEDGES

In order to obtain various fuzzy inputs A', we shall briefly review some linguistic hedges proposed by Zadeh (1975) and introduce new artificial linguistic hedges.

Let A be a fuzzy set in U. Linguistic hedges which act on the fuzzy set A are listed as follows (See Fig.2).

As a special case of A^α $(= \int_U \mu_A(u)^\alpha/u)$, we can have such linguistic hedges as

$$\text{CON}(A) = \textit{very } A = A^2 \tag{15}$$

$$\text{DIL}(A) = \textit{more or less } A = A^{0.5} \tag{16}$$

$$\textit{minus } A = A^{0.75} \tag{17}$$

$$\textit{plus } A = A^{1.25} \tag{18}$$

$$\textit{highly } A = \textit{plus very } A = A^{2.5} \tag{19}$$

where CON, DIL and the following INT stand for "concentration", "dilation" and "contrast intensification", respectively.

$$\text{INT}(A) = \int_{\mu_A(u) \leqq 0.5} 2\mu_A(u)^2/u + \int_{\mu_A(u) \geqq 0.5} 1 - 2(1 - \mu_A(u))^2/u \tag{20}$$

Using the above linguistic hedges, we can obtain the following linguistic hedges.

$$\textit{slightly } A = \text{NORM}(A \textit{ and not very } A)^{¶}$$

$$= \text{NORM}(A \cap \neg\text{CON}(A)) \tag{21}$$

$$= \frac{\sqrt{5} - 1}{2}\Big(\int_U \mu_A(u) \wedge (1 - \mu_A(u)^2)/u\Big).$$

$$\textit{sort of } A = \text{NORM}(\text{DIL}(A) \cap \neg\text{CON}(A)^2)$$

$$= \text{NORM}(\textit{more or less but not very very } A) \tag{22}$$

$$\fallingdotseq 1.232\Big(\int_U \mu_A(u)^{0.5} \wedge (1 - \mu_A(u)^4)/u\Big).$$

The above are main linguistic hedges proposed by Zadeh. It is found that linguistic hedges can be viewed as operators which act on a fuzzy set. From this point of view, we can introduce new operators on a fuzzy set. Some of these are introduced as follows.

The effect of "contrast weakening" (WEAK, for short) is the opposite of that of INT.

$$\text{WEAK}(A) \tag{23}$$

$$= \int_{\mu_A(u) \leqq 0.5} -2(\mu_A(u)^2 - \mu_A(u))/u + \int_{\mu_A(u) \geqq 0.5} 2\Big(\mu_A(u) - \frac{1}{2}\Big)^2 + \frac{1}{2}/u$$

The operator of "middle intensification" (MIDI for short) has the effect of intensifing middle grades and is defined as

$$\text{MIDI}(A) = \text{NORM}(A \cap \neg A) = 2(A \cap \neg A)$$

$$= \int_U 2\mu_A(u) \wedge 2(1 - \mu_A(u))/u. \tag{24}$$

As the opposite operator to MIDI, we can give MIDW ("middle weakening") as follows.

$$\text{MIDW}(A) = \neg\text{MIDI}(A) = \neg 2(A \cap \neg A)$$

$$= \int_U (1 - 2\mu_A(u)) \wedge (2\mu_A(u) - 1)/u. \tag{25}$$

The operator αCUT obtains the α-level set of a fuzzy set A. The operator αCUT* is the opposite operator to αCUT, that is,

$$\alpha\text{CUT}(A) = \int_{\mu_A(u) \geqq \alpha} 1/u + \int_{\mu_A(u) < \alpha} 0/u \tag{26}$$

¶ NORM (= normalization) is defined as

$$\text{NORM}(A) = \frac{1}{\mu_{A*}} A \quad \text{with } \mu_{A*} = \bigvee_u \mu_A(u)$$

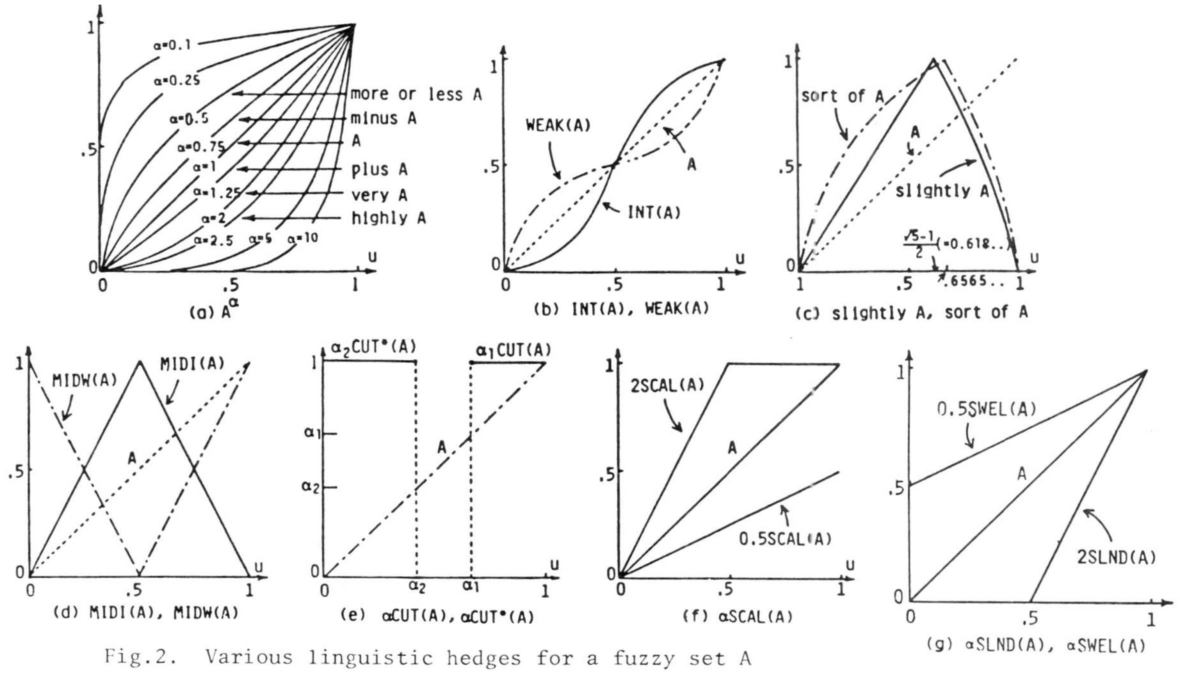

Fig.2. Various linguistic hedges for a fuzzy set A

$$\alpha\mathrm{CUT}^*(A) = \int_{\mu_A(u)\leq\alpha} 1/u \; + \int_{\mu_A(u)>\alpha} 0/u \qquad (27)$$

The operator αSCAL which gives the scalor product αA of α and A is defined as

$$\alpha\mathrm{SCAL}(A) = \int_U \alpha\mu_A(u) \wedge 1 \;/u \qquad (28)$$

Finally we shall give two operators which have the effects of "slenderizing" and "swelling" a fuzzy set A. The first is named as αSLND and the latter as αSWEL. They have the same expression but they are distinguished from the values of their parameter α. That is to say,

$$\alpha\mathrm{SLND}(A) = \int_U 0 \vee (\alpha\mu_A(u) + 1 - \alpha)/u \;\ldots\; \alpha\geq 1 \qquad (29)$$

$$\alpha\mathrm{SWEL}(A) = \int_U \alpha\mu_A(u) + 1 - \alpha \;/u \;\ldots\; \alpha\leq 1 \qquad (30)$$

Fig. 2 shows the effects of the linguistic hedges of (15)-(30) on a fuzzy set A, where A is a fuzzy set $\int_U u/u$ in U = [0,1].

INFERENCE RESULTS FOR VARIOUS FUZZY INPUTS

We shall obtain the consequence B' under each translating rule of (5)-(11) when A' is a fuzzy set given by applying linguistic hedges to A, and discuss which method can get reasonable consequences.

We shall discuss only the case of Rm (5) at $A' = A^\alpha$ (as a general case of (15)-(19)).

When $A' = A^\alpha$, the consequence Bm' is obtained as

$$\mu_{Bm'}(v) = \bigvee_u\{\mu_A(u)^\alpha \wedge [(\mu_A(u) \wedge \mu_B(v)) \vee (1-\mu_A(u))]\}$$

This expression can be rewritten as (32) by letting

$$x = \mu_A(u), \quad b = \mu_B(v), \quad bm' = \mu_{Bm'}(v) \qquad (31)$$

under the assumption that $\mu_A(u)$ takes all values in [0,1] according to u varying all over U, that is, μ_A is a function onto [0,1], i.e., x is on [0,1].

$$bm' = \bigvee_x \{x^\alpha \wedge [(x \wedge b) \vee (1-x)]\} \qquad (32)$$

$$f(x) = x^\alpha \wedge [(x \wedge b) \vee (1-x)] \qquad (33)$$

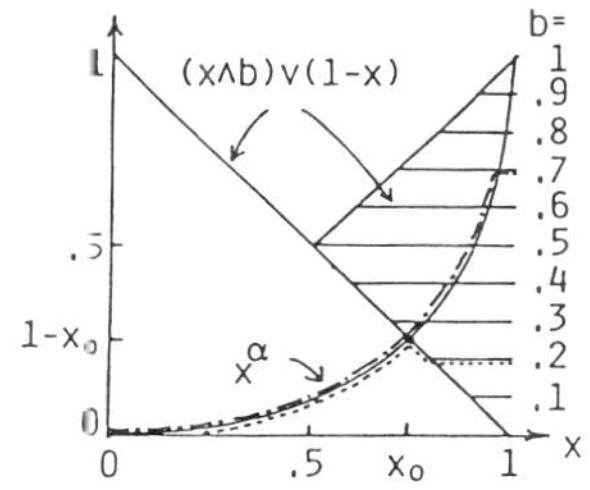

Fig.3 The way of obtaining (32)

Fig. 3 shows the expressions x^α and $(x \wedge b) \vee (1-x)$ using a parameter b. When x^α is as in this figure and b is equal to, say, 0.2, the expression f(x) is indicated by the broken line and hence bm' at b = 0.2 is the maximum value of this broken line. The value is equal to the height of the cross point of x^α and 1-x. Thus, let x_0 ($\in$ [0,1]) be the solution of $x^\alpha = 1-x$, then the height (i.e., maximum value) is given by $1 - x_0$. Therefore, we have bm' at $b \leq 1-x_0$ as

$$bm' = \bigvee_x f(x) = 1 - x_0 \;\ldots\; b \leq 1 - x_0$$

On the other hand, when b = 0.7 ($\geq 1-x_0$), f(x) is given by the dot-dash line and its maximum value is b (=0 7). Thus, bm' = b for $b \geq 1-x_0$. Therefore, we have

$$bm' = \begin{cases} 1 - x_0 & \ldots\; b \leq 1 - x_0 \\ b & \ldots\; b \geq 1 - x_0 \end{cases}$$
$$= (1 - x_0) \vee b.$$

TABLE 1 Inference Results B' = A' o R under Max-Min Composition "o"

A' \ A ⟹ B	Rm	Ra	Rc	Rs	Rg	Rb	R_Δ
A^α	$(1-x_o) \vee \mu_B$ (*)	$1-x'+\mu_B$ (**)	μ_B	μ_B^α	$\begin{cases}\mu_B^\alpha \ldots \alpha\leq 1\\ \mu_B \ldots \alpha\geq 1\end{cases}$	$(1-x_o) \vee \mu_B$	$\mu_B^{\frac{\alpha}{\alpha+1}}$
INT(A)	$0.5 \vee \mu_B$	$\frac{4\mu_B-1+\sqrt{9-8\mu_B}}{4}$	μ_B	INT(B)	$\begin{cases}\mu_B \ldots 0\leq\mu_B\leq 0.5\\ 1-2(1-\mu_B)^2 \ldots 0.5\leq\mu_B\leq 1\end{cases}$	$0.5 \vee \mu_B$	See (35)
slightly A	$(\frac{\sqrt{5}-1}{2} \vee \mu_B) \wedge m_o$ (***)	$\frac{\sqrt{5}-1}{2}(1+\mu_B) \wedge 1$	$\mu_B \wedge m_o$	$\frac{\sqrt{5}+1}{2}\mu_B \wedge 1$	$\frac{\sqrt{5}+1}{2}\mu_B \wedge 1$	$\frac{\sqrt{5}-1}{2} \vee \mu_B$	$\sqrt{\frac{\sqrt{5}+1}{2}\mu_B} \wedge 1$
sort of A	$(0.689 \vee \mu_B) \wedge 0.779$	$\frac{c(\sqrt{4(1+\mu_B)+c^2}-c)}{2} \wedge 1$ (****)	$\mu_B \wedge 0.779$	$c\sqrt{\mu_B} \wedge 1$	$c\sqrt{\mu_B} \wedge 1$	$0.689 \vee \mu_B$	$\sqrt[3]{c^2\mu_B} \wedge 1$
WEAK(A)	$0.5 \vee \mu_B$	$\frac{3+4\mu_B-\sqrt{1+8\mu_B}}{4}$	μ_B	WEAK(B)	$\begin{cases}-2(\mu_B^2-\mu_B) \ldots 0\leq\mu_B\leq 0.5\\ \mu_B \ldots 0.5\leq\mu_B\leq 1\end{cases}$	$0.5 \vee \mu_B$	See (36)
MIDI(A)	$\frac{2}{3}$	$\frac{2}{3}(1+\mu_B) \wedge 1$	$\mu_B \wedge \frac{2}{3}$	$2\mu_B \wedge 1$	$2\mu_B \wedge 1$	$\frac{2}{3} \vee \mu_B$	$\sqrt{2\mu_B} \wedge 1$
MIDW(A)	1	1	$\frac{1}{3} \vee \mu_B$	1	1	1	1
αCUT(A)	$(1-\alpha) \vee \mu_B$	$(1-\alpha+\mu_B) \wedge 1$	μ_B	αCUT(B)	$\begin{cases}1 \ldots \mu_B\geq\alpha\\ \mu_B \ldots \mu_B<\alpha\end{cases}$	$(1-\alpha) \vee \mu_B$	$\frac{\mu_B}{\alpha} \wedge 1$
αCUT*(A)	1	1	$\mu_B \wedge \alpha$	1	1	1	1
αSCAL(A)	$\begin{cases}(\frac{\alpha}{\alpha+1} \vee \mu_B) \wedge \alpha \ldots \alpha\leq 1\\ \frac{\alpha}{\alpha+1} \vee \mu_B \ldots \alpha\geq 1\end{cases}$	$\begin{cases}\frac{\alpha}{\alpha+1}(1+\mu_B) \wedge \alpha \ldots \alpha\leq 1\\ \frac{\alpha}{\alpha+1}(1+\mu_B) \wedge 1 \ldots \alpha\geq 1\end{cases}$	$\begin{cases}\mu_B \wedge \alpha \ldots \alpha\leq 1\\ \mu_B \ldots \alpha\geq 1\end{cases}$	αSCAL(B)	$\begin{cases}\mu_B \wedge \alpha \ldots \alpha\leq 1\\ \alpha\mu_B \wedge 1 \ldots \alpha\geq 1\end{cases}$	$\begin{cases}(\frac{\alpha}{\alpha+1} \vee \mu_B) \wedge \alpha \ldots \alpha\leq 1\\ \frac{\alpha}{\alpha+1} \vee \mu_B \ldots \alpha\geq 1\end{cases}$	$\begin{cases}\sqrt{\alpha\mu_B} \wedge \alpha \ldots \alpha\leq 1\\ \sqrt{\alpha\mu_B} \wedge 1 \ldots \alpha\geq 1\end{cases}$
αSLND(A) ($\alpha\geq 1$)	$\frac{1}{1+\alpha} \vee \mu_B$	$\frac{1+\alpha\mu_B}{1+\alpha}$	μ_B	αSLND(B)	μ_B	$\frac{1}{1+\alpha} \vee \mu_B$	$\frac{1-\alpha+\sqrt{(1-\alpha)^2+4\alpha\mu_B}}{2}$
αSWEL(A) ($\alpha\leq 1$)				αSWEL(B)	αSWEL(B)		

(*) x_o ($\in[0,1]$) is the solution of $x^\alpha = 1-x$

(**) x' ($\in[0,1]$) is the solution of $x^\alpha = 1-x+\mu_B$

(***) $m_o = \frac{1-\sqrt{5}+\sqrt{22-2\sqrt{5}}}{4} = 0.7376\ldots$

(****) $c = 1.232\ldots$

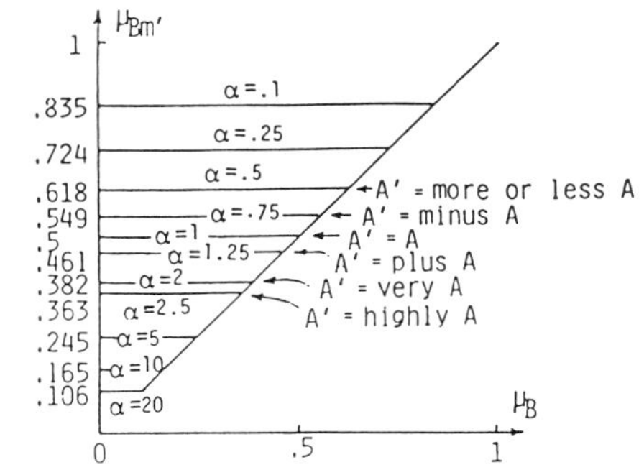

Fig.4. Bm' = A^{α} o Rm

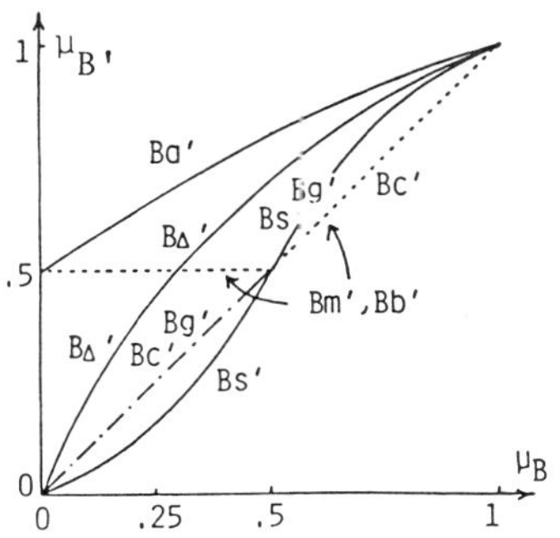

Fig.6. B' = INT(A) o R

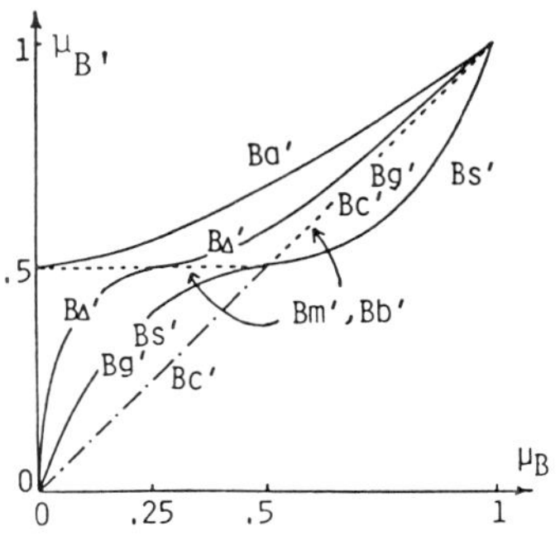

Fig.7. B' = WEAK(A) o R

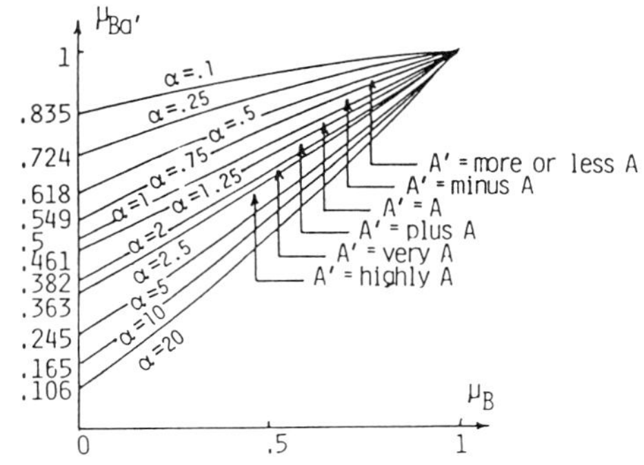

Fig.5 Ba' = A^{α} o Ra

This result is shown to hold for any α. Therefore, using the notation of (31), the consequence Bm' = A^{α} o Rm is as follows.

$$\mu_{Bm'}(u) = (1 - x_o) \vee \mu_B(u) \tag{34}$$

where x_o ($\in[0,1]$) is the solution of $x^{\alpha} = 1 - x$.

Fig.4 shows the consequence Bm' (= A^{α} o Rm) using a parameter α. Fig.5 also indicated Ba' (= A^{α} o Ra) which can be obtained in the same way as Bm'. Table 1 shows the inference results for other rules and A', where the notation μ_B stands for $\mu_B(v)$.

The consequences BΔ' for the rule R_Δ (11) at A' = INT(A) and WEAK(A) are shown in (35) and (36).

Case of BΔ' = INT(A) o R_Δ:

$$\mu_{B_\Delta'} = \begin{cases} \sqrt[3]{2\mu_B^2} & \cdots\ 0 \leqq \mu_B \leqq \frac{1}{4} \\ \frac{\mu_B}{x} & \cdots\ \frac{1}{4} \leqq \mu_B \leqq 1 \end{cases} \tag{35}$$

where

$$x = \frac{1}{3}\left(2 + \sqrt{10}\cos\frac{\theta+\pi}{3}\right), \quad \theta = \cos^{-1}\left[\frac{\sqrt{10}}{50}(27\mu_B - 14)\right].$$

Case of BΔ' = WEAK(A) o RΔ:

$$\mu_{B_\Delta'} = \begin{cases} \frac{\mu_B}{x} & \cdots\ 0 \leqq \mu_B \leqq \frac{1}{4} \\ \frac{\mu_B}{x'} & \cdots\ \frac{1}{4} \leqq \mu_B \leqq 1 \end{cases} \tag{36}$$

$$x = \frac{1}{3}\left(1 - 2\cos\frac{\psi+\pi}{3}\right), \quad \psi = \cos^{-1}\left(1 - \frac{27}{4}\mu_B\right),$$

$$x' = \frac{1}{3}\left\{1 + \sqrt[3]{\frac{1}{4}\left(27\mu_B - 5 + 9\sqrt{\frac{27\mu_B^2 - 10\mu_B + 1}{3}}\right)} + \sqrt[3]{\frac{1}{4}\left(27\mu_B - 5 - 9\sqrt{\frac{27\mu_B^2 - 10\mu_B + 1}{3}}\right)}\right\}$$

The membership function $\mu_{B\Delta'}$ of (35) and (36) are very complicated and so we shall show in Fig. 6 and 7 the diagrams of $\mu_{B\Delta'}$ together with the inference results by other methods.

In the form of the fuzzy inference of (2), it is quite natural to expect that B' = B will be obtained when A' = A (satisfaction of "modus ponens"). This criterion is satisfied by the rules Rc, Rs and Rg. See the results for A^{α} at α = 1 in Table 1. Namely, these methods obtain A o R = B. Moreover, it is also natural to expect B' ≃ B when A' ≃ A. The method Rs satisfies this criterion and obtains consequences B' = B^{α}, INT(B), WEAK(B), αCUT(B), αSCAL(B), αSLND(B) and αSWEL(B) at A' = A^{α}, INT(A), WEAK (A), αCUT(A), αSCAL(A), αSLND(A) and αSWEL(A), respectively. The other methods do not obtain such results. Note that Rc gets always B except the case of αSCAL(A), α ≦ 1. In Mizumoto (1982) we showed that all the methods except Rc satisfy the natural criterion that B' = *unknown* (= $\int_V 1/v$) is obtained at A' = *not* A. In this connection, for the criterion that B' ≃ *unknown* is obtained at A' ≃ *not* A, we may say that all the methods except Rc satisfy this criterion. In fact, these methods obtain B' = *unknown* at A' = MIDW(A) and αCUT*(A) which are similar to *not* A. Finally, it is not yet known what kinds of consequences are good for A' = *sort of* A, *slightly* A and MIDI(A) which lie between A and *not* A.

From the above considerations, we can conclude that the method Rs (8) is most suitable for the fuzzy conditional inference, though the given criteria are intuitive and rough.

INFERENCE RESULTS UNDER NEW COMPOSITION

We shall introduce new composition called "max-$\dot\wedge$ composition" for the compositional rule of inference, and show that the inference results under the new composition are better than those under the max-min composition "o" discussed above.

Introducing new operation $\dot\wedge$ (drastic product)

$$x \,\dot\wedge\, y = \begin{cases} x & \cdots\ y = 1 \\ y & \cdots\ x = 1 \\ 0 & \cdots\ \text{otherwise} \end{cases} \tag{37}$$

new composition "▲" is obtained from (13) by replacing $\wedge$ by $\dot\wedge$. For example, the consequence Ba' by Ra under ▲ is given by the following.

TABLE 2 Inference Results $B' = A' \blacktriangle R$ under Max-$\dot{\wedge}$ Composition "$\blacktriangle$"

A' \ A⟹B	Rm	Ra	Rc	Rs	Rg	Rb	R$_\Delta$
A^α	μ_B	$\begin{cases}\mu_B^\alpha \ldots \alpha \leqq 1 \\ \mu_B \ldots \alpha \geqq 1\end{cases}$	μ_B	B^α	$\begin{cases}\mu_B^\alpha \ldots \alpha \leqq 1 \\ \mu_B \ldots \alpha \geqq 1\end{cases}$	μ_B	$\begin{cases}\mu_B^\alpha \ldots \alpha \leqq 1 \\ \mu_B \ldots \alpha \geqq 1\end{cases}$
INT(A)	μ_B	$\mu_B \vee [1-2(1-\mu_B)^2]$	μ_B	INT(B)	$\mu_B \vee [1-2(1-\mu_B)^2]$	μ_B	$\mu_B \vee [1-2(1-\mu_B)^2]$
slightly A	$(\frac{3-\sqrt{5}}{2} \vee \mu_B) \wedge \frac{\sqrt{5}-1}{2}$	$(\frac{3-\sqrt{5}}{2} + \mu_B) \wedge 1$	$\mu_B \wedge \frac{\sqrt{5}-1}{2}$	$\frac{\sqrt{5}+1}{2}\mu_B \wedge 1$	$\frac{\sqrt{5}+1}{2}\mu_B \wedge 1$	$\frac{3-\sqrt{5}}{2} \vee \mu_B$	$\frac{\sqrt{5}+1}{2}\mu_B \wedge 1$
sort of A	$(0.3413 \vee \mu_B) \wedge 0.6587$	$c\sqrt{\mu_B} \wedge 1$ (*)	$\mu_B \wedge 0.6587$	$c\sqrt{\mu_B} \wedge 1$	$c\sqrt{\mu_B} \wedge 1$	$0.3413 \vee \mu_B$	$c\sqrt{\mu_B} \wedge 1$
WEAK(A)	μ_B	$-2(\mu_B^2 - \mu_B) \vee \mu_B$	μ_B	WEAK(B)	$-2(\mu_B^2 - \mu_B) \vee \mu_B$	μ_B	$-2(\mu_B^2 - \mu_B) \vee \mu_B$
MIDI(A)	0.5	$1 \wedge (\mu_B + 0.5)$	$\mu_B \wedge 0.5$	$2\mu_B \wedge 1$	$2\mu_B \wedge 1$	$0.5 \vee \mu_B$	$2\mu_B \wedge 1$
MIDW(A)	1	1	μ_B	1	1	1	1
αCUT(A)	$(1-\alpha) \vee \mu_B$	$(1-\alpha+\mu_B) \wedge 1$	μ_B	αCUT(B)	$\begin{cases}1 \ldots \mu_B \geqq \alpha \\ \mu_B \ldots \mu_B < \alpha\end{cases}$	$(1-\alpha) \vee \mu_B$	$\frac{\mu_B}{\alpha} \wedge 1$
αCUT*(A)	1	1	$\mu_B \wedge \alpha$	1	1	1	1
αSCAL(A)	$\begin{cases}0 \ldots \alpha<1 \\ (1-\frac{1}{\alpha}) \vee \mu_B \ldots \alpha \geqq 1\end{cases}$	$\begin{cases}\alpha\mu_B \ldots \alpha \leqq 1 \\ 1 \wedge (1-\frac{1}{\alpha}+\mu_B) \ldots \alpha \geqq 1\end{cases}$	$\begin{cases}0 \ldots \alpha<1 \\ \mu_B \ldots \alpha \geqq 1\end{cases}$	αSCAL(B)	αSCAL(B)	$\begin{cases}0 \ldots \alpha<1 \\ (1-\frac{1}{\alpha}) \vee \mu_B \ldots \alpha \geqq 1\end{cases}$	αSCAL(B)
αSLND(A) ($\alpha \geqq 1$)	μ_B	μ_B	μ_B	αSLND(B)	μ_B	μ_B	μ_B
αSWEL(A) ($\alpha \leqq 1$)	$(1-\alpha) \vee \mu_B$	αSWEL(B)		αSWEL(B)	αSWEL(B)	$(1-\alpha) \vee \mu_B$	αSWEL(B)

(*) c = 1.232...

$$Ba' = A' \blacktriangle Ra$$

$$\mu_{Ba'}(v) = \bigvee_u \{\mu_{A'}(u) \,\dot{\wedge}\, \mu_{Ra}(u,v)\}$$

The same way is applicable to other translating rule of (5)-(11).

Table 2 lists the inference results by all the translating rules for various fuzzy premises A' under the max-$\dot{\wedge}$ composition. It is found from the results at $A' = A^\alpha$ with $\alpha = 1$ that all the translating rules can satisfy so called modus ponens under the max-$\dot{\wedge}$ composition, though only the rules Rc, Rs and Rg satisfy the modus ponens under the max-min composition as shown in Table 1. As for other fuzzy premises A', we shall consider the case of A' = WEAK(A). The inference results for A' = WEAK(A) under the max-min composition "o" and the max-$\dot{\wedge}$ composition "$\blacktriangle$" are found in Figs 7 and 8, respectively. The rule Rs infers Bs' = WEAK(B) under each of these compositions. The other rules do not get such results. But these rules under the max-$\dot{\wedge}$ composition can infer the consequences which are very similar to WEAK(B) as in Fig.8, which leads to the satisfaction of the criterion that $B' \simeq B$ at $A' \simeq A$. Such tendency can be observed for other A'. Therefore, we may say that the max-$\dot{\wedge}$ composition is a better compositional rule of inference than the max-min composition.

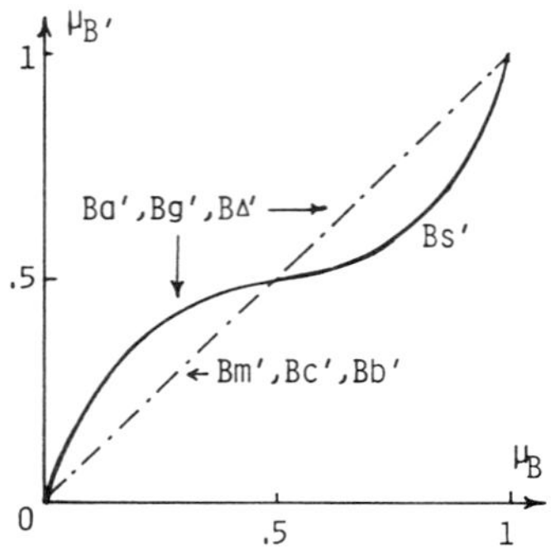

Fig.8. $B' = WEAK(A) \blacktriangle R$

CONCLUSION

Under the criterion that $B' \simeq B$ at $A' \simeq A$ and $B' \simeq$ *unknown* at $A' \simeq$ *not* A for the fuzzy modus ponens (1) and (2), it will be possible to make a quantitative analysis of the goodness of each translating rule by measuring a similarity of B' and B (or *unknown*) when A' is given which is similar to A (or *not* A).

The results of this paper will be useful to the problems such as fuzzy control, fuzzy diagnosis, fuzzy production system and so on which use fuzzy reasoning method with various fuzzy inputs.

REFERENCES

Mamdani, E.H. (1977). Application of fuzzy logic to approximate reasoning using linguistic systems. IEEE Trans. on Computer, C-26, 1182-1191.

Mizumoto, M., S. Fukami, and K. Tanaka. (1979). Some methods of fuzzy reasoning. In M.M. Gupta, et al. (Ed.), Advances in Fuzzy Set Theory and Applications, North-Holland, Amsterdam. pp. 117-136.

Mizumoto, M.,and H.J. Zimmermann. (1982). Comparison of fuzzy reasoning methods. Fuzzy Sets and Systems, 8, 253-283.

Zadeh, L.A. (1972). A fuzzy-set-theoretic interpretation of linguistic hedges. J. of Cybernetics, 2, 4-34.

Zadeh, L.A. (1975). Calculus of fuzzy restriction. In L.A. Zadeh, et al. (Ed.), Fuzzy Sets and Their Applications to Cognitive and Decision Processes, Academic Press, New York. pp. 1-39.

FUZZY INFORMATION AND FUZZY TIME

M. Vítek

Department of Economics and Management, University of Chemical Technology, Pardubice, Czechoslovakia

Abstract. A definition of fuzzy information with regard to the transition from the potential to actual, technical to sémantic, old to new information is proposed. Starting with this concept the notion of fuzzy time is developped. Then a measurement of fuzzy time by the volume of approximative knowledge is studied. The relation between fuzzy information and fuzzy time is important in socioeconomic data processing and control.

Keywords. Time-varying systems; philosophical aspects; control nonlinearities, system theory; fuzzy time; fuzzy information.

INTRODUCTION

Advances in systems theory, information theory and social planning bring an extended concept of information.
A deeper distinction of potential and actual, technical and sémantic, old and new information leeds to the consideration of fuzzy information, in which the actual aspect is enriched with potential features and possibilities.

In time-varying systems the concept of informational time is known, i.e. of time, measured by the volume of approximative knowledge. Also the dilatation /"slow-down"/ of time when the volume of knowledge is growing was studied. The concept of fuzzy time is founded on the volume of fuzzy information gained during a period of calender time.

THE CONCEPT OF FUZZY INFORMATION

Reflection and information are not only keywords in modern philosophical and system studies, but they are also general attributes of the matter. The reflection /in the philosophical sense/ means the ability of matter to record influences and change so the inner arrangement of systems. The information is one aspect of reflection, connected with its structural, formal and quantitative aspects. In a broad sense information means "arrangement", it can be also defined as 'organization", i.e. the arrangedness of a system.

We distinguish :

- technical and sémantic information /if we speak about signal transmission or about the meaning of reports/;

- actual and potential information /the first is active in the consciousness of man, the second is stored in a memory/;
- old and new information /according to the degree of originality, age etc./;
- reproduced or original information;
- condenzed or diluted information.

The border between different types of information is not sharp. It lies somewhere in the transition from the shorttime to the long-time memory /and vice versa/ and its identification needs more studies in consciousness, cognition and creativity.

The subjective approach to the both actual and potential information leads to the specification of the fuzzy information. As fuzzy information we denote an organization, to which a value of actuality for the subject is given.

More precisely, the fuzzy information I_F is a fuzzy set of elementary organizations /e.g. messages/, defined on the univerzal set of organizations :

$$I_F = \{\langle u_F(x), x\rangle\} = \bigcup_{x \in X} u_F(x)/x,$$

$x \in X$

where $u_F : X \longrightarrow [0,1]$ is the membership function for the fuzzy set I_F of elementary organizations $x \in X$.

In praxis, it is necessary to assign in an expert way the membership function values to all information quanta in a given context. The meaning of each message for the decision-maker in the view of his problems can be stated.

The fuzzy information can be expressed by means of fuzzy variables and linguistic variables. A fuzzy variable is defined as a triple

$$\langle a, X, C\rangle ,$$

where a is the name of the fuzzy variable,
X is the definition area,
C is a fuzzy set on X, interpreted as the sémantics of the fuzzy variable.

The linquistic variable is a system

$$\langle b, T, X, G, M\rangle ,$$

where b is the name of the linquistic variable,
T is the set of terms /names of fuzzy variables/,
X is the definition area for each fuzzy variable,
G is a syntactic procedure for the construction of new values of the linquistic variable,
M is a sémantic procedure for the transformation of each new value of the linquistic variable into a fuzzy variable.

As example, for the the linquistic variable "influence" fuzzy variables /terms/ are very weak, weak, middle, strong, very strong. The procedure G is an ordering of terms in the set T and the procedure M stands for an expert evaluation.

THE NOTION OF FUZZY TIME

The flow of time is incorporated in the movement and development of matter. Space and time are objective forms of the existence of matter. The motion cannot be regarded in the mechanistic sense only, but on different levels-as physical, biological and social - with various arrangement and density of events.

The notion of homogeneous, contentless, uniform and reversible time, known from the classical physics, is in the modern natural sciences overcomed by the concept of the non-homogeneous time, i.e. no uniform distribution of events is assumed. Such a heterogeneous time is not a autonomous, independent substance, but it is connected with the nature and quality of content and is not absolutely reversible /cyclic/.

In literature numerous interpretations of the nonlinear, nonhomogeneous time were given - it is considered as atomic, physiological, biological, economic, social, informational time. For example, the economic time is interpreted as an abstraction of the speed of economic processes /P. Hrubý/, the informational time is a time measured by the volume of knowledge /or by the accelerated informational surplus or by the condenzed informational stock - by the definition of J. Zeman/.

The theory of the heterogeneous time is near to the relativity theory. During high motion velocities we can describe a dilatation of time by the Lorentz transformation, known from the Relativistic Mechanics.

So, in complex processes the quality of combined physical, biological and social motion cannot be taken as simple. The fuzzy time is an interval, to which a subjective measure of heterogenity /subjectivity, interiorization, complexity/ has to be supplemented.

More exactly, the fuzzy time T_F is a fuzzy set of elementary time intervals, defined of an univerzal set of intervals Y :

$$T_F = \{\langle u_F(y), y\rangle\} =$$

$$= \bigcup_{y \in Y} u_F(y)/y, \quad y \in Y,$$

where $u_F : Y \longrightarrow [0,1]$ is the membership function to the fuzzy set T_F.

Practically, the fuzzy time can again be assesed by linquistic and fuzzy variables. For example, the linquistic variable "time of a human individuality" can be described by fuzzy variables rising, progressive, regressive and dooming, found in an expert procedure.

THE RELATION OF FUZZY INFORMATION AND FUZZY TIME

One way to measure the fuzzy time is to consider it as a form of informational time and determine the volume of fuzzy information in individual intervals.

When we take the case of accelerating growth of information during an infinite time horizont :

$$I = e^{at}, \quad a = \text{const},$$

we get after logatithming

$$t = \frac{1}{a} \ln I,$$

i.e. time t is a function of information volume I.

Taking a fuzzy function of a fuzzy argument, we obtain

$$f(I_F) = \bigcup_{x \in X} u_F(x)/f(x)$$

as fuzzification of the "normal" time t and "normal" information I.

In this way, the volume of fuzzy information I_F can be used as a measure of events, change or experience in the decision-making and control.

CONCLUSION

Such considerations are significant for studies in learning, memory and redundance in a broader sense. They can influence questions of information / decision-making systems in complex social and economic institutions and organizations. In this way, much information can be saved about subjects, their goals, motivations, wishes and interests. This information would be lost during a "hard" formalization of information and control processes.

REFERENCES

Hrubý, P. /1974/. Methods of Economic Time. Svoboda, Praha. 240 pp.

Vítek, M. /1981/. Introduction to the Modelling of Social Processes. Vysoká škola chemickotechnologická, Pardubice, 229 pp.

Zeman, J. /1978/. Theory of Reflection and Cybernetics. Academia, Praha. 251 pp.

A CLUSTERING METHOD FOR A FUZZY DIGRAPH BASED ON CONNECTEDNESS AND ITS APPLICATION TO INSTRUCTIONAL EVALUATION

M. Takeya

Computer & Communication Systems Research Laboratories, NEC Corporation Kawasaki 213, Japan

Abstract. This paper presents a formalization of cluster extraction algorithms for an asymmetric fuzzy data, i.e. a fuzzy digraph, and its applications to instructional evaluation. First, this paper classifies a fuzzy digraph into seven categories of fuzzy connectedness —1) strongly complete, 2) weakly complete, 3) bilaterally connected, 4) unilaterally connected, 5) strongly connected, 6) weakly connected, and 7) totally ordered digraphs. Second, extraction algorithms of subgraphs are presented according to individual 1)-6) connectivity categories. Then, it is shown that each cluster extraction algorithm operationally results in the clique extraction algorithm. Third, a minimum weakly complete digraph is defined and then an extraction algorithm based on a totally ordered category is presented. Fourth, these clustering procedures are applied to two kinds of the instructional evaluation problems. This paper discusses probelms on sociometry analysis and test analysis and shows usefulness of these procedures.

Keywords. Fuzzy classification; Fuzzy clustering; Fuzzy digraph; Instructional evaluation; Ordinal test theory; Sociometry.

INTRODUCTION

This paper discusses a clustering method for an asymmetric fuzzy data structure, i.e. a fuzzy digraph and its applications to instructional evaluation.

In general, investigation on the relationship among elements of a system is essential for the analysis of its behavior. The most natural measure of the relation between a pair of elements is an asymmetric similarity or a proximity similarity. The system can be represented in the form of an asymmetric data structure, which may be visualized using a fuzzy digraph. Particularly, in a complicated system, it is necessary to identify subsets of elements treated as components with common features. From a graph-theoretical point of view, these common features among items in a subset can correspond to graphical characteristics, e.g.degree of connectedness for the corresponding subgraph. The notation of connectivity is important for the application of the graph theory to social sciences.

In this paper, the concept of connectedness on a non-fuzzy graph is expanded to the connectedness concept on a fuzzy digraph. Feature extraction procedures, i.e. clustering algorithms on a fuzzy digraph, are presented according to the individual kinds of connectedness involved.

A lot of clustering procedures based on a symmetric measure are being investigated very actively, but only a few reports on clustering procedures for an asymmetric data structure have apparently been published to date. Hubert (1973) has presented hierarchical clustering procedures in such a way that the assumption of symmetric measure is unnecessary. When rearranging three procedures presented by Hubert from a connective view point, individual features of subsets, which are treated as clusters, can be categorized by three different connectivities. Hubert's algorithms have been expanded by Fujiwara (1980).

The purpose of this paper is a systematic formalization of a cluster extraction algorithms, which are little dependent upon connectivity category. First, this paper classifies a fuzzy digraph into seven categories of fuzzy connectivity — 1) strongly complete, 2) weakly complete, 3) bilaterally connected, 4) unilaterally connected, 5) strongly connected, 6) weakly connected, and 7) totally ordered digraphs on μ-level. Second, extraction algorithms of subgraphs are presented according to individual 1)-6) types of connectivity. Then, it is shown that each cluster extraction algorithm operationally results in the clique extraction algorithm. Third, a minimum weakly complete digraph is defined and then an extraction algorithm based on a totally ordered type is presented. Fourth, these clustering procedures are applied to two kinds of the instructional evaluation problems. This paper discusses problems on sociometry analysis and test analysis and shows usefulness of these procedures.

A FUZZY DIGRAPH AND ITS CONNECTEDNESS

BASIC NOTATION

Let a system be composed of n elements $v_1, v_2, \dots, v_n$. The degree of similarity measure between any pair of elements is introduced by extending the concept of a fuzzy digraph, following a fuzzy set theory which has been suggested by Zadeh (1971). The similarity measure r_{ij}, for element v_i to element v_j, is normalized such that $0 \leq r_{ij} \leq 1$. Here, $r_{ii}=1$ $(i=1,2,\dots,n)$. It is common in cluster analysis that the similarity measure r_{ij} be symmetric, namely $r_{ij}=r_{ji}$. In this paper, however, the similarity relation r_{ij} is asymmetric, i.e. in general $r_{ij} \neq r_{ji}$, because elements in the natural systems have asymmetric relation to each other.

Let $V=(v_1, v_2, \dots, v_n)$ be the set of nodes corresponding to elements in the system and let E be a fuzzy mapping function, where a membership function of E: $m_{E(v_i)}(v_j)=r_{ij}$, then, a graph $G=(V,E)$ is a fuzzy digraph. Alternatively, if $R=[r_{ij}]$ is a similarity matrix, then by defining the fuzzy relation $E(v_i, v_j)=r_{ij}$ for all $(v_i, v_j) \in V \times V$, it follows that a fuzzy digraph may be characterized by $G=(V, E)$.

First, define fundamental concepts for a fuzzy digraph.

[Definition 1] On a fuzzy digraph $G=(V, E)$:

(i) **A μ-level relation** $v_i \xrightarrow{\mu} v_j$ exists from a node v_i to a node v_j, if and only if $r_{ij} \geq \mu$, where $0<\mu<1$. If necessary, the absence of a μ-level relation $v_i \xrightarrow{\mu} v_j$ is expressed as $v_i \not\xrightarrow{\mu} v_j$. **A μ-level walk** $v_i \xrightarrow{\mu *} v_j$ exists from v_i to v_j, if and only if $v_i \xrightarrow{\mu} v_j$ or $v_i \xrightarrow{\mu} v_{i_1}, v_{i_1} \xrightarrow{\mu} v_{i_2}, \dots \xrightarrow{\mu} v_{i_k}, v_{i_k} \xrightarrow{\mu} v_j$.

(ii) **A μ-level semi-relation** $v_i \overset{\mu}{-} v_j$ exists between v_i and v_j, if and only if $v_i \xrightarrow{\mu} v_j$ or $v_j \xrightarrow{\mu} v_i$. **A μ-level semi-walk** $v_i \overset{\mu *}{-} v_j$ exists between v_i and v_j, if and only if $v_i \overset{\mu}{-} v_j$ or $v_i \overset{\mu}{-} v_{i_1}, v_{i_1} \overset{\mu}{-} v_{i_2}, \dots \overset{\mu}{-} v_{i_k}, v_{i_k} \overset{\mu}{-} v_j$.

(iii) **A μ-level closed walk** exists when a μ-level walk is $v_i \xrightarrow{\mu *} v_i$.

(iv) If a fuzzy digraph $G=(V', E')$ satisfies $V' \subseteq V$ and $E'(v_i)=E(v_i) \cap V'$ for any $v_i \in V'$, then G' is called **a fuzzy sub-digraph** of G.

(v) If a graph $G'=(V, E')$ satisfies $E'(v_i) \subset E(v_i)$ for any $v_i \in V$, then G' is called **a fuzzy partial digraph** of G.

Next, define the following operations.

[Definition 2] When two similarity matrices, $Q=[q_{ij}]$ and $R=[r_{ij}]$, exist:

(i) **A composition** of Q and R is denoted by $Q \circ R$ and is defined by a membership function $m_{Q \circ R}$:

$$m_{Q \circ R}(v_i, v_j)= \underset{v_k \in V}{\text{Max.}} \ \text{Min.} \left(m_Q(v_i, v_k), m_R(v_k, v_j)\right). \tag{1}$$

The m-fold composition $R \circ R \circ \dots \circ R$ is denoted by R^m.

(ii) Operations + and x are defined by m_{Q+R} and $m_{Q \times R}$.

$$m_{Q+R}(v_i, v_j) \equiv \text{Max.}(q_{ij}, r_{ij}).$$
$$m_{Q \times R}(v_i, v_j) \equiv \text{Min.}(q_{ij}, r_{ij}). \tag{2}$$

Especially, $R + R' \equiv S(=[s_{ij}])$, and

$$R \times R' \equiv T(=[t_{ij}]). \tag{3}$$

The transitive closure of R is defined by $\widehat{R} = R + R^2 + \dots$.

Fuzzy relations S and T, respectively, correspond to fuzzy mapping functions $E \cup E^{-1}$ and $E \cap E^{-1}$, where E^{-1} is an inverse mapping function of E. Using characteristics of similarity matrices, so that these matrices are reflective and diagonals of these matrices are 1s', $\widehat{R} = [\widehat{r}_{ij}]$, $\widehat{R}' = [\widehat{r}'_{ij}]$, $\widehat{S}=[\widehat{s}_{ij}]$ and $\widehat{T}=[\widehat{t}_{ij}]$ can be simply expressed as follows:

$$\widehat{R} = R^{n-1},\ \widehat{R}' = (R')^{n-1},\ \widehat{S}=S^{n-1} \text{ and } \widehat{T}=T^{n-1}. \tag{4}$$

CONNECTIVITY

First, introduce a concept of a clique, which is a systematization of a current notation in sociometry.

[Definition 3] On the subgraph $G'=(V', E')$ of a fuzzy digraph G, G' is **a μ-level clique** if and only if both $v_i \xrightarrow{\mu} v_j$ and $v_j \xrightarrow{\mu} v_i$ exist for any $v_i, v_j \in V'$, and no node $v_k \notin V'$ exists satisfying $v_i \xrightarrow{\mu} v_k$ and $v_k \xrightarrow{\mu} v_i$ for any $v_i \in V'$.

Harary and Ross (1957) presented a matrix method to find cliques in a non-fuzzy digraph. In this paper, the clique extraction procedure is applied to a fuzzy digraph. Let's formalize the extraction algorithm of μ-level cliques on a fuzzy digraph G by a function $f_\mu(R)$. Next, the definition of degrees of connectivity defined by Luce (1950) is expanded to other degrees of connectivity and the use of a fuzzy digraph.

[Definition 4] On the fuzzy digraph $G=(V,E)$:

(i) G is a **μ-level strongly complete** graph G_μ^1, if and only if $v_i \xrightarrow{\mu} v_j$ and $v_j \xrightarrow{\mu} v_i$ exist for any $v_i, v_j \in V$.

(ii) G is a **μ-level weakly complete** graph G_μ^2, if and only if $v_i \xrightarrow{\mu} v_j$ or $v_j \xrightarrow{\mu} v_i$ exist for any $v_i, v_j \in V$.

(iii) G is a **μ-level bilaterally connected** graph G_μ^3, if and only if $v_i \xrightarrow{\mu *} v_j$ and $v_j \xrightarrow{\mu *} v_i$ exist, going backward in the reverse order of $v_i \xrightarrow{\mu *} v_j$.

(iv) G is a **μ-level unilaterally connected** graph G_μ^4, if and only if $v_i \xrightarrow{\mu *} v_j$ or $v_j \xrightarrow{\mu *} v_i$ exist for any $v_i, v_j \in V$.

(v) G is a **μ-level strongly connected** graph G_μ^5, if and only if $v_i \xrightarrow{\mu *} v_j$ and $v_j \xrightarrow{\mu *} v_i$ exist for any $v_i, v_j \in V$.

(vi) G is a **μ-level weakly connected** graph G_μ^6, if and only if $v_i \overset{\mu *}{-} v_j$ exists for any $v_i, v_j \in V$.

(vii) G is a **μ-level totally ordered** graph G_μ^7, if and only if $v_i \xrightarrow{\mu} v_j$ exists for any v_i, v_j $(i<j) \in V$, when rearranging nodes in some adequate order.

Consider an example of a fuzzy digraph G in Fig.1. Examples of digraphs, emphasizing individual μ-level connectiveness categories, are shown in Fig.2, corresponding to Fig.1. Clearly, every μ-level strongly connected digraph is μ-level unilaterally and weakly connected. Thus, there exist inclusion relations among several connectedness categories.

[Proposition 1] Among μ-level connectivity categories for a digraph G, the following inclusion relations are established:

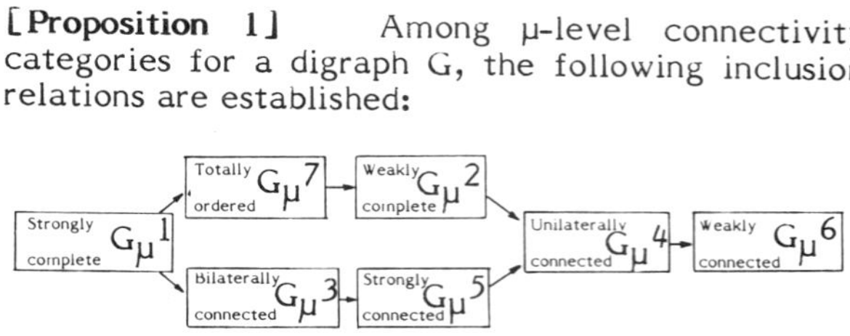

Here, an inclusion relation $G_\mu^i \longrightarrow G_\mu^j$ means that, whenever G is G_μ^i, then G is G_μ^j.

(Proof) As the proof is clear from Def.4, the proof is omitted.

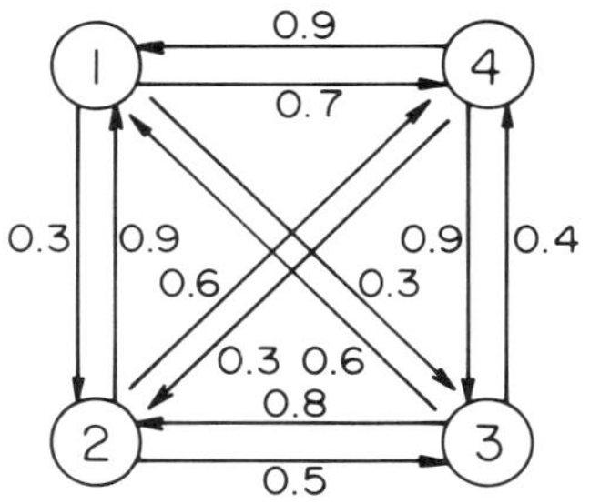

Fig. 1 An example of a fuzzy digraph.

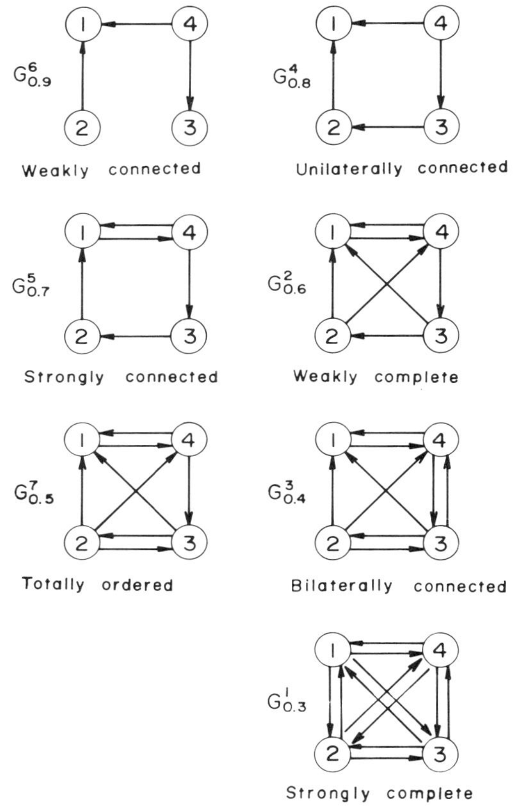

Fig. 2 Examples of digraphs, emphasizing individual connectivity categories (See the fuzzy digraph in Fig. 1).

TABLE 1 Relations between Fuzzy mapping functions and Connectivity categories

(i) Fuzzy Mapping Function:H	(ii) Fuzzy Relation Matrix	(iii) Connectivity Category
E	R	$G_\mu{}^1$
$E \cup E^{-1}$	$R + R'$	$G_\mu{}^2$
$\hat{E} \cap \hat{E}^{-1}$	$\hat{R} \times \hat{R}'$	$G_\mu{}^3$
$\hat{E} \cup \hat{E}^{-1}$	$\hat{R} + \hat{R}'$	$G_\mu{}^4$
$\hat{E}$	$\hat{R}$	$G_\mu{}^5$
$\widehat{E \cup E^{-1}}$	$\widehat{R + R'}$	$G_\mu{}^6$

CLUSTER EXTRACTION ALGORITHM ACCORDING TO CONNECTEDNESS

Represent a cluster extraction algorithm for G according to the i-th type of μ-level connectedness by a function $F_\mu{}^i(R)$. First, consider functions $F_\mu{}^i(R)$ $(i=1,2,\ldots,6)$.

[Lemma 1] On a fuzzy digraph $G=(V,E)$:
(i) $v_i \xrightarrow{\mu^*} v_j$ and $v_j \xrightarrow{\mu^*} v_i$ exist for any $v_i, v_j \in V'$ $(\subseteq V)$ and no $v_k \notin V'$ exists satisfying $v_i \xrightarrow{\mu^*} v_k$ and $v_k \xrightarrow{\mu^*} v_i$ for any $v_i \in V'$, if and only if some μ-level closed walk exists through only all $v_i \in V'$.
(ii) $v_i \xrightarrow{\mu^*} v_j$ or $v_j \xrightarrow{\mu^*} v_i$ exist for any $v_i, v_j \in V'$ $(\subseteq V)$ and no $v_k \notin V'$ exists satisfying $v_i \xrightarrow{\mu^*} v_k$ or $v_k \xrightarrow{\mu^*} v_i$ for any $v_i \in V'$, if and only if some μ-level walk exists through only all $v_i \in V'$.
(Proof) (i) First, show that some μ-level closed walk exists. On V, let U be a node set, corresponding to a μ-level closed walk with maximum numbers of nodes in V'. Show $U \supseteq V'$. At first, assume that $U \not\supseteq V'$. Then, for any $v_k \in (\bar{U} \cap V')$, $v_i \in (U \cap V')$ exists satisfying $v_k \xrightarrow{\mu^*} v_i$ and $v_i \xrightarrow{\mu^*} v_k$. As some μ-level closed walk exists through v_k, as well as in all nodes in $U \cap V'$, the assumption $U \not\supseteq V'$ cannot hold. Therefore, $U \supseteq V'$. Next, show $U=V'$. Assume that $U \supset V'$. Then, between any $v_k \in (U \cap \bar{V'})$ and any $v_i \in V'$, $v_i \xrightarrow{\mu^*} v_k$ and $v_k \xrightarrow{\mu^*} v_i$ exist. The assumption of $U \supset V'$ cannot hold. Therefore, $U=V'$. The proof of the converse is straightforward and is not presented.
(ii) Omitted, as the proof of (ii) is similar to the proof of (i).

[Lemma 2] Between a fuzzy digraph $G=(V,E)$ and its subgraph $G'=(V', E')$:
(i) G' is a μ-level strongly connected digraph, when $v_i \xrightarrow{\mu^*} v_j$ and $v_j \xrightarrow{\mu^*} v_i$ exist between v_i and $v_j \in V'$ $(\subseteq V)$ on G.
(ii) G' is a μ-level unilaterally connected digraph, when $v_i \xrightarrow{\mu^*} v_j$ or $v_j \xrightarrow{\mu^*} v_i$ exist between any v_i and $v_j \in V'$ on G.
(Proof) By Lemma 1, in both cases of (i) and (ii), if $v_i \xrightarrow{\mu^*} v_j$, for any v_i and $v_j \in V'$, then at least one μ-level walk exists through only all nodes on V'. Therefore, G' is strongly connected or unilaterally connected, respectively.

[Proposition 2] Let H be each fuzzy mapping function in Table 1 (i), and let $G''=(V', E'')$ $(V' \subseteq V)$ be a μ-level clique. Then, a subgraph $G'=(V', E')$ of G becomes a μ-level fuzzy digraph with the degree of connectivity shown in Table 1 (iii). Here, for any $v_i \in V'$, $E'(v_i)=E(v_i) \cap V'$, and $E''(v_i)=H(v_i) \cap V'$.
(Proof) In this proof, call the condition where G'' is a μ-level clique as the given condition. Where $H=E$ or $E \cup E^{-1}$, G' is a μ-level strongly complete or weakly complete graph, respectively, by the given condition, Def.2, 3 and 4 (i), (ii). In the case of $H=\widehat{E \cap E}^{-1}$, considering that T is a symmetric matrix, G' is a μ-level bilateral digraph. In the case of $H=\hat{E} \cup \hat{E}^{-1}$, by the given condition, $v_i \xrightarrow{\mu^*} v_j$ or $v_j \xrightarrow{\mu^*} v_i$ exist for any $v_i, v_j \in V'$. Therefore, G' is a μ-level unilateral digraph using Lemma 2. In the case of $H=\hat{E}$, by the given condition, $v_i \xrightarrow{\mu^*} v_j$ and $v_j \xrightarrow{\mu^*} v_i$ exist for any $v_i, v_j \in V'$. Therefore, G' is a μ-level strongly connected digraph. When $H=\widehat{E \cup E^{-1}}$, considering that S is symmetric, a fuzzy digraph $G_0=(V', E_0)$, where $E_0=(E \cup E^{-1}) \cap V'$, is a μ-level bilateral digraph, according to the given condition. Therefore, G' is a μ-level weakly connected digraph.

Proposition 2 shows that each cluster extraction algorithm can be expressed as follows:

$$F_\mu{}^1(R) = f_\mu(R), \quad F_\mu{}^2(R) = f_\mu(R + R'),$$
$$F_\mu{}^3(R) = f_\mu(\widehat{R \times R'}), \quad F_\mu{}^4(R) = f_\mu(\widehat{R} + \widehat{R'}),$$
$$F_\mu{}^5(R) = f_\mu(\widehat{R}), \text{ and } F_\mu{}^6(R) = f_\mu(\widehat{R + R'}). \quad (5)$$

That is, it is shown that individual cluster extraction algorithms according to categories 1)-6) of connectivity operationally result in the clique extraction algorithm (e.g. Okamoto (1978)).

Next, consider the cluster extraction algorithm based on μ-level totally ordered graph structure. At first, define a minimum weakly complete digraph.

[Definition 5] On a μ-level weakly complete digraph G=(V,E), only one of $v_i \xrightarrow{\mu} v_j$ and $v_j \xrightarrow{\mu} v_i$ exists for any v_i, $v_j \in V$, if and only if G is a **minimum μ-level weakly complete** digraph.

[Proposition 3] Let a partial graph G'=(V,E') of a fuzzy digraph G=(V,E) be a minimum μ-level weakly complete digraph. Then, if there is no μ-level closed walk on G', both G and G' are μ-level totally ordered digraphs.

(Proof) Assume that G' is not a μ-level totally ordered digraph. That is, no node v_i exists satisfying $v_j \xrightarrow{\mu} v_i$ for all v_j ($\in$ V) ($\neq v_i$). So, let v_i be the node to which μ-level relations exist from most of the others, and let $V_i=(v_{i_1}, v_{i_2}, \ldots, v_{i_m})$ (m< n-1) be the set of the nodes satisfying $v_{i_k} \xrightarrow{\mu} v_i$ (k=1,2,...,m) and $v_j \not\xrightarrow{\mu} v_i$, where $v_j \notin V_i$. From the condition that no μ-level closed walk exists, the relations $v_{i_k} \xrightarrow{\mu} v_j$ (k=1,2,...,m) must hold for any v_j ($\in V-V_i-(v_i)$). Moreover, $v_i \xrightarrow{\mu} v_j$. So, the assumption is not true. Therefore, a node v_i exists satisfying $v_j \xrightarrow{\mu} v_i$ for all v_j ($\in$ V) ($\neq v_i$). Next, remove v_i from V and repeat the above procedure. Both G and G' become μ-level totally ordered digraphs.

APPLICATION TO INSTRUCTIONAL EVALUATION

Last, the cluster extraction algorithms are applied to two kinds of instructional evaluation problems. Graphical representation of instructional data is useful for a teacher to analyze the data and evaluate his instructional process and his students' performance. For this purpose, a complex instructional data structure — a fuzzy digraph — should be converted into a simple non-fuzzy digraph composed of several significant subgraphs by cutting on an adequate μ-level. The cluster extraction algorithms presented in this paper is utilized in order to extract instructionally significant subgraphs.

SOCIOMETRY ANALYSIS

Sociometry, presented by Moreno (1960) is one of several measurement and evaluation methods on a group structure. In sociometry, relationships in a group are measured quantitatively and the group structure and status of each member are analyzed using a directed graphical representation, called a sociogram. In schools, sociometry is very useful to enable a teacher to understand the structure and status of students in his class. However, sociometry

TABLE 2 An Original Socio-matrix Example.

Students (columns 1–19 and 101–118; rows: Students)

Students	1234567890123456789	111111111111111111 / 000000000111111111 / 123456789012345678
1	\···112111···3··233	·····1·····1······
2	1\·1·····2·3··21·34	····11·····1······
3	14\·11····13·2··214	··················
4	21·\11···4···21··14	·······11··1······
5	1···\2···1····32221	······1····31·····
6	2·111\11·11··2··143	···········1······
7	4··1··\··3···4··1·4	·····1····11······
8	1111·11\11···111··1	··················
9	121·1···\1·2·132·12	···········1······
10	123···1··\2····1234	·····1·····1······
11	133······2\1····123	·········1·1···1··
12	1341····3··\··2·114	··················
13	1···········\······	··················
14	21··1·1···1··\2·234	·········1·2······
15	131·1231·1·1·1\·113	··················
16	21··11··422··23\·11	··················
17	12···111121·131·\14	··················
18	22311····13···12·\3	········1··1······
19	3231·····1·2·1·122\	···········1···1··
101	14·········1······2	\·11·····14···2···
102	····1····1·······12	3\2·2··211····4···
103	···················	··\12··112·1·11··1
104	1················11	32·\1··2···1···2·3
105	2······2·······1·22	2·1·\··31·21··1···
106	1················11	2··3·\·3·321··21··
107	1···1········11··13	·1····\221·11·4···
108	1··1·····1·······21	2··13··\3·1···31··
109	·················11	···1··22\332113···
110	1···1·1··2·····1·23	·····2··1\13··2···
111	·3··1······1·····14	2···31····\1··3···
112	1············1··234	·····23·31·\······
113	11·······1·······22	2·····211·13\·31··
114	···2·····1·······23	······3341·1·\····
115	1····2·······2··3·1	1····4·21·21··\···
116	21·11····1···1···21	·2··1···411··1·\··
117	1···1············13	3111····1·2··2··\·
118	···1··1······1···42	1··32···1··3··1··\

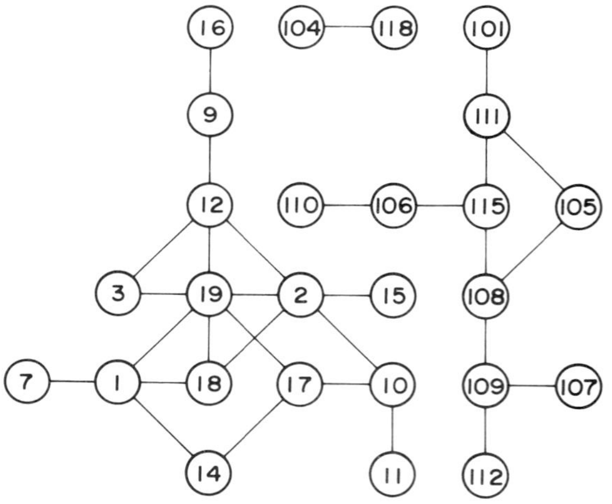

Fig. 3 A sociogram example, emphasizing 0.5-level bilaterally connected subgraphs. The connection i—j means that i→j and j→i exist.

has not spread far and wide. It is considered that the reasons are as follow:

(i) As the relation between a pair of members is represented by 1, 0 and -1, measurement reliability of sociometric data is poor.
(ii) As the structure among members more than ten is very complicated, interpretation reliability of sociogram is poor.

In order to resolve the first problem, Takeya (1981) presented multi-valued measurement methods on sociometry data. In this case, a sociogram may be expressed by a fuzzy digraph. This paper describes one solution to the second problem. To clarify the description, the following example is given. Table 2 shows an original socio-matrix of multi-valued (0,1,2,3 and 4) measurement for 37 students. This original socio-matrix is normalized by dividing by four. It is very difficult to understand features of the group structure and each student status according to the original fuzzy sociogram. In Fig.3, an example of cluster extraction result of 0.5-level bilaterally connected subgraphs is shown. Each member included in a subgroup is selecting more than one members in the subgroup with 0.5-level relation. That is, each subgroup has such a feature that every member in the group can communicate each other directly or indirectly in a friendly way. To understand the psychological structure of the group, the extracted sociogram in Fig.3 is more useful than the original sociogram. Also, as shown in Fig.3, it is very easy to understand individual student status and to find key members or isolated members.

TEST ANALYSIS (ITEM RELATIONAL STRUCTURE ANALYSIS)

The extraction method for μ-level totally ordered subgraphs is applied to ordinal test theory, i.e. an analysis method for hierarchical structuring among test items, based on student performance scores. Ordinal test theory has been presented by Airasian & Bart (1973). An IRSA (Item Relational Structure Analysis) was presented as a new system of ordinal test theory by Takeya (1980) and has been used by many schools and universities in Japan and the U.S.A., e.g. Waseda, Tamagawa (Japan), Illinois (U.S.A.) Universities (Sunouchi and Yamashita (1980), and Tatsuoka and Tatsuoka (1981)).

However, all usual ordinal test theories, including the IRSA, cannot fully satisfy the transitivity law regarding the item ordering relation. This paper presents a new method for a hierarchical item sequence satisfying the transitivity law. Test is composed of a set of n test items $V=(v_1, v_2, \ldots, v_n)$. Based on students' performance scores, the following item ordering coefficient from v_i to v_j is calculated.

$$r_{ij}* \equiv 1 - \frac{P(\bar{v}_i, v_j)}{P(\bar{v}_i)\, P(v_j)}. \qquad (6)$$

where $P(v_i)$: success probability for item v_i.
$P(\bar{v}_i)$: failure probability for item v_i.
$P(\bar{v}_i, v_j)$: failure probability for v_i and success probability for v_j.

Now, consider the positive range of $r_{ij}*$. Here, an element of R is defined as follows:

$$r_{ij} = \begin{cases} r_{ij}* & (r_{ij}* \geq 0), \\ 0 & (r_{ij}* < 0). \end{cases}$$

Then, the hierarchical structure graph (IRS graph) can be represented by a fuzzy digraph.

[Definition 6] An **ordering relationship** $v_i \xrightarrow{\mu} v_j$ exists, if and only if $r_{ij} \geq \mu$, where μ is a preassigned constant, $0 < \mu < 1$, otherwise $v_i \not\xrightarrow{\mu} v_j$.

[Proposition 4] If $v_i \xrightarrow{\mu} v_j$ and $v_j \not\xrightarrow{\mu} v_i$, then $P(v_i) > P(v_j)$.
(Proof) Omitted.

[Proposition 5] If an IRS graph $G=(V,E)$ is a μ-level weakly complete digraph, then G is a μ-level totally ordered digraph.
(Proof) As G is a μ-level weakly complete digraph, $v_i \xrightarrow{\mu} v_j$ or $v_j \xrightarrow{\mu} v_i$ exist for any $v_i, v_j \in V$. Thus, when both $v_i \xrightarrow{\mu} v_j$ and $v_j \xrightarrow{\mu} v_i$ $(i<j)$ exist, partial digraph G' of G can be obtained by converting R, according to the following conversion rules:
If $r_{ij} > r_{ji}$ (i.e. $P(v_i) > P(v_j)$), then $r_{ji}=0$.
If $r_{ij} < r_{ji}$ (i.e. $P(v_i) < P(v_j)$), then $r_{ij}=0$.
If $r_{ij} = r_{ji}$ (i.e. $P(v_i) = P(v_j)$), then $r_{ji}=0$.
Assume that a μ-level closed walk, for example $v_1 \xrightarrow{\mu} v_2$, $v_2 \xrightarrow{\mu} v_3$, ..., $v_{k-1} \xrightarrow{\mu} v_k$, $v_k \xrightarrow{\mu} v_1$, exists on G'. Then, $P(v_1)=P(v_2)=\ldots=P(v_k)$. This equation is in conflict with the conversion rule. Hence, on G', no μ-level closed walk exists. Using Proposition 3, G' is a μ-level totally ordered digraph. Therefore, G is also a μ-level totally ordered digraph.

Proposition 5 shows that cluster extraction algorithm satisfying the transitivity law of μ-level orderness results in a solution of $f_\mu(R+R')$.

First, let's show a pre-text example of the multiplication calculation in the 3rd grade. Test items consist of:

a) Calculate.
1. 213 x 3 = 2. 345 x 2 = 3. 386 x 2 =
4. 423 x 3 = 5. 872 x 9 = 6. 782 x 4 =
7. 34 x 2 x 3 =

b) Fill the following blanks.
Mr. A bought three dozen pencils. Here, the price for one dozen is 540 Yen. How much should he pay? Answer the following questions, while looking at the figure.

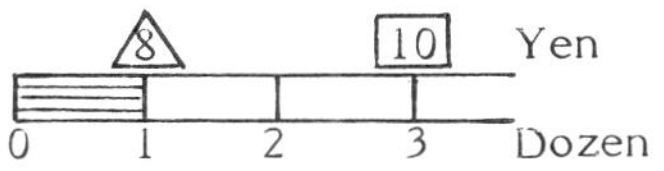

8. Write the answer in the diamond △ shaped space.
9. Write the equation resolving the question.
10. Write the answer in the square □ shaped space.

c) Practical application. One cake box contains four pieces of cake. One piece of cake costs 95 Yen. Mr. A bought two boxes.
11. How much should he pay?

Test result is shown in Table 3. In order to extract item sequences satisfying the transitivity law for item ordering on the IRS graph G, the totally ordered subgraph extraction algorithm is applied. Solutions for the function $f_{0.45}(R+R')$ are shown as follows:

1) 1 → 4 → 2 → 3 → 5 → 7 → 11
2) 1 → 4 → 2 → 3 → 5 → 6 → 11
3) 1 → 4 → 2 → 3 → 5 → 10 → 11
4) 1 → 4 → 2 → 8 → 10 → 11
5) 9 → 8 → 10 → 11 .

TABLE 3 An Example of a Test Score Matrix

Items	Students' Response Patterns	Total
1	111111111111111111101100	26
4	111111111111111110101000	24
2	111111111111101001000000	18
9	111100100000110111010110	18
3	111111101111011000000000	16
5	111111011010000000000000	12
7	111110100101000000000000	11
8	111000110000110100110000	11
10	111101010000000010010000	10
6	100011011100000000000000	9
11	110000000000000000000000	4
Freq.	311111112111111111113148	

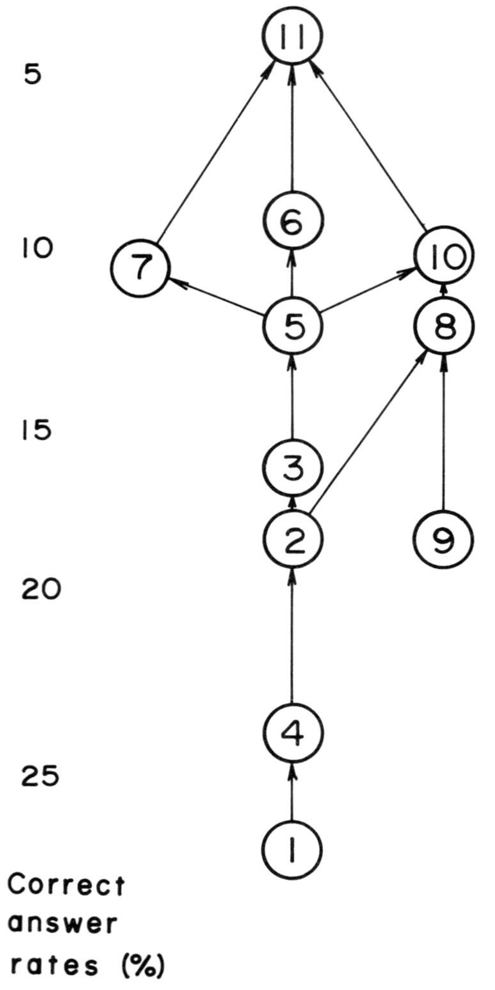

Fig. 4 An IRS digraph example, satisfying the transitive law for item ordering.

Discussion on a synthesis method for several item sequences is omitted by the space limitation. As a result of synthesis of these item sequences, an IRS graph G, satisfying the transitive law on 0.45-level ordering, can be constructed as shown in Fig. 4. The teacher can understand the relational structure among test items corresponding to every objective in an instructional unit. The use of this graph provides one kind of curriculum and instruction modification rationale. Also, the use of this graph provides the students with a diagnosis of the sources of their difficulty.

CONCLUSION

In this paper, a clustering method for a fuzzy digraph was presented in order to investigate and to analyze a complicated system. This paper classified a fuzzy digraph into seven categories of connectedness. Cluster extraction algorithms, according to individual connectedness, were formalized in the same manner, and were reduced to the operation of the clique detection procedure as a whole. Last, these algorithms were applied to two examples in the instructional evaluation field. These clustering procedures can be widely applied, not only to the instructional evaluation field, but also to other fields.

REFERENCES

Airasian, P.W. and W.M. Bart (1973). Ordering theory: A new and useful measurement model. Educational Technology, 5, 56-60.

Fujiwara, H. (1980). Methods for cluster analysis using asymmetric measures and homogeneity coefficient. Behaviometrika, 7, 12-21 (Japanese).

Harary, F. and I.C. Ross (1957). A procedure for clique detection using the group matrix. Sociometry, 20, 205-215.

Hubert, L. (1973). Min and max hierarchical clustering using asymmetric similarity measures. Psychometrika, 38, 63-72.

Luce, R.D. (1950). Connectivity and generalized cliques in sociometric group structure. Psychometrika, 15, 169-190.

Moreno, J.L. (Ed.) (1960). The Sociometry Reader. The Free press.

Okamoto, E. (1979). A cluster extraction program in sociometry. In Mathematical Models, Shinyosha, Tokyo, 160-165 (Japanese).

Sunouchi, H. and H. Yamashita (1980). Logical structure analysis of instructional programs. 4th International Conf. on Mathematical Education.

Takeya, M. (1980). Relational structure analysis among test items on performance scores. J. of Science Education, 4, 182-193.

Takeya, M. (1981). Application of a fuzzy graph theory to sociogram analysis. Trans. IECE, ET-8, 7-12 (Japanese).

Tatsuoka, K.K. and M.M. Tatsuoka (1981). Item analysis of tests designed for diagnosing bugs: Item relational structure analysis method. Computer-Based Research Lab. Research Report, 81-7, Univ. of Illinois.

Zadeh, L.A. (1971). Similarity relations and fuzzy orderings. Information Science, 8, 177-200.

THE CLINICAL DATA REDUCTION AND THE NET-MAKING FOR FUZZY CLUSTERING

Zhang Wenxiu and Zhao Ruhuai

Department of Mathematics, Xi'an Jiaotong University, Xi'an, Shaanxi Province, The People's Republic of China

Abstract. In this paper, a method for clinical data reduction is considered. It uses the concept of fuzzy clustering. A net-making method for fuzzy clustering analysis is obtained.

Keywords. Fuzzy clustering; net-making.

Suppose $D_i(i \leq n)$ denote different names of diseases and $S_j(j \leq m)$ denote different clinical symptoms. We denote a conditional probability by $P(D_i/S_j)$. Let

$$r_{ij} = \frac{\sum_{k=1}^{n} P(D_k/S_i)\ P(D_k/S_j)}{\sqrt{\sum_{k=1}^{n} P^2(D_k/S_i) \sum_{k=1}^{n} P^2(D_k/S_j)}}$$

then r_{ij} is fuzzy relation between S_i and S_j. By fuzzy clustering analysis, $S=\{S_i;\ i \leq m\}$ are clustered different classes. We select one to two symptoms in same class, so number of symptoms is reduced. It reduces not only the symptoms with less information but also the correlated symptoms. Because if S_i and S_j are correlated sympotoms, then they are clustered same classes.

For fuzzy clustering, it is defined that graph of fuzzy relation, by which the min-max transitive closure of fuzzy relation with antisimilarity is given directly, and then a net-making method for fuzzy clustering analysis can be obtained. The method has two advantages: (1) When fundamental set is finite, we may direct give the classification for any given level and do not need to obtain final form of the transitive closure, thus descreasing computational times. (2) When fundamental set is infinite, we may easily write down the classification for any increasing finite part and do not need to repeat any computation.

REFERENCES

Masahiko Okada. (1978) A method for clinical data reduction based on "Weighted Entropy" IEEE Trans. Biomed. Eng. (USA). September.

Sanchez. F. (1977) Inverses of Fuzzy relations application to possibility distributions and medical diagnosis. Proc, IEEE Conf. Decision Control.

CLUSTERING IN MIXED ENVIRONMENT (FUZZY AND NON-FUZZY) BY BRANCH AND BOUND TECHNIQUE FOR MANAGEMENT APPLICATIONS

S. Paramanick, K. S. Ray and D. Dutta-Majumder

Electronics and Communication Sciences Unit, Indian Statistical Institute, 203 B. T. Road, Calcutta-700035, India

Abstract. A resource allocation problem has been considered. The allocation algorithm has a feature of composite cost function as the convex combination of some subjective factors which are fuzzy in nature and those which are related to the resource allocation problem in the form of distance measure between people and resources (non-fuzzy part). The allocation itself is modelled as a clustering procedure. The decision involved is arrived through a branch and bound technique. The relative importance of the fuzzy and non-fuzzy criteria assigned in the decision algorithm via-a-vis and the shape of the cluster has been analysed. An attempt to optimise the coefficients of relative importance of diverse factors with the desired cluster shape has been made.

Keywords. Optimisation, Resource allocation problem, Fuzzy and Non-fuzzy environment, Clustering, Optimal Search technique, Branch and Bound method.

INTRODUCTION

Most of the decision problems in the present world involve human factors. Ultimately in any decision making system, the final decision is still made by a human being and hence involves subjective factors.

In present day management problems the human factors play a very important part and success of any decision lies in the human reaction that ensues following the decision. Apart from these there are hardcore realities in practical life.The decision has to be constrainted by the resource consideration. In the following paragraphs we have developed an algorithm which takes into account both the factors, namely those related to the resources and the human reactions which are modelled by some preference measure.

The allocation algorithm has few features (i) it combines both fuzzy and non-fuzzy criteria (ii) the algorithm is sequential in nature. This features give rise to the hierarchical decision pracedure - test if any discrepancy crops up and compensate in the subsequent stages. Hence it has an autocorrective feature (iii) the algorithm can be efficiently coded in to a computer routine, since it follows a tree structure.

The clustering introduced has also some unique features (i) the centres of the clusters have been initially known (ii) the constraint limiting the size of each cluster is linguistic.

THEORETICAL FORMULATION OF THE PROBLEM

Modelling of the Resource Allocation Cost (Non Fuzzy Part)

Let there be m groups of people denoted by the vector $\textcircled{H} = (\theta_1, \theta_2, .. \theta_m)$. Each group of people can be plotted in an n-dimensional property space. The properties are nothing but the features of the resources which will be allocated to them. Let there be p types of resources $\{R_1, R_2, .. R_p\}$. Each resource will be again an n-dimensional vector. A certain resource will be allocated to a certain group of people. The measure associated with this type of allocation will be given by a metric $d(\theta_i, R_k)$.

The form of the metric to be chosen may be a matter of further investigation. We have taken Euclidean distance for our consideration

$$d(\theta_i, R_k) = \sum_{j=1}^{n} (\theta_{ij} - R_{kj})^2 \quad (1)$$

$$\text{where} \quad \theta_i = \begin{Bmatrix} \theta_{i1} \\ \theta_{i2} \\ \cdot \\ \cdot \\ \theta_{in} \end{Bmatrix} \quad \text{and} \quad R_k = \begin{Bmatrix} R_{k1} \\ \cdot \\ \cdot \\ \cdot \\ R_{kn} \end{Bmatrix}$$

Our allocation will be such that the overall distance between the people and resources will be minimum.

MODELING OF THE SUBJECTIVE FACTORS

In this case fuzzy decision modelling has to be applied. Fuzzy decision process (FDP) is a 6 tuple $(\Theta, X, A, g_{\Theta}, \sigma_x(\cdot|\theta), J)$ where Θ is the set of clusters with approximate centre points known, in our case the group of people between whom the resources are to be distributed.

X : sample set. This will be the resource sample which will be allocated to the group of people.

g_{Θ}: This is a fuzzy measure associated with the particular classes. This measure is concerned with the groups of people whom certain discriminatory weights can be given from the management's side,- e.g. due to the urgency of needs etc.

A : Action set. It has got the membership function associated with it 'ha':A $\rightarrow [0,1]$. If 'ha' is discrete i.e. 0 or 1, the cluster will be deterministic, if it is continuous the cluster will be fuzzy. 'ha' denotes the degree of assignment to a given cluster.

$\sigma_x(\cdot|\theta)$: This is a conditional fuzzy measure. This will give rise to the preferential measure of the given group of people as to a certain resource which is given from the sample set chosen for allocation. This will model different psychological factors associated with the preference of a given group of people to a given resource.

$J(a(x),\theta)$: Loss type of function for a particular action a on x when it is classified with respect to θ.

OVERALL COST FUNCTION

The expected fuzzy cost function taking into account all the above factors is given by

$$\langle 1 \rangle = \oint_{\Theta} [\oint_x j(a(x),\theta) \circ \sigma_x(\cdot|\theta)] \circ g_{\Theta} \quad (2)$$

This is a loss type of function and in the decision process it has to be minimised.

CONSTRAINT

The constraint is mainly to limit the size of a given cluster. This is to guard that not all the resources go to one group and also that the total amount of resources is limited. The amount of resources will determine the size of the sample set.

The size of cluster will be limited by some vague linguistic decision rule like 'The number of materials acquired by the group θ_i should be roughly Z_i' etc. This can be expressed by another fuzzy integral

$$\oint_{\Theta} [\mu(n|\theta_i) \circ g_{\Theta}] < 1_c \quad (3)$$

where $\mu(n|\theta_i)$ is a membership function of a given group of people for the number of materials it has acquired and is given by a function like

$$\mu(n|\theta_i) = \frac{1}{1+(n - Z_i(\theta_i))^4} \quad (4)$$

etc. 1_c is a constraint determined from fuzzy linguistic expression.

TOTAL COST FUNCTION

$$\langle 1 \rangle_{total} = \alpha \sum_{i,k}\sum d(\theta_i, R_k) + \beta \oint_{\Theta} [\oint_x j(a(x),\theta) \circ \sigma_x(\cdot|\theta)] \circ g_{\Theta} \quad (5)$$

$$\text{s.t} \quad \alpha + \beta = 1.$$

The total cost function consists of non-fuzzy distance measure and fuzzy cost function. The fuzzy cost function is represented by fuzzy integral, meaning that having fuzzy measure it is possible to construct a functional which may be treated as a fuzzy expectation in comparison with probabilistic expectation. The decision procedure, based on this composite cost function, using a branch and bound technique is given in Fig. 1.

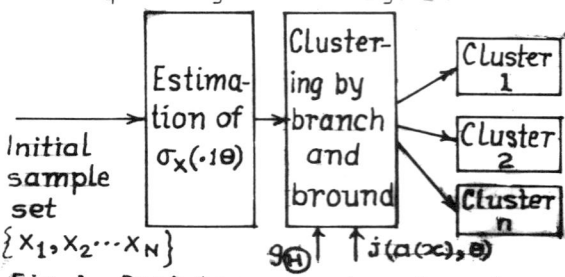

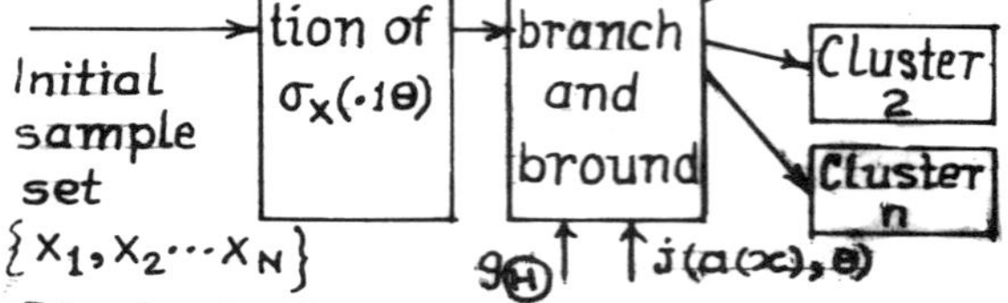

Fig.1. Decision procedure layout.

BRANCH AND BOUND TREE

Let there be 4 groups of people given by $\{\theta_1, \theta_2, \theta_3, \theta_4\}$ and a sample set $\{x_1, x_2, .. x_n\}$. The following Fig. 2, represents the solution of the objective function through branch and bound algorithm. From the starting node ① branching has been started and is directed outward. Thus at any intermidiate point in the calculations, the set of current bounding problems is identified with the set of nodes that are 'leaves' of the tree, as it has been developed at that point in the computation. Branching continues until every leaf, the nodes in the fully developed tree, can be terminated depending upon the rule that determines which of the currently active bounding problems to be chosen for branching as well as the method for solving new bounding problem.

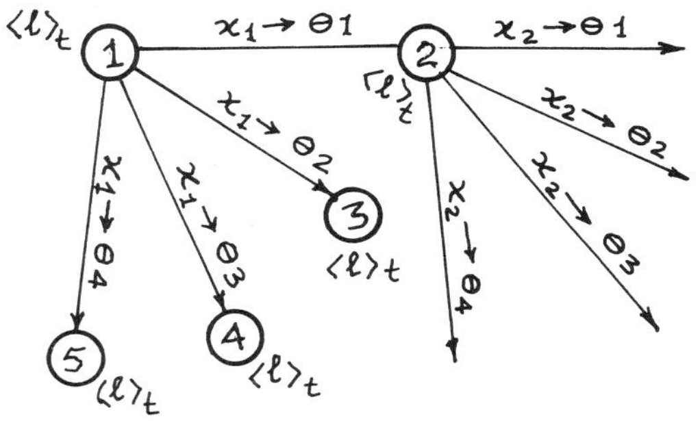

Fig.2. Optimal search through Branch and Bound technique.

where $x_i \rightarrow \theta_i$ means x_i element has been assigned to cluster θ_i and node ② is minimum of {②, ③, ④, ⑤}

For each node the following assignment matrix has to be calculated for the deterministic clustering show in Table 1. Where the elements of the matrix will be the cost function calculated while including the sample in the given cluster θ_i. The branching will occur with that node which will give the minimum cost and the node does not violate the constraint criterion.

A fessible solution for this type of allocation will look like, x_i to θ_j, and x_j to θ_k etc.

TABLE 1 Matrix Representation for Deterministic Clustering

Samples \ Clusters	θ_1	θ_2	θ_3	θ_4
x_1				
x_2				
x_3				
x_4				

As an optimum solution that path in the above tree shown in Fig. 2, will be taken which consists of the nodes of minimum cost function satisfying the constraint criterion. This solution is celled feasible optimal solution.

NUMERICAL EXAMPLE

Below we are giving a numerical example highlighting the effectiveness of the above method. For simplicity of calculation we make the following assumptions
(i) we neglet the constraint as to the size of the cluster,
(ii) the property space is two-dimensional,
(iii) the problem is essentially a two cluster one, i.e. concerns the assignment between two groups of people.

Let us take,

$$j(a(x), \theta_i) = 1 - \frac{1}{e^{n(\theta_i)}} \qquad (6)$$

$n(\theta_i)$=number of points in the cluster θ_i before the current assignment.

θ_1 has centre (5,5) and θ_2 has (-5,5).

$$\langle 1 \rangle_t = \alpha \frac{d(\theta_i, R_k)}{N} + \beta f_{\textcircled{H}} [f_x\, j(a(x), \theta) \circ \sigma_x (1\theta) \circ g_{\textcircled{H}} \qquad (7)$$

s.t. $\alpha + \beta = 1$

N=normalisation constant. N=max {intercluster distance}. For β=0 and α=1, the following points are assigned to their respective clusters.

Cluster $\theta_1(5,5)$	Cluster $\theta_2(-5,5)$
$x_1 = (4,6)$	$x_6 = (-4,4)$
$x_2 = (4,5)$	$x_7 = (-5,4)$
$x_3 = (4,4)$	$x_8 = (-6,4)$
$x_4 = (5,6)$	$x_9 = (-6,5)$
$x_5 = (6,6)$	

TABLE OF CONDITIONAL FUZZY DENSITIES (TO BE EVALUATED FROM THE FIFLD EXPERIMENTS) $(d(x_i|\theta_j))$.

TABLE 2 Data of Conditional Fuzzy Densities

Clusters \ Points	x_1	x_2	x_3	x_4	x_5	x_6	x_7	x_8	x_9
θ_1	0.1	0.3	0.5	0.4	0.8	0.3	0.2	0.6	0.9
θ_2	0.2	0.4	0.2	0.6	0.9	0.2	0.8	0.4	0.7

The entries are $d(x_i|\theta_j)$ from which $\sigma_x(.|\theta_j)$ is to be constructed.

$g_1=0.6$, $g_2=0.4$, the fuzzy densities to construct the fuzzy measure g_Θ to be introduced from the management's side.

Note : The normalisation constant N has been introduced in order that the magnitude remains less than 1 like that of the other part (i.e. the fuzzy integral part).

The form of $J(a(x),\theta)$ could have been taken any other form depending on the situation. In the present case its aim is to discourage large accumulation of resources by a single group. One could note here that though a particular group of people can express their likings or dislikings through $d(x_i|\theta_j)$ the management can regulate the assignment through a proper formulation of $J(a(x),\theta)$.

EVALUATION OF FUZZY INTEGRAL

Sugeno(1979) has shown that the integral $\int_x (j(a(x)),\theta_j) \circ \sigma_x(.|\theta_j)$ for a set $x=s_1,s_2,s_n$ is equal to

$$I(\theta_j) = \bigvee_{i=1}^{m} \left(j(a(s_i),\theta_j) \wedge \sigma_x(k_i|\theta_j)\right) \quad (8)$$

where $k_i = s_i, s_{i+1}, \ldots s_n$

s.t. $j(a(s_1),\theta_j) \leq j(a(s_2),\theta_2) \leq \ldots \leq J(a(s_n),\theta_j)$

The λ-fuzzy measure should satisfy the normalisation condition

$$\frac{1}{\lambda}\left[\prod_{i=1}^{n}(1+\lambda d(x_i|\theta_j))-1\right]=1 \quad -1\leq\lambda<\alpha \quad (9)$$

and
$$\sigma_x(k_i|\theta_j)=\frac{1}{\lambda}\left[\prod_{s_i\varepsilon k_i}(1+\lambda d(s_i|\theta_j))-1\right] \quad (10)$$

Also for a single element

$$\sigma_x(s_i|\theta_j)=d(s_i|\theta_j) \quad (11)$$

After getting $I(\theta_j)$, renumber $\theta_1,\theta_2,\ldots\theta_m$ for this step of calculation s.t. $I(\theta_1)\leq I(\theta_2)\leq\ldots\leq I(\theta_m)$ and the corresponding measure densities $g_1, g_2, \ldots g_m$ where m=no. of clusters.

Hence, the over-all fuzzy double integral is

$$\int_\Theta \left[\int_x j(a(x),\theta_j) \circ \sigma_x(.|\theta_j)\right] \circ g_\Theta = \bigvee_{i=1}^{m}\left(I(\theta_i)\wedge g(P_i)\right) \quad (12)$$

where $P_i = \{\theta_i, \theta_{i+1}, \ldots \theta_m\}$

ASSIGNEMENT CALCULATION THROUGH BRANCH AND BOUND TECHNIQUE

We shall introduce several important assumption at this point (i) the arrival and allocation of resources is sequential in nature (ii) the totality of the resources has been divided into several disjoint groups. This has been done for the simplicity of calculating the value of λ in λ-fuzzy measure. Also it is meaningful from that stand point that the choice of a certain resource may be context dependent and one phase of allocation may be assumed to be completed as soon as the assignment for one group is completed.

The finer details of this type of allocation can be worked out only with a parctical example on hand.

For our case, we have altogether 9 points partitioned into 3 groups as follows

(1) x_1, x_2, x_3 (2) x_4, x_4, x_6

(3) x_7, x_8, x_9

Now the Fig. 3 , visualise the whole matter.

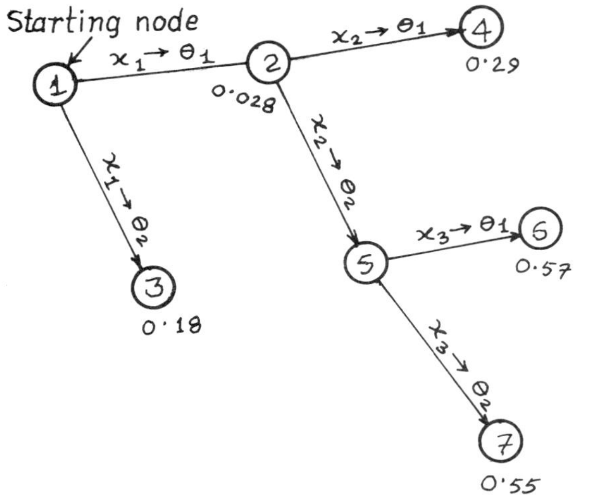

Fig.3. Assignement through Branch and Bround technique.

$N=\sqrt{(5+5)^2+0^2}=10$. We have taken N=5, since this normalised the destance to less than 1. m=2 no of clusters with $\alpha=0.1$ and $\beta=0.9$.

With the above methods of calculation and assumption, the above tree has been constructed with the values of cost function mentioned below each node. The optimal assignment will be $x_1 \to \theta_1$, $x_2 \to \theta_2$, and $x_3 \to \theta_2$.

It can be seen that with $\alpha=1$, and $\beta=0$, the points x_1, x_2, x_3, all belonged to θ_1 cluster but with the enhancement of β the assignment changed by significant amount.

Similarly the tree for other groups of resources namely that of groups (2) and (3) can be calculated.

REMARKS

An interesting study of the above algorithm can be made w.r.t. optimal decomposition of the totality of resources in independent groups and study of the overall assignment procedure.

This type of problem can be handled through a multilevel cost function.

In the present text the branch and bound technique has been adopted because it is an intelligently structured search of the space of all feasible solutions and it decomposes the main problem into a set of related sub problems which are comparatively easy to solve. The space of all feasible solutions is repeatedly partitioned into smaller and smaller subsets and a lower bound (in the case of minimisation) is calculated for the cost of the solutions within each subset. After each partitioning, those subsets with a bound that exceeds the cost of a known feasible solution satisfying its constraint (integer or mixed-integer) are excluded from all further partitionings. The partitioning continues until a feasible solution is found such that its cost is not grater than the bound for any subset. Apart from these, there is also an advantage of computing a suboptimal solution that differ from the optimum by not more than a prescribed amount subjected to management discretion. Suppose, that it is decided by management that a feasible solution whose cost exceedsthe cost of an optimal solution by not more than 10 percent would be acceptable. Then if a feasible solution is found with a cost of 150, we can terminate all problems with bounds of 137 or more $(1.10 \times 137=15o.7>150)$

The above clustering procedure is a deterministic one and is a model for the allocation of resources to different group.

The dual of this problem, namely that of resource sharing by different groups, can be done by the above procedure with the modification that the clusters will be fuzzy.

REFERENCES

Backer, E. (1978). Cluster Analysis by Optimal Decomposition of Induced Fuzzy Sets. Delft University Press.

Gillet, B.E. (1979). Introduction to Operations Research - A Computer Oriented Algorithmic Approxch. Tata Mc. GrawHill.

Halmos, P. (1954). Measure Theory Van Nostrend.

Klement, E.P. and Schwyhle, W. (1982). Correspondance between fuzzy measures and classical measures. Fuzzy Sets and Systems, Vol.7,No.1, pp.57-70.

Kacprzyk, J. (1978). A branch and bound algorithm for multistage control of a non-fuzzy system in a fuzzy environment. Control and Cybernetics, Vol.7,No.1, pp.51-64.

Sugeno, M. (1977). Fuzzy measure and fuzzy Integrals - a survey. In M.M. Gupta, G.N. Saridis and B.R. Gaines (Ed.), Fuzzy Automata and Decision Processes. North-Holland. pp. 89-102.

Wierzchon, S.T.(1982). Application of fuzzy decision making theory coping with ill defined problems. Fuzzy Sets and Systems, Vol.7, No.1, pp.1-18.

Wagner, H.M. (1969). Principles of Operations Research with Applications to Managerial Decisions, Prentice-Hall.

Walker, R.S. (1960), An enumerative technique for a class of combinatorial problems, Am. Math. SOC. Symposium on Applied Math. Proc.10. pp.91-94.

DISCRIMINANT ANALYSIS BASED ON FUZZY DISTANCE

J. Watada*, K. Montonami, H. Tanaka** and K. Asai****

**Faculty of Business Administration, Ryukoku University, Fukakusa, Fushimi, Kyoto 612 Japan*
***Department of Industrial Engineering, University of Osaka Prefecture, 4-804 Umemachi, Sakai, Osaka 591 Japan*

Abstract. It is frequently discussed in economic, social, managemental, or diagnostic problems to classify objects which are vaguely specified in a fuzzy environment. Objects in a real world have vague attributes. This paper deals with a method of discrimination between fuzzy groups with vaguely defined multi-aspect. Fuzzy distance is defined by using a fuzzy scatter matrix which consists of fuzzy covariances between aspects in fuzzy groups. The fuzzy distance plays an important and central role in this paper. This concept enables us to measure the distance of a sample from each fuzzy group in consideration of scatterness of samples in each fuzzy group.

Keywords. Discriminant analysis; fuzzy distance; fuzzy number; fuzzy mean; fuzzy variance; fuzzy scatter matrix; multivariate analysis.

INTRODUCTION

The major source of vagueness and ambiguity which are treated in economic, social, managemental and diagnostic problems should be more properly labeled fuzziness rather than randomness. In cluster analysis and pattern classification we actually encounter fuzziness of objects which comes from qualitative and quantitative properties. Bezdek (1976, 1981), Ruspini (1970) and Bezdek and Harris (1978) succeeded in pattern recognition and clustering by introducing fuzzy set theory to handle such fuzziness included in patterns and clusters. Yeh and Bang (1979) applied the theory of fuzzy graph to clustering analysis.

This paper discusses discrimination of fuzzy groups which are specified by fuzzy multivariate data. Fuzziness comes from the vagueness about groups and the possibilities (Zadeh, 1978) included in the multivariate data. Therefore, in this paper fuzzy groups and fuzzy multivariate data are treated as fuzzy sets and fuzzy numbers, respectively (Dubors and Prade, 1978; Mizumoto and Tanaka, 1979).

The concept of fuzzy distance is introduced in order to measure the proximity of a sample to each group. This fuzzy distance is based on a fuzzy mean and a fuzzy variance in a fuzzy group. Zadeh (1975b) and Watada and others (1981) deal with a variance in a fuzzy group. Kwakernaak (1978) treates a variance of fuzzy numbers. In this paper we define a variance of fuzzy numbers in a fuzzy group.

PROBLEM DESCRIPTION

Let us assume the followings: n samples i ($i=1,2,...,n$) and two groups G_j ($j=1,2$) are given, each sample i is specified by p aspects X_{ik} ($k=1,2,..,p$) and we are informed which group it belongs to. The problem in discriminant analysis is to assign a new sample to one group based on the given information. In general Mahalanobis distance is employed to measure the proximity to each group. A new sample is assigned to the group that has the smallest distance to the sample.

In a real world, however, it is not so easy and not so proper to find only one group in which a certain sample is included. It is more reasonable to describe each group as a fuzzy set due to its vagueness and also a value of an aspect as a fuzzy number due to its possibilities. According the existence of fuzziness or possibility, problems treated here can be classified into the following four cases:

(1) discrimination of non-fuzzy groups with non-fuzzy multi-aspect
(2) discrimination of non-fuzzy groups with fuzzy multi-aspect

TABLE 1 Fuzzy Multivariate Data Concerning Fuzzy Groups

Sample number	Fuzzy groups		Fuzzy multivariate		
	G_1	G_2	1	i	p
1	$\mu_{G_1}(1)$	$\mu_{G_2}(1)$	X_1^1	X_i^1	X_p^1
⋮	⋮	⋮	⋮	⋮	⋮
r	$\mu_{G_1}(r)$	$\mu_{G_2}(r)$	X_1^r	X_i^r	X_p^r
⋮	⋮	⋮	⋮	⋮	⋮
n	$\mu_{G_1}(n)$	$\mu_{G_2}(n)$	X_1^n	X_i^n	X_p^n

(3) discrimination of fuzzy groups with non-fuzzy multi-aspect, and
(4) discrimination of fuzzy groups with fuzzy multi-aspect.

The case (1) is a conventional problem treated in a discriminant analysis. The cases (1) to (3) are special problems of the case (4). Therefore, this paper discusses the general case (4) in detail.

Table 1 illustrates fuzzy multivariate data. X_i^r denotes fuzzy multivariate data, where r ($r \in S=\{1,2,...,n\}$) is a sample number and i ($i=1,2,...,p$) is a variable number. These fuzzy multivariate data X_i^r are fuzzy numbers with membership functions $\mu_{X_i^r}(x)$. Each fuzzy group G_k is characterized by a membership function $\mu_{G_k}(r)$. The fuzziness of G_k means its vague boundary. The possibilities of values of samples fuzzify a covariance through the extension principle (Zadeh,1975a; Zadeh, 1978). Let us define the size of G_k as

$$N(G_k) = \sum_{r=1}^{n} \mu_{G_k}(r).$$

A covariance V_{ij}^k between fuzzy variables X_i^r and X_j^r in a fuzzy group G_k can be defined as

$$V_{ij}^k = \frac{1}{N(G_k)} \sum_{r=1}^{n} \mu_{G_k}(r)(X_i^r - M_i^k)(X_j^r - M_j^k)$$

whose membership function is defined as

$$\mu_{V_{ij}^k}(v) = \bigvee \bigwedge_{r=1}^{n} [\bigwedge_{w=i,j} \{\mu_{X_w^r}(x_w^r) \wedge \mu_{M_w^k}(m_w)\}]$$

where $\bigvee$ and $\bigwedge$ denote max and min operations respectively and the maximum is taken over all $\{x_i^r, x_j^r, m_i, m_j\}$ which satisfy the equation

$$v = \frac{1}{N(G_k)} \sum_{r=1}^{n} \mu_{G_k}(r)(x_i^r - m_i)(x_j^r - m_j).$$

In the definition of a fuzzy variance, M_i^k denotes a fuzzy mean of a fuzzy variable X_i^r over a fuzzy group G_k:

$$M_i^k = \frac{1}{N(G_k)} \sum_{r=1}^{n} \mu_{G_k}(r) \cdot X_i^r$$

with its membership function

$$\mu_{M_i^k}(m) = \bigvee \bigwedge_{r=1}^{n} \mu_{X_i^r}(x)$$

where its maximum is taken over all x which satisfy the equation

$$m = \frac{1}{N(G_k)} \sum_{r=1}^{n} \mu_{G_k}(r) \cdot x.$$

DISCRIMINANT ANALYSIS

A variance means scatterness of sample values and divergence of its group. This concept is essential to define the measure of distance between a sample and a group. Accordingly, a fuzzy covariance matrix $\Sigma_k = (V_{ij})$ is called a fuzzy scatter matrix. Let us now use the concept of a fuzzy scatter matrix for defining a measure of fuzzy distance between a sample and each fuzzy group.

Let a fuzzy variable vector of a sample, the fuzzy mean vector of a kth fuzzy group G_k and the fuzzy scatter matrix be $X=[X_1,X_2,...,X_p]^t$, $M^k=[M_1^k,M_2^k,...,M_p^k]^t$ and Σ_k respectively, where t denotes a transpose. The fuzzy distance D_k of a sample X from a group

G_k can be described by

$$D_k(X) = (X - M_k)^t \Sigma_k^{-1} (X - M_k)$$

$$\mu_{D_k(X)}(d) = \bigvee [\mu_{M_k}(M) \wedge \mu_{\Sigma_k^{-1}}(\Sigma^{-1}) \wedge \mu_X(X)]$$

where the maximum is taken over all $\{X, M, \Sigma^{-1}\}$ which satisfy the equation

$$d = (X - M)^t \Sigma^{-1} (X - M).$$

It should be noted that this fuzzy distance is a fuzzy number due to the fuzziness of the vectors X and M_k and the fuzzy scatter matrix Σ_k.

Our problem is to decide the nearest group of a sample in the sense of fuzzy distance. Therefore, the comparison between fuzzy distances should be defined. In this paper comparison between fuzzy distances is done by the following definition in the consideration of fuzziness included in these distances.

If α-level sets of fuzzy numbers A and B satisfy the relation $a > b$ for all $a \in A^\alpha$ and all $b \in B^\alpha$, A is said to be greater in α-level than B. It is denoted by $A \overset{\alpha}{\gtrsim} B$.

If $A \overset{\alpha}{\gtrsim} B$ for all $\alpha \in [0,1]$, A is said to be greater than B with the notation $A \gtrsim B$.

Let us consider two groups for simplicity of explanation. Our approach can be applied to the case with more than two groups. Classification of a sample must be done by assigning it to the nearest group in the sense of the fuzzy distance. Mathematically speaking, the decision rule at α-level can be written as

$$G_1^\alpha = \{X \mid D_1(X) \overset{\alpha}{\lesssim} D_2(X)\},$$

$$G_2^\alpha = \{X \mid D_1(X) \overset{\alpha}{\gtrsim} D_2(X)\},$$

$$G_3 = \overline{(G_1 \cup G_2)^\alpha}.$$

This decision rule classifies a sample X into one of three fuzzy groups G_1, G_2 and the complement of $G_1 \cup G_2$. If a sample is not included in G_1 or G_2, this means that its decision should be suspended. This sample is assigned to G_3.

Let us define the degree of separation between fuzzy groups G_1 and G_2 at X, $g(X)$, as

$$g(X) = 1 - \bigvee_d \{\mu_{D_1(X)}(d) \wedge \mu_{D_2(X)}(d)\}.$$

The closer the degree of separation is to 1.0, the more reigic the classification is. Conversely, in the case that the degree of separation is 0, we can never decide the classification of its sample into one of fuzzy groups, and we must suspend our decision.

The degree of membership of a fuzzy sample X in the fuzzy groups G_1 or G_2 is estimated by the degree of separtion as

$$\mu_{G_1}(X) = \begin{cases} g(X) & : D_1(X) \overset{g(X)}{\lesssim} D_2(X) \\ 0 & : \text{otherwise} \end{cases}$$

$$\mu_{G_2}(X) = \begin{cases} g(X) & : D_1(X) \overset{g(X)}{\gtrsim} D_2(X) \\ 0 & : \text{otherwise.} \end{cases}$$

Using two numerical examples we illustrates the method. One example deals with classification of samples into one of non-fuzzy clusters with their fuzzy multi-aspects. The other one treats fuzzy clusters with their non-fuzzy multi-aspects. That is, the following cases are considered;

(i) fuzzy samples defined by fuzzy numbers and non-fuzzy clusters
(ii) non-fuzzy samples and fuzzy clusters.

<u>Numerical example 1</u>. In this case, the universal set U is discrete. Each fuzzy set consists of finite elements in U as

$$A = \mu_A(u_1)/u_1 + \mu_A(u_2)/u_2 + \ . \ . + \mu_A(u_n)/u_n$$
$$= \sum_{i=1}^{n} \mu_A(u_i)/u_i \quad : u_i \in U,$$

where + means union operation.

A worker measures wooden chips with the eye and he classifies those wooden chips into one of two non-fuzzy groups A and B. Although it seems that a wooden chip in the group A is bigger than one in the group B, his exact criterion for classifing them is not known even to him. His problem is to find his classifying criterion from the results of his classification. Assume that his classification and his measures are given in the following:

<u>Group A</u>

$${}^1X_1^1 = \text{"about 140 cm"}$$
$$= 0.5/138 \text{ cm} + 1.0/140 \text{ cm}$$

$${}^1X_1^2 = \text{"about 160 cm"}$$

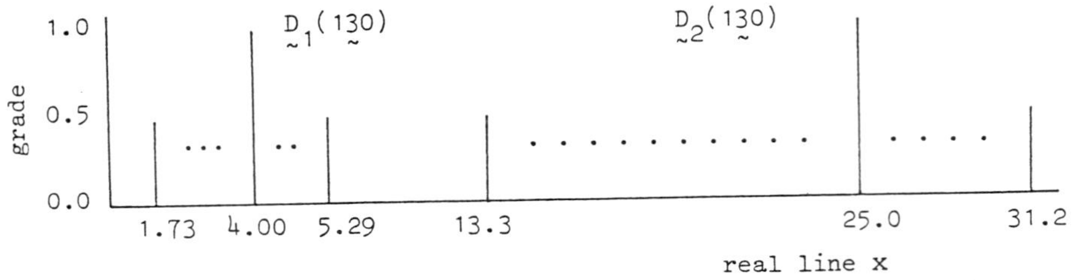

Fig. 1. Fuzzy distances $\tilde{D}_1(\tilde{130})$ and $\tilde{D}_2(\tilde{130})$.

$= 1.0/160 \text{ cm} + 0.5/162 \text{ cm}.$

Group B

$${}^2X_1^1 = \text{"about 170 cm"}$$

$$= 0.5/168 \text{ cm} + 1.0/170 \text{ cm}$$

$${}^2X_1^2 = \text{"about 190 cm"}$$

$$= 1.0/190 \text{ cm} + 0.5/192 \text{ cm}.$$

Fuzzy data are measured only by the eye. Applying the extension principle to this problem, the mean $\tilde{M}^k$ of a group G_k (k=1,2) and the variance $\tilde{V}$ can be calculated as shown in the previous section. Assuming that the variance of each group is equal, the mean and the variance are obtained as

$$\tilde{M}^1 = 0.5/149 \text{ cm} + 1.0/150 \text{ cm} + 0.5/151 \text{ cm}$$

$$= \text{"about 150 cm"}$$

$$\tilde{M}^2 = 0.5/179 \text{ cm} + 1.0/180 \text{ cm} + 0.5/181 \text{ cm}$$

$$= \text{"about 180 cm"}$$

$$\tilde{V} = 0.5/81.0 + \ldots + 1.0/100.0$$

$$+ \ldots + 0.5/169.0$$

$$= \text{"aout 100.0"}.$$

The fuzzy distances $\tilde{D}_k(\tilde{X})$ of a sample $\tilde{X}$ are calculated as

$$\tilde{D}_1(\tilde{X}) = (\tilde{X} - \tilde{150})\cdot 1/\tilde{100}\cdot(\tilde{X} - \tilde{150}),$$

$$\tilde{D}_2(\tilde{X}) = (\tilde{X} - \tilde{180})\cdot 1/\tilde{100}\cdot(\tilde{X} - \tilde{180}).$$

When a new sample $\tilde{X}$ is given by

$$\tilde{X} = \text{"about 130 cm"}$$

$$= 1.0/130 \text{ cm} + 0.5/132 \text{ cm},$$

the fuzzy distances of this sample with regard to two groups A and B can be written as

$$\tilde{D}_1(\tilde{130}) = (\tilde{130} - \tilde{150})\cdot 1/\tilde{100}\cdot(\tilde{130} - \tilde{150})$$

$$= 0.5/1.73 + \ldots + 1.0/4.00$$

$$+ \ldots + 0.5/5.29$$

$$\tilde{D}_2(\tilde{130}) = (\tilde{130} - \tilde{180})\cdot 1/\tilde{100}\cdot(\tilde{130} - \tilde{180})$$

$$= 0.5/13.3 + \ldots + 1.0/25.0$$

$$+ \ldots + 0.5/31.2.$$

Then it yiels

$$\tilde{D}_1(\tilde{130}) \tilde{\leq} \tilde{D}_2(\tilde{130})$$

which concludes that this new sample must be classified into the group A. Fig. 1 illustrates this relation.

Numerical example 2. Let us consider data of companies shown in Table 2. The first and second columns are the volumes of the employees and the capital respectively. The third column shows fuzzy clusters $\tilde{G}_1$ and $\tilde{G}_2$ of companies classified by an analyst of business management. $\tilde{G}_1$ and $\tilde{G}_2$ are labeled as "big companies" and "not big companies". The task now is to find out his classifying criterion from data shown in Table 2.
When the fuzzy distances relating to $\tilde{G}_1$ and $\tilde{G}_2$ are obtained, discriminant function Z(X) is defined by

$$Z(X) = D_1(X) - D_2(X)$$

where X is a vector $[x_1, x_2]^t$ whose elements x_1 and x_2 are the number of employees and the volume of capital, respectively.

The discriminant function is obtained from the data shown in Table 2 as

$$Z = 3.05\ x_1 + 0.30\ x_2 + 0.19.$$

In this analysis we used normalized data and assumed that each group has the same scatter matrix. This discriminant function illustrates that the judgement of the business analyst is more heavily influenced by the number of employees than the amount of capital. The discriminant function is shown in

TABLE 2 Samples and their Membership Grades to each of Fuzzy Groups

Sample number	Variables		Membership grades	
	Number of employees	Capital million yen	$\underset{\sim}{G}_1$	$\underset{\sim}{G}_2$
1	832	163.5	0.9	0.1
2	855	175.9	1.0	0.0
3	819	168.3	0.8	0.2
4	732	174.3	0.9	0.1
5	725	165.2	0.7	0.3
6	522	175.3	0.2	0.8
7	513	170.9	0.2	0.8
8	498	165.2	0.1	0.9
9	471	164.8	0.1	0.9

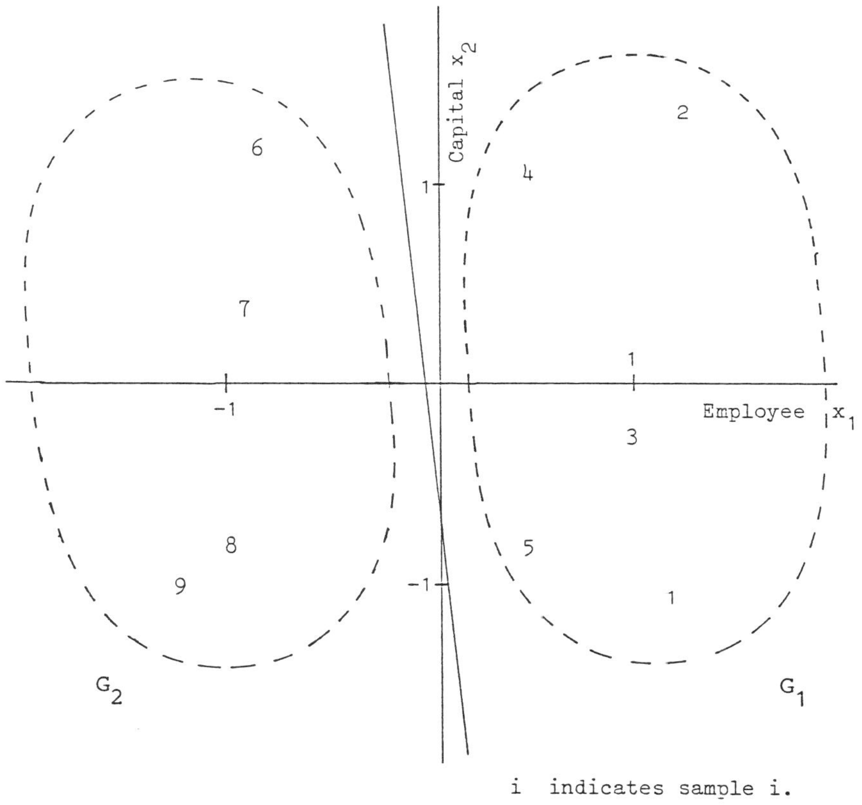

Fig. 2. Samples and discriminant function.

Fig. 2, where i denotes the noumber of a company and each of axes is scaled by normalized values. The discriminant function is obtained from the analyst's data which miror his intuitive classification of companies in a sense of size.

CONCLUDING REMARKS

We have been concerned with a method of discriminating between groups by using multivariate data, where the groups and the multivariate data are given as fuzzy sets and as fuzzy numbers, respectively. The main emphasis is that the distance of a sample from a fuzzy group is vague due to the fuzziness of a group and of measurement of its sample's aspects. Therefore, the concept of fuzzy distance is introduced to measure a fuzzy interval between a sample and a fuzzy group. This fuzzy distance is defined in consideration of the divergence of a fuzzy group and the fuzziness of sample's aspects. two fuzzy distances are compared to assign a sample to a group.

REFERENCE

Bezdek, J. C. (1976). A physical interpretation of fuzzy ISODATA. IEEE Trans. Syst. Man Cybern., SMC-6, 387-390.

Bezdek, J. C. (1981). Pattern Recognition with Fuzzy Objective Function Algorithms. Plenum Press, New York.

Bezdek, J. C., and J. Harris (1978). Fuzzy relations and partitions: and axiomatic basis for clustering. Fuzzy Sets and Systems, 1, 111-127.

Dubors, D., and H. Prade (1978). Operations on fuzzy numbers. Systems Science, 3, 613-626.

Kwakernaak, H. (1978). Fuzzy random variables I: definitions and theorems. Inf. Sci., 15, 1-29.

Mizumoto, M., and K. Tanaka (1979). Some properties of fuzzy numbers. In M. M. Gupra and others (Ed.), Advances in Fuzzy Set Theory and Applications, pp. 153-164.

Ruspini, E. (1970). Numerical methods for fuzzy clustering. Inf. Sci., 2, 319-350.

Watada, J., H. Tanaka, and K. Asai (1981). Linear representation of fuzzy groups. in Proceeding of International Conference on Policy Analysis and Information Systems in Taipei, pp. 755-763.

Yeh, R. T., and S. Y. Bang (1979). Fuzzy relations, fuzzy groups and their applications to clustering analysis. in Zadeh, Fu, Tanaka and Shimura (Ed.) Fuzzy Sets and Their Applications to Cognition and Decision Process, pp. 125-150.

Zadeh, L. A. (1975a). The concept of a linguistic variable and its application to apploximate reasoning-I. Inf. Sci., 8, 199-249.

Zadeh, L. A. (1975b). The concept of a linguistic variable and its application to apploximate reasoning-III. Inf. Sci., 9, 43-80.

Zadeh, L. A. (1978). Fuzzy sets as a basic for theory of possibility. Fuzzy Sets and Systems, 1, 3-28.

TOWARDS A MEASURE OF THE DEGREE OF SYNONYMY

R. López de Màntaras*† and E. Trillas*††

†Facultat d'Informàtica, Universitat Politècnica; Barcelona, Spain
††Dept. de Matemàtiques i Estadística, Universitat Politècnica Barcelona, Spain

Keywords: fuzzy partition, possibility distribution, semantics.

INTRODUCTION

One of the problems encountered in the study of language semantics, natural reasoning and natural or artificial language communication is that of dealing with synonyms. It seems clear that some synonyms are more adequate, in a given context, than others; therefore it is conjectured that there exists a canonical synonym which can be considered as the prototype of a class of synonyms. The usual definition of synonymy, "two concepts are synonyms if their meaning is the same or almost the same" (6), suggests that it is necessary to be able to measure degrees of synonymy.

Following (4) and (6) synonyms are represented by possibility distributions modelizing its meaning. In (1) we introduced a pattern classification algorithm to match possibility distributions. This algorithm will be used to measure the degree of synonymy between concepts and prototypes characterizing classes of synonyms in a given context. It is also introduced the use of logarithmic entropy (2) as a measure of the fuzziness of a synonym. And in each class we use the criterion of minimum entropy to select the canonical synonym or prototype of the class.

It has been shown (3) that it is possible to infer the possibility distributions representing the meaning of linguistic terms in a given context through a man-machine dialog. Therefore, the inputs of the algorithm will be concepts characterized by possibility distributions and the task of the algorithm consists in measuring the degree of synonymy between an input concept and the prototype concepts representing each one of the classes of synonyms already known. For each class the degree of synonymy will be the degree of matching or compatibility between the input concept and the prototype of that class; in pattern classification terms that means the degree of membership of the input data (concept) into each class (of synonyms). If the maximum degree of membership is greater than a given threshold $\lambda \in [0,1]$ then the algorithm decides that the input concept is another synonym of the class that gave the maximum: in this case it will compare the logarithmic entropy of the newcomer with that of the prototype and if it is lower it will become the new prototype, otherwise the prototype remains unchanged. If the maximum degree of membership is lower than the threshold then the input concept is considered as being the first instance of a new class of synonyms whose initial prototype is the newcomer itself; increasing the number of classes in such situation is justified by the assumption that we have an initial set of prototypes which form a λ-fuzzy partition (5) of the universe of objects (concepts in the given context) which has the property of the existence of new classes such that for any element there exists a class to which correspond a degree of membership greater than the level λ. Let us first set up the formal aspects on which is based our approach.

* From the 'Group of Applied Logic', Universitat Politècnica, Barcelona, Spain.

AN OUTLINE OF A NAIVE FORMALIZATION OF SYNONYMY VIA POSSIBILITY DISTRIBUTIONS

Let X be a universe of discourse and $O = \{o\}$ a family of observors which refer to the objects of X by using the linguistic terms of a predefined set T_0 of linguistic labels t. This assumes the existence of a 'logic' conveyed by the language:

$$L_0 : X \times T_0 \to [0,1] ,$$

where each value $L_0(x,t) \in [0,1]$ aggregates the particular values given by each observer.

We call $\mathcal{L}_0 = (X;T_0;L_0)$ a fuzzy relational linguistic structure; and, associated with this structure, we define the denotational meaning of each linguistic term as the mapping:

$$\sigma_t : X \to [0,1] ,$$

such that $\sigma_t(x) = L_0(x,t)$. That is, $\sigma_t \in \tilde{P}(X) = [0,1]^X$, the set of fuzzy subsets of X, and therefore, meanings are possibility distributions on X, i.e. meanings are considered graduated properties of the elements of X instead of absolute concepts. Intuitively, the meaning of each linguistic term is specified by the 'degree of applicability' of the term to the elements of X.

It is necessary to assume the existence of an 'internal logic' in T_0 which allows to relate, in meaning, primitive terms in order to form composite terms (e.g. 'young and tall', etc.). Naturally, this internal logic will be conveyed, by the meanings, to some logical structure of $\underset{\sim}{P}(X)$.

We will consider in $\underset{\sim}{P}(X)$ the usual De Morgan algebra structure given by:

$$(A \cup B)(x) = \max \{A(x), B(x)\},$$

$$(A \cap B)(x) = \min \{A(x), B(x)\},$$

$$\overline{A}(x) = 1 - A(x), \text{ and}$$

$$A \to B = \overline{A} \cup B$$

for all $x \in X$. Furthermore, we assume that the language L_0 implicitely contains the influence of the context and that this fact is conveyed to the set $\Sigma_0 = \{\sigma_t; t \in T_0\}$ of the (denotational) meanings of the terms in T_0.

Our basic conjecture is that it is possible to analyze synonymy and antonymy, in a given context, by means of operators between possibility distributions. Such operators preserve, in the set Σ_0 of meanings, the logical operations between the terms in T_0.

Formally, we assume that Σ_0 is a De Morgan subalgebra of $\underset{\sim}{P}(X)$; that is:

- $\emptyset \in \Sigma_0$, $X \in \Sigma_0$,
- if $\sigma_t, \sigma_{t'} \in \Sigma_0$, then $\sigma_t \cup \sigma_{t'} \in \Sigma_0$ and $\sigma_t \cap \sigma_{t'} \in \Sigma_0$,
- if $\sigma_t \in \Sigma_0$, then $\overline{\sigma}_t \in \Sigma_0$.

We also assume the existence of a family $\{\tau_i\}_{i \in I}$ of transformations $\tau_i: \Sigma_0 \to \Sigma_0$, such that:

- $\tau_i(\emptyset) = \emptyset$,
- $\tau_i(\overline{\sigma}_t) = \overline{\tau_i(\sigma_t)}$,
- $\tau_i(\sigma_t \cap \sigma_{t'}) = \tau_i(\sigma_t) \cap \tau_i(\sigma_{t'})$

for all $i \in I$.

Based on these hypothesis, we have the following definition that introduces the concept of synonymy as a matter of degree.

Definition. A term t' is a synonym of a term t, if there exists τ_i such that $\sigma_{t'} = \tau_i(\sigma_t)$. Similarly, a term t' is an antonym of t, if there exists τ_j such that $\overline{\sigma_{t'}} = \tau_j(\sigma_t)$.

Since we assume that the family $\{\tau_i\}_{i \in I}$ is a Group of automorphisms of Σ_0, the synonymy relation introduced by the previous definition is an equivalence relation. But in fact transitivity chains can be broken in a few steps (i.e. we can have σ_t synonym of $\sigma_{t'}$, $\sigma_{t'}$ synonym of $\sigma_{t''}$, but σ_t not synonym of $\sigma_{t''}$) in many real world situations, therefore it is convenient to be able to control the degree of synonymy between the meanings of the linguistic terms, and this is the purpose of the following algorithm.

THE ALGORITHM

Description of the set of data.

From the point of view of the algorithm, the data to classify are a sequence of samples of points X(t) in the unit cube of R^n; that is

$$X(t) = (x_1, x_2, \ldots, x_n), \text{ with } 0 \leq x_i(t) = x_i \leq 1,$$

the x_i's corresponding to a sampling of the possibility distribution $\sigma(t)$ modelizing the meaning of the input concept at a given time.

On the other hand, to each class C_k of synonyms corresponds a vector r_k of parameters characterizing the class, i.e., a sampling of the possibility distribution $\sigma(r_k)$ representing the meaning of the synonym-prototype of the class, that is

$$r_k = (r_{1,k}, \ldots, r_{n,k}), \text{ with } 0 \leq r_{i,k} \leq 1.$$

Therefore, the set R of parameters characterizing the fuzzy partition is

$$R = \{r_k \; ; \; k=1,\ldots,n\}$$

n being the number of classes of synonyms at a given time.

Membership function and class assignment.

The membership function of an element X to a class C_k is the maximum likelihood function (see (1) for a detailed study) given by

$$\mu_k(X) = \sum_{i=1}^{n} r_{i,k}^{x_i} \cdot (1 - r_{i,k})^{(1-x_i)}$$

which is maximum when $X = r_k$. This membership function can be interpreted as an "and" operation, in the sense of product, between the degree of matching of the prototype and the concept to be classified and the degree of matching of the antonym of the prototype and the antonym of the concept. This is done locally for each one of the n sampling points and finally, assuming independence among these "variables", the product gives the global degree of matching between the input concept and the prototype (degree of synonymy).

The class assignment is based on a decision rule according to the principle of maximum meaningfulness, that is

"$X \in C_j$ iff $\mu_j(X) = \max\{\mu_k(X); k=0,1,..,n\}$"

being $\mu_0(X) = \lambda$ the threshold of membership that allows to increase the number of classes.

The threshold of membership.

In general the number of classes of synonymy is, *a priori*, unknown. Therefore we must provide what we can call an "empty class" C_0 such that if the maximum degree of membership corresponds to C_0, that is if

$$\mu_0(X) = \max\{\mu_k(X); k=0,1,..,n\},$$

then the input concept becomes the first element and the prototype of a new class C_{n+1}, and C_0 will remain the "empty class" accounting for a future similar situation.

As a matter of fact $\mu_0(X)$ is independent of X and it will be written μ_0. This value is a threshold of membership such that no assignment decision to the existing classes will be made below it, instead a new class is created. The determination of the value μ_0 has to be made empirically.

Modification of the prototype.

As we have already mentioned whenever an input concept X is decided to belong to an existing class C_k of synonyms a comparison between the corresponding entropies

$$E(X) = - \sum_{i=1}^{n} x_i \log x_i$$

and

$$E(r_k) = - \sum_{i=1}^{n} r_{i,k} \log r_{i,k}$$

is made and the one that has the minimum entropy is choosen as the new prototype of the class. This implies that the concept representing the class of synonyms will always be the more specific concept already in the class. The existence of such minimal entropy synonym has been formally proved in (4) and (6).

DESCRIPTION OF AN APPLICATION EXAMPLE

Using the algorithm described in (3) with one observer, we have obtained a set of possibility distributions representing the meaning of the following (catalan) terms, corresponding to three classes of synonyms of the (english) terms: TALL, MEDIUM, SHORT, in a given context of people (students):

ALT,GRAN,LLARGARUT,LLARG (Synonyms of TALL)

PETIT,BAIX,XIC,MENUT (Synonyms of SHORT)

REGULAR,NORMAL,MIG,MITJÀ (Synonyms of MEDIUM)

We run the algorithm with a sampling of 16 points of the corresponding 12 possibility distributions. The results were the following:

CLASS 1 ← (alt,gran,llargarut,llarg)

CLASS 2 ← (petit,xic,menut,baix)

CLASS 3 ← (mitjà,mig,regular, normal)

Where the underlined term is the prototype in each class.

CONCLUDING REMARKS

The behaviour of the algorithm that we have just described has been extensively studied theoretically and experimentally in several applications to complex pattern recognition problems (1) and the results have been as good as in the previous simple example. This makes us feel that the algorithm is able to treat more complex examples than the one described. Therefore, we feel that this work is a small step towards questionning the classical dictum of Bloomfield (6) "We are unable to measure the degree of proximity of meanings". The tools provided by the theory of fuzzy sets and in particular the idea of using possibility distributions to modelize the meaninh of synonyms are the basis of our work. However, we think that a lot of work remains to be done in the theoretical side. The results described in this paper have convinced us that this work is worth being continued.

ACKNOWLEDGEMENTS

Authors are indebted to Prof. Lotfi A. Zadeh for his stimulating comments on this work. They also thank Prof. J.Aguilar-Martin (L.A.A.S., Toulouse) for his help and to the other members of the Group on Applied Logic of the "Universitat Politècnica de Barcelona".

REFERENCES

(1) Aguilar-Martin,J.; López de Màntaras,R. (1982). 'The Process of Classification and Learning the Meaning of Linguistic Descriptions of Concepts", in Approxi-

mate Reasoning in Decision Analysis, (M.M.Gupta and E. Sanchez ed.), pp. 165-177.

(2) DeLuca,A.; Termini,S. (1972). 'A definition of a Nonprobabilistic Entropy in the setting of Fuzzy Set Theory'. Information and Control, 20,pp.301-312.

(3) Freksa,C.; López de Màntaras,R. (1982). 'An adaptive Computer System for linguistic categorization of soft observations in Expert Systems and in the Social Sciences'. Proceedings of the Second World Conference on Mathematics at the service of man, pp. 288-292.

(4) Ovchinnikov,S.V. (1981). 'Representation of Synonymy and Antonymy by Automorphisms in Fuzzy Set Theory'. Stochastica, V/2, pp. 95-107.

(5) Trillas, E.; Riera, T. (1980). 'On a Special kind of variables in Fuzzy Environment', Proceedings of the X I.S.M.V.L., pp. 149-152.

(6) Trillas,E.; Riera, T. (1981). 'Towards a representation of Synonyms and Antonyms by Fuzzy Sets'. Busefal, 5, pp. 42-68.

A COMBINED FUZZY SET THEORETIC AND HEURISTIC METHOD FOR CHARACTER RECOGNITION

B. N. Chatterji

Department of Electronics and Electrical Communication Engineering, Indian Institute of Technology, Kharagpur, W. B., India

Abstract. In this paper the fuzzy set theoretic approach has been combined with a heuristic approach for the recognition of handwritten uppercase english alphabets and numerals. Linguistic description of each character is obtained in terms of the number of horizontal strokes –, vertical strokes |, right slant strokes /, left slant strokes \, A curves ∩, u curves U, c curves C, and D curves ⊃. This is done with the help of membership functions for each of the above features. For the recognition a heuristic approach is used. The recognition procedure was tested by a simulation study and a recognition accuracy better than 92 percent was achieved.

Keywords. Character recognition; classification; feature extraction; fuzzy similarity relations; linguistic description; membership function; pattern recognition.

INTRODUCTION

Zadeh (1965) introduced the concept of fuzzy set. This concept was adopted by scientists of many disciplines for theoretical work and for applications in areas like economics, management science, biology, medicine, psychology, information retrieval, artificial intelligence, linguistics, pattern recognition, social sciences and control engineering. Pattern recognition is an area where the fuzzy set theory got the attention from the very beginning of its existence. Here the fuzzy set theory has influenced tremendously in both the decision theoretic and syntactic methods. Fuzzy languages, fuzzy grammar, fuzzy automata theories and fuzzy algorithms were developed for the pattern recognition problems.

One class of pattern recognition problems is the character recognition. Most of the early works in character recognition were based on the use of correlation techniques and probabilistic concepts. Here we have a segmentation stage followed by a classification stage. Segmentation stage is based on a priori knowledge of the characters and a minimal set of features, which completely characterize the patterns of the characters, are determined. The features are determined heuristically and are used during the recognition process. The patterns of handwritten characters have biological origin and hence the variability is more appropriately represented by the concept of uncertainty than by the concent of probability. Hence it was thought that for badly written characters instead of using probability function one should use fuzzy membership function. Some of the work in character recognition using this new fuzzy set theoretic concept are those due to Tamura, Higuchi and Tanaka (1971); Siy and Chen (1974); Kickert and Koppelaar (1976); Pal, Dattamajumdar and Chaudhuri (1977), Gupta, Saridis and Gaines (1977), and Chatterji (1982).

In this paper an attempt has been made to combine fuzzy set theoretic approach and a heuristic approach for the recognition of handwritten uppercase english alphabets and numerals. Linguistic description is given to each character in terms of eight basic features. This is done with the help of the membership functions for each of the features. A heuristic approach is used for the recognition of the characters.

LINGUISTIC DESCRIPTION OF THE CHARACTERS

Using the concept of fuzzy set membership function, linguistic description of the characters are obtained in terms of eight basic features. These features are (i) horizontal stroke –, (ii) vertical stroke |, (iii) right slant stroke /, (iv) left slant stroke \, (v) A curve ∩, (vi) U curve U, (vii) c curve C and (viii) D curve ⊃. For this purpose the node points of the character are first determined. Node points are defined as those points where two or more of the above mentioned basic features meet. The strokes between the different node points are then determined. Each stroke is identified as one of the eight features with the help of the membership function of the different features (Pal, Dattamajumdar and Chaudhury, 1977) and using a heuristic approach.

The membership function for the feature to be a straight line (Fig.1) is expressed as

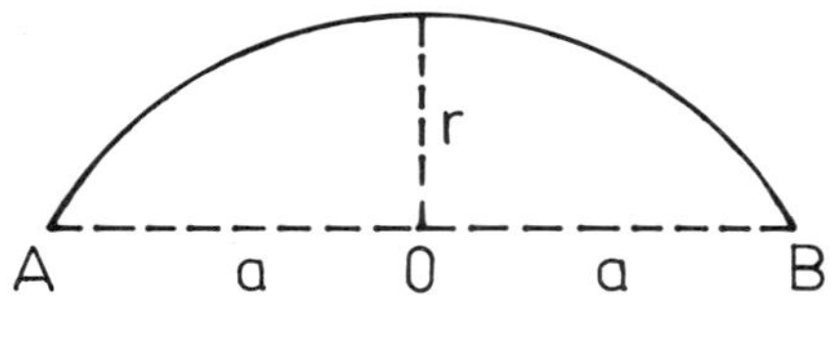

Fig.1 Deviation of arc AB from straight line

$$\mu_{SL} = (1 - \frac{r}{a}) \quad \text{if} \quad r < a$$
$$= 0 \quad \text{otherwise} \quad \ldots (1)$$

where 2a is the length of the line joining two node points of the arc (stroke between the nodes A and B). The deviation measured from the centre O is r. The membership functions μ_H and μ_V for the feature to be horizontal and vertical strokes are given by

$$\mu_H = 1 - |m_x| \quad \text{for } m_x \leq 1$$
$$= 0 \quad \text{for } m_x > 1 \quad \ldots (2)$$

and

$$\mu_V = 1 - \left|\frac{1}{m_x}\right| \quad \text{for } m_x \geq 1$$
$$= 0 \quad \text{for } m_x < 1 \quad \ldots (3)$$

where m_x is the slope of the feature.

The membership function for the feature to be oblique (i.e. either left or right slant) is expressed as

$$\mu_{ob} = 1 - \left|\frac{\theta - 45}{45}\right| \quad \text{for } 0 < |m_x| < \infty \quad \ldots (4)$$

$$\text{where} \quad \theta = \tan^{-1} m_x \quad \ldots (5)$$

The sign of m_x determines whether the feature is left slant or right slant. The membership function of the feature to be horizontal curve (i.e., either 'A' curve or 'U' curve) is given by

$$\mu_{HC} = 1 - \text{MIN}\left[\text{MIN}(|\theta|, |180-\theta|, |360-\theta|)/90, 1\right] \quad \ldots (6)$$

where $\theta = \tan^{-1}$ (slope of the straight line joining the end points of the curve) ... (7)

Similarly the membership function for the feature to be vertical curve (i.e., either 'C' or 'D' curve) is expressed as

$$\mu_{VC} = 1 - \text{MIN}\left[\text{MIN}(|90-\theta|, |270-\theta|)/90, 1\right] \quad \ldots (8)$$

Here also θ is defined by eq. (7). The identification process consists of the following eight steps :

Step 1 : Determine μ_{SL} using eq.(1). If it is less than 0.4 (which was found out by trial and error method) then go to step 5 through step 8. Otherwise go to step 2 through step 4.

Step 2: Find the slope m_x of the feature by least square error method, i.e.,

$$m_x = \frac{M \sum_{i=1}^{M} X_i Y_i - \sum_{i=1}^{M} X_i \sum_{i=1}^{N} Y_i}{M \sum_{i=1}^{M} X_i^2 - \left(\sum_{i=1}^{M} X_i\right)^2} \quad \ldots (9)$$

Step 3: Determine μ_H, μ_V and μ_{ob} using eqns. (2), (3) and (4) respectively.

Step 4: Find the maximum of μ_H, μ_V and μ_{ob}. If μ_H is maximum, the feature is a horizontal stroke. If μ_V is maximum, the feature is a vertical stroke. If μ_{ob} is maximum, the feature is right slant for positive m_x and is left slant for negative m_x.

Step 5: Determine μ_{HC} and μ_{VC} using eqns. (6) and (8) respectively.

Step 6: Determine the maximum of μ_{HC} and μ_{VC}.

Step 7: If μ_{HC} is maximum the feature is a horizontal curve. To decide whether it is 'A' curve or 'U' curve the following rule is used :

(a) For 'A' curve as shown in Fig. 2(a), the |Y| co-ordinate of any point on the curve ≤ Max [|Y| coordinate of point A, |Y| coordinate of point B].

(b) For 'U' curve as shown in Fig. 2(b), the |Y| coordinate of any point on the curve ≥ Min [|Y| coordinate of point A, |Y| coordinate of point B].

Step 8: If μ_{VC} is maximum then the feature is a vertical curve. To decide whether it is 'C' curve or 'D' curve the following rule is used.

(a) For 'C' curve, as shown in Fig.2(c), the |X| coordinate any point on the curve ≤ Max (|X| coordinate of point A, |X| coordinate of point B).

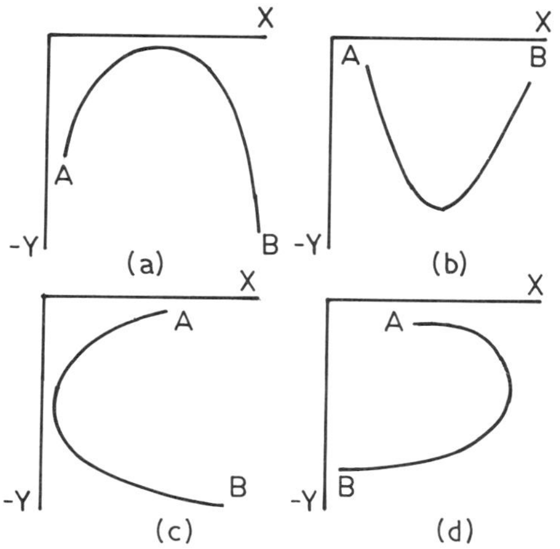

Fig.2 Representation of the different curves (a) A curve, (b) U curve, (c) C curve and (d) D curve.

(b) For 'D' curve, as shown in Fig.2(d), the |X| coordinate of any point on the curve $\geq$ MIN (|X| coordinate of paint A, |X| coordinate of point B).

CLASSIFICATION

For the extraction of the features the characters are written in a frame which is (20x20) binary matrix (Fig.3). The presence of the character is represented by the bit '1' and the absence by the bit '0'. Thereafter the features are extracted by the method discussed in the previous section and the characters are given the

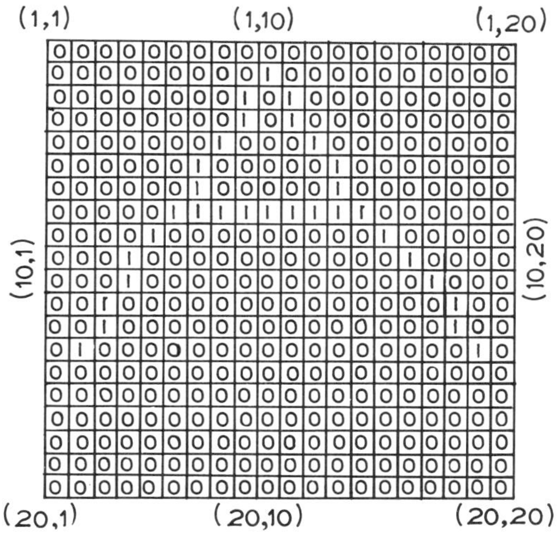

Fig.3 (20x20) binary frame for the characters.

linguistic definition in the coded form as [f_1, f_2, f_3, f_4, f_5, f_6, f_7, f_8] where f_1 and f_2 are the horizontal and vertical strokes in the character, f_3 and f_4 are the right and left slant strokes in the character and f_5, f_6, f_7 and f_8 are the 'A', 'U', 'C' and 'D' curves in the character.

For classification, the linguistic definitions are obtained in the coded form for all the 26 uppercase english alphabets and 10 numerals. This is done with the help of a training set consisting of the known characters. The linguistic definitions of uppercase english alphabets and numerals are given in Table 1. For the unknown character, which is to be recognized, the feature vector is first determined in the coded form. The feature vector is then compared with the linguistic definitions of all the 26 alphabets and numerals given in Table 1 and the unknown character is classified to be that character with which its linguistic definitions matches.

TABLE 1 Linguistic Definitions of Uppercase English Alphabets and Numerals.

ENGLISH CHARACTER	LINGUISTIC DEFINITION	CODE
A	Two left slant+two right slant+one horizontal stroke	10220000
B	Two D curve+two vertical stroke	02000002
C	One C curve	00000010
D	One vertical stroke+ one D curve	01000001
E	Three horizontal stroke+two vertical stroke	32000000
F	Two horizontal stroke+ two vertical stroke	22000000
G	One C curve+One D curve	00000011
H	One horizontal stroke +four vertical stroke	14000000
I	One vertical stroke	01000000
J	Two horizontal stroke +one vertical stroke +one U curve	21000100
K	Two vertical stroke+ one left slant+one right slant	02110000
L	One horizontal stroke +one vertical stroke	11000000
M	Two vertical stroke+ one left slant+one right slant	02110000

ENGLISH CHARACTER	LINGUISTIC DEFINITION	CODE
N	Two vertical stroke+ one left slant	02010000
O	One A curve+one U curve	00001100
P	Two vertical stroke+ one D curve	02000001
Q	Two left slant+one C curve+one D curve	00020011
R	Two vertical stroke+ one left slant+one D curve	02010001
S	One C curve+one D curve	00000011
T	Two horizontal stroke+one vertical stroke	21000000
U	One U curve	00000100
V	One right slant+one left slant	00110000
W	Two right slant+two left slant	00220000
X	Two right slant+two left slant	00220000
Y	One vertical stroke+ one left slant+one right slant	01110000
Z	Two horizontal stroke +one right slant	20100000
0	One A curve+one U curve	00001100
1	One vertical stroke	01000000
2	One horizontal stroke +one right slant+one A curve	10101000
3	Two D curve	00000002
4	One horizontal stroke +one vertical stroke +one right slant	11100000
5	One horizontal stroke +one vertical stroke +one U curve	11000100
6	Two A curve+one U curve	00002100
7	One horizontal stroke +one right slant	10100000
8	Two right slants+two left slants+one A curve+one U curve	00221100
9	One vertical stroke +one A curve+one U curve	01001100

RESULTS AND CONCLUSION

As mentioned in the previous section, a 20x20 frame was used for writing the alphabets. From this, a 20x20 binary matrix was obtained and from this matrix the features were extracted. For testing the character recognition process, 300 different persons were asked to write all the characters in the 20x20 frame. Out of the 300 sets of characters, 200 sets were used for training purpose, i.e., for the determination of the linguistic definitions of the characters. The classification procedure was tested with the remaining 100 sets of characters by the procedure outlined in the previous section. It has been found that the method gave very good results with recognition accuracy better than 92 percent. Reyad EC 1030 digital computer was used for the simulation study and the recognition time was found to be about 10 milliseconds.

The method has the advantage that it requires no threshold value for classification and has less restrictions on the shape of the alphabets and numerals. The method is simple and systematic. Using the concept of fuzzy set theory the cost of computation and hence the recognition process is considerably reduced. In this method all the characters were considered during classification process. Hence the method can probably be improved by considering the group of similar characters forming different classes and representing the recognition process to be consisting of two or more stages.

REFERENCES

Chatterji, B.N. (1982). Character recognition using fuzzy similarity relations, in Gupta, M.M. and Sanchez, E. (Ed.), Approximate Reasoning in Decision Analysis, North Holland, Amsterdam. pp. 131-137.

Gupta, M.M, G.N.Saridis, and B.R.Gaines (1977). Fuzzy Automata and Decision Processes. North Holland, Amsterdam.

Kickert, W.J.M. and H.Kopelaar (1976). Application of fuzzy set theory to syntactic pattern recognition of handwritten capitals. I.E.E.E. Trans. Syst., Man & Cybernet, 6, 148-151.

Pal, S.K., D.Duttamajumdar, and B.B.Chaudhury (1977). Fuzzy set in handwritten character recognition. In Recent Developments in Pattern Recognition and Digital Techniques, Calcutta 63-71.

Siy, P. and C.S.Chen (1974). Fuzzy logic for handwritten numerical character recognition. I.E.E.E. Trans.Syst., Man & Cybernet, 4 , 570-575.

Tamura, S., S.Higuchi, and K.Tanaka (1971). Pattern Classification based on fuzzy relations. I.E.E.E. Trans. Syst., Man & Cybernet, 1, 61-66.

Zadeh, L.A. (1965). Fuzzy sets. Inf. & Control, 8, 338-353.

ELEMENTARITY IN DESCRIBING FORMAL PROPERTIES FOR THE CLASSIFICATION OF OBJECTS

A. O. Arigoni

Department of Mathematics, University of Bologna P.zza Porta S. Donato, 5, 40127 Bologna, Italy

Abstract - In this paper, semiotic interations among descriptions of formal properties are analyzed. The main result is showing the possibility of attain_ ing to at least one elementary form in describing, formally, any given proper_ ty. Elementarity required for measuring the degree of presence, in specific ob_ jects, of simple properties which can be inferred on the basis of empirically ascertainable composite properties in the objects themselves.

Keywords - Elementarity; formal properties; classification; measure of sig_ nificativity.

1. INTRODUCTION

The subject-matter of the topic treated here_ in deals with the quantitative evaluation of to what degree specific simple properties are present in objects; this as derivable from the evidence of other properties directly shown by the objects themselves, after describ_ ing the latter properties formally.

In many attemps made to achieve the mention_ ed evaluation, the amount of evidence of the considered properties has been derived: either without taking into account the relevance of the details of the properties that said eva_ luation is relative to, or by letting the e_ valuation itself to subjective judgements (see for example Allen (1967) and, respectively, Kaufman (1973) and Zadeh (1975).

In our concern, 'relevance' is taken into ac_ count by deriving the idea of it from Carnap' s concepts (Carnap, 1956) and from those more recently elaborated by Gärdenfors (1978) and by us (Arigoni, 1974). We consider relevance itself as well as its opposite 'irrelevance' as boolean values that may be assumed by lin_ guistic variables and that result as superim_ posed on the values the same variables assume in denoting attributes of objects, i.e. in de_ noting either the presence or the absence of atomic properties in the objects themselves.

is subject is, in the goal we set for ourselves, a measure of pragmatical -finalistic- meaning. The feasibility of said measure is meanly based on the algebraic struc_ ture of the language formed by the sentences that the values of said variables give place to (Arigoni, 1980); structure that follows di_ rectly from the manner in which the same senten_ ces are semiotically related among themselves, that is, from how these may be synthetized according to the meaning of their details. The elementarity demonstrated herein permits acquiring the above mentioned feasibility, de_ finitively.

In the paper, examples as well as not strict_ ly necessary demonstrations will be omitted for brevity. These can be found in the cited references.

2. BACKGROUND AND DEFINITIONS

In this section the necessary notations and fundamental facts concerned in the developed argumentation are presented.

We define as *simple property* (SP) the formal property described by the ordered tuple of values that variables V_h of a finite set, i.e. with $h=1,2,\ldots,\ell$, assume in denoting either the possession or the non possession of one same number of corresponding *atomic properties* (AP). Such values are indicated by v_h and range on $\{0,1\}$. All the possible ℓ-tuples, indicated by x_α, form the set $\dot{X}=\prod_{h=1}^{\ell}\{0,1\}_h=\{x_\alpha: \alpha=1,2,\ldots,2^\ell\}$. Each ℓ-tuple is, thus, an ordered subset of the set $V=\{v_{h\alpha}: h=1,2,\ldots,\ell;\ \alpha=1,2,\ldots,2^\ell\}$, that is, $x_\alpha=(v_{1\alpha},v_{2\alpha},\ldots,v_{\ell\alpha})$.

One fundamental assumption is that there are *composite properties* (CP) the presence of which is empirically ascertainable and which imply the possession of one of the simple properties that interchangeably form a specific subset of X. In that manner, the degree of possession of every SP depends on which is the ascertained CP, or better, on which are the alternative SPs determining the presence of such a CP.

Thus, subsets $E,F,\ldots$ of $\dot{X}$, too, are taken into consideration. Every class resulting from one or the union of more of said subsets $E\cup F\cup\ldots$ constitutes the semantical -objective- description of one corresponding CP and is indicated by the symbol X. The elements x_α describing those EPs that are interchangeable in evidencing one X are defined as *equivalent* with respect to the CP itself; they form, in fact, an equivalence class on $\dot{X}$.

We indicate by $\Delta(\alpha,\beta)$ the *semantical distance* between x_α and x_β, i.e. the set of all indices h for which two given equivalent elements x_α and x_β, denoted by $x_\alpha\equiv x_\beta$, differ. If it so happens that $\Delta(\alpha,\beta)=\{h\}$, then $|\Delta(\alpha,\beta)|=1$ and x_α and x_β may be said to be *(semantically) adjacent elements*. The so estabilished relation is indicated by writing $\alpha|h|\beta$. Formally, if we are given the pair x_α,x_β of X, in the case that, besides being $x_\alpha\equiv x_\beta$, $\exists! h$ for which $v_{h\alpha}\neq v_{h\beta}$, then $|\Delta(\alpha,\beta)|=1$ and $\alpha|h|\beta$. If x_α and x_β are not adjacent, i.e. if $|\Delta(\alpha,\beta)|>1$, then these are said to be *distant elements*. Also pairs of E_is of $\dot{X}$ may be defined as distant; this happens when the elements of such subsets are pairwise distant.

If two elements x_α and x_β, such that relations $x_\alpha\equiv x_\beta$ and $\alpha|k|\beta$ are performed, are considered, it is axiomatic that the k-th variable, or else the values the latter may assume in said elements, are *irrelevant* in the elements themselves. Therefore, through *semiotic synthesis*, $x_\alpha \boxplus x_\beta$, an operation that may be performed on the elements in subject (Arigoni 1974), a third element x_γ is derived. This last is defined as *isosignificant* to both x_α and x_β, and also these are in the same relation among themeselves, relatively to the considered equivalence in X. The element x_γ thusly obtained will be such that $\forall h\neq k$ $v_{h\gamma}=v_{h\alpha}=v_{h\beta}$; concerning its k-th value $v_{k\gamma}$, this will be represented by one symbol which is neither '0' nor '1', it is but a third one evidencing the above defined irrelevance. Such a symbol is '$*$'. For this, x_γ is said to constitute the *pragmatic representation* of the pair x_α,x_β.

Thereupon, the values through which the SPs are denoted, $v_{h\alpha}$, shall range on $\{0,1,*\}$; further, the set of all the formable sentences describing simple properties includes $\dot{X}$ and is $\mathcal{X}=\prod_{h=1}^{\ell}\{0,1,*\}_h=\{x_\alpha: \alpha=1,2,\ldots,n=3^\ell\}$.

Example 2.1. Let us consider the set of three variables $V=\{V_1,V_2,V_3\}$. The set of SPs describable by the values the former may assume is reported in TABLE 1. TABLE 2 shows the descriptions derivable from the previous ones and forming one set indicated by $\ddot{X}$. By assembling the elements of the set $\dot{X}$ and $\ddot{X}$ together the whole set $\mathcal{X}$ is obtained.

	V_1	V_2	V_3
$x_1 =$	0	0	0
$x_2 =$	0	0	1
⋮	⋮	⋮	⋮
$x_8 =$	1	1	1

TABLE I

	v_1	v_2	v_3
x_9 =	0	0	*
x_{10} =	0	1	*
⋮	⋮	⋮	⋮
x_{15} =	1	*	0
⋮	⋮	⋮	⋮
x_{20} =	*	1	1

TABLE II

x_{21} =	0	*	*
x_{22} =	*	0	*
⋮	⋮	⋮	⋮
x_{26} =	*	*	1
x_{27} =	*	*	* = x_*

For every two elements x_α and x_β of X, if every $v_{h\beta}$ is either irrelevant or equal to $v_{h\alpha}$, then the truth of x_β derives from that of x_α and it may be said that x_α *implies semantically* x_β -Carnap' entailment (Carnap, 1956)-; we indicate this by $x_\alpha \Vdash x_\beta$ (Arigoni, 1978). It follows that every element $x_\gamma = x_\alpha \boxplus x_\beta$ is implied by both x_α and x_β, i.e. $x_\alpha \Vdash x_\gamma$ and $x_\alpha \Vdash x_\beta$. Threfore, the elements x_α of X can be distinguished in: those of $\dot{X}$, called *primary elements* -listed in TABLE I of Example 2.1-; those of $\ddot{X} = \mathcal{X} \setminus \dot{X}$, called *secondary elements* -listed in TABLE II- of the same example. Moreover, every element of $\ddot{X}$ tha is implied by two or more of those belonging to $\dot{X}$.

Every two elements x_α and x_β of $\mathcal{X}$ are called *(semantically) independent* if $\forall x_\gamma \in \mathcal{X}$ either $x_\gamma \Vdash x_\alpha$ or $x_\gamma \Vdash x_\beta$ but not both. Otherwise, x_α and x_β are said to be *dependent*.

The set of all classes of $\mathcal{X}$, i.e. the power set $\mathbb{X} = \mathcal{P}(\mathcal{X})$, is a *property set* (Allen, 1974). Every class of subsets of $\mathcal{X}$, $X = E \cup F \cup, \ldots$ of $\mathcal{X}$, constitutes a *member* of $\mathbb{X}$; each X has the general form $X = \{x_\alpha \in \mathcal{X}: x_\alpha \in \bigcup_i E_i\}$.

Given two members X and Y of $\mathbb{X}$, by 'X implies Y' is meant that $\forall x_\beta \in Y\ \exists x_\alpha \in X: x_\alpha \Vdash x_\beta$. From this follows that $\dot{X}$, by being one subset of $\mathcal{X}$, is a member of $\mathbb{X}$; we indicate it by the symbol $X_\circ$. Such a member, because of its composition, implies any other member of $\mathbb{X}$.

For every X of $\mathbb{X}$ a member X°, of $\mathbb{X}$ itself, implying it and called *maximal implicant* -of X- exists. This is such that: *i)* $\forall x_\alpha \in X° \ \exists x_\beta \in X: x_\alpha \Vdash x_\beta$; *ii)* X° is implied by itself and by $X_\circ$ exclusively.

Every member X of $\mathbb{X}$, and its X°, may be *pragmatically represented* throughout a process that develops in X^k $(k = 1, 2, \ldots, m)$ ensuing representations such that $\forall k\ X^{k-1} \Vdash X^k$. When $X^{k+1} = X^k$ is reached, then $k = m$ and X^m is the *minimal representation* of X and of X° (Arigoni 1982).

In Arigoni (1980) is shown that '$\Vdash$' on the x_αs of $\mathcal{X}$ and later generalized on those X of $\mathbb{X}$, performs the condictions listed below.
For any triplet of different members X,Y,Z of $\mathbb{X}$:

i) $X \Vdash X$ *(reflexivity)*;
ii) if $X \Vdash Y$, then $Y \nVdash X$ *(antisimmetry)*;
iii) if $X \Vdash Y$ and $Y \Vdash Z$, then $X \Vdash Z$ *(transitivity)*.

In the final analysis: 1st) '$\Vdash$' introduces a (semantical) partial ordering in $\mathbb{X}$; 2nd) the member $X_\circ$, which implies all other members of $\mathbb{X}$, is said to the *maximal member* of $\mathbb{X}$ itself; 3rd) the member $\{x_*\}$ -x_{27} of Example 2.1, (TABLE 2)-, which is implied by any other member of $\mathbb{X}$, is the *minimal member*. It follows that in $[\mathbb{X}, \Vdash]$, $X_\circ$ is the *universal upper bound;* the $[\mathbb{X}, \Vdash]$'s *universal lower bound* is $\{x_*\}$.

3. THE INTERNAL APPLICATION ON $\mathbb{X}$

In the previous section a synthetical drawing of $\mathbb{X}$'s characteristics has been given. Here a detailed analysis of the internal applications (of semantical nature) which are allowed on the members of $\mathbb{X}$ itself is performed.

The subsets of $\mathcal{X}$ forming those members X of $\mathbb{X}$ -i.e. the cantorian unions of subsets E_i of $\mathcal{X}$ itself- that are significative for the development of the subject-matter treated herein are now considered.

Since it can be proved -which is made in Arigoni (1980)- that for every X of $\mathbb{X}$ $lub(X)$, $gbl(X)$ and $\overline{X}$ are unique memebrs of $\mathbb{X}$ itself, then these memebrs constitute results of one same number of operations on $\mathbb{X}$, carried out on Xs. Said operations are respectively indicated by '$\sqcup$', '$\sqcap$', and '$-$', and are called *(semiotical) union, interserction, and complementation*. As can be derived from Definition 3.2, since operations $\sqcup$ and $\sqcap$ are considered when worked on one single member X, these might be seen as unary operations; indeed, since the Xs are formed by E_is, i.e. in general by more than one x_α, the n-arity of such operations is manifest.

In accordance with the properties of operations $\sqcup, \sqcap$, and $-$, the *semantical area* relative to whichever x_α of $\mathbb{X}$ forms one *semiotical lattice*. Definitively, because of the validity of De Morgan' laws on said structure and being $\mathbb{X}$ the power set of $\mathcal{X}$, and this last is a countable set, also $\mathbb{X}$ is countable and the system $<\mathbb{X}, \sqcup, \sqcap, X_\circ, \{x_*\}>$ is a sigma algebra. Concerning the above defined operations it is stressed that: *i)* $X \sqcup Y = (X^\circ \sqcup Y^\circ)^m$; *ii)* $X \sqcap Y = (X^\circ \sqcup Y^\circ)^m$; *iii)* $\overline{X} = (Z = \mathcal{X} \ \{x_\alpha \in X^\circ\})^m$.

3. THE ELEMENTARITY OF THE MEMBERS OF $\mathbb{X}$

The elementarity of one member X of $\mathbb{X}$ consists, like for cantorian subsets, in the *possibility of describing it, in at least one way, as a finite number of pairwise disjoint elements* x_α of $\mathcal{X}$ (see, for example, Kolmogorov and Fonin (1961)). The necessity of demonstrating such a possibility is prescribed by the transformation that the x_αs, which every X of $\mathbb{X}$ definitively is formed by, undergo to with the process through which the minimal reprentation of X itself is achieved; and this notwithstanding as well $\mathcal{X}$ as $\mathbb{X}$ are countable sets.

Definition 3.1 Any pair of equivalent elements x_α, x_β of $\mathcal{X}$ is said to be *disjoint* if 'independent' and 'distant' are its attributes.

Definition 3.2 One member X of $\mathbb{X}$ is defined as *elementary* when all the elements x_α forming it are pairwise disjoint.

4. THE ELEMENTARITY OF THE MINIMAL REPRENATION

More than one synthesis results normally as feasible on the elements x_α forming any k-th representation of a member X, X^k $k = 1, 2, \ldots, m-1$). Since each one of said x_αs can partecipate to one unique of the mentioned synthesis, every X^k becomes partitioned into groups G_i^k, each one including all elements partecipating to one same synthesis. In this manner, the elements forming the groups G° formable in X°, are synthetized to give place to those of X^1, which the elements of X° itself are represented by, and will be assembled into groups G_j^1. However the same elements will not interact semiotically with elements deriving from groups other than the one G_i° (Arigoni, 1982).

It is also noted that the set of elements x_α forming the representation of one G_i^k in X^{k+1} of anygiven X, is not the same as G_i^{k+1}. Such a set consists, in fact, in the cantoiran union $\bigcup_j (G_j^{k+1})$, i.e. in th subset of all the elements forming the k+1-th representation of G_i^k, no matter if these are adjacent or not. As a shorthand, said subset will be indicated by $(G_i^k)^1$. Therefore, it can be said that the minimal representation of whichever X, or X°, results from the cantorian union, with respect to i, of the minimal representations of the single G°s forming X°, i.e. $\bigcup_i G_i^\circ$.

lemma 4.1. *If we are given the group* G_i^k *of the k-th representation* X^k *of one member X formed by three independent elements* x_α, x_β, *and* x_γ *such that* $\alpha|h|\beta$, $\alpha|k|\gamma$ *and* $|\Delta(\beta,\gamma)| > 1$, *then* $(G_j^k)^1$ *is a disjoint element (Definition 3.1).*

Proof: We have to prove that $(G_i^k)^1$ is independent and distant. Let the two elements resulting from the feasible synthesis be $x_\chi = x_\alpha \boxplus x_\beta$ and $x_\psi = x_\alpha \boxplus x_\gamma$, respectively. The proof is given in two parts: in one, the apparent dependence between x_χ and x_ψ will be denied; in the other, $|\Delta(\chi,\psi)| > 1$ shall be shown. Such a proof can be extended to groups including any number of elements greater than three, provided these give place to a unique

Independence - Every element of $(G^k)^1$ is implied by two adjacent elements of G^k; in addition, one of these last, x_α, implies both x_χ and x_ψ. Nevertheless, such a common implication is only apparent; this in that, what really is semiotically synthetized to derive x_χ and x_ψ, and consequently what by which the latter are implied, besides x_β and x_γ, is not x_α as a whole; x_χ and x_ψ are but implied by, re_ spectively, $x_{\alpha'}$ and $x_{\alpha''}$, two virtual (distinct) elements achieved through opportunely disintegrating x_α (Arigoni, 1983). Thus, the fact that the pair x_χ, x_ψ is independent may be concluded.

Distance - Standing that

$\exists h,k$ $(h \neq k)$ for which $\Delta(\alpha,\beta) = \{h,k\}$

as well as

$\Delta(\alpha,\beta) = \{h\}$ and $\Delta(\alpha,\gamma) = \{k\}$

-which would hold even if x_α was replaced by whichever of the virtual elements x_a of x_α itself, in that the irrelevant values are the same as those of the latter- the irrelevant values in both x_χ and x_ψ, i.e, those equal to '*', are the h-th and k-th ones. Such values are, in fact, $\upsilon_{h\chi}$ and $\upsilon_{k\psi}$. This means that x_χ and x_ψ differ at least for two values, i.e. $|\Delta(\chi,\psi)| > 1$. ||

Lemma 4.2. *If we are given the group* $(G_i^k)^1$ *of the k-th representation* X^k *of one member* X, *which is formed by the four independent elements* x_α, x_β, x_γ, and x_δ *giving place to a unique synthesis, then* $(G_i^k)^1$ *is:* *i) independent;*

ii) not necessarily distant.

Assertion *ii)* means that elements susceptible of further synthesis *may* result forming the group $(G_i^k)^1$.

Proof: To prove *i)* it suffices to add one detail to the proof given for Lemma 4.1: the possibility according to which, like for one of the elements of G_i^k, in the case proven 'x_α', also the others can partecipate to sinthesis after partitioning them into virtual elements, rather than as real -unpartitioned- elements. Since this is an admitted extension of what by Lemma 4.1. has been demostrated, *i)* is proven.

ii) This consists in giving a counter example of the opposite of the statement; that is, we prove that elements representing G_i^k in more synthetical form than the one $(G_i^k)^1$ *may* exist.

Let us consider the element x_α of $(G_i^k)^1$ for which is likely to occur that

$\exists h,k$: $\upsilon_{h\alpha} = \upsilon_{k\alpha} = *$

Such an x_α results from synthetizing one pair x_β, x_γ of G_i^k such that

$\Delta(\beta,\gamma) = \{h\}$ and with $\upsilon_{k\beta} = \upsilon_{k\gamma} = *$

Each element of such a pair derives, at its turn, from another pair x_χ, x_ψ of G_i^{k-1} such that $\Delta(\alpha,\beta) = \{h\}$. In order than the four elements forming the two pairs in subject, x_β, x_γ and x_χ, x_ψ, perform the stated requirements, the distance of one from another of them can not be greater than two -as much as the '*s' appearing in x_α (Arigoni, 1974). Moreover, being the involved four elements variously adjacent among themselves, then these necessarily are included in one unique group, G_i^{k-1}. ||

By extending Lemmata 4.1 and 4.2 to groups including whichever number of elements, and on the basis of the concepts expressed earlier in this section, we attain to the possibility of analyzing the minimal representation of any group G_i° the elements of which are variously adjacent.

From the argumentation developed in proving Lemma 4.1, also the proposition that follows is estabilished with evidence.

Corollary 4.2.1. *Elements forming the k-th representation* X^k *of a member* X *may be adjacent iff they are included in the same representation* G_i^k *of one* G°

Theorem 4.3. *The minimal representation* X^m *of any given member* X *of* 𝕏 *is a disjoint member.*

What by Theorem 4.3 is stated is an alternative form of '*any member* X *of* 𝕏 *can be represented minimally by another member of* 𝕏 *itself,* X^m, *which is disjoint*'. This version yields more directly the corollary that will follow and which indicates the achievement of the goal we set curselves in the present paper. To prove the theorem, the demonstration of the following additional lemma is required.

Lemma 4.4. *Elements of two distinct* G_i^h *and* G_j^k *included in representations -eventually different- relative to one member*

X of 𝕏 are independent the ones from the others.

Proof of Theorem 4.3. Relatively to every representation X^k of one X, let us indicate by: $A^k_{i\alpha}$ every group G^k_i which, by including one unsynthetizable element exclusively, is a singleton; B^k the subset of X^k that, by including the elements x_α which, being in some way adjacent, are still susceptible to be assembled into groups and to undergo to further synthesis.

Then, by definition of k-th representation, it may be written:

$$(G^k_i)^1 = \{ \cup_\alpha A^k_{i\alpha} \} \cup \{ \cup B^k \}$$

This, since the elements x_α included in the $A^k_{i\alpha}$s are distant, can be rewritten as

$$(G^k_i)^1 = \{ \cup_\alpha A^k_{i\alpha} \} \cup \{ \cup B^k \}$$

further, due to Lemma 4.2,

$$\{ \cup_\alpha A^k_{i\alpha} \} \cup \{ \cup B^k \} = \emptyset$$

In addition, since at every ensuing representation of one group G°_i, the B^k's elements 'converge' toward other groups of type $A^k_{i\alpha}$ (Lemmata 4.1 and 4.2), the m-th representation G^m_i of G°_i can be expressed by

$$G^m_i = \cup_\alpha A^m_{i\alpha}$$

the elements appearing in which are distant, by definition, and independent, because of their derivation (Lemmata 4.1 and 4.2).

Finally, Since also the $A^m_{i\alpha}$s deriving from different G°s are distant (Corollary 4.2.1) and independent (Lemma 4.4), then:

$$X^m = \cup_i \{ \cup_\alpha A^m_{i\alpha} \} \qquad ||$$

Because of the finiteness of the values on which as well i as α range, both these indices assume, in fact, values on $\{1, 2, \ldots, 2^\ell$ (the number of primary elements)$\}$, we may conclude by stating the following proposition.

Corollary 4.3.1. *Every composite proprety -subset of interchangeable simple properties- can be pragmatically described by the finite number of disjoint elements forming the minimal representation of said composite property's description.*

7. CONCLUSION

The main result achieved in the paper is the demostration of elementarity of the synthetic description of the simple properties forming the pragmatical -minimal- description of any given composite property. Pragmatical description derivable by linguistic transformation of the one semantical, relative to the property itself, i.e. of the description in which the irrelevant atomic properties were not in to evidence.

Said result can be utilized in measuring to what degree specific formal properties are present on evidence of others, so that an efficient classification of of objects may be performed (Arigoni, 1983).

REFERENCES

Allen, A. D. (1974). Measuring the empirical properties of sets. IEEE Trans. on Systems Man, and Cybernetics, 4, 1, 66-73.

Arigoni A. O. (1974). Structure of semiotic dimension. In Proceed. of 1st Int. Congress on Semiotical Studies, Milano, Italy.

Arigoni A. O. (1978) Semiotical implication and its utilization in Fuzzy Sets Theory. In: Proced. of 4th Eur. Meet. on Cybernetics and Ststems Res., Linz, Austria.

Arigoni A. O. (1980). Math. devel. arising from 'Semantical Implication'. Fuzzy Stes and Systems. 4, 167-183.

Arigoni A. O. (1982) Transformational-generative grammar for description of formal properties. Fuzzy Stes and Systems, 8, 311-322.

Arigoni A. O. (1983) Probability of fuzzy causes. In: Sing M. Encyclopedia of Systems and Control (Pergamon Presses, N. Y.).

Arigoni A. O. (1983) Cybernetical information. (will appear on: Systmes Analysis).

Carnap R. (1956). Meaning and Necessity. Univ. of Chicago Press, N.Y.

Gärdenfors P. (1978). On the logic of relevance. Synthes, 37, 351-367.

Kaufman A. (1973). Introduction a la theory des sous-ensembles flous. Masson et C.ie, Paris.

Kolmogorov A. N. and Fonin S. V. (1961). Elements of Theory of Functions and Functional Analysis. Graylook Press, N.Y.

Zadeh L. A. (1975). Calculus of fuzzy restrictions. In: Zadeh and others, Fuzzy Sets and applications, Accademic Press, N. Y.

SOLUTION OF NONLINEAR ASSIGNMENT PROBLEMS AND STRUCTURE-BUILDING PROCEDURES ON THE BASE OF THE FUZZY APPROACH

M. Peschel, W. Mende, J. Richardt and M. Voigt

Academy of Sciences of the GDR/Scientific arec Mathematics and Cybernetics

Abstract. Assignment problems usually are np-complete and cannot be effectively solved by black&white-optimization techniques. Fuzzy clustering respective fuzzy graph decomposition offers advantages because of the graduality of membership correspondence. Two different approaches are proposed, one using fuzzy affinities to clusters in connection with nonlinear iteration and the other based on the replication form of Lotka-Volterra equations.

Keywords. Assignment problems; cluster technique; affinity measure; stationary points; replicator equation; Lotka-Volterra Equations.

INTRODUCTION

Assignment problems are of increasing importance in quite different fields, for example in industrial applications (microelectronic curcuit design, allocation problems for large manufacturing lines), in social sciences (formation of social groups, growth of settlements and their decomposition in substructures), in ecology (decomposition of a global ecosystem into biotopes).
All these problems can be formally described as cluster building processes.
Growth and structure-building are the main components of evolution processes.
Every structure is built up of elements (individuals, atoms, species, singletons etc.) on the base of a relation between pairs of elements, which is objectively given.
Depending on the concrete assignment problem we meet quite different forms of this binary relation (number of connections, competition coefficients, similarity measure, metric distance). Using this binary relation the set of all elements can be rationally decomposed into clusters not necessarily being disjunct. Thus in a natural way we meet fuzzy membership correspondences of the elements to the different clusters.
On one level of the consideration we find a fuzzy partition of the set of all elements into clusters. But we can consider the structure-building process as a hierarchical process.
On the next higher level of consideration we take the already received clusters as aggregated elements and try to build up superclusters from them.
It is important to mention that cluster-building is a dynamic process at the end of which we find a

stabilized cluster decomposition of the given set of elements.
For this process a senseful interaction of global and local performances is essential in similarity to problem-solving in general or to playing chess in special. Therefore it seems to be useful to consider the final result of a dynamic cluster-building process as an equilibrium point of a dynamic process or as a solution of a (mostly nonlinear) fixpoint equation.
In this paper two quite different fuzzy approaches to cluster-building are studied, the first is a generalisation of the cluster-building procedures from Späth resp. Bezdek and the second is an application of the replicator equation (Schuster and others /1/) for cluster building purposes.
Both approaches were checked on high-dimensional examples from microelectronics and showed quite satisfying properties.

GENERAL FEATURES OF A FUZZY CLUSTER-BUILDING PROCESS

The elements of the considered "universe" are designated with the indices 1,2,...,n, for variable elements we use the notification i and j. As names for variable clusters we use k and l and for variable superclusters $\mathcal{X}$ and λ .
The basic relation between any pair of elements i and j is given by a (deterministic) matrix of coefficients S(i,j) with real values being not necessarily a symmetric matrix and sometimes also negative values are possible (for example in the case of competition coefficients of species in an ecosystem).
The correspondence of an element i to a cluster k is modelled by membership-coefficients U(i,k) which are the state variables of the dynamic cluster-building process.
For building up superclusters we need an interaction matrix $\tilde{S}(k,l)$ between the different clusters k and l, and a fuzzy correspondence $\tilde{U}(k, \mathcal{X})$ of the clusters k to the superclusters $\mathcal{X}$. Already specific for our fuzzy approaches to cluster-building are the following notions.
We introduce an affinity measure A(i,k) for the tendency of the element i to belong to the cluster k

$$A(i,k) = \sum_j S(i,j)U(j,k)$$

The idea for this measure stems from the deterministic clustering. Here U(j,k) can only assume the values 0 or 1, 1 for all elements j belonging to the cluster k. Then A(i,k) is the number of connections of the element i into the cluster k. Our measure A(i,k) we get obviously by fuzzification of this notion from the deterministic case.
It seems to be quite natural to introduce a similarity measure for any pair of clusters k and l by

$$\tilde{S}(k,l) = \sum_{i,j} U(i,k)S(i,j)U(j,l)$$

which is obviously a fuzzy generalisation of the "number of connections between elements i belonging to the cluster k to elements j belonging to the cluster l.
During the cluster-building process we have to change in an aimoriented way the values of the memberships U(i,k) (or $\tilde{U}(k, \mathcal{X})$ in a twolevel hierarchy and so on), that means, the cluster partition is contained in the image U(i,k).
For the dynamic cluster-building process some bilance relations must always be taken into account namely

$$\sum_k U(i,k) = U(i)$$

"the mass of the particle i"

$$\sum_i U(i,k) = U(k)$$

"the intermediate mass of the cluster k"

respective in the two-level hierarchical process

$$\sum_{\varkappa} \tilde{U}(k, \varkappa) = U(k)$$

"the intermediate mass of the cluster k"

$$\sum_k \tilde{U}(k, \varkappa) = \tilde{U}(\varkappa)$$

"the intermediate mass of the supercluster "

CLUSTER-BUILDING BY A NONLINEAR ITERATION PROCEDURE

We in general assume that the masses U(i) of the elements i are given values.
Let us mention first a linear iteration formula.
We change the fuzzy memberships U(i,k) according to their affinities A(i,k) to the classes with the help of the following iterative procedure

$$F(i,k) := A(i,k)$$

We interpret F(i,k) as attraction force and by using the normalisation condition we get

$$U(i,k)=(F(i,k)/ \sum_k F(i,k)). U(i)$$

This apparently corresponds to a Markov chain procedure. It turned out to be not very successful in exposing an underlying structure.
Therefore we passed to an essential nonlinear procedure by introducing master variables and taking into account the different intermediate masses of the clusters U(k).
We introduced the following concept

$$F(i,k) := A(i,k)^{EV} / U(k)^{EG}$$

with the master variables EV and EG and then applying the normalisation condition as above

$$U(i,k) = U(i) \ F(i,k)/ \sum_k F(i,k)$$

We put EG = EV. ϱ with a slowly adapting variable ϱ, in a lot of cases it was allowed for ϱ to have a fixed value. The master variable EV is a very important parameter. We allowed for EV only nonnegative values.
Obviously a very small value EV transforms a given distribution A(i,k) in a very flat one, and a very high value of EV produces a peak-like distribution, that means by increasing the value of EV we approach more and more a deterministic correspondence of the elements i to the clusters k.

STRUCTURE-BUILDING BASED ON LOTKA-VOLTERRA EQUATIONS

For Structure-building in general we support the following system philosophy. For structure-building in real systems are besides the quantitative relations of competition, mutation and selection no special nature laws (first principles) necessary.
Therefore it must be demonstrated that on the base of the quantitative relationships structures can develop.
We concentrate on the process of structure-building arising from quantitative competition in Lotka-Volterra Equations.
As pointed out by Schuster /1/ Darwinian properties can be best studied with the replicator form of Lotka-Volterra Equations, this holds in our opinion also for the building of societies of species.

There is a standard procedure to pass from Lotka Volterra description to the replicator form

$$F\,x_i = \sum G_{ij} x_j \qquad i=1,2,\dots,n$$

We introduce as new variables socalled barycentric coordinates

$$y_o = \frac{1}{\sum x_j} \qquad y_i = y_o\, x_i$$

Then we get for the new coordinates

$$y_i \quad \text{with} \quad \sum_{i=1}^{n} y_i = 1$$

the following differential equations

$$F\,y_i = \left(\sum G_{ij} y_j - \phi \right) / \, y_o$$

$$\text{with} \quad \phi = \sum y_k G_{kj} y_j$$

For $y_o > 0$ the dynamics does not essentially depend on y_o (this is only a monotoneous transformation of time-scale along the trajectories), so that for all qualitative questions we can work with

$$F\,y_i = \left(\sum G_{ij} y_j - \phi \right)$$

This equation is called replication form of Lotka-Volterra equations /1/. This is a qualitative equivalent to the original Lotka Volterra equation.

All stationary points of the replicator equation have the form

$$y_i^* = 0 \quad \text{for} \quad i \in I \quad (1,2,\dots,n)$$

$$\sum G_{ij} y_j^* = \phi \quad \text{for} \quad i \in \bar{I}$$

We are interested in stable stationary points (sinks). Every stable stationary point contains a certain set $\bar{I}$ of not vanishing species. Every such set corresponding to a stable stationary point we consider as a possible cluster of the structure. The structure of the whole Lotka-Volterra system is then the compound of all these in general overlapping clusters.

This is a restricted structure notion, because in ecology most important clusters belong to stable limit cycles.

But nevertheless this restricted structure notion is very useful for solution of nonlinear assignment problems. In this case we introduce a control of the replication rates G_{ii} by master parameters.

REFERENCES

Hofbauer, J., Schuster, P., Sigmund, K. (1982). Game dynamics in Mendelian Populations. Biol. Cybernetics, 43, 51-57.

Maynard Smith. (1972). On Evolution. University Press, Edinburgh.

Peschel, M., Bocklisch, S.F., Meyer, W., Straube, B., Richardt, J. (1981). Daten- und Verhaltensanalyse mit Klassifikationsmodellen. Reports of the Austrian Society for Cybernetic Studies Nr. 22, 95 pp.

Peschel, M., Richardt, J. (1981). Perspektivische Möglichkeiten der Datenanalyse. Unscharfe Modellbildung und Steuerung Teil IV. Technical Highschool of Karl-Marx-Stadt, 4-18.

Peschel, M., Mende, W. (1983). Leben wir in einer Volterra-Welt? - Ein ökologischer Zugang zur angewandten Systemanalyse. Series Mathematical Research. Akademie-Verlag, Berlin. 235 pp.

Schuster, P., Sigmund, K. (1982). Coyness, Philandering and Stable Strategies in Animal Behaviour. Journ. Theoret. Biol., 29, 186-192.

Volterra, V. (1931). Lecons sur la theorie mathematique de la lutte pour la vie. Gauthier-Villars, Paris.

AN AUTOMATON FOR THE DETECTION OF VIGILANCE STATES ON LABORATORY RODENTS

J. F. Feng*, H. Emptoz**, J. L. Valatx***, G. Morin and G. Chouvet****

*Aeronautical Institute of Nanjiing (China)
**Centre de Mathématiques, Bätiment 203 INSA, 69621 Villeurbanne Cédex, France
***INSERM U 52, 8, Av. Rockefeller, 69008 Lyon, France
****INSERM U 171, Höp. Ste Eugénie, 69230 St Genis Laval, France

Abstract. In this communication we will present a summary of the definition and construction of an automaton for on-line analysis of vigilance states in laboratory rodents. A recognition system, which can be described as semi-automatic, has already been developed and currently operates successfully ; human intervention is however still needed both at the beginning of each analysis, during a learning period and, if necessary, during subsequent adaptive periods. The definition of the complete automaton thus requires the realization of an autolearning process, in which the techniques of pattern recognition are used in depth on data originating directly from preprocessing of electrical signals. Though chosen as a function of constraints linked to cost minimization, these techniques have led to the development of an original method which can be implemented on a micro-processor system.

Keywords. Sleep ; automatic classification ; micro-processor ; automaton ; articicial intelligence.

INTRODUCTION

During the last twenty years, "automatons" have invaded numerous research and industrial fields, where they have generally been designed to resolve various optimisation problems. We will first present here some studies which were the basis for the development of an automaton permitting the on-line detection of vigilance states in laboratory animals. The objective of the chronobiological approach of vigilance states is the search for structures and mechanisms responsible for the temporal organization of these states. Furthermore, within the general outline of research into the functions of paradoxical sleep (the sleep state which corresponds to dreaming in man), the genetic approach of the sleep-waking cycle, and the study of the relationships between sleep and learning, have been considerably developed in recent years. However, these different approaches demand data collection from a large number of animals using neurophysiological techniques which are difficult to implement. The practical limitation thus derives essentially from the time necessary to interpret visually the continuous polygraphic sleep recordings made over several consecutive days and simultaneously on several animals.

A recognition system, which can be described as semi-automatic, has already been developed in the laboratory of Médecine Expérimentale (Pr Michel Jouvet) in Lyon, and currently operates very successfully (Chouvet et al., 1980). Human intervention is however needed at the beginning of each analysis, during a learning period, and continuous long term analyses (for one week or more) necessitate adaptive regulation in some instances. It appeared advantageous therefore to automatise this process entirely. The objective of this study was thus the construction of an automaton which can discriminate on line the three principal vigilance states (Waking, Slow Wave Sleep and Paradoxical Sleep) in laboratory rodents (rats and inbred strains of mice). The definition of a complete automaton necessitated the realization of an autolearning process, in which the techniques of pattern recognition are used in depth on data originating directly from preprocessing of the biological signals. Though chosen as a function of constraints linked to cost minimization, these techniques have led to the development of an original method which can be implemented on a microsystem, functioning on-line and multiplexed on several animals.

THE SEMI-AUTOMATIC ANALYSER

We will first describe the semi-automatic analyser developed in 1980 (Chouvet et al., 1980). Numerous examples of this analyser are currently operational in different laboratories. Its functioning is based on interaction between the experimenter and the analyser.

Polygraphic reading of the arousal states of an animal is essentially made on the basis of electrical signals : the EEG (electroencephalogram) and the EMG (electromyogram), which result in signals which vary continuously over time (cf.below).

These polygraphic signals can be partially "summarized" by a number of indices which evolve as a function of the vigilance state of the animal. The principle of analysis is then based on the following hypothesis : each polygraphic episode is represented by a point in the space formed by the indices, and points corresponding to the same vigilance state of the animal tend to group together in this space, "Birds of a feather flock together".

In the following paragraphs we will show how the experimenter, as a function of his own interpretation criteria, supplies to the analyser the decision rules which should lead to the same result as visual interpretation.

The Reading of Polygraphic Recordings

Each vigilance state is characterized by a particular feature of one or both of the recorded signals (EEG and EMG).

Waking (W):Waking is generally characterized by strong muscular activity. This state can therefore be distinguished principally on the basis of the EMG which will have a relatively large average amplitude. The EEG trace will be rapid, desynchronized and generally of low and regular amplitude.

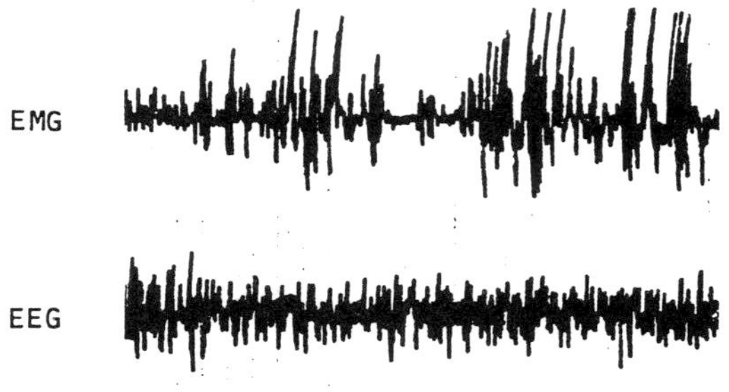

Fig. 1. Polygraphic characteristics of Waking.

Slow Wave Sleep (SWS): During periods of Slow Wave Sleep, the amplitude of the EMG is generally low ; and on the EEG this sleep state is characterized by a larger amplitude than during waking and a low average frequency (slow waves).

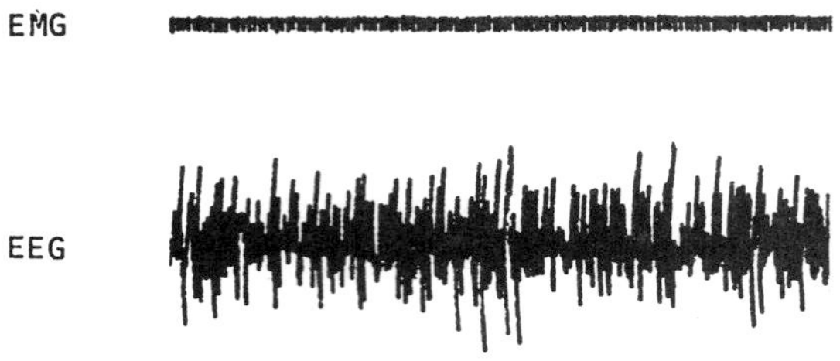

Fig. 2. Polygraphic characteristics of Slow Wave Sleep.

Paradoxical Sleep (PS): Paradoxical Sleep is characterized by a very low or zero muscular activity on the EMG. The form of the EEG signal is, at first sight, rather similar to that of waking, which leads on occasions to some uncertainty as to the state, but this can be resolved by reference to the EMG. With respect to waking, however, the EEG nevertheless presents some differences which are a signal amplitude which is sometimes greater, but above all more constant, and an average frequency which is relatively "monochromatic" and situated around the 6-8 Hz (θ) band.

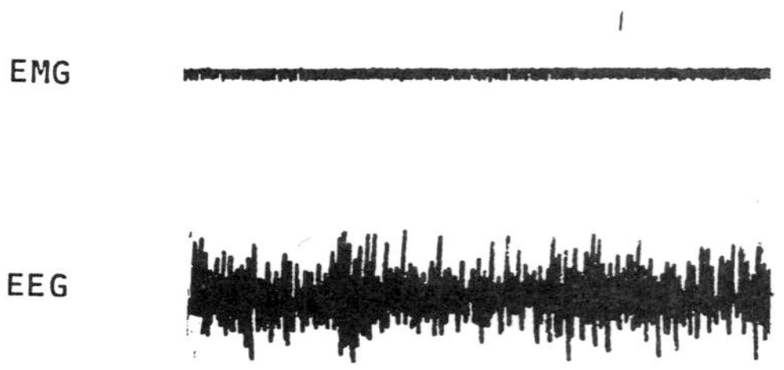

Fig. 3. Polygraphic characteristics of Paradoxical Sleep.

Definitions of the Indices and Distribution of the Episodes

Sufficient information must be extracted from the polygraphic signals, by feature extraction, to discriminate between the three states. When he examines polygraphic traces, the experimenter visualizes them over a certain time interval, for example to estimate a frequency. This led us to introduce a similar type of interval and to determine the "indices" over this time scale, which we will call an episode. The latter can vary between 10 and 60 seconds depending on the desired analysis accuracy.

The indices used for automatic analysis are the following :

a) From the EEG :
- the ratio of energy in the theta band (6-8 Hz) to energy in the delta band (1-5 Hz) (F index),
- the number of zero crossings characterizing the average EEG frequency (Z index),
- the variation of the amplitude, measuring the non-regularity of the EEG envelope over an episode (D index).

b) From the EMG :
- the average integrated value of the muscular activity (M index).

These indices are measured on-line by analogue methods and, at the end of each episode, information is available corresponding to the average values of the 4 parameters measured over that duration. The synchronisation of episodes is assured by a clock which can be regulated to give impulses separated by intervals of between 10 and 60 seconds.

As a result of both the way that vigilance states are defined polygraphically, and the above choice of indices, some "natural" relationships are established between these indices and the vigilance states.

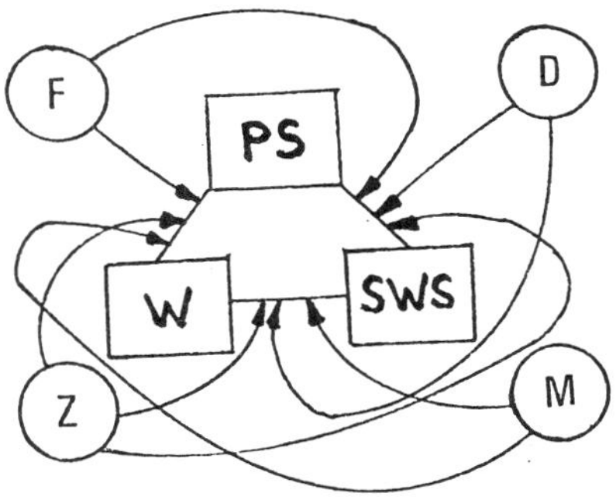

Fig. 4. The "natural" relationships between the vigilance states and indices : the arrows indicate a discriminant power.

Generally, a single index is sufficient to discriminate between one vigilance state and the other two. For example in Figure 4, the F index discriminates relatively well between PS and the other two states, W and SWS. It can therefore be used with a reasonable discriminant power to distinguish W or SWS from PS, but carries no information to help the decision "W or SWS ?".

The polygraphic signals are thus now replaced by a succession of points, each corresponding to an episode and situated in a 4-dimensional space.

Functioning of the Semi-Automatic System

We will consider in this paragraph only the principle of this system, leaving aside the physical construction which is based on standard microelectronic and electronic hardware.

The thresholds. The distribution of the index values is generally bimodal. As one of the modes usually corresponds to a single vigilance state, it is apparent that we need to convert each index into 2 "logical" values. The difficulty in this problem lies in the determination of the best threshold which will enable the index to be transformed into a useful binary signal. These thresholds determine, in the 4-dimensional space, the hyperplanes which subdivide this space into $2^4 = 16$ regions. If the thresholds have been correctly evaluated, each of the 16 regions corresponds to a vigilance state, Waking, Slow Wave Sleep or Paradoxical Sleep. This correspondence can then be established by means of a truth table. Indeed, the facility of realizing a threshold electronically was the main reason for adopting this approach.

Implementation. The truth table is hardwired by the experimenter by mean of microswitches on the basis of his own interpretational criteria, and it remains stable over time. The threshold settings are determined by the experimenter who must adopt an approach based on successive iterations to obtain a correspondence between the results from the analyser and those derived directly from visual interpretation of the EEG and EMG signals. This adjustment can take a considerable time and is dependent on the experimenter's experience. The system also does not adapt to changing thresholds caused, for example, by deterioration of the electrodes. The experimenter himself therefore must reset thresholds accordingly. This implementation of analysis can sometimes be tiresome and, in fact, corresponds to a learning process. An attempt to automatise it, therefore, by using articicial intelligence processes is difficult though necessary.

Some problems. We will discuss here only those problems related to pattern recognition, and amongst those the most important are those which

involve an episode comprising several vigilance states (for example in awakening) i.e. a change in state which can of course occur during an episode. We thus had to consider :
- simple episodes, in which the same vigilance state continues throughout the episode,
- double episodes, divided into two vigilance states of approximately the same duration,
- triple episodes, in which there is a change in state with one of them predominating in the ratio of approximately 2/3, 1/3 (two changes of state during the same episode is an extremely rare event).

The proportion of these different episodes will naturally depend on the length of the elementary analysis period. For example, using 30 second periods, the distribution of these episodes was as follows for mice of the BL/6 strain :

Episode Type		
Simple	Double	Triple
79 %	7 %	14 %

The double episodes, which correspond to transitional states, are difficult to classify, even visually. We will be concerned therefore only with simple and triple episodes ; the majority of the latter being classed in terms of the preponderant state.

TOWARDS COMPLETE AUTOMATISATION

To minimize difficulties related to :
- the length of time needed by the experimenter to set the thresholds initially,
- the stability of these thresholds over time,
- possible pharmacologically induced changes, etc.,

it seemed advantageous to automatise the analysis entirely. With this objective we set out to develop an automaton which not only recognized states but was also capable of autolearning. A priori, some questions had to be answered :
- was it necessary to keep the principle of a truth table and hence the thresholds? In other words, was it best just to add an autolearning device without change to the existing recognition principle ?
- would the percentage of correct classifications be sufficient to enable the semi-automatic system to be replaced ?
- could the cost be of the same order as that of the semi-automatic system ? (could a microprocessor be used ... ?).

Rejection of Thresholds and the Truth Table

The principle of the semi-automatic analyser is based on the complementarity between man (and his intelligence) and the machine which measures the indices. To attempt a copy of the philosophy related to thresholds and the truth table seemed impossible to us because it presupposed an intelligent system.
We thus examined ways of using the small amount of virtually stable information available. This covered :
(i) The distribution of the different types of episodes,
(ii) The fact that the projection of the signals in IR^4(the index space) is continuous ; in other words episodes corresponding to the same vigilance state project in the same zone, and
(iii) In IR^4, each zone always ocupies the same spatial position with respect to the others (given the indices used, and the polygraphic definition of vigilance states).
We would have initiated research to define the best choice of thresholds by simulation so that, over a given period of time, the results agreed with the a priori knowledge of the distribution of the different types of episodes. But a prohibitive calculation time associated with convergence problems, led us to adopt an approach based on "Data Analysis".

The Introduction of Data Analysis

The final objective for the automaton was to classify data (or to recognize form) : given the values of the indices established during an episode, to what class does a point representing these values belong ? i.e. what is the vigilance state of the animal during this episode ?
To automatize entirely this process of detecting vigilance states, it seemed necessary to construct an automaton that would not only recognize the states, but would also effect the automatic classification (or clustering) with a variable parameter during the learning period.
We therefore proposed the following procedure :
a) Definition of a method of clustering (with a variable parameter) which would enable about p% of the results, which appeared over a learning time to be grouped into three sets; the quality of the Z, D, F and M criteria ensured that the three sets would correspond well with the states of W, SWS and PS (the remaider would correspond to points which were Not Classified, and which we designated as the state NC).

b) Adoption of a discrimination method compatible with the method envisaged in (a).

Conception of the Automaton Software

The usual methods of automatic classification always enable the totality of the elements to be divided into classes, but essentially none of these methods enables some elements to be eliminated. Here however, it was necessary to separate the double states ; these states do not form a homogeneous class, because they can represent a transition period between different simple states.
We therefore had to define new methods of automatic classification, and we will present below the method used here which was that of ε -neighbours. The essential characteristic of this method is that not all the points are necessarily classified, and the classes obtained depend on the value given to a parameter ε (which is a perception parameter). We will demontrate how this value of ε is determined.
The learning period consists therefore of the recording of the sequence of several hours : this sequence, which we then classify, is the learning set. We thus look for a way of obtaining 3 principal groups which can be superimposed on the 3 vigilance states and which represent about 90 to 95% of the data (i.e. the percentage of simple and triple periods).
We can then pass to the classification (or discrimination, or recognition) phase, which is effected very simply by the method of closest neighbours. Let x be the representation of an episode in $\mathbb{R}^4$. We look for the closest neighbours to x in the learning set, and require that the state of x is the same as that of its closest neighbour (on condition that the distance between them is always less than ε); if not, this is an indetermined state, which could be a double episode, for example.
Numerous simplifications at the level of both the algorithms and the calculations were introduced : for example calculations of the distance using only additions and substractions, rapid search method for the closest neighbours, introduction of adaptive elements, etc.
The first simulations effected on the basis of the above methods gave encouraging results, and the cost of the necessary hardware seemed to be very reasonable.

AN ORIGINAL CLASSIFICATION METHOD

The Choice of a Method

It is always difficult to choose an automatic classification method to resolve a given problem. Here we wanted to set up, and above all divide classes in the space formed by the F, Z, D and M indices, which is a part of $\mathbb{R}^4$.
We wanted to determine, precisely, three homogeneous groups which corresponded to Waking, Slow Wave Sleep and Paradoxical Sleep, but among the points available (600 if the learning period lasted for 5 hours with an episode duration of 30 seconds) some would reflect other possibilities, double episodes, for example. It was necessary therefore that the latter were not taken into consideration at the time of classification.
This constraint meant that we had to eliminate all those families of methods where the number of classes are given a priori. In these methods, the totality of the points are distributed amongst these classes, followed by the optimisation of a criterion. Similarly, given that the number of elements to classify could be large, we also had to eliminate the hierarchical type of method which would be too demanding in calculation time.
We had therefore only to find a non-standard method which was not very demanding in calculation time. It was necessary that this method should be able to divide the points into homogeneous groups and to leave some points, such as those reflecting double episodes, out of the classification.

The Method of ε -Neighbours

The method of ε-neighbours, developed by Sarsoh (4) seemed to us to be the most suitable for our problem. The method is a member of a very general family of classification methods, proposed by Emptoz (2), and called propagation methods.
The basic principle is as follows. Each time that a point x in the space to classify is "observed", its ε -neighbours are also "seen". i.e. all the points in this space which are at a distance less than ε from x (ε being given).
Correlatively, we define a function, called a "structuring function", which associates with each x a numerical value. The latter reflects the concentration of the density around x; the most simple of these functions, δ_ε , consists of associating with x the cardinal of all its ε -neighbours.
Thus each x is associated with :
- A combinatorial entity, its ε-neighbours, designated by V_ε (x).
- A numerical informational entity, the structuring function, denoted by $\delta_\varepsilon(x)$.

These two concepts are complementary in the sense that :

- one of them , δ_ε , enables the points in E to be ordered, and
- the other , V_ε , enables δ_ε to be determined, and then takes part in the process of group formation.

Group formation. Three principles formed the basis for this procedure :
- only a single scan of the points in E is made, but this is done systematically by studying the points x $\in$ E, one by one, in the order of decreasing δ_ε values. The points such that $\delta_\varepsilon(x) = 1$ are called "isolated points" and are assigned to the same class, that of the points which cannot be classed.
- In fact a partial chain is introduced by assuming that x takes with it into its class its ε-neighbours (except when x is an isolated point or at the boundary between classes). x is thus called the classifier of the point which it attracts into its class.
- A division of the points will not necessarily be obtained, and a distribution with overlap is possible. Points that belong to more than one class are neutralized as far as their ability to attract other points into a class is concerned ; they are the points which are called boundary points.
(Note that for the problem under consideration here, the last principle is not of importance).

Organigram of the method.

I - Preprocessing

. For each x, $V_\varepsilon(x)$ is determined
. For each x, $\delta_\varepsilon(x)$ is determined
. Classification of x according to an order determined by the decreasing values of δ_ε in the table T
. Points x such that $\delta_\varepsilon(x) = 1$ are assigned to the ISOLATED class if U = T - ISOLATED : $\emptyset$ go to ε , if not go to α

II - Group formation

α) Initialisation — i = 0
Let e be the first element of U
$U = U - \{e\}$, go to β

β) Creating a new class — i = i + 1
e creates the class C_i and y takes with it the element of $V_\varepsilon(e)$
Go to δ

γ) Group formation — $C_k = C_k \cup V_\varepsilon(e)$

δ) Check — If $U = \emptyset$, go to ε
If not , let e be the first element remaining in U
$U = U - \{e\}$
- If there is not $k \in [1,i]$ such that $e \in C_k$, go to
- if there is only one $k \in [1,i]$ such that $e \in C_k$, go to γ
- if not, e is a boundary point, and so go to δ.

ε) Edit the classes obtained with the family of isolated points and that of boundary points.

Some remarks :
(1) Any point x can therefore be in one of the following 3 states :
- either x is an isolated point, if $\delta_\varepsilon(x) = 1$,
- or x is a boundary point between the classes to which it has been assigned,
- or x belongs to only one class.

2) The number of classes is determined by the method of ε-neighbours. In our particular problem therefore, we must, if necessary, modify in order to obtain the 3 required classes.

3) When a point is found to be common to two classes, it is certain that this point has not yet served as a classifier for other points.

Determination of ε

As we will see, the evaluation of is fundamental to our approach.
It is clear that a "correct" ε will depend on not only the size of the learning set T, but also its topology. We thus define the "critical " ε as the value of the parameter which enables our algorithm to classify 90 to 95% of the points from T into three groups, which can be superimposed on the three states.

a) The size of the learning set. It is imperative that all three vigilance states occur during the time that is set for learning. The minimum size thus corresponds to recordings over a period of three to eight hours, depending on the position in the nycthemeron when this learning period begins.
In contrast, the size does not need to be greater than 24 hours because vigilance states are under a circadian influence.

b) The topology of the learning set. Because of restraints set by the hardware, each of the F, Z, D, and M indices of can only take integral values between 0 and 127 : hence the part of $\mathbb{R}^4$ which contains T, only comprises $(128)^4$ possible values. In fact it is necessary to choose the type of metric with which to operate in $\mathbb{R}^4$ (or T). This choice will influence the way that the value of the critical ε is determined (or the zone of critical ε because we have often noted very similar results over an interval). A priori, two metrics seemed more advantageous than others :

- the classical euclidian distance :
  $d_1(x,y) = \Sigma(xi-yi)^2$, i= 1 to 4

- the "urban" distance :
  $d_2(x,y) = \Sigma|xi-yi|$, i = 1 to 4.

The particular advantage of these two metrics is principally a result of the form of the clouds that corresponds to the three states. These clouds have a form intermediate between spheres and hyper-rectangles. This phenomenon can be observed in figure 5 which shows the projection of 400 (known) episodes on a plane with the F index as abscissae and the M index as ordinates.

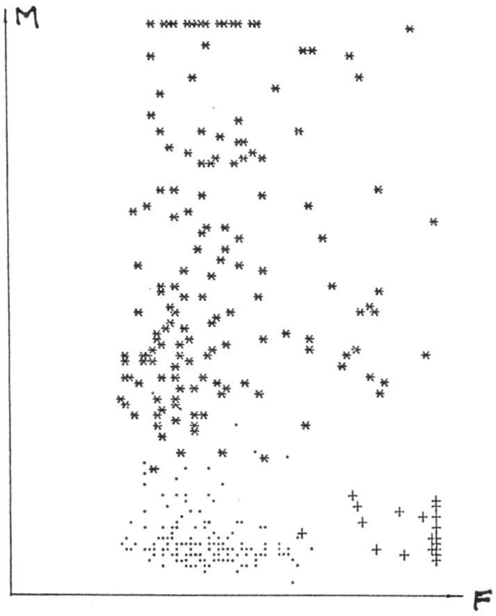

Figure 5 : Projection of 400 episodes into the F-M plane :
* represents Waking
. represents Slow Wave Sleep
\+ represents Paradoxical Sleep

Table 1 displays the complementary information to that seen in the figure, and the numerical data that it contains would be of considerable help in choosing the value of . Figures quoted in the table cover the standard statistics for 805 episodes (i.e those composed of a single arousal state). These 805 episodes were taken from a recording of 1000 consecutive episodes, comprising simple, double and triple types.

		W	S.W.S	P.S.
F	Mean	43,6	38,1	91
	σ	15,6	8,9	2,7
	Min.	22	22	77
	Max.	92	66	92
Z	Mean	89,6	42,2	40,8
	σ	8,6	6,5	7,7
	Min.	47	27	22
	Max.	101	60	58
D	Mean	36,5	81,4	49,9
	σ	6,6	9,1	6,5
	Min.	22	45	37
	Max.	68	97	65
M	Mean	67,3	8,7	7,8
	σ	26,3	2,2	2,5
	Min.	26	4	3
	Max.	106	18	16

Table 1 : Statistics concerning 805 simple episodes taken from a recording of 1000 consecutive episodes, quoted in terms of the means (x), the standard deviation ( ), the minimum value (min.), the maximum value (max.) of each of the 4 indices.

c) Evaluation of ε We have found that in $\mathbb{R}^4$ an ε of the order of 10 enables accurate enough separations to be made when working in urban distance if the size of T is about 900 to 1000 episodes.

Simulations are currently being run on data from different animals and obtained from different nycthemerons, so as to determine the range of variation of ε (around the value of 10) from which a "correct" classification, on the basis of which the classification of following episodes will correspond to physiological reality.
Results of a linear type should not be expected, however, since this would imply that the clouds corresponding to the three vigilance states are absolutely identical in all animals -or at most differing in just a few spatial positions!.

d) Adaptability introduced at the level of group formation. So as to correct variations in the forms of the "clusters" inherent in each animal, we have introduced at the level of group formation an "adaptive" ε, defined as follows : let ε_0 be the parameter estimated above ; we then determine the ε -neighbours and the δ_ε structuring function with this value ε_0

fixed. At the level of group formation we set :
$\varepsilon = \varepsilon_0$, to determine the ε-neighbours of a point which creates a class,
$\varepsilon = \varepsilon_0\, \delta(x)/\delta(y)$, to determine the ε-neighbours of every point x which by itself does not create a class ; the point y in the denominator is the element which has attracted x into the class where it is found, and thus $\delta_\varepsilon(y) > \delta_\varepsilon(x)$. ε becomes smaller as the distance from the "centre" of a class increases.
This technique of an adaptive ε has enabled us in other problems to separate a "background" and a "form", and we are in fact relatively close to this situation here if we consider that the intermediate double episodes constitute a "background". The technique of an adaptive ε has also enabled us however to reduce to some extent the importance of the choice of ε because we carry out an autocorrelation.

e) An example. Let us return to the example in figure 5 defined on the basis of 400 episodes. By setting ε = 13 we obtained three classes grouping about 370 points. In fact the set of 400 consecutives episodes is made up of

168 episodes of W (*)
161 episodes of SWS (.)
43 episodes of PS (+)

and, 28 double episodes, not represented on the figure.

The classes obtained thus give some idea of the "power" of the method of ε-neighbours, although unfortunately our illustration is limited by the fact it only displays a projection of several distinct points in $\mathbb{R}^4$.

CONCLUSION

In the particular field of the neurobiology of vigilance states, therefore, data analysis can provide an advantageous and promising contribution.

Firstly, the results obtained by the application of the method of ε-neighbours to several experimental series, enable us to envisage within the relatively near future complete automatisation of the on-line interpretation of polygraphic recordings made in laboratory animals. This method is currently being implemented on a small standard micro-processor system, which is also statistically controlling the "vigilance state" decisions supplied by the classifier multiplexed on several animals.

And secondly, the application of these classification methods can also be envisaged on data derived in the clinic from polygraphic sleep recordings of normal or pathological subjects. This would be not only to help diagnosis (by helping interpretation), but also to quantify different 'EEG" in terms of both their pathology and their pharmacological modifications.

REFERENCES

Chouvet, G., Odet, P., Valatx, J.L. and Pujol, J.F. (1980). An automatic sleep classifier for laboratory rodents. Waking and Sleeping, 4, 9-13.
Emptoz, H. (1983). Modèle prétopologique de la reconnaissance des formes. Thèse d'Etat, Université de Lyon, 300 p.
Feng, J.F. (1982). Discrimination des trois états de vigilance. Rapport de recherche, INSA de Lyon, 70 p.
Sarsoh, J. (1982). Fonctions structurantes et classification automatique. Application à des problèmes de génétique. Thèse de 3ème Cycle, Université de Lyon, 120 p.

AN EXPERIMENT WITH MULTIPLE-VALUED LOGICS IN AN EXPERT SYSTEM

R. M. Tong and D. G. Shapiro

Advanced Information & Decision Systems, 201 San Antonio Circle, Suite 286, Mountain View, CA 94040, USA

Abstract. This paper describes the first stages of an experimental investigation of the effects of using various representations of uncertainty in an interactive expert system for information retrieval. Fuzzy Set Theory is used to help analyze the results. We conclude that specification of an uncertainty calculus is a subtle problem that interacts in several ways with the scheme used to represent the expert knowledge.

Keywords. Artificial Intelligence, Expert Systems, Information Retrieval, Multiple-Valued Logic, Knowledge Representation.

INTRODUCTION

In this paper we report on an experiment intended as the first in a series designed to explore the effects of using different representations of uncertainty within a rule-based expert system. Existing systems, such as MYCIN (Shortliffe, 1976) and PROSPECTOR (Duda et al, 1979), employ a variety of uncertainty calculi which, although based on a formal theory, are usually implemented in an ad hoc way, with little or no effort expended on experimental tests of their validity.

In an attempt to clarify some of these issues, we have used RUBRIC (RUle Based Retrieval of Information by Computer), a research prototype system for rule-based information retrieval (McCune et al, 1983), as the vehicle for our investigation. Information retrieval is a good domain for such experimentation since the user is responsible for both the knowledge base and the ground truth against which performance is measured. The RUBRIC system is designed to help users by providing automated and relevant access to unformatted textual databases. A specific retrieval request is carried out by a goal-oriented inference process, in which the root node of the search tree represents a semantic concept or topic that the user wants retrieved. Nodes further down the tree represent intermediate concepts with which the root is defined, and the nodes at the leaves of the tree represent patterns of words that are to be searched for in the database. Each arc in the tree may be given a weight, which we can interpret as "truth" or "belief" or "confidence" as we wish. This allows the intermediate concepts and keyword expressions that are found to add differing amounts to our overall confidence that the root concept has indeed been retrieved. It is with the calculus by which these uncertainty values are propagated that this paper is concerned.

EXPERIMENTAL METHOD

We assume that uncertainty can be represented as a numerical value in the interval [0,1], but rather than view the uncertainty values as probabilities we view them as the truths of the associated propositions. That is, the uncertainty value attached to the proposition "x is A" is the truth of what it asserts rather than the probability that the event that it describes occurred. This being the case, we need to construct a calculus to handle these non-classical truth values. Such multi-valued logics have been studied extensively (Rescher, 1969) and we draw upon this work in what follows.

The first task is to define a set of operators for conjunction (the and connective), and disjunction (the or connective). There are many we could choose, but we shall consider four pairs as summarized in Table 1. Here v(A) and v(B) denote the truth values of the primary propositions, with v(A and B) and v(A or B) denoting the value of their conjunction and disjunction respectively. Negation (the unary operator not) is assumed always to be given by v(not A) = 1 - v(A). We can see that the conjunction operators are triangular-norms (Dubois and Prade, 1982) and the disjunction operators are triangular-conorms. The four pairs we have chosen have the ordering property that:

$$v(.\text{and}.)_1 \leq v(.\text{and}.)_2 \leq v(.\text{and}.)_3 \leq v(.\text{and}.)_4$$
$$v(.\text{or}.)_1 \geq v(.\text{or}.)_2 \geq v(.\text{or}.)_3 \geq v(.\text{or}.)_4$$

	v(A <u>and</u> B)	v(A <u>or</u> B)
1	T[v(A),v(B)]	S[v(A),v(B)]
2	max[0,v(A)+v(B)-1]	min[1,v(A)+v(B)]
3	v(A).v(B)	v(A)+v(B)-v(A).v(B)
4	min[v(A),v(B)]	max[v(A),v(B)]

$T[0,0] = 0$, $T[x,1] = T[1,x] = x$,
$T[x,y] = 0 \quad \forall x,y \in [0,1)$

$S[0,0] = 1$, $S[x,1] = S[1,x] = x$,
$S[x,y] = 1 \quad \forall x,y \in [0,1)$

Table 1. Conjunct-Disjunct Operators

	Detachment (*)	Implication (=>)
1	v(B) = min[v(A),v(A=>B)]	min[v(A),v(B)]
2	v(B) = min[v(A),v(A=>B)] if v(A)+v(A=>B) > 1 = 0 otherwise	max[1-v(A),v(B)]
3	v(B) = v(A).v(A=>B)	min[1,v(B)/v(A)]
4	v(B) = max[0,v(A)+v(A=>B)-1)]	min[1,1-v(A)+v(B)]

Table 2. Detachment Operators

The second task is to define a mechanism for performing rule-based inference. In two-valued logics the modus ponens rule allows B to be inferred from A and A=>B. In a multiple-valued logic, we need to extend this idea so that the degree-of-belief in B, denoted v(B), can be computed from any given v(A) and v(A=>B), where => is some multiple-valued implication. Functions that allow us to compute v(B) are called <u>detachment operators</u> (and are denoted *). It is usual to define them so that for a given definition of =>, v(A)*v(A=>B) is a lower bound on the value of v(B). Four of these are shown in Table 2, together with the corresponding implications.

Let us denote a particular calculus by c(i,j) where ´i` is an index over the conjunct-disjunct operators and ´j` is an index over the detachment operators. We see that some of the c(i,j) are well known; in particular, c(4,4) is Lukasiewicz`s nondenumerably infinite system (Lukasiewicz, 1930), and c(4,1) is a system proposed by Zadeh (1973). Another calculus of interest is c(3,3), which we can view as a "pseudo-probability" logic in which A and B are independent events.

Having defined the operators, our basic experiment involves the definition of a query (i.e., a set of production rules), the selection of a representative story set and the repeated application of the query to the story set. Since we have defined sixteeen separate calculi for uncertainty propagation (four pairs of conjunct-disjunct operators and four detachment operators) the experiment may result in sixteen potentially distinct orderings of the story set.

As a typical query we selected "Acts of Terrorism", and then developed a structure for this concept, part of which is shown in Figure 1. This tree of successively more precise sub-concepts, should be interpreted as a definition for a prototypical story. Thus we expect a story about TERRORISM to mention a terrorist act (EVENT), the results of this act (EFFECT), someone responsible for performing the act (ACTOR) and a reason for performing the act (REASON). Each of these sub-concepts may be defined at a further level of detail. So, a sub-concept of EVENT is KILLING which in turn can divided into SLAYING and SHOOTING. In practice this is transformed into a set of LISP production rules.

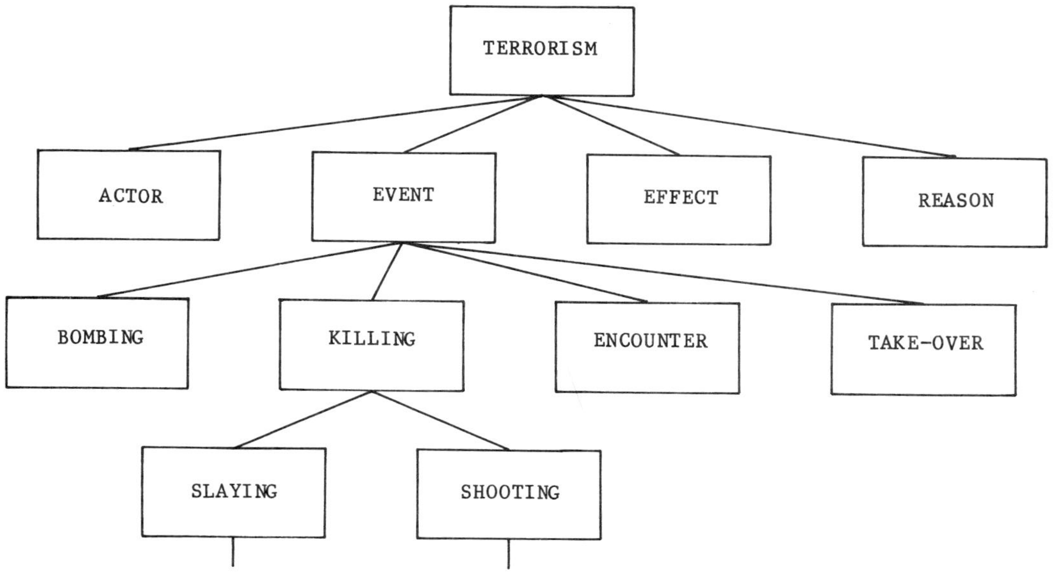

Figure 1. OR Tree of Concepts for Terrorism

1 Overview story about the war in Chad and its effects.
2 Overview story of the situation in Poland and Solidarity.
** 3 Short on car bomb in London.
* 4 US deports Palestinian terrorist to Israel.
5 FBI takes Reagan's secret code card after assassination attempt.
6 Political effects of attack on Angola's oil refinery.
** 7 Chilean secret service agent brought nerve gas into US.
** 8 Follow-on to London bombing story.
** 9 More on London bombing.
10 Boxing match - WBC featherweight champion.
11 Earthquake in Pakistan.
12 Cyclone in India.
13 Soviet reaction to Polish crisis.
14 Reaction of Soviet bloc countries to Polish crisis.
15 Spanish army officers placed under house arrest.
16 Story on Iraq-Iran conflict.
* 17 Accidental chain of explosions at Army arms dump in Zimbabwe.
18 General interest story about Napolean and Waterloo.
** 19 Bomb explosions in two Yugoslav restaurants.
20 Accidental explosion in apartment building in NE. Italy.
* 21 Italian couple freed by kidnappers after ransom paid.
** 22 Bomb explosion in central Tehran street.
** 23 Part story about murdered Italian industrialist.
24 Iranian leftists executed by firing squad in Tehran.
25 Shell exploded and killed bomb disposal experts in E. Beirut.
** 26 Mayor and seven others kidnapped and shot in Guatemala.
27 Lawyers for Sadat's assassins argue against charges.
28 Violence caused by Haitian refugees in Miami detention center.
** 29 Iranian Parliament member assassinated in Tehran.
30 Brazilian athlete recovering from auto accident.

Figure 2. Summary of Reuters Stories

We then selected a set of thirty stories taken from the Reuters wire service as representative of the data that the RUBRIC system would encounter. A one-line summary of these is given in Figure 2. Notice that they all report some kind of violent activity but not all are relevant to our query. Those marked with a double asterisk were determined to be definitely relevant and those marked with a single asterisk were determined to be marginally relevant.

MEASURES OF PERFORMANCE

RUBRIC's basic task is to assign a weight to each story in the data base. This weight is the truth of the statement "this story is relevant to the query", with its value being determined by propagating the uncertainty values through the structure defined by the query rule set. This makes the assessment of performance somewhat complicated, since we are interested in the properties of the ordering, both in absolute terms (i.e., the truth values returned) and with reference to the ordering that we determined beforehand.

A suitable mathematical framework for discussing these issues is that provided by Fuzzy Set Theory (Zadeh, 1965, 1978). Within the FST paradigm we can view the rule-based query as a <u>definitional fuzzy algorithm</u> for the retrieval concept it represents, and the system's function as one of interpreting the query, thereby generating a fuzzy sub-set of the story data base. So, if we equate the uncertainty values with membership function values then the measurement problem becomes one of constructing appropriate similarity measures between fuzzy sets.

To make these ideas more precise let us introduce some mathematical notation and basic definitions. Denote by S the set of stories in the data base: $S = \{s_1, \ldots, s_N\}$. Our a priori relevance rating of S can be described by a fuzzy set R^* where: $\mu_{R^*}(s_i)$ = "the a priori relevance of story i". Interpreting our experimental story set in this way, "definitely relevant" is assigned the membership value 1.0 and "marginally relevant" is assigned the membership value 0.5. Let the <u>support set</u> of R^* be denoted by S^*, then the number of elements in S^*, denoted N_*, is equal to the number of relevant stories in S. Obviously, N_* is the cardinality of S^* written: $N_* = \text{card}(S^*)$.

When we use RUBRIC to apply a query to the story data base we are constructing a fuzzy sub-set of S that represents the degree-of-relevance of the stories to the query. The question of how well any combination of query and representation performs is thus transformed into one of determining the "closeness" of R and R^*. Clearly, there are very many ways in which this could be done; we shall consider just a few of these.

One rather obvious method would be to form a measure such as the mean-square difference between μ_{R^*} and μ_R.

$$\sigma = \frac{1}{N} \sum_{i=1}^{N} \left(\mu_{R^*}(s_i) - \mu_R(s_i) \right)^2$$

However, such a procedure compresses all the ordering information into a scalar assessment and σ itself does not have a very intuitive interpretation. There are many variants on such functions but we have not explored these in any detail.

More interesting are measures based on assessing the cardinality of various subsets of S. The approach is to use R to define a variety of <u>α-level sets</u>, S_α, and then compute a series of counts. So, if we let

$N = \text{card}(S)$: number of stories in the database
$N_* = \text{card}(S^*)$: number of "relevant" stories
$N_\alpha = \text{card}(S_\alpha)$: number of stories in S

then the number of correctly selected stories, N_H, is given by

$$N_H = \text{card}(S_\alpha \cap S^*)$$

The number of incorrectly selected stories, N_F, is given by

$$N_F = \text{card}(S_\alpha \cap \overline{S}^*)$$

and the number of missed stories, N_M, is given by

$$N_M = \text{card}(\overline{S}_\alpha \cap S^*)$$

We are then left with the selection of S_α. Two obvious definitions are:

(1) The smallest set of stories such that N_M=0 (ie. $S_* \subseteq S_\alpha$) Remember that S_α is derived from R (the output from RUBRIC) and so the α-level is given by

$$\alpha \triangleq \min_{s_i \in S^*} \left\{ \mu_R(s_i) \right\}$$

(2) The largest set of stories such that N_F=0 (ie. $\overline{S}^* \cap S_\alpha$ =0). In this case

$$\alpha \triangleq \max_{s_i \in \overline{S}^*} \left\{ \mu_R(s_i) \right\}$$

Other choices would be: (3) the set of M-best stories, and (4) the set of stories that exceed some specific value of (in which case we would need to normalize R).

We have used all these definitions and find, in particular, that (1) gives us an insight into the system's ability to reject unwanted stories (system <u>precision</u>), whereas (2) gives us insight into the system's ability to select relevant stories (system <u>recall</u>).

The analysis of the results of our experiment is developed in terms of the measures described above. The information can be presented in several ways, of course, and we recognize that in practice the "goodness" of a system's performance will be assessed on a balance of such measures.

ANALYSIS OF RESULTS

The raw data from our experiment consists of sixteen different fuzzy sub-sets of S. In the remainder of the discussion we will distinguish between them by writing R(i,j).

Recall that we have two primary methods for determining performance. Both of these are based on the idea of using a selection threshold to partition the ordered stories so that those above it are "relevant" and those below it are "irrelevant". In the first we lower the threshold until we include all those deemed a priori relevant, and then count the number of unwanted stories that are also selected (denoted N_F). In the second we raise the threshold until we exclude all irrelevant stories, and then count the number of relevant ones that are not selected (denoted N_M). The performance of the system using these two measures is shown in Table 3. Notice, for example, that c(2,2) gave 6 false hits when we applied the precision measure and 9 missed stories when we applied the recall measure. Remembering that our story data base has 30 entries of which 12 are marked as being relevant, we see that some calculi gave good performance whichever measure we used, some performed well on one but inadequately on the other, and others did badly on both. Interestingly, no one calculus seems to be significantly better than any of the others. Indeed, there seem to be four calculi, c(2,3), c(3,2), c(3,3) and c(4,3), whose performance is practically indistinguishable.

Overall, the results are rather surprising. They show that while changing the calculus does produce performance variations, there is an apparent consistency that we might not have expected. Our initial conjecture is that the uncertainty calculus is not the major determinant of performance, with the structure of the rule-based query having at least as much impact. This implies that we need to be more subtle in our investigation. A change in calculus can certainly have a marked effect on the interpretation of a single rule, but in assessing its effects on a set of rules we need to look deeper into the role that the calculus performs.

	1	2	3	4
1	15	11	15	11
2	7	6	4	6
3	7	5	3	18
4	4	4	6	18

Precision Scores
(N_F when N_M=0)

	1	2	3	4
1	12	12	12	12
2	12	9	5	7
3	10	4	4	5
4	11	11	2	5

Recall Scores
(N_M when N_F=0)

Row indices denote disjunct-conjunct operator pairs. Column indices denote detachment operators. (See Tables 1 and 2.) Scores for the better calculi are encircled.

Table 3. System Performance

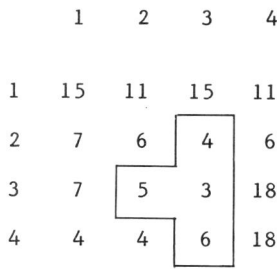

INCONSISTENCY AND DEPENDENCY

We have identified two effects that can confound our results. The first of these we have called inconsistency, which relates to the mismatch between two translations of the retrieval concept that we have used; we define both an a priori ordering of the story set, which we can view as a declarative statement of the concept, and a rule-based query, which we can view as a procedural definition of the concept. Obviously, one or the other (or both) of these may not capture exactly the user's internal model of the concept being retrieved. If that is the case, then attempts to compare them will lead to errors of assessment. As examples of mis-translation of the first kind (i.e., incorrect labelling), consider stories [24] and [25]. Since these deal with terrorist-related acts in the Middle-East, they should probably be considered at least marginally relevant to the query, yet initially we considered them to be of no interest. As examples of the second kind (i.e., inadequate specification), consider stories [9] and [10]. While story [9] is definitely relevant (a story about a car bombing), it is often not selected when the query is applied to the story data base. On the other hand, story [10] is just as clearly irrelevant (a story about a boxing match), and yet it often receives a high rating.

The second confounding effect we have called dependency. It is caused by interaction between the rule-based query and c(i,j). An implicit assumption of our experiment is that we can consider the effects of c(i,j) independently of the specification of the query. However, a particular query may rely on a particular c(i,j) for its effectiveness. Indeed, it could be that certain forms of query are basically incompatible with certain c(i,j). Thus, for example, if we wished to rely on the implicit disjunction between rules to produce a magnifying effect when there are several paths to a sub-concept, then a calculus which uses min-max as and-or connectives will never achieve the desired effect. So although calculus c(4,4) was not selected as one of the "best", this maybe because the form of query we used mediated against it. Perhaps if we had fixed c(4,4) as our uncertainty calculus, then we could have constructed an effective query around it.

In general, constructing the rule-based query gives most chance for error since it requires a translation of a vague internal concept into a formal representation using the text reference language that RUBRIC provides. Specifying R^*, on the other hand, is merely a question of assigning a subjective weighting to each of the stories, a task we have found to be much more intuitive.

CONCLUSIONS

The experiment described above shows that changing the uncertainty calculus does change the performance of RUBRIC, with some of the calculi clearly failing to produce satisfactory results. However, the variations we observed had some unexpected characteristics which led us to a consideration of some deeper issues of inconsistency and dependency.

Our conclusion is that specification of an uncertainty representation is a problem of some complexity and subtlety. We believe that there are interactions between the form of the query and the admissible representations, and will continue to explore these, and other effects, in subsequent experiments.

REFERENCES

Dubois, D., and H. Prade (1982). A Class of Fuzzy Measures Based on Triangular Norms. Int. J. General Systems, 8, 43-61.

Duda, R. O., et al (1979). A Computer Based Consultant for Mineral Exploration, Final Report AER 77-04499, SRI International, Menlo Park.

Lukasiewicz, J. (1930). Many-valued Systems of Propositional Logic. In S. McCall, Polish Logic, O.U.P, 1957.

McCune, B. P. et al (1983). Rule-Based Retrieval of Information by Computer. Final Report TR 1018-1, AI&DS, Mountain View.

Rescher, N. (1969). Many Valued Logic McGraw-Hill, New York.

Shortliffe, E. R. (1976). Computer-based Medical Consultations: MYCIN, American Elsevier, New York.

Zadeh, L. A. (1965). Fuzzy Sets. Information and Control, 8, 338-353.

Zadeh, L. A. (1978). Fuzzy Sets as a Basis for a Theory of Possibility. Fuzzy Sets and Systems, 1, 2-28.

FUZZY MATCH AND FLOATING THRESHOLD STRATEGY FOR EXPERT SYSTEM IN TRADITIONAL CHINESE MEDICINE

R. J. Guo

China Computer Technical Service Corporation Beijing, China

Abstract. The expert system has been used in the computerized diagnosis and treatment system in traditional Chinese medicine. The fuzzy match and floating threshold has been adopted in the expert system. Combination of expert system and data base technique will produce a conception of the large-scale expert system.

Keywards. Artficial intelligence; expert system; fuzzy match; floating threshold; computerized diagnosis system; traditional Chinese medicine.

INTRODUCTION

Expert system, developed concurrently with other artificial intelligence system, is regarded by many researchers today as one of the most promising topics and has proved its usefulness since its application in production.

After researches in recent years in the Computerized Diagnosis and Treatment System in Traditional Chinese Medicine(CDTSTCM) and studies of the problems arising from the production and application of the CDTSTCM, it may right to assert that the basic structure formula of the expert system is totally applicable to the diagnosing and treatment system in traditional Chinese medicine. Our conclusion is based on the following considerations:

First, given the fact that one of the major obstacles in applying the various medical diagnosing and treatment systems developed nowadays is of social and psychological, particularly so in China, a computer system aimed at simulating mature medical experience of senior medical specialists is obviously more acceptable to society at large and superior to an expert system that forms experience (or rules) by way of self-learning.

Second, the expert system is composed of three independent and yet related parts: a data base, a knowledge base or a set of production rules and a control system. Its simple and clear structure makes it easy to study each of the three parts and raise their level of efficiency accordingly. The logic the system follows in operation is almost identical with the line of thought of a physician when he treats his patient.

Third, what we have in mind is the establishment of a large-scale general data bank system for traditional Chinese medicine that is made up of a medical record data bank, a diagnosing and treatment processing bank a medicines data bank, a symptoms data bank and a bank for writings on traditional Chinese medicine etc. After being expanded, supplimented and magnified on the basis of the structure model of the present expert system, these above mentioned banks, that is, the medical record bank, diagnosing and treatment processing bank, medicines data bank, symptoms data bank and the Chinese medical writings bank will then be connected under one control system to form a large-scale expert system. This system can be used for clinical purposes, for consultations on conditions and treatment of the patient as well as for statistics and indexing. It is also capable of self-learning and self-correcting. When a patient calls for treatment, a new data base, as it is used in any expert system, is formed by combining the patient's immediate chief complaint and his laboratory reports with information sought out from related case history in the medical record bank. Then, a series of needed production rules will be selected from the diagnosing and treatment processing bank, the medicines bank etc. to form a new knowledge base (same as that in an ordinary expert system) which then proceeds to give the patient treatment under the control of a selected control system.

When set in operation, the left-part match between the data base and the production rules of all kinds of comtemporary expert systems normally adopts the method of complete or rigid match with a few exceptions of partial match. However, because of the special characteristics in traditional Chinese medicine, a new match method -- fuzzy match -- has been developed. It is different from both the complete match and partial match in that, it is, in a sense, a partial match as compared with the complete match and yet, it is not entirely a true partial match when compared with partial match. The fuzzy match takes the entirety into account with fuzzy boundaries.

Conditions of patients differ from one another. It is obviously difficult to provide a set of production rules for all possible symptoms in the course of traditional Chinese medical treatment. Besides, the left-part production rules which corresponds with typical syndrome proves more often than not unsatisfactory to the average patient. Fuzzy match is therefore especially applicable to medical treatment.

With the conflict set being already established, the scheduler of an ordinary expert system usually chooses a set of elected production rules and implement according to a previously defined conflict resolution. But in practice, the conflict set is often empty. To solve this problem, we adoped the strategy of floating threshold resolution.

FUZZY MATCH AND FLOATING THRESHOLD CONFLICT RESOLUTION

Traditional Chinese Medicine boasts of a unique theoretical system. Many noted Chinese and foreign scientists of biology, medicine and from other fields of human endeavour have attached importance to the Chinese medical theory of harmony between nature and man, adjustment of the two opposing principles of Ying and Yang (negative and positive, feminime and masculine, dark and light) as well as to the idea of taking an overall view of the patient and analysing and differentiating his symptoms in giving medical treatment.

Traditional Chinese medical treatment begins with the recognition of the overall syndrome. Because patients suffering from one illness may have only some identical symptoms, the extension of their syndrome, however, may not be the same. This shows that the extension of the syndrome is indefinite and fuzzy. Therefore, it is almost impossible to satisfy the needs of the traditional Chinese medical diagnosis and treatment system by merely adopting the complte match for production rules.

FUZZY MATCH In CDTSTCM, mark the first appearance of the data base as x_i, then all the other different possible appearances x_i ($i=1,2,\cdots\cdots, n^2$) of the data base constitutes a universe discourse. then, define a number of fuzzy subset $\underset{\sim}{A}_j$, each of which corresponds with a production rule in the data base.

When translated into geometric terms, first, define the data item of the data base as a linear aspect space of the coordinating axis on the spase base. If we construct a membership function $\mu_{\underset{\sim}{A}_j}(x_i)$, computing x_i as part of $\underset{\sim}{A}_j$ and make $\mu_{\underset{\sim}{A}_j}(x_i)$ have strong convexity, then point x_j° which corresponds with the antecedent of the production rule of the i item become the kernel of the fuzzy subset $\underset{\sim}{A}_j$.

Following are the principles and inference process for constructing $\mu_{\underset{\sim}{A}_j}(x_i)$.

Let P_j° be the set of the aspect data item of x_j° as is indicated by $P_j^\circ=\{a_1, a_2, \cdots\cdots, a_{kj}\}$ and P_i be the set of the aspect data item of x_i as is indicated by $P_i=\{b_1, b_2, \cdots\cdots, b_{qi}\}$. To correspond with all elements in P_j°, each data item is given corresponding weighting coefficients according to its value in medical diagnosis and treatment, thus forming a weighting set: $\alpha_i=\{\alpha_1, \alpha_2, \cdots\cdots, \alpha_{kj}\}$

There will also be a corresponding weighting set for each data item in the P_i set: $\beta_i=\{\beta_1, \beta_2, \cdots\cdots, \beta_{qi}\}$
If a certain b in P_i can find a certain a in P_j° that has a common meaning, then the corresponding $\beta=\alpha$. If it can not, then, there are two possilbe answers: if this b does not in any way affect the building of the production rules in the j item, $\beta=0$, if it has an unfavourable effect, then, the negative value is taken. In case, a certain b has an extraordinary negative recognizing characteristic, then β takes a very negative value.

With the above-mentioned four sets P_i, P_j°, β_i, α_i, we now proceed to delimite the following membership function:

$$\mu_{\underset{\sim}{A}_j}(x_i)=\frac{\sum_{i=1}^{q_i}\beta_i b_i}{\sum_{j=1}^{k_j}\alpha_j a_j}$$

If order $\mu_{\underset{\sim}{A}_j}(x_i)=0$ when $\sum_{i=1}^{q_i}\beta_i b_i \le 0$,

$\mu_{\underset{\sim}{A}_j}(x_i)$ then takes value between $[0,1]$, $\mu_{\underset{\sim}{A}_j}(x_i)$ then has strong convexity. When $x_i = x_j^0$, $\mu_{\underset{\sim}{A}_j}(x_i)$ takes the largest value form itself.

CONFLICT RESOLUTION -- FLOATING THRESHOLD STRATEGY

Normally, $\mu_{\underset{\sim}{A}_j}(x_i)$ can be deducted from one appearance of x_i in the data base and can take actions according to the production rule of the j item. Here, the condition for the production rule in the j item to take action must be taken into consideration. We delimite a threshold value S_j and, furthermore, $1 \leq S_j \leq 0$. When $\mu_{\underset{\sim}{A}_j}(x_i) \geqslant S_j$, allow the production rule of item j to take action.

But the correct selection of S_j is a comparatively complicated issue as conditions of patients vary from one another. We are here faced with two problems: some patients have more symtoms and relatively large P_i data set while others, with less symtoms, can only provide very small P_i data set. In the first case, the conflict set constructed by matching through production rule is very large, that is, there are many production rules that can be matched with the present data base. In the second case, the conflict set often equals to an empty set. It must be pointed out that the amount of symptoms does not necessarily manifest the seriousness of an illness. To solve these problems, we have adopted the floating threshold conflict resolution.

WAYS AND MEANS FOR FLOATING THRESHOLD STRATEGY

The range for S_j to take value is $[\underline{S}_j, \bar{S}_j]$. For each j, S_j first takes its upper limit value $\bar{S}_j$.

Arrange production rules for selsction in a priority sequence in the scheduler in accordance with the physicians experience.

When there are many production rules for selection in the conflict set, implement the scheduler.

If no production rule is selected after the first search of all the production rules, the conflict set becomes empty. Then, lower each S_j successively for research till a production set is selected. If no qualified production rule is selected after all the S_j are lowered to their lower limit $\underline{S}_j$, just stop the searching. From a medical point of view, it is either because the patient has already recovered, or the illness being sought does not belong to the range of illness the system is supposed to deal with.

SET OF PRODUCTION RULES AND CONTROL SYSTEM

CDTSTCM mainly sdopts the antecedent-driven system for its set of production rules. According to the experience of senior physicians, before giving medicaltreatment, diseases are usually differentiated and classified into main types, subtypes and individual symptoms of patient accordingly.

In keeping with these three level treatment, the structure of the production rules of the CDTSTCM also has three levels: meta meta production rule for the main types ; meta production rule for the subtypes and production rule for the adding and reducing according to the number of sumptoms.

Pre-screening is done through the three-level structure. The first level action prompted by the meta meta production rule is to direct searching into certain main types for relevant treatment, the second level of action prompted by the meta production rule is to direct searching into certain subtypes for relevant treatment while the third level action prompted by the production rule is to give special treatment to some important individual symptoms.

Here implicit control is adopted for the meta meta production rule and the meta production rule of the three level structure, whereby, the search for all types of illness that have been differentiated and classified on the basis of the cliniaal experience of senior physicians is conducted in a pre-arranged sequence.

CONCLUSION

The work are engaged in doing is to search for related data information about a visiting patient from a general medical record bank and to collect his chief complaint on the spot as well as his laboratory report so as to form the data base under an ordinary expert system .It also includes storing in the computer physicians writings about concernet diseases for medical personnel to consult. Consultation takes place in the form of man-machine communication.Man-machine communication can also be used to automatically revise the programme Therefore, it may be said that the present work is the embryonic form of the previously described large-scale expert system. Of course, to translate the ideal of a large-scale expert system into reality still more work needs to be done on the basis of what has already been achieved now.

This paper has dealt theoretically with the possibility and prespects for realizing the large-scale expert system in traditional Chinese medicine.

Our research in this respect in recent years has convinced us that our goal will be attained.

REFERENCE

Guo, R.J. (1982). Fuzzy Set Model For Computerized Diagnosis System in Traditional Chinese Medicine, Approximate Reasoning in Decision Analysis, ed, by M. M. Gupta and E. Sanchez, North Holland, Publishing Company, pp.283-287.

KNOWLEDGE DATABASES ABOUT INDUSTRIAL PLANTS

R. Badard

Département Informatique, INSA Lyon, 69621 Villeurbanne France

Abstract. The conception of a database is a fundamental point in the realisation of any decision aided system. A tendency to a rationalisation, at least at the conceptual level, has appeared. But it seems that more and more the database must capture imprecise and qualitative facts and must be able to do inferences from them. For instance our knowledge about a complex industrial plant is in great part composed of qualitative assertions from which we must be able to infere new ones and test their coherence. These inferences must be compatible with the fuzzy semantic interpretations of the qualitative constants we use, and this is a key point.

Keywords. Knowledge database, inference rules, qualitative and imprecise facts, fuzzy semantic interpretation.

INTRODUCTION

Data bases have became objects of our current environment and this principally in industrial and commercial domains. Their development shows a remarkable tendency to a rationalisation, at least at the conceptual level, without which it would not be possible to deal with the increasing complexity of the actual applications. A same care of rationalisation appeared in the software development. This tendency is not reversible and it takes its roots in the mathematical language, especially in logical theories.

Another fundamental tendency which appears is the construction of artificial systems which are much nearer of humans in the communication aspects but also in the realized functions. Artificial intelligence is an active field which has created remarkable tools. We must quoted in this area the general problem solver (Ernst and Newell, 1969) and some expert systems (Buchanan and Feigenbaum, 1978; Shortliffe, 1976).

More and more such systems must be able to accumulate informations which are expressed in qualitative and imprecise ways. The problem of dealing with qualitative and imprecise informations is crucial for the comprehension of how humans organise and use their knowledges. Important works on this subject have been done and well adapted tools have been constructed. We must quoted some of them such as extended topologies (Hammer, 1967), fuzzy sets and possibility-necessity theories (Zadeh, 1978), multivalued logics and logics with modalities (Lukaciewicz, 1970). Some of these theories have had to endure conjoint critics from both theoricians, which were impermeable with practical problems, and practicians which were impermeable with formalisations and which confound confusion with savoir-faire. These theories are complementary and have common goals. They can be bring together giving new structures well suited to deal with qualitative and imprecise aspects (Badard, 1981; Badard, 1981; Chang, 1968).

Qualitativness is not synonimous of poor quantitativness, it can be viewed as the fact of extracting a few invariants which are sufficient to describe some properties of a very complex situation. A typical example is algebraic topology. It enables us to summarize global properties of a manifold with few characteristics. These invariants being in some cases sufficient to assert that two spaces are homeomorphic or homotopic, concepts which are good abstractions of our intuitive notion of a continuous deformation of an elastic body.

About imprecision we can say it is inherent to every apprehension of complex systems. Zadeh has informally stated this point as a principle: "As the complexity of a system increases our ability to make precise and significant statements about its behaviour diminishes until a threshold is reached beyond which precision and significance become almost mutually exclusive characteristics". It is clear that a knowledge data base must integrate imprecise and qualitative informations. We will be concerned with the description of some characteristics of such a data base.

MODELS OF DATA BASE

A knowledge data base lies in the memorisa-

FIKR-H*

tion of a set of propositions we must do in order to describe a more or less great part of our real environment. The conception is then principally the study of the typology of the messages which will be allowed to be exchanged between the data base and many sources of informations. These sources of information being allowed to be composed of both human and automatic ones. The typology of the exchanged messages enables us to conceive a symbolic representation of them and this leaves us to the concept of schema. The schema is a collection of primitive objects such as individual classes, functions and relations for which we precise the associated domains and how they are related.

The classic relational model

In the relational approach, as described by Codd (Codd, 1970), we display the functional relationships existing between the entities which compose the messages. This leads us, after a decomposition of relations, to describe a standard relational version, the third normal form. The advantages of this version lie in the reduction of updating anomalies.

In this model all the messages are decomposed into messages associated with fixed size relations. It is always possible to structure an extensive set of propositions in this way, but there are situations in which it is not so easy and situations for which it is better to keep a more global form (depending of the use we want to do of them, particularly in artificial intelligence).

Many other important limitations of this model are the following :

-It only deals with extensive definitions. For example it is not possible to define an individuality by its supporting domain and limitative formal constraints. Some authors have extended the model in this sense, for example see (Phillips, Beaumont, Richardson, 1979).

- There is no systematic treatment of imprecise relations. Though it is possible to add an attribute associated with a membership degree.

- The non prime attributes must have precise values. Some extensions in this sense have been done by Codd himself (Codd, 1979), Lipski (Lipski, 1979), Farreny and Prade (Farreny and Prade, 1981).

- The memorised messages are explicit and there is no inference ability. Some authors have felt the necessity of integrating mechanisms in order to produce implicit messages from the explicit ones by the use of production like rules, see (Haar, 1977; Reiter, 1978).

We feel that these restrictions must be relaxed in order to define a model which could capture many aspects of a knowledge data base.

The assertional model

As we said in the introduction a knowledge data base is in fact composed of propositions which symbolically represent some aspects of the reality. These propositions can be viewed as expressed in a particular formal language. This language being described by its primitive constants. We may also express different constraints and links between these primitive objects. All these aspects will be described in the schema.

a) The schema

It lies in a description of the objects about which we can speak. It captures a great part of the semantic and particularly the semantical links between different levels of description. It also gives integrity constraints and mechanisms in order to generate implicit assertions by way of inference like rules.

The frame is the core of the schema, it describes the individualities, relations and functions by giving for each of them a name and attributes specifying its support. We assume we have some primitive structured sets, types of data, in order to precise the domains of the attributes.

Every individuality is described by a list of prime attributes with their corresponding domains. We obtain in this way the individuality which is the cartesian product of the domains of its attributes. We precise if the definition is extensive or not. In the first case the individuality is in fact a finite set and will be defined by a list of explicit assertions. In the second case we eventually precise some membership constraints by use of a particular computable membership predicate. We eventually attached with an individuality some parameters which will be used as links for particular transportable relations.

A relation is described by a list of prime attributes with their corresponding domains. These domains can be either structured sets or sets associated with individualities. When some attributes are defined on structured sets the relation is said to have a transportability property because it can be applied on different individualities by correctly linking its free attributes with attributes and parameters of individualities. We precise if the relation is defined extensively or as a computable predicate. A relation whose some attributes are not associated with finite sets (finite in the sense of an explicit enumeration, because in a real computer all domains are finite) is certainly defined as a computable predicate. The relations can be classic or fuzzy and when they are defined in extensive form we precise if they follow an assumption of open or closed world (do we interprete the absence of an explicit assertion as a zero membership value ?). We eventually attache with a relation a list of modifiers which change its sense, enabling us to connect relations having a commom semantic.

We finally describe functions by giving their origin and image sets. These functions are generally analogous to non prime attributes of the relational model. They can have individuality sets or structured sets as domains, in this last case they are said to verify a transportability property because we can link them with different individualities. These functions can also be defined extensively or as computable applications. We admit a list of linguistic values as images. These linguistic values will be interpreted as describing some possibility-necessity measures (as in Farreny and Prade, 1981), relating our partial ignorance about the real value. For instance the value "unknown" attached as an imprecise image value of a function will be interpreted as the possibility measure whose value is one on the all image domain (and then the corresponding necessity measure has value zero on all the image domain).

After this description of the frame we complete the description of the schema by giving constraints which relate the links existing between these constants. These constraints are analogous to the proper axioms of a formal theory. But we will describe them as production like rules (condition $\Rightarrow$ action). An action can be either the emission of a message specifying that some incoherences have been found or the generation of new assertions from the preceding ones.

b) Classification of the memorised assertions

We can classify the assertions by many criteres.
A first classification can be obtained from the synthetic and qualitative aspects of the assertions. In fact it is clear that we must classify the assertions by their synthetic level of description. For instance for a picture we can see the following levels:
A less synthetic level is concerned with a description of pixels.

We could find a topological level in which individualities such as connected components and their holes would be indexed. We would have a relative localisation relation which would tell us that a component is in a particular hole of another one, and from it a surrounding relation telling us if a component surrounds another one.

We could find a level of description by use of particular geometrical shapes and relations telling us how these shapes are positioned and how they are linked. These relations could be imprecise, for instance we could admit an assertion as: "A is near the left of B".

All these symbolic description levels must be correctly semantically linked together, in this last example we would have to create links which relate connected components and their holes with the geometrical shapes and with the pixels description.

Another critere is the stability. Some assertions are rarely modified (in order to follow the real situation), but others are frequently adjusted.

Another critere, which is practically linked with the preceding's, is the origin of the exchanged messages.

For these reasons we classify the set of the memorised assertions in categories, each one corresponding to a sub-language. We associate with every class a filter which have many functions. This filter translate assertions between their internal and external forms. He verifies some syntaxic and semantic features of input messages by use of a part of the schema. He manages the inference and coherence activation of production rules in order to detect incoherences and to generate new assertions. In this sense this concept of filter is near those of channel as defined by Rieger (Rieger, 1978), see also (Haar, 1977). The memorised assertions in a class can be, depending on the corresponding filter, simple assertions such as : "A is an occurence of individuality I", "R(A, B) has a membership value of a "... For another class it can be complex logical expressions constructed from the constants of the associated sub-language such as: "R(A,B) and H(C)] $\rightarrow$ R(A, C) has b as truth value". It is clear that filters can be subdivised into, for instance, an external part devoted to the analyse of external requests and of syntaxic validity and an internal part dealing with semantic analysis, test of coherence and implicit assertions generation.
We roughly have the following structure:

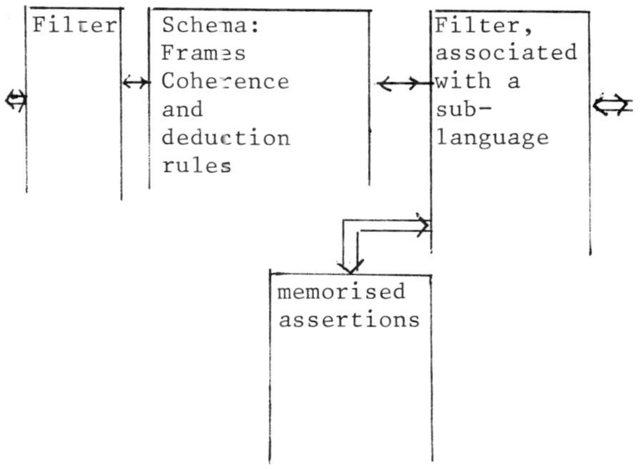

KNOWLEDGE MEMORISATION ABOUT INDUSTRIAL PROCESSES

We give in this part a brief description of the goals and constraints encountered when we want to memorise knowledge about an industrial process.

The comprehension of how an industrial plant works and how we can control it is often approximative and lies in qualitative judgements. So we will be principally concerned with the memorisation and treatment of imprecise assertions about the dynamic of the process. The building of a knowledge data base on this subject has many interesting applications:

-We can quikly obtain informations concerning a certain part of the process.
- We can confront real measures done on the variables of the process with qualitative assertions. It necessitates to have a semantic interpretation of the assertions, as those proposed by Zadeh in PRUF (Zadeh, 1978).
- We can generate new assertions by use of inference like rules and test the consistency of the data base.
- From assertions about the state of the process we can deduce what will be able to happen.
-We obtain an assistance for the conception of control algorithms. We can implement fuzzy control algorithms, but this necessitates to have a semantic interpretation in order to locate the state of the process among fuzzy functional regions of the state space.

This knowledge data base is composed of a first part in which we describe the plant. We find assertions which specify the physical elements, their location, nature,... If a variable is measured we associate with it an object of type time serie. A second part, which is more interesting for us, is composed of assertions made by experts about the dynamic of the process. These assertions are logical expressions built from logical connectives and relations about time series. These assertions always make reference to time and can specify time lags. We use qualified relations on time series such as: "to slowly increase", "to have a great value",... These relations verify a transportability property. In fact we describe, for instance, the relation "increase" as a fuzzy computable relation which associates with a time serie, an instant, a qualifier and some adjusting parameters (which enable us to take into account properties of a particular variable) a membership degree. The adjusting parameters can be numbers which can be linked with attributes values of the individuality variable. These attributes values can, for the relation "increase", specify the extrems and average speeds of variation. A correctly linking of this relation with a variable is for instance: "The temperature of the furnace slowly increases at time t" will be interpreted as: increase(time-serie(temperature-furnace),t, slowly, parameter-i(temperature-furnace),...).

The computation of "increase" can lie in an estimation, at time t, of the derivative and a confrontation of this value with the adjusting parameters and the modifier value. This construction enables us to implement an automatic semantic interpretation for a particular class of assertions. For instance we can deal with expressions such as:
"It is quite true that if the flow of oxygen is very high then the temperature of the furnace increases".

A fundamental point is that the deduction we can do from these assertions must be compatible with the semantic interpretation we have given of these constants. So, if we have an assertion like: "X is high → Y increase", with b as truth value, we must be able to compute the truth value of "X is high" when we know for instance that at a certain time we have "X >70" with a as truth value. There is no problem if we have a strict inclusion, for instance if we can assert that at a time t "X is very high" has a as truth value, then it is clear that "X is high" has a truth value, at time t, which can be renforced to a and then, at time t, the truth value of "Y increases" can be renforced to $a \wedge b$ (if we use a modus-ponens like rule: $(X)c, (X \rightarrow Y)d \Longrightarrow (Y)c \wedge d$. When we have not a strict inclusion we can use a measure of inclusion between fuzzy sets, (Badard, 1981).
A measure of inclusion is in fact a mapping I from $\mathcal{P}(E) X \mathcal{P}(E)$, $\mathcal{P}(E)$ being the set of the fuzzy subsets of E, into $[0,1]$, such that:
$-A,B \in \mathcal{P}(E)$, $A \subset B \longrightarrow I(A,B)=1$
$-A,B,C \in \mathcal{P}(E) \longrightarrow I(A,B) \wedge I(B,C) \leqslant I(A,C)$ (transitivity assumption).

For instance, from "(X has value "v1")a" we can generate "(X has value "v2")b" with the truth value $b=a \wedge I(v1, v2)$, where I(v1, v2) is the inclusion degree of the fuzzy set v1 associated to "v1" into those associated to "v2"). We must take account of many other links between the constants, for instance from "X increases" and "X is great" we can expect that in a few time X will at least be great. These links are expressed as truth value renforcement rules and incoherence detection rules. A truth value renforcement rule is a rule whose result is to increase the inferior bound of possible truth values of a proposition. An incoherence rule detects an inconsistency between inferior bound of truth values. For instance, if we have a proposition like "(A)a" then we can expect that the proposition "(not A)b" is such that $b \leqslant \phi(a)$ (where ϕ is a continuous decreasing involution), so we cannot renforce b but only detect an inconsistency as $b > \phi(a)$ (or equivalently as $\phi(b) < \phi^2(a) = a$).

CONCLUSION

A knowledge data base is a collection of assertions we must do in order to describe our knowledge about a piece of reality. We have shown that the study of these assertions enables us to extract their fundamental entities which must be described in the schema. It also integrates consistency constraints and rules for the generation of deduced assertions. The heap of the assertions must be partitioned into classes, principally depending on synthetic levels of description. With every level is associated a filter which manages the information flow between the data base and the exterior.

We have finally given some goals and problems encountered in the building of a knowledge data base about the dynamic of an industrial process.

We feel that this methodology can be applied to many domains and principally to the very captivating area of image processing. In this domain it is clear that the distinction of levels of synthetic description is crucial.

We think that our approach, which is near of the first order predicate calculus, can be extended to higher order logics. This extension would have practical purposes, it could facilitate, for instance, the treatment of problems dealing with time series.

REFERENCES

Badard, R. (1981). Fuzzy pretopological spaces and their representation. Journ. Math. Anal. Appl., 81, 378-390.
Badard, R. (1981). Fuzzy preuniform structures and the structures they induce. Report FPS-05-81, dept infor., INSA Lyon.
Buchanan, B., and F. Feigenbaum (1978). Dendral and Meta-Dendral. Art. Intell.,11, 5-24.
Chang, C.L. (1968). Fuzzy topological spaces. J. Math. Anal. Appl., 24, 182-190.
Codd, E.F. (1970). A relational model of data for large shared data banks. Comm. of the ACM, 13, 377-387.
Codd, E.F. (1979). Extending the data base relational model to capture more meaning. ACM Trans. Data base Systems, 4, 397-434.
Ernst, G.W., and A. Newell (1969). GPS: a case study in generality and problem solving. Ed. R L Ashenhurst, Acad. Press.
Farreny, H., and H. Prade (1981). The connection between Lipski's approach to incomplete information data bases and Zadeh's possibility theory. Report 154-LSI, Toulouse.
Haar, R.L. (1977). A fuzzy relational data base system. Report TR-586, Computer Science Center, University of Maryland.
Hammer, P.C. (1967). Extended topology and systems. Math. Systems Theory, 1.
Lipski, W. (1979). On semantic issues connected with incomplete information data bases. ACM Trans. Data base Systems, 4, 262-296.
Lukaciewicz, J. (1970). Selected works. Ed. L Borkowski. North Holland.
Phillips, R.J., M.J. Beaumont, and D. Richardson (1979). AESOP: An architectural relational data base. Computer-aided design, 11, 217-226.
Rieger, C. (1978). Spontaneous computation and its roles in AI modeling. Pattern-directed inferences systems, Ed. D A Waterman F Hayes-Roth, Acad. Press, 69-97.
Reiter, R. (1978). Deductive question-answering on relational data bases. Logic and Data bases, Eds H. Gallaire and J. Minker, Plenum Press.
Shortliffe, E. (1976). Computer-based medical consultations:MYCIN. American Elsevier.
Zadeh, L.A. (1978). Fuzzy sets as a basis for a theory of possibility. Fuzzy Sets and Systems, 1, 3-28.
Zadeh, L.A. (1978). PRUF-a meaning representation language for natural languages. Int. Jour. Man-Machine Studies, 10, 395-460.

STRUCTURE GENERATION ON BIBLIOGRAPHIC DATABASES WITH CITATIONS BASED ON A FUZZY SET MODEL

S. Miyamoto* T. Miyake and K. Nakayama***

**Institute of Information Sciences and Electronics, University of Tsukuba, Ibaraki 305, Japan*
***Science Information Processing Center, University of Tsukuba, Ibaraki 305, Japan*

Abstract. The present paper is concerned with data structuralization, i.e., generation of thesaurus-like structures, of bibliographic databases based on a fuzzy set model. Special attention is paid on a database with citations which have been considered to be an appropriate information for organizing bibliographic data. A finite set W of keywords, a set S of source articles, and a set C of citations are assumed to be given. Relations between W and S (resp. S and C) are represented by a matrix G (resp. H). Notions of a basis set and an objective set for the structuralization are introduced. Two kinds of fuzzy relations, that is, proximity and inclusion relations, are defined on the objective set by using these matrices and fuzzy set operations on the basis set. A thesaurus-like structure on the objective set is derived from the induced fuzzy relations by using two threshold parameters. Thesaurus-like structures on W using the basis sets C and S and that on C using the basis set S are generated. This method is applied to a bibliographic database with citations SCIENCE CITATION INDEX. Further, clustering of the bibliographic data and a graphic representation using the generated structure are discussed.

Keywords. Information retrieval; fuzzy relations; citations; modelling; set theory.

INTRODUCTION

Since problems in bibliographic information retrieval need considerations of various kinds of ambiguities, the theory of fuzzy sets should play an important role in them. Indeed, there have been researches on information retrieval using fuzzy set theory, e.g., classification in the information retrieval (Negoita, 1973) and fuzzy thesaurus (Radecki, 1976; Reisinger, 1974). However, its application and the experimental results on actual databases are insufficient.

On the other hand, recent development of on-line information retrieval system enables the use of various kinds of the bibliographic databases. Among the large scale databases there is SCIENCE CITATION INDEX (SCI) which contains the item of bibliographic citations (references) as a part of a record. SCI has been considered to be an effective tool for retrieving citations, i.e., information soucres, as well as a good data for evaluating an article and for clustering documents using the item of the citations (Small, 1973; Garfield, 1979).

Generally, advanced tools for the document retrieval are yet insufficient, although structuralization of the data such as automatic generation of a thesaurus (Salton, 1971) and citation clustering (Small, 1973;Garfield, 1979) are interesting from both theoretical and practical viewpoints. Existing studies in these subjects use heuristic approaches, therefore a basic framework for the data organization should be studied further.

A method is developed here to generate a structure on the data based on a fuzzy set model. A basis set B and a set D to be structuralized are considered. Moreover a function F of D into a family of fuzzy subsets on B is assumed to be given. Two fuzzy relations, i.e., a proximity relation and an inclusion relation between an arbitrary pair of elements in D are defined by using the fuzzy set operations on B induced by the function F.

In application with citations, a set W of keywords, a set S of source articles, and a set C of citations are considered. Each of these three sets can be taken as the basis set B and another as the set D. Here thesaurus-like structures on W and on C are considered. Moreover retrieval of citations by particular keywords is discussed which includes local clustering of the citations and a hierarchical diagram of them with the thesaurus-like relations. The last consideration is related to a structure of knowledge represented by the citations for a particular subject of research.

FUZZY SET MODEL OF THESAURUS-LIKE STRUCTURES

Let $B = \{b_1, b_2, \ldots, b_m\}$ and $D = \{d_1, d_2, \ldots, d_n\}$ be two finite sets. Assume that a function F is given which maps each element $d \in D$ onto a fuzzy set $F(\cdot, d)$ in B. The function F is represented by a matrix $A = (a_{ij})$, $1 \le i \le m$, $1 \le j \le n$, whose generic element a_{ij} satisfies $0 \le a_{ij} \le 1$. Namely, the function F maps d_j onto

$$\{F(b_i, d_j)\}_{1 \le i \le m} = \{a_{ij}\}_{1 \le i \le m} .$$

In other words, the fuzzy set corresponding to d_i is determined by the m-vector $A\,e_i$, where $e_i = (0, .., 0, 1, 0, .., 0)$ is the unit vector whose i-th element is unity and all the others are zero.

Moreover a positive measure M defined on a family of all fuzzy subsets of B is considered. We take as the measure M a simple functional of "counting numbers of elements". That is, if a fuzzy set $f = \{f_1, f_2, \ldots, f_m\}$, $0 \le f_i \le 1$, $1 \le i \le m$, is given, $M(f) = \sum_{i=1}^{m} f_i$.

Two fuzzy relations r and t on DxD are considered by using the fuzzy sets on B induced by F. That is, if A_i and A_j are two sets corresponding to d_i and d_j, respectively, then

$$r(d_i, d_j) = \frac{M(A_i \cap A_j)}{M(A_i \cup A_j)}$$

$$t(d_i, d_j) = \begin{cases} \dfrac{M(A_i \cap A_j)}{M(A_j)} , & (M(A_i) > M(A_j)) \\ 0 , & (M(A_i) \le M(A_j)) . \end{cases}$$

In terms of the matrix $A = (a_{ij})$,

$$r(d_i, d_j) = \frac{\sum_k \min(a_{ki}, a_{kj})}{\sum_k \max(a_{ki}, a_{kj})}$$

$$t(d_i, d_j) = \begin{cases} \dfrac{\sum_k \min(a_{ki}, a_{kj})}{\sum_k a_{kj}} , & (\sum_k a_{ki} > \sum_k a_{kj}) \\ 0 , & (\sum_k a_{ki} \le \sum_k a_{kj}). \end{cases}$$

Remark. The fuzzy relation r is a proximity relation (Dubois, Prade, 1980), whereas t represents an inclusion relation. It is easy to see that $t(d_i, d_j) = 1$ means inclusion of fuzzy sets $A_i \supset A_j$. If the relation between the two sets is reversible, i.e., if we consider a structure on B using the basis set D, then the function F should be reversed:

$$F^{-1}(d_j, b_i) = \{a_{ij}\} .$$

Representation by the matrix is simply A^T (transpose of A).

Traditional thesaurus for bibliographic information retrieval uses several designators of term relationships, i.e., BT (broader term), NT (narrower term), RT (related term), and UF (used for). The last designator UF concerns a particular usage in a specific database and it is not considered in a general framework of the term relations. The other three represent inclusion or similarity of terms in their meanings.

Therefore, if we suppose that D is the set of terms, B represents the set of various meanings, and F maps each term to its meanings, then we obtain a thesaurus structure by using two threshold parameters α and β $(0 < \alpha, \beta < 1)$:

d_i RT d_j (d_i is a related term of d_j)

$\Longleftrightarrow r(d_i, d_j) \ge \alpha$,

d_i NT d_j (d_j is a narrower term of d_i)

$\Longleftrightarrow d_j$ BT d_i (d_i is a broader term of d_j)

$\Longleftrightarrow t(d_i, d_j) \ge \beta$.

In application we can not easily find the set of meanings. Therefore we use actual sets of articles and citations here instead of the set of meanings. Because of this alteration of the basis set, the generated structure is called a thesaurus-like structure or a pseudo-thesaurus (Miyamoto, Miyake, Nakayama, 1983). The above model is applied here to a bibliographic database consisting of three sets of the items: $W = \{w_1, w_2, \ldots, w_\ell\}$ is the set of the keywords, $S = \{s_1, s_2, \ldots, s_m\}$ is the set of the source articles, and $C = \{c_1, c_2, \ldots, c_n\}$ is the set of the citations.

In a bibliographic database with citations, each source article (record) s_i has several keywords in W and several citations in C. The relation among W, S, and C is represented by two matrices $G = (g_{ij})$, $1 \le i \le \ell$, $1 \le j \le m$, and $H = (h_{jk})$, $1 \le j \le m$, $1 \le k \le n$. That is, the element g_{ij} represents the degree of occurrence of the keyword w_i in the source article s_j; h_{jk} means the degree of occurrence of the citation c_k in s_j. In typical cases of the occurrence, g_{ij} and h_{jk} take binary values of 0 and 1. Here, however, they need not be binary valued nor be integers. They are only assumed to be nonnegative.

Adaptation of the sets W, S, and C to the above model needs specification of the function F or the matrix A. Only one difference between A and G (or H) is that all the elements of A satisfy $0 \le a_{ij} \le 1$. Therefore each element of G is divided by a sufficiently large number K which is independent of i and j such that $0 \le g_{ij}/K \le 1$ for all i,j and the matrix $(1/K)G^T = (g_{ji}/K)$ is considered as the function F. When C is to be structured using the basis set S, the matrix $(1/L)H = (h_{jk}/L)$ such that $0 \le h_{jk}/L \le 1$ for all j,k should be considered. We need not, however, specify the numbers K and L, since the numbers disappear in the calculation of the relations r and t. Thus, the fuzzy relations of W based on S are

$$r(w_i,w_j)= \frac{\sum_k \min(g_{ik},g_{jk})}{\sum_k \max(g_{ik},g_{jk})}$$

$$t(w_i,w_j)=\begin{cases} \dfrac{\sum_k \min(g_{ik},g_{jk})}{\sum_k g_{jk}} , & (\sum_k g_{ik} > \sum_k g_{jk}) \\ 0 , & (\sum_k g_{ik} \le \sum_k g_{jk}). \end{cases}$$

If the structure of W on the basis set C should be considered, the matrix $(GH^T)^T$ is used for the fuzzy relations.

Remark. The idea of the fuzzy set model was introduced in a foregoing paper of the authors (Miyamoto, Miyake, Nakayama, 1983). Other choices of the relations r and t are given there.

Application to a practical database with citations such as SCI frequently needs processing of thousands of the elements. Therefore huge amount of the storage is needed to handle the matrices G and H. Fortunately they are sparse matrices: major part of them consists of zero elements. For example, suppose that we have 10,000 source articles and each article has 10 keywords and 20 citations on average. Then it is clear that the numbers of nonzero elements in G and in H are 10 x 10,000 and 20 x 10,000, respectively, which are not a great requirement in a present digital computer.

AN EXPERIMENTAL STUDY IN THE FIELD OF CONTROL THEORY

An experiment was performed on 190 articles in control theory and pseudo-thesauri of W and C were generated. Articles in the journal IEEE Transactions on Automatic Control in 1980 were extracted from the databases SCI and at the same time from INSPEC (INternational information Service in Physics, Electrotechnology, Computers and Control). The former includes citations and the latter contains the keywords. Citations refered only once were deleted then we have obtained 129 distinct keywords and 257 citations. Instead of the two parameters α and β, we took two thresholds to control the length of the output. That is, the average number of the related terms per one element (hereafter called AVRT) and the average number of the broader terms per one element (hereafter called AVBT) were considered. In this experiment they were fixed as AVRT = AVBT = 5.

Remark. An algorithm with sorting when AVRT and AVBT are used is given in Miyamoto, Miyake, Nakayama (1983).

Figure 1 and 2 depict the generated structures in W based on C and on S, respectively. Figure 3 illustrates that in C based on S. The former two show relations of the keywords; the latter is useful to see the structure of the information sources. If a pair has twofold relations RT and BT (or RT and NT) simultaneously, then BT (or NT) relation is adopted prior to RT in these figures.

Another use of the measures r and t is shown in Fig. 4. Here a keyword is given and the related citations are found. Two numbers after a citation are the frequency of the occurrences of the citation on the keyword and that on the entire set of the articles. Then a local clustering of the citations using the proximity r based on W was performed. Here the group average method (Anderberg, 1973) was used to generate the dendrogram. Moreover the local structure of the pseudo-thesaurus is represented as a hierarchical diagram in Fig. 5. In this figure the vertical coordinates of the elements are determined by the total frequencies of the occurrences, i.e., $M(d_i)$'s. The elements of higher frequencies are situated in the upper part and those of lower frequencies are in the lower part. The horizontal positions are determined by the orders in the dendrogram. Therefore elements of higher proximity have small distances in their x coordinates. Undirected edges in the figure show RT relations, whereas the directed edge $a \to b$ means b is a narrower term of a (a is a broader term of b). It is easy to see that all the directed edges have downward directions.

The structure of the citations as well as that of the keywords is interesting. Let us see the citations related to the keyword NUMERICAL METHOD in Fig. 4. The dendrogram shows that there are three major clusters in the citations. One cluster in the upper part of the dendrogram has seven members with ID numbers 6,7,9,12,13,14,2. This group is related to the linear system theory including the studies of eigenvalues and matrix algebra. Another cluster in the middle part consists of five members whose ID numbers are 5,8,11, 3,10. The last group has two members 1,4. The latter two groups are concerned with probability and statistics. In Fig. 2, keyword NUMERICAL METHOD has four RT terms, two NT's, and four BT's based on S. Unexpectedly, four RT terms in the upper part are related to probability or statistics, and the rest concerns linear systems. The argument here was validated by retrieving the titles of the corresponding source articles. Any researchers in this field can, however, grasp the meanings of these groups at a glance of the structures.

CONCLUSIONS

The above fuzzy set model provides proximity and inclusion relations; it is simple and applicable to a large set of the data.

In summary, advanced tools for bibliographic information retrieval is yet insufficient. Graphic methods with the relations as were given in Fig. 4 and in Fig. 5 are useful in stimulating various concerns in the retrieval of bibliography and in the studies of knowledge in publications.

REFERENCES

Anderberg, M.R. (1973). Cluster Analysis for Applications. Academic Press, New York.

Dubois, D. and H. Prade (1980). Fuzzy Sets and System, theory and applications. Academic Press, New York.

Garfield, E. (1979). Citation Indexing - its theory and application in science, technology, and humanities. Wiley, New York.

Miyamoto, S., T. Miyake, and K. Nakayama (1983) Generation of a pseudo-thesaurus for information retrieval based on co-occurrences and fuzzy set operations. IEEE Trans. Syst. Man and Cybern., Jan/Feb.

Negoita, C.V. (1973). On the application of the fuzzy sets separation theorem for automatic classification in information retrieval systems. Information Sciences, 5, 279-286.

Radecki, T. (1976). Mathematical model of information retrieval system based on the concept of fuzzy thesaurus. Inform. Proc. Manag., 12, 313-318.

Reisinger, L. (1974). On fuzzy thesauri. COMPSTAT 1974, Proc. Symp. Comput. Stat. (G. Bruckman et al. eds.), Physica-Verlag, Vienna, 119-127.

Salton G., eds. (1971). The SMART Retrieval System, experiments in automatic document processing. Prentice-Hall, Englewood Cliffs, New Jersey.

Small, H. (1973). Co-citation in the scientific literature: a new measure of the relationship between two documents. J. Am. Soc. Inform. Sci., 24, 4, 265-269.


```
NONLINEAR SYSTEMS
     RT      POLES AND ZEROS
     RT      STABILITY
     RT      STABILITY CRITERIA
     NT      BANG-BANG CONTROL
     NT      CONTROL NONLINEARITIES
     NT      FREQUENCY-DOMAIN ANALYSIS
     NT      GRAPH THEORY
     NT      HIERARCHICAL SYSTEMS
     NT      LARGE-SCALE SYSTEMS
     NT      LIE GROUPS
     NT      SWITCHING FUNCTIONS
     BT      LYAPUNOV METHODS

NUMERICAL METHODS
     RT      EIGENVALUES AND EIGENFUNCTIONS
     RT      GAME THEORY
     RT      IDENTIFICATION
     RT      MATRIX ALGEBRA
     RT      RANDOM NOISE
     RT      SPECTRAL ANALYSIS
     RT      TIME SERIES
     NT      ACCELERATION CONTROL
     NT      LINEAR ALGEBRA
     NT      MULTIDIMENSIONAL SYSTEMS
     NT      QUEUEING THEORY
     NT      ROOT LOCI
     BT      LINEAR SYSTEMS
```

Fig. 1 Structures in W based on C.

```
NONLINEAR SYSTEMS
     RT      CONTROLLABILITY
     RT      DIFFERENCE EQUATIONS
     RT      DISCRETE TIME SYSTEMS
     RT      FILTERING AND PREDICTION THEORY
     RT      LINEAR SYSTEMS
     RT      OBSERVABILITY
     RT      POLES AND ZEROS
     RT      POLYNOMIALS
     RT      STABILITY
     RT      STABILITY CRITERIA
     RT      STATE ESTIMATION
     RT      STOCHASTIC SYSTEMS
     RT      TIME-VARYING SYSTEMS
     NT      BANG-BANG CONTROL
     NT      CONTROL NONLINEARITIES
     NT      FREQUENCY-DOMAIN ANALYSIS
     NT      GRAPH THEORY
     NT      HIERARCHICAL SYSTEMS
     NT      LARGE-SCALE SYSTEMS
     NT      LIE GROUPS
     NT      LYAPUNOV METHODS
     NT      MARKOV PROCESSES
     NT      SERIES (MATHEMATICS)
     NT      SWITCHING FUNCTIONS

NUMERICAL METHODS
     RT      PARAMETER ESTIMATION
     RT      RANDOM NOISE
     RT      SPECTRAL ANALYSIS
     RT      TIME SERIES
     NT      LINEAR ALGEBRA
     NT      MULTIDIMENSIONAL SYSTEMS
     BT      DISCRETE TIME SYSTEMS
     BT      EIGENVALUES AND EIGENFUNCTIONS
     BT      LINEAR SYSTEMS
     BT      MATRIX ALGEBRA
```

Fig. 2 Structures in W based on S.

```
WILKINSON JH;1965;;;ALGEBRAIC EIGENVALUE
     RT      GARBOW BS;1977;;;MATRIX EIGENSYSTEM R
     RT      GOLUB GH;1976;V0018;P0578;SIAM REV
     RT      MACFARLANE AGJ;1977;V0025;P0081;INT J CONTROL
     RT      MARCUS M;1964;;;SURVEY MATRIX THEORY

WILLEMS JC;1971;V0009;P0105;SIAM J CONTROL
     RT      ARAKI M;1975;V0052;P0309;J MATH ANALYSIS APPL
     RT      ARAKI M;1976;V0021;P0254;IEEE T AUTOMAT CONTR
     RT      ARAKI M;1978;V0023;P0129;IEEE T AUTOMAT CONTR
     RT      HAHN W;1967;;;STABILITY MOTION
     RT      KRASOVSKII NN;1963;;;STABILITY MOTION
     RT      SAEKI M;19  ;;;UNPUBLISHED
     RT      WILLEMS JC;1976;V0064;P0024;PROC IEEE
     BT      MOYLAN PJ;1978;V0023;P0143;IEEE T AUTOMAT CONTR
     BT      NARENDRA KS;1973;;;FREQUENCY DOMAIN CRI
```

Fig. 3 Structures in C based on S.

```
ENTER DESCRIPTER
NUMERICAL METHODS

AKAIKE H;1974;V0019;P0716;IEEE T AUTOMAT CONTR      ---   1   2
ANDERSON BDO;1979;;;OPTIMAL FILTERING               ---   1   4
BOX GEP;1970;;;TIME SERIES ANAL FOR                 ---   1   3
DOOB JL;1953;;;STOCHASTIC PROCESSES                 ---   1   2
FEDOROV VV;1972;;;THEORY OPTIMAL EXPT               ---   1   2
GARBOW BS;1977;;;MATRIX EIGENSYSTEM R               ---   2   2
GOLUB GH;1976;V0018;P0578;SIAM REV                  ---   2   2
GOODWIN GC;1977;;;DYNAMIC SYSTEM IDENT              ---   1   2
GOULD LA;1969;;;CHEM PROCESS CONTROL                ---   1   2
MEHRA RK;1974;V0019;P0753;IEEE T AUTOM CONTROL      ---   1   3
PARZEN E;1961;V0032;P0951;ANN MATH STAT             ---   1   2
STEWART GW;1973;;;INTRO MATRIX COMPUTA              ---   2   3
WILKINSON JH;1965;;;ALGEBRAIC EIGENVALUE            ---   1   2
WONHAM WM;1974;;;LINEAR MULTIVARIABLE               ---   1   4

 14 DISTINCT CITATIONS FOUND
```

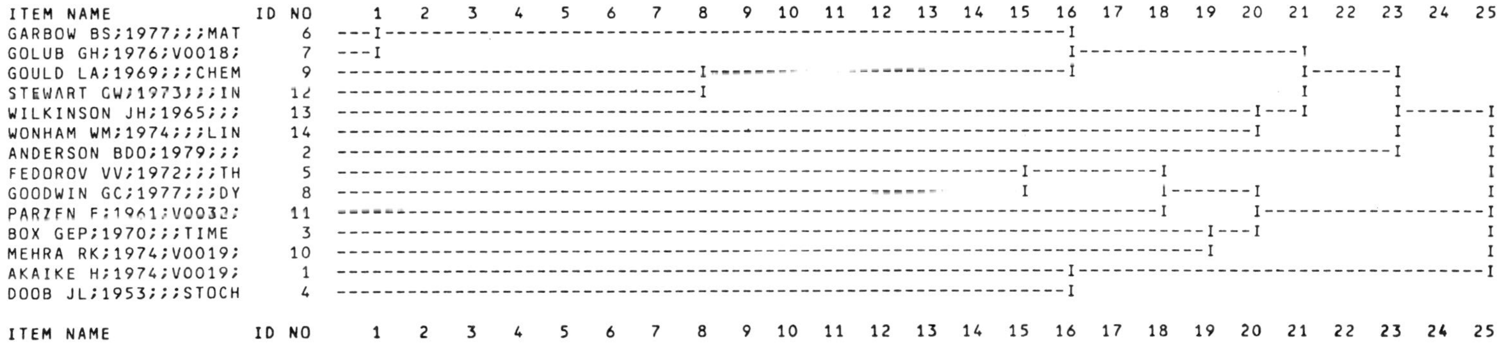

Fig. 4 Structural representation by a dendrogram of the citations related to the keyword NUMERICAL METHOD.

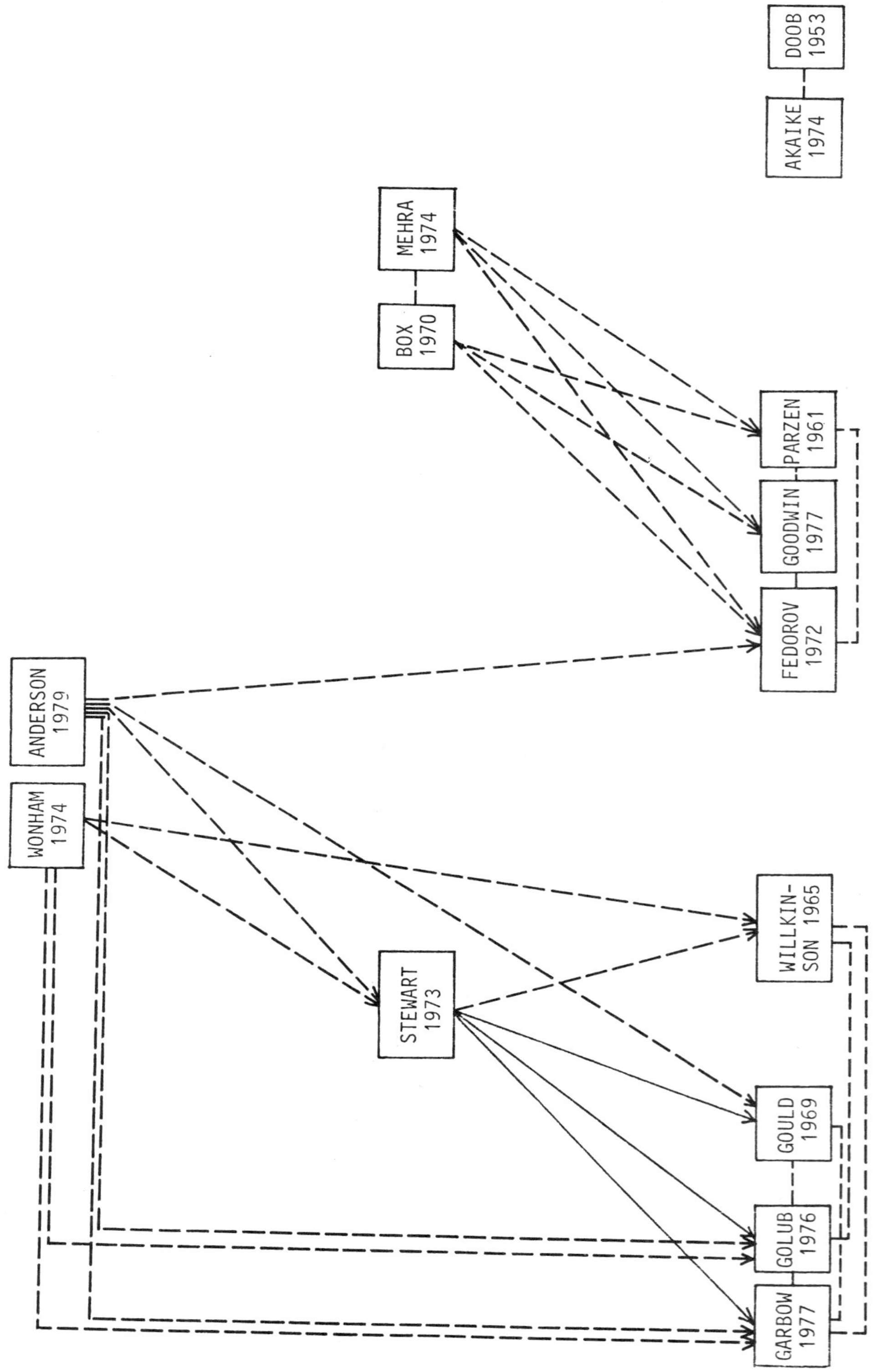

Fig. 5 A hierarchical diagram of the citations related to the keyword NUMERICAL METHOD. Solid lines are edges with AVRT=AVBT=5. Broken lines are those with AVRT=AVBT=10.

SOME CONCEPTS OF PANSYSTEMS MEDICINE

Wu Xuemou

China Digital Engineering Institute, P.O. Box 223, Wuchang, The People's Republic of China

Abstract. This paper develops some concepts of pansystems medicine which are closely connected with the interdisciplinary investigation of pansystems methodology and certain unified principles of diagnosis and treatment for general systems.

Keywords. Pansystems methodology; pansystems relations; pansystems ecology; semi-equivalence theory; bianzheng shizhi; forcing blockwise observocontrollability.

INTRODUCTION: PANSYSTEMS METHODOLOGY AND PANSYSTEMS MEDICINE

Pansystems methodology is a transfield investigation of generalized system-transformation-symmetry in things mechanism which is closely connected with mathematics-physics science, systems science, biomedicine science and thinking science. It emphasizes the mathematical research of pansystems relations —— the generalized ten-relations and their inner/outer transformations: macromicrospic, motion-rest, whole-parts, body-shadow, causality, observocontrol, series-parallel, simulation, clustering-separability, difference-identity.

Pansystems methodology was first presented in China in 1976 and from that time onwards obtained professional development which forms special investigations of pansystems logic, pansystems network analysis, pansystems recognition theory, pansystems-operations research of large scale systems, pansystems ecology and pansystems medicine. A series of concrete new results obtained (including more than 300 new theorems) are concerning cybernetics, dynamic programming, dynamic games, large scale systems, simulation theory, pattern recognition, artificial intelligence, clustering analysis, graph theory, automata, approximation transforming theory, fuzzy sets theory, universal algebra, hypercomplex variables functions, mathematical logic, economics, computer, communication, geography, agroclimatology, scientific methodology and biology-ecology-medicine, etc.

The pansystems medicine is considered as an interdisciplinary research of pansystems methodology and certain unified principles and concepts of diagnosis and treatment for general systems and for that of both Chinese traditional medicine (CTM) and western medicine. A trend of medicine science which can also be considered as the tentative ideas of development of pansystems medicine is as follows: Biology-medicine model ⟶ 1) Physics-biology-medicine model, 2) Biology-psychology-society-medicine model ⟶ Physics-biology-ecology-psychology-society-medicine model, composition and unification of CTM and western medicine, unified principles of diagnosis and treatment for general systems in combination with applications of pansystems methodology ⟶ Pansystems medicine.

PANSYSTEMS THEORY: NEW INVESTIGATIONS OF FUZZINESS, CONNECTEDNESS AND OBSERVOCONTROLLABILITY

Pansystems theory is a very good theoretical tool to analyse related problems in the diagnosis and treatment for general systems. Let's begin with the properties of semi-equivalence.

Denote $P(G)$ the power set of G, and define $I = I(G) = \{(x,x) \mid x \in G\}$.

If $f, g \in P(G^2)$, then define $f \vee g = f \cup g$, $f \wedge g = f \cap g$, $f^{-1} = \{(x,y) \mid (y,x) \in f\}$, $\bar{f} = G^2 - f$, $f \circ g = \{(x,y) \mid \exists t \in G, (x,t) \in f, (t,y) \in g\}$, and $f \leqslant g$ means $f \subset g$.

Define $f^{(n+1)} = f^{(n)} \circ f$, $f^{(0)} = I$, $f^t = \vee f^{(n)}$ $(n = 1, 2, \cdots)$, $R[G] = \{f \mid f \in P(G^2), I \leqslant f\}$, $S[G] = \{f \mid f \in P(G^2), f = f^{-1}\}$, $T[G] = \{f \mid f \in P(G^2), f^{(2)} \leqslant f\}$, $E_5[G] = R[G] \cap S[G]$, $E[G] = R[G] \cap S[G] \cap T[G]$.
We shall use the notation $(\circ)$-$\prod f_i$ for $f_1 \circ f_2 \circ f_3 \circ \cdots$ and $\overset{t}{\vee} f_i = (\vee f_i)^t$.

What are called pansystems operations are defined as the various compositions of following ones: disjunction, conjunction, union, extension, difference, confinement, intersection, complement, projection, embodiment, transformations of order of direct product, inversion and composition. And $\delta \overset{\varphi}{\longmapsto} \theta$ means the structure θ is derived from structure δ by using certain pansystems operations. Some typical operators are defined as follows.

$\varepsilon_1(g) = g \vee g^{-1} \vee I$, $\varepsilon_2(g) = \varepsilon_1(g \wedge g^{-1})$, $\varepsilon_3(g) = \varepsilon_1(g^t \wedge g^{-t})$, $\varepsilon_4(g) = \varepsilon_1(g \circ g^{-1})$, $\varepsilon_5(g) = \varepsilon_1(g^{-1} \circ g)$, $\varepsilon_{i+5}(g) = \varepsilon_i(\bar{g})$ $(i = 1, 2, \ldots, 5)$, $\delta_j(g) = (\varepsilon_j(g))^t$ $(j = 1, 2, \ldots, 10)$, $\delta_0(g) = max\{\delta \mid \delta \in E[G], \delta \leqslant g\}$ (provided it makes sense), $\delta_{11}(g) = \delta_0(\varepsilon_2(g))$, $\delta_{12}(g) = \delta_{11}(\bar{g})$. These operators are dependent on the universe G, and clearly we have $\varepsilon_i(g) \in E_5[G]$, $\delta_i(g) \in E[G]$. Generally speaking, for any given $\{g_\sigma\} \subset P(G^2)$, we can obtain various $g \in P(G^2)$ by using pansystems operations and then by the ε_i-operators and δ_i-operators they are reduced to various semi-equivalences and equivalences of G, and latter will be connected with other pansystems relations according to the pansystems methods.

If $\varepsilon \in E_5[G]$, let $G_i \subset G(d\varepsilon)$ represent $G_i = max\{Q \mid Q \in P(G), Q^2 \leqslant \varepsilon\}$, the total $\{G_i\}$ is denoted by G/ε called the quotient set of G with respect to ε, and we adopt the notation $G = \cup G_i(d\varepsilon)$ for all probable G_i. Generally speaking, $\{G_i\}$ forms a "fuzzy" classification of G, and when $\varepsilon \in E[G]$, it will be reduced to a non-fuzzy one.

If $f \subset G \times G'$ can be represented as $f = g_1 \circ g_2^{-1}$ where $g_1 : G \longrightarrow g_1(G)$, $g_2 : G' \longrightarrow g_2(G')$, $g_1(G) = g_2(G')$, then f is called a hard simulation between G and G'. If $x \circ f$, $f \circ y \neq \phi$ $(x \in G, y \in G')$, then f is called a semihard simulation, or E_5-simulation, or implicit simulation between G and G', where $x \circ f = \{y \mid (x,y) \in f\}$, $f \circ y = \{x \mid (x,y) \in f\}$.

It is easy to prove that, if f is a hard simulation, then $f \circ f^{-1} \in E[G]$, $f^{-1} \circ f \in E[G']$, and for any given $\delta \in E[G]$, we have a natural induction $f_\delta : G \longrightarrow G/\delta$ such that if $x \in G_i \subset G(d\delta)$, we have $f_\delta(x) = x \circ f_\delta = G_i$. Furthermore, if $\delta_\sigma \in E[G]$, then $\wedge \delta_\sigma$, $\overset{t}{\vee} \delta_\sigma$, δ_σ^t, δ_σ^{-1}, $\delta_\sigma^{(n)} \in E[G]$, and $(\circ)$-$\prod \delta_\sigma \in E[G]$ provided they are commutative with respect to composition.

Similarly, if f is an E_5-simulation, then $f \circ f^{-1} \in E_5[G]$, $f^{-1} \circ f \in E_5[G']$, and for any given $\varepsilon \in E_5[G]$, we have the natural induction $f_\varepsilon \subset G \times (G/\varepsilon)$ such that $x \in G_i \subset G(d\varepsilon)$ is equivalent to $(x, G_i) \in f_\varepsilon$. Furthermore, for any integers $m, n \geqslant 0$, we have $\varepsilon^{(m)} \leqslant \varepsilon^{(n+m)}$, and $\varepsilon^t \in E[G]$. Generally speaking, for given $\varepsilon_\sigma \in E_5[G]$, we have $\wedge \varepsilon_\sigma$, $\vee \varepsilon_\sigma$, ε_σ^{-1}, $\varepsilon_\sigma^{(n)} \in E_5[G]$, and when ε_σ are commutative, then $(\circ)$-$\prod \varepsilon_\sigma \in E_5[G]$.

If $f \subset G \times G'$ is an E_5-simulation, then for any given odd positive integer n we can find an E_5-simulation $g \subset G \times G'$ such that $g \circ g^{-1} = \delta^{(n)}$, $g^{-1} \circ g = \delta_*^{(n)}$, and $f^{-1} \circ \delta^{(n)} \circ f = \delta_*^{(n+1)}$, $f^{-1} \circ \delta^t \circ f = \delta_*^t$, $f \circ \delta_*^{(n)} \circ f^{-1} = \delta^{(n+1)}$, $f \circ \delta_*^t \circ f^{-1} = \delta^t$, where $\delta = f \circ f^{-1}$, $\delta_* = f^{-1} \circ f$. Furthermore, there exists an 1-1 mapping $\varphi : G/\delta^t \longrightarrow G'/\delta_*^t$, such that for $Q \subset G$, we have $(Q \circ \delta^t) \circ f = (Q \circ f) \circ \delta_*^t$, where $Q \circ f = \cup x \circ f (x \in Q)$, etc. Consequently, f can be transformed into a hard simulation $h(f) = (f_a) \circ \varphi \circ (f_b)^{-1}$, where $a = \delta^t$,

$b = \delta_*^t$; and f_a, f_b are corresponding inductions. $h(f)$ and g are called the emergence and partial emergence of respectively.

The following theorems obtained in pansystems methodology are very useful for analysis of things mechanism, including medicine problems (see references).

Theorem 1. If $f \subset G \times G'$ is a hard simulation, $\delta \in E[G)$, $\delta \leqslant f \circ f^{-1}$, or $\delta \geqslant f \circ f^{-1}$, then $f^{-1} \circ \delta \circ f \in E[G')$.

Theorem 2. If $g \in R[G)$ and f is an embodiment, then $f^{-1} \circ \delta(g) \circ f = \delta(f^{-1} \circ g \circ f)$. If f is an E_5-simulation, then $f^{-1} \circ \delta(g) \circ f \leqslant \delta(f^{-1} \circ g \circ f)$, where $\delta = \varepsilon_i, \delta_i, i = 1, 2, \ldots, 5$.

Theorem 3. If $G = \cup G_i(d\varepsilon)$, $\varepsilon \in E_5[G)$, then $\exists i(x, y \in G_i)$ is equivalent to $(x, y) \in \varepsilon$. Consequently, $\neg \exists i (x, y \in G_i)$ is equivalent to $(x, y) \in \bar{\varepsilon}$, and when $\varepsilon \in E[G)$, the latter is equivalent to $\exists ij(i \neq j, x \in G_i, y \in G_j)$.

Theorem 4. Let $G = \cup G_i(d\delta)$, $\delta \in E[G)$, $g \in P(G^2)$. If $\delta \leqslant g$, then $\exists i(x, y \in G_i)$ implies $(x, y) \in g$. If $g \leqslant \delta$, $\exists ij(i \neq j, x \in G_i, y \in G_j)$, then $(x, y) \in \bar{g}$.

Theorem 5. Let $\varepsilon \in E_5[G)$, $\varepsilon' = \varepsilon_6(\varepsilon)$, $G = \cup G_i(d\varepsilon)$, $G = \cup G'_j(d\varepsilon')$, then $\varepsilon_6(\varepsilon') = \varepsilon$, and $\exists j(x, y \in G'_j)$ is equivalent to that $x = y$ or $\neg \exists i(x, y \in G_i)$, or to that $x = y$ or $\exists ik(x \in G_i, y \in G_k, i \neq k)$ for $\varepsilon \in E[G)$.

Theorem 6. If $g \subset G^2$, $\delta \in E_5[G)$, $G = \cup G_i(d\delta)$, then between $x, y \in G_i$ there exist following corresponding connectedness-discoupling relations in respect to g. $\delta = \varepsilon_3(g)$: strongly connected; $\delta = \varepsilon_6(g^t)$: weakly discoupling; $\delta = \varepsilon_1(g^{(n)})$: n-step-connected; $\delta = \varepsilon_7(g^{(n)})$: n-step-discoupling; $\delta = \varepsilon_1(g^t)$: connected; $\delta = \varepsilon_7(g^t)$: strongly discoupling; $\delta = \varepsilon_2(g^{(n)})$: n-step-strongly connected; $\delta = \varepsilon_6(g^{(n)})$: n-step-weakly discoupling; $\delta = \varepsilon_7((I \vee g)^{(n)})$: n-step-strongly discoupling; $\delta = \delta_1(g)$: E-type weakly connected; $\delta = \delta_{12}(g)$: E-type immediately discoupling; $\varepsilon = \delta_0(\varepsilon_1(g))$: E-type immediately connected; $\delta = \delta_6(g)$: E-type weakly discoupling; $\delta = \delta_7(g)$: E-type strongly discoupling; $\delta = \varepsilon_1((I \vee g)^{(n)})$: partially n-step weakly connected; $\delta = \varepsilon_2((I \vee g)^{(n)})$: partially n-step strongly connected. Where various definitions of connectedness and discoupling are just based on corresponding $(x, y) \in \delta$.

Theorem 7. Let G, H be two given classes of sets, $E\ F \subset \cup D \uparrow H_\sigma (H_\sigma \in H)$, $D \in G$, and $G \overset{p}{\longmapsto} \varphi_i, \theta_j$ which are E_5-simulations between E and F^2 and between E and F respectively. Under the co-operation of various i and j, we can obtain the forcing blockwise observocontrollability: $A_m \subset F(d \wedge \theta_j^{-1} \circ \theta_j)$, $B_n \subset F^2(d \vee \varphi_i^{-1} \circ \varphi_i)$.

Comment. The semi-equivalence, simulation, blockwise observocontrollability are of new type fuzziness investigated by pansystems theory, and the forcingness of observocontrollability is in fact a unified pansystems model of co-diagnosis, co-operation, synthetic cure, generalized principle of CT technique, etc.

SOME CONCEPTS OF PANSYSTEMS ECOLOGY

The tentative ideas of pansystems ecology is the investigation and applications of pansystems methodology concerning the interrelations between the things and their panenvironment. Specially, the applications of pansystems network analysis, pansystems recognition theory and pansystems-operations research of large scale systems to the panenvironment relations of things, being not confined to the traditional ecosystems.

The so-called panenvironment means any things which are connected with the object investigated, including some parts or partial history of the object. If the object has the base set G, and there exists a certain binary relation $f \subset G \times G'$, then the things with the base set G' all can be considered as a panenvironment of object-things.

The interchange of structures or pan-structures between the object-things and related panenvironment enables the object-things to form a certain pansystems-holographic body —— a generalized holographic structure, including certain latent holographic information. This is a universal concept called pansystems holographic recapitulation law. Sometimes the latent holographic structure has a certain emergence, and this phenomenon or mechanism presents some new explanation for certain space-time pansymmetries of general things, from the mathematics-physics science to the systems science, and from the biomedicine -ecology science to certain social problems.

A special application of this law to the induction $\delta \in E_5[G') \longmapsto f \circ \delta \circ f^{-1} \in E_5[G)$ enables us to obtain the concept of pansystems ecotype which includes the traditional concepts (such as ecotype, biodistrict system, etc.) as certain special cases.

The structure-induction or panstructure model from the function panbox and the pansystems holographic recapitulation law (pansystems holographic body) is called pansystems zanfu jingluo. It is also a concept of universality, and generally speaking, possesses the function of relatively clustering transmision, transformation or induction for certain panstructure-information, while its structure may be fuzzy in dissection. Sometimes this concept can be described as $G = \cup G_i(d\delta(f_\sigma))$, $G/\delta(f_\sigma)$, where $\delta = \varepsilon_i$, δ_i, $\varepsilon_i(\varepsilon_j(*))$, etc.

From the viewpoint of pansystems theory, the nourishment-pyramid principle in traditional ecology is a concept derived by pansystems bird's-eye view analysis for food-energy relations, and can be represented as a sort of the form $(G \cup G')/\varepsilon_3(f)$, where f means certain food-energy relations. Similarly, we can treat the macromicroscopic system relations of other problems by using pansystems methodology. Clearly, pansystems methods and the concepts of pansystems ecology can be used to analyse the problems of every layer in fundamental investigation medicine.

PANSYSTEMS BIANZHENG SHIZHI

Pansystems methodology has developed professional observocontrol analysis and the bianzheng shizhi of CTM can be considered as a special problem of observocontrol. Bianzheng, diagnosis all are a sort of pansystems observability. The concepts such as co-diagnosis principle, panbox principle, pansystems difference principle, observocontrol level of hard simulation, pansystems emergence of E_5-simulation, pansystems sampling law and pansystems exterior-interior analysis, etc. all can be considered as some principles presented by pansystems methodology to medical diagnosis (see references). Naturally, some problems of medicine can also be reduced to a sort of dynamic gamas, which have been investigated in references.

Bianzheng shizhi of CTM, substantially, is a sort of diagnosis and treatment based on an overall analysis of the illness and the patient's condition, being nothing but some classical and naive fragrance. Consequently, the pansystematization of its substance may be extended to use to the diagnosis-treatment problems of general systems, including the conceptual unification of CTM, western medicine and artificial systems, etc.

The concepts, principles and theorems introduced in pansystems semi-equivalence theory and in pansystems ecology can be considered as a substructure of the pansystematization.

Now we present a universal pansystems analysis programme for things-treating as follows: 1) Viewpoints, source materials, criterion, primary models, intuitive consideration ($D_1 \subset \cup P(G_\sigma^2)$); 2) Object-prototype ($D_2 \subset P(G^2) \cup E_5[G) \cup E[G)$); 3) Pansystems model ($D_3 \subset P(Q^2) \cup E_5[Q) \cup E[Q)$, $Q = \cup Q_i(d\delta(f_\lambda))$, $Q/\delta(f_\lambda)$, $\delta = \varepsilon_j$, δ_j, $\varepsilon_j(\varepsilon_k(*))$, ...)

The foundation of all diagnosis and treatment is identity-seeking and difference-seeking. By using the pansystems semi-equivalence theory, we can obtain following principles concerning diagnosis, prognosis, recognition, causality (see references).

<u>Theorem 8</u>. Let $f \subset G \times G'$ be an E_5-simulation, $\theta' \in S[G')$, $x \in f \circ x'$, $y \in f \circ y'$ ($\theta \in S[G)$, $x' \in x \circ f$, $y' \in y \circ f$), $\delta \in E_5[G)$, $\delta' \in E_5[G')$, $\delta \leq \varepsilon_7(f \circ \theta' \circ f^{-1})$, $\delta' \leq \varepsilon_1(\theta')$ ($\delta \leq \varepsilon_1(\theta)$, $\delta' \leq \varepsilon_7(f^{-1} \circ \theta \circ f)$).

If $x \neq y\ (x' \neq y')$, then $\exists j(x', y' \in G_j' \subset G'(d\delta'))$, $\exists i(x, y \in G_i \subset G(d\delta))$ imply the negation of each other, and the confinement of f on $G_i \times G_j'$ is reduced to an embodiment (a projection).

<u>Theorem 9</u>. If $g_\sigma \subset F \times E_\sigma$ are some E_s-simulations, $\theta \in P(F^2)$, $\theta_\sigma \in P(E_\sigma^2)$, $x_\sigma \in x \circ g_\sigma$, $y_\sigma \in y \circ g_\sigma$, then $g_\sigma \circ \theta_\sigma \circ g_\sigma^{-1} \leqslant \theta$, $(x_\sigma, y_\sigma) \in \theta_\sigma$ imply $(x, y) \in \theta$, and $g_\sigma^{-1} \circ \theta \circ g_\sigma \lesssim \theta_\sigma$, $(x, y) \in \theta$ imply $(x_\sigma, y_\sigma) \in \theta_\sigma$.

<u>Theorem 10</u>. Let $g_1, g_2 \in P(G^2)$, $\delta \in E_s[G)$, $G = \cup G_i(d\delta)$. If $(x, y) \in g_1$ leads to $\exists i(x, y \in G_i)$, and $(x, y) \in \bar{g}_2$ leads to $\exists ij(i \neq j,\ x \in G_i,\ y \in G_j)$, then $\delta_1(g_1) \lesssim \delta \lesssim \delta_0(g_2)$. If $\exists i(x, y \in G_i)$ implies $(x, y) \in g_1$, $\exists ij(i \neq j,\ x \in G_i,\ y \in G_j)$ implies $(x, y) \in \bar{g}_2$, then $\delta_1(g_2) \lesssim \delta \lesssim \delta_0(g_1)$.

Let $\varepsilon, \delta \in E_s[G)$, and $\varepsilon \lesssim \delta$, $G = \cup G_i(d\varepsilon) = \cup G_j'(d\delta)$, define f_θ, $\theta = \delta/\varepsilon$, as that: $(G_i, G_j') \in f_\theta$ is equivalent to $G_i \subset G_j'$. And define $\theta = f_\theta \circ f_\theta^{-1}$. After having these definition, we can get another pansystems model of co-diagnosis and co-observocontrollability as follows.

<u>Theorem 11</u>. Let $g_\sigma \subset G \times G_\sigma$ be some E_s-simulations, $\delta_\sigma' \in E_s[G_\sigma)$, then $\delta_\sigma = g_\sigma \circ \delta_\sigma' \circ g_\sigma^{-1}$, $\delta = \vee \delta_\sigma$, $\varepsilon = \wedge \delta_\sigma \in E_s[G)$, $\eta = \delta/\delta_\sigma \in E_s[G/\delta_\sigma)$, $\xi = \delta_\sigma/\varepsilon \in E_s[G/\varepsilon)$, and we have E_s-simulations: $f_\theta \subset G \times (G/\theta)$, $f_\theta^{-1} \circ g_\lambda \subset (G/\theta) \times G_\lambda$, $\theta = \delta_\sigma, \delta, \varepsilon$; $f_\eta \subset (G/\delta_\sigma) \times (G/\delta)$, $f_\xi \subset (G/\varepsilon) \times (G/\delta_\sigma)$; $f_\eta \circ f_\delta^{-1} \circ g_\lambda$, $f_\xi^{-1} \circ f_\varepsilon \circ g_\lambda \subset (G/\delta_\sigma) \times G_\lambda$.

In the following we give another pansystems model of bianzheng shizhi, which seems universal and concrete.

Let G be the symptom set called symptom space, then $\varepsilon_\sigma \in E_s[G)$ will be called the bianzheng type, and G/ε_σ the syndrome space whose element G_i named syndrome, and subset $D \subset G/\varepsilon_\sigma$ the disease region. Let Q be some symptoms observed, then the decision $Q \subset G_i \in D \subset G/\varepsilon_\sigma$ through the related pan-enviroment relation $f \subset Q \times E$ and interchange $\varepsilon_\sigma \rightleftarrows \varepsilon \in E_s[G)$ by using pansystems operations is a pansystems model of bianzheng. Let G' be the drug set called drug space, then $\varepsilon_\lambda' \in E_s[G')$ are called prescription-forming types, and $G_j' \in G'/\varepsilon_\lambda'$ the prescription. If Q_0 is the normal symptom which represents the good health, then $(G_i \longmapsto G_j') \longmapsto (Q \longmapsto Q_0)$ can be considered as a model of shizhi or of treatment. Combining these two models and sufficiently using the various feedbacks among the various processes, we then obtain our pansystems model of bianzheng shizhi. It seems universal for general systems.

In western medicine, the bianzheng type can include the disease-identifying model. And in CTM, generally speaking, it includes the so-call five-systems: bagang, Liujing, Qizuejin, Zanfu, Bingyin. But in a word, the task of pansystems co-diagnosis of various bianzheng types through pansystems operations is to solve the related discriminants for the disease-property, disease-location, disease-stage, disease-cause.

CONCLUSION

Pansystems medicine mainly is a sort of mathematical metamedicine specialized by pansystems theory, within which pansystems clustering-discoupling, causality-observocontrollability and the unified inward-outward synthetic analysis play important roles. It is a newborn developing exploration and will present many transdisciplinary principles. Many works of pansystems theory originate from the modern mathematical generalization of Yijing (Book of Changes), Mojing (Book of Mozi), Neijing (Book of Medicine), Sunzi Strategy, Sunbin Strategy and Thirty-Six Stratagems, consequently they are considerably distinct from the current investigations of systems methods, general systems theory and fuzzy sets and systems. Their return to the fundamental research of medicine, perhaps, will present certain new ideas of metamedicine.

REFERENCES

Wu Xuemou (1978,79). Investigations and applications of pansystems analysis. <u>J. Wuhan Univ.</u>, <u>3</u>, 87-105; <u>1</u>, 104-117.

Wu Xuemou (1981). Pansystems analysis — a new exploration of interdisciplinary investigation (with appendixes of certain discussion of medicine, chemistry, logic and economics). Science Exploration, 1, 125-164.
Wu Xuemou (1981). Pansystems analysis and scientific methodology. Philosophical Research, 4, 30-35, 5, 24-29.
Wu Xuemou (1982, 83). Pansystems methodology: concepts, theorems and applications (I)-(IV). Science Exploration, 1, 33-56, 2, 4; 1.
Wu Xuemou (1982). Pansystems analysis and fuzzy sets. Busefal, 10, 7-16.
Wu Xuemou (1982). Pansystems methodology: a transfield investigation of generalized system-transformation-symmetry. in Gupta, M. M. and Sanchez, E. (eds.), Fuzzy Information and Decision Processes. North-Holland Publishing Co.
Wu Xuemou (1983). Pansystems metatheory of ecology, medicine and diagnostics(I). Exploration of Nature, 2.
Wu Xuemou (1983). Pansystems methodology and the pansystems medicine model of bianzheng shizhi (I). Systems Engineering of Chinese Traditional Medicine, 2.
Wu Xuemou (1983). Pansystems medicine bianzheng and pansystems recognition theory. J. Wuhan I. of Iron and Steel T., 1.
Wu Xuemou (1983). Pansystems methodology and generalized biocybernetics. J.A.I., 1.
Wu Xuemou (1983). Pansystems methodology and generalized ecology. J. West-South Agricultural College, 1.
Guo Aike (1981). Vision mechanism and pansystem recapitulation. Science Exploration, 4, 95-102.
Guo Aike (1982). A pansystem description of a sort of complex systems. Science Exploration, 1, 29-32.
Guo Aike (1982). Pansystems holographic recapitulation law. Science Exploration, 4.
Wang Pingming (1981). Biohographic recapitulation law: an analysis to the pansymmetry of bio-space -time structure by pansystems methodology. Science Exploration, 4, 103-110.

FUZZY ALGORITHMS BASED ON VAGUE PSYCHOPHYSIOLOGICAL STATEMENTS

B. Straube* and R. Schmitt**

**Central Institute of Cybernetics and Information Processes of the Academy of Sciences of the GDR*
***Humboldt University of Berlin, Department of Psychology, lab. technology*

Abstract. Polysymptomatic time series are transformed to an one-dimensional interpretation space. The influence of psychic load is to estimate. This transformation is achieved by a structured fuzzy model. Each of the submodels is a fuzzy algorithm based on vague psychological statements. Different opinions among the specialists have to be modelled. These four models were applied to the same sets of data. It results in the preference of one of them.

Keywords. Psycho-physiological models; structured fuzzy models; fuzzy algorithms; fuzzy sets.

INTRODUCTION

In psychopysiological research there is an urgent need to perform long-time experiments and field investigations with polysymptomatic measurements of time series to estimate the influence of psychic loads.

Field investigations at a monitoring centre of a modern chemical plant show clearly that too many tasks on the one hand and too easy ones on the other can cause a high staff fluctuation. In this situation the psychologist diagnoses saturation and monotony in the staff. These terms are summerizied in the expression 'negative strain effects' (see Sinz and others, 1978).

There arise several questions:

How to define these terms such that a researcher can deduce quantitative decisions ?

How to simulate a real situation to measure the relevant variables for these terms ?

How to build up a suitable transformation between measurements and definitions ?

Simulating a long-term monitoring task at our laboratory, we had to take into account three concequences:

First, it is necessary to have an adequate knowledge of the history and tendency of every particular value. Habituation, change of the strategy applied, and circadian rhythms have to be considered.

Second, the ambiguous input-output relations found in the analysis of strain effects require measurements on three levels (psychological, physiological and performance).

Third, the high dimensional model represented by these measurements has to be transformed into a one-dimensional space suitable for interpretation and decision.

It will be demonstrated that the application of fuzzy algorithms is an excellent tool for solving this problem.
To show a way of overcoming underload situations, we used two different kinds of secondary tasks during our experiment. From this follows the hypothesis that a secondary task can be used to compensate negative strain effects of the main task concomitant with an improvement of the performance-effort relation.

EXPERIMENTAL DESIGN

Six subjects selected from a total of 22 students were asked identify display configurations representing disturbances of a simulated industrial process by means of well-learnt algorithms over a period of eight hours for four days a week (see Fig. 1 - 3). In a sequence of ten-minute intervals a secondary, more demanding qualification task was interposed in the main-line experiment, and a monotoneous arithmetic task was interposed in the comparison experiment (Fig. 2).

Data were gathered on three levels:

Performance data: reaction time, short-storage performance;

Psycological data of self-judgement;

Physiological data: inter-beat intervals of heart rate, blink rate and respiratory rate (Fig. 1).

Six tonic parameters were evaluated: the heart rate measured at three different time intervals, the sinus arrhythmia, the blink rate and the respiratory rate. The phasic acceleration and deceleration of the inter-beat intervals following the cognitive demand during the identification of a simulated technical state are approximated by a model with four parameters (transfer function) that is applied each hour of the working day. The parameters were calculated by means of the method of moments. The two parameters of performance (main task reaction time and short-term storage) and eleven parameters of psychophysiological activation are represented in a thirteen-dimensional space A. To make a decision in line with the hypothesis, we found it useful to transform space A into a one-dimensional space B. According to the definitions of the four relevant terms four fuzzy algorithms has been derived to make possible this transformation.
The structure of data transformation shows Fig. 4.
If it can be shown that the structure in A is more pronounced in B by one of these algorithms the usefulness and correctness of this algorithm is accepted.

FUZZY MODEL

Let us start with some remarks to the feature extraction used.
The most important parameters are the variations of heart rate both short-term and long-term. The short-term variations are described by the estimated four parameters of the tranfer functions and by their integral. These parameters are valid for one hour only. The arithmetic mean of heart-rate, calculated over three different time-intervals in each experimental hour represents the long-term variations. Further more, the arithmetic means of sinus arrhythmia, blink rate, respiratory rate completes the features for activation of a subject. The features of the performance data are the hourly mean values.

Up to now there is not known any global deterministic or stochastic model for estimation of strain effects influenced by secondary tasks.
But, the psychophysiological knowledge achieved in other experiments or collected in literature etc. is represented in vague statements.

As an example, four vague statements are presented. They describe the relationships between performance, activation, and efficiency of a subject on an aggregated (or second) stage:

There is no change of efficiency if the variations as well activation as performance are nearly the same.

There is an opposite behaviour between the changes of efficiency and activation if the performance remains nearly the same.

Each opposite behaviour leads to extreme values of the efficiency change.

Efficiency and performance tend to the same direction if there is no activation change.

There also exist vague statements for each of the terms 'activation' and 'performance' , however, on the base of the measured variables.

It is quite obvious that a vague statement only contains information of ordinal-scale type. That is why the metric-scaled variables, like heart-rate etc. have to be transformed onto an ordinal-scaled level by means of fuzzy sets with the attributes like 'declining', 'increasing' etc. (cf. Table 1).
Now it is possible to convert a vague statement to a fuzzy conditional statement:

$$\text{IF } r_r = f_1 \vee f_2 \quad \text{AND} \quad s_s = f_6 \vee f_7 \quad \text{THEN } P = P_5,$$

with r_r reaction time, s_s human short term storage performance, P performance (f_1, ... , P_5 see Table 1).
The fuzzy set interpretation of this implication is done by the fuzzy cartesian product (Zadeh, 1973).

In doing so, fuzzy conditional statements are formulated for both the activation and performance to replace all the vague statements.

This leads to the fuzzy algorithms shown in Tables 2 and 3.

In the same way the vague statements of the second stage are considered.
A schematic depiction of the second-stage algorithm is shown in Table 4. (The arrows explain the changes).

The hierachical structure of our model is given in Fig. 5.

But there are controversal opinions among the specialists about the real influence of the second task during the experiment.
Hence, four second stage fuzzy algorithms had to be established to satisfy the diverse opinions.
These four models were applied to the same sets of data. It results in the preference of one of them.

These global models create aggregated one-dimensional fuzzy time series. The fuzzy outputs are fuzzy sets on the set of classes $\{1,2,3,4,5\}$ with its natural order.

The researcher wants to have aggregated deterministic time series with sufficient accuracy. Therefore a sharp decision rule is needed.
Because of the transformation onto an ordinal scale at the first stage of data processing (see Fig. 4) it is only possible to obtain time series on an ordinal scale. Thus the decision for the whole sample of subjects is done for each time point by using the median of the classes having the two greatest membership values.
This leads to one time series of the group.

RESULTS AND DISCUSSION

Fig. 6 shows the median of the fuzzy sets in dependence on the time intervals. These values are the results of the fuzzy algorithm called 'efficiency'. E.g., the value 5 denotes sharply increasing efficiency, and value 1 represents the most heavily declining one. The curve marked by —●— shows the median of all subjects concerning the influence of the qualifying secondary task (type I). The results of the monotonous task (type II) are demonstrated by the curve marked by —▲—. It is obvious that the two kinds of secondary tasks do not have the same influence at all time intervals. At the beginning of the experiment and after lunch break (between the fourth and the fifth hour of work), a lower efficiency is found under the influence of type I. Compared to type II, a relative and absolute improvement of efficiency occurs in the time interval between the third and the fourth hour. After the break, there are the same tendencies. A relative decline is followed by a relative improvement.

Our final conclusion is that:

> The results of fuzzy algorithms lead us to the conclusion that the fuzzy problem description is very suitable.
>
> The hypothesis, the qualifying secondary task leads to an improvement of the relationship between performance and effort, is confirmed. But it is only valid for the later time intervals, if work runs for several hours already.

This time-dependent increase in efficiency occurs only with those individuals who accept the qualification task of compensating the underload effects (Fig. 7). This conforms to questionnaire data representing higher scale values of motivation, as compared to another group of test subjects who adjust their activation level to an insufficiently demanding situation. As a result of this, their efficiency is lower in the task type I. The psychical saturation and the monotoneous scaled values obtained from questionnaires suggest that in the group in which high demands were made, an increasing efficiency is accompanied by low negative strain effects, as compared to the group with low demands.

Applicating these results to the process investigated have led to some consequences in the schedule of a normal working day in the above-mentioned monitory centre. A higher degree of availibility of the plant was obtained.

REFERENCES

Sinz, R., S. Weigel, and R. Schmitt (1978). Psychologische und psycho-physiologische Untersuchungen von Beanspruchungswirkungen bei industriellen Überwachungstätigkeiten. Symposium Optimierung kognitiver Arbeitsanforderungen, Dresden.

Straube, B., P. Richter and P.G. Richter(1978). Die Anwendung der Theorie unscharfer Mengen zur integrativen Bestimmung des Grades psychischer Beanspruchung. Informationen der TU Dresden, 22-5-78.

Zadeh, L.A. (1973). Outline of a new approach to the analysis of complex systems and decision processes. IEEE Trans. Syst., Man, and Cybern., SMC-3, 28-44.

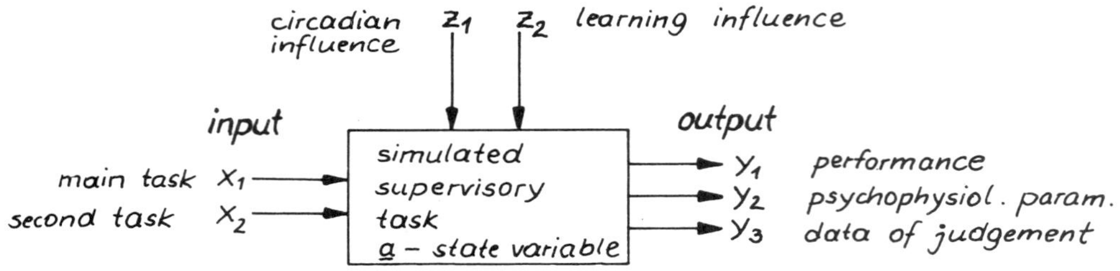

x_1 – diagnosis of simulated technical states

x_2 – two kinds (qualification task (typeI) and stupid addition task (type II))

y_1 – time for diagnosis short storage performance

y_2 – heart rate, blink rate, respiratory rate

Fig. 1 experimental design

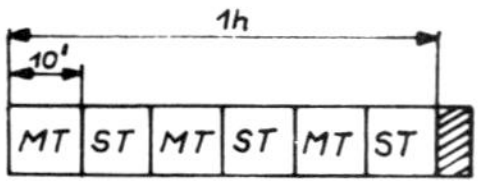

Fig. 2 structure for one hour of the eight hour experiment
MT – main task
ST – secondary task

day / subj.	1.	2.	3.	4.	day / week
1	TII	TI	TII	TI	1
2	TI	TII	TI	TII	2
3	TII	TI	TI	TII	3
4	TI	TII	TII	TI	4
5	TII	TII	TI	TI	5
6	TI	TI	TII	TII	6

Fig. 3 experimental design concerning the subjects-day-relation
TI – secondary task type I (qualification task)
TII – secondary task type II (monotonous arithmetic task)

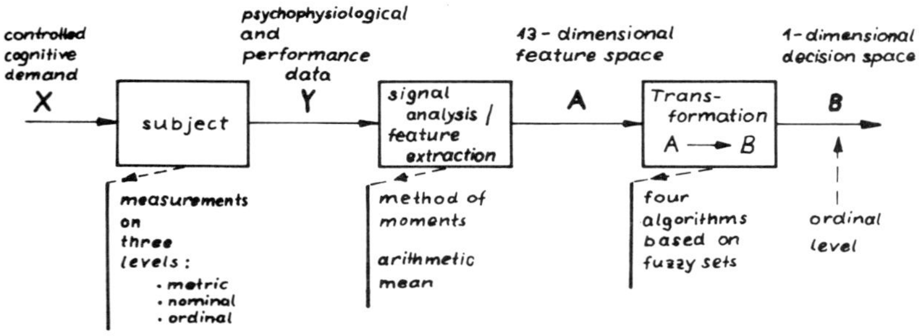

Fig. 4 Structure of data transformation

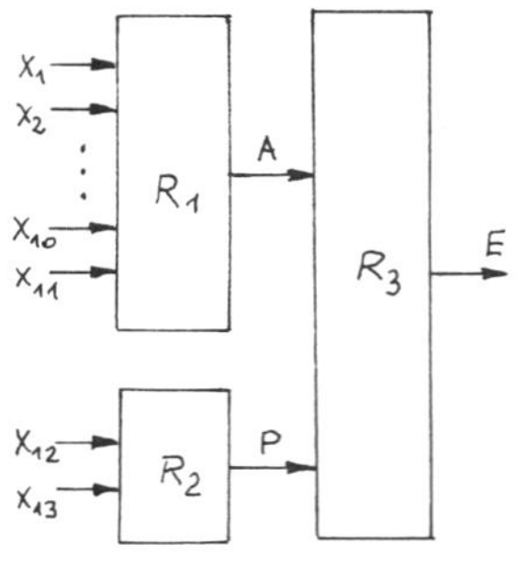

Fig. 5 Model structure

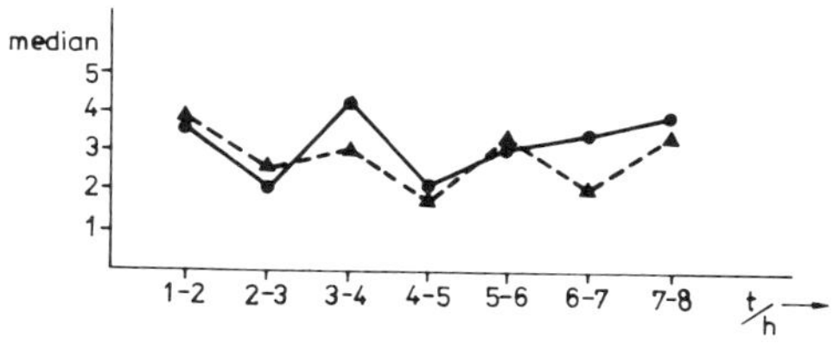

Fig. 6 Comparison: influence of both the secondary task types I ●—● and II ▲- -▲ fuzzy algorithm "efficiency"

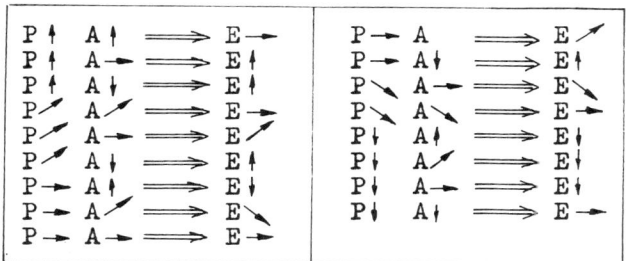

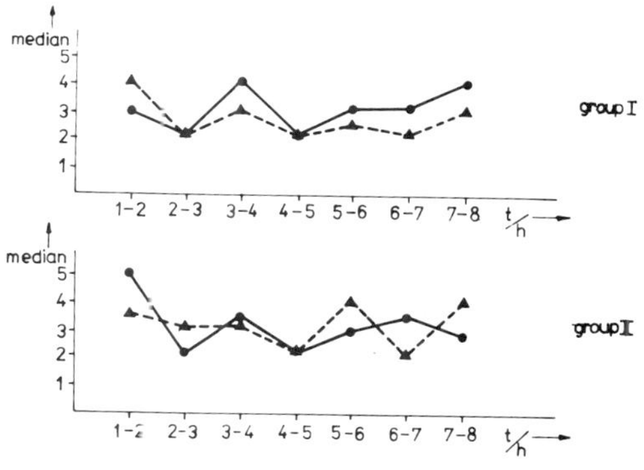

Fig. 7 Two groups of subjects: influence of both the secondary task types I ●—● and II ▲- -▲ fuzzy algorithm "efficiency"

TABLE 1 : **Linguistic variables**

f_1 , P_1 , A_1	**'heavily declining'**
f_2 , P_2 , A_2	**'decling'**
f_3	**'slightly declining'**
f_4 , P_3 , A_3	**'not changing'**
f_5	**'slightly increasing'**
f_6 , P_4 , A_4	**'increasing'**
f_7 , P_5 , A_5	**'sharply increasing'**

TABLE 2 : fuzzy 'performance' algorithm

IF $r_r=$	AND $s_s=$	THEN $P=$
$f_1 \vee f_2$	$f_7 \vee f_6$	P_5
$f_2 \vee f_3$	$f_6 \vee f_5$	P_4
$f_3 \vee f_4$	$f_5 \vee f_4$	P_3
$f_4 \vee f_5$	$f_4 \vee f_3$	P_2
$f_6 \vee f_7$	$f_2 \vee f_1$	P_1

TABLE 3 : **Fuzzy algorithm for the fuzzy notion 'activation'**

IF $h_{z1}=$	AND $h_{z2}=$	AND $h_{z3}=$	AND $s_a=$	AND $r_r=$	AND $b_r=$	AND $-M_o=$	AND $T_o=$	AND $-b_o=$	AND $D=$	AND $T_D=$	THEN A
$f_7 \vee f_6$	$f_7 \vee f_6$	$f_7 \vee f_6$	$f_1 \vee f_2$	$f_7 \vee f_6 \vee f_5$	$f_7 \vee f_6$	$f_7 \vee f_6$	$f_7 \vee f_6$	$f_7 \vee f_6$	$f_1 \vee f_2$	$f_1 \vee f_2$	A_5
$f_6 \vee f_5$	$f_6 \vee f_5$	$f_6 \vee f_5$	$f_2 \vee f_3$	$f_6 \vee f_5 \vee f_4$	$f_6 \vee f_5$	$f_6 \vee f_5$	$f_6 \vee f_5$	$f_6 \vee f_5$	$f_2 \vee f_3$	$f_2 \vee f_3$	A_4
$f_5 \vee f_4$	$f_5 \vee f_4$	$f_5 \vee f_4$	$f_3 \vee f_4$	$f_5 \vee f_4 \vee f_3$	$f_5 \vee f_4$	$f_5 \vee f_4$	$f_5 \vee f_4$	$f_5 \vee f_4$	$f_3 \vee f_4$	$f_3 \vee f_4$	A_3
$f_4 \vee f_3$	$f_4 \vee f_3$	$f_4 \vee f_3$	$f_4 \vee f_5$	$f_3 \vee f_2$	$f_4 \vee f_3$	$f_4 \vee f_3$	$f_4 \vee f_3$	$f_4 \vee f_3$	$f_4 \vee f_5$	$f_4 \vee f_5$	A_2
$f_2 \vee f_1$	$f_2 \vee f_1$	$f_2 \vee f_1$	$f_6 \vee f_7$	$f_2 \vee f_1$	$f_2 \vee f_1$	$f_2 \vee f_1$	$f_2 \vee f_1$	$f_2 \vee f_1$	$f_6 \vee f_7$	$f_6 \vee f_7$	A_1

TABLE 4

P↑	A↑	⟹	E→	P→	A	⟹	E↗
P↑	A→	⟹	E↑	P→	A↓	⟹	E↑
P↑	A↓	⟹	E↑	P↘	A→	⟹	E↘
P↗	A↗	⟹	E→	P↘	A↘	⟹	E→
P↗	A→	⟹	E↗	P↓	A↑	⟹	E↓
P↗	A↓	⟹	E↑	P↓	A↗	⟹	E↓
P→	A↑	⟹	E↓	P↓	A→	⟹	E↓
P→	A↗	⟹	E↘	P↓	A↓	⟹	E→
P→	A→	⟹	E→				

LINGUISTIC HEDGES AND REASONABLE FUZZY INFERENCES

Y. Ezawa* and M. Mizumoto**

**Faculty of Engineering, Kansai University, Suita, Osaka, 564, Japan*
***Information Science Center, Osaka Electro-Communication University Neyagawa, Osaka, 572, Japan*

Abstract. In many cases of human reasoning, the reasoning ways based on vague statements and loose concepts is often considerable. Furthermore, some vague concepts are often represented by transformations of another vague concepts. In order to make such a fuzzy inference and a transformation with fuzzy concepts, some fuzzy conditional inference rules and various linguistic hedges are proposed by Zadeh, Mamdani and Mizumoto. The relationship between the linguistic hedges and some fuzzy inference rules are reported by Mizumoto. However, the only intuitive discussion about the reasonableness for such fuzzy inference has been made. This paper introduces the concept of inference reasonableness with respect to "fuzzy modus ponens" and "fuzzy modus tollens". Then, the necessary and sufficient conditions for the reasonable fuzzy inference by the max-min composition of the fuzzy sets and the fuzzy relations is derived. Furthermore, the property of linguistic hedge which induce quite reasonable inference results for several fuzzy conditional inference rules is investigated.

Keywords. Fuzzy set; fuzzy relation; inference rule; approximate reasoning; reasonable inference; linguistic hedge; homomorphic hedge.

INTRODUCTION

Inferring an unknown fact from known facts by classical reasoning method such as syllogisms is a very interesting and attractive function of human beings, and its formal properties have been investigated by many logicians from ancient times. In the case that the known facts are loosely defined, the facts must be presented by vague statements. Then the inferred fact should also be expressed with vague concepts.

In order to formalize such a vague inference, Zadeh(1975,1979) suggested an inference rule called "compositional rule of inference with fuzzy concepts". Furthermore he introduced many linguistic hedges of translating the previously defined fuzzy concepts. The relationship between the linguistic hedges and some fuzzy inference rules are reported by Mizumoto(1979,1982). However, the only intuitive discussion about the reasonableness for such fuzzy inferences has been made.

This paper introduces the concept of reasonableness for fuzzy inference with respect to "fuzzy modus ponens" and "fuzzy modus tollens". Then, the necessary and sufficient conditions for the reasonable fuzzy inference by the max-min composition of the fuzzy sets and the fuzzy relations are derived. Furthermore, the property of linguistic hedges which induce quite reasonable inference results for several fuzzy conditional inference rules is investigated.

HOMOMORPHIC LINGUISTIC HEDGES

A simple method to deduce a fuzzy concept from another fuzzy concept is formulated as linguistic hedges by Zadeh (1972). For example, some adverbs, *very*, *more or less*, *highly*, and so forth, are introduced as linguistic hedges. In this section, we introduce formal definitions of linguistic hedges from the point of view that they are functions from a membership space to another membership space.

Def. 1. A fuzzy set A in a universe of discourse U is defined by a membership function μ_A from U into the interval [0, 1] as follows:

$$\mu_A : U \to [0, 1].$$

Def. 2. A linguistic hedge h is a function from the interval [0, 1] into the interval [0, 1].

Def. 3. For any values a_1 and a_2 in the interval [0, 1], a linguistic hedge h is called homomorphic if the semi-order relation holds such that if $a_1 \geq a_2$ then $h(a_1) \geq h(a_2)$.

Property 1. For any homomorphic linguistic hedges h_1 and h_2, there exists a homomorphic linguistic hedge h_3 such that $h_3 = h_1 h_2$ where catenation means the functional composition.

Property 2. If h_1 and h_2 are homomorphic linguistic hedges of fuzzy sets, the fuzzy set-theoretical union of h_1 and h_2 is a homomorphic linguistic hedge. And the fuzzy set-theoretical intersection is also a homomorphic linguistic hedge.

Property 3. For any homomorphic linguistic hedge h, a fuzzy set-theoretical complement (not h) is homomorphic if and only if h is a constant mapping such as $h(x) = c$ for any x in [0, 1].

Some linguistic hedges proposed by Zadeh (1972), Lakoff(1973) and Mizumoto(1982,1983) are reviewed as follows, where A is a fuzzy set in the universe of discourse U.

(a) dilation:

DIL (A) = more or less (A)

$$= \int_U \mu_A^{0.5}(u)/u.$$

(b) minus:

MINUS (A) = minus (A)

$$= \int_U \mu_A^{0.75}(u)/u.$$

(c) plus:

PLUS (A) = plus (A)

$$= \int_U \mu_A^{1.25}(u)/u.$$

(d) concentration:

CON (A) = very (A)

$$= \int_U \mu_A^{2.0}(u)/u.$$

(e) highly:

HIGH (A) = plus very (A)

$$= \int_U \mu_A^{2.5}(u)/u.$$

(f) contrast intensification:

INT (A) = int (A)

$$= \int_{\mu_A(u) \leqq 0.5} 2\,\mu_A^{2.0}(u)/u + \int_{\mu_A(u) \geqq 0.5} 1 - 2(1-\mu_A(u))^2/u.$$

(g) normalization:

NORM (A) = norm (A)

$$= \int \frac{\mu_A(u)}{\mu_A^+}/u,$$

where μ_A^+ is the maximum value of μ_A.

Using the above linguistic hedges the following linguistic hedges are obtained.

(h) slightly:

SLIGHT (A) = norm (A and not very A).

SLIGHT2(A) = int (norm (plus A and not very A)).

SLIGHT3(A) = int (SLIGHT (plus A)).

(i) sort of:

SORTOF (A) = norm (more or less A and not very very A).

SORTOF2(A) = norm (int more or less not A and int more or less A).

(j) rather:

RATHER (A) = norm (int very A and not very A).

(k) pretty:

PRETTY (A) = norm (int A and not int very A)

(l) contrast weakening:

WEAK (A) = weak (A)

$$= \int_{\mu_A(u) \leqq 0.5} 0.5\,(1-(1-2\mu_A(u))^2)/u + \int_{\mu_A(u) \geqq 0.5} 0.5\,(1+(2\mu_A(u)-1)^2)/u.$$

(m) middle intensification:

MIDI (A) = norm (A and not A).

(n) middle weakening:

MIDW (A) = norm (not MIDI (A)).

(o) α - level set:

$$\alpha\text{-}\ \mathrm{CUT}(A) = \int_{\mu_A(u) < \alpha} 0/u + \int_{\mu_A(u) \geqq \alpha} 1/u.$$

(p) scalar product:

$$\alpha\text{-}\ \mathrm{SCAL}(A) = \int_U 1 \wedge \alpha\cdot\mu_A(u)/u.$$

(q) slenderizing and swelling:

$$\alpha\text{-}\ \mathrm{SLND}(A) = \int_U 0 \vee (\alpha\,\mu_A(u) + 1 - \alpha)/u.$$

The name "slenderizing" is used in case that the parameter α is greater than one. If the parameter α is less than one, it is called "swelling" and denoted as α- SWEL(A).

The linguistic hedges (a) to (i) are proposed by Zadeh and (h) to (k) are by Lakoff, and (l) to (q) are by Mizumoto. A contrast intensification int and a contrast weakening weak are defined with quadratic formulas. In general, we can get many linguistic hedges for contrast modifications when the formula is extended from quadratic form to the form of degree α as follows:

(r) α-int (A)

$$= \int_{\mu_A(u) \leqq 0.5} 2^{\alpha-1}\,\mu_A^{\alpha}(u)/u + \int_{\mu_A(u) \geqq 0.5} 1 - 2^{\alpha-1}(1-\mu_A(u))^{\alpha}/u.$$

and

(s) α-weak (A)

$$= \int_{\mu_A(u) \leqq 0.5} 0.5\,(1-(1-2\mu_A(u))^{\alpha})/u + \int_{\mu_A(u) \geqq 0.5} 0.5\,(1+(2\mu_A(u)-1)^{\alpha})/u.$$

Example 1. 4.0-int (0.25-int A) = A.

Example 2. 4.0-weak (4.0-int A) ≠ A.

Proposition 1. The thirteen linguistic hedges from (a) to (g), (l) and from (o) to (s) are homomorphic. But the nine linguistic hedges from (h) to (k), (m) and (n) are not homomorphic.

REASONABLE FUZZY INFERENCE RULE

A fuzzy conditional inference is formulated using a compositional rule of inference by Zadeh(1975) and Mizumoto and others(1979).

Def. 4. An affirmative fuzzy inference schematic with fuzzy premises and a fuzzy consequence is formulated as follows:

Premise 1: If x is A then y is B,
Premise 2: x is C,
Consequence: y is D,

where x and y are names of objects, and A, B, C and D are fuzzy concepts which are represented by fuzzy sets in the universes of discourse U, V, U and V, respectively.

Def. 5. A refuting fuzzy inference schematic with fuzzy antecedents and a fuzzy consequence (contraposition of Def. 4) is formulated as follows:

Premise 1: If x is A then y is B,
Premise 2: y is D,
Consequence: x is C.

Def. 6. The schematic of the affirmative fuzzy inference is fuzzy modus ponens when C equals ¬ A and D equals ¬ B, which is a fuzzy extension of the classical modus ponens.

Def. 7. Two fuzzy sets A and B in the universe of discourse U are called hedge homomorphic if there exists a homomorphic linguistic hedge h such as $\mu_A(u) = h(\mu_B(u))$ for any u in U. For simplicity, it is denoted as A = h(B).

Def. 8. Given a homomorphic linguistic hedge h, the schematic of affirmative fuzzy inference is homomorphic fuzzy modus ponens when C is equal to h(A) and D is equal to h(B).

Def. 9. In the same way as in the above definitions, modus tollens may be extended to fuzzy modus tollens and homomorphic fuzzy modus tollens.

Def. 10. Let R be a fuzzy relation between the fuzzy sets A and B, and it represents a fuzzy implication "if x is A then y is B". Then, R is a fuzzy set in the universe of discourse U x V where U and V are those of A and B. Since the membership grade $\mu_R(u,v)$ of the fuzzy relation R which is a representation of the fuzzy implication depends on the membership grades of the fuzzy sets A and B that is, $\mu_A(u)$ and $\mu_B(v)$, it is denoted as follows:

$$\mu_R(u, v) = f(\mu_A(u), \mu_B(v)).$$

The relation R, the mechanism of determing the function f, is called fuzzy conditional inference rule.

Def. 11. Fuzzy consequence D is deduced from the premises (Premise 1 and Premise 2 in Def. 4) by taking the max-min composition "∘" of the fuzzy set C and the fuzzy relation R. We have

$$D = C \circ R.$$

The membership function of the consequence fuzzy set D in the universe V is given as follows:

$$\mu_D(v) = \bigvee_U \{ \mu_C(u) \wedge \mu_R(u, v) \}.$$

In case of the refuting fuzzy inference, the fuzzy consequence C is given as follows:

$$C = R \circ D,$$

and its membership function is given as

$$\mu_C(u) = \bigvee_V \{ \mu_D(v) \wedge \mu_R(u, v) \}.$$

Def. 12. Given a homomorphic linguistic hedge h, the fuzzy conditional inference rule is reasonable when the corresponding two schematics are homomorphic fuzzy modus ponens and homomorphic fuzzy modus tollens.

Theorem 1. Let R be a fuzzy relation in U x V representing the fuzzy implication "if x is A then y is B". Given a homomorphic linguistic hedge h, the affirmative fuzzy inference schematic is homomorphic fuzzy modus ponens if and only if R satisfies the following conditions:
For any u in U and v in V,

(1) if $\mu_A(u) = \mu_B(v)$
then $\mu_R(u, v) \geq \mu_{h(B)}(v)$,

(2) if $\mu_A(u) > \mu_B(v)$
then $\mu_R(u, v) \wedge \mu_{h(A)}(v) \leq \mu_{h(B)}(v)$.

(Proof) Since the fuzzy relation R is the fuzzy relation between A and B, the membership grade $\mu_R(u, v)$ is represented as a function f. The fuzzy consequence equation (Def. 11) can be partitioned into three sub expressions as follows:

$$\mu_D(v_0) = \bigvee_U \{ \mu_{h(A)}(u) \wedge \mu_R(u, v_0)\} = (a) \vee (b) \vee (c),$$

where

$$(a) = \bigvee_{\mu_A(u) < \mu_B(v_0)} \{ \mu_{h(A)}(u) \wedge f(\mu_A(u), \mu_B(v_0))\},$$

$$(b) = \bigvee_{\mu_A(u) = \mu_B(v_0)} \{ \mu_{h(A)}(u) \wedge f(\mu_A(u), \mu_B(v_0))\},$$

$$(c) = \bigvee_{\mu_A(u) > \mu_B(v_0)} \{ \mu_{h(A)}(u) \wedge f(\mu_A(u), \mu_B(v_0)) \}$$

(I) From the definition of the homomorphic linguistic hedge, i.e., for any $x \leqq y$, $h(x) \leqq h(y)$, we have

$$(a) \leqq \bigvee_{\mu_A(u) < \mu_B(v_0)} h(\mu_A(u)) \leqq h(\mu_B(v_0)).$$

(II) The second expression (b) becomes

$$(b) = \mu_{h(B)}(v_0) \wedge f(\mu_B(v_0), \mu_B(v_0))$$
$$= h(\mu_B(v_0)),$$

if and only if $h(x)$ is less than or equal to $f(x, x)$ for any x in $[0, 1]$.

(III) From the third expression (c),

$$(c) \leqq h(\mu_B(v_0)),$$

if and only if the following condition is satisfied. For any x and y in the interval [0, 1].
if $x > y$ then $h(x) \wedge f(x, y) \leqq h(y)$.

Therefore, the above three steps I, II and III conclude the proof.

Corollary 1. For any homomorphic linguistic hedge h, the affirmative inference is homomorphic fuzzy modus ponens if and only if the following conditions are satisfied.
For any u in U and v in V,
(1) if $\mu_A(u) = \mu_B(v)$ then $\mu_R(u, v) = 1$,

(2) if $\mu_A(u) > \mu_B(v)$ then $\mu_R(u, v) = 0$.

Theorem 2. Let R be a fuzzy relation representing the fuzzy implication in the universe U x V. Given a homomorphic linguistic hedge h, the refuting fuzzy inference schematic is homomorphic fuzzy modus tollens if and only if the relation R satisfies following conditions.
For any u in U and v in V,
(1) if $\mu_A(u) = \mu_B(v)$ then $\mu_R(u, v) \geqq \mu_{h(\neg A)}(u)$,

(2) if $\mu_A(u) > \mu_B(v)$ then $\mu_R(u, v) \wedge \mu_{h(\neg B)}(v) \leqq \mu_{h(\neg A)}(u)$.

The proof of this theorem is abbreviated since it is very similar to that of Theorem 1.

Corollary 2. For any homomorphic linguistic hedge h, the refuting inference is homomorphic fuzzy modus tollens if and only if the following conditions are satisfied.
For any u in U and v in V,
(1) if $\mu_A(u) = \mu_B(v)$ then $\mu_R(u, v) = 1$,

(2) if $\mu_A(u) > \mu_B(v)$ then $\mu_R(u, v) = 0$.

Theorem 3. The fuzzy conditional inference rule R is reasonable for any homomorphic linguistic hedge if and only if the following conditions are satisfied:
For any u in U and v in V,
(1) if $\mu_A(u) = \mu_B(v)$ then $\mu_R(u, v) = 1$,

(2) if $\mu_A(u) > \mu_B(v)$ then $\mu_R(u, v) = 0$.

(Proof) It is obvious from the Corollaries 1 and 2.

Corollary 3. For a constant mapping h_c, such as $h(x) = c$, the fuzzy conditional inference rule R is reasonable if and only if the maximum value of its membership value $\mu_R(u, v)$ is greater than or equal to c.

LINGUISTIC HEDGES AND REASONABLE INFERENCE RULES

This section discusses the relationship between several linguistic hedges and some fuzzy inference rules that are proposed by Zadeh(1975), Mamdani(1977) and Mizumoto and others(1979,1982).

The fuzzy conditional premise "if x is A then y is B" in the affirmative and refuting inference schematics may be represented by a certain fuzzy relation between two fuzzy sets A and B. Let A and B be fuzzy sets in the universes of discourse U and V, respectively, and let x, U, ⊕, ¬ be cartesian product, union, bounded-sum and complement, respectively.

(Rule 1) A max-min rule Rm is introduced by Zadeh as follows:

$$Rm = (A \times B) \cup (\neg A \times V),$$
$$= \int (\mu_A(u) \wedge \mu_B(v)) \vee (1 - \mu_A(u))/(u, v).$$

(Rule 2) An arithmetic rule Ra is introduced by Zadeh as follows:

$$Ra = (\neg A \times V) \oplus (U \times B),$$
$$= \int (1 - \mu_A(u) + \mu_B(v)) \wedge 1/(u, v).$$

(Rule 3) A minimum rule Rc is introduced by Mamdani as follows:

$$Rc = A \times B,$$
$$= \int \mu_A(u) \wedge \mu_B(v)/(u, v).$$

(Rule 4) A Boolean rule Rb is introduced by Mizumoto as follows:

$$Rb = (\neg A \times V) \cup (U \times B),$$
$$= \int (1 - \mu_A(u)) \vee \mu_B(v)/(u, v).$$

(Rule 5) A Gödelian rule Rg is introduced by Mizumoto as follows:

$$Rg = A \to_g B,$$
$$= \int (\mu_A(u) \to_g \mu_B(v))/(u, v),$$

where

$$\mu_A(u) \to_g \mu_B(v) = \begin{cases} 1 & \dots \mu_A(u) \leqq \mu_B(v) \\ \mu_B(v) & \dots \mu_A(u) > \mu_B(v). \end{cases}$$

(Rule 6) A standard rule Rs is introduced by Mizumoto as follows:

$$Rs = A \to_s B,$$
$$= \int (\mu_A(u) \to_s \mu_B(v)) / (u, v),$$

where

$$\mu_A(u) \to_s \mu_B(v) = \begin{cases} 1 & \dots \mu_A(u) \leqq \mu_B(v) \\ 0 & \dots \mu_A(u) > \mu_B(v). \end{cases}$$

Now, let us examine the condition of homomorphic linguistic hedges which deduce reasonable inference by the affirmative and refuting fuzzy inference schematics under these rules.

Theorem 4. A max-min rule Rm is reasonable if and only if the homomorphic linguistic hedge h is a constant mapping less than or equal to 1/2.

(Proof) The membership function of the max-min rule Rm is as follows:

$$\mu_R(u, v) = (\mu_A(u) \wedge \mu_B(v)) \vee 1 - \mu_A(u).$$

Because of the Theorem 1, the conditions that the Rm is reasonable are shown as follows:
If $\mu_A(u) = \mu_B(v)$ then

$$\mu_B(v) \vee 1 - \mu_B(v) \geqq h(\mu_B(v)). \quad \dots \quad (*1)$$

Furthermore, if $\mu_A(u) > \mu_B(v)$ then

$$(\mu_B(v) \vee 1 - \mu_A(u)) \wedge \mu_{h(A)}(u) \leqq \mu_{h(B)}(v),$$
$$(\mu_B(v) \vee 1 - \mu_A(u)) \wedge h(\mu_A(u)) \leqq h(\mu_B(v)). \quad \dots \quad (*2)$$

On the other hand, from the definition of homomorphism, $h(\mu_A(u)) \geqq h(\mu_B(v))$ if $\mu_A(u) \geqq \mu_B(v)$. So, in the case that $\mu_A(u) > \mu_B(v)$, there exists two cases that

(a) $h(\mu_A(u)) = h(\mu_B(v))$ and
(b) $h(\mu_A(u)) > h(\mu_B(v))$.

The relation (*2) holds in the former case (a). But in the latter case (b), it becomes as follows:

$$\mu_B(v) \vee 1 - \mu_A(u) \leqq h(\mu_B(v)) \quad \dots (*3)$$

Then, if $\mu_A(u)$ and $\mu_B(v)$ are both nearly equal to zero the left hand side of the relation (*3) nearly equals one and the right hand side of it nearly equals zero. Therefore, if $\mu_A(u) > \mu_B(v)$ then $h(\mu_A(u)) = h(\mu_B(v))$.

Two relations (*1) and (*3) are both true if and only if h is constant mapping less than or equal to 1/2. In this case, the conditions of the refuting fuzzy inference schematic in Theorem 2 are also satisfied.

Theorem 5. An arithmetic rule Ra is reasonable if and only if the homomorphic linguistic hedge h is a constant mapping.

Theorem 6. A minimum rule Rc is reasonable if and only if the homomorphic linguistic hedge h is a constant mapping equal to zero .

Theorem 7. A Boolean rule Rb is reasonable if and only if the homomorphic linguistic hedge h is a constant mapping less than or equal to 1/2.

Theorem 8. A Godelian rule Rg is reasonable if and only if the homomorphic linguistic hedge h is a constant mapping equal to zero .

The proofs of Theorem 5 to Theorem 8 are abbreviated because they are similar to that of Theorem 4.

Theorem 9. A standard rule Rs is reasonable for any homomorphic linguistic hedge.

(Proof) It is immediate from the Theorem 3 and definition of the rule Rg.

CONCLUSION

We have investigated the reasonableness of the fuzzy conditional inference schematics. Firstly, two criteria of reasonable fuzzy inference are introduced. Then the necessary and sufficient condition for reasonable fuzzy inference is investigated. Furthermore, the property of linguistic hedges which induce previously proposed fuzzy inference rules to be reasonable is examined.

It will be a worthy research problem to discuss the inference schematics which use new compositions other than max-min composition. These results will be presented in subsequent papers.

REFERENCES

Lakoff, G. (1973), Hedges: A study in meaning criteria and the logic of fuzzy concepts, Journal of Philosophical Logic, 2, pp. 458 - 508.

Mamdani, E. H. (1977), Application of fuzzy logic to approximate reasoning using linguistic systems. IEEE, c-26, pp. 1182 - 1191.

Mizumoto, M. (1983), Fuzzy reasoning with various fuzzy inputs, IFAC Symposium on "Fuzzy information, Knowledge representation and decision analysis", Marseille.

Mizumoto M., S. Fukami and K. Tanaka (1979), Some methods of fuzzy reasoning. In Gupta (Ed.), Advances in Fuzzy Set Theory and Application, North-Holland, pp. 117-136.

Mizumoto M. and H. J. Zimmermann (1982), Comparison of fuzzy reasoning methods. Fuzzy Sets and Systems, 8, pp. 253 - 283.

Zadeh, L. A. (1972), A fuzzy-set-theoretic interpretation of linguistic hedges, Journal of Cybernetics, 1, pp. 4 - 34.

Zadeh, L. A. (1975), Calculus of fuzzy restriction. In L.A. Zadeh (Ed.), Fuzzy Sets and Their Applications to Cognitive and Decision Processes, Academic Press, pp. 12 - 123.

Zadeh, L. A. (1979), A theory of approximate reasoning. In Horwood (Ed.), Machine Intelligence 9, Edinburgh Univ. Press, pp. 149 - 194.

BASIC OPERATIONS WITH FUZZY SETS FROM THE POINT OF FUZZY LOGIC

V. Novák* and J. Nekola**

*OKR, Automation of Control, Gregorova 3, Ostrava, Czechoslovakia
**Institute of Chemical Process Fundamentals, Czechoslovak Academy of Sciences, Prague, Czechoslovakia

Abstract. Basic operations with fuzzy sets are analyzed and some critical notes as for present time conceptions are given. Fuzzy logic on residuated lattices as the basis for the fuzzy sets theory is presented. The most important result is that it is complete. The reasons that the grades of membership should create a residuated lattice are discussed. The following basic operations with fuzzy sets are, hence, proposed: union, intersection, bald intersection and complementation. The modified compositional rule of inference is shortly discussed.

Keywords. Fuzzy logic; fuzzy set; compositional rule of inference; bald conjunction; residuated lattice.

1. INTRODUCTION

This paper reviews the basic operations with fuzzy sets though this problem has been previously discussed in many papers. The result of them is as follows (fuzzy set to be taken as a function A: U → <0, 1>): basic operations are

union $x \in A \cup B$ iff

$$(A \cup B)x = Ax \vee Bx,$$

intersection $x \in A \cap B$ iff

$$(A \cap B)x = Ax \wedge Bx,$$

complementation $x \in \bar{A}$ iff

$$\bar{A}x = 1 - x$$

where Ax, Bx are the grades of membership in the fuzzy sets A, B respectively, and $\vee$, $\wedge$ denote supremum and infimum. The operations of union and intersection have been broadly accepted. The complementation, however, is subject to some criticism since there are no fully convincing reasons for its acceptance. There exist another couples of operations which are usually presented to be alternative to the above definitions of union and intersection, namely

probabilistic sum $x \in A \;\; B$ iff

$$(A + B)x = Ax + Bx - Ax.Bx ,$$

product $x \in A \odot B$ iff

$$(A \odot B)x = Ax.Bx$$

and

bounded sum $x \in A \uplus B$ iff

$$(A \uplus B)x = 1 \wedge (Ax + Bx)$$

bold conjunction $x \in A \cap\!\!\!\!+ B$ iff

$$(A \uplus B)x = 0 \vee (Ax + Bx-1).$$

These operations are enabled because of properties of the set <0, 1>. Ambiguity in defining of the basic operations is, however, not desirable since the user can be misleaded.

This paper attempts to present tentative conclusion for the discussion on the basic operations. The conclusion is supported by the results in fuzzy logic achieved by J. Pavelka 1979.

2. FUZZY LOGIC ON RESIDUATED LATTICES

Let us recall several notions and properties of fuzzy logic on residuated lattices.

Let $\Lambda = \langle L, \vee, \wedge, \underline{1}, \underline{0}\rangle$ be a lattice with the greatest ($\underline{1}$) and smallest ($\underline{0}$) elements. Λ is called a residuated lattice if there exists a couple of binary operations $\otimes$ (product) and $\rightarrow$ (residuation) on L fulfiling the conditions:

a) $<L, \otimes, \underline{1}>$ is a commutative monoid.
b) The operation $\otimes$ is isotone.
c) The operation $\rightarrow$ is isotone in the first and antitone in the second variable.
d) Adjunction: for every α, β, γ L

$$\alpha \; x \otimes \beta \leq \gamma \quad \text{iff} \quad \alpha \leq \beta \rightarrow \gamma$$

holds.

The residuated lattice will be denoted $\Lambda = <L, \vee, \wedge, \otimes, \rightarrow, \underline{1}, \underline{O}>$. We shall further suppose that Λ is complete and infinitely distributive.

Let $L_m = \{\underline{O} = \alpha_O < \alpha_1 < \ldots < \alpha_m = 1\}$ be a finite chain. Then $\Lambda_m = <L_m, \vee, \wedge, \oplus, \rightarrow, \underline{1}, \underline{O}>$ is a residuated lattice where the operations $\oplus$, $\rightarrow$ are defined as follows:

$$\alpha_K \oplus \alpha_p = \alpha_{\max(O,\ k + p - m)} \tag{1}$$

$$\alpha_K \rightarrow \alpha_p = \alpha_{\min(m,\ m - k + p)} \tag{2}$$

Let $L = <O, 1>$. Then there can be defined two kinds of adjoint operations:

a)

$$\alpha \otimes \beta = \alpha . \beta \text{ (ordinary product of reals)}$$

$$\alpha \rightarrow \beta = \begin{cases} 1 & \text{if } \alpha \leq \beta \\ \frac{\beta}{\alpha} & \text{otherwise} \end{cases} \tag{3}$$

b)

$$\alpha \oplus \beta = O \vee (\alpha + \beta - 1) \tag{4}$$

$$\alpha \rightarrow \beta = 1 \wedge (1 - \alpha + \beta) \tag{5}$$

Theorem 1: Every residuated lattice on $<O, 1>$ whose residuation operation is continuous is isomorphic to the residuated lattice $\Lambda_I = <<O, 1>, \quad, \wedge, \oplus, \rightarrow, 1, O>$.

The language of fuzzy logic consists of
- propositional variables x, y, ...
- binary logical connectives $\underline{\vee}$, $\underline{\Delta}$, &, => called conjunction, disjunction, bald conjunction, and implication respectively,
- nullary logical connectives $\underline{\alpha}$ to all the $\alpha \in L$ called internal truth variables
- brackets (,).

Formulas:
a) The propositional variable x, y,.. ... is a formula.
b) The nullary connective $\underline{\alpha}$ is a forla.
c) If ϕ and ψ are formulas then so are $\phi \,\underline{\vee}\, \psi$, $\phi \,\underline{\Delta}\, \psi$, ϕ & ψ, $\phi \Rightarrow \psi$. We shall define also the derived connectives:
biimplication

$$\phi \Leftrightarrow \psi = (\phi \Rightarrow \psi) \wedge (\psi \Rightarrow \phi)$$

negation

$$\neg\phi = \phi \Rightarrow \underline{O}$$

Let F denote the set of all formulas. Then the truth valuation V is a fuzzy set V:F $\rightarrow$ L having the properties

$$V\underline{\alpha} = \alpha$$

$$V(\phi \,\underline{\vee}\, \psi) = V\phi \vee V\psi$$

$$V(\phi \,\underline{\Delta}\, \psi) = V\phi \wedge V\psi$$

$$V(\phi \,\&\, \psi) = V\phi \otimes V\psi$$

$$V(\phi \Rightarrow \psi) = V\phi \rightarrow V\psi$$

Let $X \subsetneq F$ be a fuzzy set of formulas. Then the semantic consequence operation is a fuzzy set

$$(\;X)\phi = \Lambda\{V\phi;\ V \text{ is a truth valuation such that } X\psi \leq V\psi \text{ for every } \psi \in F\}$$

The n-ary rule of inference r is a couple $r = <r^{syn}, r^{sem}>$ where r^{syn} is a partial n-ary operation of F and r^{sem} is n-ary operation on L preserving all the nonempty joins in L. The rule of inference r can be written

$$r: \frac{\phi_1 \ldots \phi_n}{r^{syn}(\phi_1, \ldots, \phi_n)} \left(\frac{\alpha_1 \ldots \alpha_n}{r^{sem}(\alpha_1, \ldots, \alpha_n)} \right)$$

where α_i is the truth value of the formula ϕ_i, i = 1,...,n. The rule of inference r is sound if for all the valuations V

$$Vr^{syn}(\phi_1, \ldots, \phi_n) \geq r^{sem}(V\phi_1, \ldots, V\phi_n) \tag{6}$$

holds.

Let $A \subsetneq F$ be a fuzzy set of formulas called logical axioms and R be the set of sound rules of inference. Then the couple $<A, R>$ is called fuzzy syntax.

The proof of a formula ϕ from the fuzzy set X F of nonlogical axioms is a finite sequence of formulas

$$w = <\psi_1, \psi_2, \ldots, \psi_m = \phi>$$

where each ψ_i is either logical or non-logical axiom or it is derived from n previous formulas using an n-

-ary rule of inference r R. The value of the proof is

$$\hat{w}_{(m)}X = \begin{cases} A\phi \text{ if is a logical axiom} \\ X\phi \text{ if is a non-logical axiom} \\ r^{sem}(\hat{w}_{(i_1)}X,\ldots,\hat{w}_{(i_n)}X \text{ if} \end{cases}$$

$$\phi = r^{syn}(\psi_{i_1},\ldots,\psi_{i_n})$$

provided that $\hat{w}_{i_j}$ X for j = 1,2,...,n have been determined before (in the begining they can be either $A\phi$ or $X\phi$).

The syntactic consequence operation is a fuzzy set

$$(\mathcal{C}_{A,R}X)\phi = \vee\{\hat{w}_{(m)}X;\ \ \psi_m = \phi\}$$

A fuzzy syntax $\langle A,R\rangle$ is complete if $\mathcal{C}_{A,R}X = \mathcal{C}X$ for every $X \subseteq F$.

Theorem 2: Let L = <0, 1> and let the residuation → be not continuous on L x L. Then no complete fuzzy syntax $\langle A,R\rangle$ exists on the residuated lattice $\langle\langle 0,1\rangle, \vee, \wedge, \otimes, \rightarrow, 1, 0\rangle$.

It follows from the Theorems 1 and 2, and from the fact that the residuation $\dot{\rightarrow}$ is not continuous at the point (0, 0) that the only plausible residuated lattice with L = <0, 1> to be the base for the fuzzy logic is $\Lambda_I = \langle\langle 0,1\rangle, \vee, \wedge, \oplus, \rightarrow, 1, 0\rangle$.

Theorem 3: There exists complete fuzzy syntax for the residuated lattices $\Lambda_m = \langle L_m, \vee, \wedge, \oplus, \rightarrow, \underline{1}, \underline{0}\rangle$ and $\Lambda_I = \langle\langle 0,1\rangle, \vee, \wedge, \oplus, \rightarrow 1, 0\rangle$.
For detailes see Pavelka 1979.

3. BASIC OPERATIONS WITH FUZZY SETS

It is very important that in the case of the residuated lattice L being an infinite chain, fuzzy logic is complete iff L = <0, 1>. There are good reasons for selecting <0, 1> to be the set of grades of membership. This set is:

- Natural because fuzzy sets are really the generalization of characteristic function of ordinary set. It does not predetermine any measuring units; any scale of numbers can be normalized so that its elements are elements of <0, 1>.
- Easily operable; it can be easily represented in computer memory for the case of experiments with fuzzy sets.
- Generally comprehended; a great number of various measures are set to attain values in <0, 1>.

Another important reason should be the above mentioned completness of fuzzy logic.

In the following all the fuzzy sets A, B,... $\subseteq$ U are considered to be functions U → <0, 1>. Let A be a fuzzy set in U and $x \in U$. Then $x \in A$ is a fuzzy proposition having the truth value $Ax \in \langle 0, 1\rangle$. We shall define the following abbreviations for the fuzzy propositions:

$$x \in A \cup B = x \in A \veebar x \in B \quad (7)$$

$$x \in A \cap B = x \in A \mathbin{\triangle} x \in B \quad (8)$$

$$x \in A \,\textcircled{+}\, B = x \in A \,\&\, x \in B \quad (9)$$

$$x \in A \,\textcircled{=>}\, B = x \in A \Rightarrow x \in B \quad (10)$$

Let us put for every $f, g \in \langle 0, 1\rangle^U$ ($\langle 0, 1\rangle^U$ denotes the set of all the functions U → <0, 1>)

$f \bar{\vee} g$: U → <0, 1> iff

$$(\forall x \in U)((f \bar{\vee} g)x = f(x) \vee g(x))$$

$f \bar{\wedge} g$: U → <0, 1> iff

$$(\forall x \in U)((f \quad g)x = f(x) \wedge g(x))$$

$f \bar{\oplus} g$: U → <0, 1> iff

$$(\forall x \in U)((f \bar{\oplus} g)x = f(x) \bar{\oplus} g(x))$$

$f \dot{=}\!> g$: U → <0, 1> iff

$$(\forall x \in U)((f \Rightarrow g)x = f(x) \rightarrow g(x))$$

Theorem 4: The structure $\langle\langle 0, 1\rangle^U, \bar{\vee}, \bar{\wedge}, \bar{\oplus}, \dot{=}\!>, \underline{1}, \underline{0}\rangle$ where $\underline{1}$, $\underline{0}$: U → <0, 1> are the functions $\underline{1}(x) = 1$ and $\underline{0}(x) = 0$ for every $x \in U$ is a residuated lattice.

With the help of the Theorem 4 we can find the truth valuation such that

$$[x \in A \cup B] = [x \in A] \quad [x \in B] \quad (11)$$

$$[x \in A \cap B] = [x \in A] \quad [x \in B] \quad (12)$$

$$[x \in A \,\textcircled{+}\, B] = [x \in A] + [x \in B] \quad (13)$$

$$[x \in A \,\textcircled{=>}\, B] = [x \in A] \rightarrow [x \in B] \quad (14)$$

holds for every $x \in U$, $A, B \subseteq U$ (square brackets denote truth value = grade of membership). We shall call the operations (11) - (14) the union, intersection, bald intersection and set residuation of fuzzy sets respectively. The complement of a fuzzy set $A \subseteq U$ can now be defined by $\bar{A} = \neg A = A \Rightarrow \emptyset$ with the truth value

$$[x \in \bar{A}] = [x \in A] \rightarrow 0 = 1 - [x \in A] \quad (15)$$

Writing Ax instead of $[x \in A]$ we get the ordinary formulas for the operations with fuzzy sets. Since the residuation can be replaced by negation, we can consider (11) - (13), (15) as the definitions of the <u>basic operations with fuzzy sets</u>.

Let us stop at the bald intersection. For ordinary sets this operation coincides with ordinary intersection. It appears, however, that in the fuzzy logic there should be two kinds of conjunction, and, hence, two kinds of intersection of fuzzy sets: a weak one which is computed using infimum operation, and the strong one which is computed using the bald sum operation $\oplus$.

It can be generally said that bald conjunction should be used in the case when there exists an intrinsic relation between conjuncts. For example "John is similar to Peter and is as thick as Dick" is a conjunction of two fuzzy propositions. The similarity of John and Peter includes also their thickness which is included in the second proposition. But the composite proposition brings some common information about all the three persons. Since both truth values are generally less than unit, i.e. they are uncertain, their conjunction should be more uncertain and cannot be simply equal to their infimum. It can be understood as if certain amount of the truth of one proposition were included in the second and the truth of their conjunction is reduced by it.

It is significant that there are also experimental results (Oden 1977) which support stronger rule of conjunction (e.g. ordinary product of reals) than the infimum. In Kulka, Novák it is slightly verified, too. This result can be explained in such a way that people use both kinds of conjunction dependently on their own understanding to the propositions with respect to the context.

By means of the set residuation we can define

$$A \subsetapprox_{\alpha} B \quad \text{iff} \quad A^{\alpha} \subsetapprox A \circledast B \equiv$$

$$\equiv (\forall x \in U)(\alpha \lesseqgtr Ax \rightarrow Bx) \qquad (16)$$

where A^{α} is a fuzzy set $A^{\alpha}x = \alpha$ for every $x \in U$. Note that $A \subsetapprox_{1} B$ is the usual definition of inclusion. The difference operation is:

$$B - A = B \cap\!\!\!\!\!+ \bar{A} \quad \text{iff}$$

$$(\forall x \subsetapprox U)((B - A)x = 0 \vee (Bx - Ax))$$

<u>Theorem 5</u>: Let A, B, C be fuzzy sets in U. Then

$$A \cap\!\!\!\!\!+ B = B \cap\!\!\!\!\!+ A$$

$$A \cap\!\!\!\!\!+ (B \cap\!\!\!\!\!+ C) = (A \cap\!\!\!\!\!+ B) \cap\!\!\!\!\!+ C$$

$$A \cap\!\!\!\!\!+ (B \cup C) = (A \cap\!\!\!\!\!+ B) \cup (A \cap\!\!\!\!\!+ C)$$

$$A \cap\!\!\!\!\!+ B \subsetapprox A \cap B$$

$$A \cap\!\!\!\!\!+ A \subsetapprox A \qquad A \cap\!\!\!\!\!+ U = U$$

$$A \cap\!\!\!\!\!+ \emptyset = A \qquad A \cap\!\!\!\!\!+ \bar{A} = \emptyset$$

<u>Theorem 6</u>: If $A \subsetapprox_{\alpha} \bar{A}$ for $\alpha \neq 0$ then A is subnormal and for every x U $Ax \leq 1 - \alpha/2$ holds. If $Ax \geq \beta$ for every $x \subsetapprox U$ then $\bar{A} \subsetapprox_{2\beta} A$.

4. COMPOSITIONAL RULE OF INFERENCE

Let us shortly mention the compositional rule of inference which was introduced by L.A. Zadeh and which is broadly applicated.

Originally, it was defined as follows: Let

$$B = \int_U {}^{Bx}/x \qquad C = \int_U {}^{Cx}/x$$

be fuzzy sets in U. Then the fuzzy proposition IF B THEN C (B => C) can be modelled by a fuzzy relation

$$B \Rightarrow C = \int_{U \times V} {}^{(Bx \quad Cx) \quad (1 - Bx)}/(x,y)$$

Logical rules "modus ponens" (A and A => B follow B), "modus tollens" ($\bar{B}$ and A => B follow $\bar{A}$), and logical syllogism (A => B and B => C follow A => C) can be modelled using max-min rule of composition

$$A\ (B \Rightarrow C)y = \bigvee_{x \in U} Ax \wedge (B \Rightarrow C)(x,y) \qquad (17)$$

One of the sound rules of inference defined in the section 2 is <u>modus ponens</u>

$$r_{MP}: \frac{\phi, \phi \Rightarrow \psi}{\psi} \quad \left(\frac{\alpha, \beta}{\alpha \oplus \beta}\right)$$

It suggests the following modification of the compositional rule of inference:

$$B \Rightarrow C = \int_{U \times V} {}^{1 \wedge (1 - Bx + Cy)}/(x, y)$$

$$A\ (B \Rightarrow C)y = \bigvee_{x \quad U} Ax \oplus (B \Rightarrow C)(x, y) \qquad (18)$$

Both rules of inference have been investigated by M. Mizumoto and H.J. Zimmermann (in press).

5. CONCLUSION

The aim of this paper has been to propose that bald intersection should be included among basic operations which are union, intersection, and complement. The arguments follow especially from the fuzzy logic on residuated lattices $\Lambda_m = \langle \underline{L}_m, \vee, \wedge, \oplus, \rightarrow, \underline{1}, \underline{0} \rangle$ and $\Lambda_I = \langle\langle 0, 1\rangle, \vee, \wedge, \oplus, \rightarrow, 1, 0\rangle$ which is complete and hence, should represent an appropriate basis for the fuzzy sets theory. All the operations mentioned are justified by the theoretical and experimental results. They form altogether a very transparent construction.

REFERENCES

Goguen, J. (1969). The logic of inexact concepts. Synthese, 19, 325-373.

Koczy, L., Hajnal, M. (1977). A New Attempt to Axiomatize Fuzzy Algebra with an Application Example. Problems of Contr. and Inf. Theory, 6, 47-66.

Kulka, J., Novák, V. Have Fuzzy Operations a Psychological correspondence? Studia psychologica, (to appear).

Mizumoto, M. Note on the arithmetic rule by Zadeh for fuzzy conditional inference. J. Cybernetics, (to appear).

Mizumoto, M., Zimmermann, H.J. Comparison of fuzzy reasoning methods. Fuzzy Sets and Systems, (in press).

Oden, G. (1977). Integration of Fuzzy Logical Information. J. of Exp. Psych.: Hum. Perc. and Perf., 4 565-575.

Pavelka, J. (1979). On fuzzy logic I. Zeitschr. f. math. logik und Grundlagen d. Math., 25, 45-52; 119-134; 447-464.

FUZZY MODELBUILDING FOR FORECAST TASKS OF LARGE RIVERS

B. Straube* and N. Hansel**

*Central Institute for Cybernetics and Information Processes of the Academy of Sciences of the German Democratic Republic
**Department for Hydrology and Meteorology Hydrosciences Section, Dresden Technical University, German Democratic Republic

Abstract. The development of a uniform methodology for forecasting discharge and water level of large rivers constitutes an important problem of operational hydrology. The method of fuzzy sets is suitable for modelling such large systems as a whole. For this, two fuzzy models are presented. These models setup have been successfully applied to the Elbe River in different versions for discharge and water level.

Keywords. Prediction; modelling; fuzzy models; water resources; fuzzy set theory.

INTRODUCTION

Water management mostly demands the optimal allocation of surface water for different users. This task is set more and more frequently for total large rivers (such as the Elbe). The operational management requires to forecast discharge and water level in suitable time intervals. For operational tasks, black-box methods are favoured, since they are usually quicker than other methods for these types of application. Hydrologists are interested in such methods which have regard to their available data. The following demands have to be made on these methods:

guaranteeing sufficient accuracy of forecasts,
obtining maximum period of forecast (lead time),
minimizing required expenditure.

In recent years the method of fuzzy sets has proved successful in modelling the complex hydrological character of a large river.
Fuzzy modelling allows us

to describe large rivers as a whole,
to construct the model on the basis of relatively few data,
to utilize directly the experience gained by experts and users,
to take into account topical information on the state of the system.

The forecast of discharge Q in m^3s^{-1} (controlling storages and meeting the demands made by various users) and water levels W in cm (the exeedance of bankfull levels or the stage of broadcrested weirs in polders etc.) is of prime interest. The forecast of W is important especially for very large rivers, since the technical difficulties connected with its measurement grow with the magnitude of Q.
For testing the fuzzy model in hydrological forecasting, the river system of the Elbe, including its main tributaries Mulde and Saale, was chosen. It is schematically represented in Fig. 1. This figure also explains the chosen designations W_D, W_G etc, as well as Q_D, Q_G etc. (cp. also Table 1).

FUZZY FORECAST MODELLING

The task of the forecast model is the determination of a value at the forecast station at time $t_i + T$ (lead time of forecast) by using the values measured at or before time t_i.
The relevant variables for the system behaviour have been selected from a hydrological viewpoint. Let us designate them by $X_1, X_2, \ldots, X_{10}$ and Y. Here, $X_1, \ldots, X_{10}$ represent the input sets and Y designates the output set of the models. These sets are composed of the corresponding values, the meanings of which for discharges and water levels are given in Table 1. Fig. 2 shows an example of a value triplet $(x_1,x_2,x_3) \in X_1 \times X_2 \times X_3$ of a hydrograph for W.
The connection between the input set $X=X_1 \times X_2 \times \ldots \times X_n$ and the output set

Y is given by a relation $S \subset X \times Y$. The classification property of the system can be described in the following way. For each output $y \in Y$ a set Sy of input signals exists, i.e., $Sy = \{x \in X \mid (x,y) \in S\}$ holds. The output y is assigned to this input set Sy as the class name. The relevant variables $X_1, \ldots, X_n$ may be regarded also as features of the classifier formed. Sy is also the set of all values of the relevant variables used which will lead to the same value y of the quantity to be forecast.
In forecasting, the corresponding class with $x \in Sy$ is required for each input $x=(x_1,\ldots,x_n)$ observed. With this assignment also the output signal y belonging to the input signal x is determined, i.e., thus the quantities to be forecasted are known.
In fuzzy modelling, the connection between the relevant variables is described by fuzzy relations. Therefore a fuzzy relation R on $X \times Y$ or $R : X \times Y \rightarrow [0,1]$ is required. In the case of a given fuzzy input A, the pertinent fuzzy output B with regard to the relation R will be obtained by the fuzzy composition $B=R \circ A$ with $B(y) = \bigvee_{x \in X} (R(x,y) \wedge A(x))$ for all $y \in Y$.
Non-fuzzy inputs are described by their characteristic function. With the aid of a decision rule, a non-fuzzy output y_v is to be chosen from the fuzzy forecasting B. In our investigations the following modification of a convex combination has shown good results: Let B be a fuzzy output on Y and $B(y_1) \geqslant B(y_2) \geqslant B(y)$ for all $y \in Y \setminus \{y_1,y_2\}$ (see also Fig. 3). The non-fuzzy output $y_v \in Y$ is yielded by

$$y_v = \frac{B(y_1)\cdot y_1 + a\cdot B(y_2)\cdot y_2}{B(y_1) + a\cdot B(y_2)}$$

where $a \in [0,1]$. With a = 0 the maximum decision is obtained. For a = 1 y_v is the convex combination of both the value y_1 and y_2. We used successfully

$$a = \begin{cases} 1 & \text{if } B(y_2) \geqslant 0.75\, B(y_1) \\ 0 & \text{if } B(y_2) < 0.75\, B(y_1) \end{cases}$$

that means the convex combination is applied if the fuzzy output B is not 'unique'.
The setting up or, better, the estimation of the fuzzy relation R is equivalent to the learning phase of the fuzzy model. In the following two possibilities used are given.

Model 1

We proceed from the assumption that the time series of discharge or water level or parts of them are available for the different stations. For the corresponding points of time the relevant variables are chosen from the time series. A set of input-output pairs $P = \bigcup_{i=1}^{p} \{(x,y)\}_i$ will be obtained. Typical input-output situations, as characteristic periods of flood, mean water and low water, have to be used in making up the set P.
It is assumed that the system will shown a similar behaviour in the neighbourhood of a pair $(x,i)_i$. This similarity is described by an unimodal fuzzy set N_i in such a way that its maximum lies at the point $(x,y)_i$. These N_i can be regarded as fuzzy subrelations on $X \times Y$. Their fuzzy union constitutes an estimation for the fuzzy relation $R = \bigvee_{i=1}^{p} N_i$ required.
In Fig. 3 a fuzzy relation for a single input variable and an output variable is schematically represented (n=1,p=3). In a simplified form, the calculation of the pertinent fuzzy output B is indicated also for a non-fuzzy input x_1.
Put $Z=X \times Y$ and $z=(x,y)=(x_1,x_2,\ldots,x_n,y)$. The property of the points to lie in the neighbourhood of the measured point $z_i=(x,y)_i$ is coordinatewise described by the fuzzy set N_{ij} with $N_{ij}(z_{ij})=1/(1+a_j d_j(z_j,z_{ij}))$, where $d_j(z_j,z_{ij})$ is a suitable metric and a_j determines the influence of the j-th relevant variable. By selecting suitable a_j, the frequencies and random phenomena occuring in the process can be weighted. Also an 'optimization' can be performed by means of the a_j values.
Now N_i can be represented as the fuzzy Cartesian product: $N_i = \bigtimes_{j=1}^{n+1} N_{ij}$.
This fuzzy subrelation, however, can be interpreted also as a fuzzy conditional statement; $x_1,x_2,\ldots$ and y now designating fuzzy variables. E. g., for n=2 holds

'if' x_1='nearly x_{i1}' 'and' x_2= 'nearly x_{i2}' 'then' x ='nearly y_i',

where the fuzzy notion 'nearly x_{ij}' is described by N_{ij}. With the assumption of similarity the set P can be clustered. We have this done for subsets of P. The established clusters are described by unimodal fuzzy sets. A further data reduction is achieved if local linear regression models are used (Straube and Arendt, 1978).

Model 2

In modelling complex systems there will be often possible to uncover a structure of coupled subsystems. This is an additional information for the process of modelbuilding. This fact is used in the following model. Fig. 4 shows the schematic representation of the forecast model. It consists of the five fuzzy relations $R_1, R_2, \ldots, R_5$. First, the fuzzy relation R_1 is considered. It is a fuzzy classifier. It establishes the basic shape of the hydrograph measured at the station 1; the dynamic situation of this hydrograph is summarized by the triplet $(x_1, x_2, x_3) \in X_1 \times X_2 \times X_3$ (see Fig. 2), and a set of hydrologically significant hydrograph-interval types U_1. The fuzzy classifier (fuzzy relation) R_1 is designed as a fuzzy algorithm in form of fuzzy conditional statements (Zadeh, 1973), the beginning of the algorithm being shown, for station 1, in Table 2. The pertinent fuzzy sets are given in Table 3. They are defined on a normalized set for computational reasons and therefore, they are used for all variables. The symbols F1, F2, F3 etc. are fuzzy descriptions of typical sections of the hydrographs.
The relations R_2 and R_3 are constructed in the same way. They are used for the fuzzy classification of the dynamic situation at the station 2 and 3. R_4 classifies x_{10} in a fuzzy mode into 'high' and 'low'.
The connection between the sets of hydrograph-interval types U_1 to U_3 of the individual stations, the set of heights H and the set of forecast values Y is established by the fuzzy relation R_5. The pertinent fuzzy algorithm is indicated in Table 4.

The model might be extended in two ways:

- to build in more relations or classifiers R_j with $j > 3$ (e.g., more stations) or more hierarchical levels (e.g., more complex river networks)
- to include more relevant variables in R_j.

APPLICATIONS

In model 1 there were used 512 fuzzy conditional statements which were formed by a set of input-output pairs to build-up the fuzzy relation.
In model 2 the numbers of fuzzy conditional statements which comprise the individual relations $R_1, \ldots, R_5$ are given for the model versions of Q and W in Table 5.
Due to the structured setup of the forecast model, each relation leads to a data reduction. By means of fuzziness, however, additional information on a similar behaviour of the system is always taken into account.
In the case of a practical forecast, with regard to the measured values $x_j \in X_j$ on R_1 to R_4, the respective fuzzy hydrograph-interval types are assigned. These fuzzy sets on the U_1, U_2, U_3 and H are the inputs of relation R_5 which gives the fuzzy forecast B. The forecast value $y_v \in Y$ is determined by means of the above mentioned decision rule.
In this structural approach it is possible to combine the two described modelling approaches.

Fig. 5 shows a part of a listing from a Q-forecast at the station Wittenberge by means of model 1. Fig. 6 is a detailed example of a W-forecast at the station Barby by model 2. More information about hydrological aspects of applications for model 1 see (Arendt, Straube and Hansel, 1979; Hansel and Straube, 1980) and for model 2 (Hansel, Oppermann and Straube, 1982; Hansel and Straube, 1982). In (Hansel, Oppermann and Straube, 1982) we estimate not only values y_v but short sections of forecast time series.

REFERENCES

Arendt, F., and B. Straube (1977). Zur Anwendung der Theorie unscharfer Mengen auf Vorhersage- und Steuerungsprobleme. Material zur Vorlesung 'Kennwertermittlung und Modellbildung' TH Karl-Marx-Stadt, Teil 1.

Arendt, F., B. Straube, and N. Hansel (1979). Durchflußvorhersage für Flüsse mittels unscharfer Mengen. Acta Hydrophysica, Berlin, XXIV (4), 221-240.

Hansel, N., and B. Straube (1980). Application of the method of fuzzy sets to the discharge forecast for large rivers. J. of Hydrological Sciences, 7 no 1-2, 64-76.

Hansel, N., and B. Straube (1982). Fuzzy modelling for forecasting discharge and water level of large rivers. Optimal Allocation of Water Resources (Proc. of the Exeter Symp., July 1982). IAHS Publ. no. 135.

Hansel, N., R. Oppermann, and B. Straube (1982). Vorhersage von Wasserstand und Durchfluß für die Elbe mit Hilfe einer unscharfen Modellierung. Wasserwirtschaft-Wassertechnik, 32, No. 12.

Straube, B., and F. Arendt (1978). Unscharfe Algorithmen mit lokalen linearen Regressionsmodellen. (Internal report).

Zadeh, L.A. (1973). Outline of a new approach to the analysis of complex systems and decision processes. IEEE Trans. Syst., Man and Cybern., SMC-3, 28-44.

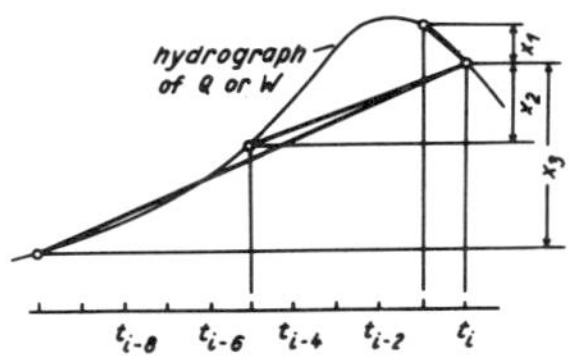

Fig. 2.

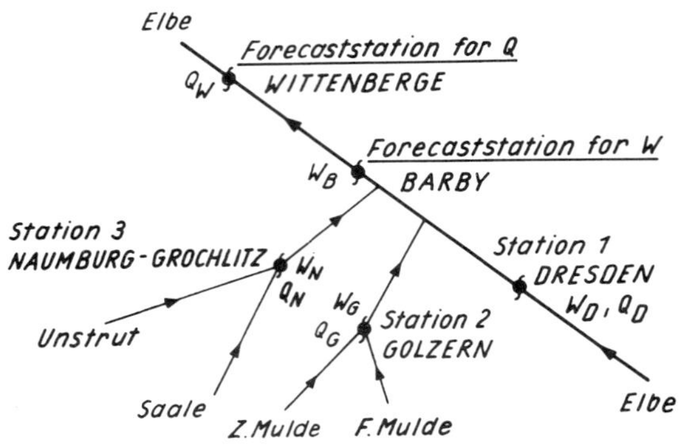

Fig. 1.

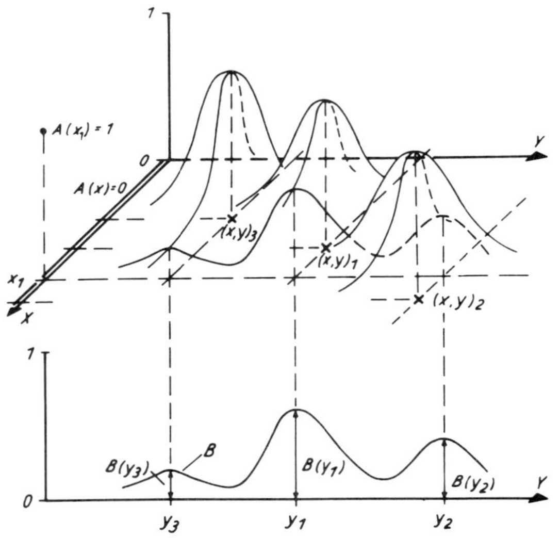

Fig. 3.

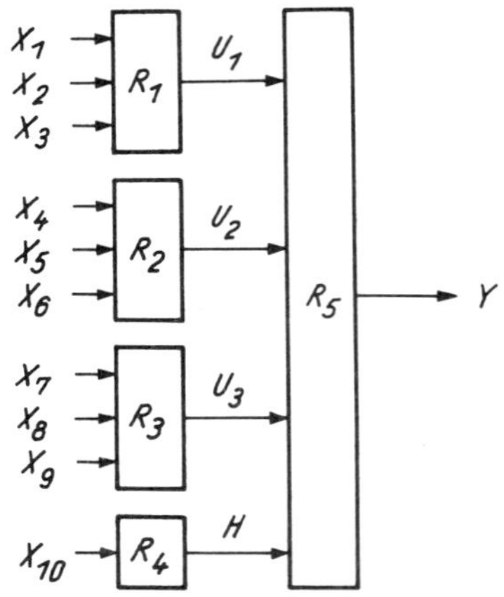

Fig. 4.

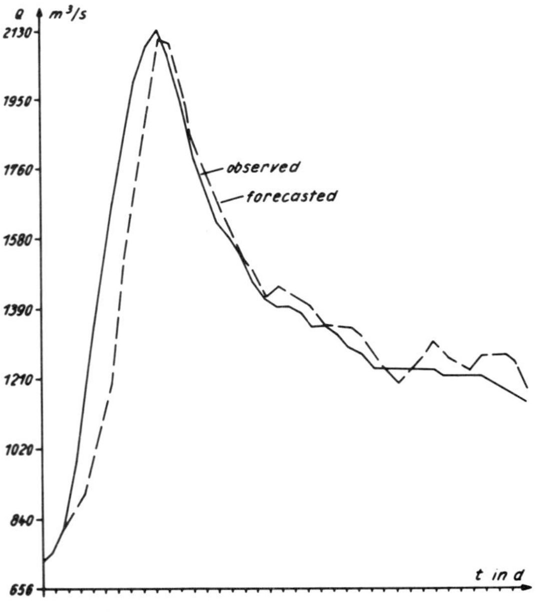

Fig. 5.

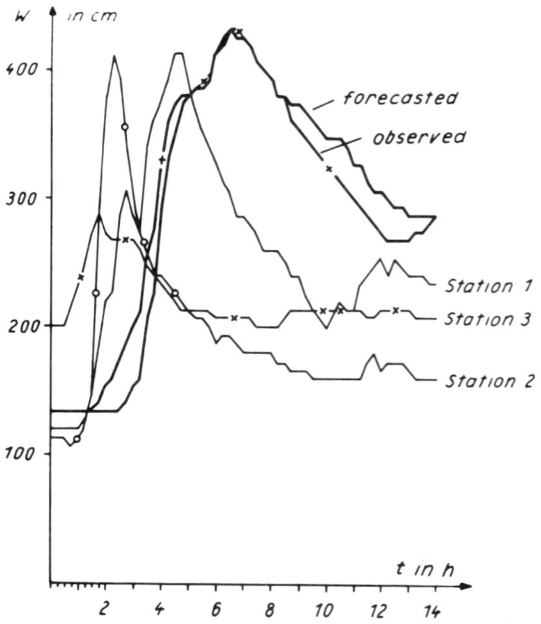

Fig. 6.

TABLE 1

	discharge t = 1d	water-level t = 6h	station
x_1	$Q_D(t_i) - Q_D(t_{i-1})$	$W_D(t_i) - W_D(t_{i-1})$	station 1
x_2	$Q_D(t_i) - Q_D(t_{i-4})$	$W_D(t_i) - W_D(t_{i-5})$	Dresden
x_3	---	$W_D(t_i) - W_D(t_{i-10})$	(Elbe)
x_4	$Q_G(t_{i+1})-Q_G(t_i)$	$W_G(t_{i+4})-W_G(t_{i+3})$	station 2
x_5	$Q_G(t_{i+1})-Q_G(t_{i-4})$	$W_G(t_{i+4})-W_G(t_i)$	Golzern
x_6	---	$W_G(t_{i+4})-W_G(t_{i-4})$	(Mulde)
x_7	$Q_N(t_i) - Q_N(t_{i-1})$	$W_N(t_i) - W_N(t_{i-1})$	station 3
x_8	$Q_N(t_i) - Q_N(t_{i-4})$	$W_N(t_i) - W_N(t_{i-4})$	Naumburg-Grochlitz
x_9	---	$W_N(t_i) - W_N(t_{i-8})$	(Saale)
x_{10}	$Q_W(t_i)$	$W_B(t_i)$	Wittenberge / Barby
y	$Q_W(t_{i+4})-Q_W(t_{i+3})$	$W_B(t_{i+9})-W_B(t_{i+8})$	Wittenberge / Barby

TABLE 2

IF x_1 = c AND x_2 = c AND x_3 = c THEN u_1 = F1
IF x_1 = wi AND x_2 = c AND x_3 = c THEN u_1 = F2
IF x_1 = i AND x_2 = wi AND x_3 = wi THEN u_1 = F3
etc.

TABLE 4

IF u_1 = F1 AND u_2 = F1 AND u_3= F1 AND h = 'high' THEN y = 0
IF u_1 = F1 AND u_2 = F1 AND u_3= F3 AND h = 'high' THEN y = 4
IF u_1 = F3 AND u_2 = F2 AND u_3= F1 AND h = 'high' THEN y = 10
IF u_1 = F4 AND u_2 = F3 AND u_3= F2 AND h = 'low' THEN y = -5
etc.

TABLE 3

si: 'strongly increasing'
i : 'increasing'
wi: 'weakly increasing'
c : 'constant'
wd: 'weakly decreasing'
d : 'decreasing'
sd: 'strongly decreasing'

TABLE 5

	Number of fuzzy conditional statements of model 2	
	Q-version	W-version
R_1	27	61
R_2	15	54
R_3	17	54
R_4	2	2
R_5	133	332

THE APPLICATION OF FUZZY SETS THEORY IN INNOVATION PROCESS MODELLING

J. Nekola* and V. Novák**

**Institute of Chemical Process Fundamentals, Czechoslovak Academy of Sciences, Praha, Czechoslovakia*
***OKR, Automation of control, k.ú.o., Ostrava, Czechoslovakia*

Abstract. The requirements for innovation process model construction are characterized with respect of uncertain factors influencing the innovations. The possibilities of modelling under uncertainty and the applications of fuzzy sets theory is analysed. The general fuzzy model of the innovation is proposed and shortly discussed.

Keywords. Innovation process, fuzzy modelling, fuzzy sets, linguistic modelling, modelling under uncertainty.

1. THE PROBLEM CHARACTERISTICS

In the recent past research of possibilities of mathematical modelling of the process of innovation when methods as parametric and nonparametric correlation analysis were used (Nekola, Vrba 1971, 1972) or the theory of production function and factor analysis was applied (Nekola, Vrba 1973 and Kratochvíl, Nekola 1979) brought many useful results. Many limitations, however, had to be formulated rooted in the rigid requirement on the quantifiability of all the parameters involved.

Results reached so far enabled to formulate basic presumptions for the innovation process model construction and for innovation management and its final effect estimation.

The principles for the construction of the model should respect the following general prerequisities:

1. The fact should be versatively respected that the process of innovation performs the combination of immense amount of factors and is so complex that for its governing neither pure verbal analysis nor deterministic programs are sufficient. The reality requires to search for operations with hardly quantifiable, undirectly or approximately expressed phenomena and to apply the approaches for apprehension of the uncertainty of all components involved.

2. External and internal conditions influencing the relation of factors of innovation process and those of socioeconomic development will never be fully recognized. This fact follows both from the character of the creation of new knowledge (R+D system) and complex relations and situations when realizing the results gained in R+D in industrial and socioeconomic systems. Construction of the model has therefore to enable step by step programming with satisfactory space for operations with uncertain factors.

3. The model has to respect the fact that the methods and algorithms applicable for the judgement of the functions and relations have to be permanently completed.

4. The model must enable to apprehend the characteristics of the final effect of innovation process in the most complete form, i.e. it has to enable the quantitative and qualitative analysis of maximum factors influencing inputs, internal functioning and output of R+D and innovation process systems. The conditions for the realization and utilization of its results in industry and other socio-economic systems should be taken into consideration.

5. The data basis corresponding to the needs of new methods has to be formed and gradually enriched. The fact has to be carefully considered that the information needed will be mostly nontraditional, i.e. not only in statistical form, principally new,

and sometimes extraordinal. From the character of the data available follows that linguistic variables are supposed to be applied in a great scale. The model has to comprehend all phases of innovation process. From the point of view of control, it is most important to recognize the real weight of particular factors and to reach their operational utilization on every level of innovation process. The general data basis, the methods and quantification procedures have to be understood in mutual accordance and unity. The preceeding publications (Nekola, Kratochvíl 1978 and Nekola, Kratochvíl 1979) describe the problematics of the orders of variables in more details.

6. The data and information objectively available are of different orders of quantifiability and certainty. In the first step the endeavour for comprehension of greatest amount of factors possible regardless the degree of their quantification leads to the models based primarily on the verbal evaluations of factors involved. If we consider at the same time the uncertainty in the estimation both of the intensity degree of factors influencing the effect and of the effect expected then it is evident that the construction of the model has to be based more on the verbal than on pure quantitative variables. The model based on the theory and technics of fuzzy sets represents one of the promising ways how to comprehend decisive factors and to estimate the real effect of innovation process.

2. GENERAL SOLUTION

The model can be characterized by a general dependence

$$Y = F(X_1,\dots,X_N) \qquad (1)$$

where X_i, $i = 1,2,\dots,N$ represent independent variables (factors), Y is dependent variable (effect) and F is the symbol of explicit dependence. We shall suppose that every variable X_i, Y, $i = 1,\dots,N$ attains values from the universes X_i, Y, $i = 1,\dots,N$. In our case, however, the values of majority of factors and the effect itself are not determined exactly but approximately only using verbal notions such as "high", "low", "average" etc. As for the dependence, we can say only that it exists and is so far intricate that it is not possible to express it exactly.

The evaluation of factors by verbal labels is dependent more or less on the level of skills, opinion, knowledge, and experiences of experts or on subjective judgements. The application of the fuzzy sets theory seems therefore to be appropriate.

Let us suppose that each factor X_i, $i = 1,\dots,N$ can be evaluated by a verbal expression H^i_j having been chosen from the scale

$$\varepsilon^i = \langle H^i_1,\dots,H^i_{J_i}\rangle$$

and similarly for the effect:

$$\varepsilon^Y = \langle H^Y_1,\dots,H^Y_{J_Y}\rangle \quad .$$

We shall further suppose that a semantic rule μ is given according to which a fuzzy set

$$H^i_j \subsetneq X_i$$

is adjoined to every evaluating verbal expression H^i_j, $i = 1,\dots,N$, $j = 1,\dots,J_i$, and similarly for the effect Y:

$$H^Y_j \subsetneq Y,\ J = 1,\dots,J_Y \quad .$$

The logical connectives AND and OR are modelled by intersection and union of fuzzy sets as usual.

Hence, for each factor X_i, Y, $i = 1,\dots,N$ there exist fuzzy relations

$$M^i \subsetneq \varepsilon^i \times X_i,\ i = 1,\dots,N$$

$$M^Y \subsetneq \varepsilon^Y \times Y$$

with membership functions

$$M^i\langle H^i_j, x_i\rangle = H^i_j x_i,\ x_i \in X_i,\ H^i_j \in \varepsilon^i$$

$$M^Y\langle H^Y_j, y\rangle = H^Y_j y,\ y \in Y,\ H^Y_j \in \varepsilon^Y \quad .$$

Each $M^i\langle H^i_j, x_i\rangle$ expresses the degree to which the element $x_i \in X_i$ corresponds to the verbal evaluation H^i_j.

The functioning of the system in a given moment can be expressed by linguistic proposition:

$$\text{IF } X_1 \text{ is } H^1_j \in \varepsilon^1 \text{ AND } \dots \text{ AND } X_N \text{ is } H^N_j \in \varepsilon_N \text{ THEN } Y \text{ is } H^Y_j \in \varepsilon^Y \qquad (2)$$

This statement can be modelled with respect to (1) using fuzzy sets as follows.

Let the factor X_i be evaluated by

$H_j^i \in \varepsilon^i$. Then there exists a set (possibly empty) $\varepsilon_j^{i,Y} = \langle H^Y_{j_{i_1}}, \ldots, H^Y_{j_i} \rangle \subset \varepsilon^Y$ of effect evaluations which is induced fixing the evaluation H_j^i of X_i. This situation can be expressed by a fuzzy statement

IF X_i is H_j^i ε^i THEN Y is $H_j^{i,Y}$ (3)

where $H_j^{i,Y}$ is the fuzzy statement

$H^Y_{j_{i_1}}$ OR $H^Y_{j_{i_2}}$ OR ... OR $H^Y_{j_{i_s}}$.

Using the semantic rule μ, the fuzzy statement $H_j^{i,Y}$ is adjoined a fuzzy set

$$H_j^{i,Y} = \bigcup_{k=j_{i_1}}^{j_{i_s}} H_k^Y$$

Hence, each evaluating scale ε^i induces the evaluating scale $\varepsilon^{i,Y} = \langle H_1^{i,Y}, \ldots, H_J^{i,Y} \rangle$ having the same number of i elements as ε^1 for each i = 1,...N. We shall define the fuzzy relation

$$M^{i,Y} \subsetneqq \varepsilon^i \times Y$$

by the membership function

$$M^{i,Y} \langle H_j^i, y \rangle = H_j^{i,Y} y \quad .$$

Each $M^{i,Y} \langle H_j^i, y \rangle$ expresses the degree to which the element y corresponds to the evaluation H_j^i through the dependence (1). The relation between $x_i \in X_i$ and $y \in Y$ with respect to the dependence (1) is now given by the relation

$$R^i = (M^i)^{-1} \circ M^{i,Y} \subsetneqq X_i \times Y \quad . \qquad (4)$$

Suppose now that each factor X_i, i = 1,...,N is evaluated by a verbal expression K_i to which a fuzzy set $K_i \subsetneqq X_i$ is adjoined. Then the final effect $K_Y \subsetneqq Y$ is given by the formula

$$K_Y = \bigcap_{i=1}^{N} (K_i \circ R^i) \quad . \qquad (5)$$

The formulae (4), (5) represent a general fuzzy modelling approach to the innovation process.

Using the linguistic approximation we can adjoin to the fuzzy set K_Y the closest verbal term $H_j^Y \in \varepsilon^Y$. The best way seems to be in computing the number

$$h_j = \sum_{y \in Y} |K_Y y - H_j^Y y| \quad .$$

Then we can adjoin a term $H_j^Y \in \varepsilon^Y$ to K_Y for which h_j is minimum. This approach is based on the procedure presented in more details in publications (Vrba, Nekola 1982).

Certain internal limitations of independent functioning of particular factors can exist in some cases. For instance in the case of innovation process, the full exploitation of top research equipment can be conditioned by skills of research workers, by the level of organization of research team work, and by other dependent factors respecting general conditions for creative and innovation activity.

The problem of conditioned and independent functioning of factors is closely connected with the possibilities of substitution. If the model is constructed correctly, the substitution will be only partial and it is rooted in the dynamics of the system.

The application of the above procedures is being verified on particular problems in chemistry, water economy, energy balances and on specific problems of innovation in electronics and electrotechnics. Some results will be mentioned in the symposium discussion.

3. CONCLUSION

The aim of this contribution is to show that fuzzy sets theory is suitable for the modelling of innovation process. The requirements for its management are ever increasing. Present time quantitative methods become rather obsolete as they do not enable to reflect uncertain factors substantially influencing the process of innovation. Linguistic variables and modelling with the help of fuzzy technics seem to represent a remarkable step forward in innovation process management.

REFERENCES

Nekola, J., and J. Vrba (1971). Study on the Problems of the Relations of R+D Potential and the factors of Socioeconomic Development. Research Paper 28, Ekonomicko-matematická laboratoř EÚ ČSAV, Praha.

Nekola, J., and J. Vrba (1972). Ekonomicko-matematický obzor, 8, 1.

Nekola, J., and J. Vrba (1973). Ekonomicko-matematický obzor, 9, 1.

Kratochvíl, P., and J. Nekola (1979). Ekonomicko-matematický obzor, 15, 3.

Nekola, J., and P. Kratochvíl (1978). Systémy ekonomicko-matematických modelů a informační systémy. Sborník ze symposia, Part I, p. 131, Praha.

Vrba, J., and J. Nekola (1982). Hodnocení efektivnosti projektů vědecko-technického rozvoje s použitím techniky fuzzy množin. Chemický průmysl, 32/57, 6.

Zadeh, L.A. (ed.) (1975). Fuzzy Sets and their Application to Cognitive and Decision Processes. Academic Press, New York Inc.

MATHEMATICAL MODELS OF MULTIFACTORIAL DECISION AND WEATHER FORECAST

Wenqian Zhang* and Yongyi Chen**

**Academy of Meteorological Science National Meteorological Bureau, Beijing, China*
***Beijing Meteorological Institute, Beijing, China*

Abstract. In this paper, we show several improved models of multifactorial decision on the foundation of the primary model, and point out the application of them to the weather forecast. Lastly, we give a realistic example of the forecast of precipitation trend of flood season in Tian Jin. For three years running (1979-1981), the results of forecast are satisfactory.

Keywords. Set theory; decision theory; multivariable systems; fuzzy transformation; fuzzy operator.

INTRODUCTION

The problem of multifactorial decision presents in various domains. As everyone knows, a great majority of things results from multifactor, not results from single factor in natural world. When people make any decision, at first, they take the effect of every factor into account respectively, and then they make a synthetic decision of all factors again. For example, when we want to buy clothes, we usually take price, durableness, style, design and colour, etc. into account; after we weigh these, we take one's choice which is satisfactory.

When we make weather forecast, we often think about the various omens; after synthesizing these, we make out a weather forecast.

A mathematical model of multifactorial decision on the bases of fuzzy set theory has been given in China[4]. It has been applied to environmental protection[6], weather forecast[7], agricultural climatic divisions[8], education[9] etc.; and well received by the men. We call it here after model I — Primary Model.

PRIMARY MODEL

Let $Y=\{y_1, y_2, \cdots, y_m\}$ denote the set of the objects, and $X=\{x_1, x_2, \cdots, x_n\}$ denote the factor set. Let R_i be the decision matrix of the ith factor,

$$R_i = (r_{i1} \;\; r_{i2} \cdots r_{im})$$

R_i is a fuzzy set on Y, but r_{ij} denotes the membership grade of the ith factor to the jth object of decision.

Obviously, the total decision matrix of all single factor is expressed as

$$R = \begin{pmatrix} R_1 \\ R_2 \\ \vdots \\ R_n \end{pmatrix} = \begin{pmatrix} r_{11} & r_{12} & \cdots & r_{1m} \\ r_{21} & r_{22} & \cdots & r_{2m} \\ \cdots & \cdots & \cdots & \cdots \\ r_{n1} & r_{n2} & \cdots & r_{nm} \end{pmatrix} \qquad (1)$$

Simultaneously, we may estabilish a matrix of weigh distribution by statistical experiment or expert evaluation etc. The matrix of weigh distribution is written as

$$A = (a_1 \;\; a_2 \cdots a_n)$$

where $a_i \geqslant 0$, and $\sum_{i=1}^{n} a_i = 1$.

We obtain primary model I by the use of composite operation of fuzzy matrix, that is

$$A \circ R = Y = (y_1 \;\; y_2 \cdots y_m) \qquad (2)$$

where

$$y_i = \bigvee_{i=1}^{n} (a_i \wedge r_{ij}) \qquad (3)$$

R is a fuzzy relation from X to Y. It determines a fuzzy transformation. In reality, a set of weigh A is transformed into a synthetic decision through fuzzy relation R.

Model I is a very simple and fundamental one. It only contains "$\wedge$" and "$\vee$" operations, so it is too rough. Hence the result is often undetermined, that is, the model actually does not have any effect.

For this reason, we introduce the extended fuzzy operation into model I.

EXTENDED MULTIFACTORIAL DECISION MODEL

Definition. A generalized fuzzy "and" operator (expression in $*$) is a mapping

$$*: \; I \times I \longrightarrow I \qquad I = [0, 1]$$

satisfying

1) Every variable is continuous and monotone increasing in I;

2) For $\forall a,b \in I$ $\quad a*b = b*a$

3) For $\forall a \in I$ $\quad a*0=0, \quad a*1=a$

4) For $\forall a,b,c \in I$ $\quad (a*b)*c = a*(b*c)$

(Note that this $*$ is a continuous triangular norm, see [2])

In addition, we define a generalized fuzzy "or" operator. All we need to do is to change $*$ to $\overset{+}{*}$, and $a*0=0$, $a*1=1$ to $a\overset{+}{*}0=a$, $a\overset{+}{*}1=1$ in the above definition, that is a generalized fuzzy "or" [5] . $*$ and $\overset{+}{*}$ are the extension of $\wedge$ and $\vee$ respectively.

Replacing $\wedge$ and $\vee$ into $*$ and $\overset{+}{*}$ respectively in primary model, we have extended multifactorial decision model:

$$A \circ R = Y = (y_1\ y_2 \cdots y_m) \qquad (4)$$

where

$$y_j = (a_1 * r_{1j}) \overset{+}{*} (a_2 * r_{2j}) \overset{+}{*} \cdots \overset{+}{*} (a_n * r_{nj}) \qquad (5)$$

it is written as $M(*,\overset{+}{*})$ in simplified form.

The primary model is a special case of $M(*,\overset{+}{*})$, it may be written as $M(\wedge,\vee)$.

We usually take a few of fuzzy operations to make up several multifactorial decision model, as follows

1. Model II

$$A \circ R = Y = (y_1\ y_2 \cdots y_m)$$

where

$$y_j = \bigvee_{i=1}^{n} a_i r_{ij} \qquad (6)$$

In formala (5) we change the "multiplication" for $*$ and "logical sum" for $\overset{+}{*}$, then we obtain Model II, it is written as $M(\cdot,\vee)$.

2. Model III

$$A \circ R = Y = (y_1\ y_2 \cdots y_m)$$

where

$$y_j = \sum_{i=1}^{n} {}^{\oplus}\, a_i \wedge r_{ij} \qquad (7)$$

After changing $\wedge$ for $*$ and $\oplus$ for $\overset{+}{*}$ in formula (5), we obtain Model III , it is written as $M(\wedge,\oplus)$.

Where operator $\oplus$ means $a \oplus b = min(1, a+b)$, and $\Sigma^{\oplus}$ expresses the sum in $\oplus$ operation.

3. Model IV

$$A \circ R = Y = (y_1\ y_2 \cdots y_m)$$

where

$$y_j = \sum_{i=1}^{n} {}^{\oplus}\, a_i r_{ij} \qquad (8)$$

After changing "multiplication" and $\oplus$ for $*$ and $\overset{+}{*}$ respectively in formula (5), we obtain Model IV, it is written as $M(\cdot,\oplus)$.

The above several models mirror the difference in decision making. we call $M(\wedge,\vee)$ after the type determining by main factor. Its result is principally determined by the main factor; it isn't affected by the another factors even variating in some region. It is applicable to this case that the result is principally determined by main factor. $M(\cdot,\vee)$ and $M(\wedge,\oplus)$ are called the conspicuous type of main factor. They are all about the same as $M(\wedge,\vee)$, but their operation are more cafeful than $M(\wedge,\vee)$. These two models are applicable to the case that $M(\wedge,\vee)$ no longer has any effect . $M(\cdot,\oplus)$ is called weighted sum type,it take all factors into consideration in accordance with weight . It is applicable to this case in consideration of all factors.

MULTISTAGE DECISION MODEL

In a complex system, we must consider many factors at a time. If we use above models over again, determining the weight distribution is difficult . Even if the weight distribution has been determined,because of $\sum_{i=1}^{n} a_i = 1$, the weights a_i are very small surely.Thus, all decision of single factor are denied easily . We can not obtain any significant result. When we meet with this situation , we may divide factor set into several parts according to their attribute. First of all,we make the decision for every part and call the first stage decision; and then we make the decision again for the above results and call the secondary decision, the rest may be deduced by analogy . At last, a multistage decision model is constituted. The steps are follows

step 1. Let factor set X be divided into N subsets in P-Partition and written as

$$X/P = \{X_1, X_2, \cdots, X_N\}$$

X_i consists of n_i factors.

$$X_i = \{x_{i1}, x_{i2}, \cdots, x_{in_i}\} \qquad i = 1,2,\cdots,N$$

$$\sum_{i=1}^{n} n_i = n$$

step 2. Make first decision for n_i factors in X_i respectively. Assuming the weight distribution of X_i is

$$A_i = (a_{i1}\ a_{i2} \cdots a_{in_i})$$

The total decision matrix of single factor in X_i is

$$R_i = \begin{pmatrix} R_{i1} \\ R_{i2} \\ \vdots \\ R_{in_i} \end{pmatrix} = \left(r_{jl}^{(i)} \right)_{n_i \times m} \qquad (9)$$

then the first decision is

$$Y_i = A_i \cdot R_i = (y_{i1}\ y_{i2} \cdots y_{im}) \qquad (10)$$

step 3. Make the secondary decision. Y_i is either the result of the first decision or the secondary decision of single factors. The secondary total decision matrix of single factors is

$$R^* = \begin{pmatrix} Y_1 \\ Y_2 \\ \vdots \\ Y_N \end{pmatrix} = (y_{ij})_{N \times m} \qquad (11)$$

The weight distribution is

$$A^* = (a_1\ a_2 \cdots a_N)$$

The secondary decision is

$$Y = A^* \cdot R^* = (y_1\ y_2 \cdots y_m)$$

$$= A^* \circ \begin{pmatrix} A_1 \circ R_1 \\ A_2 \circ R_2 \\ \vdots \\ A_N \circ R_N \end{pmatrix} \qquad (12)$$

If X/P consists excessive numbers of factor, we divide X/P again and proceed with step 2,3, and so on and so forth.

This model reflects the various attributes having an influence upon the object. Simultaneously, the disadvantage of difficult making of weight division is avoided. This model is applicable to the synthetic weather forecast of the multifactor.

A REALISTIC EXAMPLE

We select eight factors on the basis of correlation analysis for the sea surface temperature field and the precipitation trend of the flood season (June-September) in Tian Jin, 1951—1978. We have made the precipitation trend forecast of the flood season in Tian Jin for three years(1979--1981). Data are in table.

	x_1	x_2	x_3	x_4	x_5	x_6	x_7	x_8	y
1952	-16.2	-15.8	11.9	27.1	25.0	-21.4	-22.2	22.8	283
1953	-15.8	-15.5	-9.9	-25.9	-23.2	22.7	-21.9	22.2	647
1954	-15.0	-14.7	-10.7	-25.8	-23.4	23.2	23.1	23.1	731
1955	-15.7	-14.0	-10.4	-25.6	-23.3	-21.7	23.0	22.3	561
1956	-15.7	17.5	12.5	27.0	-23.7	-21.2	-22.3	22.3	467
1957	19.3	17.5	12.5	-26.0	-23.1	-20.5	-22.3	21.9	399
1958	17.4	17.2	-11.1	27.2	24.1	22.3	-22.4	-21.6	315
1959	16.7	-16.3	-9.1	27.1	24.8	22.5	22.7	22.5	521
1960	16.7	16.7	-11.7	-25.2	24.1	22.3	22.8	21.9	472
1961	16.8	17.1	-11.0	-25.7	24.6	22.0	-21.8	-21.0	536
1962	17.2	16.7	-11.6	26.9	25.5	-21.6	-22.1	-21.3	385
1963	17.0	17.4	12.2	27.1	24.5	-21.3	-21.4	-20.3	259
1964	16.7	16.5	-11.4	27.3	-23.2	-21.3	22.9	-21.8	657
1965	17.5	17.0	-11.3	27.6	-23.3	22.0	-22.0	-21.8	348
1966	16.9	-15.1	-10.9	26.7	-22.7	-21.8	23.1	22.3	644
1967	16.9	-15.9	-11.8	27.4	24.0	22.0	22.7	22.0	431
1968	16.8	17.1	13.3	27.1	24.0	-21.0	-21.5	-20.0	179
1969	16.7	-15.2	13.9	-26.1	-22.8	22.3	23.3	22.9	615
1970	16.7	16.4	13.2	-26.4	-23.0	22.1	-22.0	-21.4	433
1971	-15.9	-16.2	-11.4	27.1	24.0	-21.9	-22.2	-20.1	401
1972	16.9	16.4	12.4	27.1	24.2	-21.0	-21.9	22.0	206
1973	-16.5	-15.8	-11.7	-26.4	-23.8	22.2	23.4	22.1	639
1974	17.5	17.5	12.3	26.9	24.2	-21.8	-21.8	21.9	418
1975	16.7	-15.8	11.9	-26.0	-23.3	-21.9	22.6	22.4	570
1976	-15.6	16.7	12.5	27.5	-23.3	-21.5	23.0	-21.5	415
1977	-15.8	16.6	-11.3	-25.1	-23.6	22.6	22.5	22.2	796
1978	-16.0	-15.6	13.5	-26.6	24.0	24.2	23.0	22.2	626
1979	-16.4	-15.9	-11.4	27.4	24.9	-21.1	22.7	23.0	422
1980	17.8	17.4	-11.0	-25.7	24.6	-21.2	23.0	22.3	318
1981	17.3	17.6	12.2	26.8	24.3	22.7	-22.2	-21.8	501

where

x_1: sea surface temperature at (27) point in Jan. last year,
x_2: sea surface temperature at (25) point in Feb. last year,
x_3: sea surface temperature at (32) point in May last year,
x_4: sea surface temperature at (25) point in the last ten days of Aug. last year,
x_5: sea surface temperature at (25) point in the last ten days of Oct. last year,
x_6: sea surface temperature at (15) point in the first ten days of Dec. last year,
x_7: sea surface temperature at (9) point in the last ten days of Jan. this year,
x_8: sea surface temperature at (9) point in the second ten days of Feb. this year,
y : precipitation of flood season (June-Sept.) in Tian Jin.

The steps are,

step 1. Partition. Let the precipitation of the flood season in Tian Jin be $Y=\{y_1, y_2, y_3\}$, the norm of the partition as follows

$y \leq 300$ mm is y_1 to express "drought",
300 mm $< y < 600$ mm is y_2 to express "normol",
$y \geq 600$ mm is y_3 to express "flood".

Let $X=\{x_1, x_2, x_3, x_4, x_5, x_6, x_7, x_8\}$ be the factor set.

In order of time, we divide X into three subsets, that is

$$X = \{X_1, X_2, X_3\}$$

X_i , $i=1,2,3$ denotes the factor set of the three time intervals respectively (the first half year of last year, the last half year of last year and the first half year of this year), then

$$X_1 = \{x_1, x_2, x_3\}$$

$$X_2 = \{x_4, x_5, x_6\}$$
$$X_3 = \{x_7, x_8\}$$

This partition tells us the process of multistage decision to wit the process of stepwis amendement forecast.

step 2.Make the first forcast decision for X_1 at the end of the first time interval.

First of all, we establish the total decision matrix of single factors for X_1 according to formula (9), as

$$R_1 = \begin{pmatrix} R_{11} \\ R_{12} \\ R_{13} \end{pmatrix} = (\gamma_{j\ell}^{(1)})_{3\times3}$$

Let $(\gamma_{j\ell}^{(1)})$ denote the frequency of occurrence of the ℓth forecast object in the jth factor of the first partition.But the representative of frequency of one value "j" is no good for lack of data.Hence,We take five values around "j" factor;then assign various weight, the weight distribution takes (0.2 0.5 1 0.5 0.2). We use five weighted frequency sum (normalization) to represent $\gamma_{j\ell}^{(1)}$.

For example: 1981 forecast.

In the first partition, the first factor is x_1=17.3. The five values are 17.1, 17.2, 17.3, 17.4 and 17.5 around x_1 .We calculate their weighted frequency sum(normalization), that is

$$R_{11} = (0\ 1\ 0)$$

We may use the same way to calculate R_{12} and R_{13} .

We obtain the total decision matrix of single factors for X_1 , to be

$$R = \begin{pmatrix} 0 & 1 & 0 \\ 0.12 & 0.88 & 0 \\ 0.70 & 0.30 & 0 \end{pmatrix}$$

Secondly,establish the weight distribution matrix of X_1,to be

$$A_1 = (a_{11}\ a_{12}\ a_{13})$$

We define the absolute value of the correlation coefficint of sea surface temperature and precipitation of flood season at the rate of percentageas weight. For example, for the first partition, the absolute value of the correlation coefficient of x_1 and y

$$|\gamma_1| = 0.38$$

the absolute value of the correlation coefficient of x_2 and y

$$|\gamma_2| = 0.43$$

the absolute value of the correlation coefficient of x_3 and y

$$|\gamma_3| = 0.41$$
$$|\gamma_1| + |\gamma_2| + |\gamma_3| = 1.22$$

We obtain

$$A_1 = (0.38/1.22\ \ 0.43/1.22\ \ 0.41/1.22) = (0.31\ 0.35\ 0.34)$$

Lastly, calculate the first forecast decision. From formula (10), we obtain

$$Y_1 = A_1 \circ R_1 = (0.31\ 0.35\ 0.34) \cdot \begin{pmatrix} 0 & 1 & 0 \\ 0.12 & 0.88 & 0 \\ 0.70 & 0.30 & 0 \end{pmatrix}$$

by $M(\wedge, \vee)$, the result is

$$Y_1 = (0.49\ 0.51\ 0)$$

It may be seen that by $M(\wedge, \vee)$, the precipitation of the flood season in 1981 should belong to y_2 (normal),but we are not quite sure.

We make the supplementary forcast again by means of the other model. The results are

$$M(\cdot, \vee): Y_1 = (0.44\ 0.56\ 0)$$
$$M(\wedge, \oplus): Y_1 = (0.32\ 0.68\ 0)$$
$$M(\cdot, \oplus): Y_1 = (0.26\ 0.74\ 0)$$

From the supplementary forecast, we say that the precitation of flood season in 1981 belongs to "normal" with assurance.

step 3. Make the secondary forecast decision at the end of the second time interval (i.e. amendement forecast).

In the first place, we make the first forecast decition of X_2 by the same way as step 2. we obtain

$$Y_2 = A_2 \circ R_2 = (0.38\ 0.36\ 0.26) \cdot \begin{pmatrix} 0 & 0.86 & 0.14 \\ 0.44 & 0.56 & 0 \\ 0 & 0.12 & 0.88 \end{pmatrix}$$

According to above models, the results are

$$M(\wedge, \vee): Y_2 = (0.36\ 0.38\ 0.26)$$
$$M(\cdot, \vee): Y_2 = (0.22\ 0.46\ 0.32)$$
$$M(\wedge, \oplus): Y_2 = (0.22\ 0.53\ 0.25)$$
$$M(\cdot, \oplus): Y_2 = (0.16\ 0.56\ 0.28)$$

In the second place, we make the secondary forecast decision for Y_1 and Y_2 from formulae (11), (12), as

$$Y^* = A^* \circ \begin{pmatrix} Y_1 \\ Y_2 \end{pmatrix}$$

where

$$A^* = \left(\sum_{i=1}^{3} |\gamma_i| / \sum_{i=1}^{6} |\gamma_i| \quad \sum_{i=4}^{6} |\gamma_i| / \sum_{i=1}^{6} |\gamma_i| \right)$$

through calculation, we obtain

$$A^* = (0.44\ 0.56)$$

By the above models,the secondary forecast decision are

$M(\wedge,\vee)$: $Y^*=(0.39\ 0.39\ 0.22)$
$M(\cdot,\vee)$: $Y^*=(0.30\ 0.41\ 0.29)$
$M(\wedge,\oplus)$: $Y^*=(0.31\ 0.55\ 0.14)$
$M(\cdot,\oplus)$: $Y^*=(0.20\ 0.64\ 0.16)$

From above, it may be seen that $M(\wedge,\vee)$ has lost effect, but the other models are distingishable. The results of the above three models are the same,so the secondary forecast decision is still "normal".

step 4.Make the third (i.e. last) forecast decision at the end of the third time interval.

First of all, make the first forecast decision of X_3 by the same way. We obtain

$$Y_3 = A_3 \circ R_3 = (0.62\ 0.38) \circ \begin{pmatrix} 0.25 & 0.75 & 0 \\ 0 & 0.55 & 0.45 \end{pmatrix}$$

For all models, the results are

$M(\wedge,\vee)$: $Y_3=(0.20\ 0.50\ 0.30)$
$M(\cdot,\vee)$: $Y_3=(0.20\ 0.59\ 0.21)$
$M(\wedge,\oplus)$: $Y_3=(0.15\ 0.62\ 0.23)$
$M(\cdot,\oplus)$: $Y_3=(0.16\ 0.67\ 0.17)$

After that , we make the last forecast decision for Y^* and Y_3 , as

$$Y^{**} = A^{**} \cdot \begin{pmatrix} Y^* \\ Y_3 \end{pmatrix}$$

where

$$A^{**} = (0.62\ 0.38)$$

The last forecast decisions are

$M(\wedge,\vee)$: $Y^{**}=(0.37\ 0.37\ 0.26)$
$M(\cdot,\vee)$: $Y^{**}=(0.30\ 0.41\ 0.29)$
$M(\wedge,\oplus)$: $Y^{**}=(0.26\ 0.53\ 0.21)$
$M(\cdot,\oplus)$: $Y^{**}=(0.18\ 0.65\ 0.17)$

From above, we may seen that,$M(\wedge,\vee)$ still losses effect. According to the decisions of the other three models, the last forecast decision should be "normal". It is identical with the real value.

In addition, we also make the precipitation trend forecasts of the flood season in Tian Jin, 1978 and 1980. As for 1979, the first forecast decision is "flood",after thorough amendement,the secondary forecast decision is "normal", the last forecast decision is still "normal".For 1980,the forecast decisions of every stage are all "normal". The results of this two years are identical with the real values.

In a word, the results are satisfactory for three years running (1979—1981).Especially, the multistage decision model $M(\cdot,\oplus)$ is better for the synthetic forecast of the multifactor.

We sincere thanks are also due to Professor Peizhuang Wang of Beijing Normal University.

REFERENCES

[1] Zadeh,L.A. (1965). Fuzzy sets. Inf. Control, 8, 338-353.
[2] Sohweizer,B., and Sklar,A. (1963). Associative functions and abstract semi-groups. Publ.Math.Debrecen, 10, 69-81.
[3] Capocelli,R.M., and De Luca,A. (1973). Fuzzy sets and decision theory. Inf. Control, 23, 446-473.
[4] Peizhuang Wang. Introduction of fuzzy mathematics. Practice and Knowledge on Mathematics, 2,3,1980 (China).
[5] Yongyi Chen. An approach to fuzzy operators, BUSEFAL, 9,1982, 59-65.
[6] Rong Yue, Jian Jin, (1981). Accounting Synthetic Evaluation Index of Water Circumstance with Fuzzy Set Theory. Study of Water Circumstance, 2,(China).
[7] Youkun Kong, (1981). Fuzzy Transformation and Synthetic Evalution, The Collection of Applications of Fuzzy Math. in Meteorology, 67-79,(China).
[8] Suhua Gao, (1982). Primary Applications of Fuzzy Math. in Agricultural Meteorology, The Collection of Meterological Science and Technology (in Agricultural Meteorology), 3,(China).
[9] Han Wu, (1982). The Questions of Synthetic Evalution in Teaching Process, Fuzzy Math. 1,(China).

FUZZY RELATIONAL EQUATIONS WITH TRIANGULAR NORMS IN MODELLING OF DECISION-MAKING PROCESSES

W. Pedrycz

Department of Automatic Control and Computer Science Silesian Technical University, 44-100 Gliwice Poland

Abstract. The paper deals with fuzzy relational equations with triangular norms and their application in modelling of decision-making processes. Generally speaking, in any decision-making problem we face with a set of constraints and a set of goals with various grades of importance, very often conflicting in their nature. A final decision D which is made upon on available information can be formally written as a result of composition of constraints C_i, i=1,2,...,n, and goals G_j, j=1,2,...,m by means of a relation R expressing the relationship between goals, constraints and decisions. We shall propose a fuzzy relational equation as a model of the process of decision-making. Analytical and numerical methods of solving this class of equations are provided and discussed in detail.

Keywords. Decision theory; fuzzy set theory; fuzzy relational equations.

INTRODUCTION

The problem of description and analysis of decision-making processes in a presence of a large variety of goals and constraints, sometimes conflicting in their nature, has been attacked by means of diverse mathematical techniques and a large group of various methods was established. While dealing with a human being playing a significant role in any decision-making process, we have to underline a need of a clear understanding the mechanism involved in such an activity and a form of uncertainty existing in the decision-aking process. Very often this form of uncertainty cannot be satisfactorily modelled in a stochastic way (in a real life nobody can repeat his decision-making process many times, which implies you cannot collect a large set of data needed for an appropiate application of the probabilistic methods). Fuzzy set theory (Zadeh, 1976) creates a natural and an acceptable tool for handling this form of uncertainty, especially in situations where the goals and constraints are rather fuzzy than defined in a precise way. Thus there is no doubt that this attractive tool was widely discussed. Starting from the fundamental paper of Bellman and Zadeh (1970) we can enumerate a sequence of works devoted to the problems of fuzzy sets in decision-making (Carlsson, 1982; Czogała, Pedrycz, 1981; Kickert, 1978; Pedrycz, 1982; Yager, 1977). A lot of interesting algorithms were established and discussed for different decision-making situations (Dubois, Prade, 1980). Unfortunatelly there is only a relatively small number of works oriented towards experimental verification of the assumptions, which are usually made almost in every paper on fuzzy decision-making; for instance "and" is modelled by an intersection of fuzzy sets, while "or" is treated as their union. Underlying the fact of the local properties of fuzzy logic (Bellman, Zadeh, 1977) we should look for new models for "and" and "or" connectives existing in the decision-making process (Hamacher, 1976; Pedrycz; 1982; Rödder, 1975).

In this paper we shall present a class of fuzzy relational equations with triangular norms, presenting their analytical and numerical solutions and discussing the models of decision-making with the help of the equations introduced. The paper is structured as follows. In Section 2 analytical and numerical methods of solving fuzzy relational equations are given. The model of the decision-making process formulated on the basis of this class of equations is discussed in Section 3. The numerical example is given in Section 4. Concluding remarks are contained in Section 5.

FUZZY RELATIONAL EQUATIONS WITH TRIANGULAR NORMS

Triangular norms, t-, and s-norms, which play a significant role in further investigations, are defined as follows (Frank, 1979; Menger, 1942; Pedrycz, 1982)

Def.1.

A t-norm is a function of two variables

$$t: [0,1]\times[0,1] \rightarrow [0,1] \quad (1)$$

such that

(i) $0tx=0, 1tx=x$ (2)

(ii) $xty \le zty$ if $x \le z$ (3)

(iii) $xty=ytx$ (4)

(iv) $(xty)tz=xt(ytz)$ (5)

$x,y,z \in [0,1]$.

Def.2.

A s-norm is a function of two variables

$$s: [0,1]\times[0,1] \rightarrow [0,1] \quad (6)$$

such that

(i) $0sx=x, 1sx=1$ (7)

(ii) $xsy \le zsy$ if $x \le z$ (8)

(iii) $xsy=ysx$ (9)

(iv) $(xsy)sz=xs(ysz)$ (10)

$x,y,z \in [0,1]$. For each t-norm defined as above the function expressed as

$$xfy=1-(1-x)\,t\,(1-y), \quad (11)$$

$x,y \in [0,1]$ is a s-norm.

Some of t- and s-norms are listed below,

$$xt_1y=1-\min(1,((1-x)^p+(1-y)^p)^{1/p}) \quad p\geq 1 \quad (12)$$

$$xt_2y=xy \quad (13)$$

$$xt_3y=xy/(\gamma+(1-\gamma)(x+y-xy)) \quad \gamma\geq 0 \quad (14)$$

and

$$xs_1y=\min(1,(x^p+y^p)^{1/p}), p\geq 1 \quad (15)$$

$$xs_2y=x+y-xy \quad (16)$$

$$xs_3y=(xy(\gamma-2)+x+y)/(xy(\gamma-1)+1) \quad \gamma\geq 0 \quad (17)$$

$x,y \in [0,1]$. The values of some triangular norms depend on parameters e.g. p,γ. The triangular norms can be treated as the models for the intersection of fuzzy sets t-norms and the union s-norms. The choice of the value of the parameter standing in the formula of the triangular norm provides an opportunity to fit the operator to the collected empirical data. Moreover the following holds,

$$\lim_{p\to\infty}(xt_1y)=\min(x,y), \quad (18)$$

$$\lim_{p\to\infty}(xs_1y)=\max(x,y), \quad (19)$$

so the classical connectives are contained in a class of the triangular norms.

SOLUTION OF THE FUZZY RELATIONAL EQUATIONS Y=X∘R AND DUAL EQUATIONS

We consider the fuzzy relational equation of the type,

$$Y = X \square R \tag{20}$$

and a dual fuzzy relational equation

$$Y = X \Delta R \tag{21}$$

where X, Y, R are fuzzy sets defined on the spaces $\mathbb{X}$ and $\mathbb{Y}$ and a fuzzy relation defined on the cartesian product $\mathbb{X}\times\mathbb{Y}$. We shall adopt a notation $X\in F(\mathbb{X})$, $Y\in F(\mathbb{Y})$, $R\in F(\mathbb{X}\times\mathbb{Y})$. "$\square$" and "$\Delta$" are sup-t and inf-s composition operators. Thus formulas (20) -(21) are read as follows,

$$Y(y) = \sup_{x\in\mathbb{X}}[X(x)\ t\ R(x,y)], \tag{22}$$

$$Y(y) = \inf_{x\in\mathbb{X}}[X(x)\ s\ R(x,y)], \tag{23}$$

$y\in\mathbb{Y}$. The problem of solving fuzzy relational equations as given by (20)-(21) can be formulated as follows,

- X, Y are given, determine R,
- Y, R are given, determine X.

The solution of the abovestated problems can be found elsewhere (Pedrycz, 1982a; Pedrycz 1982; Pedrycz, 1983). We shall remember the main results

Theorem 1.

(i) If $X\in F(\mathbb{X})$, $Y\in F(\mathbb{Y})$ satisfy (20), then the greatest fuzzy relation $\hat{R}\in F(\mathbb{X}\times\mathbb{Y})$ such that $X\square\hat{R} = Y$ holds, is equal to

$$\hat{R} = X \circledcirc\!\!\!\!\varphi\ Y \tag{24}$$

$$\hat{R}(x,y) = (X \circledcirc\!\!\!\!\varphi\ Y)(x,y) = X(x)\,\varphi\, Y(y) \tag{25}$$

(ii) If $Y\in F(\mathbb{Y})$ and $R\in F(\mathbb{X}\times\mathbb{Y})$ satisfy (20), then the greatest fuzzy set $\hat{X}\in F(\mathbb{X})$, such that $\hat{X}\square R = Y$, is equal to

$$\hat{X} = R \circledcirc\!\!\!\!\varphi\ Y \tag{25}$$

$$\hat{X}(x) = (R \circledcirc\!\!\!\!\varphi\ Y)(x) = \inf_{y\in\mathbb{Y}}[R(x,y)\,\varphi\, Y(y)]$$

where φ is a function

$$\varphi : [0,1]\times[0,1] \to [0,1]$$

such that

$$x\,\varphi\,\max(y,z) \geq \max(x\varphi y, x\varphi z) \tag{26}$$

$$x\,t\,(x\varphi y) \leq y \tag{27}$$

$$x\varphi(x t y) \geq y, \tag{28}$$

$x, y, z \in [0,1]$.

Theorem 2.

(i) If $X\in F(\mathbb{X})$, $Y\in F(\mathbb{Y})$ satisfy (21), then the least fuzzy relation $\check{R}\in F(\mathbb{X}\times\mathbb{Y})$ such that $X\Delta\check{R} = Y$ holds, is equal to

$$\check{R} = X \circledcirc\!\!\!\!\beta\ Y, \tag{29}$$

$$\check{R}(x,y) = (X \circledcirc\!\!\!\!\beta\ Y)(x,y) = X(x)\,\beta\, Y(y) \tag{30}$$

(ii) If $Y\in F(\mathbb{Y})$ and $R\in F(\mathbb{X}\times\mathbb{Y})$ satisfy (21) then the least fuzzy set $\check{X}\in F(\mathbb{X})$ ($\check{X}\Delta R = Y$) is equal to

$$\check{X} = R \circledcirc\!\!\!\!\beta\ Y, \tag{31}$$

$$\check{X}(x) = (R \circledcirc\!\!\!\!\beta\ Y)(x) = \sup_{y\in\mathbb{Y}}[R(x,y)\,\beta\, Y(y)] \tag{32}$$

$x\in\mathbb{X}$, where β is a function

$$\beta : [0,1]\times[0,1] \to [0,1] \tag{33}$$

such that

$$x\beta\min(y,z) \leq \min(x\beta y, x\beta z) \tag{34}$$

$$x s (x\beta y) \geq y \tag{35}$$

$$x\beta(x s y) \leq y \tag{36}$$

$x, y, z \in [0,1]$. These results could be extended for the fuzzy relational equations of the type,

$$Y = (X_1 \times X_2 \times \ldots \times X_n) \square R \tag{36}$$

$$Y(y) = \sup[X_1(x_1)\ t\ X_2(x_2)\ t \ldots t X_n(x_n)\ t\ R(x_1, x_2, \ldots, x_n, y)] \tag{37}$$

(supremum is taken over all $x_i \in \mathbb{X}_i$, $i=1,2,\ldots,n$), and dual fuzzy relational equations. Now we get

Theorem 3.

(i) If $X_i\in F(\mathbb{X}_i)$ and $Y\in F(\mathbb{Y})$ fulfil (36) then the greatest fuzzy relation $\hat{R}\in F(\mathbb{X}_1\times\mathbb{X}_2\times\ldots\times\mathbb{X}_n\times\mathbb{Y})$ satisfying (36) is equal to

$$\hat{R} = (\mathop{T}_{i=1}^{n} X_i) \circledcirc\!\!\!\!\varphi\ Y \tag{38}$$

$\mathop{T}_{i=1}^{n} X_i$ is the fuzzy relation (cartesian product with respect to the t-norm).

(ii) If $X_i\in F(\mathbb{X}_i)$ and $Y\in F(\mathbb{Y})$ fulfil dual fuzzy relational equation then the least fuzzy relation $\check{R}\in F(\mathbb{X}_1\times\mathbb{X}_2\times\ldots\times\mathbb{X}_n\times\mathbb{Y})$ which fulfils dual equation, is equal to

$$\check{R} = (\mathop{S}_{i=1}^{n} X_i) \circledcirc\!\!\!\!\beta\ Y, \tag{39}$$

$\mathop{S}_{i=1}^{n} X_i$ is the fuzzy relation (cartesian product with respect to the s-norm).

Note that the results presented here form a generalisation of the results derived for sup-min and inf-max fuzzy relational equations (Sanchez, 1976).

The problem of the determination of the fuzzy sets (relations) standing in the equations and discussed above can be also solved in a numerical manner. Consider the case of determination of R for X and Y given. We get the following task,

-for X,Y given find such a relation R which optimizes minimizes a performance index

$$Q = \sum_{j=1}^{m} \{Y(y_j) - \max_{1\le i\le n}[X(x_i)\ t\ R(x_i,y_j)]\}^2 \qquad (40)$$

we have assumed the universes of discourse $\mathbb{X}$ and $\mathbb{Y}$ are finite, card($\mathbb{X}$)= n, card($\mathbb{Y}$)= m. We propose the following iteration scheme,

$$R^{(K+1)}(x_s,y_t) = R^{(K)}(x_s,y_t) - a_K\, \partial Q/\partial R \qquad (41)$$

$s=1,2,\dots,n$, $t=1,2,\dots,m$. K denotes the number of iteration, $K=0,1,\dots$ a_K is a nonincreasing sequence of numbers chosen in such a way to avoid oscillations and to assure a reasonable convergence rate. Good results can be obtained putting down a_K equal to

$$a_K = 1/(K^w + b)\ , \qquad (42)$$

$w \geq 0$,and $b = \sup_{x,y\in[0,1]} |\partial/\partial x(xty)|$

THE MODEL OF DECISION-MAKING PROCESS

Now we present an approach to those decision-making processes that can be represented using fuzzy relational equations. Usually in any decision-making process in a fuzzy environment we deal with a collection of goals G_i, $i=1,2,\dots,n$, and constraints C_j, $j=1,2,\dots,m$ expressed as fuzzy sets on the respective spaces of goals $\mathbb{G}_i$ and constraints $\mathbb{C}_j$ viz. $G_i \in F(\mathbb{G}_i)$, $C_j \in F(\mathbb{C}_j)$. A decision D is expressed as a fuzzy set on the space $\mathbb{D}$ and resulting from the aggregation of all the goals and constraints,

$$D = f(G_1,G_2,\dots,G_n,C_1,C_2,\dots,C_m) \qquad (43)$$

where "f" denotes the operation of aggregating of the G_i's and C_j's . If all the goals and constraints are defined on the same space $\mathbb{G}_i = \mathbb{C}_j = \mathbb{X}$ the fuzzy decision is calculated as (Bellman, Zadeh, 1970) ,

$$D(x) = \min[G_1(x), G_2(x),\dots,G_n(x), C_1(x), C_2(x),\dots,C_m(x)] \qquad (44)$$

or

$$D(x) = \prod_{i=1}^{n} G_i(x) \prod_{j=1}^{m} C_j(x) \qquad (45)$$

$x\in\mathbb{X}$. If the space of goals and constraints are not the same, then the determination of the decision can be performed via the fuzzy relational equation of the type,

$$D = (G_1 \times G_2 \times \dots \times G_n \times C_1 \times C_2 \times \dots \times C_m) \circ R \qquad (46)$$

where the cartesian product is treated in sense of the triangular norm. Note that the model presented by Bellman and Zadeh is embedded in the scheme provided above (see (46)). If goals and constraints are defined on the same space $\mathbb{X}$, the relation R is a diagonal one defined on the cartesian product $\underbrace{\mathbb{X}\times\mathbb{X}\times\dots\times\mathbb{X}}_{n+m+1\text{ -times}}$ with the membership function

$$R(x_1,x_2,\dots,x_{n+m+1}) = \begin{cases} 1, & \text{if } x_1=x_2=\dots=x_{n+m+1} \\ 0, & \text{otherwise} \end{cases} \qquad (47)$$

$x_i \in \mathbb{X}$, $Ii=1,2,\dots,n+m+1$ and the triangular norm is the min operator, we have,

$D(x_{n+m+1}) = \sup[\min(G_1(x_1), G_2(x_2),\dots,$

$$\ldots G_n(x), C_1(x_{n+1}), C_2(x_{n+2}), \ldots, C_m(x_{n+m}), R(x_1, x_2, \ldots, x_{n+m+1})] \quad (48)$$

where sup is taken over all $x_i \in \mathcal{X}$. The abovestated assumptions lead to the formula (44).

NUMERICAL EXAMPLE

In order to illustrate our considerations let us discuss a numerical example. It should be treated as a means for the explanaition of the method introduced, rather than a solution of a concrete decision-making problem.

The determination of the fuzzy relation R standing in the equation (45) is one of the crucial points of this approach. Note that dealing with N_1 goals and N_2 constraints the determination of R according to the formulas provided in the previous section can be a tedious task. We arrive to the following iteration scheme,

$$R^{(K+1)}(\underline{s},\underline{w},d) = R^{(K)}(\underline{s},\underline{w},d) - a_K \partial Q/\partial R(\underline{s},\underline{w},d)$$
$$\underline{s} = (s_1, s_2, \ldots, s_{N_1}) \quad , \underline{w} = (w_1, w_2, \ldots, w_{N_2}) \quad (49)$$

where Q is equal to,

$$Q = \sum_{j=1}^{card(\mathbb{D})} (D(d_j) - (G_1 \times G_2 \times \ldots \times G_{N_1} \times C_1 \times C_2 \times \ldots \times C_{N_2} \square R)(d_j))^2 \quad (50)$$

The calculations of the relation R can be performed on the basis of a family of fuzzy sets and goals and respective decisions,

$$G_1^\ell, G_2^\ell, \ldots, G_{N_1}^\ell, C_1^\ell, C_2^\ell, \ldots, C_{N_2}^\ell, D^\ell$$

$\ell = 1, 2, \ldots, L$. The choice of the number of elements of the abovestated family of fuzzy sets and their elements is strictly tied with an identification algorithm in fuzzy relational systems (Pedrycz, 1982). The performance index takes a form,

$$Q = \sum_{\ell=1}^{L} \sum_{j=1}^{card(\mathbb{D})} (D^\ell(d_j) - (G_1^\ell \times G_2^\ell \times \ldots \times G_{N_1}^\ell \times C_1^\ell \times C_2^\ell \times \ldots \times C_{N_2}^\ell \square R)(d_j))^2 \quad (51)$$

Now the optimization problem becomes much more complicated than given before. Therefore we shall propose a suboptimal solution, which essence lies in replacing the multivariable optimization problem by the optimization problem with one variable. Fixing the triangular norm with the parameter e.g. p, γ, Q will be minimized over the range of this parameter with the relation R calculated as,

$$R = \bigcap_{\ell=1}^{L} [G_1^\ell \times G_2^\ell \times \ldots \times G_{N_1}^\ell \times C_1^\ell \times C_2^\ell \times \ldots \times C_{N_2}^\ell) \varphi D^\ell] \quad (52)$$

Let a collected fuzzy data for the decision-making problem be given below (L=5),

G_1				G_2			
0.3	0.6	0.7	0.4	0.5	0.2	0.9	1.0
1.0	0.5	0.5	0.0	0.2	1.0	0.9	0.8
0.2	1.0	0.8	0.7	0.3	0.4	1.0	1.0
1.0	1.0	0.8	0.7	0.3	0.4	0.6	0.9
0.2	1.0	0.8	0.7	0.7	1.0	0.9	0.4

G_3				C_1			
1.0	0.0	0.4	0.5	0.7	0.8	0.5	0.2
0.7	1.0	1.0	0.6	0.2	0.4	0.8	1.0
1.0	1.0	0.9	0.4	0.0	0.2	1.0	0.3
0.5	0.6	0.8	0.4	1.0	0.9	0.9	0.8
0.3	0.9	1.0	0.7	1.0	1.0	0.9	0.5

D			
0.8	1.0	0.5	0.1
0.9	0.2	0.1	0.0
0.8	1.0	0.4	0.2
0.3	0.4	0.6	0.4
0.2	0.9	0.8	0.4

The equation under consideration takes a form,

$$D = (G_1 \times G_2 \times G_3 \times C_1) \square R \quad . \quad (53)$$

The triangular norm t which stands in

(53) is t_3 given before. Additionally we consider the corresponding min-s_3 fuzzy relational equation,

$$D = (G_1 \times G_2 \times G_3 \times C_1) \Delta R \qquad (54)$$

"x" is treated in sense of s_3-norm. Calculating the values of the performance index (51) for different values of γ it was shown that max-t_3 composition operator is preferred with $\gamma = 1.3$. Then Q attains its minimal value equal to 0.27.

CONCLUSIONS

We have proposed a model of decision-making process based on the fuzzy relational equation with the triangular norms. The triangular norms can be treated as a flexible tool for modelling the aggregation operators of fuzzy goals and constraints. It was clearly demonstrated that the decision-making model introduced by Bellman and Zadeh can be derived from the equations discussed here. The basic task concerned with the determination of the relation between goals, constraints and decisions was presented in detail. Several analytical and numerical methods of solving this problem were also considered. Further investigations dealing e.g. with the analysis of the fuzzy relations standing in the equation which can help in understanding the grades of importance of various goals constraints and interrelationships between them are really needed. They could create topics of further researches.

REFERENCES

Bellman R., and L.A. Zadeh (1970). Decision making in a fuzzy environment. Management Sci., 17, 141-164.

Bellman R., and L.A. Zadeh (1977). Local and fuzzy logics. In J.M. Dunn, and D. Epstein Eds., Modern Uses of Multiple Valued Logic, D. Reidel, Dordrecht, pp. 103-165.

Carlsson Ch. (1982). Tackling in MCDM-problem with the help of some results from fuzzy set theory. J. Oper. Res., 10, 270-281.

Czogała E., and W. Pedrycz (1981). Some problems concerning the construction of algorithms of decision-making in fuzzy systems, Int. J. Man-Mach. Stud., 15 201-211.

Dubois D., and H. Prade (1980). Fuzzy Sets and Systems: Theory and Applications. Academic Press, New York.

Frank H.J. (1979). On the simultaneous associativity of F(x,y) and x+y-F(x,y). Aequationes Mathematicae, 19, 194-226.

Hamacher H. (1976). On logical connectives of fuzzy statements and their affilated truth-functions, presented at 3rd European Meeting on Cybernetics and Systems Research, Vienna, Austria.

Kickert W.J.M. (1978). Fuzzy Theories on Decision-Making, N. Nijhoff, Leiden.

Menger K. (1942). Statistical metric spaces. Proc. Nat. Acad. Sci. USA., 28, 535-537.

Pedrycz W. (1982a). Fuzzy relational equations with triangular norms and their resolutions, BUSEFAL, 11, 24-32.

Pedrycz W. (1982b). Fuzzy control and fuzzy systems. Rep. 82 14 Delft Univ. of Technology, Delft, the Netherlands.

Pedrycz W. (1983). On generalized fuzzy relational equations and their applications, J. Math. Anal. and Appl., to appear.

Sanchez E. (1976). Resolution of composite fuzzy relation equations. Inf. Control, 30, 38-48.

Rödder R. (1975). On "and" and "or" connectives in fuzzy set theory. Rep. 75/07. RWTH Aachen.

Yager R.R. (1977). Multiple objective decision-making using fuzzy sets, Int. J. Man-Mach. Stud., 9, 375-382.

Zadeh L.A. (1976). A fuzzy algorithmic approach to the definition of complex or imprecise concepts, Int. J. Man-Mach. Stud. 8, 249-291.

ON MEASURES OF FUZZINESS OF SOLUTIONS OF COMPOSITE FUZZY RELATION EQUATIONS

A. Di Nola and S. Sessa

Universita' di Napoli Facolta' di Architettura, Istituto di Matematica, Via Monteoliveto 3, 80134 Napoli, Italy

Abstract . In this paper we study the solutions of a max-min fuzzy relation equation of Sanchez (1976) by determining ones which have the greatest energy measure and the smallest entropy measure of fuzziness.

Keywords. Fuzzy relation, fuzzy equation, fuzziness measures.

INTRODUCTION

Let X,Y,Z be finite sets and $F(X) = \{A: X \to [0,1]\}$ the class of all fuzzy sets of X. Let be $Q \in F(X\times Y)$, $R \in F(Y\times Z)$ and $T \in F(X\times Z)$ fuzzy relations correlated in such a way the following equation holds:

$$R \circ Q = T \qquad (1)$$

where '$\circ$' is the max-min composition and R is unknown. Let's denote with $\mathcal{R}$ the set of solutions R of Eq.(1). Sanchez (1976) gives a fundamental theorem for the existence and the determination of the greatest element of $\mathcal{R}$ and points out that the study of $\mathcal{R}$ can be enlarged by investigating the entropy measures of fuzziness of the relations $R \in \mathcal{R}$. Following this suggestion Di Nola and Sessa (1983) give algorithms which minimize the entropy measure of a given element of $\mathcal{R}$, whereas Di Nola and Pedrycz(1982) establish a necessary and sufficient condition for existence of boolean solution $R \in \mathcal{R}$ (viz. $R: X\times Y \to \{0,1\}$) which has entropy measure equal to zero. In this note we discuss about the existence of the relation R which have

(E) the greatest energy measure of fuzziness and

(H) the smallest entropy measure of fuzziness.

In order to attain this aim, in section 2 we recall the definitions and notations of composite max-min fuzzy relation equations theory.

Section 3 contains the resolution of problem (E). In section 4 we have furtherly developped the theory of Sanchez (1977). By using some results of Di Nola and Ventre (1980) in section 5 we give the resolution of the problem (H). For semplicity of exposition, we identify fuzzy sets and fuzzy relations with their membership functions. Furthermore, we represent the fuzzy relations as real matrices and we assume $\mathcal{R} \neq \emptyset$.

2. PRELIMINARIES

Let $X \neq \emptyset$ and let's denote by $a \in [0,1]$ the fuzzy set constantly equal to a. It is well known that the following operations defined pointwise for x X: $(A \wedge B)(x) = \min\{A(x),B(x)\}$, $(A \vee B)(x) = \max\{A(x),B(x)\}$, where $A,B \in F(X)$, induce in F(X) a complete lattice structure with universal bounds 0,1 and a partial ordering given by

$A \leq B$ iff $A(x) \leq B(x)$ for any $x \in X$.
From now on we suppose that $X = \{x_1, x_2, \ldots, x_n\}$, $Y = \{y_1, \ldots, y_m\}$, $Z = \{z_1, \ldots, z_p\}$ are finite sets and let be $I_n = \{1, 2, \ldots, n\}$ the set of first n natural numbers. Now we recall the following definitions (Sanchez 1976):

Definition 1. A fuzzy relation Q between X and Y is an element of $F(X \times Y)$. We call inverse of Q the fuzzy relation $Q^{-1} \in F(Y \times X)$ as $Q^{-1}(y_j, x_i) = Q(x_i, y_j)$ for every $i \in I_n, j \in I_m$.

Let be $Q \in F(X \times Y)$ and $R \in F(Y \times Z)$ two fuzzy relations. Now we put: $Q(x_i, y_j) = Q_{ij}$ for every $i \in I_n, j \in I_m$ and $R(y_j, z_k) = R_{jk}$ for every $j \in I_m, k \in I_p$.

Definition 2. We define $RoQ = T$, $T \in F(X \times Z)$, the max-min composite fuzzy relation of R and Q as

$$T_{ik} = T(x_i, z_k) = \bigvee_{j=1}^{m} (Q_{ij} \wedge R_{jk})$$

for every $i \in I_n, k \in I_p$.

Definition 3. We define $S = Q^{-1} \circledast T$, $S \in F(Y \times Z)$, the @-composite fuzzy relation of Q^{-1} and T, as

$$S_{jk} = \bigvee_{i=1}^{n} (Q_{ij} \circledast T_{ik})$$

for every $j \in I_m, k \in I_p$ where

$$Q_{ij} \circledast T_{ik} = \begin{cases} 1 & \text{if } Q_{ij} \leq T_{ik} \\ T_{ik} & \text{if } Q_{ij} > T_{ik} \end{cases}.$$

Definition 4. L is called lower solution of Eq.(1) if for every $R \in \mathcal{R}$ such that $R \leq L$, we have $R = L$.

We introduce further definitions (De Luca and Termini 1979):

Definition 5. Let $f: [0,1] \to [0,1]$ be a real function such that $f(0) = 0$, $f(1) = 1$ and strictly increasing in $[0,1]$. We define energy measure of fuzziness of $R \in F(Y \times Z)$, the number

$$E(R) = \frac{1}{p \cdot m} \sum_{k=1}^{p} \sum_{j=1}^{m} f(R_{jk}).$$

Definition 6. Let $g: [0,1] \to [0,1]$ be a real function such that $g(0) = g(1) = 0$, $g(a) = g(1-a)$ for any $a \in [0,1]$ and strictly increasing in $[0, \frac{1}{2}]$. We define entropy measure of fuzziness of $R \in F(Y \times Z)$ the number

$$H(R) = \frac{1}{p \cdot m} \sum_{k=1}^{p} \sum_{j=1}^{m} g(R_{jk}).$$

3. RESOLUTION OF PROBLEM (E).

By considering the Eq.(1), Sanchez(1976) proved that $\mathcal{R}$ is non-empty iff S is the greatest element of $\mathcal{R}$. Then the problem (E) is immediately solved by knowledge of S.

4. FURTHER RESULTS

The following lemma will be useful in the sequel.

Lemma. Let be $R', R'' \in \mathcal{R}$ and $R \in F(Y \times Z)$ such that $R' \leq R \leq R''$. Then $R \in \mathcal{R}$.

Proof. By def.2, one can easily deduce that $R' o Q \leq RoQ \leq R'' oQ$ and then $R \in \mathcal{R}$ since $R'oQ = R''oQ = T$.

Let's define, for every $i \in I_n$, the fuzzy sets $Q_i \in F(\{x_i\} \times Y)$ and $T_i \in F(\{x_i\} \times Z)$ as $Q_i(x_i, y_j) = Q_{ij}$, $T_i(x_i, z_k) = T_{ik}$ for every $j \in I_m, k \in I_p$.

Then the study of Eq.(1) is led on the fuzzy system of n equations:

$$RoQ_i = T_i \quad . \qquad (2)$$

For any $i \in I_n$, let be $\mathcal{R}_i$ the set of solutions of Eq.(2). Clearly we have

$$\mathcal{R} = \bigcap_{i=1}^{n} \mathcal{R}_i \qquad (3)$$

The equation (2) is called σ-fuzzy relation equation. We recall some results of Sanchez (1977):

- If $\mathcal{R}_i \neq \emptyset$, the fuzzy relation $W_i = Q_i \circledS T_i \in F(Y \times Z)$ defined by $W_i(y_j, z_k) = Q_{ij} \sigma T_{ik}$ for every $i \in I_n$ $j \in I_m, k \in I_p$ where

$$Q_{ij} \sigma T_{ik} = \begin{cases} 0 & \text{if } Q_{ij} < T_{ik} \\ T_{ik} & \text{if } Q_{ij} \geq T_{ik}, \end{cases}$$

belongs to $\mathcal{R}_i$.

- If $\mathcal{R}_i \neq \emptyset$, then $\mathcal{R}_i$ has lower solutions $L_i \neq 0$, whose fuzzy union is just W_i. In order to determine such relations L_i it suffices to keep a non-zero element in each column of W_i.

We illustrate the following example quoted by Sanchez (1976):
let be $X = \{x_1, x_2, x_3\}$, $Y = \{y_1, y_2, y_3, y_4\}$, $Z = \{z_1, z_2, z_3\}$ and let's consider Q, T, S, W_i, $i \in I_3$, as

Q =

	y_1	y_2	y_3	y_4
x_1	0.2	0.0	0.8	1.0
x_2	0.4	0.3	0.0	0.7
x_3	0.5	0.9	0.2	0.0

T =

	z_1	z_2	z_3
x_1	0.7	0.3	1.0
x_2	0.6	0.4	0.7
x_3	0.8	0.9	0.2

S =

	z_1	z_2	z_3
y_1	1.0	1.0	0.2
y_2	0.8	1.0	0.2
y_3	0.7	0.3	1.0
y_4	0.6	0.3	1.0

W_1 =

	z_1	z_2	z_3
y_1	0.0	0.0	0.0
y_2	0.0	0.0	0.0
y_3	0.7	0.3	0.0
y_4	0.7	0.3	1.0

$W_2 =$

	z_1	z_2	z_3
y_1	0.0	0.4	0.0
y_2	0.0	0.0	0.0
y_3	0.0	0.0	0.0
y_4	0.6	0.4	0.7

$W_3 =$

	z_1	z_2	z_3
y_1	0.0	0.0	0.2
y_2	0.8	0.9	0.2
y_3	0.0	0.0	0.2
y_4	0.0	0.0	0.0

The lower solutions of $\mathcal{R}_i$, $i \in I_3$, are the following relations:

$L_1 =$

	z_1	z_2	z_3
y_1	0.0	0.0	0.0
y_2	0.0	0.0	0.0
y_3	0.7	0.3	0.0
y_4	0.0	0.0	1.0

$L'_1 =$

	z_1	z_2	z_3
y_1	0.0	0.0	0.0
y_2	0.0	0.0	0.0
y_3	0.0	0.0	0.0
y_4	0.7	0.3	1.0

$L''_1 =$

	z_1	z_2	z_3
y_1	0.0	0.0	0.0
y_2	0.0	0.0	0.0
y_3	0.7	0.0	0.0
y_4	0.0	0.3	1.0

$L'''_1 =$

	z_1	z_2	z_3
y_1	0.0	0.0	0.0
y_2	0.0	NO	0.0
y_3	0.0	0.3	0.0
y_4	0.7	0.0	1.0

$L_2 =$

	z_1	z_2	z_3
y_1	0.0	0.4	0.0
y_2	0.0	0.0	0.0
y_3	0.0	0.0	0.0
y_4	0.6	0.0	0.7

$L'_2 =$

	z_1	z_2	z_3
y_1	0.0	0.0	0.0
y_2	0.0	0.0	0.0
y_3	0.0	0.0	0.0
y_4	0.6	0.4	0.7

$L_3 =$

	z_1	z_2	z_3
y_1	0.0	0.0	0.0
y_2	0.8	0.9	0.0
y_3	0.0	0.0	0.2
y_4	0.0	0.0	0.0

$L'_3 =$

	z_1	z_2	z_3
y_1	0.0	0.0	0.0
y_2	0.8	0.9	0.2
y_3	0.0	0.0	0.0
y_4	0.0	0.0	0.0

$L''_3 =$

	z_1	z_2	z_3
y_1	0.0	0.0	0.2
y_2	0.8	0.9	0.0
y_3	0.0	0.0	0.0
y_4	0.0	0.0	0.0

The following theorems hold:

<u>Theorem 1</u>. If $\mathcal{R}_i \neq \emptyset$, for any $R \in \mathcal{R}_i$ there exists at least a lower solution L_i such that $L_i \leq R$.

<u>Proof</u>. Let be $R \in \mathcal{R}_i$ and for any $k \in I_p$ we have:

$$\bigvee_{j=1}^{m} (Q_{ij} \wedge R_{jk}) = T_{ik} .$$

For any $k \in I_p$, then there exists an index $j^* \in I_m$ such that $Q_{ij^*} \geq Q_{ij^*} \wedge R_{j^*k} = T_{ik}$. This implies $W_i(y_{j^*}, z_k) = T_{ik} \leq R_{j^*k}$. Therefore we define for any $k \in I_p$, $j \in I_m$:

$$L_i(y_j, z_k) = \begin{cases} 0 & \text{if } j \neq j^* \\ T_{ik} & \text{if } j = j^* \end{cases}$$

which is a lower solution of $\mathcal{R}_i$ such that $L_i \leq R$.

<u>Theorem 2</u>. If $\mathcal{R}_i \neq \emptyset$ for every $i \in I_n$, then the set $\mathcal{L}_i = \{ L_i \in \mathcal{R}_i$ such that $L_i \leq S \}$ is non-empty.

<u>Proof</u>. By theorem 5 of Sanchez (1976), $S \in \mathcal{R}$ and by (3) we have $S \in \mathcal{R}_i$ for every $i \in I_n$. By theorem 1, then follows the thesis.

<u>Theorem 3</u>. If $\mathcal{R} \neq \emptyset$, the fuzzy relation

$$L = \bigvee_{i=1}^{n} L_i$$

with $L_i \in \mathcal{L}_i$ for every $i \in I_n$, belongs to $\mathcal{R}$.

<u>Proof</u>. Since $\mathcal{R} \neq \emptyset$, then $\mathcal{R}_i \neq \emptyset$ for every $i \in I_n$. By theorem 2, $\mathcal{R}_i$ has lower solutions $L_i \in \mathcal{L}_i$. Furthermore we have for every $i \in I_n$:

$$L_i \leq \bigvee_{i=1}^{n} L_i \leq S .$$

By lemma, the fuzzy relation L belongs to $\mathcal{R}_i$ for every $i \in I_n$ and therefore it is an element of $\mathcal{R}$ because (3) holds.

In according to the theorem 3, now we define the following set:

$$\mathcal{L} = \left\{ L \in \mathcal{R} \text{ such that } L = \bigvee_{i=1}^{n} L_i,\ L_i \in \mathcal{L}_i \text{ for any } i \in I_n \right\} .$$

By returning to the foregoing example we have $\mathcal{L}_1 = \{L_1, L''_1\}$, $\mathcal{L}_2 = \{L_2\}$, $\mathcal{L}_3 = \{L_3, L'_3, L''_3\}$ and consequently $\mathcal{L}$ has the following six relations:

$L =$

	z_1	z_2	z_3
y_1	0.0	0.4	0.0
y_2	0.8	0.9	0.0
y_3	0.7	0.0	0.2
y_4	0.6	0.3	1.0

$M =$

	z_1	z_2	z_3
y_1	0.0	0.4	0.0
y_2	0.8	0.9	0.0
y_3	0.7	0.3	0.2
y_4	0.6	0.0	1.0

N=

	z_1	z_2	z_3
y_1	0.0	0.4	0.0
y_2	0.8	0.9	0.2
y_3	0.7	0.0	0.0
y_4	0.6	0.3	1.0

P=

	z_1	z_2	z_3
y_1	0.0	0.4	0.0
y_2	0.8	0.9	0.2
y_3	0.7	0.3	0.0
y_4	0.6	0.0	1.0

U=

	z_1	z_2	z_3
y_1	0.0	0.4	0.2
y_2	0.8	0.9	0.0
y_3	0.7	0.0	0.0
y_4	0.6	0.3	1.0

V=

	z_1	z_2	z_3
y_1	0.0	0.4	0.2
y_2	0.8	0.9	0.0
y_3	0.7	0.3	0.0
y_4	0.6	0.0	1.0

5. RESOLUTION OF PROBLEM (H)

By using a result of Di Nola and Ventre (1980) we prove that there exists at least an element of $\mathcal{R}$ which has the smallest fuzzy entropy measure, basing on the knowledge of S and on the elements of $\mathcal{L}$. Now we recall (Di Nola and Sessa, to appear) the following:

Definition 7. Let $g:[0,1]\to[0,1]$ as in def.6. Let be $a,b\in[0,1]$, we define a partial ordering as: $a \sqsubseteq b$ if and only if $g(a)\leq g(b)$, and a meet operation:

$$a \sqcap b = \begin{cases} a & \text{if } g(a)< g(b) \\ a\wedge b & \text{if } g(a)= g(b). \end{cases}$$

We can prove that the structure $([0,1], \sqsubseteq, \sqcap)$ is a semilattice which no depends by the choice of the function g (Di Nola and Sessa, to appear). Then we consider the function $g(a)=a\wedge\bar{a}$ where $\bar{a}=1-a$ for any $a\in[0,1]$. We will use later the following:

Theorem 4. Let be $a,b,c\in[0,1]$ such that $a\leq b\leq c$. Then $a\sqcap c=(a\sqcap c)\sqcap b$.

Proof. We distinguish two cases:

(i) $a\sqcap c=a$. Then

$$a\wedge\bar{a}\leq c\wedge\bar{c}. \qquad (4)$$

We must prove $a\wedge\bar{a}\leq b\wedge\bar{b}$. For absurd, let be

$$a\wedge\bar{a}> b\wedge\bar{b}. \qquad (5)$$

From (4), we have:

$$c\wedge\bar{c}> a\wedge\bar{a}> b\wedge\bar{b}. \qquad (6)$$

We claim $b\wedge\bar{b}=b$ otherwise from (6) it follows that $\bar{c}\geq c\wedge\bar{c}>b\wedge\bar{b}=\bar{b}$, which implies $c<b$, a contradiction to the hypothesis $b\leq c$. So from (5), we deduce $a\geq a\wedge\bar{a}>b$, a contradiction to the hypothesis $a\leq b$.

(ii) $a\sqcap c=c$. Then

$$c\wedge\bar{c}\leq a\wedge\bar{a}. \qquad (7)$$

We must prove $c\wedge\bar{c}\leq b\wedge\bar{b}$. For absurd, let be

$$c\wedge\bar{c}>b\wedge\bar{b}. \qquad (8)$$

From (7), we have:

$$a\wedge\bar{a}>c\wedge\bar{c}>b\wedge\bar{b}. \qquad (9)$$

(9) implies $b\wedge\bar{b}=\bar{b}$ otherwise we obtain $a\geq a\wedge\bar{a}>b\wedge\bar{b}=b$, a contradiction to the assumption $a\leq b$. Then (8) implies $\bar{c}\geq c\wedge\bar{c}>\bar{b}$, that is $c<b$, a contradiction to the hypothesis $b\leq c$.

Now we recall the following result of Di Nola and Sessa (1983):

Theorem 5. Let be $\mathcal{R}\neq\emptyset$ and $R,R'\in\mathcal{R}$ such that $R\leq R'$. Then the relation $R\sqcap R'\in F(Y\times Z)$ so defined for every $j\in I_m$, $k\in I_p$: $(R\sqcap R')(y_j,z_k)=R_{jk}\sqcap R'_{jk}$ is an element of $\mathcal{R}$.

Finally we have the following:

Theorem 6. Let be $\mathcal{R}\neq\emptyset$. For any $L\in\mathcal{L}$, let be $L^*\in F(Y\times Z)$ defined for every $j\in I_m$, $k\in I_p$ as $L^*_{jk}=L_{jk}\sqcap S_{jk}$. In the set $\mathcal{L}^*=\{L^*, L\in\mathcal{L}\}$ there is at least one element $\tilde{L}^*$ of $\mathcal{R}$ such that $H(\tilde{L}^*)\leq H(R)$ for any $R\in\mathcal{R}$.

Proof. By theorem 5 L^* belongs to $\mathcal{R}$ for any $L\in\mathcal{L}$. Since $\mathcal{L}^*$ is a finite set, let be $\tilde{L}^*$ such that

$$H(\tilde{L}^*) = \min_{L^*\in\mathcal{L}^*} H(L^*) \qquad (10)$$

Let be $R\in\mathcal{R}$. By th.5 of Sanchez (1976) we have $R\leq S$ and being $R\in\mathcal{R}_i$ for any $i\in I_n$, by theorem 1, there exists a lower solution $L_i\leq R$. Then $L_i\leq S$ and therefore $L_i\in\mathcal{L}_i$ for any $i\in I_n$. By theorem 3 the fuzzy relation

$$L = \bigvee_{i=1}^{n} L_i$$

is an element of $\mathcal{L}$ such that $L\leq R\leq S$. By theorem 4, we have:

$$L^*= L\sqcap S= (L\sqcap S)\sqcap R= L^*\sqcap R. \qquad (11)$$

(11) implies $H(L^*)\leq H(R)$ and therefore the thesis because we obtain from (10), for any $R\in\mathcal{R}$, $H(\tilde{L}^*)\leq H(L^*)\leq H(R)$.

Bearing in mind the above example, we have that $\mathcal{L}^*$ contains the following six relations:

L^*=

	z_1	z_2	z_3
y_1	0.0	1.0	0.0
y_2	0.8	1.0	0.0
y_3	0.7	0.0	1.0
y_4	0.6	0.3	1.0

M^*=

	z_1	z_2	z_3
y_1	0.0	1.0	0.0
y_2	0.8	1.0	0.0
y_3	0.7	0.3	1.0
y_4	0.6	0.0	1.0

N^*=

	z_1	z_2	z_3
y_1	0.0	1.0	0.0
y_2	0.8	1.0	0.2
y_3	0.7	0.0	0.0
y_4	0.6	0.3	1.0

P^*=

	z_1	z_2	z_3
y_1	0.0	1.0	0.0
y_2	0.8	1.0	0.2
y_3	0.7	0.3	0.0
y_4	0.6	0.0	1.0

U^*=

	z_1	z_2	z_3
y_1	0.0	1.0	0.2
y_2	0.8	1.0	0.0
y_3	0.7	0.0	0.0
y_4	0.6	0.3	1.0

V^*=

	z_1	z_2	z_3
y_1	0.0	1.0	0.2
y_2	0.8	1.0	0.0
y_3	0.7	0.3	0.0
y_4	0.6	0.0	1.0

which have respectively fuzzy entropy measures:

$H(L^*) = H(M^*) = 0.1$ and $H(N^*) = H(P^*) = H(U^*) = H(V^*) = 0.116$.

6. CONCLUDING COMMENTS

The problem (H) was anticipated by Di Nola and Sessa (to appear) and there solved for Ⓞ-fuzzy relation equations. For a wide class of fuzzy relation equations, which have like limit case the Ⓞ-fuzzy equations of Sanchez (1977), the problems (E) and (H) have been solved by Di Nola, Pedrycz and Sessa (to appear). In future papers, we carefully shall study further algebraic properties of the set $\mathcal{R}$ and applications to fuzzy systems shall be attempted too.

7. REFERENCES

De Luca.A. and Termini.S.(1979). Entropy and energy measures of fuzzy sets, in "Advances in fuzzy sets theory and applications", Eds.M. M.Gupta, and R.K.Ragade and R.R. Yager, North-Holland, Amsterdam, 321-328.

Di Nola.A. and Pedrycz.W.(1982). Entropy and energy measure characterization of resolutions of some fuzzy relational equations, Busefal, 10.

Di Nola.A., Pedrycz.W. and Sessa.S. (to appear). On measures of fuzziness of solutions of fuzzy relation equations with generalized connectives, J.Math.Anal.Appl.

Di Nola.A. and Sessa.S. (to appear). On the fuzziness of solutions of Ⓞ-fuzzy relation equations on finite spaces, Fuzzy Sets and Systems.

Di Nola.A. and Ventre,A. (1980). On some chains of fuzzy sets, Fuzzy Sets and Systems, 4, 185-191.

Sanchez.E. (1977). Solutions in composite fuzzy relation equations: applications to medical diagnosis in brouwerian logic, in "Fuzzy Automata and Decision Processes" Eds. M.M.Gupta, G.N.Saridis, B.R. Gaines, North-Holland, 221-234.

Zadeh.L.A. (1965). Fuzzy Sets, Inform. and Control, 8, 338-353.

E. Sanchez, Resolution of composite fuzzy relation equations, Inform. and Control 30, 1976, 38-49.

A CHARACTERISTIC OPTIMISM FACTOR IN FUZZY DECISION-MAKING

H. R. van Nauta Lemke*, T. G. Dijkman, H. van Haeringen** and M. Pleeging***

**Department of Electrical Engineering, Delft University of Technology, Delft, The Netherlands*

***Department of Mathematics, Delft University of Technology, Delft, The Netherlands*

Abstract. In multi-objective attribute fuzzy decision making one wishes to select the best alternative. Such a selection is based on many different aspects of varying degrees of importance. To solve the problem of selection one has to find at least approximate values that represent the satisfaction of each alternative to all objectives and also values that represent the relative importance of each objective. Moreover, the operation necessary to combine the values corresponding to the objectives is not uniquely defined and has to be chosen. This paper presents a general formulation of different operators for the computation of these values and provides some insight into the effect of applying a particular operation. This can be considered as a subjective optimism index of the decision maker.

Keywords. Decision making; fuzzy; multi-objective; subjective appreciation.

INTRODUCTION

In real situations decisions have to be made mostly on the basis of vague, imprecise and uncertain information. For decision making in the presence of uncertainties one may rely on the abundance of literature on statistical decision theory. Sensitivity analysis can be used to determine the sensitivity of the decision for variations in the different values of the variables. Moreover, fuzzy sets, introduced by Zadeh (1965), offer a wonderful possibility to handle imprecise data and to adapt decision making to subjective considerations that are inherently connected with the way humans think of real-life problems.

Many researchers have investigated the application of fuzzy sets in decision making (Bellman and Zadeh 1970, Zadeh 1973, Jain 1977, Efstathion and Rajkovic 1979, 1980, Freeling 1980, Eshragh 1980, Yager 1978, 1980), as well as in related fields, e.g. control engineering (Mamdani 1977, Assilian, Gaines, Kickert, Van Nauta Lemke). Up to now much research has been done in three important areas.

The first important field of research comprises the evaluation of the decision maker's preferences, which are sometimes available only in ordinal information.

Assume a finite set of alternatives

$$A = \{a_1, a_2, \ldots, a_n\}$$

and a finite set of objectives

$$C = \{c_1, c_2, \ldots, c_m\}$$

We assume that it is possible to associate a number $\mu_{ij} \in [0,1]$ with each pair (a_i, c_j) $i=1,2,\ldots,n$; $j=1,2,\ldots,m$, such that μ_{ij} represents its degree of appreciation of the alternative a_i with respect to the objective c_j.
In the application of fuzzy sets it is possible to cope with vague or imprecise fuzzy objectives. The evaluation itself can be of an objective or subjective nature. The introduction of subjective values can also be based on linguistic expressions or interrogations.

The second important field involves the evaluation of the relative importance of the different objectives. To be more specific, it is necessary to associate a value w_j $(j=1,2,\ldots,m)$ expressing the relative importance or weight of the different objectives. In the literature there are various possibilities for combining these weights with the values of appreciation of the objectives.

The third field of research concerns the final stage in the decision process : the choice of the best, optimal alternative. It is customary to define a decision function on the set of alternatives such that it takes its highest (sometimes: lowest) value for the or an optimal alternative. In the literature

one can find different approaches leading to a variety of operators to form a decision function.
In this paper we present a simple general formula which comprises many different operators. More specifically, we introduce a family of operators that are characterized by a parameter or index s. This index s can take any real value, which means that we have a continuous, gradual change in these operators. The decision maker has to choose subjectively a particular value for s. We consider the introduction of this class of operators as a valuable extension to the theory of the processing of subjective information in decision making.

2. OPERATIONS FOR COMBINING OBJECTIVES IN DECISION MAKING

In section 1 we introduced the set of alternatives $A=\{a_i\}_{i=1}^{n}$, the set of objectives $C=\{c_j\}_{j=1}^{m}$, and the degree of appreciation of a_i with respect to c_j, $\mu_{ij}=\mu(a_i,c_j)\in[0,1]$. A decision function D should give a measure of the overall appreciation of each of these alternatives, respectively. Let

$$D_i = D(a_i)$$

be the degree to which the alternative a_i meets all objectives $c_1,c_2,\ldots,c_m$. Then D will take its maximal value D_{i0} for the or a 'best' alternative a_{i0}. Each value D_i has to be composed of the numbers μ_{ij}, $j=1,2\ldots,m$. A large variety of methods for forming such a composition exists.

The reasoning underlying these methods is so logical that selecting one particular, i.e. the most appropriate, combination for a specific application is a decision problem in itself. According to a linguistic connection between the objectives, one can argue that the decision function D depends on the satisfaction of an alternative to all objectives, that is to objective c_1 and to objective c_2, etc. In that case one should use an intersection-type operator, e.g. the minimum, or non-interactive 'and' operator, given by

$$D_i = \mu_{i1} \wedge \mu_{i2} \wedge \ldots \wedge \mu_{im} = \bigwedge_{j=1}^{m} \mu_{ij} \qquad (1)$$

or the product, or interactive 'and' operator, modified to

$$D_i = (\mu_{i1},\, \mu_{i2} \ldots \mu_{im})^{\frac{1}{m}} = (\prod_{j=1}^{m} \mu_{ij})^{\frac{1}{m}} \qquad (2)$$

It is also reasonable to argue that it is important to judge the satisfaction of all alternatives on a more balancing, comparative or averaging basis. In this case an averaging operation seems more appropriate, e.g.

$$D_i = \frac{1}{m} \sum_{j=1}^{m} \mu_{ij} \qquad (3)$$

These three operations are special cases of one general formula:

$$D_i^{(s)} = [\frac{1}{m} \sum_{j=1}^{m} (\mu_{ij})^s]^{\frac{1}{s}} \quad , s\neq 0 \qquad (4)$$

It is well known (cf. Hardy et al. 1978, Van Haeringen 1980) that for

$s\to-\infty$	$D_i^{(-\infty)} = \bigwedge_{j=1}^{m} \mu_{ij}$	:	minimum non-interactive 'and'
$s=-1$	$D_i^{(-1)} = [\frac{1}{m} \sum_{j=1}^{m} \frac{1}{\mu_{ij}}]^{-1}$	:	harmonic mean
$s\to0$	$D_i^{(0)} = [\prod_{j=1}^{m} \mu_{ij}]^{\frac{1}{m}}$	:	modified product, interactive 'and' geometric mean
$s=1$	$D_i^{(1)} = \frac{1}{m} \sum_{j=1}^{m} \mu_{ij}$	:	arithmetic mean
$s=2$	$D_i^{(2)} = [\frac{1}{m} \sum_{j=1}^{m} (\mu_{ij})^2]^{\frac{1}{2}}$	:	quadratic mean
$s\to+\infty$	$D_i^{(\infty)} = \bigvee_{j=1}^{m} \mu_{ij}$	:	maximum, exclusive 'or'

The operation (4) therefore includes the different combinations of the objectives in a very general form. As we stated before, the index s allows the operation to change gradually, from a minimum operator into a maximum operator, hereby passing the distinct operations harmonic mean (s=-1), geometric mean (s=0), arithmetic mean (s=1) and quadratic mean (s=2).
This is represented in Fig. 1.

It is known that $D_i^{(s)}$ is monotonically increasing in s ,

$$s < t \Rightarrow D_i^{(s)} < D_i^{(t)}$$

Considering that fuzzy sets are symbolized as sets with vague, cloudy boundaries, it is not surprising that there is a gradual change from the intersection into the union operation.
In analogy to the previous section, a generalized operator can be defined which includes the weighting factors that take into consideration the fact that the objectives are not equally important.

$$D = [\sum_{j=1}^{m} w_j \mu_{ij}^{s}]^{\frac{1}{s}} , \qquad (5)$$

$$\sum_{j=1}^{m} w_j = 1, \; w_j \in [0,1]$$

3. THE 'OPTIMISM' INDEX

In the foregoing it has been shown that in passing all values from minus infinity to plus infinity the index s changes the decision operator gradually from the minimum operator into the maximum operator. For simplicity we shall in this section take all weighting factors equal to one another, and hence equal to 1/m, cf. Eq. (4). In this case it is easier to clarify the effect of a gradual change of s. First we shall consider $D_i^{(s)}$ for some special values of s.

$s \to -\infty$ gives the minimum operator. So for this value the decision function D for any alternative depends only on the minimum of its objective values μ_{ij} and is insensitive to a small change in the other objective values.
$s \to \infty$ gives the maximum operator. Therefore D is insensitive to changes in all values of μ except when the maximum value is exceeded.
s=1 gives the arithmetic-mean operator. For this value D is equally sensitive to an absolute change in each value of μ_{ij} and D depends on all values of μ_{ij}.
$s \to 0$ gives the geometric-mean operator. For this value of s, D is equally sensitive to relative changes in each value of μ_{ij} and D also depends on all values of μ_{ij}.
In other words: by varying the value of s we get different operators for combining the objective values. This means that a different emphasis is given to lower or higher values of s.

For $s \to \infty$ the lowest value of μ_{ij} is dominant,
for $s \to \infty$ the highest value of μ_{ij} is dominant,
for $s \to 0$ all values of μ_{ij} are of equal relative importance, and
for s=1 all values of μ_{ij} are of equal absolute importance.

In general for low values of s (large negative values) the lower values of μ_{ij} are more important, whereas for high values of s (large positive values) the higher values of μ_{ij} play a more dominant role.

In reconsidering a real-life decision problem a relation can be made to a certain degree of pessimism or optimism in the character of the decision maker. A very pessimistic person will lay all the emphasis on those objectives that badly meet his desires or requirements, whereas a very optimistic person will stress those objectives that best meet his wishes, to the point of almost neglecting those with a poor performance.

To express this linguistic relation concerning optimism and pessimism in terms more suitable for use in the theory of fuzzy sets the range of s has to be transformed into [0,1]. This can be achieved in several ways, e.g.

$$r_1 = (1 + e^{-as})^{-1} \quad , \; a > 0 \qquad (6)$$

$$r_2 = \frac{1}{\pi} (\frac{\pi}{2} + \text{arctg } bs) \quad , \; b > 0 \qquad (7)$$

$$r_3 = \frac{1}{2} (1 + \frac{s}{c+|s|}) \quad , \; c > 0 \qquad (8)$$

Little experience has been obtained with different transformations. As an example the following transformation is chosen:

$$r_3 = \frac{1}{2} (1 + \frac{s}{1+|s|}) \qquad (9)$$

In Fig. 2 examples of r_1, r_2 and r_3 are shown with a=.2, b=1 and c=1, respectively.

The degree of optimism can be represented by a fuzzy set. The decision maker can introduce his view of the problem by the choice of the membership function, see Fig. 3.

4. EXAMPLE

Assume a decision problem with five alternatives a_1, a_2, a_3, a_4, a_5 and six objectives c_1, c_2, c_3, c_4, c_5, c_6. For reasons of simplicity the importance of the different objectives are considered to be equal; therefore the weighting factors are 1/6. We assume the relation according to table 1.

TABLE 1

Objectives	1	2	3	4	5	6
Alternatives						
a_1	0.5	0.5	0.5	0.5	0.6	0.6
a_2	0.4	0.5	0.7	0.7	0.7	0.7
a_3	0.4	0.4	0.6	0.8	0.8	0.8
a_4	0.4	0.4	0.4	0.9	0.9	0.9
a_5	0.4	0.4	0.4	0.5	0.5	1.0

According to section 2 the following solutions are in general obtained:
When the minimum operator is applied, $s \to -\infty$:
D_1=0.5000, D_2=0.4000, D_3=0.4000, D_4=0.4000, D_5=0.4000.
Therefore when the maximum value is selected a_1 is given as the best solution.
When the harmonic-mean operator is applied, s=-1:
D_1=0.5294, D_2=0.5874, D_3=0.5760, D_4=0.5538, D_5=0.4800.
Therefore a_2 is the best solution.
When the geometric-mean operator is applied, $s \to 0$:
D_1=0.5313, D_2=0.6029, D_3=0.6052, D_4=0.6000, D_5=0.5020.
Therefore a_3 is the best solution.
When the arithmetic-mean operator is applied, s=1:
D_1=0.5333, D_2=0.6167, D_3=0.6333, D_4=0.6500, D_5=0.5333.
Therefore a_4 is the best solution.
When the maximum operator is applied, $s \to +\infty$:
D_1=0.6000, D_2=0.7000, D_3=0.8000, D_4=0.9000, D_5=1.0000.
Therefore a_5 is the best solution.

Better insight into the solution can be obtained with the help of Eq. (4) where s runs from $-\infty$ to $+\infty$. This is shown in Figs. 4a and 4b where for the five alternatives D is represented graphically as a function of s and of r, respectively.

It is clear that the decision maker's choice of the alternative will depend on his characteristic degree of optimism. If he is very optimistic he will choose alternative a_5, if he is optimistic alternative a_4, if he is indifferent alternative a_3, if he is pessimistic alternative a_2 and if he is very pessimistic alternative a_1. This particular example is chosen for this particular purpose. In real-life decision problems the results will not be so pronounced, but the introduction of the optimism index will provide a solution in cases in which no choice could otherwise be made.

5. CONCLUSION

We have introduced a general formula to replace and generalize decision rules that are usually applied. This formula contains a parameter - introduced in this paper as an optimism index - which is related to the intuitive idea of optimism or pessimism of the decision maker. If a degree of optimism can be defined, this approach offers a practical method to determine the best alternative.
An example has been given in which different choices between alternatives are made for different degrees of optimism. More research is needed to give a better evaluation of the optimism index.

6. REFERENCES

Yager, R.R. (1978). Fuzzy decision making including unequal objectives. Fuzzy sets and systems, 1, no. 2, 87-95.

Freeling, A.N.S. (1980). Fuzzy sets and decision analysis. IEEE Trans. Svst., Man & Cybern., SMC-10, no. 7, 341-355.

Yager, R.R. (1980). Finite linearly ordered fuzzy sets with application to decisions. Int. J. Man-Mach. Stud., 12, 299-322.

Jain, R. (1977). A procedure for multiple-aspect decision making using fuzzy sets. Int. J. Syst. Sci., 8, no. 1, 1-7.

Zadeh, L. (1973). Output of a new approach to the analysis of complex systems and decision processes, IEEE Trans. Syst., Man & Cybern., SMC-3, no. 1, 28-44.

Eshragh, F. (1980). Subjective multi-criteria decision making. Int. J. Man-Mach. Stud., 13, 117-141.

Efstathion J., and V. Rajkovic (1980). Multi-attribute decision making using a fuzzy, heuristic approach. Int. J. Man-Mach. Stud., 12, 141-156.

Efstathion J., and V. Rajkovic (1979). Multi-attribute decision making using a fuzzy heuristic approach. IEEE Trans. Syst., Man & Cybern., SMC-9, no. 6, 326-333.

Mamdani, E. (1977). Application of fuzzy theory to control. In Fuzzy Automation and Decision Processes, North Holland.

Hardy, G.H., J.E. Littlewood, and G. Polya (1978). Inequalities, Cambridge U.P., London, 2nd ed.

Haeringen, H. van (1980). Mean values, inequalities and fuzzy sets. Report 80.05, Delft University of Technology

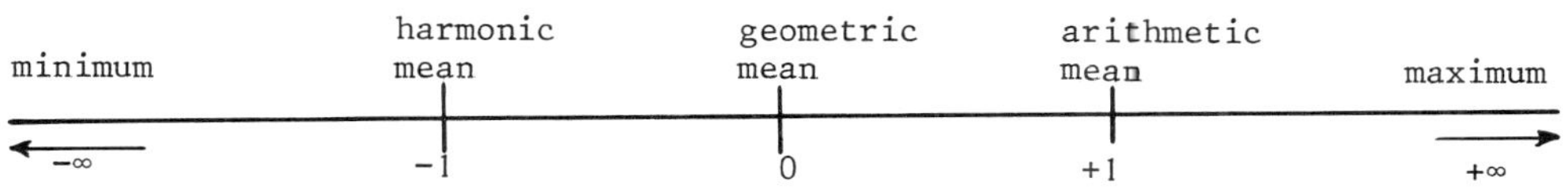

Fig. 1 Special cases of the general formula (4) as a function of the index s.

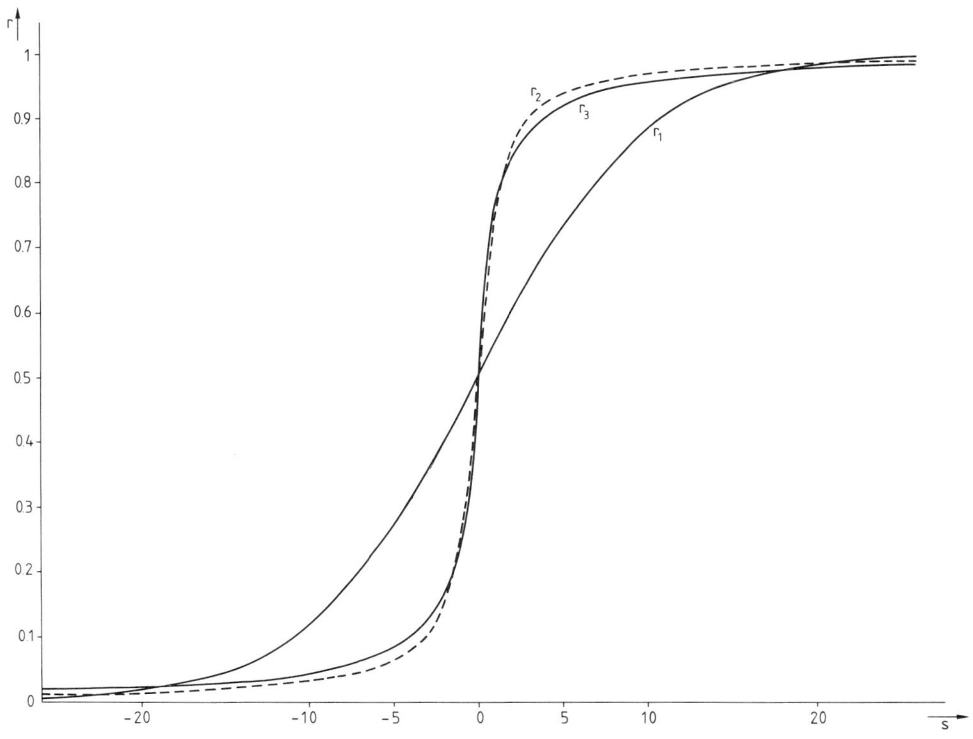

Fig.2 The transformation of the index s into a more suitable index r.

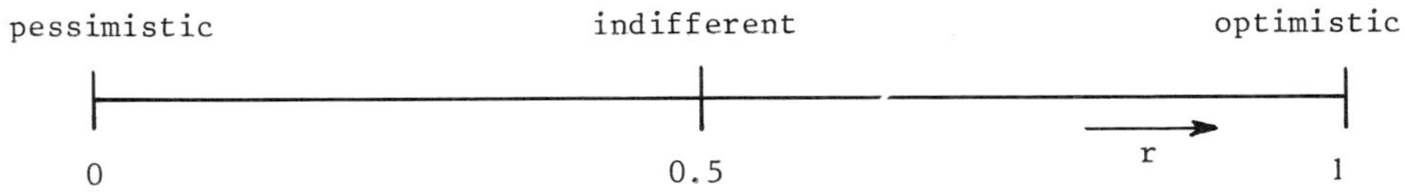

Fig. 3 The degree of optimism as a function of the index r.

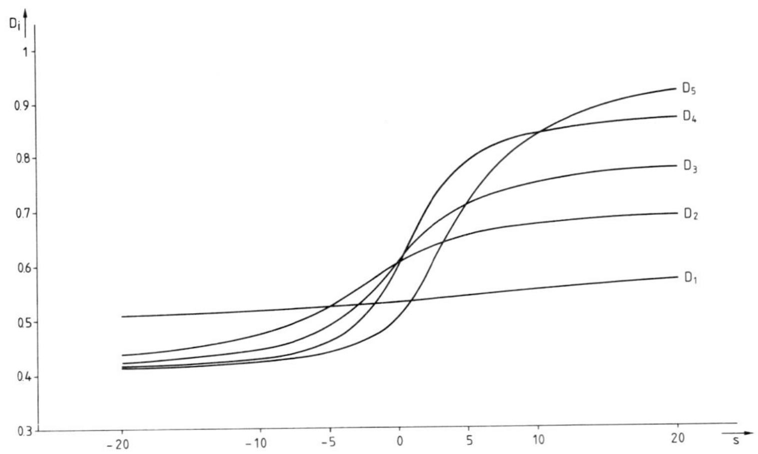

Fig. 4a The decisionfunction D_i as a function of the index s.

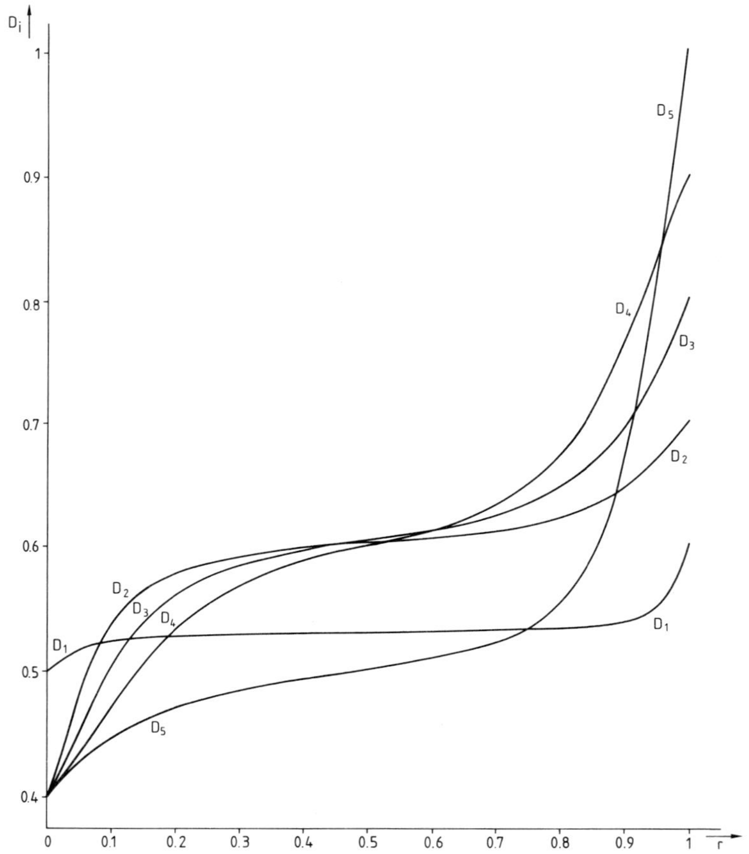

Fig. 4b The decisionfunction D_i as a function of the index r.

FUZZY DECISION ANALYSIS IN THE TASKS OF ELECTRIC NETWORK DEVELOPMENT

A. N. Borisov, Y. Y. Luns and V. A. Popov

Department of Automatized Control Systems, Riga Polytechnical Institute, 1 Lenin Street, Riga, Latvian SSR, USSR

Abstract. The choice task of optimal project decision of electric distribution network development under fuzzy initial information and dynamic preferences of decision maker is considered. Fuzzy set theory and value theory are used. The concept of fuzzy stream of consequences of the alternative (project decision) realization is introduced. The method and algorithm of the construction of the choice model under dependence of objectives and/or elementary consequences of the stream are described. In this case the decision maker's preferences elucidation is carried out on the basis of the procedure of fuzzy tradeoffs and the construction of the value function using Taylor's formula. The results and some stages of the choice task decision of optimal variant of three-time period development of the electric distribution network in some region are described.

Keywords. Decision-making theory; fuzzy set theory; fuzzy stream of consequences; fuzzy tradeoff; value function.

INTRODUCTION

The problem of the large scale systems development such as power systems and, in particular, electric distribution networks (EDN) is very important today and is worthy of serious attention in the decision-making theory and practice. The sought variant of the development of EDN is a sequence of actions introducing new or reconstructing the existing lines of electrotransmission and feeding substations being carried out in the network. Each of these actions leads to a quite certain consequence, which will be realized at some time instant according to the network development. The sequence of such elementary consequences form the time stream of consequences, which is the main characteristic of the project decision. The comparison of time streams of consequences over number of objectives, taken into account by the decision maker (DM), and determination of the preference relation are the main points of multiobjective evaluation and choice task decision on the set of alternatives (project decisions).

The choice task of optimal EDN development may be considered as a perspective decision analysis, when streams of consequences of realization of concrete project decision will come only in the future and the DM has no exact information about them. Thus, the shown class of decision making tasks is characterized by the following peculiarities:

- multiobjective evaluation of the development consequences;
- the presence of fuzzy initial information about technical and economic data of decisions over time and about elements of the DM's preference structure.
- the presence of time factor which is taken into account for exposing and modelling the DM's preferences.

The first of the above peculiarities stipulates the necessity of using the theory of multiobjective evaluation and choice (Keeney, 1975; Larichev, 1979; Fishburn, 1970; Vilkas, 1981), in particular, value theory (Keeney, 1975). The second peculiarity demands the apparatus of fuzzy set theory, a concept of fuzzy and linguistic variables as obligatory conditions (Zadeh, 1975). In confirmation of the above said we may recall, for example, the work by Dubois (1982). The third peculiarity touches the aspects of formalization of the so called dynamic pre-

preferences or, otherwise, preferences over time.

STATEMENT OF THE PROBLEM

Twenty nine variants of possible three-stage development of electric distribution network for some region were proposed by the project department for the manager's consideration. Diversion of the project decisions is provided by adding new branches to the scheme, by switching on branches to other junctions, by removing branches from the scheme or by changing their parameters. The basis for preparing initial data about technical and economic qualitative indices of EDN is a calculated algorithm (Dale, 1979) which includes existing and possible perspective elements of the network - electrical transmission lines and substations.

In some project decision the new equipment was introduced, the technical characateristics of which were not sufficient in the course of electric systems' work. So, it was necessary to get the meanings of experts for these variants of EDN as values of corresponding linguistic variables-objectives. For each of three stages and for each development alternative the following objective values were determined: the sum of capital investments, the losses of power, expenditures for regulation (i.e. additional expenditures to eliminate the exceeding of tension losses over permissible 5 per cent degree) and engineering reliability. The last objective, RELIABILITY characterized the EDN functioning on the whole and dealt only with the final moment of development. The primary analysis of twenty nine initial variants of project decisions allowed to reduce the whole set of alternatives to ten Pareto-optimal alternatives with the help of nonfuzzy dominance's condition. The set of these project decisions (alternatives) is shown in Table 1.

For all this, the following formalization of fuzzy meanings x'_1 = SMALL, x''_1 = MIDDLE, x'''_1 = BIG capital investment satisfied the DM (see Fig.1).

The form of membership functions for the meanings of linguistic objective RELIABILITY x'_4 - SMALL, x''_4 - MIDDLE, x'''_4 - BIG is shown in Fig.2.

Table 1
Pareto-optimal variants of EDN development

Variants / Objectives and periods	a_1	a_2	a_3	a_4	a_5	a_6	a_7	a_8	a_9	a_{10}
X_{11}-Capital inv. first period, (ths.rbl.)	67	77	MDL	MDL	67	77	67	MDL	MDL	MDL
X_{12}-Capital inv. second period, (ths.rbl.)	I25	II5	BIG	MDL	97	88	88	BIG	MDL	BIG
X_{13}-Capital inv. third period (ths.rbl.)	20	20	SML	2I0	2I0	2I0	2I0	SML	SML	SML
X_{21}-Power loss. first period(kw)	360	350	360	350	360	350	360	360	360	440
X_{22}-Power loss. second period (kw)	730	730	730	730	780	780	790	740	740	740
X_{23}-Power loss. third period (kw)	920	920	960	960	920	920	930	930	930	930
X_{31}-Expend. for regul., first per. (ths.rbl.)	0	0	0	0	0	0	0	0	0	0
X_{32}-Expend. for regul., sec. per. (ths.rbl.)	6	6	6	6	I7	I7	26	I3	I3	II
X_{33}-Expend. for regul., third per. (ths.rbl.)	95	95	99	99	II	II	I9	I03	I03	I03
X_4-Reliability, (Prob. of effective work)	MDL	MDL	MDL	MDL	BIG	BIG	BIG	SML	SML	SML

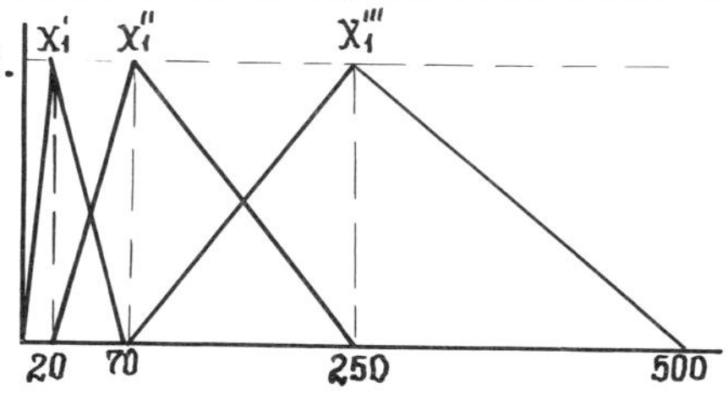

Fig.1. Membership functions of fuzzy meanings x'_1, x''_1, x'''_1

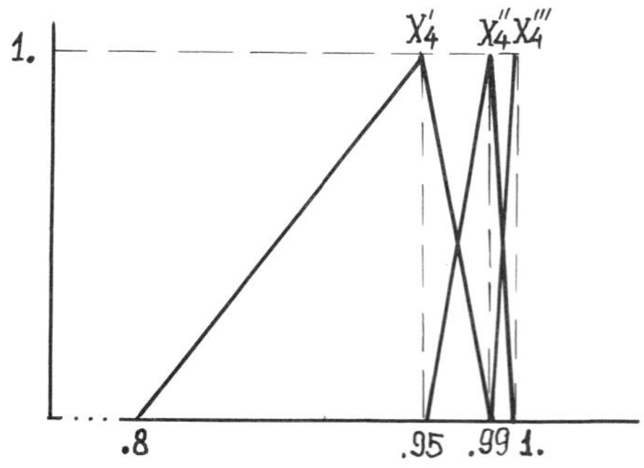

Fig.2. Membership functions of fuzzy meanings x'_4, x''_4, x'''_4

Thus, for the DM a set of alternatives (project decisions) $A = \{a_i\}$, $i \in N_{10}$, $N_{10} = \{1, 2, \ldots, 10\}$ was proposed for the subsequent analysis, corresponding to Table 1. The choice and realization of each $a_i \in A$ gives a stream of consequences $C_i = (C_{i1}, C_{i2}, C_{i3})$ and each elementary consequence of the stream is evaluated over three objectives X_1, X_2, X_3, respectively for three periods $X_{11}, X_{12}, X_{13}, X_{21}, X_{22}, X_{23}, X_{31}, X_{32}, X_{33}$, but the whole stream is evaluated over the objective X_4 without devision into periods. The following problems were necessary to solve:

1. To determine the relations MORE and EQUAL for fuzzy meanings of objectives.
2. To determine the dominance over fuzzy streams of consequences $\tilde{c} = (\tilde{c}_1, \ldots, \tilde{c}_f)$.

Under $\tilde{c}$ let's understand a stream of consequences, the components of which are fuzzy sets and their objective values $\tilde{x}_k$ are linguistic (fuzzy numbers).

3. To formalize the DM's preference structure in order to receive a scalar-value of preferability, or "value" for all $a \in A$. Formally, this is adequate to specify a scalar-valued function V defined on the space of streams of consequences with the property that

$$a' \succsim a'' \Leftrightarrow c' \succsim c'' \Leftrightarrow x' \succsim x'' \Leftrightarrow V(x') \geqslant V(x''), \tag{1}$$

where $c = (c_1, c_2, c_3)$, $x = (x_1, x_2, x_3)$; the function $V : \{x\} \rightarrow [0, 1]$ is a value function in accordance with Keeney (1976), Meyer (1977), Barrager (1980). Then, if V reflects the DM's preferences as in (1), the choice of the best $a^* \in A$ can be put into the format of standard optimization problem:

$$a^* = \arg \max_{a \in A} V[X(C_a)], \tag{2}$$

where, in the case of fuzzy meanings of objectives X, a value function V is a nonfuzzy function of fuzzy attributes, but in the case of fuzzy meanings of objectives X and fuzzy parameters determined for the value function, we have a fuzzy function of fuzzy attributes (Borisov, 1982).

Note, that the choice task of optimal variant of EDN development is formulated under certainty, as follows from the statement, i.e., the certain stream of consequences and their formalization in the sense of values on objectives correspond to each alternative.

THE PROBLEM OF DOMINANCE OVER FUZZY STREAMS OF CONSEQUENCES

It is natural to use the well-known concept of dominance in nonfuzzy case for the definition of dominance over fuzzy streams of consequences. But then it is necessary to define relations MORE and EQUAL between two fuzzy numbers. Let V and W be fuzzy numbers defined on the positive real interval.

The solution of this task is proposed in many well-known works, for example, by Dubois (1980, 1982a, 1982b), Yager (1981). But it is obvious that the choice of the procedure of fuzzy number ranking is a subjective process just like presentation and formalization of fuzzy information by the DM. If the DM is a pessimist, our experiment with him resultes in the following simple scheme

$$V \geqslant W \Leftrightarrow d_{VW} = \int_0^{.5} [1 - \mu_{P_{VW}}(z)]\,dz + \int_{.5}^{1} \mu_{P_{VW}}(z)\,dz \geqslant .5, \tag{3}$$

where $\mu_{P_{VW}}$ is a membership function of fuzzy relation $P_{VW} = \bigcup_{z \in [0,1]} \mu_{P_{VW}}(z)/z$; $(\forall z : z = \frac{v}{v + w})$

$$(\mu_{P_{VW}}(z) = \min[\mu_V(v), \mu_W(w)]). \tag{4}$$

For discrete fuzzy numbers V and W the integration in (3) is changed to summing up the corresponding values of base variables.

Definition 1. Fuzzy stream of consequences x dominates fuzzy stream x" whenever

$$d_{\tilde{x}'_{jk} \tilde{x}''_{jk}} \geqslant .5, \text{ for all j and k,}$$

and

$$d_{\tilde{x}'_{jk} \tilde{x}''_{jk}} > .5, \text{ for some j or k,} \tag{5}$$

where $d_{\tilde{x}'_{jk} \tilde{x}''_{jk}}$ are calculated according to Eq.(3) for fuzzy values of the j-th objective at time instant k. This situation will be also denoted as $\tilde{x}' \succ \tilde{x}''$. The use of definition 1 in our problem allows us to conclude that $a_9 \succ a_8$ and $a_4 \succ a_3$. Thus, only eight alternatives remain for the further analysis.

FUZZY TRADEOFFS AND CONSTRUCTING A VALUE FUNCTION

We may get practically full information about the DM's preferences on the basis of the well-known procedure of value tradeoffs (Keeney, 1976) between objectives' values and/or moments of time for streams of consequences. It is true since on the basis of analysis of marginal rate of substitution, we may determine the character of preferential independence for objectives and time instants and obtain a priori the value function with accuracy to the parameters. But the experiment with the DM has shown that it is very difficult for him to designate an exact compensating meaning of the objective at one time instant for the changes at the other one. For the DM it is easier to operate with fuzzy meanings of changes.

For simplicity of the opening analysis we consider the single-objective and two-period stream of consequences (x_1, x_2), where $x_1, x_2 \in \{x\}$ and $\{x\}$ is a set of possible objective values.

Let $V : \{x\} \rightarrow R^1$ be a value function which is to order the streams of consequences x' and x" according to the DM's preferences. It is necessary to determine its form with accuracy to arbitrary monotonous transformation. For the purpose of alternative ordering it is enough, since the evaluation of alternatives will result in an interval scale.

Assumption 1. Let V be such that on the interval of permissible meanings of x_1 and x_2 there are m derivatives over x_1 and x_2, that is the preferability is fluently changed with changes on each objective meanings. Then, in accordance with Taylor's formula a value function V may be presented as follows

$$V(x_1+\Delta x_1, x_2+\Delta x_2) = V(x_1,x_2) + \left(\frac{\partial V}{\partial x_1}\Delta x_1 + \frac{\partial V}{\partial x_2}\Delta x_2\right) + \dots + \frac{1}{m!}\left(\frac{\partial V}{\partial x_1}\Delta x_1 + \frac{\partial V}{\partial x_2}\Delta x_2\right)^m + Rm,$$

where Rm is a residual member in Taylor's formula.

Let (k + 1) indifferent points be obtained on the basis of nonfuzzy tradeoff procedure for some point (x_1, x_2). Then it enables us to write a linear equation system (Eq.6) relative to unknown meanings of derivatives

$$\left(\frac{\partial V}{\partial x_1}\Delta x_{1i} + \frac{\partial V}{\partial x_2}\Delta x_{2i}\right) + \dots + \frac{1}{m!}\left(\frac{\partial V}{\partial x_1}\Delta x_{1i} + \frac{\partial V}{\partial x_2}\Delta x_{2i}\right)^m = -R_{mi} \quad (6)$$

where the i-th point $(x_1 + \Delta x_{1i}, x_2 + \Delta x_{2i})$ is indifferent to the point (x_1, x_2) and an index $i \in N_k$, is a number of derivatives; if $m = 1 \Rightarrow k = 2$, if $m = 2 \Rightarrow k = 5$ and so on; R_{mi} is a residual member under determination of a value at i-th point.

If we assume strictly that $R_{mi} = 0$ for all i N_{k+1}, then it is true if the determinant of equation system (6) equals 0, i.e. $\Delta = 0$. This is a necessary condition for the check, whether or not $R_{mi} = 0$; but taking into account assumption 1 the condition becomes a sufficient one, too. Hence, if the determinant of system (6) in Taylor's formula is not equal to zero, i.e. $R_{mi} \neq 0$, under the chosen order m, then it is worth to increase m and set the necessary complementary insufficient number of indifferent point. When, at certain m the equality $R_{mi} = 0$ for all i is achieved, to solve system (6) we are to complement it by the condition of value function's equality to one (under the best meanings of criteria (i.e. the best possible stream of consequences x*), and its equality to zero under the worst ones (x^0). Then it is possible to determine all the partial derivatives at point (x_1, x_2).

If index is an arbitrary (m + m'), where $m' = \{1,2,3, \dots\}$, the equality $R_{m+m'} = 0$ is not achieved or it becomes impossible to obtain the required number of indifferent points, then it's worth either to refuse from the decomposition into a series, or evaluate interval meanings of partial derivatives of all m orders provided that m is a priori fixed.

Assumption 2. A value function, formalizing the DM preferences is monotonous and convex or concave on each of criteria x_1 and x_2.

If assumption 2 is true, that is easily proved by interrogating the DM, than maximal or minimal value of the partial derivative V, provided that the order is less or equal to m + 1, is achieved at points (x_1^I, x_2^I), (x_1^S, x_2^I), (x_1^I, x_2^S) or (x_1^S, x_2^S), where $x_1 \in [x_1^I, x_1^S]$ and $x_2 \in [x_2^I, x_2^S]$. The latter follows from the condition that the derivative of monotonous (convex or concave) function is a monotonous (convex or concave) function, too. Then, taking into account assumption 2, we can write

$$R_{mi} \in [0, R_{mi}^{max}],$$

where

$$R_{mi}^{max} = \frac{1}{(m+1)!} \max_{\substack{\varepsilon_{1i} \in \{x_{1i}, x_{1i}+\Delta x_{1i}\} \\ \varepsilon_{1i} \in \{x_{1i}, x_{1i}+\Delta x_{1i}\}}} \left(\frac{\partial V(\varepsilon_{1i})}{\partial x_1} \Delta x_{1i} + \frac{\partial V(\varepsilon_{2i})}{\partial x_2} \Delta x_{2i} \right)^{m+1}$$

Given an interval estimate of the residual member in Taylor's formula, while solving system (6) it's possible to get interval estimates for all partial derivatives at the point being determined. Then the calculated value of alternatives will also be an interval estimate.

Let's consider the case when information about tradeoffs is fuzzy. It's natural to suppose that the scheme of value function construction remains analogous to the one discussed above. Then the solution of the system of given fuzzy factors (tradeoff estimates) is made on the basis of the reverse procedure of fuzzy numbers arithmetic, proposed in (Dubois, 1980) and definitively stated in (Alexeyev, 1982). It results in fuzzy meanings of partial derivatives and a residual member, single for all equations. On the basis of expression (4) the residual member's equality to zero is checked and if it's true, the meanings of derivatives are considered to be determined. Otherwise, the order in Taylor's formula should be raised. Theoretically, the complete impossibility of the residual member's equality to zero, in this case, leads to interval fuzzy estimate of partial derivatives. Practical work with such a value function, that contains fuzzy meanings of partial derivatives is reduced to the usual operating with fuzzy function of nonfuzzy streams of consequences and fuzzy function of fuzzy argument of given fuzzy streams of consequences.

It does't cause difficulties to generalize the proposed method of value function construction with the help of Taylor's formula, for f-period stream of consequences and j-th attributes. It should be taken into account that this generalization causes transformations in Taylor's formula and in the corresponding system of equations which is to be solved.

Thereby, an algorithm for value function construction by the proposed method can be presented as follows:

Step 1. Set the minimal order in Taylor's formula (m = 1).

Step 2. Question the DM in order to determine the necessary number of indifferent points in the set of estimates over attributes and time periods.

Step.3. Calculate the determinant of system(6)

Step.4. If $\Delta = 0$, then $R_m = 0$; and Step 6 follows.
If $\Delta \neq 0$, then $R_m \neq 0$
The DM's possibilities in increasing the number of indifferent points are revealed. If the DM agrees, then return to Step 2. Otherwise, go to Step 5.

Step.5. Evaluate $R_m \in [0, R_m^{max}]$.
Solve equation system (6)
Step 7 follows.

Step.6. Complement system (6) with condition $V(x^*) - V(x^o) = 1$.
Solve the system and determine all derivatives. The value function then is constructed with accuracy to monotonous transformation and parameters meanings. Step 8 follows.

Step.7. The value function is constructed with accuracy to monotonous transformation and interval meanings of parameters.

Step 8. Stop.

THE RESULTS OF VALUE FUNCTION CONSTRUCTION FOR THE CHOICE OF EDN DEVELOPMENT VARIANT

We tried to get a value function for ranking eight Pareto-optimal EDN development variants under the first order of the derivatives in Taylor's formula. For this, it was necessary to get (1 + 10) indifferent points. The stream of consequences, corresponding to the alternative a_1 in Table 1 was assumed to be the basic one. For that, we got nonfuzzy values of tradeoffs from the DM. This resulted in $R_1 \neq 0$. So we have increased an order in Taylor's formula. Then it became necessary to get the (1 + 10 + 55) indifferent streams of consequences. In this case the experiment with the DM was carried out in accordance with a special table.

The information contained in our table allowed us to get sixty five indifferent streams of consequences and to write the required equation system. The analysis of the indicated system led to the result $R_2 = 0$. So, the conditions of value function equality to one given the best meanings of objectives (best stream of consequences) and to zero at the worst meanings were completed. After all that the derivatives were found, from which only twenty nine were not null. Taking into account that the calculation of value of the best stream of consequences on the basis of alternative a_1 (our base point) must give a unit, we determine a value of a_1. When a value of alternative a_1 was found, we got the meanings of a value function for other alternatives as shown in Fig.3 with accuracy to figure.

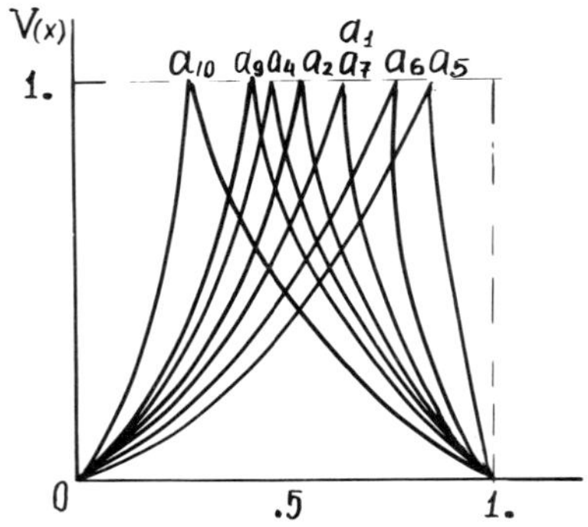

Fig.3. Membership functions for fuzzy meanings of ordered alternatives value

With reference to Eq.(4) we got the following ranking of alternatives

$a_5 \succ a_6 \succ a_1 \sim a_7 \succ a_2 \succ a_4 \succ a_9 \succ a_{10}$.

CONCLUSION

It is necessary to note that the proposed procedure of a value function construction, formalizing the DM's preferences and automating the hard process of streams of consequences analysis is sufficiently universal. But it requires much expertal information and large computer power. In some cases it leads to receiving only interval estimation of alternatives value, that doesn't always result in a single decision. There is a very interesting fact that under fuzzy tradeoffs the subjectivity in the definition of equality between fuzzy meanings plays a very important part as the finding of derivative's fuzzy meanings is based on this subjectivity. In this case the question of sensitivity analysis for the proposed choice model is not carried out by the authors.

All the software for the proposed approach computer realization was written in FORTRAN for Simulation and Decision-Making Interactive System-SDMIS.

REFERENCES

Alexeyev, A.V. (1982). Application of fuzzy mathematics in decision analysis. Practical Problems of Decision Analysis in Large Scale Systems. Riga Polytechn. Inst., Riga, pp. 33-39 (in Russian).

Barrager, S. (1980). Assessment of simple joint time/risk preference function. Management Science, Vol. 26, pp. 620-632.

Borisov, A.N., and V.A.Popov (1981). Decision analysis of dynamic alternatives with account of truth. Decision Making Methods and Systems. Riga Polytechn. Inst., Rigã, pp. 11-23 (in Russian).

Borisov, A.N. (Ed.) (1982). Decision Making Models Based on Linguistic Variable. Zinatne, Riga. 256 pp. (in Russian).

Dale, V.A., Krishan, Z.P., and O.H.Piegle (1979). Dynamic Methods of Analysis of Energetic Network Development. Zinatne, Riga. 259 pp. (in Russian).

Dubois D., and H. Prade (1980). Fuzzy Sets and Systems. Theory and Applications. Academic Press, New York. 393 pp.

Dubois D., and H. Prade (1982). The use of fuzzy numbers in decision analysis. In: M.M. Gupta and E. Sanches (Eds.), Fuzzy Information and Decision Processes. North-Holland, p. 301-309.

Fishburn P. (1970). Utility Theory for Decision Making. Wiley, New York. 352 pp.

Keeney R.L., and H. Raiffa (1976). Decisions with Multiple Objectives: Preferences and Value Tradeoffs. Wiley, New York. 569 pp.

Larichev O.I. (1979) Science and Art of Decision Making. Nauka, Moscow, 200 pp. (in Russian).

Meyer R.F. (1977). Preferences over time. In: D. Bell, R. Keeney, and H. Raiffa (Eds.), Conflicting Objectives. Wiley, New York, p. 232-246.

Vilkas E.Y., and E.Z. Maiminas (1981). Decisions: Theory, Information, Modelling. Radio i Svyaz, Moscow. 328 pp. (in Russian).

Yager R. (1981). A procedure for ordering fuzzy subset of the unit interval. Information Sciences, Vol. 24, pp. 142-161.

Zadeh L. (1975). The concept of a linguistic variable and its application to approximate reasoning. I, II, III. Information Sciences, Vols. 8, 9, pp. 199-249, 301-357, 43-80.

INTERACTIVE FUZZY DECISION MAKING FOR MULTIOBJECTIVE LINEAR PROGRAMMING PROBLEMS AND ITS APPLICATION

M. Sakawa

Department of Systems Engineering, Faculty of Engineering, Kobe University, Rokko, Nada, Kobe 657, Japan

Abstract. In this paper, we present an interactive fuzzy decisionmaking method for the solution of multiobjective linear programming problems. By considering the imprecise nature of decision maker's (DM) judgements, we assume that he has fuzzy or imprecise goals for each of the objective functions. Through the use of five types of membership functions including nonlinear functions, the fuzzy or imprecise goals of the DM are quantified. Although the formulated problem becomes a nonlinear programming problem, it can be reduced to a set of linear inequalities if some variable is fixed. Based on this idea, we propose a new method by combined use of bisection method and linear programming method. On the basis of the proposed method, FORTRAN programs are developed to implement man-machine interactive procedures. An application to an optimal operation problem in packaging system in automated warehouses is demonstrated together with the computer outputs.

Keywords. Multiobjective linear programming, Fuzzy decisionmaking, Linear and nonlinear membership functions, Interactive computer program.

INTRODUCTION

In his 1978 paper "Fuzzy programming and linear programming with several objective functions," Zimmermann first presented an application of fuzzy approaches to multiobjective linear programming (MOLP) problems. By adopting the maximizing decision proposed by Bellman and Zadeh (1970) and introducing linear membership functions, he showed that the compromise solution of the decision maker (DM) could be obtained through linear programming technique.

In 1981, Leberling proposed a special nonlinear membership function described with a hyperbolic function in MOLP problems by considering that the rate of increase in membership of satisfaction must not always be constant as in case of a linear membership function. Following the maximizing decision together with a hyperbolic membership function, he proved that there exists an equivalent linear programming problem.

On the other hand, in 1981, Hannan demonstrated how fuzzy or imprecise goals of the DM can be quantified through the use of piecewise linear membership function and proposed three different approaches to linear programming with multiple fuzzy goals.

However, suppose that the interaction with the DM establishes that the first membership function should be linear, the second hyperbolic, the third piecewise linear and so forth. In such a situation, following the maximizing decision the resulting problem becomes a nonlinear programming problem and cannot be solved by a linear programming technique.

In this paper, by considering the imprecise nature of DM's judgements, we assume that he has fuzzy or imprecise goals for each of the objective functions in MOLP problems. In order to quantify the fuzzy goals of the DM by eliciting the corresponding membership functions, we propose five types of membership functions; linear, exponential, hyperbolic, hyperbolic inverse and piecewise linear functions. Through the use of these membership functions including nonlinear ones, the fuzzy or imprecise goals of the DM are quantified. Then following the maximizing decision, the formulated problem becomes a nonlinear programming problem. However, it can be reduced to a set of linear inequalities if some variable is fixed. Based on this idea, we propose a new method by combined use of bisection method and linear programming method. On the basis of the proposed method, FORTRAN programs are developed to implement man-machine interactive procedures. An application to an optimal operation problem in packaging systems in automated warehouses is demonstrated along with the corresponding computer outputs.

INTERACTIVE FUZZY DECISIONMAKING

In general, the multiobjective linear programming (MOLP) problem is represented as

MOLP

$$\begin{array}{ll} \text{maximize} & \left\{ \begin{array}{l} z_1(x) = c_1 x \\ \cdots\cdots\cdots\cdots \\ z_k(x) = c_k x \end{array} \right. \\ \text{subject to} & x \in X = \{x \in R^n \mid Ax = b, \\ & \qquad x \geq 0\} \end{array} \quad (1)$$

where $c_i \in R^n$ $(i=1,2,\ldots,k)$, A is an $m \times n$ matrix, $b \in R^m$.

Fundamental to the MOLP is the Pareto optimal concept, also known as a noninferior solution. Qualitatively, a Pareto optimal solution of the MOLP is one where any improvement of one objective function can be achieved only at the expense of another. Mathematically, two slightly different notions of Pareto optimality are defined.

Definition 1. A decision $x^* \in X$ is said to be a strong Pareto optimal solution to the MOLP, if and only if there does not exist another $\bar{x} \in X$ so that $z_i(\bar{x}) \geq z_i(x^*)$ for all i with strict inequality holding for at least one i.

Definition 2. A decision $x^* \in X$ is said to be a weak Pareto optimal solution to the MOLP, if and only if there does not exist another $\bar{x} \in X$ so that $z_i(\bar{x}) > z_i(x^*)$ for all i.

Note that the set of strong Pareto optimal solutions is a subset of the set of weak Pareto optimal solutions. Usually, strong Pareto optimal solutions consist of an infinite number of points, and some kinds of subjective judgement should be added to the quantitative analyses by the DM. The DM must select his compromise solution from among strong Pareto optimal solutions.

In order to determine the compromise solution of the DM, there are three major approaches:
(1) goal programming (e.g. Charnes and Cooper (1977)).
(2) interactive approach (e.g. Geoffrion et al. (1972), Sakawa (1982)).
(3) fuzzy approach (e.g. Zimmerman (1978), Leberling (1981), Hannan (1981)).
Each of these approaches has its own advantages and disadvantages relative to the other approaches.

However, considering the imprecise nature of the DM's judgements, it is natural to assume that the DM may have fuzzy or imprecise goals for each of the objective functions. For example, a goal stated by the DM may be to achieve "somewhat larger" than A. This type of statement can be quantified by eliciting a corresponding membership function.

In order to elicit a membership function $\mu_{z_i}(x)$ from the DM for each of the objective functions $z_i(x)$, $i=1,2,\ldots,k$, we first calculate the individual minimum z_i^{min} and maximum z_i^{max} of each objective function $z_i(x)$ under given constraints. By taking account of the calculated individual minimum and maximum of each objective function, the DM can select his membership function in a subjective manner from among the following five types of functions; linear, exponential, hyperbolic, hyperbolic inverse and piecewise linear functions. Then the parameter values are determined through the interaction with the DM.

(1) Linear membership function (TYPE 1)
For each objective function, the corresponding linear membership function is defined as follows:

$$\mu_{z_i}(x) = [z_i(x) - z_i^0]/[z_i^1 - z_i^0] \quad (2)$$

The linear membership function can be determined by asking the DM to specify the two points z_i^0 and z_i^1 within z_i^{max} and z_i^{min} where

z_i^0: worst acceptable level for $z_i(x)$

z_i^1: totally desirable level for $z_i(x)$.

Fig. 1 illustrates the graph of the linear membership function.

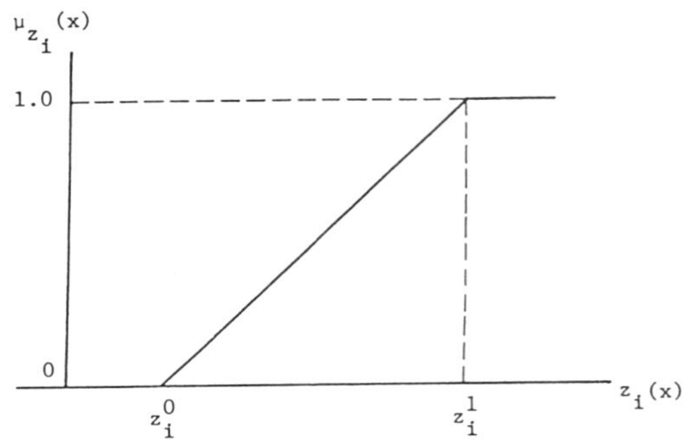

Fig. 1. Linear membership function.

(2) Exponential membership function (TYPE 2)
For each objective function, the corresponding exponential membership function is defined by

$$\mu_{z_i}(x) = a_i[1 - \exp\{- b_i(z_i(x) - z_i^0)/(z_i^1 - z_i^0)\}] \quad (3)$$

where $a_i > 1$, $b_i > 0$ or $a_i < 0$, $b_i < 0$.
The exponential membership function can be determined by asking the DM to specify the three points z_i^0, $z_i^{0.5}$ and z_i^1 within z_i^{max} and z_i^{min}, where z_i^a represents the value of $z_i(x)$ such that the degree of membership function $\mu_{z_i}(x)$ is a.

Fig. 2 illustrates the graph of the exponential membership function.

(3) Hyperbolic membership function (TYPE 3)
For each objective function, the corresponding hyperbolic membership function is defined by

$$\mu_{z_i}(x) = \frac{1}{2} \tanh ((z_i(x) - b_i)\alpha_i) + \frac{1}{2} \quad (4)$$

where $\alpha_i > 0$.

The hyperbolic membership function can be determined by asking the DM to specify the two points $z_i^{0.25}$ and $z_i^{0.5}$ within z_i^{max} and z_i^{min}.

Fig. 3 illustrates the graph of the hyperbolic membership function.

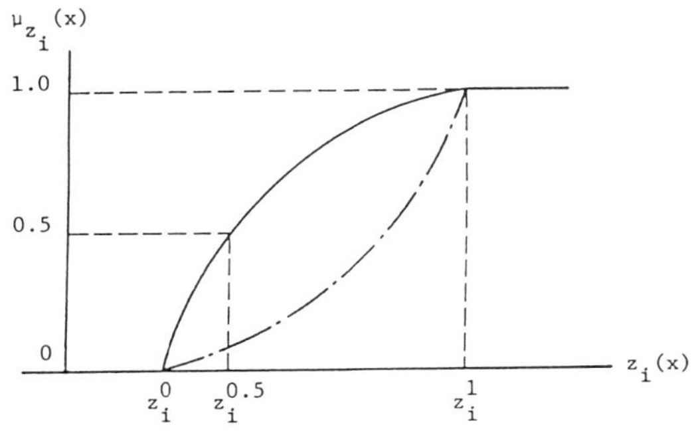

Fig. 2. Exponential membership function.

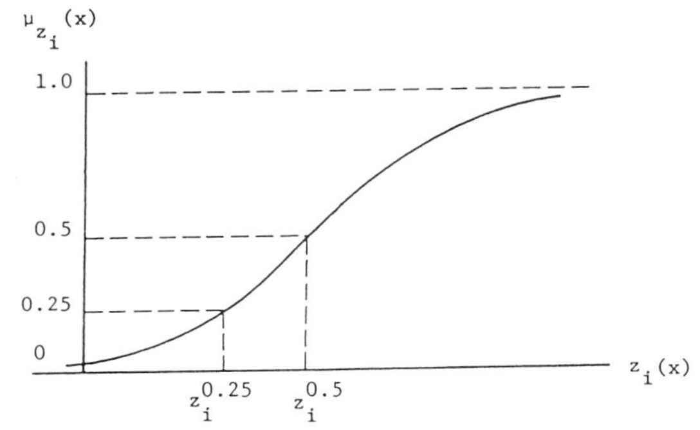

Fig. 3. Hyperbolic membership function.

(4) Hyperbolic inverse membership function (TYPE 4)

For each objective function, the corresponding hyperbolic inverse membership function is defined by

$$\mu_{z_i}(x) = a_i \tanh^{-1}((z_i(x) - b_i)\alpha_i) + \frac{1}{2} \quad (5)$$

where $a_i > 0$; $\alpha_i > 0$.

The hyperbolic inverse membership function can be determined by asking the DM to specify the three points z_i^0, $z_i^{0.25}$ and $z_i^{0.5}$ within z_i^{max} and z_i^{min}.

Fig. 4 illustrates the graph of the hyperbolic inverse membership function.

(5) Piecewise linear membership function (TYPE 5)

For each objective function, the corresponding piecewise linear membership function is defined by

$$\mu_{z_i}(x) = t_{ir} z_i(x) + s_{ir} \quad \text{for}$$

$$f(g_{ir-1}) \leqq \lambda \leqq f(g_{ir}) \quad (6)$$

Here, it is assumed that $\mu_{z_i}(x) = t_{ir} z_i(x) + s_{ir}$ for each segment $g_{ir-1} \leqq z_i(x) \leqq g_{ir}$. That is, t_{ir} is the slope and s_{ir} is the y-intercept for the section of the curve initiated at g_{ir-1} and terminated at g_{ir}.

The piecewise linear membership function can be determined by asking the DM to specify the degree of membership in each of several values of objective functions within z_i^{max} and z_i^{min}.

Fig. 5 illustrates the graph of the piecewise linear membership function.

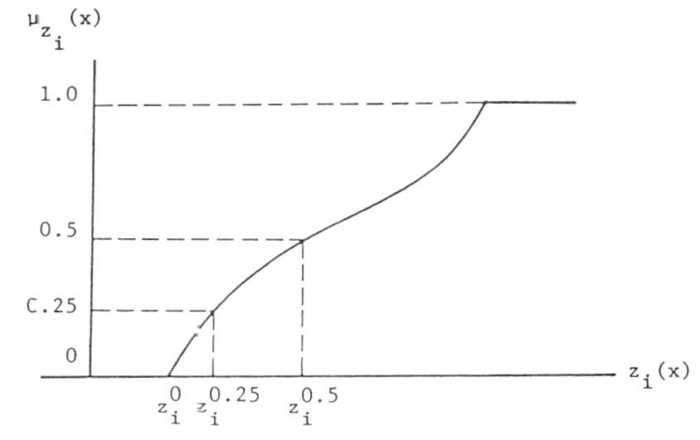

Fig. 4. Hyperbolic inverse membership function.

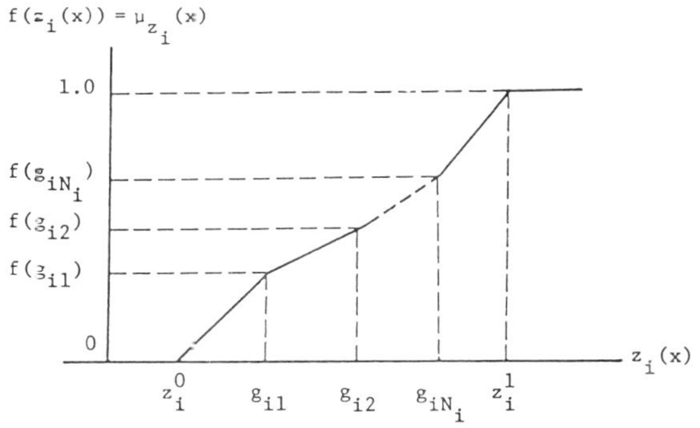

Fig. 5. Piecewise linear membership function.

After determining the membership functions for each of the objective functions, if we adopt the maximizing decision proposed by Bellman and Zadeh (1970) the resulting problem to be solved is

$$\max_{x \in X} \{\min_i (\mu_{z_i}(x))\} \quad (7)$$

Namely, the problem is to maximize the minimum membership function value. As is well known, this problem is equivalent to solving the following problem:

$$\max \quad \lambda$$

$$\text{subject to } \lambda \leq \mu_{z_i}(x) \quad i=1,2,\ldots,k \qquad Ax = b, \quad x, \lambda \geq 0 \tag{8}$$

However, with the five types of membership functions given by (2)-(6), the resulting problem is a nonlinear programming problem. In order to solve the formulated problem on the basis of linear programming method, if we convert each constraint $\lambda \leq \mu_{z_i}(x)$ into the form like $z_i(x) \geq D_i(\lambda)$, we arrive at the following problem-formulation.

$$\max \ \lambda$$

$$\left.\begin{array}{ll} \text{subject to} & \\ z_i(x) \geq \lambda(z_i^1 - z_i^0) + z_i^0 & (i \in \text{TYPE 1}) \\ z_i(x) \geq z_i^0 - ((z_i^1 - z_i^0)/b_i) \cdot \log((a_i - \lambda)/a_i) & (i \in \text{TYPE 2}) \\ z_i(x) \geq b_i + (1/\alpha_i) \tanh^{-1}(2\lambda - 1) & (i \in \text{TYPE 3}) \\ z_i(x) \geq b_i + (1/\alpha_i) \tanh((\lambda - 1/2)/a_i) & (i \in \text{TYPE 4}) \\ z_i(x) \geq (\lambda - s_{ir})/t_{ir} \ \text{ for } f(g_{ir-1}) \leq \lambda \leq f(g_{ir}) & (i \in \text{TYPE 5}) \\ Ax = b, \quad 0 \leq \lambda \leq 1, \ x \geq 0. & \end{array}\right\} \tag{9}$$

It is important to note that in this formulation, if the value of λ is fixed, it can be reduced to a set of linear inequalities. Obtaining the optimal solution λ^* to the above problem is equivalent to determining the maximum value of λ so that there exists an admissible set satisfying the constraints of (9). Since λ satisfies $0 \leq \lambda \leq 1$, we have the following method for solving this problem by combined use of the bisection method and the phase one of linear programming technique.

Step 1. Set $\lambda = 0$ and test whether an admissible set satisfying the constraints of (9) exists or not by making use of the phase one of the simplex method. If an admissible set exists, proceed. Otherwise, the DM must reassess his membership function.

Step 2. Set $\lambda = 1$ and test whether an admissible set satisfying the constraints of (9) exists or not using the simplex method. If an admissible set exists, the maximum degree of membership $\lambda^* = 1$ is achieved. Otherwise go to the next step, since the maximum λ which satisfies the constraints of (9) exists between 0 and 1.

Step 3. For the initial value of $\lambda_1 = 0.5$, update the value of λ using the bisection method as follows:

$$\begin{cases} \lambda_{n+1} = \lambda_n + 1/2^{n+1} & \text{if admissible set exists for } \lambda_n \\ \lambda_{n+1} = \lambda_n - 1/2^{n+1} & \text{if no admissible set exists for } \lambda_n. \end{cases}$$

Namely, for each λ_n (n=1,2,...), test whether an admissible set of (9) exists or not using the simplex method and determine the maximum value of λ satisfying the constraints of (9).

In this way, we can determine the optimal solution λ^*. Then, the DM selects his most important objective from among the objectives $(z_i(x))$. If it is the jth objectives, the following linear programming problem is solved for $\lambda = \lambda^*$.

$$\left.\begin{array}{l} \max \ z_j(x) \\ \text{subject to the constraints of (9)} \end{array}\right\} \tag{10}$$

It can be easily shown that each optimal solution to (10), say x^*, satisfies weak Pareto optimality.

In order to test the strong Pareto optimality, we solve the following linear programming problem:

$$\left.\begin{array}{l} \bar{V} = \max \sum_{j=1}^{k} \varepsilon_j \\ \text{subject to} \\ z_j(x) - \varepsilon_j = z_j(x^*) \\ x \in X, \quad \varepsilon > 0 \end{array}\right\} \tag{11}$$

Let $\bar{x}$ be an optimal solution to (11). If $\bar{V} = 0$ and all $\varepsilon_j = 0$, then x^* is a strong Pareto optimal solution. In case of $\bar{V} > 0$ and consequently at least one $\varepsilon_j > 0$, we adopt the solution $\bar{x}$ as the compromise solution of the DM, because it can easily be shown that $\bar{x}$ is a strong Pareto optimal solution.

AN INTERACTIVE COMPUTER PROGRAM AND ITS APPLICATION

Fuzzy decisionmaking processes for multiobjective linear programming problems include eliciting a membership function from the DM for each of the objective functions. Thus, mitigation and speed-up of computation works are indispensable to this approach, and interactive utilization of computer facilities is highly recommended. Based on the method described above, we have developed a new interactive computer program. Our new package includes graphical representations by which the DM can figure the shape of his membership functions, and he can find incorrect assessments or inconsistent evaluations promptly, revise them immediately and proceed to the next stage more easily.

Our program is composed of one main program and several subroutines. The main program calls in and runs the subprograms with commands indicated by the user (DM). Here we give a brief explanation of the commands prepared in our program.

(1) MINMAX: Displays the calculated individual minimum and maximum of each of the objective functions under the given constraints.
(2) MF: Elicits a membership function from the DM for each of the objective functions.
(3) GRAPH: Depicts graphically the shape of the membership function for each of the objective function.
(4) GO: Solves the maximization problem (9) by the proposed method.
(5) STOP: Exists from the program.

Now consider an application of our interactive computer program to an optimal operation of a packaging system in automated warehouses. The problem formulation was originally given by Hamada (1976). Then three objective linear programming problem formulation was developed by Nakayama et al.(1980) and the following numerical example was solved by the interactive multiobjective decisionmaking method called IRM (Interactive Relaxation Method).

$$\max z_1 = x_1 + x_2 + 5x_3 + 5x_4 + 4x_5 + 4x_6 + 2x_7 + 2x_8$$

$$\min z_2 = x_3 + x_4 + x_5 + x_6$$

$$\min z_3 = 5.6x_1 - 5.6x_2 - 2.4x_3 + 2.4x_4 - 5.25x_5 + 5.25x_6 - 8.4x_7 + 8.4x_8$$

subject to

$$x_1 + x_2 \leqslant 1000,\ x_3 + x_4 \leqslant 200,\ x_5 + x_6 \leqslant 250,\ x_7 + x_8 \leqslant 500$$

$$-x_1 + x_2 \leqslant 300,\ x_3 - x_4 \leqslant 100,\ x_5 - x_6 \leqslant 200,\ x_7 - x_8 \leqslant 200$$

$$x_1 + x_3 + x_5 + x_7 \leqslant 1500,\ x_2 + x_4 + x_6 + x_8 \leqslant 1500$$

$$500 \leqslant x_1 + x_2 + x_5 + x_6 \leqslant 1500,\ 300 \leqslant x_3 + x_4 + x_7 + x_8 \leqslant 1000$$

$$2x_1 + 2x_2 + x_3 + x_4 \leqslant 3000,\ x_1 + x_2 + 6x_3 + 6x_4 \leqslant 1000$$

$$3x_5 + 3x_6 + x_7 + x_8 \leqslant 800,\ x_5 + x_6 + 2x_7 + 2x_8 \leqslant 1400$$

In applying our computer program to this problem, suppose that interaction with the hypothetical DM establishes the following membership functions and corresponding assessment values for the three objective functions.

z_1: exponential membership function

$$(z_1^0, z_1^{0.5}, z_1^1) = (1820, 2260, 2400)$$

z_2: linear membership function

$$(z_2^0, z_2^1) = (195, 0)$$

z_3: hyperbolic inverse membership function

$$(z_3^0, z_3^{0.25}, z_3^{0.5}) = (-1350, -1602, -2950)$$

In Appendix, the interaction processes using our computer program are shown with the aid of some of the computer outputs. The obtained compromise solutions achieve nearly equal membership function values for the three objective functions. Furthermore, the strong Pareto optimality test yields

$$\varepsilon_1 + \varepsilon_2 + \varepsilon_3 \cong 0$$

which means the obtained solutions approximately satisfy the strong Pareto optimality.

CONCLUSION

In this paper, by adopting the maximizing decision together with the five types of membership functions including nonlinear functions, we have proposed an interactive method by combined use of bisection method and linear proramming method in order to deal with the fuzzy goals of the DM in multiobjective linear programming problems. By performing a strong Pareto optimality test, strong Pareto optimality of the compromise solution of the DM is also guaranteed in our method. Based on the proposed method, the time-sharing computer program has been written to facilitate the interactive processes.

An application to an optimal operation problem in packaging systems in automated warehouses demonstrates the feasibility and efficiency of both the proposed method and its interactive computer program under the hypothetical DM. Although the actual DM for the formulated problem would of course compromise other values of the three objectives than the ones which were compromised by the hypothetical DM used in this paper, the way to iterate and calculate is essentially the same.

REFERENCES

Bellman, R.E. and L.A. Zedah (1970). Decision making in a fuzzy environment. Management Sci., 17, 141-164.
Charnes, A. and W.W. Cooper (1977). Goal programming and multiple objective optimizations. European J. Operational Res., 1, 39-54.
Geoffrion, A.M., J.S. Dyer and A. Feinberg

(1972). An interactive approach for multi-criteria optimization, with an application to the operation of an academic department. Management Sci., 19, 357-368.

Hamada, H. (1976). Studies on optimal operations in automated warehouses, Doctoral Dissertation, Osaka University.

Hannan, E.L. (1981). Linear programming with multiple fuzzy goals, Fuzzy Sets and Systems, 6, 235-248.

Leberling, H. (1981). On finding compromise solution in multicriteria problems using the fuzzy min-operator, Fuzzy Sets and Systems, 6, 105-118.

Nakayama, H., Y. Karasawa and S. Dohi (1980). Subjective programming applied to optimal operation in automated warehouses. Int. J. Systems Sci., 11, 513-525.

Sakawa, M. (1982). Interactive multiobjective decision making by the sequential proxy optimization technique: SPOT. European J. Operational Res., 9, 386-396.

Zimmermann, H.J. (1978). Fuzzy programming and linear programming with several objective functions. Fuzzy Sets and Systems, 1, 45-55.

APPENDIX

INTERACTIVE FUZZY DECISIONMAKING PROCESSES

```
COMMAND:
=MINMAX

           INDIVIDUAL MINIMUM AND MAXIMUM

          I        MINIMUM          I        MAXIMUM
 ---------+-------------------------+-------------------------
 Z( 1)    I    0.1100000000D+04     I    0.2400000000D+04
 Z( 2)    I    0.                   I    0.3235294118D+03
 Z( 3)    I   -0.4650000000D+04     I    0.1032500000D+05
```

```
COMMAND:
=MF
INPUT THE OBJECTIVE FUNCTION NUMBER:
=1
DO YOU WANT LIST OF MEMBERSHIP FUNCTION TYPE ?
=YES
LIST OF MEMBERSHIP FUNCTION TYPE
 (1) LINEAR
 (2) EXPONENTIAL
 (3) HYPERBOLIC
 (4) HYPERBOLIC INVERSE
 (5) PIECEWISE LINEAR
INPUT MEMBERSHIP FUNCTION TYPE:
=2
INPUT THREE POINTS(Z1,Z2,Z3) SUCH THAT
          M(Z1)=0.0 ( Z1 : WORST ACCEPTABLE LEVEL )
          M(Z2)=0.5
          M(Z3)=1.0 ( Z3 : TOTALLY DESIRABLE LEVEL )
=1820 2260 2400
ANOTHER MSET ?
=YES
INPUT THE OBJECTIVE FUNCTION NUMBER:
=2
DO YOU WANT LIST OF MEMBERSHIP FUNCTION TYPE ?
=NO
INPUT MEMBERSHIP FUNCTION TYPE:
=1
INPUT TWO POINTS(Z1,Z2) SUCH THAT
          M(Z1)=0.0 ( Z1 : WORST ACCEPTABLE LEVEL )
          M(Z2)=1.0 ( Z2 : TOTALLY DESIRABLE LEVEL )
=195 0
ANOTHER MSET ?
=YES
INPUT THE OBJECTIVE FUNCTION NUMBER:
=3
DO YOU WANT LIST OF MEMBERSHIP FUNCTION TYPE ?
=NO
INPUT MEMBERSHIP FUNCTION TYPE:
=4
INPUT THREE POINTS(Z1,Z2,Z3) SUCH THAT
          M(Z1)=0.00 ( Z1 : WORST ACCEPTABLE LEVEL )
          M(Z2)=0.25
          M(Z3)=0.50
=-1350 -1602 -2950
ANOTHER MSET ?
=NO
```

```
COMMAND:
=GRAPH
INPUT THE MEMBERSHIP FUNCTION NUMBER:
=1
GRAPH OF THE MEMBERSHIP FUNCTION ( NO. 1)
  MEMBERSHIP FUNCTION TYPE --- EXPONENTIAL
```

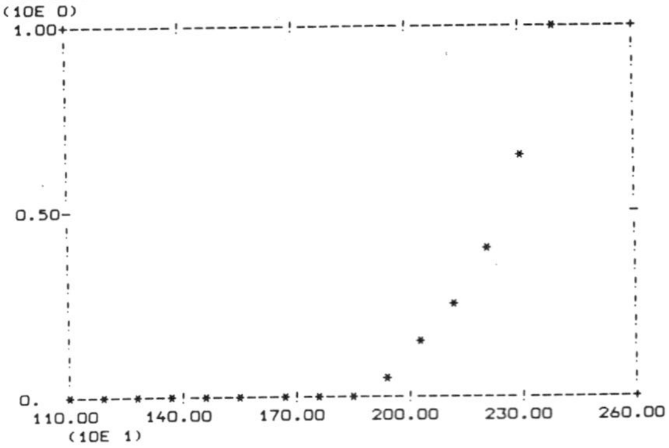

. . . .

```
COMMAND:
=GO

LAMBDA-MAX CALCULATION BY BISECTION METHOD

       LAMBDA                LAMBDA-MAX             FEASIBILITY
  1    0.                    0.                         F
  2    0.1000000000D+01      0.                         I
 ----------------------------------------------------------
  1    0.5000000000D+00      0.5000000000D+00           F
  2    0.7500000000D+00      0.5000000000D+00           I
  3    0.6250000000D+00      0.5000000000D+00           I
  4    0.5625000000D+00      0.5625000000D+00           F
  5    0.5937500000D+00      0.5937500000D+00           F
  6    0.6093750000D+00      0.6093750000D+00           F
  7    0.6171875000D+00      0.6093750000D+00           I
  8    0.6132812500D+00      0.6132812500D+00           F
  9    0.6152343750D+00      0.6132812500D+00           I
 10    0.6142578125D+00      0.6142578125D+00           F
 11    0.6147460938D+00      0.6142578125D+00           I
 12    0.6145019531D+00      0.6145019531D+00           F
 13    0.6146240234D+00      0.6145019531D+00           I
 14    0.6145629883D+00      0.6145629883D+00           F
 15    0.6145935059D+00      0.6145935059D+00           F
 16    0.6146087646D+00      0.6146087646D+00           F
 17    0.6146163940D+00      0.6146163940D+00           F
 18    0.6146202087D+00      0.6146202087D+00           F
 19    0.6146221161D+00      0.6146202087D+00           I
 20    0.6146211624D+00      0.6146202087D+00           I

OPTIMAL LAMBDA = 0.6146202087D+00


MAXIMIZE Z( 1)

   -------------------------------------------------------------
     X( 1)= 0.3500000000D+03
     X( 2)= 0.6500000000D+03
     X( 3)= 0.
     X( 4)= 0.
     X( 5)= 0.7514861263D+02
     X( 6)= 0.4466654608D-03
     X( 7)= 0.3500000000D+03
     X( 8)= 0.1500000000D+03
   -------------------------------------------------------------
             I   OBJECTIVE FUNCTION    I        MEMBERSHIP
   ----------+-------------------------+------------------------
     Z( 1)   I     0.2300596237D+04    I     0.6146993378D+00
     Z( 2)   I     0.7514905930D+02    I     0.6146202087D+00
     Z( 3)   I    -0.3754527871D+04    I     0.6146202087D+00

PARETO OPTIMALITY TEST

MAXIMIZE SUM(EPS)

   -------------------------------------------------------------
       X( 1)= 0.3500000000D+03
       X( 2)= 0.6500000000D+03
       X( 3)= 0.
       X( 4)= 0.
       X( 5)= 0.7514905930D+02
       X( 6)= 0.
       X( 7)= 0.3500000000D+03
       X( 8)= 0.1500000000D+03
     EPS( 1)= 0.
     EPS( 2)= 0.
     EPS( 3)= 0.4689987339D-02
   -------------------------------------------------------------
    SUM(EPS)= 0.4689987339D-02
   -------------------------------------------------------------
             I     PARETO OPTIMUM      I        MEMBERSHIP
   ----------+-------------------------+------------------------
     Z( 1)   I     0.2300596237D+04    I     0.6146993378D+00
     Z( 2)   I     0.7514905930D+02    I     0.6146202087D+00
     Z( 3)   I    -0.3754532561D+04    I     0.6146210153D+00
```

OPTIMIZATION IN FUZZY ENVIRONMENT

D. Ralescu

Department of Mathematics, University of Cincinnati, Cincinnati, Ohio 45221 U.S.A.

Abstract. In this paper we study the possibility of representing an optimization problem with inexact constraints, as a fuzzy integral. This integral is to be taken with respect to a capacity (or to an outer measure) rather than to a fuzzy measure.

We prove a mean-value theorem for the fuzzy integral, which has as a consequence the possibility of reducing optimization with inexact constraints to classical optimization. We also give some sufficient conditions such that this reduction holds.

Keywords. Fuzzy constraint; fuzzy measure, fuzzy integral.

1. INTRODUCTION

The concepts of decision-making in a fuzzy environment were defined by Bellman and Zadeh [1]. To recall briefly this approach, let X be a set, $P : X \to [0,1]$ a cost function, and $u : X \to [0,1]$ a fuzzy constraint. Note that P is a fuzzy set (fuzzy goal, in [1]) which can be derived from a positive, bounded function on X, by a simple normalization.

The optimization problem $\sup_u P$ is defined by

$$\sup_u P = \sup_{x \in X} [P(x) \wedge u(x)] \tag{1.1}$$

where $\wedge = \min$.

The dynamic programming approach was used in [1] to solve optimization problems with inexact constraints, mainly where X is a finite set.

In [12], Tanaka, Okuda and Asai are led to a reformulation of the problem (1.1), under the name fuzzy mathematical programming. Namely, it can be shown (see the short proof in [5, p. 160]) that

$$\sup_u P = \sup_{\alpha > 0} [\alpha \wedge \sup_{u(x) \geq \alpha} P(x)] \tag{1.2}$$

Written under this form, it was observed that the right-hand side in (1.2) is related to the fuzzy integral [11].

To explain this point let $\mathcal{A}$ be a σ-algebra of subsets of X. A fuzzy measure is a positive, extended real-valued set function $\mu : \mathcal{A} \to [0,\infty]$, with the properties:

(FM1) $\mu(\emptyset) = 0$.

(FM2) $A \subset B \Rightarrow \mu(A) \leq \mu(B)$.

(FM3) $A_1 \subset A_2 \subset \ldots \Rightarrow \mu(\bigcup_{n=1}^{\infty} A_n) = \lim_{n\to\infty} \mu(A_n)$.

(FM4) $A_1 \supset A_2 \supset \ldots,\ \mu(A_1) < \infty \Rightarrow \mu(\bigcap_{n=1}^{\infty} A_n) = \lim_{n\to\infty} \mu(A_n)$.

Let $f : X \to [0,\infty)$ be a measurable function (i.e. $\{f \geq \alpha\} \in \mathcal{A}$ for any $\alpha \geq 0$). The fuzzy integral of f with respect to μ is

$$\int_X f\, d\mu = \sup_{\alpha \geq 0} (\alpha \wedge \mu\{f \geq \alpha\}) \tag{1.3}$$

where $\{f \geq \alpha\} = \{x \in X | f(x) \geq \alpha\}$. This definition was given in [11] in a more restrictive context. Properties of this integral, especially convergence theorems, are studied in [9].

It is now clear the similarity between (1.2) and (1.3). If, for a cost function $P : X \to [0,1]$, we define the set function $\mu_P : \mathcal{P}(X) \to [0,1]$ by

$$\mu_P(A) = \sup_{x \in A} P(x) \tag{1.4}$$

where $\sup_{\emptyset} P = 0$, then (1.2) can be written under the form:

$$\sup_u P = \int_X u\, d\mu_P \tag{1.5}$$

It was pointed out in [4, p. 1629], [6, p. 665], [11, p. 12] that the set function μ_P is a fuzzy measure.

As we shall see in Section 2, μ_P always has properties (FM1) - (FM3), for any set X and any cost function P. In which property (FM4) is concerned, we show that it only holds for μ_P in trivial cases (see also [8]).

We study in Section 2 the set function μ_P; we prove that μ_P should be viewed as an outer measure, rather than as a fuzzy measure. In a topological setting, we show that μ_P is a capacity (as it was also mentioned in [7]).

In Section 3 we give a general mean-value theorem for the fuzzy integral, and see its relevance to optimization with inexact constraints. One of the main result in [12] is a consequence of this mean-value theorem.

In Section 4 we give sufficient conditions that optimization on fuzzy sets can be reduced to classical optimization (this will be called the "reduction problem"). These conditions are more general than those in [12].

Related work on fuzzy mathematical programming was recently reported in [3].

2. THE SET FUNCTION μ_P

Let us consider first X to be an arbitrary set, and $P:X \to [0,\infty)$ a positive function. We can prove that the set function μ_P as defined by (1.4) satisfies properties (FM1)-(FM3) of a fuzzy measure:

PROPOSITION 1. (a) $A \subset B \Rightarrow \mu_P(A) \leq \mu_P(B)$.

(b) $A_1 \subset A_2 \subset \ldots \Rightarrow \mu_P(\bigcup_{n=1}^{\infty} A_n) = \lim_{n\to\infty} \mu_P(A_n)$.

(c) μ_P is an outer-measure.

Proof. (a) is obvious.

(b) Denote by $A = \bigcup_{n=1}^{\infty} A_n$; we have $\mu_P(A_1) \leq \mu_P(A_2) \leq \ldots \leq \mu_P(A)$ thus $\lim_{n\to\infty} \mu_P(A_n) = \sup_{n\geq 1} \mu_P(A_n)$ exists, and

$$\sup_{n\geq 1} \mu_P(A_n) \leq \mu_P(A) \tag{2.1}$$

Let us suppose strict inequality in (2.1); then there is an $a \in A$ with $\sup_{n\geq 1} \mu_P(A_n) < P(a)$. But $a \in A_{n_0}$ for some $n_0 \geq 1$. We get:

$\mu_P(A_{n_0}) = \sup_{x\in A_{n_0}} P(x) < P(a)$, a contradiction.

(c) It only remains to show that μ_P is subadditive, i.e.

$$\mu_P(\bigcup_{n=1}^{\infty} A_n) \leq \sum_{n=1}^{\infty} \mu_P(A_n) \tag{2.2}$$

for any sequence $\{A_n\}$ of subsets of X.

In the same way as in (b), we can show that

$$\mu_P(\bigcup_{n=1}^{\infty} A_n) = \sup_{n\geq 1} \mu_P(A_n) \tag{2.3}$$

and (2.2) immediately follows.

Let us show now that property (FM4) is not generally true for μ_P:

Example 1. Take $X = \mathbb{R}$, $P(x) = 1$, and the sequences of sets $A_n = (n,\infty)$. We observe that $\mu_P(\bigcap_{n=1}^{\infty} A_n) = \mu_P(\emptyset) = 0$, while

$\lim_{n\to\infty} \mu_P(A_n) = 1.$

This simple example shows that property (FM4) is much too strong to be satisfied by μ_P.

In fact, it can be shown that, under fairly general hypotheses, if μ_P satisfies (FM4), then P is a "trivial" function.

To be more precise, let us consider X to be a metric space without isolated points. If we take the cost function $P(x) = 0$, then μ_P satisfies (FM4). Under the assumption of continuity, the converse is also true:

PROPOSITION 2. If the function P is continuous and μ_P satisfies (FM4), then $P(x) = 0$, for any $x \in X$.

Proof. Let us take $x_0 \in X$, and let $B(x_0,1/n)$ be the ball with center x_0 and radius $1/n$. Since x_0 is not an isolated point, it follows that $A_n = B(x_0,1/n) \setminus \{x_0\}$ is nonempty, for all $n \geq 1$. Also $A_1 \supset A_2 \supset \ldots$ and $\bigcap_{n=1}^{\infty} A_n = \emptyset$. Since μ_P satisfies (FM4), it follows that $\mu_P(A_n) \to 0$.

Let $\{x_j\}_j \subset X$ be a sequence, such that $x_j \to x_0$, $x_j \neq x_0$. Let $n \geq 1$ be fixed; then $x_j \in B(x_0,1/n) \setminus \{x_0\}$ for all $j \geq j_n$. Thus: $P(x_j) \leq \sup_{x\in A_n} P(x)$ for $j \geq j_n$. It follows that

$0 \leq \limsup_{j\to\infty} P(x_j) \leq \sup_{x\in A_n} P(x) = \mu_P(A_n)$. Since this is true for any $n \geq 1$, we conclude that $\limsup_{j\to\infty} P(x_j) = 0$, thus $\lim_{j\to\infty} P(x_j) = 0$. Finally: $\lim_{y\to x_0} P(y) = 0$; since P is continuous, it follows that $P(x_0) = 0$.

We see now that, except for functions which are "almost" zero (P = 0 at any point of continuity), it is not to be expected that μ_P is a fuzzy measure.

By a suitable change of (FM4) we can, however, prove that μ_P is a capacity in the sense of Choquet (see [2]).

Suppose that X is a topological space. A capacity is a set function $C: P(X) \to [0,\infty]$ with properties:

(C1) $A \subset B \Rightarrow C(A) \leq C(B)$.

(C2) $A_1 \subset A_2 \subset \ldots \Rightarrow C(\bigcup_{n=1}^{\infty} A_n) = \lim_{n\to\infty} C(A_n)$.

(C3) $K_1 \supset K_2 \supset \ldots, K_n$ compact $\Rightarrow C(\bigcap_{n=1}^{\infty} K_n) = \lim_{n\to\infty} C(K_n)$.

Proposition 3. If P is upper semi-continuous, then μ_P is a capacity. Moreover $\mu_P(K) < \infty$, for any compact subset K of X.

Proof. Since P is upper semi-continuous, then $\sup_{x\in K} P(x) < \infty$ for any compact set K; therefore the last assertion follows.

Properties (C1) and (C2) were already proved in Proposition 1. It only remains to prove (C3); let $K_1 \supset K_2 \supset \ldots, K_n$ compact. We always have

$$\lim_{n\to\infty} \mu_P(K_n) = \inf_{n\geq 1} \mu_P(K_n) \geq \mu_P(K) \qquad (2.4)$$

where $K = \bigcap_{n=1}^{\infty} K_n$. Suppose that strict inequality holds in (2.4). Then: $\inf_{n\geq 1} \mu_P(K_n) > c > \mu_P(K)$, for some $c > 0$. Thus $\mu_P(K_n) = \sup_{K_n} P > c > \mu_P(K)$, for any $n \geq 1$.

It follows: $(\exists)\ x_n \in K_n$, such that $P(x_n) > c > \mu_P(K)$. Since $\{x_n\} \subset K_1$, there is a subsequence $x_{n_j} \to x_0 \in K$. From $P(x_{n_j}) > c > \mu_P(K)$ and the upper semi-continuity of P, we conclude that $P(x_0) \geq c > \mu_P(K)$, a contradiction.

It is clear that the integral representation (1.5) still holds, where μ_P is viewed either as an outer measure, or as a capacity.

3. THE MEAN-VALUE THEOREM

In this section we give a mean-value theorem for the fuzzy integral. Its applicability will be in an immediate proof of a basic result in fuzzy mathematical programming.

Let us consider X a set, A a σ-algebra of subsets, and $\mu: A \to [0,\infty)$ a set function, with the only requirement of being non-decreasing:

$$A, B \in A,\ A \subset B \Rightarrow \mu(A) \leq \mu(B) \qquad (3.1)$$

If $f: X \to [0,\infty)$ is a measurable function, the fuzzy integral can still be defined by:

$$\int_X f\, d\mu = \sup_{\alpha\geq 0} (\alpha \wedge \mu\{f \geq \alpha\}) \qquad (3.2)$$

The next result is a counterpart of the mean-value theorem in classical measure theory:

THEOREM 1. If the function $F(\alpha) = \mu\{f \geq \alpha\}$ is continuous, then there exists an $\bar{\alpha} \geq 0$, such that

$$\int_X f\, d\mu = \mu\{f \geq \alpha\} \qquad (3.3)$$

Proof. Consider the function $F: [0,\infty) \to [0,\infty)$; clearly it is impossible that $F(\alpha) > \alpha$ for any $\alpha \geq 0$. From this, and the fact that F is continuous, it follows that F has a fixed point: $(\exists)\ \bar{\alpha} \geq 0$, $F(\bar{\alpha}) = \bar{\alpha}$. We show that

$$\int_X f\, d\mu = \bar{\alpha} = F(\alpha).$$

Suppose $\alpha < \bar{\alpha}$; then $F(\alpha) \geq F(\bar{\alpha}) = \bar{\alpha} > \alpha$. Thus $\alpha \wedge F(\alpha) = \alpha < \bar{\alpha}$. If $\alpha > \bar{\alpha}$, then $F(\alpha) \leq F(\bar{\alpha}) = \bar{\alpha} < \alpha$. Then $\alpha \wedge F(\alpha) = F(\alpha) \leq \bar{\alpha}$. Thus $\bar{\alpha} \wedge F(\alpha) \leq \bar{\alpha}$, for any $\alpha \geq 0$. Therefore

$$\int_X f\, d\mu = \sup_{\alpha\geq 0} [\alpha \wedge F(\alpha)] \leq \bar{\alpha} \qquad (3.4)$$

But $\bar{\alpha} = \bar{\alpha} \wedge F(\bar{\alpha}) \leq \int_X f\, d\mu$, which ends the proof.

Applied to a cost function $P: X \to [0,1]$ and to the set function μ_P, Theorem 1 yields the following

COROLLARY 1. ([12]). If the function $F(\alpha) = \sup_{u(x)\geq\alpha} P(x)$ is continuous, then there exists an $\bar{\alpha} \in [0,1]$, such that

$$\sup_u P = \sup_{u(x)\geq\bar{\alpha}} P(x).$$

The meaning of this result is the following: if F is continuous, then the fuzzy mathematical programming problem $\sup_u P$ can be reduced to the ordinary optimization problem $\sup_{u(x)\geq\bar{\alpha}} P(x)$, for some $\bar{\alpha} \in (0,1]$ (if $P \neq 0$).
In other words, for a given cost function P and fuzzy constraint u, only part of the information contained in u is needed.

We shall call the possibility of writing $\sup_u P = \sup_{u(x)\geq\bar{\alpha}} P(x)$, the reduction problem.
In the next section we discuss some sufficient conditions for this reduction problem.

4. SUFFICIENT CONDITIONS FOR THE REDUCTION PROBLEM

The hypothesis of continuity for F is complicated from the practical point of view. This was also observed in [12], where a sufficient condition is given in terms of fuzzy convexity. We shall given here a new proof of this result, under more general assumptions. Our conditions will also include, as a particular case, classical optimization (i.e. $u = \chi_A$, the characteristic function of a compact set A).

Let us consider X a topological space. The fuzzy constraint $u : X \to [0,1]$ is supposed to be upper semi-continuous (u.s.c.), with a compact support supp u.

The cost function $P : X \to [0,1]$ is also assumed to be u.s.c.

LEMMA 1. The function $F(\alpha) = \sup_{u(x)\geq\alpha} P(x)$ is left continuous.

Proof. Consider $\{\alpha_n\}_n \subset [0,1]$, $\alpha_1 \leq \alpha_2 \leq \dots$, $\alpha_n \to \alpha_0$. Then $\{u\geq\alpha_1\} \supset \{u\geq\alpha_2\} \supset \dots$ and $\bigcap_{n=1}^{\infty} \{u\geq\alpha_n\} = \{u\geq\alpha_0\}$.
Also $\{u\geq\alpha_n\}$ are compact, since u is u.s.c. and supp u is compact. Then, by Proposition 3, we get $F(\alpha_0) = \lim_{n\to\infty} F(\alpha_n)$.

Suppose now X a topological vector space. A fuzzy set $u:X \to [0,1]$ is called strictly convex if

$$u(\lambda x + (1-\lambda)y) > u(x) \wedge u(y) \qquad (4.1)$$

for any $x,y \in X$, $x \neq y$, and $\lambda \in (0,1)$.

THEOREM 2 ([12]). If P is continuous, u is u.s.c., strictly convex, and supp u is compact, there exists an $\bar{\alpha} \in (0,1]$ such that

$$\sup_u P = \sup_{u(x)\geq\bar{\alpha}} P(x).$$

Proof. We only have to check the right-continuity $F(\alpha) = \sup_{u(x)\geq\alpha} P(x)$. Take $\alpha_1 \geq \alpha_2 \geq \dots$, $\alpha_n \to \bar{\alpha}_0$, $\alpha_0 \neq 1$; then $\{u\geq\alpha_1\} \subset \{u\geq\alpha_2\} \subset \dots$ and $\bigcup_{n=1}^{\infty} \{u\geq\alpha_n\} = \{u>\alpha_0\}$.
By Proposition 1, then $F(\alpha_n) \to \sup_{u(x)>\alpha_0} P(x)$.

If we show that $\sup_{u(x)>\alpha_0} P(x) = \sup_{u(x)\geq\alpha_0} P(x) = F(\alpha_0)$, we are done. Suppose that strict inequality $\sup_{u(x)>\alpha_0} P(x) < \sup_{u(x)\geq\alpha_0} P(x)$ holds. Then ($\exists$) $a \in \{u \geq \alpha_0\}$ with $\sup_{u(x)>\alpha_0} P(x) < P(a)$. In fact $u(a) = \alpha_0$, obviously. Let us fix an $x \in \{u>\alpha_0\}$; clearly $x \neq a$ and, by the strict convexity of u, we can write:

$$u((1-1/n)a + (1/n)x) > u(a) \wedge u(x) = \alpha_0 \qquad (4.2)$$

Thus $(1-1/n)a + (1/n)x \in \{u>\alpha_0\}$ and $(1-1/n)a+(1/n)x \to a$. Therefore:
$P((1-1/n)a + (1/n)x \leq \sup_{u(x)>\alpha_0} P(x) < P(a)$.

By the continuity of P we get $P(a) \leq \sup_{u(x)>\alpha_0} P(x) < P(a)$, a contradiction which ends the proof.

Other sufficient conditions are given in the next theorem.

Recall that a function $u:X \to [0,1]$ is called concave, if:

$$u(\lambda x + (1-\lambda)y) \geq \lambda u(x) + (1-\lambda)u(y) \qquad (4.3)$$

for any $x,y \in X$, $\lambda \in [0,1]$.

Observe that u concave implies u quasi-concave (i.e. $u(\lambda x+(1-\lambda)y) \geq u(x) \wedge u(y)$, see [10]), but, generally, u concave does not imply u strictly fuzzy convex.

THEOREM 3. If P is continuous, u is u.s.c., concave, and supp u is compact, then there exists an $\bar{\alpha} \in (0,1]$ such that

$$\sup_u P = \sup_{u(x)\geq\bar{\alpha}} P(x).$$

Proof. The proof is carried out in much the same way as in the previous theorem.

Instead of (4.2), we have:

$$u((1-1/n)a+(1/n)x) \geq (1-1/n)u(a)+(1/n)u(x)>\alpha \qquad (4.4)$$

and the rest is identical to the proof above.

Both Theorem 2 and Theorem 3 give sufficient conditions for the reduction problem, when X is assumed to have a linear structure. The typical example in applications is $X = \mathbb{R}^n$.

If X is not assumed to be a vector space, the following topological sufficient conditions can be useful:

THEOREM 4. If P is continuous, u is u.s.c., supp u is compact, and if, for any $x_0 \in X$, $u(x)_0 \neq 1$, there exists a sequence $\{x_n\}_n$, $x_n \to x_0$, $u(x_n) > u(x_0)$, then the reduction problem holds.

Proof. Is obtained by examining the proof of Theorem 2.

Informally, the last condition in Theorem 4 can be stated in the following way: any point $x_0 \in X$ with membership degree $\neq 1$, can be approximated arbitrarily close by points $x \in X$ with strictly greater membership degrees.

REFERENCES

1. R. Bellman and L. A. Zadeh, Decision-making in a fuzzy environment, Management Sci. 17 (1970), B141-B146.

2. Cl. Dellacherie, Capacités et Processus Stochastiques, Springer-Verlag, Berlin, 1972.

3. J. Flachs and M.A. Pollatschek, Further results on fuzzy-mathematical programming, Inf. and Control 38 (1978), 241-257.

4. A. Kandel, Fuzzy sets, fuzzy algebra, and fuzzy statistics, Proc. of the IEEE, 66 (1978), 1619-1639.

5. C.V. Negoita and D.A. Ralescu, Applications of Fuzzy Sets to Systems Analysis, Wiley, New York, 1975.

6. H.T. Nguyen, On fuzziness and linguistic probabilities, J. Math. Anal. Appl. 61 (1977), 658-671.

7. H.T. Nguyen, Some mathematical tools for linguistic probabilities, Fuzzy Sets and Syst. 2 (1979), 53-65.

8. M.L. Puri and D. Ralescu, A possibility measure is not a fuzzy measure, Fuzzy Sets and Systems 7 (1982), 311-311.

9. D. Ralescu and G. Adams, The fuzzy integral, Journal of Math. Analysis and Applic. 75 (1980), 562-570.

10. A.W. Roberts and D.E. Varberg, Convex Functions, Academic Press, New York, 1973.

11. M. Sugeno, Theory of fuzzy integrals and its applications, Ph.D. Dissertation, Tokyo Inst. of Technology, 1974.

12. H. Tanaka, T. Okuda and K. Asai, On fuzzy-mathematical programming, J. of Cybernetics 3 (1974), 37-46.

METHODS OF UTILITY EVALUATION IN DECISION-MAKING PROBLEMS UNDER FUZZINESS AND RANDOMNESS

A. N. Borisov and G. V. Merkuryeva

Department of Automatized Control Systems, Riga Polytechnical Institute, 1 Lenin Street, Riga, Latvian SSR, USSR

Abstract. The problem of decision-making under fuzziness and randomness is considered. The concept of linguistic preference relation is introduced allowing decision-making models to contain the decision maker's uncertainty about preferences and his opinions on the strength of these preferences. The L-lottery is used providing the description of underterministic consequences with fuzzy components. The axiomatic definition of utility function in the decision-making problems under fuzziness and randomness is given. Methods of linguistic utility evaluation are presented. The illustrative examples are given.

Keywords. Decision theory; fuzzy set theory; linguistic lottery; linguistic preference relations; linguistic utility evaluation.

INTRODUCTION

Application of the known models of decision making under undeterministic consequences and fuzzy description of decision problem elements such as criterial consequences estimates and their probabilities meets difficulties in practice because methods of linguistic utility evaluation don't exist. These latter, as a rule, are assumed to be set up. In the report some methods of utility evaluation in problems of decision making under fuzziness and randomness are presented and their analysis is given. Some examples of direct, indirect, deterministic, possibilistic as well as combined methods of utility evaluation are described.

BASIC CONCEPTS

Linguistic Preference Relation

The decision maker's preference relation is treated as an initial and basic concept, providing an opportunity to structurize and formalize the decision maker's preference system. The fact that fuzzy description of decision problem elements leads to fuzzy preference relations is taken into account. Linguistic preference relations are introduced, allowing decision making models to contain both the description of the decision maker's uncertainty about preferences and his approximate opinions on the strength of these preferences.

The linguistic preference relation is defined (Borisov, 1982a) as a linguistic variable PREFERENCE $= \langle S, T(S), U_s, G_s, M \rangle$. There are the following elements of the term-set $T(S)$: no preference, approximately equivalent, slightly preferable, considerably preferable, etc. The basic variable u_s from universe of discource U_s is interpreted as a ration u_i/u_j of numeric utilities of the objects to be compared (criterial estimates, consequences, alternatives, etc.).

Methods of the membership functions construction for fuzzy sets, which allow to describe linguistic preference relations numerically, are considered in (Borisov, 1982c). Some properties of the linguistic preference relations (such as transitivity, symmetry, etc.) are discussed by Borisov (1982a), Efstathiou (1980), Zadeh (1977).

Linguistic Lottery

To describe real-world situations in decision-making and construct hypothetical ones under fuzziness and randomness methods of L-lotteries construction are used.

Let a random variable X, corresponding to some undeterministic consequence of decision problem assume meanings X_i from the set of possible linguistic meanings $\{X_1, \ldots, X_i, \ldots, X_r\}$, each of them being a fuzzy variable $\langle X_i, U_X, \tilde{X}_i \rangle$ and being formalized by fuzzy set $\tilde{X}_i = \bigcup_{x \in U_X} \mu_{X_i}(x)/x$. The probability of a variable X to assume linguistic meaning X_i is characterized by fuzzy number P_i or linguistic probability of X_i: $P_i \in P_o$, $P_o = \{\langle P_i, [0,1], \tilde{P}_i \rangle\}$, $P_i = \bigcup_{p \in [0,1]} \mu_{P_i}(p)/p$. The list of linguistic probability meanings $(P_1, \ldots, P_i, \ldots, P_r)$, corresponding to the set of possible meanings $(X_1, \ldots, X_i, \ldots, X_r)$ of a random variable X, is called linguistic probability distribution for this random variable, and the variable X is called a linguistic fuzzy variable.

By analogy with the nonfuzzy case linguistic lottery, or L-lottery, is defined (Borisov, 1982b) as a linguistic random variable with known linguistic probability distribution and presented by vector L:

$$L = (P_1, X_1; \ldots; P_i, X_i; \ldots P_r, X_r).$$

Lottery L_2 with two consequences is an example of linguistic lottery:

L_2 = (probability Large, profit increment LOW; probability SMALL, profit increment HIGH).

Methods of L-lotteries construction, where their components X_i are described by simple or composite fuzzy variables are discussed by Borisov (1982b). Here basic meanings p in $\mu_{P_i}(p)$ are treated as numerical meanings of a fuzzy event probability, when it is caused by consequence X_i :

$$p_i = \int_{U_X} \mu_{X_i}(x) f(x) dx,$$

where f(x) is a function of probability density on numerical domain of consequence definition U_X. Let's also note that linguistic lottery definition, introduced by Dubois and Prade (1982), doesn't assume fuzzy description of its components X_i and correponds to the decision-making situations, where only consequences probabilities $P_1, \ldots, P_i, \ldots, P_r$ are fuzzy.

AXIOMATIC DEFINITION OF UTILITY

Let's formulate the system of postulates providing numerical decision maker's preferences representation and leading to some rule of risky alternatives ranking in decision making problems under fuzziness and randomness. Let's take as a principle the system of axioms of utility function existence in the nonfuzzy case and nonformal recommendations about choosing these axioms (Keeney, 1976; Luce, 1957; Neumann, 1947). The most meaningful from the practical point of view is the recommendation, which states that each of postulates has to have an intuitively clear meaning and interpretation.

Axiomatic definition of utility function in decision making problems under fuzziness and randomness is based on the justice of the following assumptions.

<u>Axiom 1. The completeness of the decision maker's preference system</u>

For any two alternatives A_k and A_l the decision maker can always set some preference or equivalence relation R :

$$(\forall A_k, A_l)\ (\exists S \in T(S))\ [R(A_k, A_l) = S].$$

Let's define some properties of the preference relation R. Let $T(S) = S_o \cup S_e \cup \{S_i\}$, where S_o, S_e correspond to the terms: no preference, approximately equivalent; $S_i \in T(S) \cap \overline{S_o \cup S_e}$.

Let also $R = R_e \cup \bar{R}_e$, where R_e is an equivalence relation, but $\bar{R}_e$ is its complement.

The relation R_e is symmetric:

$$R_e(A_k, A_l) = S_e \Rightarrow R_e(A_l, A_k) = S_e.$$

The relation $\bar{R}_e$ is antisymmetric and antireflexive:

$$\bar{R}_e(A_k, A_l) = S_i \Rightarrow \bar{R}_e(A_l, A_k) = S_o,$$

$$\bar{R}_e(A, A) = S_o.$$

<u>Axiom 2. Transitivity of decision maker's preferences</u>

The preference relation is transitive (Zadeh, 1977):

$$R \supset R \circ R \text{ or } \tilde{S}_{ij} \geqslant \bigcup_k (\tilde{S}_{ik} \cap \tilde{S}_{kj}),$$

where $\tilde{S}_{ij}$ denotes fuzzy set, describing corresponding linguistic preference relation $R(A_i, A_j) = S_{ij}$, $S_{ij} \in T(S)$.

In particular

$(A_i S_{ij} A_j) \wedge (A_j S_{jk} A_k) \Rightarrow$
$(A_i S_{ik} A_k), \tilde{S}_{ik} \succ \tilde{S}_{ij} \cap \tilde{S}_{jk},$
$(A_i S_e A_j) \wedge (A_j S_e A_k) \Rightarrow (A_i S_e A_k).$

Axiom 3. Reduction of composite lotteries

A composite lottery $L = (P_1, A_1; \ldots; P_i, A_i; \ldots; P_r, A_r)$, whose components A_i are simple lotteries of the form $A_i = (P_{ix}, X; P_{iy}, Y)$, is equivalent to the simple lottery $L' = (P'_1, X; P'_2, Y)$ with consequences X and Y. Probabilities of these consequences are calculated by the following expressions:

$$\tilde{P}'_1 = \tilde{P}(\tilde{A}_1\tilde{X} \oplus \ldots \oplus \tilde{A}_i\tilde{X} \oplus \ldots \oplus \tilde{A}_r\tilde{X}),$$
$$\tilde{P}'_2 = \tilde{P}(\tilde{A}_1\tilde{Y} \oplus \ldots \oplus \tilde{A}_i\tilde{Y} \oplus \ldots \oplus \tilde{A}_r\tilde{Y}) -$$

in accordance with operation rules for fuzzy event probabilities and minimax principle of extension.

Remark 3.1. If $L = (P_1, A_1; P_2, A_2)$, then the basic meanings of linguistic probability $\tilde{P}(\tilde{A}_1\tilde{X} \oplus \tilde{A}_2\tilde{X})$, taking into account the independence of fuzzy events $\tilde{A}_i$ and $\tilde{X}$, are calculated by the following expressions:

$$p(\tilde{A}_1\tilde{X} \oplus \tilde{A}_2\tilde{X}) = p(\tilde{A}_1)\cdot p(\tilde{X}/\tilde{A}_1) + p(\tilde{A}_2)\cdot p(\tilde{X}/\tilde{A}_2) - p(\tilde{A}_1\tilde{A}_2)\, p(\tilde{X}/\tilde{A}_1\tilde{A}_2) = p_1 p_x^1 + p_2 p_x^2 - p_{12} p_x^1 p_x^2.$$

For the case of lottery L with r consequence we have:

$$p(\tilde{A}_1\tilde{X} \oplus \ldots \oplus \tilde{A}_i\tilde{X} \oplus \ldots \oplus \tilde{A}_r\tilde{X}) = \sum_i p_i p_x^i - \sum_{i,j} p_{ij} p_x^i p_x^j + \sum_{i,j,k} p_{ijk} p_x^i p_x^j p_x^k - \ldots \ldots - (-1)^r p_{12\ldots r} p_x^1 p_x^2 \ldots p_x^r,$$

where $p_i, p_{ij}, p_{ijk}, \ldots$ are interpreted by numerical probabilities of fuzzy events, caused by consequences $A_i, (A_i A_j), (A_i A_j A_k), \ldots$ respectively.

Axiom 4. Continuity. If $X_1 S_{1j} X_j$ and $X_j S_{jk} X_k$, $S_{1j} \in T(S) \cap \overline{S_o \cup S_e}$, $S_{jk} \in T(S) \cap \overline{S_o \cup S_1}$, then an L-lottery $L_j = (P_e X_1; P_k, X_k)$ with consequences X_1 and X_k exists, such that $X_j \sim A_j$ or $X_j S_e A_j$, where $\sim$ denotes the relation of strong equivalence.

Axiom 5. Equivalence. Let $L = (P_i, X_i; P_j X_j)$, then

if $X_i S_e X'_i$, then $L S_e (P_i, X'_i; P_j, X_j)$;
if $X_i \sim X'_i$, then $L \sim (P_i, X'_i; P_j, X_j)$.

The axiom of equivalence supposes that alternatives remain equally valuable (or approximately equally valuable) not only when comparing them alone, but also when substituting them into lotteries.

Axiom 6. Comparison of fuzzy numbers. The largest of fuzzy numbers $\tilde{V}_i$ and $\tilde{V}_j$ being described by membership functions $\mu_{\tilde{V}_i}(u_i)$ and $\mu_{\tilde{V}_j}(u_j)$ is a fuzzy set

$$\tilde{V}_{max} = \{(\tilde{V}_i, \mu_{\tilde{V}_{max}}(\tilde{V}_i))\},$$

in which the degree of fuzzy number's $\tilde{V}_i$ membership to this set, $\mu_{\tilde{V}_{max}}(\tilde{V}_i)$, is interpreted as a degree of relation "V_i is greater than V_j" truth and calculated by the expression:

$$\mu_{\tilde{V}_{max}}(\tilde{V}_i) = \sup_{u_i, u_j} \min \{\mu_{\tilde{V}_i}(u_i), \mu_{\tilde{V}_j}(u_j), \mu_R(u_i, u_j)\}.$$

The introduction (Baldwin, 1979) of fuzzy relation R (for example, "Is significantly greater") with membership function $\mu_R(u_i, u_j)$ in every particular decision making problem allows to take into account and describe the particular decision maker's notions about the essentiality of distinctions u_i and u_j for ascertaining superiority relation for $\tilde{V}_i$ concerning $\tilde{V}_j$.

Axiom 7. Comparison of L-lotteries. Let $L_1 = (P_i^1, X_i; P_j^1, X_j)$ and $L_m = (P_i^m, X_i; P_j^m, X_j)$, $X_i S_{ij} X_j$, $S_{ij} \in T(S) \cap \overline{S_o \cup S_e}$.

The best is a lottery with the greatest consequence X_i probability:

$$\mu(L_1 \succ L_m) = \mu(\tilde{P}_i^1 > \tilde{P}_i^m).$$

An effective lottery is described by fuzzy set L_o:

$$L_o = \{(L_1, \mu_{L_o}(L_1))\},$$

in which $\mu_{L_o}(L_1)$ is defined as a degree of

the relation $\tilde{P}_i^l > \tilde{P}_i^m$ truth, $\mu_{L_o}(L_l) = \mu_{\tilde{P}^i_{max}}(\tilde{P}_i^l)$, and interpreted by the degree of the decision maker's confidence in preference of L_l concerning L_m.

If the assumptions given above are true, we come to the following statement.

Statement I.

I. If preference relations meet axioms 1 - 7, then fuzzy numbers $\tilde{V}_i$, connected with consequences X_i exist, such that for L-lotteries

$$L_l = (P_1^l, X_1; \ldots; P_i^l, X_i; \ldots; P_r^l, X_r) \text{ and}$$
$$L_m = (P_1^m, X_1; \ldots; P_i^m, X_i; \ldots; P_r^m, X_r)$$

correlation of fuzzy expected utilities $\tilde{V}^l_{exp}$ and $\tilde{V}^m_{exp}$ determines which of the alternatives is preferable. The basic meanings of the utilities are expressed as (I) and membership degrees are calculated taking into account expression (I) on the basis of minimax extension principle

$$u_{exp} = \sum_i p_i u_i - \sum_{i,j} p_{ij} u_i u_j + \sum_{i,j,k} p_{ijk} u_i u_j u_k - \ldots - (-1)^r p_{12\ldots r} u_1 u_2 \ldots u_r.$$

II. Comparison of fuzzy estimates of the expected utilities $\tilde{V}^1_{exp}, \ldots, \tilde{V}^l_{exp}, \ldots, \tilde{V}^n_{exp}$ of alternatives $L_1, \ldots, L_l, \ldots, L_n$ leads to fuzzy set L_o of the most preferable alternative:

$$L_o = \{(L_l, \mu_{L_o}(L_l))\}.$$

The degree of the decision maker's confidence in that alternative L_l is more preferable in comparison with other alternatives is calculated by the following formula:

$$\mu_{L_o}(L_l) = \sup_{u_1,\ldots,u_n} \min\{\mu_{\tilde{V}^1_{exp}}(u_1), \ldots, \mu_{\tilde{V}^l_{exp}}(u_2), \ldots, \mu_{\tilde{V}^n_{exp}}(u_n), \mu_R(u_1, \ldots, u_l, \ldots, u_n)\},$$

where $\mu_R(\cdot)$ is a membership function for fuzzy relation R, describing how essential are the distinctions in numerical utilities when determining the superiority relation.

METHODS OF UTILITY EVALUATION

Statement I gives an opportunity to analyse formally alternative decisions under risk and fuzzy information. The component of the analysis is the process of determination of consequences utilities $\tilde{V}_i$. The methods of utility evaluation proposed in the given paper are methods of empirical utility analysis under fuzziness and known approaches are used (Fishburn, 1967; Keeney, 1976; Kozeletsky, 1979): determination of utility on the basis of the analysis of the choices and series of decisions, being made by a subject in the process of problem solving (indirect methods, axiomatically based) or on the basis of direct judgements about utilities or correlations of consequences utilities (direct methods); application in the process of utility evaluation probabilistic or only deterministic judgements of the decision maker (probabilistic and deterministic methods).

Let's give some methods of linguistic utilities evaluation. We shall denote the kind of method by the list $\langle T_1, T_2 \rangle$:

T_1 - direct (M_1) or indirect (M_2),

T_2 - probabilistic (P) or deterministic (D).

Method of Direct Utility Evaluation $\langle M_2, D \rangle$.

We suppose that from the set of possible consequences the decision maker may choose three of them - the least preferable $\tilde{X}^o$, the most preferable $\tilde{X}^*$ and an average as to preference, $\tilde{X}^+$, which correspond to the following utility meanings - the lowest $\tilde{V}^o$ (about 0), the highest $\tilde{V}^*$ (say, near 1) and average $\tilde{V}^+$ (approximately 0.5):

$$\text{utility } (\tilde{X}^o) = \tilde{V}^o,$$
$$\text{utility } (\tilde{X}^*) = \tilde{V}^*,$$
$$\text{utility } (\tilde{X}^+) = \tilde{V}^+.$$

On the basis of the equations given above fuzzy restriction R_g on numerical meanings of utility u and criterial consequences estimates x is constructed:

$$R_g(x,u) = \tilde{X}^o \times \tilde{V}^o + \tilde{X}^+ \times \tilde{V}^+ + \tilde{X}^* \times \tilde{V}^*,$$

where x and + denote Cartesian product and the union of fuzzy sets.

The determination of unknown utility $\tilde{V}_j$ of a consequence $\tilde{X}_j$ is reduced to the composition of R_g and $\tilde{X}_j$:

$$V_j = \tilde{X}_j \circ R_g -$$

and is made in accordance with the following expression:

$$\mu_{\tilde{V}_j}(u_j) = \max_{x \in U_X} \min\{\mu_{X_i}(x_i), \mu_{R_g}(x_i, u_j)\}.$$

The method of direct evaluation assumes that membership functions for fuzzy sets, describing the semantics of criterial consequences estimates, are known.

Method of Linguistic Relations Evaluation $\langle M_2, D \rangle$.

Let the utility of a consequence $\tilde{X}_i$, equal to $\tilde{V}_i$ with membership function $\mu_{\tilde{V}_i}(u)$ and linguistic relation S_{ij} between $\tilde{X}_i$ and consequence $\tilde{X}_j$ with unknown utility be known: $S_{ij} \in T(S)$ and is formalized by fuzzy set with membership function $\mu_{S_{ij}}(s)$.
Taking into account the interpretation of basic variable s, the ratio between possible numerical utilities of consequences $\tilde{X}_i$, $\tilde{X}_j$ is expressed as follows: $u_i = s \cdot u_j$. Then the reconstruction of membership function of the unknown utility of consequence $\tilde{X}_j$ is made in accordance with the following expression:

$$\mu_{\tilde{V}_j}(u_j) = \max_{s \in U_S} \min\{\mu_{S_{ij}}(s), \mu_{\tilde{V}_i}(s \cdot u_j)\}. \quad (4)$$

If we treat $\mu_{S_{ij}}(s)$, where $s = u_i/u_j$ as a nonevident representation of fuzzy relation $R_g(u_i, u_j)$, then the latter expression is reduced to the form:

$$\mu_{\tilde{V}_j}(u_j) = \tilde{V}_i \circ S_{ij}.$$

The method of linguistic relations evaluation assumes the linguistic preference variable to be given.

Combined Method of the Kind $\langle M_2, D \rangle$.

The decision maker fixes up two characteristic consequences - the least preferable $\tilde{X}^o$ and the most preferable $\tilde{X}^*$, corresponding to the utility meanings $\tilde{V}^o$ and $\tilde{V}^*$. For some $\tilde{X}_i$ linguistic relations S such as $\tilde{X}_o S \tilde{X}_i$ or $\tilde{X}_i S \tilde{X}^*$ are determined and the corresponding membership functions for utilities are calculated taking into account expression (4).
For the other consequences $\tilde{X}_j$, utilities $\tilde{V}_j$ are reconstructed according to expression (2) and (3).

The joint application of the mentioned methods for determination of linguistic consequences utilities allows to vary the size of information being required from the decision maker. At the same time the interpolation of utility $\tilde{V}_j$ on the basis of expression (2) and (3) is probably more correct in the sense that it is based on a greater amount of information in comparison with the algorithms constructed on the method of direct utility evaluation only.

The Method of Choice Analysis $\langle M_1, P \rangle$.

This method is an extension of the method of standard game (Fishburn, 1967) for the case of fuzzy decision analysis. There are two modifications of the method, depending on the type of information, being required from the decision maker:

1) The decision maker is asked to set such a meaning of linguistic probability $\tilde{P}_j^o$, which creates for him approximately equivalent decision-making situations and the corresponding gains, caused by lottery $L_j = (P_j^o, X^o; P_j^*, X^*)$ and consequence $\tilde{X}_j$ with unknown utility;

2) The decision maker is asked to set a preference relation, S between lottery L_j with consequences $\tilde{X}^o$, $\tilde{X}^*$ and fixed probabilities $\tilde{P}_j^o$, $\tilde{P}_j^*$ and consequence $\tilde{X}_j$ with unknown utility.

In both cases the determination of unknown utility $\tilde{V}_j$ of consequence $\tilde{X}_j$ includes:

1) determination of the expected linguistic utility of lottery L_j on the basis of Statement I;

2) reconstruction of utility $\tilde{V}_j$ taking into consideration utility of lottery L_j and preference relation S on the basis of expression (4).

The method of choice analysis (as well as other indirect methods of utility evaluation) is based on the accepted axiomatic concerning the decision maker's preference system and in that sense is more grounded. At the same time from the computational point of view it turns out more complicated and more difficult in comparison with other above mentioned methods.

Application of the method of choice analysis is worth while in the cases, when in practice a number of decisions are known - alternative choice, made by an individual in the given class of problems.

EXAMPLE

Let's give an example, which illustrates construction of utility function with the help of applied program package, METHOD, which is written in FORTRAN and realizes the above mentioned methods of utility evaluation in decision making problems under fuzziness.

Let's show a fragment of the dialogue between the decision maker (DM) and a computer (C) when determining utility by the method of linguistic relations evaluation:

C. Tell a consequence, the least preferable for you

DM. PROFIT INCREMENT. LOW

C. Tell a consequence, the most preferable for you

DM. PROFIT INCREMENT. HIGH

C. Tell a consequence, which utility it's necessary to determine

DM. PROFIT INCREMENT. LARGE

C. Tell the preference of the given consequence concerning the least preferable consequence

DM. PREFERENCE. CONSIDERABLY PREFERABLE

C. Tell the preference of the most preferable consequence concerning the given consequence

DM. PREFERENCE. SLIGHTLY PREFERABLE

C. Give the next meaning of the consequence which utility it's necessary to determine

DM. PROFIT INCREMENT. AVERAGE

C. Tell the preference of the given consequence concerning the least preferable consequence

DM. PREFERENCE. SLIGHTLY PREFERABLE

The membership functions of linguistic utilities for the consequences LARGE PROFIT INCREMENT and AVERAGE PROFIT INCREMENT, obtained with the help of the method of linguistic relations evaluation, are approximated by the terms - AVERAGE and SMALL.

CONCLUSION

The methods of linguistic utility evaluation described in this paper give an opportunity to determine utility function also in the decision making problems under fuzziness and randomness with multidimensional consequences.

REFERENCES

Baldwin, J.F., and N.C. Guild (1979). Comparison of fuzzy sets on the same decision space. Fuzzy Sets & Systems, 2, 213-231.

Borisov, A.N. (Ed.) (1982a). Decision Making Models Based on Linguistic Variable. Zinatne, Riga. 256 pp.

Borisov, A.N., and G.V. Merkuryeva (1982b). Linguistic Lotteries - Construction and Properties. Busefal, II, 39-46.

Borisov, A.N., and G.V. Merkuryeva (1982c). Linguistic Preference Relations Modeling in the Decision Making Problems. In R.Trapple, N.P., Findler, and W. Korn (Eds.), Progress in Cybernetics and Systems Research, vol. XI. Hemisphere Publishing Corporation, Washington. pp. 269-274.

Dubois, D., and H. Prade (1982). The use of fuzzy numbers in decision analysis. In M.M. Gupta, and E. Sanchez (Eds.), Fuzzy Information and Decision Processes. North-Holland.

Efstathiou, J., and R. Tong (1980). Ranking Fuzzy Sets Using Linguistic Preference Relations. Proc. 10th Int. Symp. Multiple-Valued Logic. Evanston, 1980, N.Y. pp. 137-142.

Fishburn, P.C. (1967). Methods of estimating additive utilities. Management Science, 13, 435-453.

Keeney, R.L., and H. Raiffa (1976). Decisions with Multiple Objectives: Preferences and Value Tradeoffs. Wiley & Sons, New York. 569 pp.

Kozeletsky, Y. (1979). Psychological Decision Theory. Progress, Moscow. 504 pp.

Kuzmin, V.B. (1982). Construction of Group Decisions in the Spaces of Nonfuzzy and Fuzzy Binary Relations. Nauka, Moscow. 168 pp.

Luce, R.D., and H. Raiffa (1957). Games and Decisions. Wiley, New York.

Neumann J., and O. Morgenstern (1947). Theory of Games and Economical Behaviour. 2 nd. ed. Princeton University Press, Princeton.

Zadeh, L.A. (1977). Linguistic Characterization of Preference Relations as a Basis of Choice in Social Systems. Mem. No. UCB/ERL M77/24. University of California, Berkeley. 37 pp.

A FUZZY CONCEPT IN THE THEORY OF STRATEGIC DECISIONS WHERE SEVERAL OBJECTIVES EXIST

B. Urban and V. Hänsel

Zittau College of Engineering, German Democratic Republic

Keywords. Game theory; Decision theory; Probability; Fuzzy sets; Multiobjective decision making.

PROBLEMS IN THE APPLICATION OF THE GAME THEORY IN TECHNOLOGICAL-ECONOMIC DECISION SITUATIONS

As a rule technological-economic decisions are linked with problems of lack of information. This means that too little information about a part or even all the state variables of the model to describe the decision situation is available.

The causes of this lack of information are frequently of an objective nature. For long-term decisions in particular it is mostly impossible to make precise predictions on the state variables at the moment when the decision to be taken is made.

Often the consequences of this lack of information are so great that from the outset it cannot be disregarded. At the Zittau College of Engineering the possibilities of utilizing the game theory, or more precisely the theory of games against nature, were examined for decision-making taking into account the lack of information.

At first it seems reasonable to represent the lack of information on the state variables of the situation model by permitting the use of several possible values for each variable. The combinations of these possible values of the state variables constitute different situation model states, in the following referred to as environmental conditions or environmental strategies. These environmental strategies are compared with various alternatives for action to be examined, i. e., different combinations of values of the decision variables. In practice there is no difficulty in considering all variables of the model as discrete values, so that the number of environmental strategies and alternatives for action can be regarded as finite.

Forming a decision matrix by fitting the respective decision and state variables into a given objective function results in the establishment of a game situation. This procedure is accompanied, above all, by three problems, which are described briefly represented here.

Firstly, in contrast to the game theory (Neumann, 1953) the player "environment", "technological-economic system" does not pursue an objective. Thus the minimax criterion loses the basic position it has in the game theory as a decision matrix evaluation procedure because of its precondition that one player pursues the objective of maximising his gain at the cost of his opponent. For the utilization range of technological-economic decisions numerous rational criteria like the minimax criterion have been developed (Savage, 1951; Hurwicz, 1951; Hodges, 1952).

All these criteria assume that deviation from the minimax strategy is linked with a risk, but that this risk will not necessarily be taken advantage of by the unconsciously acting opponent "evironment". Thus the running of a limited risk is unquestionably justifiable, if a gain exceeding the minimax gain can be expected.

In addition, these criteria start from different decision standpoints and can therefore show different alternatives for action as being optimum for the same decision situation. Each of these standpoints is only valid for spezific idealized situations. However, most ways of looking at technological-economic problems will not be adequately represented by any of these decision standpoints. Thus the person in charge of such decision situations is permanently confronted with the problem of a subjective criterion choice.

Secondly, with technological-economic problems it is frequently the case that the decision has to be taken not just once but several times. The formulation of mixed strategies established in the game theory, albeit for other reasons, is unsuitable for dealing with this aspect, as it postulates the technically irrelevant infinite taking of the decision, again in connection with the lack of an objective for the player "environment".

Lastly, the existence postulated in the game theory of only one objective function for technological-economic tasks seldom applies. Usually several objectives will be pursued simultaneously with one decision. These can mostly be explicitly presented as objective functions. Therefore, it seems useful to consider further all these objectives as being equivalent instead of a subjectively established utility function. Therefore, for the moment the existence of several decision matrices is linked.

Based on this brief analysis, a new concept for elaborating technological-economic decision situations is intended to be introduced in the following. Although this concept is based on the game theory, it is free of the faults mentioned here. This concept has to permit the display of the lack of information and the frequency of decision-taking, the taking into account of several decision objectives, and a caleulable risk in the decision-taking. Furthermore, the concept is meant to exclude subjectivity in the selection of a decision criterion.

RATIONAL EVALUTION OF THE CONSEQUENCES OF A DECISION

Representation of the Lack of Information

The first problem to be more closely investigated is that of the rational evaluation of the consequences of a single-objective-oriented decision for an alternative for action Y_h. Later the linking of several decision objectives will be analysed. The procedure to be developed here is based on a representation of lack of information with respect to a state variable x_i in the form of several feasible discrete values of this variable provided with probability intervals for the occurence of the individual values:

$$x_i = \begin{pmatrix} x_{i1}, x_{i2}, \ldots, x_{in_i} \\ p_{iu1}, p_{iu2}, \ldots, p_{iun_i} \\ p_{io1}, p_{io2}, \ldots, p_{ion_i} \end{pmatrix} \qquad (1)$$

By means of technical considerations it is usually possible to define the discrete values $\{x_{ir}\}$ without serious difficulties. The specification of the probability intervals is more complicated and a basic distinction must be made between two examples:

A) For the variable x_i no experimental values are available. This example is characterized either by a determined variable x_i (if experimental values were available, the only possible value of the variable would be known, i.e., the variable would be free of lack of information) or a stochastic variable x_i.

B) The variable x_i is represented by a number of NSP_i experimental sample values. This can only be the case for a stochastic variable.

As any reliable statement on the real value and the probabilities of the variable x_i is lacking the following relations only can apply for example A

$$p_{iur} = 0; \; p_{ior} = 1; \; r = 1,2,\ldots,n_i \qquad (2)$$

In contrast to this in example B) the probability intervals obviously depend on the sample available. If $(h_{i1}, h_{i2}, \ldots, h_{in_i})$ are the relative frequencies of $(x_{i1}, x_{i2}, \ldots, x_{in_i})$ of the sample variable x_i, the confidence intervals of the n_i alternative distributions can be used for the n_i variable values as the probability intervals sought. These can be calculated for a error probability using

$$\left\{\begin{matrix} p_{iur}(NSP_i) \\ p_{ior}(NSP_i) \end{matrix}\right\} = \frac{1}{NSP_i + z_\alpha^2} \left(h_{ir}NSP_i + \frac{z_\alpha^2}{2} \mp z_\alpha \sqrt{h_{ir}NSP_i(1 - h_{ir}) + \frac{z_\alpha^2}{4}}\right) \qquad (3)$$

In this case z_α is the error probability quantile α calculated by using the standardized normal distribution.

These confidence intervals ensure that the real probabilities of the values $\{x_{ir}\}$ lie in these intervals with the probability $(1 - \alpha)$:

$$P(p^*_{ir} \in [p_{iur}(NSP_i); p_{ior}(NSP_i)]) = 1 - \alpha \qquad (4)$$

Thus all probability distributions $(p_{i1}, p_{i2}, \ldots, p_{in_i})$ where

$$p_{ir} \in [p_{iur}(NSP_i); p_{ior}(NSP_i)] \qquad (5)$$

can be regarded as a possible origin of the sample studied. These probability intervals constitute quite a useful presentation of the existing lack of information. Where a sample is lacking ($NSP_i = 0$, $h_{ir} = 0$, $r = 1,2,\ldots,n$) (actually separately denoted as example A) relation (3) gives

$$p_{iur}(0) = 0 \; ; \; p_{ior}(0) = 1 \qquad (6)$$

On diminishing the lack of information, i.e., on increasing the sample size an isotonic function develops for $p_{iur}(NSP_i)$ and an antitonic function for $p_{ior}(NSP_i)$. The common limit of these two functions becomes

$$\lim_{NSP\to\infty} p_{iur}(NSP) = \\ = \lim_{NSP\to\infty} p_{ior}(NSP) = p^{*}_{ir} \quad (7)$$

because

$$\lim_{NSP\to\infty} h_{ir} = p^{*}_{ir} \quad (8)$$

holds.

In this way the set of probability distributions coming into consideration as the origin of the sample studied diminishes with a decreasing lack of information. If several state variables exist the environmental conditions will be established by a combination of the discrete possible values of all variables, i.e., each environmental strategy can be represented by

$$X_U = (x_{1U}, x_{2U}, \ldots, x_{vU}) \; ;$$
$$U = 1,2,\ldots,NU \; ;$$
$$NU = \prod_{i=1}^{v} n_i \quad (9)$$

The probability intervals for these environmental strategies are established analogously by multiplying the probability interval limits of the combined variable values calculated by use of (3)

$$p_{uU} = \prod_{i=1}^{v} p_{iuU}(NSP_i)$$
$$p_{oU} = \prod_{i=1}^{v} p_{ioU}(NSP_i) \quad (10)$$

Because of the existing lack of information all probability distributions on the established environmental strategy with

$$p_{uU} \leq p_U \leq p_{oU} \quad (11)$$

cannot be eliminated with the error probability α. Here it is sufficient to know both the distributions $\{p^-_U\}$ and $\{p^+_U\}$ giving the smallest and the greatest expected objective function values $Z(Y_h,X)$, respectively.

For this purpose both problems

$$\min \mu(Z(Y_h,X)) = \min \sum_{U=1}^{NU} p^-_U \, Z(Y_h,X_U)$$
$$\sum_{U=1}^{NU} p^-_U = 1 \quad (12)$$
$$p_{uU} \leq p^-_U \leq p_{oU}$$

and

$$\max \mu(Z(Y_h,X)) = \max \sum_{U=1}^{NU} p^+_U \, Z(Y_h,X_U)$$
$$\sum_{U=1}^{NU} p^+_U = 1 \quad (13)$$
$$p_{uU} \leq p^+_U \leq p_{oU}$$

have to be solved and have a single solution, under the insignificant restriction

$$Z(Y_h,X_{U1}) \neq Z(Y_h,X_{U2}) \; ;$$
$$U1 = 1,2,\ldots,NU \; ; \; U2 = 1,2,\ldots,NU$$
$$U1 \neq U2 \quad (14)$$

Provided that the indexing with U has been done in such a way that $Z(Y_h,X_U) > Z(Y_h,X_{U+1})$; $U = 1,2,\ldots, NU-1$ the solution can be given directly by the recursion relation

$$p^-_U = \min\left\{1 - \sum_{l=1}^{U-1} p^-_l - \sum_{l=U+1}^{NU} p_{ul} \; ; \; \max p_{oU}\right\}$$
$$p^+_U = \min\left\{1 - \sum_{l=U+1}^{NU} p^+_l - \sum_{l=1}^{U-1} p_{ul} \; ; \; \max p_{oU}\right\} \quad (15)$$

As a counterpart of the existing lack of information the most unfavourable probability distribution of the environmental strategies, i.e., a maximizing problem

$$\{p_{IU}\} = \{p^-_U\} \quad (16)$$

and a minimizing problem

$$\{p_{IU}\} = \{p_U^+\} \tag{17}$$

should be used further.

It was thus possible to find occurence probabilities for all possible environmental strategies permitting a consideration of the lack of information.

Evaluation of the Consequences

The procedure presented above can also provide in a second step an evaluation of the consequences to the application of the alternative for action Y_h taking into account the number of its applications. The most unfavourable probabilities of the formulated environmental strategies $\{p_{IU}\}$ preriously determined taking into account the lack of information and the realization number of the decision NRZ constitute the starting point.

The NRZ decision realizations are again characterized by confidence intervals for the particular environmental strategy probabilities which can similarly (1) be calculated:

$$\left\{\begin{matrix} p_{uU}(NRZ) \\ p_{oU}(NRZ) \end{matrix}\right\} = \frac{1}{NRZ + z_\alpha^2}\left(p_{IU}NRZ + \frac{z_\alpha^2}{2} \mp \mp z_\alpha \sqrt{p_{IU}NRZ(1-p_{IU}) + \frac{z_\alpha^2}{4}}\right) \tag{18}$$

A solution of (12) and (13) using these probability intervals again results in distributions $\{p_U^-(NRZ)\}$ and $\{p_U^+(NRZ)\}$ giving the confidence interval limits of the expected value of the decision result $\mu_u(Z(Y_h,X))$ and $\mu_o(Z(Y_h,X))$.

The decision result actually obtained will be found with

$$P(\mu(Z(Y_h,X)) \in [\mu_u(Z(Y_h,X));\mu_o(Z(Y_h,X))]) = 1 - \alpha \tag{19}$$

in this confidence interval.

The final evaluation of the consequences is made by selecting the most unfavourable result from this interval expressed by

$$\mu_B(Z(Y_h,X)) = \begin{cases} \sum_{U=1}^{NU} p_U^-(NRZ)\, Z(Y_h,X_U) & \text{for a maximizing problem} \\ \sum_{U=1}^{NU} p_U^+(NRZ)\, Z(Y_h,X_U) & \text{for a minimizing problem} \end{cases} \tag{20}$$

since here again no result can be excluded from the confidence interval calculated owing to the limited realization number.

The limiting processes already mentioned illustrate that assuming an increasing realization number in the case of maximization $\mu_B(Z(Y_h,X))$ results in a monotonically increasing sequence and in the case of minimization in a monotonically decreasing sequence with a joint limit

$$\lim_{NRZ \to \infty} \mu_B(Z(Y_h,X)) = \sum_{U=1}^{NU} p_{IU}\, Z(Y_h,X_U) \tag{21}$$

Thus a relevant evaluation of the consequences of a decision is established, processing the lack of information associated with the decision as well as the number of decision realizations without any subjective interference. This evaluation closes the gap between the decision standpoints of the minimax and Bayes principles, which has been relevant to technological-economic tasks, but so far not utilizable. From now on both these standpoints are merely peripheral cases of the procedure presented here.

PROCESSING OF SEVERAL DECISION OBJECTIVES

The above developed evaluation $\mu_B(Z(Y_h,X))$ of an alternative for action Y_h allows an ordering of the action alternatives to be studied with respect to the objective $Z(Y,X)$ considered, but no consistent ordering if several objectives $Z_k(Y,X)$, $k = 1, 2,\ldots,K$ exist. This intention demands the transformation of the evaluations with regard to the individual objectives to a homogeneous, comparable basis.

Such a basis is the utility ϱ_{hk} of the alternatives for action with respect to the individual objectives, determined by a utility function

$$\varrho_{hk} = \varrho_{hk}(\mu_B(Y_h,X)) \qquad (22)$$

The utility function can be any functional connectivity complying with the utility theory conditions (Fishburn, 1970).

The conversion of the alternative evaluations into objective levels of the form

$$\varrho_{hk} = \frac{\mu_B(Z_k(Y_h,X)) - \mu_B(Z_k^-)}{\mu_B(Z_k^+) - \mu_B(Z_k^-)} \qquad (23)$$

has proved a success,

where $\mu_B(Z_k^-)$ is the most unfavourable evaluation of all alternatives for action studied in respect of the objective $Z_k(Y,X)$,

and $\mu_B(Z_k^+)$ is the most favourable evaluation of all alternatives for action in respect of the objective $Z_k(Y,X)$.

This conversion has the advatage that all alternatives for action evaluations can be related to the /0,1/ region.

This makes it possible to interpret the objective levels ϱ_{hk} as interpolation nodes of membershipfunctions of the fuzzy sets "set of the optimum alternatives for action with regard to the objective $Z_k(Y,X)$". Thus the problem of the evaluation of alternatives for action with regard to several objectives becomes a common problem of fuzzy set combination.

In a concrete situation the character of this combination is determined by concrete additional information. This additional information can develop in two different ways.
On the one hand the decision-maker is obviously only forced to consider several decision objectives by that fact that he is able to formulate his global objective "optimum decision", but is unable to evaluate its fulfilment. Thus he breaks this global objective down into several subobjectives. The fulfilment of these subobjectives is open to a one-dimensional evaluation. However, it has to be taken into account that the global objective is generally not covered by each subobjective to the same degree. The importance of the fulfilment of the individual subobjectives for the fulfilment of the global overall objective is characterized by certain relations. These relations, in the following denoted as objective weighting g_k, provide the first important additional information for the evaluation of the alternatives for action with regard to the global objective.

Breaking the global objective down

into subobjectives and the logical combination of the subobjectives to a global objective will on the other hand differ from case to case depending upon the specific situation.
A conjunctive and a disjunctive subobjective combination constitute the basic combination possibilities. The disjunctive combination presupposes certain balancing effects between the fulfilment of the respective subobjectives, whereas the conjunctive combination allows the conclusion to be drawn that the global objective will not be fulfilled if one of the respective subobjectives is not be fulfilled. Usually both the combination modes will be required simultaneously to break the global objective down into subobjectives. This breaking down subobjectives provides further important additional information for the evaluation of the alternatives for action.

This additional information is meant to be used for the final evaluation of the alternatives for action as follows: The objective weightings are already included in the calculation of the objective levels by a generalisation of equation (23) of the sort

$$\rho^*_{hk} = 1 - (1 - \rho_{hk})^{1/g_k} \qquad (24)$$

A large objective weighting ($g_k > 1$) results in a downward alternative for action adjustment, whereas an upward alternative for action adjustment is caused by a low objective weighting ($g_k < 1$). The second information logically combining the subobjectives to the global objective eventually determines the combination operator.

Conjunctively combined subobjectives demand a conjunctive operator and disjunctively combined subobjectives a disjunctive one.
Used here as conjunctive operator is a min-operator

$$\rho_{Mh} = \min_k \rho_{hk} \qquad (25)$$

and as disjunctive operator a summation operator

$$\rho_{\Sigma h} = \frac{1}{K} \sum_{k=1}^{K} \rho_{hk} \qquad (26)$$

Any global objective structure can be processed by these two operators by integrating uniformly-combined subobjectives step-by-step.
The combination of six subobjectives to a global objective according to

$$Z = \left[(Z_1 \vee Z_2) \wedge (Z_3 \vee Z_4 \vee Z_5)\right] \vee Z_6 \qquad (27)$$

results in this multiplication law using equations (25), (26):

$$\rho_h = \frac{1}{2}\left[\min\left\{\frac{1}{2}(\rho_{h1} + \rho_{h2}) \; ; \; \frac{1}{3}(\rho_{h3} + \rho_{h4} + \rho_{h5})\right\} + \rho_{h6}\right] \qquad (28)$$

It is thus possible to evaluate the alternative for action optimality for any given subobjective structure. However, this evaluation should be only regarded as a guideline for the decision-maker because the starting points of the final evaluation
- utility function
- objective weightings
- logical subobjective combination

can be only subjectively determined and can therefore be influenced by the respective processor only to a certain extent. Nevertheless the priority of alternatives for action thus obtained offers an important decision aid for the decision processor.

REFERENCES

Fishburn, P.C. (1970). Utility Theory for Decision Making. Wiley, New York.

Hodges, Jr.J.L., E.L.Lehmann (1952). The use of previous experience in reaching statistical decision.

Ann. Math. Statistics, 23, 396-407.

Hurwicz, L. (1951). Optimality criteria for decisionmaking under ignorance. Cowles Commission Discussion Paper, Statistics, 370.

Neumann, J.v., O. Morgenstern (1953). Theory of Games and Economic Bahavior. Princeton University Press, Princeton.

Savage, L.J. (1951). The theory of statistical decision. J. Am. Statistic Ass., 46, 55-67.

DECOMPOSABLE MEASURES AND MEASURES OF INFORMATION FOR CRISP AND FUZZY SETS

S. Weber

Fachbereich Mathematik, Johannes Gutenberg-Universität, Mainz, Federal Republic of Germany

Abstract. There exist bijections between the decomposable informations of Kampé de Fériet and Forte (1967a) and the decomposable measures of Weber (1982). Using integrals for Archimedean decomposable operations, introduced by Weber (1982), informations and measures of this type are extended from crisp to fuzzy sets. For $\vee$-decomposable measures, Sugeno's (1974) integral is used. For $\wedge$-decomposable informations, Nguyen's (1977) construction and a modification are discussed.

Keywords. Decomposable information, decomposable measure, t-conorm, Archimedean semigroup, fuzzy measure, fuzzy integral.

1. INTRODUCTION

Let $(\Omega, \mathcal{B})$ be a measurable space. The concept of information $J(A)$ of (crisp) events $A \in \mathcal{B}$ was generalized by Kampé de Fériet and Forte (1967a). The crucial axiom is the decomposability of J,

$$J(A \cup B) = J(A) \top J(B) \quad ,$$

with respect to an appropriate semigroup operation $\top$ on $[0,\infty]$. In subsequent papers there have been considered the special types M (Kampé de Fériet and Forte, 1967b), M' (Kampé de Fériet, Forte and Benvenutti, 1969), inf (Kampé de Fériet and Benvenutti, 1969). The first two correspond to Archimedean operations $\top$, which are characterized by Ling's (1965) representation,

$$a \top b = f^{(-1)}(\ f(a) + f(b)\ ) \quad ,$$

by additive generators f and there pseudo-inverses $f^{(-1)}$. In a previous paper Weber (1982) has considered decomposable measures m,

$$m(A \cup B) = m(A) \perp m(B) \quad ,$$

with respect to t-conorms $\perp$ as appropriate semigroup operations on $[0,1]$, including also the Archimedean types with representation

$$a \perp b = g^{(-1)}(\ g(a) + g(b)\ ) \quad .$$

In the classical case there exists a bijection between the probabilities $m=P$ and the Wiener-Shannon-informations $J= -\ln \circ P$. The same holds for decomposable measures and informations, more generally,

$$J = l \circ m \quad ,$$

with appropriate functions $l : [0,1] \rightarrow [0,\infty]$. In the Archimedean cases this reduces to the relationship

$$f = g \circ l^{-1} \quad \text{with} \quad f^{(-1)} = l \circ g^{(-1)} \quad .$$

Extending now measures and informations to fuzzy events, i. e. to measurable fuzzy sets (Zadeh, 1965) $\varphi : \Omega \rightarrow [0,1]$, the first step was done by Zadeh (1968) for $m=P$ using Lebesgue's integral,

$$\tilde{P}(\varphi) = \int_{\Omega} \varphi \, dP \quad .$$

This idea was adopted by Sugeno (1974), who introduced an integral for measures m which are only monotone instead of additive, the so called "fuzzy integral",

$$\check{m}(\varphi) = \oint_{\Omega} \varphi \circ m \quad .$$

For $\vee$-decomposable measures m, also called "F-additive", the integral $\check{m}$ is $\vee$-decomposable. This does not hold for $\perp$-decomposable measures with respect to an Archimedean t-conorm $\perp$ and, what is more, $\check{m}$ does not reduce to Lebesgue's integral for $m=P$. These two drawbacks have been overcome by Weber (1982), who introduced an appropriate $\perp$-decomposable extension of Lebegue's integral for $\perp$-decomposable measures in the Archimedean cases, which can be used to define $\tilde{m}$,

$$\tilde{m}(\varphi) = \hat{\int_{\Omega}} \varphi \perp m \quad .$$

Applying l to those measures we obtain informations for fuzzy events,

$$l \circ \tilde{P} \quad , \quad l \circ \check{m} \quad , \quad l \circ \tilde{m} \quad .$$

Another construction, $\hat{J}$, similar to that of Sugeno, was presented by Nguyen (1977) for $\wedge$-decomposable informations, but in general

$$\hat{J} \neq l \circ \check{m} \quad ,$$

and $\hat{J}(\varphi)$ assumes the trivial value o for many nontrivial fuzzy sets.

Section 2 introduces decomposable informations J and measures m for crisp events. The basic results concerning the Archimedean cases are presented in section 3. Section 4 deals with the measures $\tilde{m}$ and informations $\tilde{J} = 1 \circ \tilde{m}$ for fuzzy events with their basic properties. In section 5 the other constructions of measures $\check{m}$ and informations $1 \circ \check{m}$ and $\hat{J}$ are discussed.

2. DECOMPOSABLE MEASURES AND INFORMATIONS FOR CRISP SETS: THE GENERAL CASE

Let $(\Omega, \mathcal{B})$ be a measurable space.

2.1 Definition (Kampé de Fêriet and Forte, 1967a): A function $J : \mathcal{B} \to [o,\infty]$ with $J(\emptyset) = \infty$ and $J(\Omega) = o$ will be called a Π- or σ-Π-decomposable information :$\Leftrightarrow$

$J(A \dot{\cup} B) = J(A) \,\Pi\, J(B)$ or

$J(\dot{\bigcup}_{k=1}^{\infty} A_k) = \prod_{k=1}^{\infty} J(A_k)$ resp.,

where Π is a binary operation on $[o,\infty]$, which is
(1) non decreasing in each argument,
(2) commutative,
(3) associative, and has
(4Π) ∞ as unit.

We use the notations $\prod_{k=1}^{\infty} a_k := \lim_{n\to\infty} \prod_{k=1}^{n} a_k$, $\prod_{k=1}^{n} a_k := (\prod_{k=1}^{n-1} a_k) \Pi a_n$, $\prod_{k=1}^{1} a_k := a_1$, and $(\sigma\text{-})\,\Pi$-decomposable for Π- or σ-Π-decomposable. For simplicity we consider Π as a semigroup operation on $[o,\infty]$, clearly Π has to be defined at least for all $(J(A),J(B))$ with $A \cap B = \emptyset$.

2.2 Definition (Weber, 1982): A function $m : \mathcal{B} \to [o,1]$ with $m(\emptyset) = o$ and $m(\Omega) = 1$ will be called a $\perp$- or σ-$\perp$-decomposable measure :$\Leftrightarrow$

$m(A \dot{\cup} B) = m(A) \perp m(B)$ or

$m(\dot{\bigcup}_{k=1}^{\infty} A_k) = \mathop{\perp}_{k=1}^{\infty} m(A_k)$ resp.,

where $\perp$ is a binary operation on $[o,1]$ with properties (1)(2)(3) from 2.1 and (4$\perp$) o as unit.

The decomposable operations $\perp$ with the properties from 2.2 are called t-conorms.

2.3 Theorem: Each decreasing continuous function $1 : [o,1] \to [o,\infty]$ with $1(1) = o$ and $1(o) = \infty$ determines a bijection, given by

$J = 1 \circ m$,

between the $\perp$-decomposable measures m and the Π-decomposable informations J, where

$1(a \perp b) = 1(a) \,\Pi\, 1(b)$.

Proof: Immediate.

The result of 2.3 permits us to simplify formulations and proofs in the following.

2.4 Example (Kampê de Fêriet and Benvenutti, 1969; Sugeno, 1974): Take $\Pi = \wedge :=$ min and $\perp = \vee :=$ max, then each 1 from 2.3 gives a bijection between the $\vee$-decomposable measures m (also called F-additive) and the $\wedge$-decomposable informations J (also called type inf). E.g. take any function $\Phi_J : \Omega \to [o,\infty]$ with $\bigwedge_{\omega\in\Omega} \Phi_J(\omega) = o$ and set $J(A) := \bigwedge_{\omega\in A} \Phi_J(\omega)$ or take any $\Phi_m : \Omega \to [o,1]$ with $\bigvee_{\omega\in\Omega} \Phi_m(\omega) = 1$ and set $m(A) := \bigvee_{\omega\in A} \Phi_m(\omega)$.
Other examples will follow in the next section or can be found in the references.

Some of the properties of classical measures (Lebesgue) or informations (Wiener-Shannon) remain valid in the general decomposable cases, e.g. the following ones.

2.5 Theorem (Kampe de Fêriet, 1970; Weber, 1982):
(i) a) J decomposable $\Rightarrow$ J non increasing,
b) m decomposable $\Rightarrow$ m non decreasing,
(ii) J (or m resp.) decomposable and continuous from below $\Leftrightarrow$ J (or m resp.) σ-decomposable,
(iii) a) J Π-decomposable $\Leftrightarrow$ $J(A\cup B) \,\Pi\, J(A\cap B) = J(A) \,\Pi\, J(B)$,
b) m $\perp$-decomposable $\Leftrightarrow$ $m(A\cup B) \perp m(A\cap B) = m(A) \perp m(B)$.

As a corollary of 2.5(iii), the $\wedge$-decomposability of J is equivalent to $J(A\cup B) = J(A) \wedge J(B)$ for all $A,B \in \mathcal{B}$, and the $\vee$-decomposability of m is equivalent to $m(A\cup B) = m(A) \vee m(B)$ for all $A,B \in \mathcal{B}$.

To derive more properties we need a more specific structure of Π and $\perp$, namely the Archimedean property.

3. DECOMPOSABLE MEASURES AND INFORMATIONS FOR CRISP SETS: THE ARCHIMEDEAN CASE

The basic tool for Archimedean decomposable operations of 3.1 is Ling's representation theorem 3.2.

3.1 Definition (Ling, 1965): A binary operation Π on $[o,\infty]$ or $\perp$ on $[o,1]$ resp. with properties (1)(2)(3)(4) from 2.1 or 2.2 resp. will be called Archimedean :$\Leftrightarrow$

(5) $\top$ ($\perp$) is continuous,
(6 $\top$) $\top(a,a) < a$ for all $a \in (0,\infty)$ or
(6 $\perp$) $\perp(a,a) > a$ for all $a \in (0,1)$ resp.

An Archimedean operation $\top$ ($\perp$) will be called strict :$\Leftrightarrow$

(7) $\top$ ($\perp$) is increasing.

3.2 Theorem (Ling, 1965):
a) A binary operation $\top$ on $[0,\infty]$ is Archimedean $\Leftrightarrow$
There exists a decreasing and continuous function $f : [0,\infty] \to [0,\infty]$ with $f(\infty) = 0$ so that

$$a \top b = f^{(-1)}(\, f(a) + f(b)\,) \quad ,$$

where $f^{(-1)}$ is the pseudoinverse of f, given by

$$f^{(-1)}(y) := f^{-1}(\, \min(y, f(0))\,) \; .$$

Moreover: $\top$ strict $\Leftrightarrow$ $f(0) = \infty$.

b) A binary operation $\perp$ on $[0,1]$ is Archimedean $\Leftrightarrow$
There exists an increasing and continuous function $g : [0,1] \to [0,\infty]$ with $g(0) = 0$ so that

$$a \perp b = g^{(-1)}(\, g(a) + g(b)\,) \quad ,$$

where $g^{(-1)}$ is the pseudoinverse of g, given by

$$g^{(-1)}(y) := g^{-1}(\, \min(y, g(1))\,) \; .$$

Moreover: $\perp$ strict $\Leftrightarrow$ $g(1) = \infty$.

The functions f and g from 3.2 are called (additive) generators. They are unique except for multiplication with positive numbers. In the non strict case we will call the generators with $f(0) = 1$ or $g(1) = 1$ resp. the normed generators.

For the Archimedean case, 2.3 leads to

3.3 Theorem: In the situation of 2.3, if $\top$ is continuous or Archimedean or strict, then $\perp$ has the same properties and vice versa. In the Archimedean case of 3.2 ,

$$f = g \circ l^{-1} \;,\; f^{(-1)} = l \circ g^{(-1)} \;,\; f(0) = g(1).$$

Proof: Using 2.3 and 3.2 .

The crucial step for section 4 will be noted in

3.4 Theorem (For step b) see Weber, 1982):
a) Let J be a $(\sigma\text{-})\top$-decomposable information with respect to an Archimedean operation $\top$ with generator f and let $\mu := f \circ J$, or

b) Let m be a $(\sigma\text{-})\perp$-decomposable measure with respect to an Archimedean operation $\perp$ with generator g and let $\mu := g \circ m$.

$$\Rightarrow \mu(\dot{\bigcup_{k \in K}} A_k) = \min\,(\sum_{k \in K} \mu(A_k), \mu(\Omega)\,),$$

where K is finite or countable resp. in case of decomposable or σ-decomposable measure (information).

Note that for $J = l \circ m$ the two μ from 3.4 are equal.

In view of the result 3.4 we can distinguish between three cases:

(S) $\top$ ($\perp$) strict and therefore μ an infinite (σ-) additive measure ,

(NSA) $\top$($\perp$) non strict and μ a finite (σ-) additive measure ,

(NSP) $\top$($\perp$) non strict and μ pseudo(σ-) additive, i.e. where

$$\mu(\dot{\bigcup_k} A_k) = \mu(\Omega) < \sum_k \mu(A_k)$$

is possible.

We can state a result reverse to 3.4.

3.5 Theorem: Let μ be a (σ-) additive measure. Then

a) Let f be some generator of an Archimedean operation $\top$ with $f(0) \leq \mu(\Omega)$
$\Rightarrow$ $J := f^{(-1)} \circ \mu$ is a $(\sigma\text{-})\top$-decomposable information of type

(S) if $\top$ strict and $f(0) = \mu(\Omega) = \infty$,
(NSA) if $\top$ non strict and $f(0) = \mu(\Omega) < \infty$,
(NSP) if $\top$ non strict and $f(0) < \mu(\Omega)$.

b) Let g be some generator of an Archimedean operation $\perp$ with $g(1) \leq \mu(\Omega)$
$\Rightarrow$ $m := g^{(-1)} \circ \mu$ is a $(\sigma\text{-})\perp$-decomposable measure, for which the same as in a) is valid replacing $\top$ by $\perp$ and $f(0)$ by $g(1)$.

Proof: a) By definition of $f^{(-1)}$ we have

$$(f \circ J)(A) = \min\,(\mu(A),\, f(0)\,) \; .$$

Applying 3.4 a) to $\bar{\mu} := f \circ J$ we obtain the result. Note that $\bar{\mu} = \mu$ in cases (S) and (NSA), but $\bar{\mu} \lneq \mu$ in case (NSP) . Clearly,

$\bar{\mu}(\Omega) = f(0)$.

b) Here $\bar{\mu} := g \circ m$ applies.

3.6 Remark: For informations, the cases (S) and (NSA) in 3.5 a) correspond to the type M from Kampé de Fériet and Forte (1967b), while (NSP) corresponds to the proper type M' from Kampé de Fériet, Forte and Benvenutti (1969), where "proper" refers to "$<$" in $f(0) < \mu(\Omega)$.

One of the properties for informations and

measures of Archimedean type is what can be seen as "subtractivity". Other properties can be found in the references.

3.7 Theorem (For step b) see Weber, 1982):

a) Let J be a $\top$-decomposable information with respect to an Archimedean operation $\top$ with generator f. Let $A \subseteq B$ with the additional assumption that $J(A) > o$ for (S) or $J(B) > o$ for (NSP)

$$\Rightarrow \quad J(B \setminus A) = f^{-1}(f(J(B)) - f(J(A))) .$$

b) Let m be a $\perp$-decomposable measure with respect to an Archimedean operation $\perp$ with generator g. Let $A \subseteq B$ with $m(A) < 1$ for (S) or $m(B) < 1$ for (NSP)

$$\Rightarrow \quad m(B \setminus A) = g^{-1}(g(m(B)) - g(m(A))) .$$

Typical examples shall illustrate this section and will be continued in the following sections.

3.8 Example (Kampé de Fériet and Forte, 1967a; Weber, 1982):

Let Ω be countable, $\mathcal{B} = \mathcal{P}(\Omega)$, $\mu(A) = |A|$ and $f(x) = x^{-1}$. Then 3.5 a) implies that $J(A) := |A|^{-1}$ gives a σ-$\top$-decomposable information of type (S) with

$a \top b = (a^{-1} + b^{-1})^{-1}$.

For all $A \subseteq B$ with $|A| < \infty$, 3.7 a) gives

$J(B \setminus A) = (|B| - |A|)^{-1}$.

We can apply 3.3 in two ways, given l or given g. Let e.g. $l(x) = -c \cdot \ln x$. Then $m(A) := l^{-1}(J(A)) = \exp (-c |A|)^{-1}$ is a σ-$\perp$-decomposable measure of type (S), where $\perp$ is generated by $g(x) = (-c \cdot \ln x)^{-1}$. Or let e.g. $g(x) = -\ln (1-x)$. Then $l^{-1}(x) = 1-\exp(-x)^{-1}$ and $m(A) := l^{-1}(J(A)) = 1-\exp(-|A|)$. This last measure is that of example (S) from Weber (1982).

3.9 Example (Kampé de Fériet, Forte and Benvenutti, 1969; Weber, 1982):

Let $(\Omega, \mathcal{B}, \mu)$ be a measure space and $f(x) = f(o) \cdot \exp(- x/c)$ with $c > o$, $o < f(o) < \infty$, $f(o) \leq \mu(\Omega)$. Then

$$J(A) = \begin{cases} -c \cdot \ln \dfrac{\mu(A)}{f(o)} & \text{if} \quad \mu(A) \leq f(o) \\ o & \text{if} \quad \mu(A) \geq f(o) \end{cases}$$

gives a σ-$\top$-decomposable information of type (NSA) if $f(o) = \mu(\Omega) < \infty$,

(NSP) if $f(o) < \mu(\Omega)$, with

$a \top b = -c \cdot \ln(\min[\exp(-a/c) + \exp(-b/c), 1])$.

The (NSA) type leads to the Wiener-Shannon-information

$J(A) = -c \cdot \ln P(A)$ with $P(A) = \dfrac{\mu(A)}{\mu(\Omega)}$.

The (NSP) type leads to a generalized W-S-information

$J(A) = -c \cdot \ln m(A)$ wih $m(A) = \min(\dfrac{\mu(A)}{f(o)}, 1)$.

Here m is only $\perp_m$-decomposable, $a \perp_m b := \min(a+b, 1)$, but in general not a probability because of $f(o) < \mu(\Omega)$.

For $\Omega = \{ s_1, \dots, s_N \}$, $\mathcal{B} = \mathcal{P}(\Omega)$, $\mu(A) = \dfrac{|A| \cdot \ln(1+\lambda)}{N \cdot \lambda}$ and $g(x) = \lambda^{-1} \ln(1+\lambda \cdot x)$ with $\lambda > -1$ we are led to the measure of example (NSA) from Weber (1982),

$m(A) = \lambda^{-1} \cdot [\exp(\lambda \cdot \min(\mu(A), f(o))) - 1]$.

For $\Omega = \mathbb{N}$, $\mathcal{B} = \mathcal{P}(\Omega)$, $\mu(A) = \dfrac{|A| \cdot f(o)}{N}$ with $N \geq 2$ and $l(x) = -c \cdot \ln x$ we are led to the measure of example (NSP) from Weber (1982),

$m(A) = \min (\dfrac{|A|}{N}, 1)$.

4. MEASURES AND INFORMATIONS FOR FUZZY SETS: THE ARCHIMEDEAN CASE

By means of the property derived in 3.4 for the Archimedean case, we can extend measures and informations from crisp to fuzzy sets, using Lebesgue's integral.

In this section let $(\Omega, \mathcal{B})$ be a measurable space, J a σ-$\top$-decomposable information and m a σ-$\perp$-decomposable measure, with respect to Archimedean operations $\top$ and $\perp$ resp. generated by f and g resp. The fuzzy sets $\varphi, \psi, \dots$ are supposed to be measurable functions from Ω to $[0,1]$, i.e. fuzzy events (Zadeh, 1968).

4.1 Definition (For step b) see Weber, 1982):

$$\text{a)} \quad \tilde{J}(\varphi) := \begin{cases} f^{-1}(\int_\Omega \varphi \, d(f \circ J)) & \text{for (S)(NSA)} \\ f^{(-1)}[\sum_{k=1}^{\infty} \int_{A_k} \varphi \, d(f \circ J)] & \text{for (NSP) and } \Omega \text{ J-achievable} \end{cases}$$

$$\text{b)} \quad \tilde{m}(\varphi) := \begin{cases} g^{-1}(\int_\Omega \varphi \, d(g \circ m)) & \text{for (S)(NSA)} \\ g^{(-1)}[\sum_{k=1}^{\infty} \int_{A_k} \varphi \, d(g \circ m)] & \text{for (NSP) and } \Omega \text{ m-achievable} \end{cases}$$

where Ω will be called J- (m-)achievable $:\Leftrightarrow$ there exists $\Omega = \dot{\bigcup}_{k=1}^{\infty} A_k$ with $J(A_k) > o$ ($m(A_k) < 1$).

The integrals $\int_{\Omega}$ and $\int_{A_k}$ in 4.1 are of Lebesgue type, because $f \circ J$, $g \circ m$ in cases (S) and (NSA) and their restrictions to each A_k in case (NSP) are σ-additive measures. The definitions are independent from the special choices of generators and sequence $\{A_k\}$ which makes Ω achievable. For more details see Weber (1982), where instead of $\tilde{m}(\varphi)$ symbols $\int_{\Omega} \varphi \perp m$, $I_{(m)}\varphi$ or $I\varphi$ are employed.

4.2 Theorem: The following diagram is commutative:

$$\begin{array}{ccc} m & \longleftrightarrow & J = l \circ m \\ \downarrow & & \downarrow \\ \tilde{m} & \longleftrightarrow & \tilde{J} = l \circ \tilde{m} \end{array} \quad .$$

Proof: Using 2.3, 3.3, 3.4, there is $(l \circ \tilde{m})(\varphi) = \tilde{J}(\varphi)$.

Clearly, the classical cases $m = P$ with $g(x) = x$ and $J = -\ln \circ P$ with $f(x)=\exp(-x)$ lead to Lebesgue's integral $\tilde{m}(\varphi) = \tilde{P}(\varphi) = \int_{\Omega} \varphi \, dP$ and $\tilde{J} = -\ln \circ \tilde{P}$ resp.

All characteristic properties of J and m are inherited by $\tilde{J}$ and $\tilde{m}$, as will be noted in the following.

4.3 Theorem (For $\tilde{m}$ see Weber, 1982): The information $\tilde{J}$ and measure $\tilde{m}$ for fuzzy events are

extensions of J and m : $\tilde{J}(1_A) = J(A)$ and $\tilde{m}(1_A) = m(A)$ for $A \in \mathcal{B}$,

normed : $\tilde{J}(1) = o$, $\tilde{J}(o) = \infty$ and $\tilde{m}(1) = 1$, $\tilde{m}(o) = o$,

decomposable : $\tilde{J}(\varphi + \psi) = \tilde{J}\varphi \,\pi\, \tilde{J}\psi$ and $\tilde{m}(\varphi + \psi) = \tilde{m}\varphi \perp \tilde{m}\psi$ for $\varphi + \psi \leq 1$,

continuous from below : $\tilde{J}\varphi_n \downarrow \tilde{J}\varphi$ and $\tilde{m}\varphi_n \uparrow \tilde{m}\varphi$ for $\varphi_n \uparrow \varphi$ a.e.

Proof: For $\tilde{J}$ by 2.3 and 4.2.

4.4 Theorem:

(i) a) $\tilde{J}$ decomposable $\Rightarrow \tilde{J}$ non increasing,
b) $\tilde{m}$ decomposable $\Rightarrow \tilde{m}$ non decreasing,

(ii) $\tilde{J}$ (or $\tilde{m}$ resp.) decomposable and continuous from below $\Leftrightarrow$ $\tilde{J}$ (or $\tilde{m}$ resp.) σ-decomposable.

Proof: For $\tilde{m}$ see Weber (1982) except the step (ii) "$\Leftarrow$" , proved here: $\tilde{m}$ σ-decomposable means naturally that for all sequences $\{\psi_k\}$ with $\sum_{k=1}^{\infty} \psi_k \leq 1$,

$\tilde{m}(\sum_{k=1}^{\infty} \psi_k) = \perp_{k=1}^{\infty} \tilde{m}(\psi_k)$. Trivially $\tilde{m}$ is also decomposable. Now, $\varphi_n \uparrow \varphi$ and set $\psi_1 := \varphi_1$, $\psi_k := \varphi_k - \varphi_{k-1}$ for $k \geq 2$. Then the result follows,

$$\tilde{m}(\varphi) = \tilde{m}(\lim_{n\to\infty} \varphi_n) = \tilde{m}(\lim_{n\to\infty} \sum_{k=1}^{n} \psi_k) =$$

$$\tilde{m}(\sum_{k=1}^{\infty} \psi_k) = \perp_{k=1}^{\infty} \tilde{m}(\psi_k) = \lim_{n\to\infty} \perp_{k=1}^{n} \tilde{m}(\psi_k) =$$

$$\lim_{n\to\infty} \tilde{m}(\sum_{k=1}^{n} \psi_k) = \lim_{n\to\infty} \tilde{m}(\varphi_n) \ .$$

4.5 Remark: For types (S) and (NSA), $\tilde{J}$ and $\tilde{m}$ are positively homogeneous :

$\tilde{J}(c \cdot \varphi) = c *_f \tilde{J}\varphi$ and $\tilde{m}(c \cdot \varphi) = c *_g \tilde{m}\varphi$

for $c \in (o,1)$, where $c *_f x := f^{-1}(c \cdot f(x))$ and $c *_g x$ by analogy. Clearly, $1 * x = x$. For any fixed $c \in (o,1)$: $x \mapsto c *_f x$ is a bijective unitary operation in (o,∞), $x \mapsto c *_g x$ the same in $(o,1)$. But

$$c *_f o \begin{cases} > o \text{ for (NSA)} \\ = o \text{ for (S)} \end{cases} \quad \text{and}$$

$$c *_g 1 \begin{cases} < 1 \text{ for (NSA)} \\ = 1 \text{ for (S)} \end{cases} \quad ,$$

which means that for type (S) all fuzzy events $\varphi = c$ with $c \in (o,1)$ have equal $\tilde{J}(c) = \tilde{J}(1) = o$ and $\tilde{m}(c) = \tilde{m}(1) = 1$.
This would be a drawback in situations, where is difficult to distinguish between "$\varphi = o$" and "$\varphi = c$ with c near o" . Then, better we choose non strict Archimedean operations π and $\perp$, because for type (NSA) we have "continuity in c" , i.e.

$\tilde{J}(c) = c *_f o \rightarrow \tilde{J}(o) = \infty$ and

$\tilde{m}(c) = c *_g 1 \rightarrow \tilde{m}(o) = o$, if $c \rightarrow o$.

4.6 Example (continuation of 3.8):

$\tilde{J}\varphi = (\sum_{k=1}^{\infty} \varphi(k))^{-1}$ and $\tilde{m}\varphi = l^{-1}(\tilde{J}\varphi)$

assume non trivial values $\tilde{J}\varphi > o$ and $\tilde{m}\varphi < 1$ iff $\sum_{k=1}^{\infty} \varphi(k) < \infty$, typically for type (S).

4.7 Example (continuation of 3.9):

Type (NSA): $\tilde{J} = -c\cdot(\ln\circ\tilde{P})$ with $\tilde{P}\varphi = \int_\Omega \varphi\, dP$.

Type (NSP): $\tilde{J} = -c\cdot(\ln\circ\tilde{m})$ with

$$\tilde{m}\varphi = \min\left(\frac{1}{f(o)}\cdot\sum_{k=1}^{\infty}\int_{A_k}\varphi\, d\mu\ ,\ 1\right)$$

and achievable $\Omega = \dot{\bigcup}_{k=1}^{\infty} A_k$. The special measure space from 3.9 leads to

$$\tilde{m}\varphi = \min\left(\frac{1}{N}\cdot\sum_{k=1}^{\infty}\varphi(k)\ ,\ 1\right) .$$

5. OTHER CONSTRUCTIONS FOR MEASURES AND INFORMATIONS FOR FUZZY SETS

For measures and informations other than the decomposable ones of Archimedean type we need other constructions. I will discuss here the two ones used in fuzzy set theory.

5.1 Definition (Sugeno, 1974): For any function $m : \mathcal{B} \to [o,1]$ let

$$\check{m}\varphi := \bigvee_{t\in(o,1]} (\,t \wedge m\{\varphi \geq t\}\,) .$$

5.2 Remark (Sugeno, 1974): $\check{m}$ is an extension of m , and $\check{m}$ is normed, non decreasing, continuous from below if m has the corresponding properties. Furthermore, $\check{m}(c) = c$ and $\check{m}(\varphi\vee\psi) = \check{m}\varphi \vee \check{m}\psi$ if m is $\vee$-decomposable. But $\check{P}\varphi$ is not Lebesgue's integral if P is a probability. For more details I refer to Sugeno (1974), where $\check{m}\varphi$ is denoted by $\oint_\Omega \varphi\circ m$ and called fuzzy integral.

5.3 Definition (Nguyen, 1977): For any function $J : \mathcal{B} \to [o,\infty]$ let

$$\hat{J}\varphi := \bigwedge_{t\in(o,1]} (\,t \vee J\{\varphi \geq t\}\,) .$$

5.4 Remark: $\hat{J}$ is an extension of J , and $\hat{J}$ is normed, non increasing, continuous from below if J has the corresponding properties. Furthermore, $\hat{J}(\varphi\vee\psi) = \hat{J}\varphi \wedge \hat{J}\psi$ if J is $\wedge$-decomposable. But $\hat{J}(c) = \hat{J}(1) = o$ for all $c \in (o,1)$, the same situation as for $\tilde{J}$ of type (S), see 4.5.

5.5 Remark: The following diagram is not commutative:

$$\begin{array}{ccc} m & \longleftrightarrow & l\circ m = J \\ \downarrow & & \downarrow \\ \check{m} & \longleftrightarrow & l\circ\check{m} \neq \hat{J} \end{array} .$$

Proof: $(l\circ\check{m})(\varphi) = \bigwedge_{t\in(o,1]} (\,l(t) \vee J\{\varphi\geq t\}\,) \neq \hat{J}\varphi$, e.g. $\varphi = c \in (o,1)$ has $(l\circ\check{m})(c) = l(c) > o = \hat{J}(c)$.

In view of 5.5 I suggest to use the informations $l\circ\check{m}$ instead of $\hat{J}$ if m is $\vee$-decomposable or J is $\wedge$-decomposable.

5.6 Example (continuation of 2.4):
Take $\Omega = [o,1]$,

$$\Phi_J(w) = \begin{cases} \frac{1}{3w} - 1 & \text{if } w \leq \frac{1}{3} \\ o & \text{if } \frac{1}{3} \leq w \leq \frac{2}{3} \\ \frac{3w-2}{3-3w} & \text{if } \frac{2}{3} \leq w \end{cases} \quad \text{and}$$

$\varphi = c_1\cdot 1_{[o,3/4)} + c_2\cdot 1_{[3/4,1]}$ with $o \leq c_1 < c_2 \leq 1$. Then

$$\hat{J}\varphi = \begin{cases} o & \text{if } c_1 > o \\ \frac{1}{3} & \text{if } c_1 = o \end{cases} ,$$

$$(l\circ\check{m})(\varphi) = \begin{cases} l(c_2) & \text{if } \frac{1}{3} \leq l(c_2) \\ \frac{1}{3} & \text{if } l(c_2) \leq \frac{1}{3} \leq l(c_1) \\ l(c_1) & \text{if } l(c_1) \leq \frac{1}{3} \end{cases} .$$

5.7 Example (continuation of 4.7):
Take $\Omega = [o,1]$, $\mu = m = P$ = Lebesgue probability, $f(o) = 1$, i.e. (NSA) type, $c = 1$, and φ from 5.6. Then
$\tilde{P}\varphi = \frac{3}{4}\cdot c_1 + \frac{1}{4}\cdot c_2$, $\tilde{J}\varphi = -\ln(\tilde{P}\varphi)$, compared with $\check{P}\varphi = c_1 \vee \frac{1}{4}$, $J(A) = -\ln P(A)$,

$$\hat{J}\varphi = \begin{cases} o & \text{if } c_1 > o \\ \ln 4 & \text{if } c_1 = o \end{cases} .$$

In both examples it does not import the value c_2 for $\hat{J}\varphi$.

CONCLUSION

In this paper I have presented the concepts of decomposable informations and measures for crisp events and suggested how to extend these to fuzzy events.

For the Archimedean cases, $\tilde{J}$ and $\tilde{m}$ from

4.1 are prefered to Nguyen's $\hat{J}$ from 5.3 and Sugeno's $\check{m}$ from 5.1, because $\tilde{J}$ and $\tilde{m}$ extend Wiener-Shannon's and Lebesgue's concepts and are decomposable. Only for type (S) all non μ-integrable fuzzy sets have the same trivial value as the "full (fuzzy)set" 1 .

For the case of $\vee$-decomposable measures, Sugeno's $\check{m}$ will be used, whereas for $\wedge$-decomposable informations, $1 \circ \check{m}$ from 5.5 has advantage over Nguyen's $\hat{J}$.

REFERENCES

Kampé de Fériet, J., B. Forte (1967a). Information et probabilité. C.R. Acad. Sc. Paris, 265, 110-114.

Kampé de Fériet, J., B. Forte (1967b). Information et probabilité. C.R. Acad. Sc. Paris, 265, 142-146.

Kampé de Fériet, J., P. Benvenutti (1969). Sur une classe d'informations. C.R.Acad. Sc. Paris, 269, 97-101.

Kampé de Fériet, J., B. Forte, and P. Benvenutti (1969). Forme générale de l'opération de composition continue d'une information. C.R.Acad. Sc. Paris, 269, 529-534.

Kampê de Fériet, J. (1970). Mesure de l'information fournie par un évènement. In: Les probabilités sur les structures algébrique. Centre Nacional de la Recherche Scientifique, Paris. pp. 191-221.

Ling, C.H. (1965). Representation of associative functions. Publ. Math. Debrecen, 12, 189-212.

Nguyen, H.T. (1977). On fuzziness and linguistic probabilities. J. Math. Anal. Appl., 61, 658-671.

Sugeno, M. (1974). Theory of fuzzy integrals and its applications. Doctoral dissertation, Tokyo Institute of Technology.

Weber, S. (1982). $\perp$-decomposable measures and integrals for Archimedean t-conorms $\perp$. Preprint, to appear in J. Math. Anal. Appl.

Zadeh, L.A. (1965). Fuzzy sets. Inf. and Control, 8, 338-353.

Zadeh, L.A. (1968). Probability measures of fuzzy events. J. Math. Anal. Appl., 23, 421-427.

ASYMPTOTIC STRUCTURAL CHARACTERISTICS OF FUZZY MEASURE AND THEIR APPLICATIONS

Wang Zhenyuan

Department of Mathematics, Hebei University, Baoding, Hebei, China

Abstract. In the fuzzy measure theory, as Sugeno's fuzzy measures lose the additivity in general, the concept "almost", which is well known in the classical measure theory, splits into two different concepts "almost" and "pseudo-almost". It complicates the relations among the convergences of the sequence of measurable functions on the fuzzy measure space, and increases the content of the theory of convergences of the sequence of fuzzy integrals. In order to replace the additivity, it is quite necessary to investigate some asymptotic behaviors of a fuzzy measure at the sequences of sets which are called "waxing" and "waning", and to introduce some new concepts, such as "autocontinuity", "converse-autocontinuity" and "pseudo-autocontinuity". They describe some asymptotic structural characteristics of a fuzzy measure. By means of them, we give four forms of generalization for each of the Egoroff's theorem, the Liesz's theorem and the Lebesgue's theorem respectively, and prove a lot of convergence theorems on the sequence of fuzzy integrals.

Keywords. Set theory; measure theory; fuzzy measure; fuzzy integral.

INTRODUCTION

Sugeno (1974), Ralescu (1980) and Batle (1979) began the researches of fuzzy measures and fuzzy integrals. Wang (1981b) introduced the concept of "autocontinuity of a set function", used it in the above-mentioned researches, and obtained a series of new results. the autocontinuity is one of the asymptotic structural characteristics of set function. As that is well known, Sugeno's fuzzy measures lose the additivity in general. Therefore, it looks very important to study its some other structural characteristics.

In this paper, we shall study some asymptotic structural characteristics of a fuzzy measure, introduce the concepts of "almost" and "pseudo-almost" on the fuzzy measure space, and give their applications in the theory of convergences of the sequence of measurable functions and of fuzzy integrals. We shall discover that a fuzzy measure always possesses some good asymptotic structural characteristic when it is F-additive, subadditive, superadditive, quasi-additive, or satisfies the λ-rule respectively, and the condition given by means of these asymptotic structural characteristics is more suitable than other for a lot of important theorems of convergence.

All concepts and signs not defined in this paper may be found in the references (Halmos, 1967; Wang, 1981b).

ASYMPTOTIC STRUCTURAL CHARACTERISTICS OF FUZZY MEASURE

Let X be a nonempty set, $\mathcal{F}$ be a σ-algebra of subsets of X, $\mu : \mathcal{F} \longrightarrow [0,\infty]$ be a fuzzy measure on $(X, \mathcal{F})$.

Definition 2.1. Let $A \in \mathcal{F}$, $\{B_n\} \subset \mathcal{F}$. $\{B_n\}$ is called μ-waning outside A, if

$$\mu(B_n \cup A) \longrightarrow \mu(A);$$

$\{B_n\}$ is called μ-waxing inside A, if

$$\mu(B_n \cap A) \longrightarrow \mu(A).$$

If it does not bring a confusion, we omit "μ-" from "μ-waning" and from "μ-waxing", and "waning outside ϕ" is simply said "waning". Evidently,

$$\{B_n\} \text{ is waning} \Longleftrightarrow \mu(B_n) \longrightarrow 0.$$

Definition 2.2. μ is called autocontinuous from above, denoted by autoc.↓ (resp. autocontinuous from below, denoted by autoc.↑), if $\forall A \in \mathcal{F}$, $\forall \{B_n\} \subset \mathcal{F}$,

$\{B_n\}$ is waning $\Longrightarrow$

$\{B_n\}$ is waning outside A

(resp. $\{B_n^c\}$ is waxing inside A);

μ is called autocontinuous, denoted by autoc., if it is both autoc.↓ and autoc.↑.

Definition 2.3. μ is called converse-autocontinuous from above, denoted by c.autoc.↓ (resp. converse-autocontinuous from below, denoted by c.autoc.↑), if $\forall A \in \mathcal{F}$ with $\mu(A) < \infty$, $\forall \{B_n\} \subset \mathcal{F}$,

$\{B_n\}$ is waning outside $A \Longrightarrow$

$\{B_n - A\}$ is waning

(resp. $\{B_n\}$ is waxing inside A $\Longrightarrow$

$\{A - B_n\}$ is waning);

μ is called converse-autocontinuous, denoted by c.autoc., if it is both c.autoc.↓ and c.autoc.↑.

Definition 2.4. Let $A \in \mathcal{F}$, $\mu(A) < \infty$. μ is called pseudo-autocontinuous from above with respect to A, denoted by p.autoc.↓/A (resp. pseudo-autocontinuous from below with respect to A, denoted by p.autoc.↑/A), if $\forall \{B_n\} \subset \mathcal{F}$,

$\{B_n\}$ is waxing inside A $\Longrightarrow$

$\{A - B_n\}$ is waning outside C

(resp. $\{B_n\}$ is waxing inside C)

for every $C \in A \cap \mathcal{F}$; μ is called pseudo-autocontinuous with respect to A, denoted by p.autoc./A, if it is both p.autoc.↓/A and p.autoc.↑/A; μ is called pseudo-autocontinuous from above, denoted by p.autoc.↓ (resp. pseudo-autocontinuous from below, denoted by p.autoc.↑; and pseudo-autocontonuous, denoted by p.autoc.), if it is p.autoc.↓/A (resp. p.autoc.↑/A; and p.autoc./A), whenever $A \in \mathcal{F}$, $\mu(A) < \infty$.

The autocontinuity, converse-autocontinuity and pseudo-autocontinuity describe some asymptotic structural behaviors of a fuzzy measure. And we have the following propositions for them.

Proposition 2.1. If μ is p.autoc.↓, then it is c.autoc.↑.

Proposition 2.2. If μ is autoc.↓ (resp. autoc.↑) and c.autoc.↑, then it is p.autoc.↓ (resp. p.autoc.↑).

In the current reseaches on the fuzzy measure theory, the concepts of "subadditivity", "F-additivity", "superadditivity", "quasi-additivity" and "λ-rule" are often used (cf. Ralescu, 1980; Sugeno, 1974; Wang, 1981a, 1981b; Zhao, 1981; Zheng, 1981). Now, we study the relations among the above-mentioned asymptotic structural characteristics and them.

Proposition 2.3. If μ is subadditive, then it is autoc.; Particularly, if μ is F-additive, then it is subadditive, and therefore, it is autoc..

Definition 2.5. μ is called superadditive, if we have

$$\mu(A \cup B) \geq \mu(A) + \mu(B),$$

whenever $A \in \mathcal{F}$, $B \in \mathcal{F}$, and $A \cap B = \phi$.

Proposition 2.4. If μ is superadditive, then it is c.autoc..

Definition 2.6. μ is called quasi-additive, if there exists a strictly monotone, continuous function θ: $[0, \mu(X)] \longrightarrow [0, \infty]$, with $\theta(0) = 0$ and $\theta^{-1}(\{\infty\}) \subset \{\infty\}$, such that $\theta \circ \mu$ is additive.

This definition is a little stronger than that one given by Wang (1981a, 1981b), in which the condition "$\theta^{-1}(\{\infty\}) \subset \{\infty\}$" was not required.

Proposition 2.5. If μ is quasi-additive, then it is both autoc. and c.autoc..

By using Proposition 2.2. and Proposition 2.5., we have

Proposition 2.6. If μ is quasi-additive, then it is p.autoc..

Definition 2.7. μ is called a g_λ-fuzzy measure, if there exists

$$\lambda \in (-\frac{1}{\mu(X)}, \infty) \cup \{0\},$$

such that

$$\mu(A \cup B) = \mu(A) + \mu(B) + \lambda \cdot \mu(A) \cdot \mu(B) \quad (*)$$

whenever $A \in \mathcal{F}$, $B \in \mathcal{F}$, $A \cap B = \phi$. And (*) is called λ-rule.

Proposition 2.7. Any g_λ-fuzzy measure is quasi-additive (Wang, 1981a), and therefore, it is autoc., c.autoc. and p.autoc..

Furthermore, we introduce some other concepts which are useful in the fuzzy measure theory.

Definition 2.8. μ is called null-additive, denoted by 0-add., if we have $\mu(A \cup B) = \mu(A)$, whenever $A \in \mathcal{F}$, $B \in \mathcal{F}$, $\mu(B) = 0$.

Definition 2.9. μ is called converse-null-additive, denoted by c.0-add., if we have $\mu(A - B) = 0$, whenever $A \in \mathcal{F}$, $B \in A \cap \mathcal{F}$, $\mu(B) = \mu(A) < \infty$.

Definition 2.10. Let $A \in \mathcal{F}$, $\mu(A) < \infty$. μ is called pseudo-null-additive with respect to A, denoted by p.0-add./A, if we have $\mu(B \cup C) = \mu(C)$, whenever $B \in A \cap \mathcal{F}$, $C \in A \cap \mathcal{F}$, $\mu(A - B) = \mu(A)$; μ is called pseudo-null-additive, denoted by p.o-add., if it is p.o-add./A, whenever $A \in \mathcal{F}$, $\mu(A) < \infty$.

Evidently, if $\mu(A) \neq 0$ whenever $A \in \mathcal{F}$, $A \neq \phi$, then μ is 0-add.; if $\mu(A) \neq \mu(B)$ whenever $A \in \mathcal{F}$, $B \in \mathcal{F}$, $A \neq B$, then μ is c.0-add., p.0-add. and 0-add.. And if X is finite, then 0-add. $\Longleftrightarrow$ autoc., c.0-add. $\Longleftrightarrow$ c.autoc., and p.0-add./A $\Longleftrightarrow$ p.autoc./A.

In general, we have

Proposition 2.8. If μ is p.0-add., then it is c.0-add..

Proposition 2.9. If μ is both c.0-add. and 0-add., then it is p.0-add..

Proposition 2.10. If μ is autoc.↓ or autoc.↑, then it is 0-add..

Proposition 2.11. If μ is c.autoc.↓ or c.autoc.↑, then it is c.0-add..

Proposition 2.12. Let $A \in \mathcal{F}$, $\mu(A) < \infty$. If μ is p.autoc.↓/A or p.autoc.↑/A, then it is p.0-add./A.

In the following, we denote "increasing" by "↗", "decreasing" by "↘".

Theorem 2.1. (1) Let μ be 0-add.. $\forall A \in \mathcal{F}$, $\forall B_n \in \mathcal{F}$, n=1,2,···, if $B_n \searrow$, $\mu(B_n) \longrightarrow 0$, then $\mu(A - B_n) \longrightarrow \mu(A)$; Moreover, if there exists at least one n_0, such that $\mu(A \cup B_{n_0}) < \infty$ as $\mu(A) < \infty$, then $\mu(A \cup B_n) \longrightarrow \mu(A)$.

(2) Let μ be c.0-add.. $\forall A \in \mathcal{F}$, $\forall B_n \in \mathcal{F}$, n=1,2,···, if $A \supset B_n \nearrow$ (resp. $A \subset B_n \searrow$), $\mu(B_n) \longrightarrow \mu(A) < \infty$, then $\mu(A - B_n) \longrightarrow 0$ (resp. $\mu(B_n - A) \longrightarrow 0$).

(3) Let μ be p.0-add./A, $A \in \mathcal{F}$, $\mu(A) < \infty$.

$\forall B_n \in \mathcal{F}$, $n=1,2,\cdots$, if $A \supset B_n \nearrow$, $\mu(B_n) \to \mu(A)$, then $\mu(B_n \cap C) \to \mu(C)$ and $\mu(C \cup (A-B_n)) \to \mu(C)$ for all $C \in A \cap \mathcal{F}$.

Proof. We only give the proof for one of these conclusions, since the rest is similar. If μ is c.0-add., $A \subset B_n \searrow$, $\mu(B_n) \to \mu(A) < \infty$, then there exists n_0, such that $\mu(B_{n_0}) < \infty$. Denote $\bigcap_{n=1}^{\infty} B_n$ by B, then $B \supset A$, and by using the continuity of μ, $\mu(B_n) \to \mu(B) = \mu(A)$. As μ is c.0-add., we have $\mu(B - A) = 0$. From $B_n - A \searrow$, $\bigcap_{n=1}^{\infty}(B_n - A) = B - A$, and $\mu(B_{n_0} - A) \le \mu(B_{n_0}) < \infty$, using the continuity of μ again, we obtain $\mu(B_n - A) \to \mu(B-A)=0$. ∎

It is interesting to contrast the conclusions of Theorem 2.1. with the definitions of autoc., c.autoc. and p.autoc.. We can discover that the autoc., c.autoc. and p.autoc. are only a little stronger than the 0-add., c.0-add. and p.0-add. respectively.

"ALMOST" AND "PSEUDO-ALMOST"

Since the fuzzy measures lose the additivity in general, it is necessary to introduce two different concepts of "almost" and "pseudo-almost" on the fuzzy measure space.

In the following, let $A \in \mathcal{F}$, P be a proposition.

Definition 3.1. If there exists $E \in \mathcal{F}$ with $\mu(E)=0$, such that P is true on A-E, then we say "P is almost everywhere true on A"; If there exists $F \in \mathcal{F}$ with $\mu(A-F)=\mu(A)$, such that P is true on A-F, then we say "P is pseudo-almost everywhere true on A"; And if P is pseudo-almost everywhere true on C for all $C \in A \cap \mathcal{F}$, then we say "P is pseudo-almost everywhere true in A".

Definition 3.2. If there exists $\{E_n\} \subset \mathcal{F}$, with $\mu(E_n) \to 0$, such that P is true on $A-E_n$, $n=1,2,\cdots$, then we say "P is almost true on A"; If there exists $\{F_n\} \subset \mathcal{F}$, with $\mu(A-F_n) \to \mu(A)$, such that P is true on $A-F_n$, $n=1,2,\cdots$, then we say "P is pseudo-almost true on A"; And if P is pseudo-almost true on C for all $C \in A \cap \mathcal{F}$, then we say "P is pseudo-almost true in A".

We denote "almost everywhere", "pseudo-almost everywhere", "almost" and "pseudo-almost" by "a.e.", "p.a.e.", "a." and "p.a." respectively.

It is important to point out that: If P is a.e. (resp. a.) true on A, then it is also a.e. (resp. a.) true on C for all $C \in A \cap \mathcal{F}$; But, the similar conclusions for p.a.e. and for p.a. do not hold (see Example 3.1.). Therefore, the concepts of "a.e. on A" (resp. "a. on A") and of "p.a.e. on A" (resp. "p.a. on A") are not symmetric.

Example 3.1. Let $X = \{a,b,c\}$, $\mathcal{F} = \mathcal{P}(X)$,

$$\mu(A) = \begin{cases} \text{the cardinal number of } A, & A \ne \{a,b\}, \\ 3, & A=\{a,b\}, \end{cases}$$

$$f(X) = \begin{cases} 0, & x = a, b, \\ 1, & x = c, \end{cases}$$

then μ is a fuzzy measure on $(X, \mathcal{F})$ and $f(x)=0$ p.a.e. on X, but it does not hold that $f(x)=0$ p.a.e. on $\{a,c\}$.

Now, let $f: X \to (-\infty, \infty)$ be a real-valued measurable function and $\{f_n\}$ be a sequence of real-valued measurable functions.

Definition 3.3. If $\mu(\{|f_n - f| \ge \varepsilon\} \cap A) \to 0$ for any given $\varepsilon > 0$, then we say $\{f_n\}$ converge in fuzzy measure μ to f on A (if it is without confusion, we say $\{f_n\}$ converge in measure to f on A), and denote it by $f_n \xrightarrow{\mu} f$ on A; If $\mu(\{|f_n - f| < \varepsilon\} \cap A) \to \mu(A)$ for any given $\varepsilon>0$, then we say $\{f_n\}$ converge pseudo-in fuzzy measure μ to f on A (or, for short, $\{f_n\}$ converge pseudo-in measure to f on A), and denote it by $f_n \xrightarrow{p.\mu} f$ on A; And if $f_n \xrightarrow{p.\mu} f$ on C for all $C \in A \cap \mathcal{F}$, then we say $\{f_n\}$ converge pseudo-in measure to f in A, and denote it by $f_n \xrightarrow{p.\mu} f$ in A.

When μ is a measure (i.e. it is additive), on $A \in \mathcal{F}$,

P is a.e. true $\Longrightarrow$ P is p.a.e. true;
P is a. true $\Longrightarrow$ P is p.a. true;
$f_n \xrightarrow{\mu} f \Longrightarrow f_n \xrightarrow{p.\mu} f$.

Furthermore, if μ is a measure with $\mu(A)<\infty$, then the converse implication relations hold too.

Proposition 3.1. If P is a.e. true on A, μ is 0-add., then P is p.a.e. true in A; If P is p.a.e. true on A, μ is c.0-add. and $\mu(A)<\infty$, then P is a.e. true on A; If P is p.a.e. true on A, μ is p.0-add./A and $\mu(A)<\infty$, then P is p.a.e. true in A.

Proposition 3.2. If P is a. true on A, μ is autoc.↑, then P is p.a. true in A; If P is p.a. true on A, μ is c.autoc.↑ and $\mu(A)<\infty$, then P is a. true on A; If P is p.a. true on A, μ is p.autoc.↑/A and $\mu(A)<\infty$, then P is p.a. true in A.

Proposition 3.3. If $f_n \xrightarrow{\mu} f$ on A, μ is autoc.↑, then $f_n \xrightarrow{p.\mu} f$ in A; If $f_n \xrightarrow{p.\mu} f$ on A, μ is c.autoc.↑ and $\mu(A)<\infty$, then $f_n \xrightarrow{\mu} f$ on A; If $f_n \xrightarrow{p.\mu} f$ on A, μ is p.autoc.↑/A and $\mu(A)<\infty$, then $f_n \xrightarrow{p.\mu} f$ in A.

The above definitions and propositions, and the results given in the following sections can be extended to the case of f and $f_n: X \to [-\infty, \infty]$ without any essential difficulty.

APPLICATIONS ON CONVERGENCES OF SEQUENCE OF MEASURABLE FUNCTIONS

Wang (1981b, 1982) gave some generalizations on the fuzzy measure space for the Liesz's theorem, the Egoroff's theorem and the Lebesgue's theorem well known in the classical measure theory. Now, by means of the asymptotic structural characteristics of fuzzy measure and the concepts of "almost" and "pseudo-almost", we give all of four forms of generalization for each of the three theorems respectively.

Lemma 4.1. Let $\{E_n\} \subset \mathcal{F}$, $A \in \mathcal{F}$. If $\mu(E_n) \to 0$, and μ is autoc.↓ (resp. autoc.↑), then there exists some sequence $\{\mathcal{E}_j\}$ of subsequences of $\{E_n\}$, where $\mathcal{E}_j = \{E_{n_i^{(j)}}\}$, such that

$$\lim_{j\to\infty} \mu(\bigcup_{i=1}^{\infty} E_{n_i^{(j)}}) = 0$$

$$(\text{resp. } \lim_{j\to\infty} \mu(A - \bigcup_{i=1}^{\infty} E_{n_i^{(j)}}) = \mu(A));$$

Furthermore, there exists some subsequence $\{E_{n_i}\}$ of $\{E_n\}$, such that

$$\mu(\bigcap_{j=1}^{\infty} \bigcup_{i=j}^{\infty} E_{n_i}) = 0$$

$$(\text{resp. } \mu(A - \bigcap_{j=1}^{\infty} \bigcup_{i=j}^{\infty} E_{n_i}) = \mu(A)).$$

If $E_n \subset A$, $\mu(E_n) \to \mu(A) < \infty$, and μ is p.autoc.↓/A (resp. p.autoc.↑/A), then there some sequence $\{\mathcal{E}_j\}$ of subsequences of $\{E_n\}$, where $\mathcal{E}_j = \{E_{n_i^{(j)}}\}$, such that

$$\lim_{j\to\infty} \mu(A - \bigcap_{i=1}^{\infty} E_{n_i^{(j)}}) = 0$$

$$(\text{resp. } \lim_{j\to\infty} \mu(\bigcap_{i=1}^{\infty} E_{n_i^{(j)}}) = \mu(A));$$

Furthermore, there exists some subsequence $\{E_{n_i}\}$ of $\{E_n\}$, such that

$$\mu(A - \bigcup_{j=1}^{\infty} \bigcap_{i=j}^{\infty} E_{n_i}) = 0$$

$$(\text{resp. } \mu(\bigcup_{j=1}^{\infty} \bigcap_{i=j}^{\infty} E_{n_i}) = \mu(A)).$$

Proof. We only prove the last one of the four conclusions of this lemma, the rest is similar to it.

Let $E_n \subset A$, $\mu(E_n) \to \mu(A) < \infty$. For arbitrarily given $\varepsilon > 0$, there exists $E_{n_1(\varepsilon)}$, such that

$$\mu(E_{n_1(\varepsilon)}) \geq \mu(A) - \frac{\varepsilon}{2};$$

And for $E_{n_1(\varepsilon)} \subset A$, as μ is p.autoc.↑/A, there exists $E_{n_2(\varepsilon)}$, such that

$$\begin{aligned}\mu(E_{n_1(\varepsilon)} \cap E_{n_2(\varepsilon)}) &= \mu(E_{n_1(\varepsilon)} - (A - E_{n_2(\varepsilon)}))\\ &\geq \mu(A) - \frac{\varepsilon}{2} - \frac{\varepsilon}{4}\\ &= \mu(A) - \frac{3}{4}\varepsilon; \quad \cdots\end{aligned}$$

and so on. Finally, we obtain a subsequence $\{E_{n_i(\varepsilon)}\}$, such that

$$\mu(\bigcap_{i=1}^{\infty} E_{n_i(\varepsilon)}) \geq \mu(A) - \varepsilon.$$

Furthermore, we take a subsequence $\{E_{n_i^{(1)}}\}$ such that

$$\mu(\bigcap_{i=1}^{\infty} E_{n_i^{(1)}}) \geq \mu(A) - 1;$$

And as $\mu(E_{n_i^{(1)}}) \to \mu(A)$ too, there exists a subsequence $\{E_{n_i^{(2)}}\}$ of $\{E_{n_i^{(1)}}\}$, such that

$$\mu(\bigcap_{i=1}^{\infty} E_{n_i^{(2)}}) \geq \mu(A) - \frac{1}{2};$$

$\cdots$. In general, there exists a subsequence $\{E_{n_i^{(j)}}\}$ of $\{E_{n_i^{(j-1)}}\}$, such that

$$\mu(\bigcap_{i=1}^{\infty} E_{n_i^{(j)}}) \geq \mu(A) - \frac{1}{j}, \quad j=2,3,\cdots.$$

If we take $n_i = n_i^{(i)}$, then $\{E_{n_i}\}$ is a subsequence of $\{E_n\}$, and

$$\bigcap_{i=j}^{\infty} E_{n_i} \supset \bigcap_{i=1}^{\infty} E_{n_i^{(j)}}, \quad j=1,2,\cdots.$$

Consequently,

$$\begin{aligned}\mu(A) &\geq \mu(\bigcup_{j=1}^{\infty} \bigcap_{i=j}^{\infty} E_{n_i}) \geq \mu(\bigcap_{i=j}^{\infty} E_{n_i})\\ &\geq \mu(\bigcap_{i=1}^{\infty} E_{n_i^{(j)}}) \geq \mu(A) - \frac{1}{j}\end{aligned}$$

for all $j=1,2,\cdots$. And therefore,

$$\mu(\bigcup_{j=1}^{\infty} \bigcap_{i=j}^{\infty} E_{n_i}) = \mu(A). \quad ▌$$

The following theorem is a generalization of the Liesz's theorem.

Theorem 4.1. (1) If μ is autoc.↓, $f_n \xrightarrow{\mu} f$ on A, then there exists some subsequence $\{f_{n_i}\}$ of $\{f_n\}$, $f_{n_i} \xrightarrow{a.e.} f$ on A, and $f_{n_i} \xrightarrow{p.a.e.} f$ in A.

(2) If μ is autoc.↑, $f_n \xrightarrow{\mu} f$ on A, then there exists some subsequence $\{f_{n_i}\}$ of $\{f_n\}$, $f_{n_i} \xrightarrow{p.a.e.} f$ on A.

(3) If μ is p.autoc.↓/A, $\mu(A) < \infty$, $f_n \xrightarrow{p.\mu} f$ on A, then there exists some subsequence $\{f_{n_i}\}$ of $\{f_n\}$, $f_{n_i} \xrightarrow{a.e.} f$ on A.

(4) If μ is p.autoc.↑/A, $\mu(A) < \infty$, $f_n \xrightarrow{p.\mu} f$ on A, then there exists some subsequence $\{f_{n_i}\}$ of $\{f_n\}$, $f_{n_i} \xrightarrow{p.a.e.} f$ in A, and $f_{n_i} \xrightarrow{a.e.} f$ on A.

Proof. We only prove the fourth conclusion, the rest is similar.

There is no harm in assuming $A = X$. As $f_n \xrightarrow{p.\mu} f$, for every $k=1,2,\cdots$, there exists n_k respectively, such that

$$\mu(\{\,|f_{n_k} - f| < \frac{1}{k}\,\}) > \mu(X) - \frac{1}{k}.$$

without any loss of generality, we suppose $n_{k+1} > n_k$, $k=1,2,\cdots$. If we denote

$$E_k = \{\,|f_{n_k} - f| < \frac{1}{k}\,\},$$

then

$$\lim_{k\to\infty} \mu(E_k) = \mu(X) < \infty.$$

By using Lemma 4.1., there exists a subsequence $\{E_{k_i}\}$ of $\{E_k\}$, such that

$$\mu(\bigcup_{j=1}^{\infty} \bigcap_{i=j}^{\infty} E_{k_i}) = \mu(X).$$

For every $x \in \bigcup_{j=1}^{\infty} \bigcap_{i=j}^{\infty} E_{k_i}$, there exists $j(x)$, such that $x \in \bigcap_{i=j(x)}^{\infty} E_{k_i}$, namely

$$|f_{n_{k_i}}(x) - f(x)| < \frac{1}{k_i}$$

as $i \geq j(x)$. Thus, for arbitrarily given $\varepsilon > 0$, if we take i_0 such that $\frac{1}{k_{i_0}} < \varepsilon$, then

$$|f_{n_{k_i}}(x) - f(x)| < \frac{1}{k_i} < \frac{1}{k_{i_0}} < \varepsilon$$

as $i \geq j(x) \vee i_0$. That is to say, $\{f_{n_{k_i}}\}$ converges to f on $\bigcup_{j=1}^{\infty} \bigcap_{i=j}^{\infty} E_{k_i}$.

Furthermore, by using Proposition 2.12. and Proposition 3.1., we have $f_{n_{k_i}} \xrightarrow{p.a.e.} f$ in A; And by a similar reason, we can obtain $f_{n_{k_i}} \xrightarrow{a.e.} f$ on A. ▌

If $\{f_n\}$ converges almost (resp. pseudo-almost) uniformly to f on A, then we denote it by $f_n \xrightarrow{a.u.} f$ (resp. $f_n \xrightarrow{p.a.u.} f$) on A.

Now, we give four forms of the generalization of the Egoroff's theorem.

Theorem 4.2. (1) If μ is autoc.$\downarrow$, $\mu(A)<\infty$, then on A,

$$f_n \xrightarrow{a.e.} f \Longrightarrow f_n \xrightarrow{a.u.} f .$$

(2) If μ is autoc.$\uparrow$, $\mu(A)<\infty$, then on A,

$$f_n \xrightarrow{a.e.} f \Longrightarrow f_n \xrightarrow{p.a.u.} f .$$

(3) If μ is p.autoc.$\downarrow$/A, $\mu(A)<\infty$, then on A,

$$f_n \xrightarrow{p.a.e.} f \Longrightarrow f_n \xrightarrow{a.u.} f .$$

(4) If μ is p.autoc.$\uparrow$/A, $\mu(A)<\infty$, then

$$f_n \xrightarrow{p.a.e.} f \text{ on A} \Longrightarrow f_n \xrightarrow{p.a.u.} f \text{ in A} .$$

Proof. (1) has been proved by Wang (1981b). Using the corresponding conclusions in Lemma 4.1., the proofs of (2), (3) and (4) are similar to one of (1). ∎

Finally, we study the Lebesgue's theorem. Ralescu (1980) gave the first conclusion of the following theorem.

Theorem 4.3. (1) $\mu(A)<\infty$, $f_n \xrightarrow{a.e.} f$ on A $\Longrightarrow$

$f_n \xrightarrow{\mu} f$ on A.

(2) $f_n \xrightarrow{p.a.e.} f$ on A $\Longrightarrow f_n \xrightarrow{p.\mu} f$ on A.

(3) $f_n \xrightarrow{a.e.} f$ on A, μ is 0-add. $\Longrightarrow f_n \xrightarrow{p.\mu} f$ in A.

(4) $f_n \xrightarrow{p.a.e.} f$ on A, μ is c.0-add., $\mu(A)<\infty \Longrightarrow$

$f_n \xrightarrow{\mu} f$ on A.

Proof. For (1), the classical proof works. And by using Proposition 3.1., we can obtain (3) and (4) from (2) and (1) respectively. Now, we prove (2).
As $f_n \xrightarrow{p.a.e.} f$ on A, there exists $B \in \mathcal{F}$ with $B \subset A$, $\mu(B) = \mu(A)$, such that $f_n \longrightarrow f$ on B. For arbitrarily given $\varepsilon>0$ and $x \in B$, there exists N(x), such that

$$|f_n(x) - f(x)| < \varepsilon$$

as $n \geq N(x)$. Denote

$$A_n = \{N(x) \leq n\} \cap B,$$

we have $A_n \nearrow$, $\bigcup_{n=1}^{\infty} A_n = B$. Since

$$\{|f_n - f| < \varepsilon\} \cap A \supset A_n,$$

$$\mu(A) \geq \mu(\{|f_n - f| < \varepsilon\} \cap A)$$
$$\geq \mu(A_n) \longrightarrow \mu(B) = \mu(A),$$

and therefore

$$\mu(\{|f_n - f| < \varepsilon\} \cap A) \longrightarrow \mu(A).$$

That is to say, $f_n \xrightarrow{p.\mu} f$ on A. ∎

APPLICATIONS ON FUZZY INTEGRAL

In this section, we give the a.e. (resp. p.a.e.) convergence theorem and the convergence in measure (resp. pseudo-in measure) theorem of the sequence of fuzzy integrals.

Theorem 5.1. (1) Whenever $f_1 = f_2$ a.e. on A, it holds $\int_A f_1 \, d\mu = \int_A f_2 \, d\mu$, if and only if μ is 0-add..

(2) Whenever $f_1 = f_2$ p.a.e. on A, it holds $\int_A f_1 \, d\mu = \int_A f_2 \, d\mu$, if and only if μ is p.0-add./A and $\mu(A) < \infty$.

Proof. The proof of (1) was given by Wang (1981b). Now, we prove (2).

Starting from the definition of fuzzy integral directly, it is easy to prove the sufficiency.

For the necessity, $\mu(A)<\infty$ is evident, Now, let $B \in A \cap \mathcal{F}$, $\mu(A) = \mu(A-B)$ and $C \in A \cap \mathcal{F}$. We take $N > \mu(A)$, and

$$f_1(x) = \begin{cases} N, & x \in B \cup C, \\ 0, & x \notin B \cup C, \end{cases}$$

$$f_2(x) = \begin{cases} N, & x \in C, \\ 0, & x \notin C. \end{cases}$$

Then, $f_1 = f_2$ p.a.e. on A. Therefore,

$$\mu(C \cup B) = \int_A f_1 \, d\mu = \int_A f_2 \, d\mu = \mu(C),$$

That is, μ is p.0-add./A. ∎

By a theorem given by Wang (1981b) (Theorem 15) and Theorem 5.1., we can give the following statement.

Theorem 5.2. Whenever $f_n \xrightarrow{a.e.} f$ (resp. $f_n \xrightarrow{p.a.e.} f$) on A, and there exist n_0 and a constant $c \leq \int_A f \, d\mu$, such that

$$\mu(\{\sup_{n \geq n_0} f_n > c\} \cap A) < \infty,$$

then

$$\int_A f_n \, d\mu \longrightarrow \int_A f \, d\mu,$$

if and only if μ is 0-add. (resp. μ is p.0-add./A and $\mu(A)<\infty$).

In order to prove the convergence in measure (pseudo-in measure) theorem, we need the following lemma proved by Wang (1981b).

Lemma 5.1. For $\alpha \in [0,\infty)$,

(1) $\int_X f \, d\mu = \alpha \Longrightarrow \mu(F_\alpha) \geq \alpha \geq \mu(F_{\alpha+\varepsilon})$, for $\varepsilon>0$.

(2) $\int_X f \, d\mu \geq \alpha \Longleftrightarrow \mu(F_\alpha) \geq \alpha$.

Where $F_\alpha = \{f \geq \alpha\}$.

Theorem 5.3. Whenever $f_n \xrightarrow{\mu} f$ (resp. $f_n \xrightarrow{p.\mu} f$) on A, then

$$\int_A f_n \, d\mu \longrightarrow \int_A f \, d\mu ,$$

if and only if μ is autoc. (resp. μ is p.autoc./A and $\mu(A)<\infty$).
Proof. We only prove the second conclusion, the other is similar.

Sufficiency: There is no harm in assuming $A = X$ and $\int_X f \, d\mu > 0$. Let $F_\alpha^n = \{f_n \geq \alpha\}$ and $c = \int_X f \, d\mu$. As $c \leq \mu(X)<\infty$, by Lemma 5.1., we have $\mu(F_c) \geq c$. For arbitrarily given $\varepsilon>0$ with $\varepsilon<c$, since $f_n \xrightarrow{p.\mu} f$, we have

$$\mu(\{|f_n - f| < \varepsilon\}) \longrightarrow \mu(X).$$

From

$$F_{c-\varepsilon}^n \supset F_c - \{|f_n - f| \geq \varepsilon\} = F_c \cap \{|f_n - f| < \varepsilon\},$$

since μ is p.autoc.$\uparrow$/X and $\mu(X)<\infty$, there exists n_0, such that

$$\mu(F_{c-\varepsilon}^n) \geq \mu(F_c) - \varepsilon \geq c - \varepsilon$$

as $n \geq n_0$. Using Lemma 5.1. again, we obtain

$$\int_X f_n \, d\mu \geq c - \varepsilon$$

as $n \geq n_0$.

Analogously, from

$$F^n_{c+2\varepsilon} \subset F_{c+\varepsilon} \cup \{ |f_n - f| \geq \varepsilon \}$$
$$= F_{c+\varepsilon} \cup (X - \{|f_n - f| < \varepsilon\}),$$

since μ is p.autoc.$\downarrow$/X, there exists n'_o, such that

$$\mu(F^n_{c+2\varepsilon}) \leq \mu(F_{c+\varepsilon}) + \varepsilon < c + 2\varepsilon,$$

and therefore,

$$\int_X f_n \, d\mu < c + 2\varepsilon$$

as $n \geq n'_o$.

Consequently, $\int_X f_n \, d\mu \longrightarrow c$.

Necessity: The conclusion $\mu(A) < \infty$ is evident. Let $\{B_n\} \subset \mathcal{F}$, $B_n \subset A$, $n=1,2,\cdots$, $\mu(A - B_n) \longrightarrow \mu(A) < \infty$. For $C \in A \cap \mathcal{F}$, if we take N with $\mu(A) < N < \infty$, and

$$f(x) = \begin{cases} N, & x \in C, \\ 0, & x \notin C, \end{cases}$$

$$f(x) = \begin{cases} N, & x \in C - B_n, \\ 0, & x \notin C - B_n, \end{cases} \quad n=1,2,\cdots,$$

then $f_n = f$ on $A - B_n$, and therefore $f_n \xrightarrow{p.\mu} f$ on A. Consequently,

$$\mu(C - B_n) = \int_A f_n \, d\mu \longrightarrow \int_A f \, d\mu = \mu(C),$$

that is, μ is p.autoc.$\uparrow$/A.

Analogously, if we take

$$f_n(x) = \begin{cases} N, & x \in C \cup B_n, \\ 0, & x \notin C \cup B_n, \end{cases} \quad n=1,2,\cdots,$$

then it may be obtained that

$$\mu(C \cup B_n) \longrightarrow \mu(C),$$

that is, μ is p.autoc.$\downarrow$/A. ▮

The first conclusion of the above theorem was given by Wang (1981b) with a quite long proof, and a prolix concept of "local-uniformly autocontinuity" was used there. The present proof is much briefer than that.

REFERENCES

Batle, N. and E. Trillas (1979). Entropy and fuzzy integral. J. Math. Anal. Appl., 69, 469-474.

Halmos, P. R. (1967). Measure Theory. Van Nostrand, New York.

Ralescu, D and G. Adams (1980). The fuzzy integral. J. Math. Anal. Appl., 75, 562-570.

Sugeno, M. (1974). Theory of fuzzy integrals and its applications. Ph. D. dissertation, Tokyo Institute of Technology.

Wang Zhenyuan (1981a). Une classe de mesures floues — les quasi-mesures. BUSEFAL, 6, 28-37.

Wang Zhenyuan (1981b). The autocontinuity of set function and the fuzzy integral. 12th meeting of the EWG on fuzzy sets, Hamburg (in J. Math. Anal. Appl. to appear).

Wang Zhenyuan (1982). On the convergence of sequences of measurable functions on the fuzzy measure space. BUSEFAL, 9, 27-30.

Zhao Ruhuai (1981). (N) fuzzy integral. Mathematical Research and Exposition (in Chinese), Vol. 1, No. 2, 55-72.

Zheng Daopeng and Huang Jinli (1981). A series of convergence theorems for fuzzy integral and the neighbourhood property of a pseudo-fixed point. Fuzzy Mathematics (in Chinese), Vol. 1, No. 2, 37-44.

HYPERFIELDS AND RANDOM SETS

Wang Pei-Zhuang* and E. Sanchez**

**Department of Mathematics, Beijing Normal University, Beijing, China*
***Service Universitaire de Biomathématiques, Faculté de Médecine, Marseille, France*

Abstract. Basically, σ- hyperfields on σ- complete De Morgan algebras are introduced and explored. Measurable spaces (topological spaces) are extended to hypermeasurable spaces (hypertopological spaces) constructed considering σ- hyperfields. Random sets are then given a pure measure theoretical definition and, finally, random fuzzy sets are characterized.

Keywords. Hyperfield ; random set ; random fuzzy set.

INTRODUCTION.

Recently, Zhang Nan-Lun (1982), Wang Pei-Zhuang and Sanchez (1982) investigated a model of fuzzy-statistical experiments, illustrated by "fix the element, move the set". It is important to note that, analogously to probability-statistical experiments, there exists a stability in the observation of the frequency of a movable interval (expressing a fuzzy concept) that covers a fixed point in the real line $\mathbb{R}$. These results provided an objective method for the determination of grades of membership. The model of fuzzy-statistical experiments prompted us to describe fuzzy sets as a "cumulus" of random sets and, following the works of Nguyen (1978) and Goodman (1982), in (Wang Pei-Zhuang and Sanchez, 1982) we proposed a unified framework to treat a fuzzy set as a random set. Depending on topological structures, random sets may be defined in several ways. Here, we give a pure measure theoretical definition of random sets, in which the notion of hyperfiels we propose, plays an important role. The random fuzzy sets of this paper can be useful in the study of fuzzy point processes.

σ- HYPERFIELDS ON A DE MORGAN ALGEBRA.

Let $B = (B,\vee,\wedge,c)$ be a σ- complete, De Morgan algebra (a complemented distributive lattice holding De Morgan's laws) with greatest element, 1, and least element, 0. For any family $(D_n)_{n\in\mathbb{N}}$, of subsets of B, we denote the following :

$$\dot{\bigcup_{n=1}^{\infty}} D_n \triangleq \{\alpha \mid \exists\, n,\ \alpha \in D_n\} \quad ; \qquad (1)$$

$$\dot{\bigcap_{n=1}^{\infty}} D_n \triangleq \{\alpha \mid \forall\, n,\ \alpha \in D_n\} \quad ; \qquad (2)$$

$$D^{c} \triangleq \{\alpha \mid \alpha \in B,\ \alpha \notin D\} \ ; \qquad (3)$$

$$D^{\backslash c} \triangleq \{\alpha \mid \alpha^{c} \in D\}. \qquad (4)$$

Let

$$\dot{\alpha} \triangleq \{\beta \mid \beta \in B,\ \beta \geqslant \alpha\} \ ; \qquad (5)$$

$$\underset{\cdot}{\alpha} \triangleq \{\beta \mid \beta \in B,\ \beta \leqslant \alpha\} \ , \qquad (6)$$

where $\leqslant$ is the order relation in B.
Let us now denote

$$\dot{D} \triangleq D^{\cdot} \triangleq \{\dot{\alpha} \mid \alpha \in D\} \quad ; \qquad (7)$$

$$\underset{\cdot}{D} \triangleq D_{\cdot} \triangleq \{\underset{\cdot}{\alpha} \mid \alpha \in D\} \quad ; \qquad (8)$$

$$\underset{\cdot}{\dot{D}} \triangleq D^{\cdot}_{\cdot} \triangleq \dot{D} \cup \underset{\cdot}{D} \qquad (D \in \mathcal{P}(B)) \ ; \qquad (9)$$

$$\dot{\downarrow}\mathcal{D} \triangleq \{\alpha \mid \dot{\alpha} \in \mathcal{D}\} \qquad ; \qquad (10)$$

$$\underset{\cdot}{\downarrow}\mathcal{D} \triangleq \{\alpha \mid \underset{\cdot}{\alpha} \in \mathcal{D}\} \qquad (D \subset \mathcal{P}(B)) \ . \qquad (11)$$

DEFINITION 1.

$\mathcal{H} \subset \mathcal{P}(B)$ is called a σ- hyperfield on B iff

i) $\mathcal{H} \supset \{B,\phi\}$ is a σ- field on $\mathcal{P}(B)$, that is, it is closed under

$$\dot{\bigcup_{n=1}^{\infty}}\ ,\ \dot{\bigcap_{n=1}^{\infty}}\ ,\ c.$$

ii) $\mathcal{H}$ is closed under $\backslash c$ (see (4)).

iii) $\dot{\downarrow}\mathcal{H} = \underset{\cdot}{\downarrow}\mathcal{H} = M$, and M is a σ- subalgebra of B, i.e., M is closed under

$$\dot{\bigvee_{n=1}^{\infty}}\ ,\ \dot{\bigwedge_{n=1}^{\infty}}\ ,\ c.$$

For a given $D \in P(B)$, we denote

$$[D] \triangleq \cap \{E \mid D \subset E \subset B, E \text{ is closed under } \overset{\infty}{\underset{n=1}{\vee}}, \overset{\infty}{\underset{n=1}{\wedge}}, c\}. \quad (12)$$

For a given $\mathcal{D} \subset P(B)$, we denote

$$[\mathcal{D}] \triangleq \cap \{\mathcal{E} \mid \mathcal{D} \subset \mathcal{E} \subset P(B), \mathcal{E} \text{ is closed under } \overset{\infty}{\underset{n=1}{\cup}}, \overset{\infty}{\underset{n=1}{\cap}}, c\}. \quad (13)$$

Generally, $[\underset{.}{D}]$ is not a σ- hyperfield on B. The following theorem gives conditions for it.

THEOREM 1.

$[\underset{.}{D}]$ is a σ-hyperfield on B if

i) $D^{\setminus c} = D$; (14)

ii) $\downarrow^{\cdot}[\underset{.}{D}] = \downarrow_{\cdot}[\underset{.}{D}] \triangleq M$, (15)

and in this case, we have

$M \supset D$. (16)

To prove Theorem 1, we need the following lemmas. The proofs are clear.

LEMMA 1.

$$(D^c)^{\setminus c} = (D^{\setminus c})^c \quad (\forall D \in P(B)) . \quad (17)$$

LEMMA 2.

$$C \subset D \Rightarrow C^{\setminus c} \subset D^{\setminus c} \quad (\forall C, D \in P(B)); \quad (18)$$

$$(\overset{\infty}{\underset{n=1}{\cup}} D_n)^{\setminus c} = \overset{\infty}{\underset{n=1}{\cup}} D_n^{\setminus c},$$

$$(\overset{\infty}{\underset{n=1}{\cap}} D_n)^{\setminus c} = \overset{\infty}{\underset{n=1}{\cap}} D_n^{\setminus c},$$

$$(\forall D_n \in P(B)) . \quad (19)$$

LEMMA 3.

$$\widehat{\dot{\alpha}^c} = (\underset{.}{\alpha})^{\setminus c}, \ \underset{.}{\underbrace{\alpha^c}} = (\dot{\alpha})^{\setminus c}$$

$$(\forall \alpha \in B) . \quad (20)$$

LEMMA 4.

$$\overset{\infty}{\underset{n=1}{\cap}} \dot{\alpha}_n = \overbrace{\overset{\infty}{\underset{n=1}{\vee}} \alpha_n}^{\cdot}, \ \overset{\infty}{\underset{n=1}{\cap}} \underset{.}{\alpha}_n = \underbrace{\overset{\infty}{\underset{n=1}{\wedge}} \alpha_n}_{\cdot}$$

$$(\forall \alpha_n \in B) . \quad (21)$$

For $C \subset P(B)$, set

$$C^{\setminus c} \triangleq \{c \mid c^{\setminus c} \in C\} . \quad (22)$$

LEMMA 5.

For all $C \subset P(B)$, we have

$$[C]^{\setminus c} \supset [C^{\setminus c}] . \quad (23)$$

Proof.

Set $\mathcal{D} = [C]$.

1) From $C \subset \mathcal{D}$, it follows $C^{\setminus c} \subset \mathcal{D}^{\setminus c}$.

2) Suppose that $D_n \in \mathcal{D}^{\setminus c}$ $(n \in \mathbb{N})$, i.e., $D_n^{\setminus c} \in \mathcal{D}$, we have $(\overset{\infty}{\underset{n=1}{\cup}} D_n)^{\setminus c} = \overset{\infty}{\underset{n=1}{\cup}} D_n^{\setminus c} \in \mathcal{D}$, so that $\overset{\infty}{\underset{n=1}{\cup}} D_n \in \mathcal{D}^{\setminus c}$.

3) If $D \in \mathcal{D}^{\setminus c}$ then $D^{\setminus c} \in \mathcal{D}$, and then $(D^{\setminus c})^c \in \mathcal{D}$. But, $(D^{\setminus c})^c = (D^c)^{\setminus c}$ (from lemma 1) yields $(D^c)^{\setminus c} \in \mathcal{D}$, therefore $D^c \in \mathcal{D}^{\setminus c}$.

From 1), 2), 3), we see that $\mathcal{D}^{\setminus c}$ is a σ-field containing $C^{\setminus c}$, so that (23) is true. ∎

LEMMA 6.

$$[\underset{.}{D}]^{\setminus c} = [(\underset{.}{D})^{\setminus c}] . \quad (24)$$

Proof.

With $C = \underset{.}{D}$, in lemma 5, it follows

$$[\underset{.}{D}]^{\setminus c} \supset [(\underset{.}{D})^{\setminus c}] .$$

With $C = (\underset{.}{D})^{\setminus c}$, in lemma 5, it follows

$$[(\underset{.}{D})^{\setminus c}]^{\setminus c} \supset [((\underset{.}{D})^{\setminus c})^{\setminus c}] .$$

But, $((C)^{\setminus c})^{\setminus c} = C$,

so that $[(\underset{.}{D})^{\setminus c}] \supset [\underset{.}{D}]^{\setminus c}$, and (24) is true. ∎

Proof of theorem 1.

We only need to prove that

1) $[\underset{.}{D}]$ is closed under $\setminus c$;
2) M is a σ- subalgebra of B ;
3) $M \supset [D]$.

1) Since $D^{\setminus c} = D$, it is clear that $(\underset{.}{D})^{\setminus c} = \underset{.}{D}$. From lemma 6, we have

$$[\underset{.}{D}] = [\underset{.}{D}]^{\setminus c}, \quad (25)$$

which means that $[\underset{.}{D}]$ is closed under $\setminus c$.

2) If $\alpha \in M$, then $\underset{.}{\alpha} \in [\underset{.}{D}]$, but (from lemma 3),

$$\widehat{(\dot{\alpha}^c)} = (\underset{.}{\alpha})^{\setminus c} . \quad (26)$$

From (25) we have

$$\alpha^c \in \downarrow^{\cdot}[\underset{.}{D}] = M .$$

If $\alpha_n \in M$, then $\dot{\alpha}_n \in \underset{.}{D}$ and from lemma 4,

$$\overbrace{(\overset{\infty}{\underset{n=1}{\vee}} \alpha_n)}^{\cdot} = \overset{\infty}{\underset{n=1}{\cap}} \dot{\alpha}_n \in [\underset{.}{D}] .$$

Hence,

$$\overset{\infty}{\underset{n=1}{\vee}} \alpha_n \in \downarrow^{\cdot}[\underset{.}{D}] = M.$$

We have proved that M is a σ- subalgebra of B.

3) Obviously, $M \supset [D]$. ∎

HYPERMEASURABLE SPACE $(X, B, \check{B})$ EXTENDED FROM A MEASURABLE SPACE (X,B).

Let (X,B) be a measurable space, i.e., B is a σ- field on X, so that, B is a σ- complete Boolean subalgebra of $(P(X), \cup, \cap, c)$. Refering to the notations of last section, we denote here

$$\dot{\alpha} = \{\beta \mid \beta \in B, \beta \supset \alpha\} ,$$

$$\underset{\cdot}{\alpha} = \{\beta \mid \beta \in B, \beta \subset \alpha\} ; \quad (27)$$

$$\dot{B} = \{\dot{\alpha} \mid \alpha \in B\} ; \quad (28)$$

$$\underset{\cdot}{B} = \{\underset{\cdot}{\alpha} \mid \alpha \in B\} ; \quad (29)$$

$$\dot{\underset{\cdot}{B}} = \dot{B} \cup \underset{\cdot}{B} . \quad (30)$$

THEOREM 2.

If we denote $\check{B} \triangleq [\dot{\underset{\cdot}{B}}]$, the smallest σ- field containing $\dot{\underset{\cdot}{B}}$ on $P(B)$, then $\check{B}$ is a σ- hyperfield on B.

Proof.

Take D = B in theorem 1; obviously, we have

i) $B^{\setminus c} = B$; (31)

ii) $\downarrow^{\cdot}[\dot{\underset{\cdot}{B}}] = \downarrow_{\cdot}[\dot{\underset{\cdot}{B}}] = B.$ (32)

i.e. (14) and (15) are satisfied. ∎

DEFINITION 2.

We call $(X, B, \check{B})$ the hypermeasurable space extended from (X,B).

TOPOLOGICAL HYPERMEASURABLE SPACES.

Let $B = (B, \vee, \wedge, c)$ be a σ- complete De Morgan algebra with greatest element 1 and least element 0.

Let G be a given topology in B, that is, G is such that

1) $1,0 \in G$; (33)

2) For any index set T

$\alpha_t \in G \ (t \in T) \Rightarrow \bigvee_{t\in T} \alpha_t \in G$; (34)

3) $\alpha_i \in G \ (t \in T,\ T \text{ finite}) \Rightarrow \bigwedge_{t\in T} \alpha_t \in G$. (35)

Let $F = G^{\setminus c}$.

ASSUMPTION A :

Given $\{(\alpha_t,\beta_t) \mid \alpha_t \in G, B_t \in F, \alpha_t \leqslant \beta_t (t \in T)\}$

there exists a countable subset T' of T, such that

$$(\forall t \in T)\ (\exists t' \in T' : \alpha_{t'} \leqslant \alpha_t \leqslant \beta_t \leqslant \beta_{t'}) . \quad (36)$$

Set $\mathcal{G}_1 = \{D \mid D^c \in \dot{G}\}$;

$\mathcal{G}_2 = \{D \mid D^c \in \underset{\cdot}{F}\}$;

$\mathcal{G} = \bigcup_{i=1}^{2} \mathcal{G}_i$.

Set $T(\mathcal{G}) \triangleq$

$$\{ \bigcup_{s\in S} \bigcap_{i_s=1}^{n_s} D_{si_s} \mid (\forall S)((\forall s \in S)((\forall n_s \in N)((\forall i_s \leqslant n_s)(D_{si_s} \in \mathcal{G}))))\} . \quad (37)$$

Obviously, $T(\mathcal{G}) \cup \{B\}$ is a topology on $P(B)$.

Let $[T(\mathcal{G})]$ be the smallest σ- field containing $T(\mathcal{G})$:

$$[T(\mathcal{G})] \triangleq \{L \mid T(\mathcal{G}) \subset L \subset P(B), L \text{ is a } \sigma\text{-field}\}. \quad (38)$$

DEFINITION 3.

We call $T(\mathcal{G})$ an hyper-topology, and $[T(\mathcal{G})]$ a topological hyperfield.

If B is a σ- field on X and (X,G) is a topological space, we call $(X,G,T(\mathcal{G}))$ the hyper-topological space extended from (X,G), and $(X,[G],T(\mathcal{G}))$ the topological hypermeasurable space extended from $(X,[G])$.

There is an important consequence expressed as follows.

THEOREM 3.

Under Assumption (A), we have

$$[T(\mathcal{G})] = [\dot{G} \cup \underset{\cdot}{F}] . \quad (39)$$

In order to prove theorem 3, we need the following lemma.

LEMMA 7.

Under Assumption (A) we have

$$[T(\mathcal{G})] = [\mathcal{G}] . \quad (40)$$

Proof.

From (37), we consider

$$I \triangleq \bigcup_{s\in S} \bigcap_{i_s=1}^{n_s} D_{si_s} \quad (D_{si_s} \in \mathcal{G})$$

which can be expressed as

$$I^c = \bigcap_{s\in S} \bigcup_{i_s=1}^{n_s} D^c_{si_s} = \bigcup_{\substack{\varphi: S\to \mathbb{N} \\ \varphi(s)\leqslant n_s}} \bigcap_{s\in S} D^c_{s\varphi(s)} . \quad (41)$$

Furthermore, formula

$$\bigcap_{s\in S} D^c_{s\varphi(s)} = \bigcap_{i=1}^{2} \bigcap_{\substack{s\in S \\ D_{s\varphi(s)}\in\mathcal{G}_i}} D^c_{s\varphi(s)} \quad (42)$$

holds if it is added that :

If $\{s \mid D_{s\varphi(s)} \in \mathcal{G}_i\} \neq \phi$ then

$$\bigcap_{\substack{s\in S \\ D_{s\varphi(s)}\in\mathcal{G}_i}} D^c_{s\varphi(s)} \triangleq B \ (i=1,2) \quad (43)$$

Thus, we have

$$I^c = \bigcup_{\substack{\varphi: S\to \mathbb{N}\\ \varphi(s)\leq n_s}} \bigcap_{i=1}^{2} \bigcap_{\substack{s\in S\\ D_{s\varphi(s)}\in \mathcal{G}_i}} D^c_{s\varphi(s)}$$

$$\triangleq \bigcup_{\substack{\varphi: S\to \mathbb{N}\\ \varphi(s)\leq n_s}} \bigcap_{i=1}^{2} J_{\varphi_i} \quad , \qquad (44)$$

$$J_{\varphi_1} = \bigcap_{\substack{s\in S\\ D_{s\varphi(s)}\in\mathcal{G}_1}} D^c_{s\varphi(s)} = \bigcap_{\substack{s\in S\\ D^c_{s\varphi(s)}=\dot{\alpha}_{s\varphi(s)}\in\dot{G}}} \dot{\alpha}_{s\varphi(s)}$$

$$= \overbrace{\bigvee_{s\in S'} \alpha_{s\varphi(s)}}^{\bullet} .$$

Because $\alpha_{s\varphi(s)} \in G$, we have

$$\bigvee_{s\in S'} \alpha_{s\varphi(s)} \triangleq \alpha_\varphi \in G \quad ,$$

and then $J_{\varphi_1} = \dot{\alpha}_\varphi \in \dot{G}$.

$$J_{\varphi_2} = \bigcap_{\substack{s\in S\\ D_{s\varphi(s)}\in\mathcal{G}_2}} D^c_{s\varphi(s)} = \bigcap_{\substack{s\in S\\ D^c_{s\varphi(s)}=\underset{\bullet}{\beta}_{s\varphi(s)}\in \underset{\bullet}{F}}} \underset{\bullet}{\beta}_{s\varphi(s)} = \underset{\bullet}{\bigwedge_{s\in S'} \beta_{s\varphi(s)}} .$$

Because $\beta_{s\varphi(s)} \in F$, we have

$$\bigwedge_{s\in S'} \beta_{s\varphi(s)} \triangleq \beta_\varphi \in F$$

and then

$$J_{\varphi_2} = \underset{\bullet}{\beta}_\varphi \in \underset{\bullet}{F} \quad .$$

Let

$$\Phi \triangleq \{\varphi \mid \varphi : S \to \mathbb{N},\ \varphi(s) \leq n_s\ (s\in S)\} \quad , \qquad (45)$$

then,

$$I^c = \bigcup_{\varphi\in\Phi} (\dot{\alpha}_\varphi \cap \underset{\bullet}{\beta}_\varphi)$$

$$= \{\delta \mid \delta \in B,\ \exists\, \varphi \in \Phi : \alpha_\varphi \leq \delta \leq \beta_\varphi\}. \qquad (46)$$

Take $T = \Phi$ in Assumption (A), there exists a countable subset Φ' of Φ, such that

$$(\forall \varphi \in \Phi)\ \ (\exists\, \varphi' \in \Phi' : \alpha_{\varphi'} \leq \alpha_\varphi \leq \beta_\varphi \leq \beta_{\varphi'}) \quad (47)$$

so that

$$I^c = \{\delta \mid \delta \in B,\ \exists\, \varphi' \in \Phi' : \alpha_{\varphi'} \leq \delta \leq \beta_{\varphi'}\}$$

$$= \bigcup_{\varphi\in\Phi'} (\dot{\alpha}_\varphi \cap \underset{\bullet}{\beta}_\varphi) \in [\mathcal{G}]. \qquad (48)$$

Hence, $I \in [\mathcal{G}]$, which means that $T(\mathcal{G}) \subset [\mathcal{G}]$ and (40) is satisfied. ∎

Proof of theorem 3.

From lemma 7, we have $[T(\mathcal{G})] = [\mathcal{G}]$, so that we only need to prove that

$$[\mathcal{G}] = [\dot{G} \cup \underset{\bullet}{F}]. \qquad (49)$$

$$D \in \mathcal{G}_1 \Rightarrow D^c \in \dot{G} \Rightarrow D = (D^c)^c \in [\dot{G} \cup \underset{\bullet}{F}] \ ;$$

$$D \in \mathcal{G}_2 \Rightarrow D^c \in \underset{\bullet}{F} \Rightarrow D = (D^c)^c \in [\dot{G} \cup \underset{\bullet}{F}] \ ;$$

$$B = \dot{0} \in \dot{G} \subset [\dot{G} \cup \underset{\bullet}{F}] \quad ,$$

so that $\mathcal{G} \subset [\dot{G} \cup \underset{\bullet}{F}]$, and then $[\mathcal{G}] \subset [\dot{G} \cup \underset{\bullet}{F}]$. Analogously,

$$D \in \dot{G} \Rightarrow D^c \in \mathcal{G}_1 \Rightarrow D = (D^c)^c \in [\mathcal{G}] \quad ;$$

$$D \in \underset{\bullet}{F} \Rightarrow D^c \in \mathcal{G}_2 \Rightarrow D = (D^c)^c \in [\mathcal{G}] \quad ,$$

so that $\dot{G} \cup \underset{\bullet}{F} \subset [\mathcal{G}]$, and then $[\dot{G} \cup \underset{\bullet}{F}] \subset [\mathcal{G}]$. ∎
The following proposition is not difficult to prove.

PROPOSITION 1.

$T(\mathcal{G})$ defines the convergence in B in the sense that : δ_n converges to δ iff

i) $(\forall \alpha \in G)(\alpha \not\leq \delta \Rightarrow \exists N : \alpha \not\leq \delta_n\ (n \geq N))$;(50)

ii) $(\forall \beta \in F)(\beta \not\geq \delta \Rightarrow \exists N : \beta \not\geq \delta_n\ (n \geq N))$.(51)

Let $(X,[G],[T(\mathcal{G})])$ be the topological hypermeasurable space extended from the topological space (X,G), $\alpha \in G$ is an open set in X, and $\beta \in F$ is a closed set in X. Taking $B = [G]$, we have the following proposition.

PROPOSITION 2.

Suppose that G has a countable base, then $T(\mathcal{G})$ defines the convergence in B in the sense that δ_n ($\in$ B) converges to δ ($\in$ B) iff

i) For any $x \in \delta$, there exists $\{x_n\}$ ($n \in \mathbb{N}$) such that $x_n \in \delta_n$ and x_n converges to x;

ii) For any $x \in \delta^c$, there exists $\{x_n\}$ ($n \in \mathbb{N}$) such that $x_n \in \delta^c_n$ and x_n converges to x.

Proof.

Given $x \in \delta^c$, let $\{\alpha_k\}$ ($k \in \mathbb{N}$) be the neighbourhood base of x such that $\alpha_k \in G$,

$$\alpha_1 \supset \alpha_2 \supset \ldots \supset \alpha_k \supset \ldots$$

Obviously, $\alpha_k \not\subset \delta$ ($k \in \mathbb{N}$). According to (50), we have

$$(\forall k)\ \ (\exists N_k)\ \ (n \geq N_k \Rightarrow \alpha_k \cap \delta^c_n \neq \phi) \quad . \qquad (52)$$

Choose N_k such that $N_1 \leq N_2 \leq \ldots \leq N_k \leq \ldots$

and choose $x_n \in \delta^c_n$ such that

$$x_n \in \alpha_{f(n)} \qquad (53)$$

where

$$f(n) \triangleq k \qquad (N \leq n < N_{k+1}) \quad ,$$

then x_n converges to x. By analogy, the rest of the "only if" part can be proved. The "if" part is clear.

When $X = \mathbb{R}^n$, G, the ordinary topology in $\mathbb{R}^n$, and B, the Borel field on $\mathbb{R}^n$, then the restriction of $\mathcal{T}(\mathcal{G})$ to co$\mathcal{C}$ — the collection of convex hulls of all nonempty, compact subsets of $\mathbb{R}^n$ — is equivalent to the topology induced by the Hausdorff distance

$$d(\delta_1,\delta_2) = \max(\sup_{x\in\delta_1} d(x,\delta_2), \sup_{y\in\delta_2} d(\delta_1,y))$$

$$(\delta_i \subset X) \quad . \qquad (54)$$

The detailed proof is not given here.

RANDOM SETS

A random set on X is a measurable mapping defined on an abstract probability space $(\Omega,\mathcal{F},P)$ and taking values in a measurable space $(D,\mathcal{B})$, where $D \subset \mathcal{P}(X)$, and $\mathcal{B}$ is a σ- field on D.

There are several possible choices for the definition of $(D,\mathcal{B})$. For exaple, D may be the collection of all nonempty compact subsets or $\mathbb{R}^n$ or the Borel field, $\mathcal{B}$, generated by the Hausdorff distance defined in (54). All these choices depend on metric or topological structures.

There is a pure measure theoretical definition of random sets :

DEFINITION 4.

Let $(\Omega,\mathcal{F},P)$ be a probability space and let $(X,B,\check{B})$ be an hypermeasurable space, a mapping

$$S : \Omega \rightarrow B$$

is called a random set on X if it is $\mathcal{F}$ - $\check{B}$ measurable, i.e., for any $E \in \check{B}$, one has

$$S^{-1}(E) = \{\omega \mid S(\omega) \in E\} \in \mathcal{F}. \qquad (55)$$

THEOREM 4.

Let $\mathcal{T}(\mathcal{G})$ be a hyper-topology extended from the topological space (X,G), $B \triangleq [G]$, and $(X,B,\check{B})$ be the hypermeasurable space extended from (X,B). If Assumption (A) holds for G, then each random set on X is always $\mathcal{F}$ - $[\mathcal{T}(\mathcal{G})]$ measurable.

Proof.

Under Assumption (A) we have $\mathcal{T}(\mathcal{G}) = [\dot{G} \cup \underset{\bullet}{F}]$, but $\check{B} = [\underset{\bullet}{\dot{B}}] \supset [\dot{G} \cup \underset{\bullet}{F}]$, so that the theorem is true.

DEFINITION 5 .

A mapping

$$S : \Omega \rightarrow B$$

is called basically measurable if for all $\alpha \in B$, one has

$$S^{-1}(\dot{\alpha}) = \{\omega \mid S(\omega) \supset \alpha\} \in \mathcal{F}, \qquad (56)$$

and

$$S^{-1}(\underset{\bullet}{\alpha}) = \{\omega \mid S(\omega) \subset \alpha\} \in \mathcal{F}. \qquad (57)$$

THEOREM 5 .

S is a random set on X iff it is basically measurable.

The proof is trivial.

RANDOM FUZZY SETS

Let $\mathcal{F}(X)$ be the class of all fuzzy subsets of X. According to L.A. Zadeh's usual definitions, $(\mathcal{F}(X),\cup,\cap,c)$ is a complete De Morgan algebra. If B is a subalgebra of $\mathcal{F}(X)$, the previous results apply to B.

DEFINITION 6.

Let B be a subalgebra of $(\mathcal{F}(X),\cup,\cap,c)$, we call $(x,B,\check{B})$ defined in definition 2, the fuzzy hypermeasurable field.

DEFINITION 7 .

Let $G \subset B$ be a fuzzy topology, we call $\mathcal{T}(\mathcal{G})$, defined in definition 3, the hyper-topology extended from G.

DEFINITION 8 .

Let $(\Omega,\mathcal{F},P)$ be a probability space, and $(X,B,\check{B})$ be a fuzzy hypermeasurable field, a mapping

$$S : \Omega \rightarrow B$$

is called a random fuzzy set iff it is $\mathcal{F}$ - $\check{B}$ measurable. One can finally restate the previous theorems and propositions, which apply here.

REFERENCES

Gaines, B.R. (1978). Fuzzy and Probability uncertainty logics. Information and Control, 38, 154-169.

Goodman, I.R. (1982). Fuzzy Sets as equivalence classes of random sets. In R.R. Yager (Ed), Recent Developments in Fuzzy Set and Possibility Theory. Pergamon Press.

Nguyen, H.T. (1978). On random sets and belief functions, J. Math. Anal. and Applic., 65, 531-542.

Matheron, G. (1975). Random sets and integral geometry, John Wiley, New York.

Wang Pei-Zhuang and Sanchez, E. (1982). Treating a fuzzy subset as a projectable random subset. In M.M. Gupta and E. Sanchez (Eds), Fuzzy Information and Decision Process, North-Holland, Amsterdam.

Zadeh, L.A. (1978). Fuzzy sets as a basis for a theory of possibility. Fuzzy Sets and Systems, 1, 3-28.

Zhang Nan-Lun. (1982). A preliminary study to the theoretical bases of fuzzy sets. In P.P. Wang (Ed), Advances in Fuzzy Set Theory and Application. Pergamon Press.

ON THE FUZZY MEASURES AND THE MEASURES OF FUZZINESS FOR L-FUZZY SETS

Zi-Xiao Wang

Department of Mathematics, Northeast Normal University, Changchun, Jilin, People's Republic of China

Abstract. This paper is based on the results of De Luca and Termini [1,2] and Wang [6,7,8] . It includes the following three parts. (1) We extend the fuzzy integrals taking value in the unit inteval [0, 1] to a fuzzy integrals taking value in a lattice. When the lattice is total order, it is given that one group equivalent conditions such that the L-fuzzy measure can be represented as a L-fuzzy integral. (2) We extend the measure of fuzziness for general fuzzy sets to the measure of fuzziness for L-fuzzy sets, where L is a complement lattice. (3) We investigate the L-fuzzy integral representation for the measure of fuzziness on the L-fuzzy sets.

Keywords L-Fuzzy Measure ; L-Fuzzy Integrals ; L-Measure of Fuzziness.

1 INTRODUCTION

In 1968, Zadeh [11] first suggested the fuzzy probability on fuzzy events and peoneered the consideration of fuzzy measure. Klement et al. [4] worked important results at this direction. Their work's characteristic is that their tool is the classical Lebesgqe integral. In 1974, Sugeno [5] suggested a complete theory of fuzzy integral which is independent of Lebesgue measure and integral. Generally, Sugeno's fuzzy integral is extendation of Lebesgue integral. Therefore, in [6,7], we supplemented the fuzzy additivity for the fuzzy measure such that the measure is really a extendation of a " similarity " At the same time, we gave a new definition which is equivalent with Sugeno's fuzzy integral and is more convenient. This is anather deriction on the investigation of fuzzy measure, it is more intuitive than the first deriction on fuzzy measure.

By a measure of fuzziness, we mean a measurment of the fuzzy grade for a fuzzy set. It is similar to the entropy in the theory of probability. In 1972, De Luca and Termini first provided the concept of fuzziness and gave three principles as a measure of fuzziness. This paper has itself generated a number of papers in this area. Yager [9] suggested an intuitive definition of the fuzziness and Yager [10] and De Luca and Termini [2] investigated the measure of fuzziness on lattice. These authors consider the fuzziness of a lattice becouse of the L-fuzzy sets on the universal space is a lattice. Thus their rather consider the fuzziness of a lattice with the betweenness of elements of a lattice than investigate the fuzziness on L-fuzzy sets.

In this paper, first of all, we consider the fuzzy integral taking value in a lattice and we call it as L-fuzzy integral, Next, we investigate immediatly the fuzziness of L-fuzzy sets. Here, we require the lattice L is a metric lattice. At the same time, an axiomatic definition of the measure of fuzziness for L-fuzzy sets is given, Finally, we prove an integral representation of the measure of fuzziness.

2 L-FUZZY INTEGRAL

In [5], Sugeno first suggested complete theory of fuzzy integral. In [6,7], we gave a new definition of the fuzzy integral which is equivalent with Sugeno's integral. Now, we shall consider the fuzzy integral taking value in a lattice.

Throughout this paper, we suppose that L denotes a complete distributive metric lattice with the maximum element I and the minimum element 0. The universal space is a measurable space ($X, \mathfrak{B}$), where $\mathfrak{B}$ is ordinary Borel field on X.

I L-fuzzy measure

Definition 2.1 By a L-fuzzy measure $m(\cdot)$: $\mathfrak{B} \longrightarrow L$, we mean that $m(\cdot)$ satisfies the following conditions

(1) $0 \leqslant m(A) \leqslant I$ for all $A \in \mathfrak{B}$, and $m(\phi)=0$

(2) If $A, B \in \mathfrak{B}$, then $m(A \cup B)=m(A) \vee m(B)$

(3) If $\{A_n\}_{n \geqslant 1} \subset \mathfrak{B}$ is a monotone sequence, then

$$m(\lim A_n) = \lim m(A_n) .$$

Remark. (i) The condition (2) implies the monotonicity of $m(\cdot)$, i.e. if $A \subset B$, then $m(A) \leqslant m(A) \vee m(B) = m(B)$.

(ii) The condition (2) is equivalent to the following condition that if $A, B \in \mathfrak{B}$ and $A \cap B = \phi$, then $m(A \cup B) = m(A) \vee m(B)$

II L-fuzzy integral

In this section, we suppose that $D(\cdot)\cdot L \longrightarrow [0, M]$ is a distance on L, where M is a finite positive real number, is the metric topology and $\mathfrak{B}_0$ is the Borel field on L.

A function $h(x)$ is called measurable iff $h^{-1}(c) \in \mathfrak{B}$ for all $c \in \mathfrak{B}_0$.

Proposition 2.1 If $h(x)$ is measurable, then there exists a sequence $\{h_n(x)\}_{n \geqslant 1}$ of simple functions such that

$$\lim h_n(x) = h(x) \text{ (uniformly)}$$

Definition 2.2 Let $(X, \mathfrak{B}, m)$ be a fuzzy measure space, $h(x)$ a measurable function

(i) $h(x)$ is a simple function

$$h(x) = \bigvee_{i=1}^{n} [\alpha_i \wedge \chi_{E_i}(x)] \quad E_i \cap E_j = \phi \quad (i \neq j)$$

then

$$\oint_E h(x)\, dm = \bigvee_{i=1}^{n} [\alpha_i \wedge m(E_i \cap E)]$$

is called L-fuzzy integral of $h(x)$ on $E \in \mathfrak{B}$.

(ii) $h(x)$ is a general function, we take a monotone increasing sequence of simple functions such that $h_n(x) \longrightarrow h(x)$ (uniformly) then

$$\oint_E h(x)\, dm = \lim \oint_E h_n(x)\, dm$$

is called L-fuzzy integral of $h(x)$ on $E \in \mathfrak{B}$.

Proposition 2.1 The L-fuzzy integral defined in the Def. 2.2 equivalent with the fuzzy integral defined by Sugeno, $\mathfrak{B}_0$ is the Borel field of the unit inteval [0, 1] .

Proof. Sugeno's fuzzy integral is defined

$$(S) \oint_E h(x)\, dm = \sup[\alpha \wedge m(F_\alpha \cap E)]$$

where $F_\alpha = \{x : h(x) \geqslant \alpha\}$, $0 \leqslant \alpha \leqslant 1$.

If $h(x)$ is a simple function, say,

$$h(x) = \bigvee_{i=1}^{n} [\alpha_i \wedge \chi_{E_i}(x)] \quad E_i \cap E_j = \phi \quad (i \neq j)$$

No lest generality, we assume that

$$\alpha_1 \geqslant \alpha_2 \geqslant \cdots \geqslant \alpha_n .$$

So that

$$(S) \oint_E h(x)\, dm = \bigvee_{i=1}^{n} [\alpha_i \wedge m(F_i \cap E)]$$

where $F_i = \bigcup_{k=1}^{i} E_k$. It follows that

$$(S) \oint_E h(x)\, dm = \bigvee_{i=1}^{n} [\alpha_i \wedge m(E_i \cap E)] = \oint_E h(x)\, dm$$

If $h(x)$ is a general function, the proposition is immediate result of the monotonicity of Sugeno's fuzzy integral and L-fuzzy integrals.

Now $\tilde{\mathfrak{B}}$ denote the set of all measurable L-fuzzy sets, and define that

$$\tilde{m}(A) = \oint A(x)\, dm \quad \text{for all } A \in \tilde{\mathfrak{B}}$$

It is clear that $\tilde{m}(\cdot)$ is a fuzzy measure on $\tilde{\mathfrak{B}}$. If L is also total order lattice, then we have

Theorem 2.1 Let $\tilde{m}(\cdot): \tilde{\mathfrak{B}} \rightarrow L$ be a L-fuzzy measure on $\tilde{\mathfrak{B}}$, then the following statements are equivalent.

(1) $\tilde{m}(\cdot)$ is represented as a L-fuzzy

integral, i.e. there exists a L-fuzzy measure $m(\cdot): \mathfrak{B} \longrightarrow L$ such that

$$m(A) = \int A(x)\, dm \quad \text{for all } A \in \widetilde{\mathfrak{B}},$$

(ii) $\widetilde{m}((\alpha \wedge A(x)) \vee (\beta \wedge B(x))) =$

$$= (\alpha \wedge \widetilde{m}(A(x))) \vee (\beta \wedge \widetilde{m}(B(x)))$$

for all $\alpha, \beta \in L$ and $A, B \in \mathfrak{B}$,

(iii) $\widetilde{m}(\alpha \wedge A(x)) = \alpha \wedge \widetilde{m}(A(x))$ for all $\alpha \in L$ and $A \in \mathfrak{B}$,

(iv) For all α, $a \in L$ and $A \in \mathfrak{B}$,

$$\lim \frac{D(\widetilde{m}(\alpha \wedge (a \wedge A)),\ \widetilde{m}(\beta \wedge (a \wedge A)))}{D(\alpha, \beta)} = K((a \wedge A), \alpha) \begin{cases} = 0 & \widetilde{m}(a \wedge A) \leqslant \alpha \\ > 0 & \widetilde{m}(a \wedge A) > \alpha \end{cases}$$

Proof. The procedure of the proof is the following that

$$(i) \Longrightarrow (ii) \Longrightarrow (iii) \Longrightarrow (iv) \Longrightarrow (iii) \Longrightarrow (i)$$

(iv) $\Longrightarrow$ (iii). Let α, $a \in L$ and $A \in \mathfrak{B}$.

1) $\alpha \geqslant \widetilde{m}(A)$. By the condition (iv) and the connectivity of the inteval $[\widetilde{m}(a \wedge A(x)), 1]$ it follows that $\widetilde{m}(\alpha \wedge (a \wedge A(x)))$ is a constant function on the inteval $[\widetilde{m}(a \wedge A(x)), 1]$ and $\widetilde{m}(\alpha \wedge (a \wedge A(x))) = \widetilde{m}(a \wedge A(x))$, thus, $\widetilde{m}(\widetilde{m}(a \wedge A(x)) \wedge (a \wedge A(x))) = \widetilde{m}(a \wedge A(x))$.

(2) $\alpha < \widetilde{m}(A)$. If $a < b < \widetilde{m}(A)$, then $\widetilde{m}(a \wedge A(x)) < \widetilde{m}(b \wedge A(x))$.

Indeed, if $\widetilde{m}(a \wedge A(x)) = \widetilde{m}(b \wedge A(x))$, then

$$\widetilde{m}(a \wedge A(x)) = \widetilde{m}(\alpha \wedge A(x)) = \widetilde{m}(b \wedge A(x)),$$

so that $K(A, \alpha) = 0$ for all $\alpha \in [a, b]$. Therefore, if $\widetilde{m}(\alpha \wedge A(x)) < \alpha$, then

$$\widetilde{m}(\alpha \wedge A(x)) = \widetilde{m}(\widetilde{m}(\alpha \wedge A(x)) \wedge (\alpha \wedge A(x))) < \widetilde{m}(\alpha \wedge A(x))$$

this is impossible. Thus, we have

$$\widetilde{m}(\alpha \wedge A(x)) = \alpha \wedge \widetilde{m}(A)$$

(iii) $\Longrightarrow$ (i). For any $A \in \mathfrak{B}$, define $m(A) = \widetilde{m}(A)$, then $m(\cdot): \mathfrak{B} \longrightarrow L$ is a fuzzy measure on $\mathfrak{B}$ such that

$$m(A) = \int A(x)\, dm \qquad \text{for all } A \in \widetilde{\mathfrak{B}}.$$

3 AXIOMATIC DEFINITION OF FUZZINESS

In [1], De Luca and Termini first suggested the measure of fuzziness and they gave three conditions what the measure of fuzziness should satisfy at least, His intial paper has generated a lot of papers on this area. In particular, in [9], Yager suggested the intuitive definition of fuzziness, he associates fuzziness with the lack of distintion between a proposition and its negation. In [10], Yager devoloped his concept to L-fuzzy sets.

In [7,8], we have pointed out that some faults of De Luca and Termini's conditions of fuzziness measurement and have retieved it.

In this section, first of all, we shall give an axiomatic definition for measure of fuzziness on ordinary fuzzy sets that the grades of menbership lie the unit inteval [0, 1], then we shall extend this definition to L-fuzzy sets.

The conditions for measure of fuzziness which De Luca and Termini suggested in [1] are the following that

Let X be an universal set, $\mathfrak{F}(X)$ the set of all ordinary fuzzy sets on X, $P(X)$ the power set of X.

A mapping $d(\cdot): \mathfrak{F}(X) \longrightarrow [0, 1]$, then $d(\cdot)$ satisfies at least the following conditions that

(D_1) $d(A) = 0$ for all $A \in P(X)$,

(D_2) $d(A)$ is maximum of $d(\cdot)$ if $A(x) = \frac{1}{2}$ for all $x \in X$,

(D_3) $d(A) \geqslant d(B)$ if

$$B(x) \geqslant A(x) \quad \text{when} \quad A(x) \geqslant \tfrac{1}{2}$$
$$B(x) \leqslant A(x) \quad \text{when} \quad A(x) \leqslant \tfrac{1}{2}$$

In [2], De Luca and Termini suggested further the following definition,

Let us introduce in the inteval $I = [0, 1]$ the partial order relation '$\geqslant$' defined, for all $x, y \in I$, as

$$x \,'\!\geqslant y \quad = \quad x \leqslant y \leqslant \tfrac{1}{2} \quad \text{or} \quad x \geqslant y \geqslant \tfrac{1}{2}$$

So that

$$A \leqslant' B \Longleftrightarrow A(x) \leqslant' B(x) \qquad \text{for all } x \in X$$

for all $A, B \in \mathfrak{F}(X)$ and a real value mapping $d(\cdot)$ is called a measure of fuzziness, if

(D_1') $d(A) = 0$ iff A is a crisp set,

(D_2') $d(A)$ reaches its maximum value iff $A = A'$,

(D_3') $d(A) \geqslant d(B)$ if $A \,'\!\geqslant B$.

Remark. (1) It is obverous that (D_1) includes (D_1'), (D_2) includes (D_2') and (D_3) includes (D_3'). We think that the condition (D_1), (D_2) and (D_3) are more reasonable than the conditions (D_1'), (D_2') and (D_3') are.

(2) $d(\cdot)$ satisfying the condition (D_1), (D_2) and (D_3) may be not a voluation on $\mathcal{F}(X)$.

Counterexample. Let $X = [0, 1]$, $\mathfrak{B}$ is the Borel field of X. $\mathfrak{B}_1 = \mathfrak{B} \cap [0, \frac{1}{2}] =$

$= \{ A \cap [0, \frac{1}{2}] : A \in \mathfrak{B} \}$, $\mathfrak{B}_2 = \mathfrak{B} \cap (\frac{1}{2}, 1] =$

$= \{ A \cap (\frac{1}{2}, 1] : A \in \mathfrak{B} \}$, then $\mathfrak{B}_1$ and $\mathfrak{B}_2$ are Borel fields on $[0, \frac{1}{2}]$ and $(\frac{1}{2}, 1]$ respectively.

Let m_1 and m_2 are fuzzy measures on $\mathfrak{B}_1$ and $\mathfrak{B}_2$ respectively, so that a mapping

$$m(\cdot) : m(A) = m_1(A) \vee m_2(A) \text{ for all } A \in \mathfrak{B}$$

is a measure on $\mathfrak{B}$. The following mapping

$$d(A) = 2 \int |A(x) - \underline{A}(x)| \, dm$$

where $\underline{A} = \{ x : A(x) > \frac{1}{2} \}$,

is a measure of fuzziness on $\tilde{\mathfrak{B}}$. But $d(\cdot)$ is not a valuation on $\tilde{\mathfrak{B}}$. Indeed, let

$$A : A(x) = \tfrac{1}{2} \quad \text{when } x \in [0, \tfrac{1}{2}]$$
$$A(x) = 0 \quad \text{when } x \in (\tfrac{1}{2}, 1]$$
$$B : B(x) = 0 \quad \text{when } x \in [0, \tfrac{1}{2}]$$
$$B(x) = \tfrac{1}{2} \quad \text{when } x \in (\tfrac{1}{2}, 1],$$

So that

$$d(A \cup B) = 1, \; d(A \cap B) = 0,$$

and $d(A) = d(B) = 1$, this is to say $d(\cdot)$ is not a valuation on $\tilde{\mathfrak{B}}$.

In [8], we suggested the concept of measure of fuzziness as the following that Assume that the universal set is a fuzzy measure space $(X, \mathfrak{B}, m)$, $\tilde{\mathfrak{B}}$ denote the set of all $\mathfrak{B}$ - measurable fuzzy sets on X.

Definition 3.1 A mapping $d(\cdot) : \tilde{\mathfrak{B}} \to [0, 1]$ is called a measure of fuzziness on $\tilde{\mathfrak{B}}$ iff $d(\cdot)$ satisfies the following conditions

(F_1) $d(A) = 0$ if $A \in \mathfrak{B}$,

(F_2) $d(A) \geq d(B)$ if

$$B(x) \geq A(x) \vee A'(x) \quad \text{if } A(x) > \tfrac{1}{2} \text{ or } B(x) > \tfrac{1}{2}$$
$$B(x) \leq A(x) \quad \text{if } B(x) \leq \tfrac{1}{2}$$

(F_3) $d(A) = d(A')$ for all $A \in \tilde{\mathfrak{B}}$,

(F_4) If $\{A_n\}_{n \geq 1} \subset \tilde{\mathfrak{B}}$ is a monotone sequence, then

$$d(\lim A_n) = \lim d(A_n).$$

In order to extend the above mentioned concept to L-fuzzy sets, let us to suppose that the lattice L is complementory, Here, the complementory operation is defined by Dubois in [3]

Definition 3.2 A mapping $c(\cdot) : L \to L$ is called a complementory operation, if $c(\cdot) : L \to L$ satisfies the following conditions

(1) $c(0) = I$,

(2) c is strictly decreasing,

(3) c is involutive ($c \circ c$ = identify)

Proposition 3.1 Let L be a complete complementory metric lattice. Any element x and its complementory element x' is comparable. To exist uniquely an element a such that $a = a'$ it is necessary and sufficient conditions that if $x < x'$, then there exist an element $e \neq x$ and x' such that (x, e, x') and (x, e', x') (see [10])

Proof. Put

$$L_1 = \{ x : x \leq x' \}$$
$$L_2 = \{ x : x \geq x' \}$$

L_1 and L_2 are non-empty. By the completeness of L, it follows that $\vee \{ x : x \in L_1 \}$ and $\vee \{ x : x \in L_2 \}$ exist in L and they are denoted by a and b respectively. It is easily to prove $a \in L_1$ and $b \in L_2$, thus, $a \leq a'$ and $b \geq b'$. It is sufficient to show $a = b$. By the definitions of L_1 and L_2, we have

$$x \to x' : L_1 \to L_2 \; ; \quad x \to x' : L_2 \to L_1$$

Indeed, if $x \in L_1$, then

$$x \to x' \Rightarrow x = (x')' \leq x' \Rightarrow x' \in L_2,$$

similarly, if $x \in L_2$, then $x' \in L_1$, Thus,

$$a \geq b' \Rightarrow a' \leq b \text{ and } a' \in L_2 \Rightarrow b \leq a' \Rightarrow b = a'$$
$$b \leq a' \Rightarrow b' \geq a \text{ and } b \in L_1 \Rightarrow b' \leq a \Rightarrow a = b'$$

so that $a \leq a' = b$, and $a = b$ since $a < b$ is impossible.

To prove that the condition is necessary, we assume that any element x such that $x < x'$, then the element a is an element satisfying the condition.

Assume that L denote the lattice satisfying the condition of the proposition 3.1, then we have

Definition 3.4 By a measure of fuzziness

$d(\cdot) : \mathfrak{F}(X) \longrightarrow L$ we mean that $d(\cdot)$ satisfies the following conditions

(L_1) $d(A) = 0$ if A is a crisp set,

(L_2) $d(A) \geqslant d(B)$ if

$B(x) \vee B'(x) \geqslant A(x) \vee A'(x)$ if $A(x) > a$ or $B(x) > a$

$B(x) \leqslant A(x)$ if $A(x) \leqslant a$ and $B(x) \leqslant a$

(L_3) $d(A) = d(A')$ for all $A \in \mathfrak{F}(X)$,

(L_4) If $\{A_n\}_{n \geqslant 1} \subset \mathfrak{F}(X)$ is a monotone sequence of L-fuzzy sets, then

$d(\lim A_n) = \lim d(A_n)$.

To illustrate the above mentioned definition is reasonable, we have

Proposition 3.2 The element a is comparable with any element $x \in L$.

In order to compare the fuzziness grades between two elements in the lattice L , in [10] , Yager suggested the following concept.

Definition 3.5 Given two elements x and y in a lattice, we shall say that element x is at least as fuzzy as y, denoted xfy, if the following two conditions hold.

(1) (y, x, y^*)

(2) (y, x^*, y^*)

where x^* and y^* are the negations of x and y respectively.

Proposition 3.3 Assume that A and B are two L-fuzzy sets on X. Then AfB iff A and B satisfy the condition (L_2), where A^* is the complement of A.

4 THE INTEGRAL REPRESENTATION OF THE MEASURE OF FUZZINESS FOR L-FUZZY SETS

In this section, we shall consider the relation between the fuzzy integral and the measure of fuzziness.

For all $A \in \widetilde{\mathfrak{B}}$, we define that

$$d(A) = ⨏ h(A(x))\, dm : \widetilde{\mathfrak{B}} \longrightarrow L,$$

where the continuous function $h(x) : L \longrightarrow L$ satisfies the following conditions

(i) $h(0) = 0$

(ii) $h(u) = h(u')$ for all $u \in L$

(iii) $h(\cdot)$ is strictly monotone increasing on $[0, a]$.

Proposition 4.1 The mapping $d(\cdot)$ is a measure of fuzziness.

Proof. By $h(0) = h(I) = 0$, therefore, it is $h(A(x)) = 0$ for all $A \in \widetilde{\mathfrak{B}}$.

By A and B satisfy the condition (L_2), therefore, $h(A(x)) \geqslant h(B(x))$, so that $d(A) \geqslant d(B)$. Other conditions are clearly satisfied.

The measure of fuzziness defined in the proposition 4.1 possesses the following property.

If $d(\cdot)$ satisfies fuzzy additivity, i.e.

$$d((\alpha \wedge A(x)) \vee (\beta \wedge B(x))) = d(\alpha \wedge A(x)) \vee d(\beta \wedge B(x))$$

for all $\alpha, \beta \in L$ and A and $B \in \widetilde{\mathfrak{B}}$, then

(L_5) For all $E \in \mathfrak{B}$ and $\alpha \in L$

(i) If $\alpha \geqslant a$, then $d(\alpha \wedge E(x)) = d(\alpha' \wedge E(x))$

(ii) If $\alpha < a$, then there exists an element $\lambda(E)$ such that

$$d(\alpha \wedge E(x)) = \begin{cases} \lambda(E) & \text{when } h(\alpha) \geqslant \lambda(E) \\ h(\alpha) & \text{when } h(\alpha) < \lambda(E) \end{cases}$$

where $\lambda(\cdot) : \mathfrak{B} \longrightarrow L$ and

$$\lambda(A \cup B) = \lambda(A) \cup \lambda(B)$$

for all $A, B \in \mathfrak{B}$,

$h(\cdot)$ satisfies that

$$h(\alpha \vee \beta) = h(\alpha) \vee h(\beta)$$

for all $\alpha, \beta \leqslant a$.

Conversely, we have

Theorem 4.1 Let $d(\cdot)$ be a measure of fuzziness satisfying the fuzzy additivity and the condition (L_5), then there exists a fuzzy measure $m(\cdot) : \mathfrak{B} \rightarrow L$ such that

$$d(A) = ⨏ h(A)\, dm \qquad (*)$$

for all $A \in \widetilde{\mathfrak{B}}$.

Proof. First of all, we define the following mapping $m(\cdot) : \mathfrak{B} \longrightarrow L$ that

$$m(A) = \lambda(A) \quad \text{for all } A \in \mathfrak{B} .$$

It is clear that the mapping $d(\cdot)$ is a fuzzy measure on $\mathfrak{B}$.

Now, we shall show that

$$d(A) = ⨏ h(A(x))\, dm \quad \text{for all } A \in \widetilde{\mathfrak{B}} .$$

i) $A = \alpha \wedge E(x)$ where $E \in \mathfrak{B}$,

$$I_A = ⨏ h(A(x))\, dm = \begin{cases} h(\alpha') \wedge m(E) & \alpha \geqslant a \\ h(\alpha) \wedge m(E) & \alpha < a \end{cases}$$

Therefore, it is obverously that $d(A) = I_A$

by the condition (L_5)

ii) $A(x) = \bigvee \alpha_i \wedge E_i(x)$ is a simple function. The formula (*) holds since $d(\cdot)$ possesses the fuzzy additivity.

iii) $A(x)$ is a general function. We take a sequence $\{A_n\}$ of simple functions such that $A_n(x) \longrightarrow A(x)$ (uniformly). By the continuity of the function $h(\cdot)$, it follows that $h(A_n(x)) \longrightarrow h(x)$ and

$$\int h(A(x))\, dm = \lim \int h(A_n(x))\, dm = \lim d(A_n) = d(A).$$

Remark. The function $h(\cdot)$ may be regarded as a subjective principle of the measurement of fuzziness, thus, it is very interesting problem how $h(\cdot)$ is determined.

REFERENCES

[1] A. De Luca and Termini, A definition of a non-probabilistic entropy in the setting of fuzzy sets theory, Inform. and Contr., 20 (1972) 301—312.

[2] ———, On some algebraic aspects of the measure of fuzziness, " Fuzzy Information and Decision Processes " M.M.Gupta and E.Sanchez (Editors) North-Holland Publishing Company, 1982.

[3] D. Dubois, Uncertainty in transportation network design : some new techniques, Presented at the IVth European congress on operation reseach.

[4] E.P.Klement et al., Fuzzy probability measures, J. Int. Fuzzy Sets and Systems 5(1981) 21—30.

[5] M.Sugeno, Theory of fuzzy integrals and its applications, Ph.D.Dissertation, Tokyo Institute of Tech. , 1974.

[6] Zi-Xiao Wang, Note on the fuzzy measure, A Monthly J. of Science, Vol. 26, No. 11 (Beijing, China) 961— 964.

[7] ———, The structure of fuzzy Lebesgue measures, " Fuzzy Information and Decision Processes ", M.M.Gupta and E.Sanchez (Editors), North-Holland Publishing Company, 1982. 71—78

[8] ———, Fuzzy measures and measures of fuzziness, Presented at " Procedings of the Second World Conference on Mathematics at Severce of Man ", June 28th to July 3th, 1982, Canary Islands, Spain.

[9] R.R.Yager, On the measure of fuzziness and negation part I : Membership in the unit inteval, Int. J. General Systems 5, 221—229.

[10] ———. On the measure of fuzziness and negation part II : Lattice. Information and Control, 44(1980) 236—260

[11] L.A.Zadeh, Probability measure on fuzzy events. J.Math.Anal.Appl. (1968).

A NEW APPROACH TO SYNTHESIS IN SYSTEM THEORY

A. Di Nola and A. G. S. Ventre

Universita' di Napoli, Facolta' di Architettura, Istituto di Matematica, Via Monteoliveto 3, 80134, Napoli, Italy

Abstract. A new approach to synthesis of a multiple description in system theory is presented. The process is described in an anagogical context.

Keywords: Synthesis, Analogy, Anagogy, Pullback, Entropy.

INTRODUCTION

Analogy and induction play a fundamental role in the search. The creative impulse in mathematics and, more generally in science is often supported by the reasoning by analogy or induction. The classical book of Polya' (1954) deeply illustrates and analyzes the meaning of both concepts more specifically in mathematical search and the growth of the mathematical thinking. Also in cybernetics the concepts of analogy and induction consist. It is well known that the first one represents a method and a classical behaviour for the cybernetician, also it makes sense in the treatment of the exceedingly complex systems, e.g. societal systems, the anagogical approach (Negoita 1982).

A situation which occurs in system theory is that a partial information gives rise to a multiplicity of perspectives about an object: an anagogycal process reduces manifoldness to the unity, and then creates a higher level of knowledge from a fragmentary information. A multiplicity of description of a single object is based on a dialectical context which has theoretical, descriptive, and analytical support into fuzzy sets theory (Bellman and Giertz, 1973) and (Negoita, 1981).

In this framework Negoita (1981) introduced the pullback principle, by means a synthesis operation on fuzzy sets as anagogical process. In the present paper we will present another aspect of pullback principle when the reduction to the unity is built in terms of reduction of fuzzy entropy (De Luca and Termini 1974; Knopfmacher 1975) and then in terms of amount of information.

SYNTHESIS OPERATION

Let X a non-empty set and $F(X)=\{f: X \to [0,1]=I\}$ denote the set of all fuzzy sets defined on X. The lattice on F(X) defined by Zadeh (1965) will bee denoted by L_1. A function φ : $x \to (f_1(x),\dots,f_n(x))$, $n \geq 1$, where $f_i \in F(X)$, $i=1,\dots,n$, is said to be a multiple description of the element x (Capocelli and De Luca, 1973). We assume that the function φ is the fuzzy set theoretical counterpart of a description of a single object by means of more than a single property.

We recall that another ordering δ in F(X) is defined (Di Nola and Ventre, 1980) as follows. Let t a real function defined in I such that $t(0)=t(1)=0$, $t(\alpha)=t(1-\alpha)$ for $\alpha \in I$, which is strictly increasing in $[0,\frac{1}{2}]$. For any $f_1, f_2 \in F(X)$ it is $f_1 \delta f_2$ if $t(f_1(x)) > t(f_2(x))$ or $t(f_1(x))=t(f_2(x))$ implies $f_1(x) \geq f_2(x)$, for any $x \in X$.

A lattice structure is induced in the poset $(F(X), \delta)$ by the operations

$f_1(x) \wedge f_2(x)=f_i(x)$ such that either $t(f_i(x)) < t(f_j(x))$ or $t(f_i(x))=t(f_j(x))$ implies $f_i(x) \leq f_j(x)$;

$f_1(x) \vee f_2(x)= f_i(x)$ such that either $t(f_i(x)) > t(f_j(x))$ or $t(f_i(x))=t(f_j(x))$ implies $f_i(x) \geq f_j(x)$, $i \neq j$, $i,j=1,2$.

We set $glb(f_1,f_2)=f_1 \wedge f_2$ and $lub(f_1, f_2)=f_1 \vee f_2$. The structure $L_2=(F(X),\wedge, \vee)$ is a lattice. Let us define synthesis in the lattice L_2 the function $s: F^n(X) \longrightarrow F(X)$:

$$s(f_1(x),\ldots,f_n(x)) = s(f_i(x)) = \bigcap_{i=1}^{n} f_i(x) =$$

$f_j(x)$, such that $t(f_j(x)) < t(f_h(x))$, for any $h \neq j$, or if $t(f_j(x)) = t(f_h(x))$, for some h, then $f_j(x) \leq f_h(x)$. Function s defines a synthesis operation in the sense of Negoita (1978, 1981). Synthesis s is correlated with synthesis with respect to operation "min" in the lattice L_1 (Di Nola and Ventre, to appear). We just mention that the two synthesis processes coincide when the object is described by means of orthogonal properties, i.e. $\sum f_i(x) = 1$.

When s acts on measurable fuzzy sets, then the increase of the amount of information can be measured. Indeed let us assume the definition of entropy given by Knopfmacher (1975):

$$d(f) = \frac{1}{\mu(X)} \int_X t(f(x))\, d\mu(x).$$

Then $d(s) \leq d(f_i)$, for any i. Thus the new knowledge corresponds to the new amount of information measured by

$$|d(s) - \min d(f_i)| .$$

In particular, for a fuzzy system (Negoita and Ralescu,1975) with standard state equation

$$x_{k+1} = \eta\,(x_k, u_k)$$

if the dynamic η is expressed in terms of $\cap$ operation, i.e.

$$x_{k+1} = x_k \cap p(u_k),$$

then a decreasing of the entropy in the states is a chain in the lattice L_2 (Di Nola and Ventre,1980), then the sequence (x_k) is convergent and therefore the system is stable.

PULLBACK PRINCIPLE

The pullback principle, expressed by a categorical model depends on the latticeal structures on F(X). We are now able to extend some considerations of Negoita (1981, 1982). If δ is viewed as a morphism, the following diagram can be constructed. For every pair A, B of morphisms having the same domain g, $f_1 \xrightarrow{A} g \xleftarrow{B} f_2$, there is a fuzzy set p, the pullback, such that the square on p commutes (AP=BQ, in the diagram of Fig.1), and the universal property holds: for any commutative square BK=AH, on $z \in F(X)$ and on the arrows A and B, there exists a unique morphism $S: z \longrightarrow p$ such that the whole diagram in Fig.1 commutes. Therefore the process of reduction of a multiplicity of perspectives to the unity has a rigorous setting.

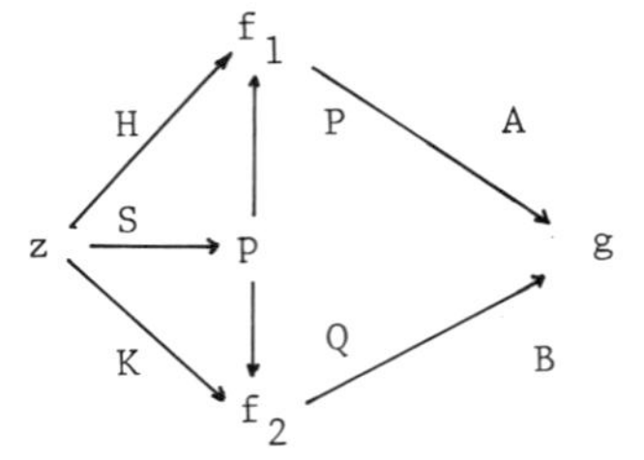

Fig.1 The pullback scheme.

REFERENCES

Bellman,R.E. and Giertz,M.(1973). On the analytic formalism of the theory of fuzzy sets. Information Science,5, 149-157.

Capocelli,R. and De Luca,A. (1973). Fuzzy sets and decision theory. Information and Control, 23, 446-473.

De Luca,A. and Termini,S.(1974). Entropy of L-fuzzy sets. Information and Control,24, 55-73.

Di Nola,A. and Ventre,A.(1980). On some chains of fuzzy sets. Int. Journal Fuzzy Sets and Systems,4, 185-191.

Di Nola,A. and Ventre,A.(to appear) Synthesis with respect fuzzy entropy. Kybernetes.

Knopfmacher,J. (1975). On measure of fuzziness. J.Math.Anal.Appl., 49, 529-534.

Negoita,C.(1978). On the stability of fuzzy systems. Symp. on Systems Science and Cybernetics, Int.Conf. on Cyb. and Soc., Tokio,Nov.3-5.

Negoita,C. (1982). Cybernetics and Society. Kybernetes, 11, 97-101.

Negoita,C. and Ralescu,A.(1975). Applications of fuzzy sets to system Analysis (Birkhauser Verlag, Basel).

Polya,G. (1954). Induction and Analogy in Mathematics, Princeton University Press. Princeton, N.J.

Zadeh,L. (1965). Fuzzy Sets. Information and Control, 8, 338-353.

METHOD OF FUZZY STATISTICS AND APPLICATIONS

Heng-ling Hong, Lin-Ge Sang, Zhong-Wu Mei

Computer Station, Changchun Institute of Geology, Changchun, Jilin, People's Republic of China

Abstract. This paper is based on the theory of knowledge, the process forming a conception is discussed. One of statistical methods to establish membership function is given. The theorem of stability of the membership function is proved. The fuzzy statistical model is established. These methods were employed in petroleum prospecting and geological prospecting.

Key words. Fuzzy statistics; fuzzy conception; multi-dimension attribute space; stability; computer organization; geophysics.

INTRODUCTION

The essential attribute of a conception is called the intention of the conception. The total objects conformed with a conception is called the extension of the conception. Therefore the extension of a conception is a set of a certain space. In other words, a conception may be expressed by a set on a certain space. A conception is called a clear conception if its extension exists a clear boundary, otherwise it is called a fuzzy. Since a clear set is special one of fuzzy sets, a clear conception is special one of fuzzy conceptions. Thus, clear and fuzzy conceptions are called by a joint name, the fuzzy conception.

To determine a fuzzy set, it is important to determine its membership funcition. Without a good method of establishing the membership function it would be limited for developing and applying the theory of the fuzzy set.

Based on the theory of knowledge, the process forming a conception were discussed and a mathematical description of the process was given. In this paper a statistical method of imitating human's process of cognition was developed to establish the membership function, and the stablity of the membership function was proved. The method was used to give a fuzzy statistical recognition model. These methods were employed in petroleum prospecting and geological prospecting, and good results are achived.

FUZZY CONCEPTION AND FUZZY SETS

A conception is a reflection of objective things in a humen's subjective consciousness. The process of the reflection consists of four stages.

The first is the sense perception. People have senses of sight, hearing, taste, touch etc, and by these instincts individual attributes of each object, such as colour, tone, taste etc, will be obtained directly or indirectly.

At this stage, people have not obtained a complete knowledge of the object except for individual attribute and have not known what is the object. But people can obtain "a quantity" of the individual attribute, such as hardness etc. People can say that it is large or small, strong or weak and so on, although the accurate values of these guantities is still unknown. This kind of guantity is called a fuzzy guantity.

A single attribute is described by an one-dimension real space which is called an one-dimension attribute space. A fuzzy quantity is described by a fuzzy number on an one-dimension attribute space.

The second stage is consciousness. It is the reflection of an object as a whole in the man's mind. At this stage, not only each single attribute of an object is reflected, but also they are united to form a whole image of an object.

If a individual attribute can be described by an one-dimension attribute space, then many attributes can be described by a multi-dimension space. Consciousness is a reflection of an object on the space. It makes people obtain a multi-dimensional quantity which may be clear or fuzzy, so that it is called multi-dimensional fuzzy quantity. It will be described by a fuzzy number on a multi-dimensional space.

The third stage is idea. The stage is based on sense perception and consciousness. The distinction between idea and sense perception, consciousness is that the latter is an immediate reflection of an objective thing in man's mind and the former is formed after many times of reflection. At this stage, the image of the object is still presented in man's mind, even if at time when man don't get in touch with the object. Once an object is sensed, a fuzzy quantity is obtained on the attribute space. If the object is sensed many times, then many fuzzy quantities are received. However, these quantities are neither simply overlap, nor unlimitedly scattered. So long as the objects contain common attributes or approximate attributes, the fuzzy quantities will relatively concentrate on attribute space. Thus, the fuzzy set which expresses kind of the objects is formed. It is the mathematical model of a kind of the objects.

The fourth stage is the conception. A object contains many attributes. If the attributes can distinguish a object from another, these attributes are essential, otherwise, they are non-essential. On the basis of the above mentioned process, by the logical methods of compare, analysis, synthesis, abstracte and generalization man investigating model in man's mind. The non-essential attributes will be rejected. while the essential attributes will be retained. The number of the dimension of attribute space is decreased as low as possible. Thus the conception is the thinking form which is a reflection of essential attributes of a objective thing.

We may notice from above mentioned process that a fuzzy conception may be expressed by a fuzzy set. If a fuzzy conception has n essential attributes, then a n-dimension Euclidean space R^n is adopted, where every facter space R_i $(i = 1, 2, \ldots, n)$ is the representation of a essential attribute so that $R^n = R_1 \times R_2 \times \ldots \times R_n$. A fuzzy conception may be expressed by a fuzzy set on the space R^n. A fuzzy conception is formed after a kind of object is sensed many times. Every time once a object is sensed, a reflected image is presented in man's mind and these is a corresponding fuzzy mumber (a fuzzy set) on a n-dimension space R^n.

The first fuaay number is a original model of a conception. When the second fuzzy number is obtained, two fuzzy numbers are synthesised to produce a new fuzzy set which is a improved model of the original model. In this way, a man sense the objective thing continualy and improve the old model in the mind. This process will be given with the method of fuzzy statistics in latter part of this paper.

The objective world is infinite, the number of dimension of the space to investigate things is infinite too. With development of the science and technology the people's ability of understanding the world becomes more and more intensiting. The number of dimension of the attribute space for investigating objective things is increased. In a long period a great deal of scintific data are accumlated at production and scintific experiment, and many important conceptions and principls are contained in these data. In this case the dimensions of the attribute space are so many that nobody will be able to recognize all of them. Now, however, the modern computer are used to imitate the cognition process of a people and to extablish a model of a complex conception, i.e a fuzzy set on the multi-dimensional space.

THE STATISTICAL METHOD MAKING THE MEMBERSHIP FUNCTION OF A FUZZY SET

One-Dimension Fuzzy Number Method

We call a conception (or a law) that will be determined a parent population which is denoted by $\underset{\sim}{A}$. Assume that a parent population $\underset{\sim}{A}$ has n essential attributes and each of them is expressed by a one-dimension space X_j $(j = 1, 2, \ldots, n)$. Then a parent population $\underset{\sim}{A}$ is a fuzzy set on a n-dimension space $X = \prod_{j=1}^{n} X_j$. And assume that some fuzzy quantities of each essential attribate of $\underset{\sim}{A}$ have been obtained and they are expressed by the fuzzy number $\underset{\sim}{x}_{ij}$,

$$\underset{\sim}{x}_{ij} = \int_{x_j \in X_j} M_{\underset{\sim}{x}_{ij}}(x_j)/x_j \quad (i=1,2,\ldots,m_j) \qquad (1)$$

Where $\underset{\sim}{X}_{ij}$ is the ith fuzzy guantity of the jth attribute, m_j is the number of the fuzzy quantities of the jth attribute.

Let $\underset{\sim}{A}_j$ is the fuzzy set that express $\underset{\sim}{A}$ only by the jth attribute, then the membership function of $\underset{\sim}{A}_j$ is as following

$$M_{\underset{\sim}{A}_j}(x_j) = \sum_{i=1}^{m_j} M_{\underset{\sim}{x}_{ij}}(x_j)/m_j \quad \forall x_j \in X_j \qquad (2)$$

$M_{\underset{\sim}{A}_j}(x_j)$ is called one-dimension statistical membership function or one-demension experienced membership function. Then n fuzzy sets on one-dimension space will be

$$\underset{\sim}{A}_j = \int_{x_j \in X_j} M_{\underset{\sim}{A}_j}(x_j)/x_j \quad (j=1,2,\ldots,n) \qquad (3)$$

By making cylindrical extensions of $\underset{\sim}{A}_1$, $\underset{\sim}{A}_2$, ..., $\underset{\sim}{A}_n$ in the X respectively, we get

$$\underset{\sim}{A}^*_j = \int_{\bar{x} \in X} M_{\underset{\sim}{A}_j}(x_j)/\bar{x}, \quad (j=1,2,\ldots,n) \qquad (4)$$

Where $\bar{x} = (x_1, x_2, \ldots x_n) \in X$, $x_j \in X_j$.

The intersection of them will be

$$\underset{\sim}{A}^* = \bigcap_{j=1}^{n} \underset{\sim}{A}_j^* , \tag{5}$$

and the membership function of $\underset{\sim}{A}^*$ is

$$M_{\underset{\sim}{A}^*}(\bar{x}) = \bigwedge_{j=1}^{n} M_{\underset{\sim}{A}_j^*}(x_j). \tag{6}$$

$M_{\underset{\sim}{A}^*}(\bar{x})$ was formed using the fuzzy information of $\underset{\sim}{A}$, thus $M_{\underset{\sim}{A}^*}(\bar{x})$ is an approximate function or experienced membership function of $M_{\underset{\sim}{A}}(\bar{x})$. $\underset{\sim}{A}^*$ is called the statistical model.

It will be proved that the larger the m_j $(j=1,2,\ldots,n)$ is, the stabler the $M_{\underset{\sim}{A}^*}(\bar{x})$ is. Thus, if m_j is large enough then $M_{\underset{\sim}{A}^*}(\bar{x})$ may replace $M_{\underset{\sim}{A}}(\bar{x})$ and model $\underset{\sim}{A}^*$ may replace $\underset{\sim}{A}$.

Multi-Dimension Fuzzy Number Method

Let $X = \prod_{j=1}^{n} X_j$ is a n-dimension real Euclidean space and X_j $(j=1,2,\ldots,n)$ is a real line. It is defined that a convex fuzzy set in X is a n-dimension fuzzy number,

$$\underset{\sim}{a} = \int_{\bar{x}\in X} M_{\underset{\sim}{a}}(\bar{x})/\bar{x} \qquad \forall\ \bar{x}\in X \tag{7}$$

Assume that a parent population $\underset{\sim}{A}$ in X contains m fuzzy quantities $\underset{\sim}{\bar{x}}_i$ $(i=1,2,\ldots,m)$ and they are given by multi-dimension fuzzy numbers

$$\underset{\sim}{\bar{x}}_i = \int_{\bar{x}\in X} M_{\underset{\sim}{\bar{x}}_i}(\bar{x})/\bar{x} , \qquad (i=1,2,\ldots,m) \tag{8}$$

They are called a fuzzy sample. The fuzzy set is defined as

$$\underset{\sim}{A}^* = \int_{\bar{x}\in X} M_{\underset{\sim}{A}^*}(\bar{x})/\bar{x} \tag{9}$$

and its membership function is

$$M_{\underset{\sim}{A}^*}(\bar{x}) = \sum_{i=1}^{m} M_{\underset{\sim}{\bar{x}}_i}(\bar{x}) \tag{10}$$

$M_{\underset{\sim}{A}^*}(\bar{x})$ is called the statistical membership function or the experienced membership function of $\underset{\sim}{A}$. $\underset{\sim}{A}^*$ is called the statistical model of $\underset{\sim}{A}$.
It can be proved that larger the m is, the stabler the $M_{\underset{\sim}{A}^*}(\bar{x})$ is, and $M_{\underset{\sim}{A}^*}(\bar{x})$ will more approximate the membership function $M_{\underset{\sim}{A}}(\bar{x})$ of $\underset{\sim}{A}$. Therefore $M_{\underset{\sim}{A}^*}(\bar{x})$ may replace $M_{\underset{\sim}{A}}(\bar{x})$ and the model $\underset{\sim}{A}^*$ may replace $\underset{\sim}{A}$ if m is large enough.

THE THEOREM OF STABILITY

The membership function formed by statistical method possesses the stability in a sense. Assume that a fuzzy sample of a parent population $\underset{\sim}{A}$ is $\{\underset{\sim}{\bar{x}}_1, \underset{\sim}{\bar{x}}_2, \ldots, \underset{\sim}{\bar{x}}_m\}$ and they are the fuzzy results of m observations of $\underset{\sim}{A}$. The fuzzy statistical model $\underset{\sim}{A}_m$ of $\underset{\sim}{A}$ is

$$\underset{\sim}{A}_m = \int_{\bar{x}\in X} \frac{1}{m} \sum_{i=1}^{m} M_{\underset{\sim}{\bar{x}}_i}(\bar{x})/\bar{x} . \tag{11}$$

Let $\{\underset{\sim}{\bar{x}}_{m+1}, \underset{\sim}{\bar{x}}_{m+2}, \ldots, \underset{\sim}{\bar{x}}_{m+\ell}\}$ be another ℓ observations of $\underset{\sim}{A}$ on the basis of former m observations, then a new model of $\underset{\sim}{A}$ will be

$$\underset{\sim}{A}_{m+\ell} = \int_{\bar{x}\in X} \frac{1}{m+\ell} \sum_{i=1}^{m+\ell} M_{\underset{\sim}{\bar{x}}}(\bar{x})/\bar{x} . \tag{12}$$

The model $\underset{\sim}{A}_{m-\ell}$ may be different from $\underset{\sim}{A}_m$. Let

$$D_{m.\ell} = \max_{\bar{x}\in X} \left| M_{\underset{\sim}{A}_{m+\ell}}(\bar{x}) - M_{\underset{\sim}{A}_m}(\bar{x}) \right| . \tag{13}$$

we have the following theorem about relation of m, ℓ and $D_{m.\ell}$.

Theorem of Stability

For a given $\varepsilon\in(0,1)$ and a positive integer L there will be a positive integer M, when $m\geq M$ and $\forall\ \ell\leq L$, $D_{m.\ell}<\varepsilon$ or

$$\left| M_{\underset{\sim}{A}_{m+\ell}}(\bar{x}) - M_{\underset{\sim}{A}_m}(\bar{x}) \right| < \varepsilon \quad \forall \bar{x}\in X \tag{14}$$

Proof

$$\begin{aligned} D_{m.\ell} &= \max_{\bar{x}\in X} \left| \frac{1}{m+\ell}\sum_{i=1}^{m+\ell} M_{\underset{\sim}{\bar{x}}_i}(\bar{x}) - \frac{1}{m}\sum_{i=1}^{m} M_{\underset{\sim}{\bar{x}}_i}(\bar{x}) \right| \\ &= \max_{\bar{x}\in X} \left| \frac{1}{m+\ell}\sum_{i=m+1}^{m+\ell} M_{\underset{\sim}{\bar{x}}_i}(\bar{x}) - \frac{\ell}{m(m+\ell)}\sum_{i=1}^{m} M_{\underset{\sim}{\bar{x}}_i}(\bar{x}) \right| \\ &\leq \max\left\{ \max_{\bar{x}\in X} \frac{1}{m+\ell}\sum_{i=m+1}^{m+\ell} M_{\underset{\sim}{\bar{x}}_i}(\bar{x}) ,\ \max_{\bar{x}\in X} \frac{\ell}{m(m+\ell)}\sum_{i=1}^{m} M_{\underset{\sim}{\bar{x}}_i}(\bar{x}) \right\} \\ &\leq \max\left\{ \frac{\ell}{m+\ell}, \frac{\ell}{m+\ell} \right\} \\ &= \frac{\ell}{m+\ell} \end{aligned} \tag{15}$$

and if $\ell\leq L$ then $\ell/(m+\ell)\leq L/(m+L)$. Thus for given $\varepsilon\in(0,1)$ and positive integer L, let

$M \geq L/\varepsilon - L$, so long as $m \geq M$ and $\ell \leq L$,

$$D_{m,\ell} < \varepsilon \qquad \forall\ \bar{x} \in X \qquad (16)$$

(Q.E.D.)

Corollary 1. For given $\varepsilon \in (o,1)$, there is an integer $L \geq o$ which makes $D_{m.\ell} < \varepsilon$ for any ℓ satisfying $o \leq \ell \leq L$.

The supremum of L is called the stable life of m observation-fuzzy model for $\varepsilon > o$ and it is denoted by

$$L_s(m.\varepsilon) = \text{Sup}\ (L) \qquad (17)$$

Corollary 2. For given $\varepsilon \in (o,1), L_s(m,\varepsilon) \to \infty$ when $m \to \infty$

This corollary shows that the bigger the number m of observations is, the longer the stable life $L_s(m,\varepsilon)$ becomes. In other words, the more observed data are, the more stable the statistical membership function of the $\underset{\sim}{A}$ becomes. It explains that why it is hard to change a prejudice formed in man's mind and why the older a doctor is, the reliabler his experience will be. the conclusion is of great value for us.

It should be noticed that for given $\varepsilon \in (o,1)$ and fixed m, if $\ell \to \infty$, it will be possible that $D_{m.\ell} > \varepsilon$. It indicates that the stability of the model is relative. It is possible to change the conception formed in man's mind with the lapse of time ceaselessly

THE DIFFERENCE BETWEEN THE FUZZY STATISTICS AND THE PROBABILITY STATISTICS

In the productive practice and the scientific experimente, in order to find a law or to detemine a conception, people have to make statistical analysis to the data after their experiment. Depending on the aim of the analysis, people use different statistical method.

Fuzzy statistics is quite different from probability statistics. The probability statistics is a method to consider the randomness, while the fuzzy statistics is a method to research the fuzziness. It is well known that the randomness and the fuzziness are two completely different kinds of the uncertainties. The randomness is to destroy the law of causation and the fuzziness is to destroy the law of excluded middle. An event may possess the two uncertainties.

The probability can be described by the stability of the frequency which emerged from the random experiment. There are four main factors on the randomness experiments.
1. The set Ω of the foundamental events.
2. The event A which is a set in Ω.
3. A variable ω in Ω.
4. A condition S. It is a restriction on the action of the variable ω.

The event A is called occurrence if $\omega \in S \cap A$ and the event A is called non-occurrence if $\omega \notin S \cap A$. The event A is a random event under the condition S. The event A is fixed in every experiment. $\omega \in S$ and it is changeable. After n experiments the frequency W(A) of A can be computed. i.e $W(A) = m/n$, where m is a number of times of $\omega \in A$. In practice, as n increases, W(A) will get more and more stable. Thus we will be get the probability of the event A under the conditions S.

There are four main factors for the fuzzy statistics too.
1. The universal set U.
2. A variable $u \in U$.
3. A changeable fuzzy subset $\underset{\sim}{A}_i \subset U$.

It is a representation of a certain fuzzy conception $\underset{\sim}{A}$.
4. A condition F, it is a power set which consists of some fuzzy sets on U. F restrict the change of $\underset{\sim}{A}_i \subset F$.

The characteristic of the experiment is that u is any fixed member belonging to U and $\underset{\sim}{A}_i$ is changeable. A result of every experiment is $M_{\underset{\sim}{A}_i}(u)$ which is a possible value that how a variable u belong $\underset{\sim}{A}$. Thus after n experiments the statistical membership function is defined by

$$M_{\underset{\sim}{A}*}(u) = \sum_{i=1}^{n} M_{\underset{\sim}{A}_i}(u)/n \ \forall u \in U \qquad (18)$$

The theorem of stability has shown that the stability of $M_{\underset{\sim}{A}*}(u)$ will become stronger and stronger as n increases. Therefore the the statistical membership function $M_{\underset{\sim}{A}*}(u)$ may be regarded as a approximate function of $M_{\underset{\sim}{A}}(u)$.

APPLICATION - THE FUZZY MODEL RECOGNITION

The Membership Principle

If there are enough fuzzy information and the fuzzy sets expressing each fuzzy conception are established on n-dimension attribute space, then there will be a model to recognize every object.

Assume that there are fuzzy sets $\underset{\sim}{A}_1, \underset{\sim}{A}_2, \ldots, \underset{\sim}{A}_n$ Which have established in the universal set U. The fixed point $\bar{x} \in U$ is subordinated to $\underset{\sim}{A}_i$ relatively if there is a positive number $i \in \{1,2,\ldots,n\}$ which makes

$$M_{\underset{\sim}{A}_i}(\bar{x}) = \max\left\{M_{\underset{\sim}{A}_1}(\bar{x}), M_{\underset{\sim}{A}_2}(\bar{x}), \ldots, M_{\underset{\sim}{A}_n}(\bar{x})\right\} \qquad (19)$$

This principle is called the maximum membership principle.

The Application in the Interpretation of Petroleum Logging.

Distinguishing the water-bearing stratum and the oil-bearing stratum is one of the main tasks of the interpretation of logging in petroleum prospection. It is impossible to see an underground stratum immediately, but people can measure various physical parameters which reflect the properties of a stratum. Since each parameter is a one-dimension attribute space expressing the stratum, the attribute space discribing the stratum is a high dimension space and a certain stratum, for example an oil-bearing stratum, is a set in the attribute space. Usualy a stratum is so complex that the conception "oil-bearing stratum" hasn't exact boundary in the attribute space. Thus "oil-bearing stratum" is a fuzzy set in the space.

Assume we have obtained a lot of data about the oil-bearing stratum and the water-bearing stratum and these data have n parameters which form a n-dimension parameter space. Since the water-bearing stratum is different from the oil-bearing stratum, the data will form two different fuzzy sets in the space. It should be noticed that every datum expresses a fuzzy information of the parent population. For instance, resistivity of a oil-bearing stratum is 4.5Ω, it shows that the resistivity of the parent population of the oil-bearing stratum is about 4.5Ω which is a fuzzy information, and is denoted by a fuzzy number. The fuzzy number is once indication of possibility distribution of the fuzzy conception "oil-bearing stratum" on the resistivity attribute space. Using previous fuzzy statistical mathod we may make a fuzzy model distinguishing the oil-bearing stratum and the water-bearing stratum.

The Living Example

In a oil field, 77 samples of oil-bearing strata and 94 samples of water-bearing strata were taken and 9 physical properties were selected as the parameters. These parameters are R_{mf}(resistivity of mud filtrate), R_w(resistivity of formation water), $R_{a.s}$(apparente resistivity with short electrod spacing), $R_{a.\ell}$(apparente resistivity with long electrod spacing), C(electrical conductivity), SP(Self-potention), SSP(Static self-potention), GR(natural radioactive) and ∇T(interval transit time). The models of the oil-bearing stratum and the water-bearing stratum were made by using one-dimension fuzzy number and multi-dimension fuzzy number respectively.

The statistical models of one-dimension fuzzy number are as following.

__The model of the oil-bearing stratum__

$$M_{\underset{\sim}{o}_j}(x_j) = \sum_{i=1}^{77} e^{-\left(\frac{x_j - x_{ij}}{\sigma_{o_j}}\right)^2} /77 \quad (j=1,2,\ldots,9) \tag{20}$$

$$M_{\underset{\sim}{o}}(\bar{x}) = \bigwedge_{j=1}^{9} M_{\underset{\sim}{o}_j}(x_j) \tag{21}$$

__The model of water-bearing stratum__

$$M_{\underset{\sim}{w}_j}(x_j) = \sum_{i=1}^{94} e^{-\left(\frac{x_j - x_{ij}}{\sigma_{w_j}}\right)^2} /94 \quad (j=1,2,\ldots,9) \tag{22}$$

$$M_{\underset{\sim}{w}}(\bar{x}) = \bigwedge_{j=1}^{9} M_{\underset{\sim}{w}_j}(x_j) \tag{23}$$

where $\bar{x} = (x_1, x_2, \ldots, x_9)$ and $x_j(j=1,2,\ldots,9)$ denotes variables Rmf, Rw, Ra.s, Ra.ℓ , C, SP, SSP, GR, ∇T respectively, X_{ij} denotes the value of jth variable of ith sample, σ_{oj}, σ_{wj} are the mean-variances of the oil-bearing stratum and the water-bearing stratum samples respectively.

By using previous method the curves of the membership functions of the oil-bearing stratum and the water-bearing stratum were obtained (See Fig.1 - Fig.9)

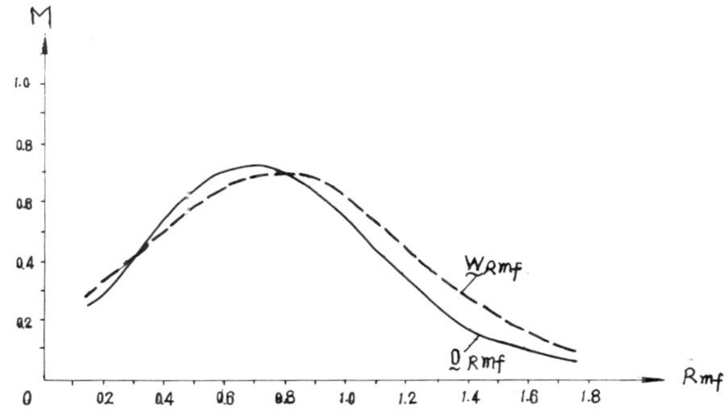

Fig. 1.

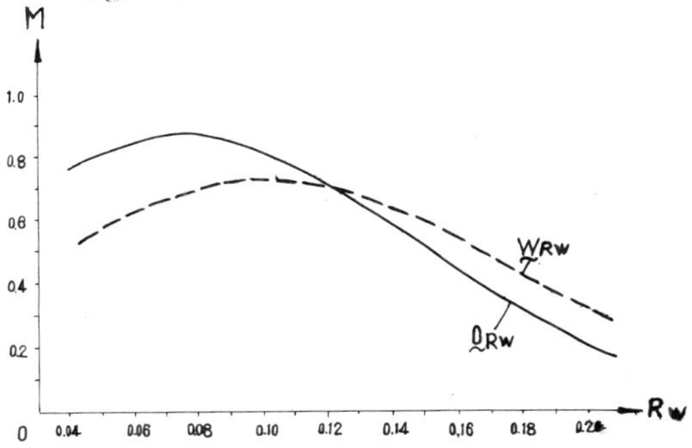

Fig. 2.

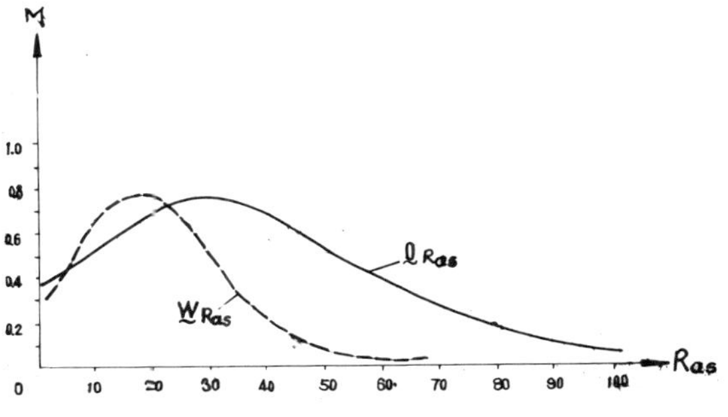

Fig. 3

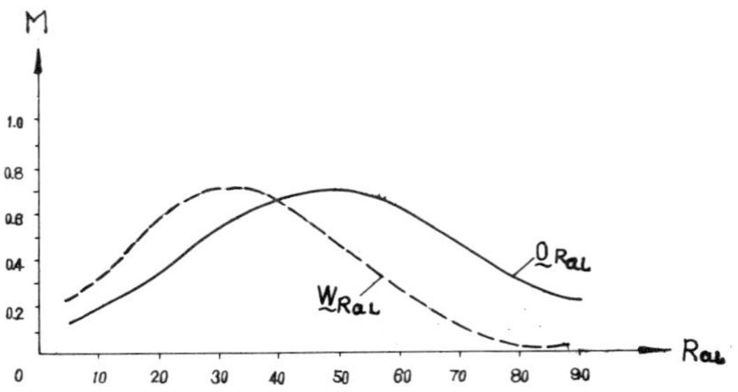

Fig. 4.

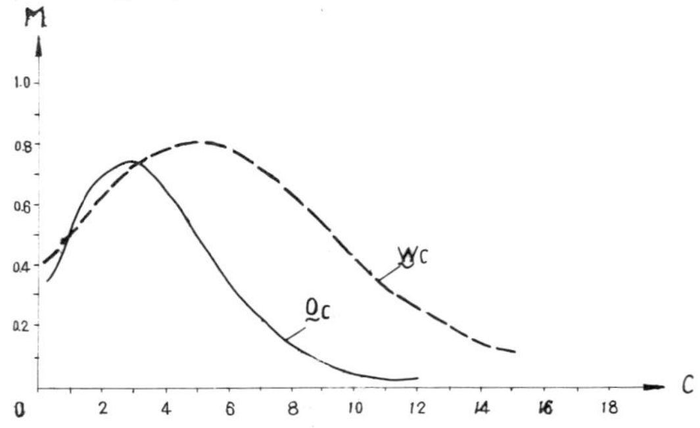

Fig. 5.

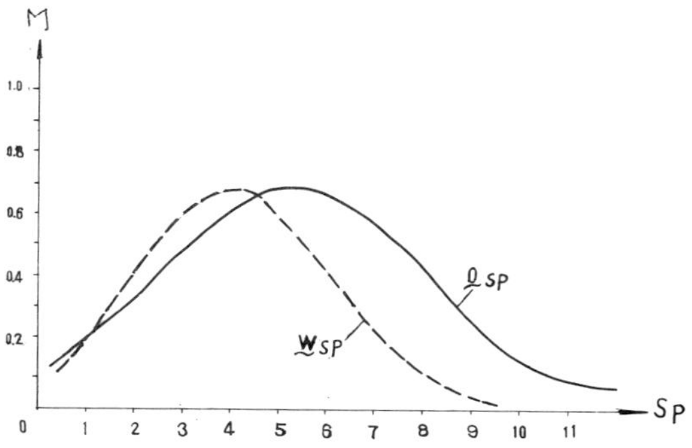

Fig. 6.

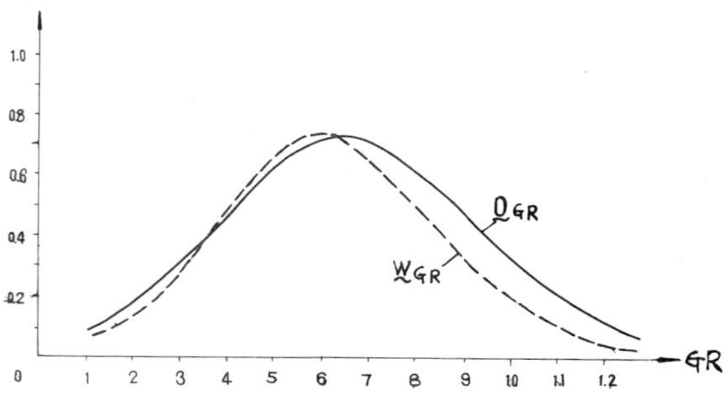

Fig. 7.

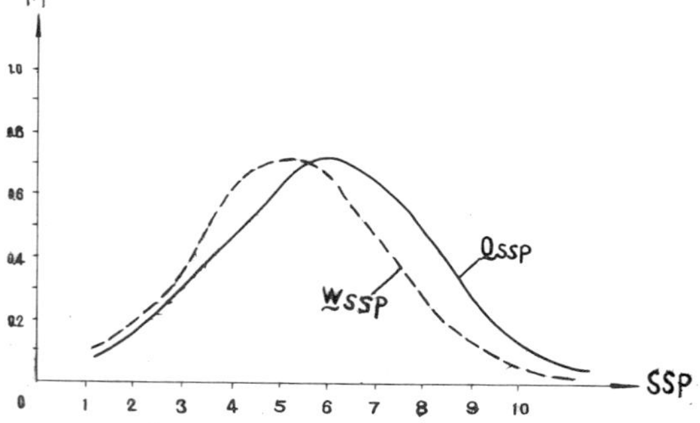

Fig. 8.

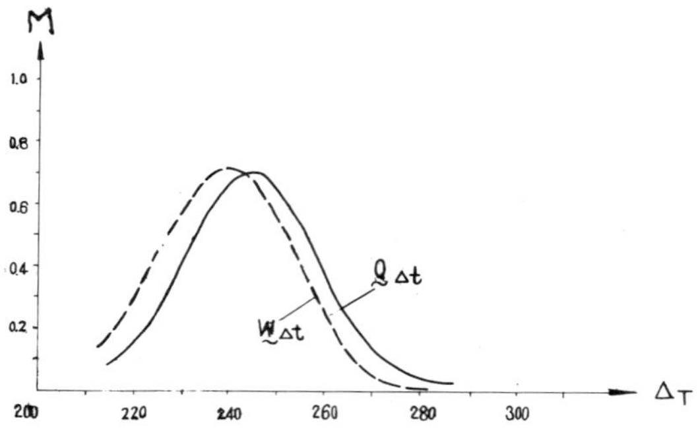

Fig. 9.

The statistical model of multi-dimension fuzzy number is as following.

<u>The model of the oil-bearing stratum</u>

$$M_{\underset{\sim}{O}}(\bar{x}) = \sum_{i=1}^{77} M_{\underset{\sim}{\bar{x}}_i}(\bar{x})/77$$

$$= \sum_{i=1}^{77} \left[\bigwedge_{j=1}^{9} e^{-\left(\frac{x_j - x_{ij}}{\sigma_{o_j}}\right)^2} \right] /77 \qquad (24)$$

<u>The model of the water-bearing stratum</u>

$$M_{\underset{\sim}{W}}(\bar{x}) = \sum_{i=1}^{94} M_{\underset{\sim}{\bar{x}}_i}(\bar{x})/94$$

$$= \sum_{i=1}^{94} \left[\bigwedge_{j=1}^{9} e^{-\left(\frac{x_j - x_{ij}}{\sigma_{w_j}}\right)^2} \right] /94 \qquad (25)$$

where $\underset{\sim}{\bar{x}}_i$ denotes the 9-dimension fuzzy number of the samples.

Using two fuzzy statistical model we did a feedback recognition on the known stratums. Using the one-dimension fuzzy model, we got the rate of right of 88%. Using the 9-dimension fuzzy model, we got the rate of right of 93%. It shows that the discernment of the multi-dimensin model is greater than the one-dimension model.

REFERENCES

Zadeh, I. A. (1978). Fuzzy Sets as a Basis for a Theory of Possibility, Fuzzy Sets and Systems, 1, 3-28.

Zadeh, L. A. (1965). Fuzzy Set, Information and Control, 8, P.338-353.

Pei-Zhang Wang. (1980). The Synopsis of Fuzzy Mathematics, The practice and knowledge of Mathematica, 2. 3.

IMPACT OF NEW MEXICAN INDUSTRIAL PORTS ON TRANSPORTATION NETWORKS: A FUZZY SETS APPROACH

J. P. Antún

Coordinación de Sistemas B101, Instituto de Ingenieria, National University of Mexico, Apdo Postal 70-472, Mexico DF (04510), Mexico

Abstract. An alternative approach based on fuzzy sets concepts is suggest to model the impact of industrial ports on plurimodal land transportation.

Keywords. Ports & Plurimodal Land Transportation Planning, Modelling, Fuzzy Sets, Graph Theory, Reliability Theory.

1. MEXICO: PROGRAM OF INDUSTRIAL PORTS

1.1 Objectives

An industrial port is a port conceived as a convenient location for industries with requirements of a waterfront. Usually an industrial port is placed in national plans of industrial and regional development, and is the result of a normative framework of territorial ordering. In general, the industries located in it have markets of products and output factors that can be reached by means of an aquatic transportation; those markets are usually of international nature.

In developed countries [1] the industrial ports have been associated to the evolution of the petrochemistry industry and in some ways to the basic chemistry and steel industry. In developing countries not only the industrial development strategies are associated, but it also includes the fostering of modern ports.

In 1979, the Mexican Government established a Program of Industrial Ports [2], its main objectives are:
1) to ease the foreign trade by sea with an endowment of an adecuate port infrastructure,
2) to back the ruling national setting of land regulation which underline the coastline as one of the areas with the greatest potential for socioeconomic development,
3) to establish the convenient infrastructure for the development of strategic industrial sectors, specially: basic metal industries (iron, steel, aluminium), basic petrochemistry(ethylene, high density polyethylene, propylene, polipropylene, amoniac, methanol, etc.) and secondary petrochemistry (resins, plastics, textil fibers, etc.), paper and cellulose paste manufacturing, making up capital goods of great volume, specially for the oil industry, and shipyards.

1.2 Locations

The mexican industrial ports program points out five locations:

1) *OSTION*, in a spot south of the Ostión Lagoon, a few kilometers north of Coatzacoalcos (Veracruz),
2) *ALTAMIRA*, in the south end of the San Andrés Lagoon, nearby Altamira (Tamaulipas), a few kilometers north of Tampico (Tamaulipas),
3) *LAZARO CARDENAS*, in the Balsas river outlet, at the opposite side border of the Sicartsa steel plant, in the town of the same name (Michoacán),
4) *SALINA CRUZ*, in a place nearby and north of the present Salina Cruz Port (Oaxaca),
5) *DOS BOCAS*, nearby Paraíso (Tabasco).

1.3 General Contents of the Strategy

The Industrial Ports Program was created as an action for the integral development and it ponders on:
- the port infrastructure project, construction and operations [3],
- the fitting out and the development of an incustrial park destined to strategic industries, and with requirements of a waterfront,
- the fitting out and the development of an industrial park destined to the local small and medium size industries,
- the urban development of new population centers and the regularization of the growth of the present ones,
- the fostering of the activities of the port services, and
- a set of subprograms for social development (housing, education, laboral training, communitary development, etc.) and complementary actions (local agroindustries development, environment control quality, etc).

2. PORTS AND PLURIMODAL LAND TRANSPORTATION

2.1 Hinterland, foreing-trade and cabotage basin

The port cities are powerful generators of land transportation because:
- the port is subjected to transit traffic from foreign and coastal (caotage) trade,
-the markets of goods and output products of industrial activities carried out in this port cities can be found inside the territory.

The set of nodes of the land transportation network(highway and railroad modes) that form up the chains of transportation that include a port as an origin or destination, is called the *Hinterland* of the port.

On the other hand, the *foreign-trade region* of a port is the set of foreign countries or regions of these countries with the ones it has connections by means of regular maritime lines.
The features of the economic activities and the political frameworks in the hinterland as well as in the foreign trade region, not only define the profile of the traffic of foreign trade but they also give form to the same.
Finally, the *cabotage basin* of a port is the set of ports that are related to it by means of a regular line of the coastal trading network.

2.2 Subnetwork of land links of primary impact in a port (SPI)

With the aim of planning the evolution of lands links related to a maritime port the following links must be taken into consideration:
-the links in the hinterland of the port,
- those links of the hinterland of the ports in its cabotage basin(considering the existing and potential traffic by coastal trading).

This set es denominated *subnetwork of land links of primary impact* of the port(SPI).

2.3 Impact of the industrial ports on land transportation

The industrial ports repercussion on the transportation in México is due to its double role as
-transit center of foreign commerce, and as
- strategic industrial activity center.

Based upon these facts, it may be identified the following effects[4]:
a. potential changes in the *transit patterns* of foreign trade,
b. change in the *cargo handling technology* in the local transportation,
c. a new demand of *strategic freights* transportation involved with the ports production itself,
d. impulse for the development of coastal trading transportation (*cabotage*),
e. planning needs of the development of some *railroads* connectios,
f. planning needs of the development of *highways* in the hinterland,
g. planning needs of the development of *regional plurimodal logistics centers*.

3. MODELLING APPROACHS

In the present section, some outlooks are presented to elaborate a mathematical model in order to build a quantitative analysis of the influence of the industrial ports on the development of the overland transportation, based on:
- *the definition of SPI using fuzzy sets,*
- *the assignation of traffic by competence criteria,* and
- *the identification of critics archs by means of an analysis of reliability.*

The objective of the model is to back the making of decisions in the planning of the mainland transportation links related to the ports, that is, in the SPI.

3.1 Previous considerations

a. *Service quality and capacity of a mainland link*
In general, one of the products expected from the analysis of the influence of the future ports on the mainland links, is the determination of the service quality that can be expected, according archs(modal and pluri modal). It is common that the quality is measured agreeing with the concept of "service levels"[5]. Nevertheless, this has been greatly criticized. It must be remembered that in the land links planning, the most important fact is to get a mesure that should br translated, in a proposal of new works and new transportation services.This measure result from the diagnostic of the present situation and from the analysis of the perspective of the traffic demand.

Thus, if the model is built according to concepts of reliability, ti might be a metric space of the service quality; and, it might be useful for planning purposes if it is able to transform reliability into capacity values in specific archs.

b. *Model asignation in the plurimodal mainland transportation system*
To assign the cargo freight in transit by port and that generated by industrial production in the port area, to the highway and railroad, the following hypothesis will be adopted:
-the products of the extractive industry will be shifted preferably by railroad
-the imports of agricultural products will be shifted to the consumption centers of the high plateau preferably by railroad
- the huge and heavy importated capital goods will be gradually transfered to the railroad as some of its connections are modernized
-the perishable freights, adn specially those under refrigeration as fishing products, will be shifted preferably by truck
- the non-imported agricultural products in whose shifting the cabotage transportation is involved, will be send by trailer (they get in and out of the port in mainland using

truck-tractors; on cabotage lines, the operation will be roll-on/roll-off), in some cases fi there is a railroad, and is a long distance and to high plateau, the might to use piggy-back system

- the imported containered fright will be shifted by trucks and railroad; but the participation fo a railways will be growing in case there exist both modes and when the railroads are modernized

-the strategic industrial production at industrial ports will be shifted by trucks and railroads, adn the cabotage will be encouraged; the railways will have a growing participation in basic petrochemistry(products that can not be delivered by pipelines) and secondary petrochemistry, fertilizers (specially in the links with the high plateau), steel sheet and aluminum lingots.

c. *Alternative methodologies*

It is common to use demand models of traffic of the gravitatory type[6] in transportation planning, and the consequent derivation of need capacity on an arch of the network for the defined flow by the model and a set of hypothesis associated to some metric of service quality. The gravitatory models use the physics analogy of gravity, traslating mass into gross product per capita, population or some other variable derived to the central place theory, by means of constants more or les sophisticated.

In the case of a port city, the freights-in-transit (foreign trade and cabotage) is more important than the freights derived from local production; in this case the gravitatoty model is deficient: the analogic variable to mass necer has the importance that allows to derive a traffic demand by transit, unless using some patchwork.

The use of a model of gravitatory type(without a "patchwork") to networks thar include tha ports, allows the identification a traffic demands which do not include neither the cabotage neither the foreign trade. It is proper to incorporate these, with simplicity(another "patchwork"), only to the mainland liks involved with the ports, this is, to the SPI. In order to achieve this, the subnetwork must be defined and the main traffic too, according to their origin-destination and modal assignation.

The SPI might be modeled either as an interconnected queues network or as a flow-network, accepting the hypothesis that it would not be a great mistake to "inyect" the estimated traffics (for example by some market model) through the origin-destination nodes, the foreign trade and potential coastal trading, and the involved port In the case of interconnected queues network[7], their length may be translated into arch capacity by means of an association of the service time to the speed; alternately might be consider the track (railroad and highway) as a service module.

In the model of flow-networks[8], the traffic demand between two nodes (one the port) is satisfy by means of a path(combination of some archs) according some matrix of conductive type (operation cost, journey time, etc.); an analysis of reliability of the network may be achieved (for a flow-traffic),and a posteriori to allow a translation of reliability of archs into capacity of link.

3.2 <u>An outline to build a model of analysis of reliability in flow-networks applied to mainland transportation links influenced by a port</u>

a. *Hinterland definition*

The definition of a hinterland is a problem of "regionalisation";that is,of classificatory analysis in regional economy or economic geography.

A way to solve this is to use fuzzy sets-sous ensembles flous[9],according the experience of the Institute de Mathématiques Economiques of the University of Dijon[10].

Two alternatives to be explored:

1) to define the hinterland of each port p as a fuzzy set $\tilde{H}_p$ of the non fuzzy set of the traffic generators centers C,whith a grade of membership function μ_{H_p} (c) defined for each element c of C using certain ad-hoc coefficient that will measure its connection through foreign trade,cabotage and/or industrial production market(input factors & products) via/realized in the specific port.

2) to define the hinterland of each port p as a fuzzy set $\tilde{H}_p$ of non fuzzy set of traffic generator centers C based on a numeric taxonomy settled on distance minimization (Hamming,Euclidean, Max-min relation)between values which take the attributes in association with the foreign trade,cabotage and industrial production market in each center c of C.

Notice that 2) allows the simultaneous definition of all $\tilde{H}_p$;p=1,2,...,p.

b. *Determination of the SPI*

The definition of the hinterland of each port as a fuzzy set $\tilde{H}_p$ allows to determine a fuzzy network $\tilde{R}_p$ of the non fuzzy network of mainland transportation links with fuzzy nodes[11].The fuzziness of a node is related to the possibility[12] of socioeconomic links with an industrial port; SPI is time variant[13] fuzzy systems[14].In addition for each arch (non fuzzy) it can be defined a set of non fuzzy attributes (operation cost,capacity,etc.) and a set of fuzzy attributes (journey time,etc.).

b. *Identification of the strategic nodes*

By a.2) $\tilde{H}_p$;p=1,2,...,p. is defined;a fuzzy set of strategic nodes $\tilde{H}^*$ may be defined on $\tilde{H}_p$ using a new set of non fuzzy atributes derived from ad-hoc conception of the strategic character(for example:a fraction of foreign commerce in relation to the total commerce in transit generated by p;input factors & products markets measured as a fraction in relation to the total in transit by p;etc.)or fuzzy attributes(for example: relative importance in national plans of development).That is producing a "regionalization" into the hinterland(fuzzy).

c. *Adoption of a model of analysis of reliability in flow-networks*

In flow-networks (production,communication,transportation)the relative reliability of a given flow F is defined as the probability that the network might deliver a flow equal or greater than the specified F value,from the point of origin to its destination one.This probability is a function of four factors:

-magnitude F of the flow
-capacity of each of the network branches
-network topology
-reliability of each one of the network branches

In general,when the flow F is grater,the reliability of the network is smaller[15].Another research show a topologycal analysis of reliability in flow-networks with capacity restrictions[16]; it is proposed as a background reference,taking into consideration that:

1) Due to the fact that the methodology was developed for two-terminal-nodes network,one of this will be the port and the other one,a strategic node.
2) Due to the fact that our problem is a fuzzy one,the matrix calculation methodology shoul be extended to the control of fuzzy sets[17].
3) The arch capacity k_{ij}(non fuzzy) is adopted as independently of time.
4) The reliability r_{ij}(non fuzzy) of an arch is defined as the probability of being sucessful when trying to make the arch deliver a flow value f_{ij}: $0<f_{ij}<k_{ij}$.
5) An arch reliability is adopted according to the flow function that travels through it as $r_{ij}=r_{ij}(f_{ij})$.
6) The matrix of branch conductivity G,for the flow assignation in the model,might be defined by non fuzzy parameters as operation cost,or by fuzzy parameters as journet time.
7) The failure process of the network,a markovianic process[18],allows to identify the archs whose critic capacity should be modified(planning outputs:modernization,new works,new services).
8) Indeed,according 1),it can be iterated for different strategic nodes;nevertheless it is advisable to find a generator equivalent to the generators analogous to the traffic inyection among nodes and ports,by means the usage of network theorems[19].

Nowdays,Port Hinterland Models are being developed,based on the present outline,to facilitate the planning of the highways development in a multimodal perspective specially for the Altamira Industrial Port in Tamaulipas,and Lazaro Cardenas Industrial Port in Michoacan.

4. ACKNOWLEDGMENT

This work was presented some results from a research contract between the INSTITUTO DE INGENIERIA-UNAM(The Institute of Engineering of the National University of Mexico) and the Secretaria de Asentamientos Humanos y Obras Públicas -SAHOP (The Mexican Federal Ministry of Human Settlements and Public Works).

The author wishes to acknowledge Mr C V Negoita,Visiting Professor from Bucarest University, for providing time,discussion and encouragement.

5. REFERENCES

1 HANAPPE,P;SAVY,M (1980) Industrial Ports and Economic Transformations,The International Association of Ports and Harbors,Tokyo.

2 GOBIERNO DE MEXICO (1979)Programa de Puertos Industriales,Coordinación de Proyectos de Desarrollo,Presidencia de la República,México.

3 THE JAPAN OVERSEAS COASTAL AREA DEVELOPMENT INSTITUTE-OCDI (1981) Terminal de Usos Múltiples y Módulo Polivalente de Lázaro Cárdenas,Altamira, Ostión,Salina Cruz y Dos Bocas,Coordinación de Proyectos de Desarrollo,Presidencia de la República,México.

4 ANTUN,JP (1982) Determinación de la Influencia de los Puertos Marítimos sobre la Evolución de los Enlaces Terrestres de México,Instituto de Ingeniería-SAHOP,Proy 1507,México.

5 GOBIERNO DE MEXICO (1976) Manual de Proyecto Geométrico de Carreteras,SAHOP,México.

7 αHO,YC (1981)"Optimization analysis of discret event stochastic systems with application to manufacturing automation",in Proceedings of the 1981 IFAC World Conference (Kyoto), Pergamon Press,Tokyo,PS-40/50.

βCHANDY,KM;SAUER,Ch "Approximate solution of queueing models",Computer(IEEE) ,April 1980,pp. 25-32.

ψPRADE,H (1980)"An outline of fuzzy possibilistic models for queueing systems",in WANG,P;CHANG, S(eds)Fuzzy Sets:Theory and Applications to Policy Analysis and Information Systems,Plenum Press, New York,pp.147-153.

6 HABERMANN,R(1977) Mathematical Models:Mechanical Vibration,Population Dynamics and Traffic Flow Prentice Hall,New Jersey.

8 HARARY,F(1969)Graph Theory,Addison Wesley,New York.

9 αZADEH,L(1965) "Fuzzy Sets",Infomation and Control,8,pp.338-353.
βKAUFFMAN,A Introduction a la théorie des sous-ensembles flous,Vol 1 (1973),Vol 2 (1975),Vol 3 (1975),Vol 4 (1977),Masson,Paris.
ψDUBOIS,D;PRADE,H (1980)Fuzzy Sets and Systems: theory and applications,Academic Press,New York.
φNECOITA,CV (1981) Fuzzy Systems,Abacus Press, Turnbridge Wells.

10 αDELOCHE,R (1975) Theorie des sous-ensembles flous et clasification economique spatial,Document de Travail N°11,Institute de Mathematiques Economiques,Universite de Dijon,31p.,Dijon.
βPONSARD,C (1977) La region en analyse spatial, Document de Travail N°21,Institute de Mathematiques Economiques,Universite de Dijon,25p.,Dijon.

11 αROSENFELD,A (1975) "Fuzzy Graphs",in ZADEH,L et al. (1975)Fuzzy Sets and their applications to cognitive and decision processes,Academic Press, New York,pp.77-95.
βYEH,R;BANG,S (1975) "Fuzzy relations,fuzzy graphs and their applications to clustering analysis",in ZADEH,L et al (1975) Fuzzy Sets and heir applications to cognitive and decision process,Academic Press,New York,pp.125-149.

12 ZADEH,L (1978) "Fuzzy sets as a basis for a theory of possibility",Fuzzy Sets and Systems, 1,pp.3-28.

13 NOWAKOWSKA,M (1981) "A new theory of time:generation of time from fuzzy temporal relations",in

LASKER,GE (ed) (1981) Applied Systems and Cybernetics,Vol VI,pp.2742-2747,Pergamon Press,New York.

[14] BERNARD,N (1982) Multiensembles,Thése,Université de Lyon I,Villeurbaine,Lyon.

[15] LARA,F (1980) Análisis de Confiabilidad en Redes de Flujo,Serie Azul N°422,Instituto de Ingeniería,UNAM,México.

[16] LARA,F (1973) Anális Topológico de la Confiabilidad de Redes de Flujo sujetas a Restricciones de Capacidad,Tesis Dr Ing ,Universidad de México,México.

[17] HALPERN,J (1975) "Set-adjency measures in fuzzy graphs",Journal of Cybernetics,5,N°4,pp.77-87.

[18] LARA,F (1973),op.cit.,p.82 and ss.

[19] SESEHU,S;BALABANIAN,N (1967)Linear Network Analysis,Wiley,New York.

Fig 1 . MEXICO: Location of new industrial ports
Source: Programa de Puertos Industriales
Coordinación de Proyectos de Desarrollo, Presidencia de la República,México,1979.

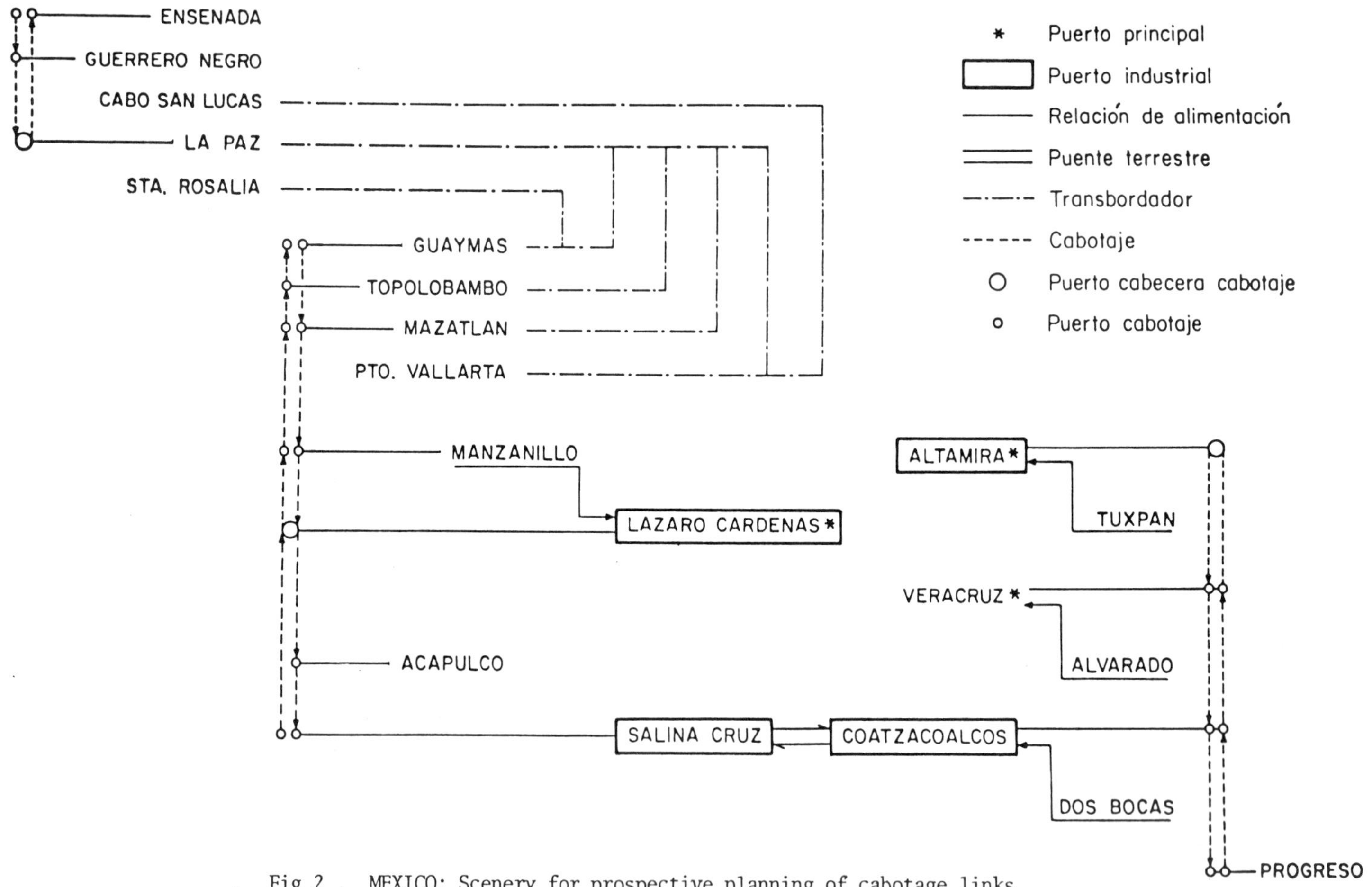

Fig 2 . MEXICO: Scenery for prospective planning of cabotage links
Source: Antún,JP Determinación de la influencia de los puertos marítimos en la evolución de los enlaces terrestres de México,Proy 1507,Instituto de Ingeniería-UNAM,México,1982

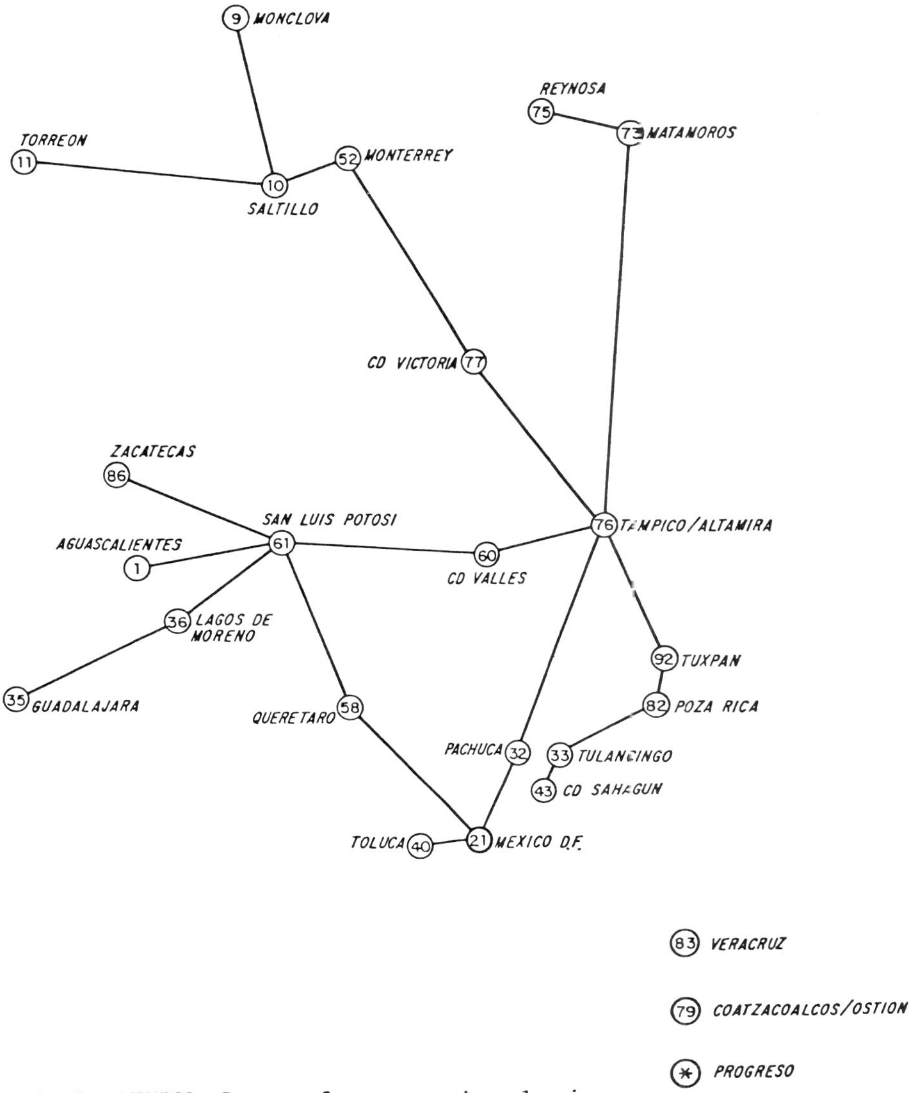

Fig 3. MEXICO: Scenery for prospective planning of SPI related to Altamira Industrial Port
Source: Antún,JP Determinación de la influencia de los puertos marítimos en la evolución de los enlaces terrestres de México, Instituto de Ingeniería-UNAM, Proy 1507, México, 1982.

FUZZY DECISION ANALYSIS ON THE DEVELOPMENT OF CENTRALIZED REGIONAL ENERGY CONTROL SYSTEM

S. Murakami*, H. Maeda* and S. Imamura**

**Department of Computer Science, Kyushu Institute of Technology 1-1 Sensuicho, Tobata, Kitakyushu 804, Japan*
***Ohtsuka Assay Laboratory, 463-10 Kagasuno, Kawauchicho Tokushima 771-01, Japan*

Abstract. In this paper, we develop the methodology to analyse the problem which contains fuzziness included the decision maker's judgement on subjective probabilities and subjective trade-offs among attributes. Also, we propose two methods (α-cut method and centroid method) to evaluate alternatives by using fuzzy expected utility curves calculated from fuzzy probabilities and fuzzy utilities. Next, we apply the above methods to a fuzzy decision analysis on the development of a Centralized Regional Energy Control System in Tobata Ward of Kitakyushu City. In this system, energy abandoned from heavy industries, power generation plants and refuse disposal plants is collected at the Heat Supply Center and exchanged into hot water; the hot water is then supplied to the regional air conditioning systems and the regional hot water supply systems through the pipe line networks. We analyse whether, and for what part of the region, such a system should be contructed in order to save energy consumption in the region and to supply easy-to-use energy under the fuzzy environments.

Keyward. Fuzzy decision analysis; Multiattribute utility function; Fuzzy expected utility; Energy system; Local energy saving system.

INTRODUCTION

The theory of multiattribute utility function is usable as a methodology for evaluating the alternatives of large scale systems such as a social or an energy system. With his trade-offs among the system objectives, the decision maker could select the optimal alternative on the basis of this methodology. The multiattribute utility function method, which was developed by Keeney (1976), decomposes a multiattribute utility function into unidimensional utility functions. After the decision maker structures his multiattribute utility function and assigns probabilities to the possible outcomes of alternatives, he can calculate the optimal alternative which maximizes expected utility.

When a decision maker or an expert evaluates attribute values for alternatives, particularly, in the form of assigning subjective probabilities to them, he may describe these probabilities in a fuzzy expression. For example, he might say, "the probability that the project will cost 100 million yen is about 0.3," or, "the probability that the 95 percent level of reliability could be retain seems to be in the range around between 0.1 and 0.2."

Fuzziness could be included in the values of certainty equivalents and lottery probabilities which will be given by the decision maker when his unidimensional utility function is to be identified on the basis of the 50-50 lottery. In another case, when he is asked to take the trade-offs among attributes to evaluate weights in an additive utility function or to form a multiplicative utility function, he would express some fuzziness in his answers.

The purpose of this paper is to develop a method to analyse these fuzzy decision problems and to apply the method to a case study: the development of a Centralized Regional Energy Control System (CRECS) in the city of Kitakyushu. The function of this system, a form of local energy saving system, is to collect and control waste heat released from industries and refuse disposal plants in order to utilize it for such uses as regional air conditioning, hot water supply, heated swimming pools and cultivation. There are numerous factories, including large scale iron mills and power generating plants, in Kitakyushu, which produce the huge amount of waste energy altogether. By recycling this energy effectively, the CRECS aims to save regional energy consumption, to lessen air pollution and heat contamination, and to provide the citizens with easy-to-use energy, all of which will lead to the realization of the more comfortable environments for them. Among several wards of Kitakyushu, we shall exclusively examine in this paper the feasibility to contruct a CRECS in Tobata Ward which is especially adjacent to the factory area.

FUZZY DECISION ANALYSIS

Fuzzy Expected Utility

Fuzziness, more or less, slips into the evaluation of uncertainty which haunts the possible concequences of an alternative. In particular, when the decision maker assigns probabilities to the possible events, some fuzziness is inevitable in estimating the probabilities. This seems to be the case when the decision maker has to determine the indifferent points (trade-offs) between couples of attributes in order to identify scaling constants in a multiattribute utility function. A fuzzy decision analysis method proposed in this paper can treat these cases: more specifically, the probability distribution of the outcomes of an alternative and their utility values are all fuzzy.

Now let us suppose a decision problem whose overall goal is specified in measurable n objectives, that is, n attributes: $X_1, X_2, \ldots, X_n$. These attributes take values $x_1, x_2, \ldots, x_n$, respectively. In the decision tree in Fig. 1, there are supposed to be m outcomes for an alternative, $A_i (i = 1,2,..,l)$, and the joint fuzzy probability of a possible outcome, $(x_{1j}, x_{2j}, .., x_{nj} : j = 1,2,..,m)$, is designated $\tilde{p}_{ij}$. $\tilde{u}_{ij}$ is the fuzzy utility which will be provided by the outcome. Using these symbols, we define fuzzy expected utility for alternative A_i, $\tilde{U}_i$, to be

$$\tilde{U}_i = [\tilde{p}_{i1} \otimes \tilde{u}_{i1}] \oplus [\tilde{p}_{i2} \otimes \tilde{u}_{i2}] \oplus \ldots \oplus [\tilde{p}_{im} \otimes \tilde{u}_{im}] \quad (1)$$

where sum and product of fuzzy variables, $\oplus$ and $\otimes$, are defined as follows:

$$\text{Sum} : \tilde{z} = \{(z/\mu(z))\} = \tilde{x} \oplus \tilde{y}$$

$$\mu(z) = \max_{z=x+y} [\min(\mu(x), \mu(y))] \quad (2)$$

$$\text{Product} : \tilde{z} = \{(z/\mu(z))\} = \tilde{x} \otimes \tilde{y}$$

$$\mu(z) = \max_{z=x \times y} [\min(\mu(x), \mu(y))] \quad (3)$$

In Eq. (2) and (3), $\tilde{x} = \{x/\mu(x)\}$, $\tilde{y} = \{y/\mu(y)\}$, and $\tilde{z} = \{z/\mu(z)\}$ are fuzzy variables and $\mu(x)$, $\mu(y)$, and $\mu(z)$ are membership functions. As to the form of a fuzzy multiattribute utility function employed in his paper, we assume the following multiplicative form:

$$[1 \oplus \tilde{k} \otimes u(x_1, x_2, \ldots, x_n)] = [1 \oplus \tilde{k} \otimes \tilde{k}_1 \otimes u_1(x_1)] \otimes [1 \oplus \tilde{k} \otimes \tilde{k}_2 \otimes u_2(x_2)] \otimes \ldots \otimes [1 \oplus \tilde{k} \otimes \tilde{k}_n \otimes u_n(x_n)] \quad (4)$$

where $\tilde{k}_1, \tilde{k}_2, .., \tilde{k}_n$, and $\tilde{k}$ are fuzzy scaling constants and $u_1(x_1)$, $u_2(x_2), .., u_n(x_n)$ are unidimensional utility functions corresponding to the n attributes and are assumed to be non-fuzzy functions. By Eq. (1) and (4), we can derive fuzzy expected utility for each alternative.

Fuzzy Scaling Constants

In this place, we show how to assess the fuzzy scaling constants. Without the loss of generality, we proceed our discussion about the case of the two-attribute utility function with the multiplicative form. Let the two-attribute utility function be

$$\tilde{u}(x_1, x_2) = \tilde{k}_1 \otimes u_1(x_1) \oplus \tilde{k}_2 \otimes u_2(x_2) \oplus \tilde{k} \otimes \tilde{k}_1 \otimes \tilde{k}_2 \otimes u_1(x_1) \otimes u_2(x_2) \quad (5)$$

To assess the scaling constants, it is necessary to get the specific values of trade-offs between two attributes answered by the decision maker. In Fig. 2, for example, fuzzy indifferent points, $(\tilde{a}, x_2^\circ) \sim (x_1^\circ, x_2^*)$ and $(\tilde{b}, x_2^*)$ (x_1^*, x_2°), are shown, where x_1^* and x_2^* imply the most prefered concequences, and x_1° and x_2° imply the least prefered ones. $\tilde{a}$ and $\tilde{b}$ represent the fuzzy answers by the decision maker. Strictly speaking, the point that is indifferent to (x_1°, x_2^*) should be the unique point for the sake of consistency. Nevertheless, it seems to be a very difficult problem for the decision maker to identify this indifferent point. However, it will be easy for him to give, instead, a rough value around a.

The meaning of "indifferent (~)" is interpreted as that a concequence (x_1°, x_2^*) must be substituted for (a, x_2°) in any situations. Thus, a membership $\mu(a)$ of fuzzy set $\tilde{a}$ will be interpreted as the degree of conviction with which a certain concequence, (a, x_2°), is substituted for (x_1°, x_2^*). Hence, (x_1°, x_2^*) can be considered to be indifferent to $(\tilde{a}, x_2^\circ)$ with, the degree of conviction, $\mu(a)$.

Substituting the indifferent relations as shown in Fig. 2 into Eq. (5), we get the following equations,

$$\tilde{k}_2 = \tilde{k}_1 \otimes u_1(\tilde{a}) \quad (6)$$

$$\tilde{k}_1 = \tilde{k}_1 \otimes u_1(\tilde{b}) \oplus \tilde{k}_2 \oplus \tilde{k} \otimes \tilde{k}_1 \otimes \tilde{k}_2 \otimes u_2(\tilde{b}) \quad (7)$$

And from the property of scaling constants, we have another equation, that is,

$$\tilde{k}_1 \oplus \tilde{k}_2 \oplus \tilde{k} \otimes \tilde{k}_1 \otimes \tilde{k}_2 = 1 \quad (8)$$

Given fuzzy estimates, that is, $\tilde{a}$ and $\tilde{b}$, for a set of specific values of $\tilde{a}$ and $\tilde{b}$, we can obtain a set of solutions of scaling constants from Eq. (6)-(8). Let define this set of solutiond as follows,

$$\tilde{K} = \{((k_1, k_2, k)/\mu(k_1, k_2, k))\}$$
$$\mu(k_1, k_2, k) = \min(\mu(a), \mu(b)) \quad (9)$$

In the case of a n-attribute utility function, we can easily devive the fuzzy scaling constants from the fuzzy trade-offs among attributes in the same manner as the case of the two-attribute utility function.

Selecting Criteria for Alternatives

Now that fuzzy expected utility has been derived for each alternative, we proceed in the present section to introducing some criteria for selecting the optimal alternative.The methodology to evaluate fuzzy expected utility has been studied by several authors, Baas (1977), Jain (1976), Baltwin (1979), Watoson (1979), Freeling (1980). In this paper, however, we propose 1) α -cut method and 2) centroid method.

1) α-cut method

For alternative A_i, we let f_i^α be the maximum of the fuzzy expected utility values whose membership values are not less than a critical level, say, α. That is,

$$f_i^\alpha = \max\{U_i \,|\, \mu(U_i) \geq \alpha\} \quad (i=1\sim l) \tag{10}$$

Then an alternative with the maximum f_i^α will be selected as the optimal solution. This procedure can be restated as follows: select an alternative with the maximum expected utility out of the qualified candidates in that their expected utility values have degrees of conviction not less than α.

2) centroid method

When we evaluate alternatives on the basis of fuzzy expected utility and its membership function (degree of conviction), we should select in principle those which are dominant in terms of the both measurements. In order to combine the two into one measure, we have contrived the centroid of fuzzy expected utility, (U_0,μ_0), that is,

$$U_0 = \frac{\int_0^1 U\mu(U)\,dU}{\int_0^1 \mu(U)\,dU} \tag{11}$$

$$\mu_0 = \frac{\int_0^1 U\mu(U)\,d\mu(U)}{\int_0^1 U\,d\mu(U)} \tag{12}$$

where U is the fuzzy expected utility with the membership $\mu(U)$. By calculating the centroid of fuzzy expected utility for each alternative, we can select an alternative as being optimal which attains the maximum value on either of the U and μ axes. In other words, this method seeks an alternative which has the maximum expected degree of conviction on the expected utility. Note here that there is not always the unique optimal solution: that is, Pareto solutions (alternatives not dominated by others) are possible to be found. Still, in this situation, the final decision to select only one alternative could be made by introducing the decision maker's judgement on which of expected utility and conviction may be more important for him.

FUZZY DECISION ANALYSIS ON CENTRALIZED REGIONAL ENERGY CONTROL SYSTEM

As we mentioned earlier, we confine the CRECS application to Tobata Ward in Kitakyushu. Considering the geographical conditions, we suppose that the sources of waste heat are 200 °C steam from Hiagari Refuse Disposal Plant, digestive gas from Hiagari and Nishiminato Waste Disposal Plants, and exhaust gas as hot as 200 °C - 400 °C from the coke and combustion furnaces at Tobata Work of Nippon Steel Corporation. The waste energy from these sources is to be collected and then transformed into high-heated water (130 °C) at the Heat Supply Center (HSC), which in turn distributes it to Subcenters through Primary Pipelines.

At Subcenters, equipped with such facilities as heat exchangers, refrigerating machines, hot water storage tanks, and pumps, the high-heated water from the HSC is changed to three kinds of water: 80 °C-hot water for general use, 80 °C-hot water for air-heating, and 5 °C-cool water for air-cooling. The hot or cool water is supplied to users through Secondary Pipelines. The water which has been used for air conditioning is pumped back to Subcenters again. We have assumed heretofore that regional air conditioning and hot water supply are all the services which the users can directly take from the CRECS and that for residences the both are accomodated but for stores only air conditioning service is supplied.

System Objectives and Their Attributes

We propose the following three as the objectives which a CRECS should attain:

(1) maximizing energy savings in the target area,
(2) minimizing the total cost to construct the CRECS,
(3) realizing the more comfortable city life by the brord distribution of easy-to-use energy

Next, for each of these, we have to define an attribute to measure its attainment. Attribute X_1 for the first objective is defined to be the amount of energy (Gcal/year) which will be saved by the CRECS out of the average amount of energy the area residences and stores are now consuming for air conditioning and hot water use. Attribute X_2 for the second objective is the total construction cost (in 100 million yen) of the CRECS, which includes the expenditures for construcing the HSC, Primary and Secondary Pipelines, and Subcenters. Attribute X_3 for the last objective is the coverage ratio (%) of the area served by the system to the whale area in Tobata. These areas are measured by the total floor space of residences and stores in the respective domains. We assume here that a residence corresponds to a household.

CRECS Alternatives

After investigating the density of residences and stores in Tobata and the planned location for the HSC, we have projected the following four alternatives of CRECS:

A_1 : not constructing any CRECS,
A_2 : constructing a small-scale CRECS serving the nearby area to the HSC from Nakabaru-nishi to Nakabaru-higashi to Tenjin,
A_3 : constructing a medium-scale CRECS serving the broader area, including the A_2 coverage, down to the vicinity of Tobata Station,
A_4 : constructing a large-scale CRECS serving the whole area of Tobata.

Fig. 3 shows the serving areas of the respective alternatives and the planned locations for the HSC, Subcenters and the sources of waste heat. Table 1 shows the numbers of residences and stores and their respective total floor spaces in the area covered by each alternative. Out of these total numbers, however, how many residences and stores would actually be willing to take the service of regional air conditioning and hot water supply by the four CRECS alternatives has not yet been investigated. Therefore, in this paper, we use fuzzy estimation for this, instead, which will be made subjectively by the decision maker.

For each alternative, the probability that the number of residences willing to take the CRECS service would amount to 10 percent of the total area residences is evaluated to be "roughly 0.6"; for the coverage of 15 percent, the probability will be "a little smaller than 0.4". On the other hand, the probability that 30 percent of the area stores would take the service of regional air conditioning is evaluated to be "a little larger than 0.5"; for the coverage of 50 percent, the probability will be "roughly between 0.4 and 0.5". In Fig. 4, membership functions are shown which correspond to these fuzzy probabilities, $\tilde{p}_{10\%}$ and $\tilde{p}_{15\%}$ for the served residences and $\tilde{p}_{30\%}$ and $\tilde{p}_{50\%}$ for the served stores.

Evaluation of Alternatives

For each alternative, the amount of energy savings, the total construction cost, and the coverage ratio were estimated as shown in Table 2. The maximum energy demand estimates in the fourth column were calculated for the extreme situation that all the users maximize their use of the available CRECS service simultaneously. The estimating methods for these values are described in detail in Imamura (1982).

Now, let us determine the optimal CRECS alternative. First, we define unidimensional utility functions for three attributes as $u_1(x_1)=x_1^{0.64213}$, $u_2(x_1)=(1-x_2/150)^{0.71724}$, and $u_3(x_3)=(x_3/50)^{0.64332}$, in which we set up the most preferred concequences of attributes as $x_1^* = 1.0$ (10^5Gcal/year), $x_2^* = 0$ (yen), and $x_3^* = 50$ (%), the least preferred ones as $x_1^\circ = 0$ (10^5Gcal/year), $x_2^\circ = 150$ (100 million yen), and $x_3^\circ = 0$ (%). And then, we assume the following multiplicative form of fuzzy multiattribute utility function:

$$[1\oplus\tilde{k}\otimes\tilde{u}(x_1,x_2,x_3)] = [1\oplus\tilde{k}\otimes\tilde{k}_1\otimes u_1(x_1)]\otimes[1\oplus\tilde{k}\otimes\tilde{k}_2\otimes u_2(x_2)]\otimes[1\oplus\tilde{k}\otimes\tilde{k}_3\otimes u_3(x_3)] \quad (13)$$

where $u_1(x_1)$, $u_2(x_2)$ and $u_3(x_3)$ are non-fuzzy unidimensional utility functions for attribute X_1, X_2 and X_3, respectively, and $\tilde{k}_1$, $\tilde{k}_2$, $\tilde{k}_3$, and $\tilde{k}$ are fuzzy scaling constants.

To obtain fuzzy scaling constants, the decision maker was asked indefferent points to A, B and C shown in Fig. 5. His answers wereas follows,

$\tilde{a}$ = "about 75 (100 million yen)"
$\tilde{b}$ = "roughly between 120 and 125 (100 million yen)",
$\tilde{c}$ = "about 110 (100 million yen)".

In Fig. 6, the membership functions of these answers are shown.

From these fuzzy answers, we get

$$\tilde{k}_1 = \tilde{k}_2 \otimes u_2(\tilde{a}) \quad (14)$$

$$\tilde{k}_3 = \tilde{k}_2 \otimes u_2(\tilde{b}) \quad (15)$$

$$\tilde{k}_2 = \tilde{k}_1 \oplus \tilde{k}_2 \otimes u_2(\tilde{c}) \oplus \tilde{k} \otimes \tilde{k}_1 \otimes \tilde{k}_2 \otimes u_2(\tilde{c}) \quad (16)$$

$$(1 \oplus \tilde{k}) = (1 \oplus \tilde{k} \otimes \tilde{k}_1) \otimes (1 \oplus \tilde{k} \otimes \tilde{k}_2) \otimes (1 \oplus \tilde{k} \otimes \tilde{k}_3) \quad (17)$$

From Eq. (14) - (17) and the membership functions of $\tilde{a}$, $\tilde{b}$, and $\tilde{c}$, we can calculate the fuzzy scaling constants.

For each alternative, the values of fuzzy expected utility to the varying degree of conviction of the decision maker, μ's, are plotted in Fig. 7.

Finally, we proceed to evaluating the alternatives by the two methods introduced in the previous section. First, let us employ the α-cut method. With $\alpha = 0.5$, that is, the degree of conviction not less than 0.5, the decision maker will rank the four alternatives as A_2, A_1, A_3, and A_4, in order of dominance, with A_2 being optimal. Next, from the results of the centroid method shown in Fig. 8, we can find that the alternative A_2 and A_1 are Pareto solutions. A_2 has the highest expected utility and the second-ranked expected degree of conviction. On the other hand, A_1 has the third-ranked expected utility and the highest expected degree of conviction. Since the decision maker regards the expected utility more important than the expected degree of conviction, A_2 is apparently the optimal solution.

As the conclusion of the CRECS decision problem, we could say that, by either of α-cut and centroid methods, the optimal alternative is to construct a small-scale CRECS which serves the area from Nakabaru-nishi to Nakabaru-higashi to Tenjin. From this conclusion, we would like to propose the feasibility to construct a CRECS for the limited area in Tobata which is both adjacent to the HSC and of the high density of residences and stores.

CONCLUDING REMARKS

From the fuzzy decision analysis on the CRECS alternatives, it turned out that the small-scale system would be feasible. In this paper, fuzzy variables introduced into the decision analysis were restricted to fuzzy trade-offs and percentages of the residences and stores which are supposed to take air conditioning service and hot water supply from the CRECS. All these variables were evaluated on fuzzy judgement by the decision maker.

To make the analysis more practical, however, we should further take into account the uncertainty of attribute values (probability distribution) and the fuzziness with evaluating it. For still further development in the future, we will proceed to the development of a general fuzzy decision analysis method which could treat fuzziness in the identification process of the unidimensional utility function for an attribute and in assessing process of concequences for alternatives.

REFERENCES

Keeney, R.L. and H. Raiffa (1976). Decision with Multiple Objectives, John Wiley.

Murakami, S., et al. (1979). Decision analysis on the development of regional energy saving system, Kyushu Kogyo Daigaku Kenkyu Hokoku (Kogaku), No.38, 89/97 (in Japanese).

Zadeh, L.A. (1965). Fuzzy sets, Information and Control, Vol.8, 338/353.

Adamo, J.M. (1980). Fuzzy decision trees, Fuzzy Sets and Systems, Vol.4, 207/219.

Baas, S.M., et al. (1977). Rating and ranking of multi-aspect alternatives using fuzzy sets, Automatica, Vol.13, 47/58.

Jain, R. (1976). Decision-making in the presence of fuzzy variables, IEEE, SMC-6, 698/703.

Baldwin, J.F., et al. (1979). Comparison of fuzzy sets on the same decision space, Fuzzy Sets and Systems, Vol.2, 213/231.

Watson, S.R., et al. (1979). Fuzzy decision analysis, IEEE, SMC-9, No.1, 1/9.

Freeling, A.N.S. (1980). Fuzzy sets and decision analysis, IEEE, SMC-10, No.7.

Imamura, S. (1982). Studies on the fuzzy decision analysis, Master's dissertation of Kyushu Institute of Technology (in Japanese).

TABLE 1 Number and Total Floor space of Residences and Stores

	Number of residences	Number of stores	Total floor space of residences (m^2)	Total floor space of stores (m^2)
A_2	9,955	832	530,992	122,154
A_3	16,317	1,762	922,589	280,027
A_4	27,847	2,761	1,521,607	433,469

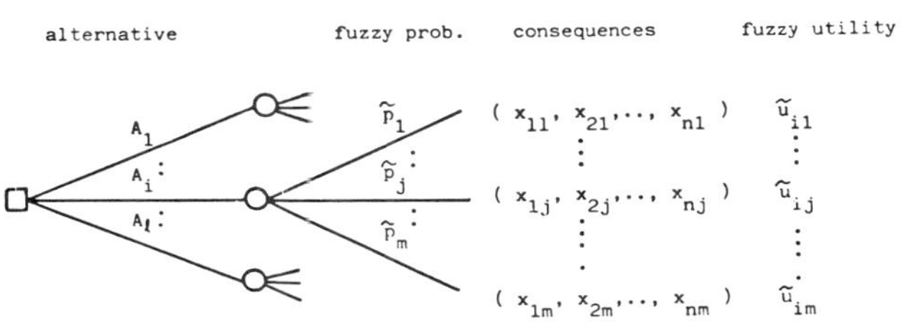

Fig. 1. Fuzzy decision tree.

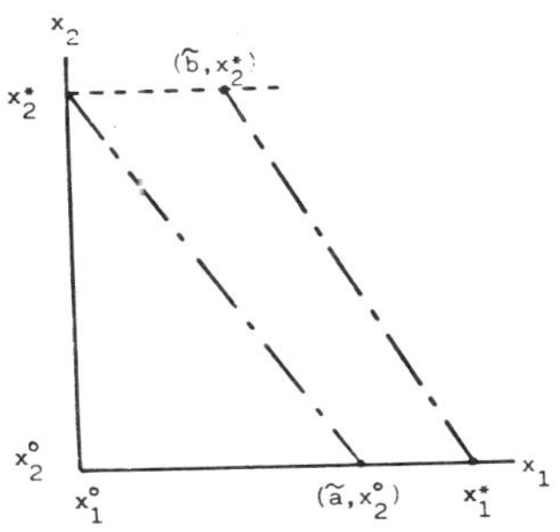

Fig. 2. Fuzzy indifferent points.

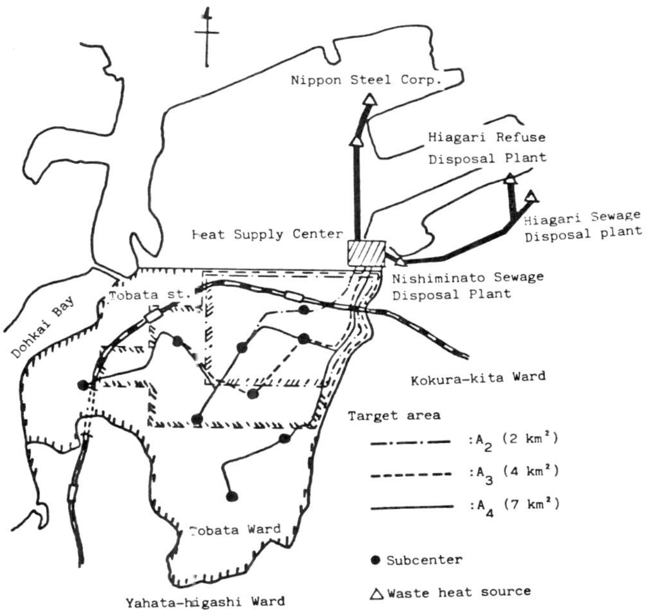

Fig. 3. Target area of CRECS and planned locations of HSC, Subcenter and waste heat sources.

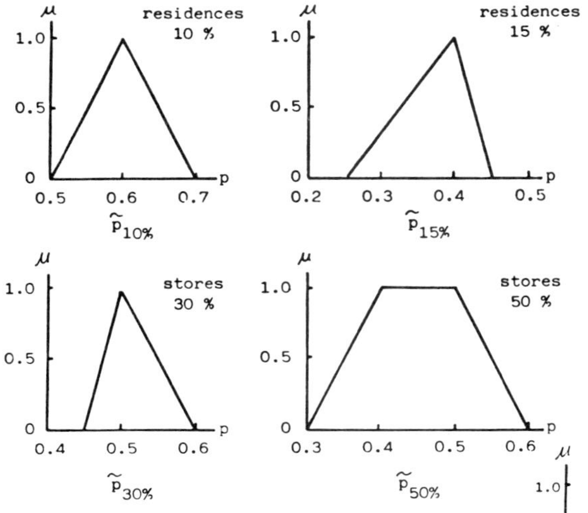

Fig. 4. Fuzzy probabilities of residences and stores served by CRECS.

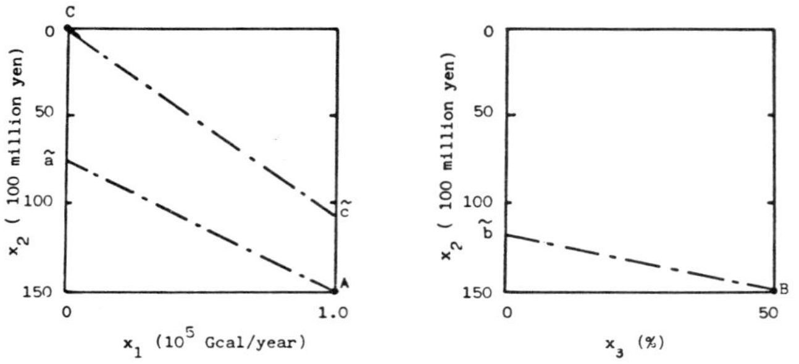

Fig. 5. Fuzzy indifferent points answered by decision maker.

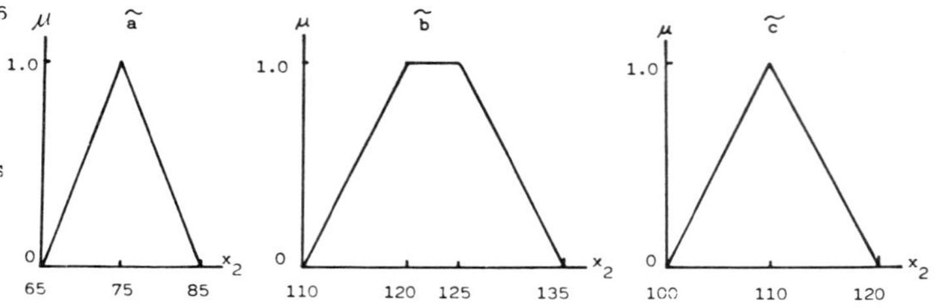

Fig. 6. Membership function of fuzzy indifferent points.

TABLE 2 Estimates of Attributes for CRECS Alternatives

Alternative	Residences (%)	Stores (%)	Maximum energy demand (Gcal/h)	Secondary pipeline (km)	Amount of energy savings (Gcal/y)	Total construction cost (10^8 yen)	Coverage ratio (%)
A_2	10	30	17.2	6.0	11,190	29.8	4.7
	10	50	21.0	7.0	13,786	35.9	6.0
	15	30	22.0	8.5	14,824	40.0	6.2
	15	50	26.8	9.5	17,420	46.4	7.4
A_3	10	30	32.1	10.5	20,202	54.1	9.0
	10	50	41.0	12.5	25,706	67.8	11.9
	15	30	41.5	14.0	26,165	70.4	11.4
	15	50	50.4	16.5	31,665	86.6	14.2
A_4	10	30	52.1	18.0	33,307	80.2	14.4
	10	50	65.7	20.5	41,954	100.0	18.9
	15	30	68.0	25.0	43,483	107.3	18.3
	15	50	81.5	27.5	52,130	128.7	22.8

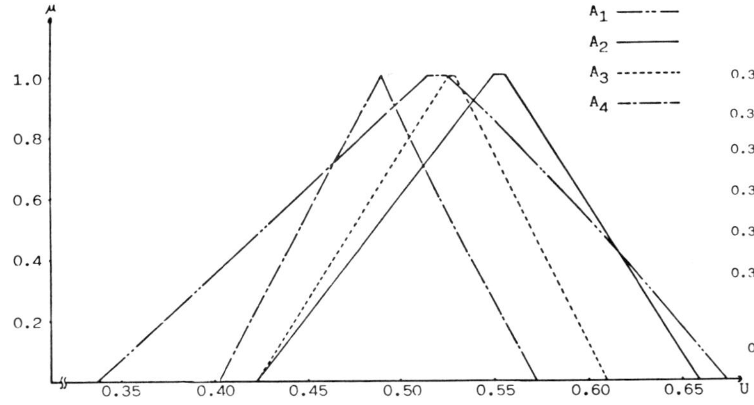

Fig. 7. Fuzzy expected utility for CRECS alternatives.

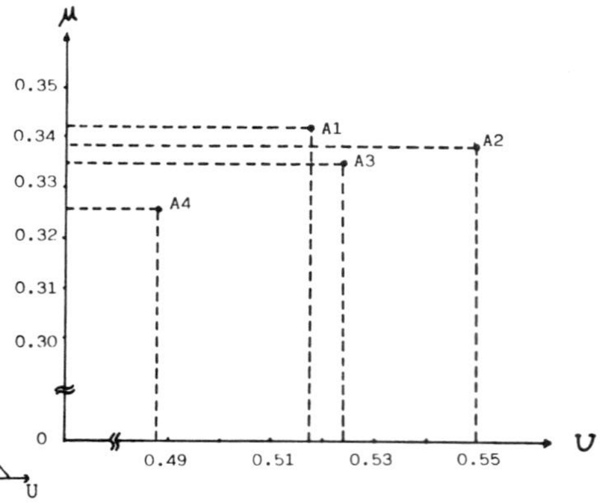

Fig. 8. Centroids of fuzzy expected utility for CRECS alternatives.

FUZZY CHOICE FUNCTIONS

S. V. Ovchinnikov

Department of Mathematics, San Francisco State University, San Francisco, CA 94132, U.S.A.

Abstract. An approach to choice function theory based on fuzzy set theory is suggested. A number of necessary and sufficient conditions on a fuzzy choice function to be a fuzzy rational choice function of a certain type are established.

Keywords. Choice function; mechanism of choice; fuzzy binary relation; fuzzy choice function.

INTRODUCTION

In general, choice function theory considers the following model (see, for example, Aizerman,Malishevski,I98I). Let A be a fixed finite set of alternatives. For each nonempty subset $X \subseteq A$ a nonempty subset $Y \subseteq X$ is chosen in accordance with some rule. In such a manner a choice function $Y = C(X)$ is given, which associates with each $X \subseteq A$ its subset $Y \subseteq X$. There are two different methods to describe the "entire choice" defined in this way. The first one points out a mechanism of choice whereby part Y is found from X. This method can be called an internal method. The second method indicates the set of all pairs (X,Y) and is called an external method.

Mostly, classical choice mechanisms are "pair-dominant" ones. This means that the choice of element $y \in X$ is made as a result of comparisons of this element with all $x \in X$. Some given structure on the set A is utilized to make these comparisons, for instance, a binary relation. The choice function thus arised has various attractive properties. One of the main problems in choice theory is a description of characteristic properties of choice functions. These properties, known as "choice axioms", separate the functions which have an equivalent description in pair-dominance optimization terms.

A framework of the choice function theory just described is an algebraic or, better to say, a non-probabilistic one. Moreover, it is non-deterministic in the sense that a subset C(X) chosen from X is assumed to be any subset, not necessarily a singleton.

There is another an approach to choice theory which may be described as follows. Let P(x,X) be the probability of choosing an element x from a set X. It is supposed that $P(x,X) \geqslant 0$, and

$$\sum_{x \in X} P(x,X) = I$$

i.e. P(x,X) defines a probability distribution on X. In practice, the choice probabilities P(x,X) are estimated by the relative frequences of choosing x. It is important to note, that on each trial one and only one alternative is chosen; this means deterministic character of choice mechanism involved. A good deal of research has been done in the area of probabilistic choice theory; the reader is refered to the Luce"s (1977) survey for further details and references.
In this paper an approach to choice theory is suggested which is probabilistic and non-deterministic. Empirically speaking, non-deterministic character of this approach means that the relative frequences of choosing x from X are not, already, estimations of a probability distribution on X. Let us consider, for example, the following data, where I indicates that a given alternative is chosen in a given trial, and 0 indicates that an alternative is rejected.

	Alternative				
Trial	I	2	3	4	5
I	0	0	I	I	0
2	0	I	0	I	0
3	0	0	I	I	0
Relative frequency	0	I/3	2/3	I	0

Obviously, any data of this kind is not consistent with the general probabilistic model described above: the relative frequences do not form a probability distribution on X. Nevertheless, we still have a nice probabilistic interpretation of these frequences: they are estimations of probabilities $P(x \in \underline{X})$ where $\underline{X}$ is a certain random subset of X. Since there is a one-to-one correspondence between classes of random subsets and fuzzy sets (see Orlov (I975) and Goodman (I982) for details), one may regard relative frequences obtained from non-deterministic choice experiments as estimations of membership function values of a fuzzy subset of X.

Generalizing this observations, we study fuzzy choice functions in this paper; these functions assign to each fuzzy set X a fuzzy subset $C_X \subseteq X$. Note again that a membership function is estimated by relative frequences obtained by averaging experimental data in non-deterministic choice experiments (for nonfuzzy X's). Hence, fuzzy set theory is used here as a mathematical tool providing consistent representation of empirical data in a stochastic model of choice.

BEST VARIANT CHOICE

It was mentioned above that there are two alternative approaches to the algebraic theory of choice. The first one is concerned with a mechanism of choice. In this paper only mechanisms based upon binary relations are considered. Let R be a binary relation on the set A. We read xRy as "x is preferred to or indifferent with y", i.e. R is a preference relation. A choice function based upon R is defined by

$$Y = C(X) = \{ x \in X | xRy \text{ for all } y \in X \} . \quad (I)$$

This mechanism is based on comparisons in pairs of variants (alternatives). Such "pair-dominance" mechanisms can be regarded as abstract forms of classical optimization mechanisms based on scalar and vector criteria. Various types of binary relations, such as partial orderings, weak orderings, etc., define, by (I), classes of choice functions possessing specific "rational" properties.

An alternative approach to choice theory considers "characteristic properties" of choice functions and the main problem is to describe combinations of characteristic properties which separate exactly the same classes as given by pair-dominant choice mechanisms.

Following Aizerman, Malishevski (I98I) we define main characteristic properties as follows:

I. Heritage (H): if $X' \subseteq X$, then $C(X') \supseteq C(X) \cap X'$.

2. Strict heritage (K) : if $X' \subseteq X$, and $X' \cap C(X) \neq \emptyset$, then $C(X') = C(X) \cap X'$.

3. Concordance (C) : if $X = X' \cup X''$, then $C(X) \supseteq C(X') \cap C(X'')$.

4. Independence (I) : if $C(X) \subseteq X' \subseteq X$, then $C(X') = C(X)$.

Most works concerned with choice theory unanimously declare these properties as clearly representing the idea of what is "better".

The following proposition represents main classical statements on correspondence between pair-dominant mechanisms and choice functions.

Proposition I. (Aizerman, Malishevski, I98I) For a choice function to be generated by choice mechanism (I) of i) an arbitrary binary relation, ii) a weak ordering, and iii) a quasi-transitive binary relation it is necessary and sufficient that it satisfies a condition i) H&C, ii) K, and iii) H&C&O.

This proposition is extended on fuzzy choice function theory in this paper.

FUZZY PREFERENCES

Recall that a fuzzy binary relation R on a set A is a fuzzy set with universe A×A and is defined by its membership function R(x,y).

Definition I. Fuzzy binary relation R is said to be
reflexive if $R(x,x) = I$ for all $x \in A$;
antireflexive if $R(x,x) = 0$ for all $x \in A$;
symmetric if $R(x,y) = R(y,x)$ for all $x,y \in A$;
antisymmetric if $R(x,y) > 0$ implies $R(y,x)=0$ for all $x \neq y$;
complete if $R(x,y) = 0$ implies $R(y,x) > 0$ for all $x,y \in A$;
acyclic if $R(x_i,x_{i+I}) > 0$ for $i = I,..,k-I$, implies $R(x_k,x_I) = 0$ for any sequence $x_I,...,x_k$.
transitive if $R(x,y) > 0$ and $R(y,z) > 0$ imply $R(x,z) > 0$ for all $x,y,z \in A$.

It should be noted that our definition of transitivity is different from standart ones (see Zadeh (I97I)). Fuzzy set theory admits various definitions of transitivity (Goguen, I967). Transitivity defined above may be regarded as the weakest one.

The notion of a strict preference plays a significant role in choice theory.

Definition 2. Let R be a fuzzy binary relation (a preference). A fuzzy binary relation P_R defined by

$$P_R(x,y) = \begin{cases} R(x,y), & \text{if } R(y,x) = 0, \\ 0 & \text{otherwise,} \end{cases}$$

is said to be a strict preference.

In choice theory aP_Rb if and only if not bRa, i.e. $P_R = R \cap \overline{R}^{-1}$ where $\overline{R}^{-1}$ is a complement of the converse relation R^{-1} considered as a subset of A×A. Definition 2 is based upon the intuitionistic negation

(Ovchinnikov, 1983) for which we have

$$P_R = R \cap \bar{R}^{-1}.$$

Definition 3. A fuzzy binary relation R is said to be

1) partial ordering if it is reflexive, antisymmetric, transitive fuzzy binary relation;
2) chain if it is a complete partial ordering;
3) ordering if it is reflexive, complete and transitive fuzzy binary relation;
4) quasi-transitive relation if it is reflexive, complete relation and P_R is a transitive relation.

The following structural properties of fuzzy binary relations are used in the paper.

Proposition 2. Let R be an ordering. Then
1) R is a qusi-transitive relation, and
2) $P_R(x,y) > 0$ and $R(y,z) > 0$ imply $P_R(x,z) > 0$, and $R(x,y) > 0$ and $P_R(y,z) > 0$ imply $P_R(x,z) > 0$.

We omit proofs of these statements which are similar to the crisp ones.

FUZZY PAIR-DOMINANT CHOICE FUNCTIONS

The following definition is an immediate extension of definition (1) and is in accordance with a general approach to a fuzzy decision-making developed by Bellman and Zadeh (1970).

Definition 4. A pair-dominant choice function based on a fuzzy binary relation R is a mapping which, to each fuzzy set X, assigns a fuzzy subset with a membership function

$$C_X^R(x) = \bigwedge_{y \in X} R(x,y) \wedge X(x) \qquad (2)$$

(We denote $y \in X$ iff $X(y) > 0$.)

One can also compare (2) with the definition of a fuzzy upper bound due to Zadeh (1971). Note, that if R and X are crisp sets then (2) is equivalent to (1).

Lemmas 1-3 establish general properties of fuzzy pair-dominant mechanisms. We denote $\text{car}X = \{x \in A \mid X(x) > 0\}$.

Lemma 1. $C_X^R \subseteq C_{\text{car}X}^R$ (3)

Proof follows immediately from (2).

Lemma 2. C_X^R fulfills the heritage property (H) $X' \subseteq X$ implies $C_{X'}^R \supseteq C_X^R \cap X$.

Proof. $C_{X'}^R(x) = \bigwedge_{y \in X'} R(x,y) \wedge X(x) \geqslant$

$$\bigwedge_{y \in X} R(x,y) \wedge X'(x) \wedge X(x) = C_X^R(x) \wedge X'(x).$$

■

Lemma 3. C_X^R satisfies the concordance property (C): $C_{X \cup Y}^R \supseteq C_X^R \cap C_Y^R$.

Proof. $C_X^R(x) \wedge C_Y^R(x) =$

$$[\bigwedge_{u \in X} R(x,u) \wedge X(x)] \wedge [\bigwedge_{v \in Y} R(x,v) \wedge Y(x)] =$$

$$\bigwedge_{y \in X \cup Y} R(x,y) \wedge X(x) \wedge Y(x) \leqslant$$

$$\bigwedge_{y \in X \cup Y} R(x,y) \wedge (X(x) \vee Y(x)) = C_{X \cup Y}^R(x).$$

■

Two last lemmas show that a fuzzy pair-dominant choice function satisfies the same properties as a crisp one (cf. Proposition 1) In addition to these properties, it satisfies very important property (3) which has no crisp analog. The role of this property will be clarified in the last section of the paper. We only note here that (3) has a quite clear interpretation: if x is chosen from a fuzzy set X then it should be chosen from the carrier of X and the degree of its belongness to C_X does not exceed of that to $C_{\text{car}X}$.

Lemma 4 gives some properties of fuzzy pair-dominant choice mechanisms following from main properties of fuzzy preferences.

Lemma 4. 1) Let R be a reflexive relation. Then

$$C_{\{x\}}^R = \{x\}. \qquad (4)$$

2) Let R be a reflexive and complete relation. Then

$$C_{\{x,y\}}^R = \begin{cases} R(x,y), & \text{if } u = x, \\ R(y,x), & \text{if } u = y, \\ 0 & \text{otherwise.} \end{cases}$$

3) Let R be reflexive, complete relation and P_R is an acyclic relation. Then

$$C_X^R \neq \emptyset, \text{ for any } X \neq \emptyset. \qquad (5)$$

Proof. 1) and 2) follow immediately from (2). 3) The proof is quite similar to the crisp one. ■

Corollary. $C_X^R \neq \emptyset$ for any nonempty X if R is an ordering or quasi-transitive relation.

The main properties of fuzzy pair-dominant choice functions based on particular types of fuzzy binary relations are given below.

Lemma 5. Let R be an ordering. Then $X' \subseteq X$ and $C_X^R \cap X' \neq \emptyset$ imply toghether

$$\text{car}C_{X'}^R = \text{car}C_X^R \cap \text{car}X'. \qquad (6)$$

Proof. By lemma 2 it is sufficient to prove that

$$\text{car}C_{X'}^R \subseteq \text{car}C_X^R \cap \text{car}X'.$$

Let $x \in \text{car}C_{X'}^R$ and $x \notin \text{car}C_X^R \cap \text{car}X'$, i.e.

$x \notin carC_X^R$. Then there is y such that $R(x,y)=0$, i.e. $P_R(y,x) > 0$. On the other hand, x belongs to $carC_{X'}^R$, which impies $y \in carX \setminus carX'$. Since $C_X^R \cap X' \neq \emptyset$ there is z such that $z \in carX'$ and $z \in carC_X^R$. Hence, $R(X,z) > 0$. By Proposition 2 we have $P_R(y,z) > 0$ which implies $R(z,y) = 0$. But $z \in carC_X^R$ which implies $R(z,y) > 0$. The contradiction completes the proof. ∎

Lemma 6. Let R be a quasi-transitive relation. Then

$$C_X^R \subseteq X' \subseteq X \text{ implies } carC_{X'}^R = carC_X^R \ . \quad (7)$$

Proof. By lemma 2 it is sufficient to prove that $carC_{X'}^R \subseteq carC_X^R$. Let us suppose that $x \in carC_{X'}^R$ and $x \notin carC_X^R$. Then there is y_I in X such that $P_R(y_I,x) > 0$. We have $y_I \notin carC_X^R$ implying $R(x,y) > 0$ for all $y \in X'$. Since $y_I \in carC_X^R$ there is y_2 such that $P_R(y_2,y_I) > 0$ which implies $P_R(y_2,x) > 0$, by transitivity of P_R. If y_2 does not belong to C_X^R then there is y_3 different from y_I and y_2 such that $P(y_3,x) > 0$ and so on. By finiteness of A we find y such that $P_R(y,x) > 0$ and $y \in carC_X^R$.

But $x \in carC_{X'}^R$ and $y \in carX'$, i.e. $R(x,y) > 0$ which contradicts to $P_R(y,x) > 0$. ∎

Note, that any fuzzy pair-dominant choice function fulfills properties (H) and (C). On the other hand, properties (6) and (7) are weaker than (K) and (O), respectively, although for crisp sets and preferences they coincide with them. Simple examples demonstrate that there are fuzzy orderings and quasi-transitive relations which do not satisfy (K) and (O). As it is mentioned in (Ovchinnikov, I98I) it is stipulated by pair-dominance choice function structure (2), "since this function considers not only the ties between alternatives but also their 'power'. Having excluded certain alternatives from consideration, we have naturally increased the degree of membership to the fuzzy set C_X for other alternatives."

CHARACTERIZATIONS OF FUZZY CHOICE FUNCTIONS

In this section characteristic properties of fuzzy choice fuctions are introduced. Various conjunctions of these properties define classes of choice functions based upon pair-dominant choice mechanisms.

Definition 5. The following -roperties of fuzzy choice functions are said to be characteristic properties:

I) Boundedness (B): $C_X \subseteq C_{carX}$;
2) Heritage (H) : $X' \subseteq X$ implies $C_{X'} \supseteq C_X \cap X'$;
3) Concordance (C) : $X = X' \cup X''$ implies $C_X \supset C_{X'} \cap C_{X''}$;
4) Fuzzy strict heritage (FK) : if $X' \subseteq X$ and $C_X \cap X' \neq \emptyset$, then $carC_{X'} = carC_X \cap carX'$;
5) Fuzzy independence (FO) : $C_X \subseteq X' \subseteq X$ implies $carC_{X'} = carC_X$;
6) Singleton law (S): $C_{\{x\}} = \{x\}$;
7) Nonvoideness (N) : $X \neq \emptyset$ implies $C_X \neq \emptyset$.

Note, that properties (B), (K), (O), (S), and (N) are the same as 3, 6, 7, 4, and 5, respectively.

Lemma 7. Conjunction (B)&(H) implies

$$C_X = C_{carX} \cap X \quad (8)$$

Proof. We have $C_X \supseteq C_{carX} \cap X$, by (H), and $C_X \subseteq C_{carX} \cap X$, by (B). ∎

Theorem I. A fuzzy choice function C_X is a fuzzy pair-dominant choice fuction for some fuzzy preference R iff C_X satisfies properties (B), (H) and (C).

Proof. Necessity follows from lemmas I-3. To prove sufficiency let us define R by

$R(x,y) = C_{\{x,y\}}(x)$.

By (H) and (8), $X \cap \{x,y\} \subseteq X$ implies

$$C_X \cap X \cap \{x,y\} \subseteq C_{X \cap \{x,y\}} = C_{\{x,y\}} \cap X \cap \{x,y\}$$

or $C_X \cap \{x,y\} \subseteq C_{\{x,y\}} \cap X$.

Hence,

$$C_X(x) \leq C_{\{x,y\}}(x) \wedge X(x),$$

which implies

$$C_X(x) \leq \bigwedge_{y \in X} C_{\{x,y\}} \wedge X(x) = \bigwedge_{y \in X} R(x,y) \wedge X(x) = C_X^R(x).$$

On the other hand,

$$X = \bigcup_{y \in X} [X \cap \{x,y\}]$$

which implies, by (8) and (C),

$$C_X \supseteq \bigcap_{y \in X} C_{X \cap \{x,y\}} = \bigcap_{y \in X} [C_{\{x,y\}} \cap \{x,y\} \cap X]$$

or

$$C_X(x) \geq \bigwedge_{y \in X} [C_{\{x,y\}}(x) \wedge X(x)] = C_X^R(x). \quad ∎$$

One can compare the statement of this theorem with statement I) of proposition I.

By theorem I the mapping F: $R \to C_X^R$ is a surjection of the set of all fuzzy binary relations onto the set of all fuzzy choice functions satisfying properties (B), (H) and (C). This mapping is not a bijection since there are binary relations with the same image under F. Let us define $R_I \sim R_2$ if and only if $C_X^{R_I} = C_X^{R_2}$. Then $\sim$ is an equivalence relation on the set of all fuzzy preferences. Each fuzzy choice function satisfying (B), (H) and (C) is an image of some class of the relation $\sim$ under the mapping F. Theorem 2 describes all fuzzy preferences $R \in F^{-1}(C_X)$ for a given C_X.

Theorem 2. Let us define R_c for any given R by

$$R_c(x,y) = R(x,y) \wedge R(x,x).$$

Then $R \sim R_c$, and $R' \sim R''$ iff $R'_c = R''_c$.

Proof. We have

$$C_X^{R_c}(x) = \bigwedge_{y \in X} [R_c(x,y) \wedge X(x)] =$$

$$\bigwedge_{y \in X} [R(x,x) \wedge R(x,y) \wedge X(x)] =$$

$$\bigwedge_{y \in X} [R(x,y) \wedge X(x)] = C_X^R(x)$$, i.e $R \sim R_c$.

Let $R'_c = R''_c$; then $R' \sim R'_c = R''_c \sim R''$ which implies $R' \sim R''$, by transitivity of $\sim$.

Let $R' \sim R''$, i.e. $C_X^{R''} = C_X^{R'}$. Then

$$R'_c(x,y) = R'(x,y) \wedge R'(x,x) = C^{R'}_{\{x,y\}}(x) =$$

$$C^{R''}_{\{x,y\}}(x) = R''(x,y) \wedge R''(x,x) = R''_c(x,y).$$ ∎

One can consider R_c as a "cannonical representative" in the class of $\sim$ containing R. These cannonocal representatives are completeley characterized by the property

$$R_c(x,y) \le R_c(x,x).$$

Note, that this condition is always stisfied for reflexive binary relations. We have the following

Corollary. The mapping F is a bijection of the set of all reflexive fuzzy binary relations onto the set of all fuzzy choice functions satisfying properties (B), (H), (C) and (S).

In general, it is possible for C_X^R to be an empty set for some nonempty X. It was mentioned already that acyclicity implies nonvoideness of a choice from nonempty sets. The converse is also true.

Theorem 3. Let R be a reflexive and complete fuzzy binary relation. Then C_X^R is a nonempty fuzzy set for all nonempty X iff P_R is an acyclic fuzzy binary relation.

Proof. The necessity follows from lemma 4. Let $C_X^R \neq \emptyset$ for all $X \neq \emptyset$. Suppose that P_R is not an acyclic relation. Then there is a sequence $x_I,\ldots,x_n$ such that

$$P_R(x_i,x_{i+I}) > 0 \text{ for } i = I,\ldots,n-I, \text{ and}$$
$$P_R(x_n,x_I) > 0. \tag{9}$$

By definition 4, we have

$$C^R_{x_I,\ldots,x_n}(x) = \bigwedge_{i=I}^{n} R(x,x_i)$$

for $x \in \{x_I,\ldots,x_n\}$.

We have $R(x_{i+I},x_i) = 0$ for $i = I,\ldots,n-I$ and $R(x_I,x_n) = 0$, by (9). Hence, $C^R_{\{x_I,\ldots,x_n\}} = \emptyset$. This contradiction completes the proof. ∎

We will study now conditions determining the class of fuzzy choice functions having an equivalent description in terms of a pair-dominant choice mechanism based on a quasi-transitive fuzzy binary relation.

Theorem 4. A fuzzy choice function C_X is a pair-dominant choice function C_X^R based on a fuzzy quasi-transitive relation R iff it satisfies conditions (B), (H), (C), (FO), (N) and (S).

Proof. The necessity follows from lemmas I-4 and 6. Let C_X satisfies the conditions listed in the theorem. By theorem I, we have $C_X = C_X^R$ for some R. By (S) and (N), R is a reflexive complete fuzzy binary relation. It is sufficient now to prove that P_R is a transitive relation. Let $X = \{x,y,z\}$. Then

$$C_X^R(t) = R(t,x) \wedge R(t,y) \wedge R(t,z) \quad \text{for } t \in X.$$

Hence,

$$C_X^R(x) = R(x,y) \wedge R(x,z),$$

$$C_X^R(y) = 0, \text{ if } P_R(x,y) > 0, \text{ and}$$

$$C_X^R(z) = 0, \text{ if } P_R(y,z) > 0.$$

By (N), $C_X^R(x) > 0$ which implies $R(x,z) > 0$ and $\text{car}C_X^R = \{x\}$. Let now $X' = \{x,z\}$. Then $C_{X'}^R(t) = R(t,x) \wedge R(t,z)$ for $t \in \{x,z\}$.

Hence, $C_{X'}^R(x) = R(x,z)$ and $C_{X'}^R(z) = R(z,x)$.

Now, by (FO), $C_X^R \subseteq X' \subseteq X$ implies $\text{car}C_{X'}^R = \text{car}C_X^R = \{x\}$ which implies $R(z,x)=0$.

Hence, $P_R(x,z) > 0$. ∎

As it follows from proposition I in the crisp case the property (K) is a very strong one. The power of this property shows itself very clearly in the fuzzy case too. Let C_X fulfils conditions (K) and (S). Let X be a crisp set and $x \in C_X$. Then $\{x\} \subseteq X$ and $C_X \cap \{x\} \neq \emptyset$. By (K) we have $C_{\{x\}} = C_X \cap \{x\}$, or $C_X(x) = C_{\{x\}}(x) = I$, by (S), i.e. C_X is a crisp set. From (K) it also follows immediatelly that $C_X = C_{carX} \cap X$. (I0)

Hence, C_X is, essentially, a crisp choice function which coincides with C_X^R for some crisp ordering R, by proposition I and (I0).

On the other hand it was proven in lemma 5 that fuzzy orderings satisfy the condition (FK) which coincides with (K) for crisp sets and orderings.

We complete this section by the following

Theorem 5. A fuzzy choice function C_X is a pair-dominant choice function based on a fuzzy ordering iff it satisfies conditions (B), (H), (C), (FK), (N), and (S).

Proof. The necessity follows from lemmas I-5. By (B), (H), (C), (N), and (S) we have $C_X = C_X^R$ where R is fuzzy reflexive complete relation. Let us prove that R is a transitive relation, i.e. that $R(x,y) > 0$ and $R(y,z) > 0$ imply $R(x,z) > 0$. Let $X = \{x,y,z\}$. Then

$C_X^R(x) = R(x,y) \wedge R(x,z)$,

$C_X^R(y) = R(y,x) \wedge R(y,z)$, and

$C_X^R(z) = R(z,x) \wedge R(z,y)$.

Let now $X' = \{x,y\}$. Then

$C_{X'}^R(x) = R(x,y)$ and $C_{X'}^R(y) = R(y,x)$.

If $C_X^R \cap X' = \emptyset$, i.e. $carC_X^R = \{z\}$,

then $R(x,z) = R(y,x) = 0$. By (N),

$C_X^R(z) > 0$ which implies $R(z,x) > 0$ and

$R(z,y) > 0$. Let $X'' = \{y,z\}$. We have

$C_{X''}^R(y) = R(y,z) > 0$, and

$C_{X''}^R(z) = R(z,y) > 0$.

Since $C_X^R \cap X'' \neq \emptyset$, then, by (FK),

$carC_{X''}^R = carC_X^R \cap \{y,z\}$. But $carC_{X''}^R = X''$

which contradicts $carC_X^R = \{z\}$.

Hence, $C_X^R \cap X' \neq \emptyset$. Then, by (FK), $carC_{X'}^R = carC_X^R \cap X'$. We have $x \in carC_{X'}^R$ since $R(x,y) > 0$. Hence, $x \in carC_X^R$ which implies $R(x,z)$. ∎

CONCLUSION

Fuzzy choice theory described above has some characteristic features which distinguish it from the crisp one. The difference is mainly stipulated by the purely fuzzy property (B). For instance, there are many "pathological" fuzzy choice functions which satisfy (H) and (C) and do not satisfy (B).

From fuzzy set theory point of view there is a significant difference between classical properties (H), (C) (O) and (K). The properties (H) and (C) play the same role in both fuzzy and crisp cases. It seems that various transitivity properties which play a great role in classical choice theory, are not so important in general fuzzy choice theory.

Only pair-dominant mechanisms based on fuzzy preferences are considered in this paper. It is an interesting problem to study different mechanisms of choice based, for example, upon fuzzy utility functions. We leave this study for further publications.

REFERENCES

Aizerman, M.A. and Malishevski, A.V. (I98I) General theory of best variant choice: some aspects. IEEE Transactions on Aut. Control, vol.AC-26, no.5, I030-I040.

Bellman, R.E. and Zadeh, L.A. (I970). Decision-making in a fuzzy environment. Management Sci., I7, 8I4I-8I64.

Goguen, J.A. (I967). L-fuzzy sets. J.Math.Anal.Appl., I8, I45-I74.

Goodman, I.R. (I982). Fuzzy sets as equivalence classes of random sets. In R.R.Yager(ed.), Recent Developments in Fuzzy Set and Possibility Theory. Pergamon Press, New York.

Luce, R.D. (I977). The choice axiom after twenty years. J.of Math.Psych. I5, 2I5-233.

Orlov, A.I. (I975). Foundations of fuzzy set theory. In Algorithms of Multivariate Statistical Analysis and Applications. Central Econom.Math. Institute, pp.I69-I75 (in Russian)

Ovchinnikov, S.V. (I98I). Structure of fuzzy binary relations. Fuzzy Sets and Systems, 6, I69-I95.

Ovchinnikov, S.V. (I983). General negations in fuzzy set theory. J.Math.Anal.Appl., 9I (in print)

Zadeh, L.A. (I97I). Similarity relations and fuzzy orderings. Inf.Sci., 3, I77-200.

A VALUATION MODEL OF SUBJECTIVE SPACES

C. Rolland-May

Department of Geography — University of Metz, Ile du Saulcy, Metz, France

Abstract. In the following work we propose an application of the fuzzy subset theory to the study of subjective urban spaces. So A valuation model of mental maps is presented. Its term measure for each spatial component the localization wich can be more or less fuzzy, the eventual distorsion of area, length or orientation relatively to the geographical space. Then we define the fuzzy spatial information matrix of a set of subjects. Terms for valuation of the individual spatial perception process are proposed.

Keywords. Fuzzy subsets theory ; urban systems ; mental maps ; geographical spatial analysis

INTRODUCTION

In this research that we define as subjective space, is Geographical space such as perceived by a special subject and restored by itself in a oral document - as oral replies to a questionnaire - or as written replies like literary descriptions of a given space - or geographical descriptions, like mental maps.
The subjective space consequently - through definition - is imprecise, fuzzy, nay, erroneous with respect to geographical space. On that account, the action of valuation of a subjective space consists in proposing a likely measure in order to seize this fundamental character of imprecision.

In the following matter we present a valuation model of urban subjective space represented by a mental map, using the fuzzy subsets theory. At once, we define this model and the array that we name the fuzzy spatial information matrix of a set of subjects. Next we present several indexes adaptable to measure the process of spatial perception of each subject and consequently to allow comparative studies and studies of fuzzy spatial behaviours.

VALUATION MODEL OF SUBJECTIVE SPACE

Method of Approach

Let $\underline{E}_G$: be a given geographical space [1]

E_{S_i} : the subjective space of the subject number i

c_j : a spatial component of the geographical space.

we can define $\underline{E}_G = \{c_j\}$ $j \in \underline{P}$ where $\underline{P}$ is a set of indices, as a finite set of spatial components.
Table 1 describes the different components of urban space.
The question is here to define an application μ from a geographical space $\underline{E}_G$ to $\underline{M}$, where $\underline{M} = [0,1]$ and E_S a fuzzy subset of $\underline{E}_G$, such that :

$$\mu : \underline{E}_G \longmapsto \underline{M}$$
$$\forall c_j \in \underline{E}_G \longmapsto \mu_{E_S}(c_j) \in \underline{M}$$

So we can formally define a subjective space E_S as a fuzzy subset of $\underline{E}_G$

$$E_S = \{(c_j, \mu_{E_S}(c_j)); \forall c_j \in \underline{E}_G : \mu_{E_S}(c_j) \in \underline{M}\}$$

It means that one supposes that a spatial component belongs more or less to a subjective space. The degree of membership of each spatial component C_j to the subjective space E_S can be posed as following:

$\mu_{E_S}(c_j) = 0$ if the c_j component hasn't bee represented in the subjective space.

$\mu_{E_S}(c_j) = 1$ if the c_j component has been represented in the subjective space without localisation error or distorsion in respect to the geographical space.

$0 < \mu_{E_S}(c_j) < 1$ if c_j has been represented : wether with localisation error or distorsion of shape, area or orientation in respect to geographical space, wether not precisely, but in a fuzzy zone where it should be found. In this case it is possible to studie the case of fuzzy knowledge or fuzzy perception of geographical space.

Now we have to determine the value of $\mu_{E_S}(c_j)$ in the terms of the model wich is below proposed. We elabore for this a checkerboard locating map wich is applied on the geographical and on the mental map and wich allows us to measure the error or fuzziness of loca-

1 It will be underlined in the study the non fuzzy sets : Ex : $\underline{E}_G$ and it will be written such as fuzzy subsets : Ex : E_S

degree of membership of the c_j's component to the E_{s_i}'s subjective space.

So $m_{ij} = \mu_{E_{si}}(c_j)$

and $0 \leqslant m_{ij} \leqslant 1$, according to the valuation model. We call M the fuzzy spatial information matrix. We can used it to define the three following spatial perception indexes.

Spatial Integration Coefficient

It means a valuation of an individual subjectif space in respect to that we call a basis subjective space wich is defined by the researcher.

Basic subjective space. We define a basic subjective space E_{s_b} as a fuzzy subset of $\underline{E}_G$ where the degree of membership of each spatial component is fixed according to a level wich can be for example the degree minimum of spatial perception. We define an application

$$f : \underline{E}_G \longmapsto \underline{M}$$
$$\forall c_j \in E_G \longmapsto f_{E_{s_b}}(c_j) \in \underline{M}$$

f (c_j) represents the degree of membership of each spatial component to the basic subjective space. It is estimated by the researcher according to his own preoccupations. So he can estimate fundamental the perception of spatial landmarks like a cathedral, a monument... and make heavier their basis function, or - as in geography - he can estimate that spatial components which explain the urban dynamic, as streets, renovation zones..., are more important and make heavier their basis function. So the basic subjective space can be considered as a basis reference in a given research.

The Spatial integration coefficient of a i subject corresponds to following valuation :

$$C_i = \bigvee_{j=1}^{p} \left(\mu_{E_{s_i}}(c_j) \wedge f(c_j)\right)$$

with $0 \leqslant C_i \leqslant 1$

This coefficient explains spatial perception of a subject relatively the basis perception defined previously. It allows to quantify the individual integration of a subject to a geographical space and, contengently, to study quantitatively the evolution of this integration.

MIN and MAX Relative Position Index

It explains the position of an i subject, particularly its spatial perception in respect to spatial perception of the set of subjects. Min and max spatial structures. We define as weak subjective space, or MIN spatial structure the fuzzy subset :

$$E_f = \left\{(c_j, \mu_{E_f}(c_j))/\mu_{E_f}(c_j) = \bigwedge_{i=1}^{n} \mu_{E_{s_i}}(c_j)\right\} \quad (6)$$

E_f is such as each spatial component is associated the MIN of the degree of membership in the column number j in the fuzzy spatial information matrix. It is the urban space wich "everybody knows", with the important precision that it is differencied : each component is knew with a given degree by the group, wich is different from another degree.

In the same way we define strong subjective space or MAX spatial structure, the fuzzy subset :

$$E_F = \left\{(c_j, \mu_{E_F}(c_j))/\mu_{E_F}(c_j) = \bigvee_{i=1}^{n} \mu_{E_{s_i}}(c_j)\right\} \quad (7)$$

It explains the MAX of the degree of perception of the spatial components. We suppose that it explains the MAX level of perception that the studied group can have from the urban space.

The Min relative position index of a i subject corresponds to the distance - defined here as a Hamming generalized relative distance as explained by Kaufmann (1977) between the fuzzy subset E_{s_i} and the MIN spatial structure E_f.

$$I_{Min_i} = \frac{1}{p} \sum_{j=1}^{p} \left| \mu_{E_{s_f}}(c_j) - \mu_{E_{s_i}}(c_j) \right| \quad (8)$$

The MAX relative position index of a i subject corresponds to the Hamming generalized relative distance between the fuzzy subset E_{s_i} and the MAX spatial structure E_F

$$I_{Max_i} = \frac{1}{p} \sum_{j=1}^{p} \left| \mu_{E_{s_F}}(c_j) - \mu_{E_{s_i}}(c_j) \right| \quad (9)$$

This two relative position indexes precise the position of a subject relatively to the min and max spatial perception of the set of subjects.

Spatial Conquest Coefficient

It precises for a i subject, the spatial perception process in respect to the one of entire group.

Fuzzy belt of spatial integration. Let's realise on E_{S_F} the α-cut (α has been chosen by the researcher); we define so the ordinary subset of $\underline{E}_G$

$$\underline{E}_{F_\alpha} = \left\{c_j \,/\, \mu_{E_{S_F}}(c_j) \geqslant \alpha\right\} \quad (10)$$

and its complementary in $\underline{E}_G$

$$\overline{\underline{E}_{F_\alpha}} = \left\{c_j \,/\, \mu_{E_{S_F}}(c_j) < \alpha\right\} \quad (11)$$

$\underline{E}_{F_\alpha}$ definies a strong spatial structure on threshold α.

Let's realise now on E_s in the same way a β-cut and define the ordinary subset of $\underline{E}_G$ called a weak spatial structure on β threshold

$$\underline{E}_{f_\beta} = \left\{c_j \,/\, \mu_{E_f}(c_j) \geqslant \beta \text{ with } \beta < \alpha\right\} \quad (12)$$

we define a fuzzy belt of spatial integration on $\alpha\beta$ level, the ordinary subset of $\underline{E}_G$:

$$\underline{F}_{\alpha\beta} = \left\{c_j \,/\, \mu_{E_F}(c_j) < \alpha \text{ and } \mu_{E_f}(c_j) \geqslant \beta\right\} \quad (13)$$

Consequently :

$$c_j \in \underline{F}_{\alpha\beta} \Longleftrightarrow c_j \in \overline{\underline{E}_{F_\alpha}} \cap \underline{E}_{f_\beta}$$

lisation, configuration... of each spatial component.

The Valuation Model of subjective space.

Its general expression is :
$\forall j, j \in P$, P being a set of indices

$$\mu_{E_s}(c_j) = L_f * D_f * O_r / D \quad (1)$$

with
L_f : fuzzy locating index ; it values the character of fuzziness of the component's localisation. It means its possible localisation in a fuzzy zone of the subjective space.
D_f : distorsion index; it values the deformation of the surface of an area component or of the length of an axial component.
O_r : orientation index, values the modification by the subject of the orientation of area or axial spatial component in respect to its "geographical" orientation.

Consequently, the developped model is written as following

$$\mu_{E_s}(c_j) = \frac{\left(1 - \frac{H_L}{H_G}\right)\left(1 - \frac{|m_d - m_r|}{m_d}\right)\left(|\cos\alpha|\right)}{D} \quad (2)$$

with $D > 0$

The Terms of the Valuation Model.

Fuzzy locating index. The developped expression is

$$L_f = 1 - \frac{H_L}{H_G} \quad (3)$$

where H_G ist defined as general entropy
H_L ist defined as local entropy
according to Walliser (1977).
The localization of punctual spatial component (= the component which the area doesn't exceed 1 in checkerboard locating map), of the gravity center of area component or of the midst of axial component :

- is equiprobable between the n squares of the checkerboard locating map if the component is unkwoned to the subject. Then we definie for this component, the notion of general entropy wich is maximal $H_L = -\log_n$

If the spatial component isn't fuzzy, it means, if a punctual component, or the center of gravity of an area one, or the midst of an axial one, are localized in one precise square of the checkerboard locating map, then we define the local entropy, wich is minimal.
$H_L = -\log 1 = 0$

If the component exists but is represented in a fuzzy zone, the m squares of this zone have an equal probability to contain the gravity center or the midst of the component, or the punctual component.
Then

$$H_L = -\log m$$

The possible values of the fuzzy locating index are summarized in table 2.

Distorsion index. The developped expression is

$$D_f = 1 - \frac{|m_d - m_r|}{m_d}$$

with m_d : "subjective" length or area (measured in number of squares of the checkerboard locating map) of an axial or an area component or the mental map,
and m_r : "geographical" length or area of the Same component on the geographical map. This terms come to seize distorsions of the areas or of the lengths of non fuzzy components.

In case of fuzzy representation, it is naturally impossible to take this measures. The loss of information is then sanctioned in this way : m_d becomes the measure of the area of the fuzzy zone of localization in case of area component or of the length of the longest axis of the fuzzy zone in case of axial component.

The possible values of distorsion index are summarized in table 3.

Orientation index : the developped expression is :

$$O_r = |\cos\alpha|$$

It explains the modification of orientation of the subjective component in respect of geographical ore. α is the angle wich explains this modification. In case of fuzzy representation, the angle is determined between geographical orientation of the component and the general orientation of the fuzzy zone.
In case of the greatest possible distorsion of orientation, α = 90 deg, we propose $O_r = 0.01$

Distance : D is the distance between subjective and geographical spatial component (indeed the distance between the representative points of these component as showned abowe). In this example we have chosen the rectilinear distance, because its fully adapted to orthogonal plan of city center wich we studie, but it's obvious that all other definition of a distance is possible in another research - In any case $D \geqslant 1$.

Summary Table of Valuation Model of Subjective Space.

Table 4 summaryzes the different terms of the valuation model and deduces the degree of membership of a spatial component to a subjective space .

PROPOSITION OF APPLICATION SPATIAL PERCEPTION INDEXES

Definition of the Fuzzy Spatial Information Matrix.

This work of valuation concerns in fact a set of n subjects wich have realized n mental maps of an urban space.
Let $E_G = \{c_j\}$, be the geographical space,
E_G is a set of p spatial components,
so we have M, a matrix consisting on n rows and p columns, were the (i, j) element is the

with Card ($\underline{F}_{\alpha\beta}$) = K < Card ($\underline{E}_G$).

Geographicaly, $\underline{F}_{\alpha\beta}$ represents a more or less large spatial belt, continuous or not, where the process of perception of the whole group has overtaken a β threshold and reached a α threshold for same subjects. It represents a space which is in the cours of integration by the studied group.

The spatial conquest index of a i subject precises his MAX degree of perception relatively of the fuzzy edge of spatial integration of $\alpha\beta$ level.

$$S_i = \mu_{E_{s_i}}(c_j)/c_j \in \underline{F}_{\alpha\beta}$$

$$\text{and } \mu_{E_{s_i}}(c_j) = \bigvee_{j=1}^{K} \mu_{E_{s_i}}(c_j) \qquad (14)$$

So, relatively to $\underline{F}_{\alpha\beta}$, each interrogated subject is defined by its own spatial position value. This measure precises the position of each subject in a sub-space itself in course of integration by the group.

CONCLUSION

The presented valuation model of subjective space is certainly perfectible by multiplying or improving of the terms. So it will be preciser but will loose its handiness. The examples of application are diversified. We have presented three terms as may be of interest to economical, geographical or psychological studies : they try to value the individual or collective process of spatial perception trough quantification of individual of collective subjective spaces.

REFERENCES

Dubois, D., Prade, H. (1980). Fuzzy sets and Systems theory and Applications. Academic Press, New York.

Kaufmann, A. (1977). Introduction à la theorie des sous-ensembles flous. Masson, Paris.

Bailly, A., Beguin, H. (1982). Introduction à la géographie humaine. Masson, Paris

Beguin, H., Thisse, J. (1979). An Axiomatic Approach to geographical Space. Geographical Analysis, 11, 325-341.

Ponsard, C. (1980). Fuzzy Economic Spaces. I.M.E. document de travail, Dijon.

TABLE 1 Components of Urban Space

Components	Geographical meaning	Geomatrical formalization	Examples
Static spatial components	Elementary component	Punctual comp. Area component Axial component	Buildings, churches, monuments Places, parks, parkings Barrier, rail track, stream
Spatial components of flows	Component of geographical structure	Punctual c. Area component Axial component	Cross roads, markets, forum Streets
Components of spatial dynamic	Geographical process	Area component Axial component	Urban renovation zones, urbanized zones Links with sub-urban zones

TABLE 2 Values of the Fuzzy Locating Index

	H_G	H_L	L_f
Non fuzzy component	Log_n	$m = 1$ $Log_m = 0$	1
Fuzzy component	Log_n	$1 < m < n$ $H_L = log_m$	$0 < L_f < 1$
Non existent component	Log_n	Log_n	0

TABLE 3. Values of the Distorsion Index

Type of Component		m_d	m_r	D_f
Area Comp.	Non Fuzzy	area of subj. comp.	real area	$D_f=1$ if $m_d=m_r$ $0<D_f<1$ if not
	Fuzzy	area of fuzzy zone	real area	$0<D_f<1$
	Non Existent	0		D_f is not seized in the model
Axial Comp.	Non Fuzzy	length of subj. comp.	real length	$D_f=1$ if $m_d=m_r$ $0<D_f<1$ if not
	Fuzzy	Max. length of fuzzy zone	real length	$0<D_f<1$
	Non Existent	0		D_f ist not seized in the model

TABLE 4 Valuation Model of Subjective Space

Expression of the model : $\mu_{E_S}(c_j) = L_f * D_f * O_r / D$ – where $D > 0$
– in the cases 11, 12, 13, 21, 31, D_f and O_r are not valued.

Components \ Terms		Fuzzy Locating Index L_f	Distorsion Index D_f	Orientation Index O_r	$V = \mu_{E_S}(c_j)$
Punctual Component	11	$L_f = 0$	—	—	V=0 No existent component.
	12	$L_f = 1$ no fuzzy locating			V = 1/D : if D=1 V=1 if D > 1 0 < V < 1
	13	$0 < L_f < 1$ fuzzy locating			0 < V < 1
Area Component (Taking into account of gravity center)	21	$L_f = 0$	—	—	V=0 No existent component.
	22	$L_f = 1$ no fuzzy locating	$0 < D_f \leqslant 1$ (areas relation)	$0 < O_r \leqslant 1$	$0 < V \leqslant 1$
	23	$0 < L_f < 1$ fuzzy gravity center	$0 < D_f < 1$ (Relation between geographical and fuzzy area)	$0 < O_r \leqslant 1$	0 < v < 1
Axial Component	31	$L_f = 0$	—	—	V=0 No existent component
	32	$L_f = 1$ no fuzzy locating	$0 < D_f \leqslant 1$ (lengths relation)	$0 < O_r \leqslant 1$	$0 < V \leqslant 1$
	33	$0 < L_f < 1$ fuzzy midst of axis	$0 < D_f < 1$ (Relation between geogr. and fuzzy axis)	$0 < O_r \leqslant 1$ Extension axis of fuzzy zone	0 < V < 1

PLANNING IN MANAGEMENT BY FUZZY DYNAMIC PROGRAMMING

T. Terano*, M. Sugeno and Y. Tsukamoto****

**Faculty of Engineering, Hosei University, Koganei, Tokyo, Japan*
***Department of Systems Science, Tokyo Institute of Technology, Ohokayama, Meguro, Tokyo, Japan*

Abstract. Mathematical Programming is a useful tool for the planning in management. However, management is an ill-defined system and the building of its mathematical model is difficult. Since "The Principle of Optimality" in Dynamic Programming is still effective in such an ill-defined system under some assumptions, fuzzy dynamic programming can be applied to management. In this paper, the authors discuss basic problems in its applications to real planning and suggest a general idea of fuzzy dynamic programming.

Keywords. Dynamic programming; Management system; Man-machine system; Operations research; Set theory.

INTRODUCTION

Mathematical programming is a powerful tool for managerial problems in industry and business. Especially dynamic programming (D.P.) is effective for an optimal control and also for an optimum planning of dynamical systems. In order to apply D.P., a complete mathematical model is needed, that is, the objective system should be a white-box for designer. However, in case of the execution of managerial plan, a manager makes a rough plan and gives instructions linguistically to clerks who carry out those by their own judgement. Since the instruction, its executing process and the result are not defined exactly nor expressed numeically, the construction of a mathematical model is infeasible. So, management is an ill-defined system, and is a gray box which is very difficult to be handled with ordinary D.P.. Since the principle of optimality of D.P. is essentially described in words, it is not difficult to combine both theories. Also we should note that the exact optimal solution is not favored in real management. The substantially optimal plan obtained by fuzzy dynamic programming is very suitable, because it admits human interpretation and interference. In this aspect, fuzzy dynamic programming (F.D.P.) is not a mathematical tool but a decison-support-system in management.

PROBLEMS IN FUZZY DYNAMIC PROGRAMMING

Bellman and Zadeh [2] first studied fuzzy dynamic programming as to the case when the objective system is deterministic but its goal and constraints are fuzzy. Since then, some other studies about fuzzy automata have been published [3] in which the state is finite and deterministic but the transition of state is fuzzified. Recently many papers [1] deal with the case where the state, transition, control, goal and constraint are all fuzzified. However, if we try to apply these theories to some real problems, there still remain many difficulties we must overcome, and there are also many different ways of modelling which puzzle us.

Before discussing difficulties, we point out some expectable merits of F.D.P.. (i) Planning or policy-making in management can be nearly optimized. (ii) The solution is qualitative, flexible, robust and easily interfered. (iii) The exact model of an objective system is not needed. (iv) Even when the rules describing the system are incomplete, they can be supplemented by inference. (v) Multi-variable schedule which is hardly planned by human intuition can be made easily.

Next, we mention the very important assumptions which must be satisfied in order that the principle of optimality holds well. (1) The state-transition should be a Markovian Process. In other words, the future state of the system depends only on the present state and the control policies adopted from now. (2) The performance-index should be additive, i.e., the sum of the evaluations in each step. Fortunately these premises are satisfied qualitatively in most of the managerial problems.

Now, we turn to discuss the problems that are necessary for the formulation of F.D.P.. They are classified into three categories; definition, modelling and discretization. These are rather conceptual and philosophical than mathematical. In the ordinary D.P., there are six important items; Step, State, Control, Constraint, Transition and Performance. Among them, State and Transition are expressed discretely, for example, with a directed graph. This is not changed in F.D.P., but six items are related so closely to each other that it is difficult to define them independently.

First, we consider fuzzy state. State variables should be chosen so as to effectively represent the progress of a plan of which measure is usually qualitative and linguistic. The fuzziness of a state is due to two reasons, one of which is the lack of a numerical measure and the other is the vagueness of measuring time. The latter is closely related to the definition of fuzzy step.

The number of states of a system should be finite in D.P. regardless they are fuzzy or not. However, new states are generated during D.P. calculation. These must be classified into some pre-determined states.

The definition and the evaluation of fuzzy control (strategy) is similar to those of fuzzy state, that is, a control variable is a feasible policy and its measure is its effect on the next state in a step. This means that fuzzy control is related to fuzzy transition. If an initial fuzzy state and a fuzzy control are given at the beginning of a step, the next fuzzy state is determined according to fuzzy transition rules. These rules are usually given in natural language.

Fuzzy constraint can be defined in the same way as fuzzy state or fuzzy control. However, we suggest here that fuzzy constraints had better be included in fuzzy transition, that is, the controls or the states which contradict the constraints should be eliminated beforehand from the transition rules.
The last problem is fuzzy step (phase). The selection of steps in ordinary D.P. is only a technique of calculation. However, this is very essential in F.D.P. because it has an effect on the definition of fuzzy state, fuzzy control and fuzzy transition. Besides, fuzzy step can be considered as a time-measure in fuzzy dynamic process which has not been discussed so far.

As mentioned before, fuzzy transition is expressed by linguistic rules but it does not tell about the running-time explicitly. Nevertheless, we understand that these rules contain some concept of time tacitly. For example, if we put a control to system, its effect will be revealed after a while. This period can be considered as a time-unit of fuzzy dynamic process. If we choose the period of a step in F.D.P. too much longer or shorter than this tacit-unit, that rule must become invalid.

Another problem we must consider is that the fuzziness of the final state is effected by the number of steps in F.D.P.. Everytime the calculation fuzzy reasoning is repeated, the fuzziness of a state is increased, whatever model of implication is used. That is, the fuzziness of a plan is increased if we choose a large number of steps in order to obtain a more precise time-schedule. This is unrealistic. Therefore, we must study what are the causes of increasing fuzziness in a real process, before applying F.D.P..

We discuss the above points as to some examples which are shown in TABLE 1.

Multi-Stage Heat Exchangers

Fig. 1 shows the scheme of multi-stage heat exchanger. Though this is an engineering system and its performance is exactly represented with a mathematical model, we consider the case when the state and the performance are described in natural language. The fluid temperature at the inlet and the outlet of each stage is expressed as "low", "medium" and "high", and the steam amount distributed to them is also expressed as "much", "normal" and "little". The process of a heat exchanger is described as follows.

> "If the inlet temperature is low and the steam amount is little, then the increase of temperature in this stage is small."

The goal of this problem is to find the optimum steam distribution which minimizes the total steam consumption and makes the final temperature near the given value. We can see that additivity of the performance index is approximately satisfied.

Plan of Golf Play

Before selecting a club at teeing ground, a golfer images whole shots until green-on and composes the order of clubs. Each club has its character but the result of shot is fuzzy, because it is effected by many uncertain factors. Generally speaking, a wood club hits a ball very far but the direction is not so exact as an iron club. Since we have no scale with which the location of ball is measured exactly, the location of ball can be considered as fuzzy state in this process. Then, the selection of a club is fuzzy control. The performance index is the total number of shot. The aim of this fuzzy dynamic programming is to find the best coposition of clubs which compromizes the total number of shots and the certainty of approaching green. This is similar to the shortest-path problem in a fuzzy graph.

Systems Engineering

One of the important points of systems engineering is a reasonable way of designing of a very complex system. The final goal may be given when a new large system is developed, but this is usually abstract and gives us no clue to realize the system. According to the methodology of systems engineering, we must set up some intermeadiate sub-goals between the initial state and the final goal. The selection of steps and states in Systems Engineering is a very serious problem. We first choose some sub-goals as fuzzy states of which measure is the degree of achievement of the final goal. Next, the number and the location of sub-goals should be set. The alternatives how to attain the sub-goals may be created by considering the routes connecting the initial state and the sub-goals.

Puzzle or Game

Usually a game solving process is structured with a tree graph. The thinking process of human is different from it. He images some patterns of its solution and sets up some intermediate sub-goals before the imaged goal. Then he tries to find ways which connect the initial state and sub-goal, and so on. This two-way thinking is the feature of human thought. In order to simulate this process with fuzzy dynamic programming, we consider a network type maze [4]. It is shown in Fig. 2. The purpose is to find the shortest path from the entrance to the exit.

In order to solve it, the maze is divided into some blocks of which boundaries are fuzzy. These blocks are considered as the sub-goals, and the possibility of transition from one block to another is fuzzily estimated. By using fuzzy dynamic programming, we can find a macroscopically optimal route. If we search precisely in this macro-route, the real optimal path may be found.

We can see from these examples that fuzzy dynamic programming is useful in the cases, when we have no quantitative measure for the state of object, when the result of control is uncertain, and also when the intermediate sub-goals are not clearly known to us.

MODELS OF FUZZY TRANSITION

One of the key points of F.D.P. is the choice of mathematical models for fuzzy transition which is given as a conditional statement linguistically as follows.

"If A then B"

, where A, B are the cause and the result respectively described in natural language. This conditional statement is represented with four types of fuzzy models.

(a) Fuzzy implication with fuzzy truth value

$$A \rightarrow B, \ \tau \qquad (1)$$

(b) Fuzzy relation

$$A \circ R = B \qquad (2)$$

(c) Fuzzy number

$$f(A\ ,\ C) = B \qquad (3)$$

(d) Conditional fuzzy measure and fuzzy integral [5]

$$\int \sigma(\,\cdot\,|x) \circ A = B(\,\cdot\,) \qquad (4)$$

Model (c) is an equation in which state A, control C and next state B are all fuzzy numbers. Here f is an extended function that models a fuzzy transition. This model is very simple and general among the four types especially when all the fuzzy numbers are normalized.
In the sequel we shall use this type of model in this paper.

ALGORITHMS

There are three problems in calculating F.D.P.. First one is discretization. If two of three variables (initial state, control, next state) are given, the rest is determined by the transition model. This must be discretized by classifying it into the pre-determined fuzzy sets. Second, the calculation of fuzzy transition forward and backward not always coincide each other, because Extension Principle is applied to calculate f in Eq. (3). The third problem is how to compare two fuzzy sets in order to obtain better control. There have been many ideas of prefering fuzzy sets, but none of them is always effective. Human judgement is most excellent for the first and the third problems. In this sense, F.D.P. has to be considered as a supporting system of human decision making.

For the first problem, two types of discretization are considered. One of them is the dicretization of controls. After all the fuzzy states are set up, the controls derived from a model are not well-formed. Those are classified into a finite number of pre-determined controls. The other is concerned with states. If a class of controls is given, a new state generated by a control is also classified into admissible states. This is done by human according to the similarity of membership functions.

For the third problem, we suggest the separation of fuzziness from the performance in the solution. The preference of fuzzy sets requires the trade-off between performance and fuzziness, but this is too much troublesome in F.D.P.. Therefore, we omit it in each step and judge the trade-off only at the final solutions.

This separation is very easy when the fuzzy number model Eq.(3) is adopted to express a fuzzy transition. It is assumed that the membership function of fuzzy number is an isosceles triangle of which smallest-, center- and largest-values are denoted by L, M, U respectively. The results of addition and subtraction of some fuzzy numbers are shown as follow.

$$M = \sum_{i=1}^{m} M_i - \sum_{j=1}^{n} M_j$$
$$U = \sum_{i=1}^{m} U_i - \sum_{j=1}^{n} L_j \qquad (5)$$
$$L = \sum_{i=1}^{m} L_i - \sum_{j=1}^{n} U_j$$

For example to obtain $(L, M, U) = (L_1, M_1, U_1) + (L_2, M_2, U_2)$, one sets $m = 2$ and $n = 0$. We can see that the center value is not different from the calculation of ordinary number but the grade of fuzziness (U-L) is changed by operations. Therefore, the solutions like those of ordinary D.P. are obtained by using center values, and after that, the grade of fuzziness is calculated as to the candidate solutions. Thus, the result of F.D.P. is usually of Pareto Type, then human decision maker should find a compromizing point on it.

NUMERICAL EXAMPLES

Resource Allocation in Scheduling

We suppose a job which consists of three sub-jobs. Its time limit is 20 days. The aim of fuzzy dynamic programming is to find the optimal plan of resource allocation which minimizes the total consumption of resources and completes the job as soon as possible. The starting date of each sub-job is chosen as fuzzy state. The state-transition is formulated approximately as follows.

(starting date of a sub-job) + (period of sub-job) = (completing date of sub-job) (6)

We suppose that the period of sub-job depends on the allocation policy of resources but this relation is qualitative. It may be described as follows.

(i) if labor and payment allocated to sub-jobs are normal, the period of each job is about one week,

(ii) if payment is increased, the period will be shorten by two days,

(iii) if labor is increased, shortenning of one or two days may be expected,

(iv) if both resources are increased together, the effect will be additive.

(v) any increase of resources are not favored.

(vi) in the first and the third sub-job, they prefer the increase of payment to that of labor. In the second sub-job, it is contrary,

(vii) when the job is advanced, the increase of resources is difficult.

Now let x_0, x_1, x_2 denote the starting date of sub-jobs, and y_1, y_2, y_3, y_4 denote the period of sub-jobs corresponding to the four policies mentioned from (i) to (iv). Since the state and the transition rules are described in words, x_i and y_j should be all fuzzy numbers. The membership functions of y_j are given from (i) - (iv) as Fig. 3.

We suppose that the cost is not fuzzy but fixed values depending on the allocation policy. This is shown in TABLE 2. Of course, it is possible to fuzzify the cost, but the final trade-off becomes too difficult.

The fuzzy transition is written as follows.

$$x_{n-1} + y_n = x_n \qquad (7)$$

Among sixteen feasible solutions, six do not satisfy the time limit, and the remaining ten are clustered into seven. The final candidate solutions are shown in TABLE 3. Considering both the cost and the fuzziness of completion date, the second solution may be best for us.

Soft-Landing of Rocket

We discuss here the problem of fuel control in case of soft-landing of a rocket. This is a pure physical problem formulated as follows.

$$M \frac{d^2x}{dt^2} = u - g \qquad x \geqslant 0 \qquad (8)$$

$$x = x_0, \quad \frac{dx}{dt} = y_0 \quad \text{for } t = 0 \qquad (9)$$

$$x = 0, \quad \frac{dx}{dt} = 0, \quad \text{for } t = T \qquad (10)$$

$$J = \int_0^T u \, dt \rightarrow \min \qquad (11)$$

, where x denotes the height of the rocket from surface, u is thrust (control variable), and t is time. Our goal is to find u(t) which satisfies Eqs. (8) - (11).

If we suppose that we have no quantitative measure for x and u, this problem can be considered as an F.D.P. problem by using fuzzy number model (Eq.3) instead of Eq.(8). Now, above equations are modified as follow.

$$x_{n+2} - x_{n+1} = y_{n+1},$$
$$y_{n+1} - y_n = v_n, \qquad (8')$$

$$x_0 = 5,\ y_0 = -1, \quad \text{for } n = 0 \qquad (9')$$

$$x_5 = 0,\ y_5 = 0, \quad \text{for } n = 5 \qquad (10')$$

$$J = \sum_{n=1}^{5} C_n(v) \rightarrow \text{min.} \qquad (11')$$

For simplification, it is assumed that x, y and v are all fuzzy numbers of which center value is integer and membership funciton is of the same form. Another constraint is added to v_n, that is, the center value of v_n is limited to [-2, -1, 0, 1].

As far as the center value is considered, this is an ordinary D.P. problem and solved when control cost is given as a function of v and n as shown in Table 4. All feasible solutions are shown in Table 5. There is no fuzziness in the cost value in Table 4. However, since x and v are fuzzy numbers, the final state must have some fuzziness which depends on the number of operations. This is rather unrealistic as we discussed in previously. It seems more natural in this case to assume that the fuzziness of the final state depends on the frequency of control switching, because the fuzziness of control-effect is accumulated everytime when a new control is put. Under these considerations, we have trade-off curves as shown in Fig. 4 where the axes are cost and fuzziness. We can see the control policy #1, #2 and #8 are indifference and best.

CONCLUSION

The paper states that fuzzy dynamic programming is an effective tool for improving managerial plans. The authors pointed out some problems to overcome before applying F.D.P. to a real management process and discussed the countermeasures. They hope many applications will be developed in this field.

The authors thank Mr. D. Tong and Mr. F. Taguchi who studied the examples of F.D.P. in Tokyo Institute of Technology.

REFERENCES

[1] Baldwin, J. F., and B. W. Pilsworth (1982). Dynamic Programming for Fuzzy Systems with Fuzzy Environment, J.Math. Anal. and Appl. 85, 1-23

[2] Bellman, R.E., and L. A. Zadeh (1970). Decision Making in a Fuzzy Enfironment, Management Science, 17, 141-164 or

Esogbue, A. O., and R.E. Bellman (1982). Contributions to Fuzzy Eynamic Programming, 2nd World Conf. on Mathematics at the Service of Man, Las Palmas

[3] Santos, E. S. (1968). Maximin Automata, Information & Control, 13

[4] Taguchi, F. and T. Terano (1981). Macroscopic Optimization of Network Problem, Summary of Papers on General Fuzzy Problems, 7 (Tokyo Inst. of Tech.)

[5] Terano, T., and M. Sugeno (1975). Conditional Fuzzy Measure and its Application, in Fuzzy Sets and Their Applications to Cognitive & Decision Process (ed. by L. A. Zadeh, et al), Academic Press

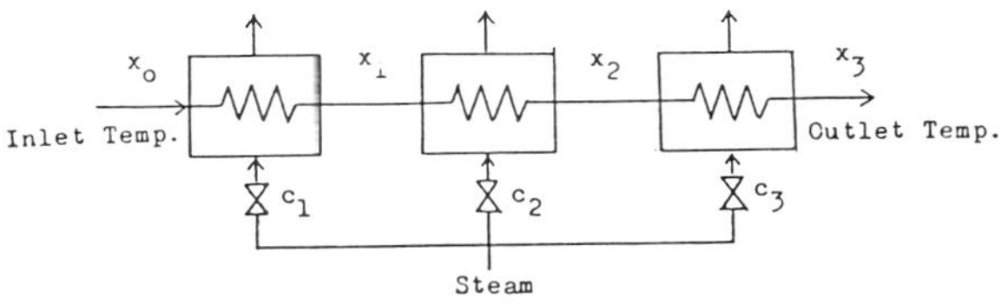

Fig. 1 Multi-state Heat Exchanger

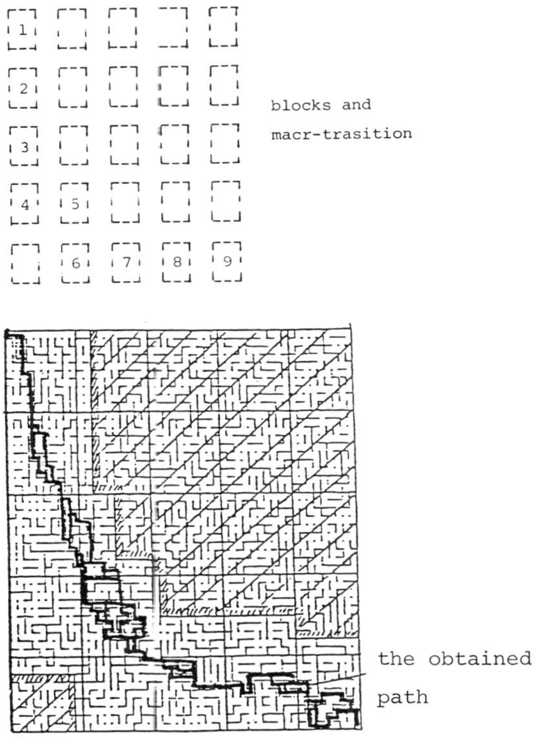

Fig. 2 Maze

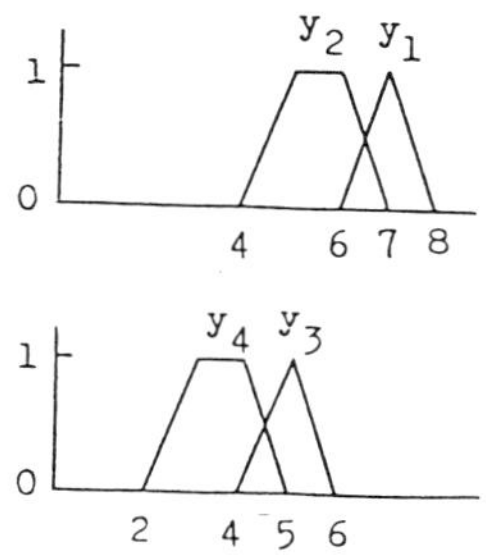

Fig. 3 Effect of Policy

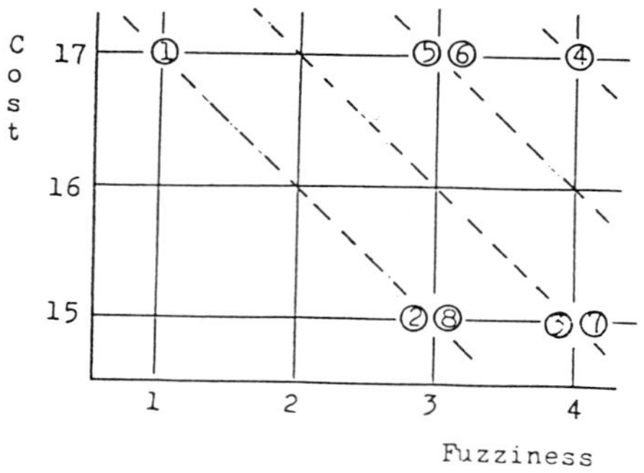

Fig. 4 Trade-off between Cost & Fuzziness

TABLE 1 Applications of Fuzzy Dynamic Programming

Topics	State	Transition	Control	Evaluation	Goal
Multi-Stage Heat Exchanger	Temperature	(inlet temp.) + (incr. of temp.) = (outlet temp.)	Steam Distribution	Steam Consumption	Outlet Temperature
Golf Play	Distance	(initial position) + (flight) = (new position)	Club Selection	Total Shot	Green on
Scheduling	Starting Date	(starting date) + (period) = (completion date)	Allocation of Resources	Total Cost	Time Limit
Systems Engineering	Progress	(initial state) + (advance) = (new state)	Alternatives	Degree of Success	Completion
Puzzle & Game	Goal Attainability	(initial state) + (play) = (new state)	Strategy	Time or Efficiency	Solution

TABLE 2 Cost of Resource

	1st subjob	2nd subjob	3rd subjob
(i) normal	0	0	0
(ii) increase payment	3	6	4
(iii) increase labor	6	4	7
(iv) increase both	8	9	10

TABLE 4 Cost of Control

v \ n	1	2	3	4	5
1 (-2)	0	0	0	0	0
2 (-1)	1	1	1	1	1
3 (0)	2	4	4	4	2
4 (1)	3	5	5	5	3

TABLE 3 Candidate Solutions of Scheduling

No.	Policy	Cost	Completion
I	(ii) - (iii) - (i)	10	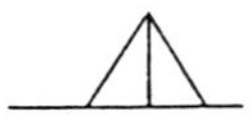
II	(ii) - (iv) - (i)	12	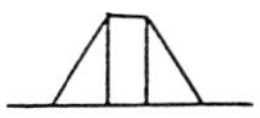
III	(ii) - (iv) - (ii)	16	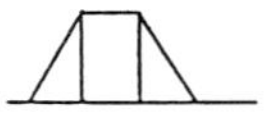
IV	(iii) - (iii) - (iii)	17	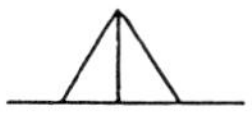

TABLE 5 Feasible Solutions

No.	Control Policy	Cost	Control Freq.
①	3 - 3 - 3 - 3 - 4	17	1
②	3 - 2 - 4 - 4 - 3	15	3
③	3 - 2 - 4 - 3 - 4	15	4
④	2 - 4 - 3 - 3 - 4	17	3
⑤	2 - 4 - 3 - 4 - 3	17	4
⑥	2 - 3 - 4 - 4 - 3	17	3
⑦	4 - 1 - 4 - 3 - 4	15	4
⑧	2 - 4 - 4 - 2 - 4	15	3

YAGER'S PROBABILITY OF A FUZZY EVENT IN STOCHASTIC CONTROL UNDER FUZZINESS

J. Kacprzk*

Machine Intelligence Institute, Hagan School of Business, Iona College New Rochelle, New York 10801, U.S.A

Abstract. Multistage control of a stochastic system (Markov chain) in a fuzzy environment is considered. Fuzzy constraints are imposed on the subsequent controls and a fuzzy goal is imposed on the final state (output). The fuzzy decision is of intersection type. An optimal sequence of controls is sought which maximizes the probability of attainment of the fuzzy goal subject to the fuzzy constraints. The fuzzy goal is viewed as a fuzzy event and its probability is assumed in the sense of Yager, i.e. a fuzzy set in [0, 1]. . Dynamic programming is used to solve the control problems formulated.

Keywords. Stochastic control; fuzzy control; fuzzy event; probability of a fuzzy event.

INTRODUCTION

Control models provide powerful tools to deal with problems in which basically a directed (goal-oriented) influence is exercised. Such problems prevail in virtually all the fields, hence the relevance of control. An important issue is the uncertainty which is usually equated with the randomness in the behavior of the system under control. Stochastic control theory provides means to formulate and solve the resulting control problems.

Moreover, some "softness"(fuzziness) may exist, e.g., due to the lack of clear-cut goals and constraints. It is not of randomness type, but rather of vagueness and ambiguity type. Probabilistic tools are therefore inadequate. Adequate and efficient means may be provided by fuzzy sets and possibility theory (Zadeh, 1965; 1978).

A general framework within which to deal with control under fuzziness was proposed by Bellman & Zadeh(1970). It triggered a huge research effort, including that in stochastic control under fuzziness (e.g., Jacobson, 1976; Kacprzyk, 1978a, 1978b, 1982b, 1983b; Kacprzyk & Staniewski, 1980; Sladky, Kacprzyk & Staniewski, 1982). Various problem classes were considered; their classification was mainly based on the type of termination time(fixed and specified, fuzzy, and infinite). For details, see Kacprzyk(1982a, 1983a).

The main underlying concept to deal with stochastic control under fuzziness is the probability of a fuzzy event. In all works mentioned, the probability of a fuzzy event was assumed as originally proposed by Zadeh(1968), i.e. a real number from [0, 1] . Another approach was recently proposed by Yager(1979) in which the probability of a fuzzy event was defined as a fuzzy set in [0, 1] . This may be viewed as more intuitively appealing, because the probability of a fuzzy event is now fuzzy. The purpose of this work is to propose the use of Yager's(fuzzy) probability of a fuzzy event in stochastic control under fuzziness.

*on leave from: Systems Research Institute
Polish Academy of Sciences
ul.Newelska 6
01-447 Warsaw, Poland

PRELIMINARIES

To make the work self-contained, let us review some needed elements of fuzzy sets and control under fuzziness.

Fuzzy sets and related topics

A fuzzy set in $X = \{x_1, \ldots, x_n\}$, $A \subseteq X$, is characterized by its membership function μ_A: $X \longrightarrow [0, 1]$, and written as $A = \mu_A(x_1)/x_1 + \ldots + \mu_A(x_n)/x_n$.

A crucial concept is a fuzzy event. For simplicity, let $X = R^m$. If now (R^m, B, p) is a probability space in which B is a Borel field in R^m, and p: $B \longrightarrow [0, 1]$ is a probability measure, then a fuzzy event in R^m is a fuzzy set $A \subseteq R^m$ whose membership function is Borel measurable.

The probability of a fuzzy event $A \subseteq R^m$ is now defined by Zadeh(1968) as

$$p(A) = E\mu_A(x) = \int_{R^m} \mu_A(x)\,dp \tag{1}$$

or if $A \subseteq X = \{x_1, \ldots, x_n\}$, then

$$p(A) = \sum_{i=1}^{n} \mu_A(x_i) \cdot p(x_i) \tag{2}$$

For an analysis of p(A) and its properties, see Smets(1982) and Zadeh(1968).

Example 1. Let $X = \{x_1, x_2, x_3\}$ be a set of people and $A = 0.1/x_1 + 0.7/x_2 + 1/x_3$ be a fuzzy set of "strong people". If $p(x_1)=0.6$, $p(x_2)=0.3$ and $p(x_3)=0.1$, then by randomly selecting $x \in X$ the probability of choosing a "strong person is

$$p(A)=\mu_A(x_1)p(x_1) + \mu_A(x_2)p(x_2) + \mu_A(x_3)p(x_3)= =0.1\cdot 0.6 + 0.7\cdot 0.3 + 1\cdot 0.1=0.37$$

Thus, $p(A) \in [0, 1]$ which may seem to be somewhat counter-intuitive, because the event is fuzzy, but its probability is not. To overcome this, another approach was proposed by Yager (1979). It assumes the classic Zadeh's definition of a fuzzy event, but its probability is defined as a fuzzy set in $[0, 1]$. To present this definition, let us start with some notions.

The α-level set (α-cut) of a fuzzy set $A \subseteq X$, A_α, is

$$A_\alpha = \{ x \in X: \mu_A(x) \geqslant \alpha \} \subseteq X \quad \text{for each } \alpha \in (0,1] \qquad (3)$$

A fuzzy set $A \subseteq X$ may be represented by its A_α's as follows

$$A = \bigcup_{\alpha=0}^{1} \alpha A_\alpha \qquad (4)$$

Let Y and W be some nonfuzzy sets, G be a family of all nonfuzzy subsets of Y, and $A \subseteq Y$. Now, if there is a mapping $f: G \longrightarrow W$, then the extension principle makes it possible to extend f to act on fuzzy sets in Y as follows

$$f(A) = \bigcup_{\alpha=0}^{1} \alpha f(A_\alpha) \qquad (5)$$

Now, if $A \subseteq X$ is a fuzzy event and p is a probability function, then the (fuzzy) probability of a fuzzy event is defined by Yager(1979) as

$$P(A) = E_f \mu_A(x) = \bigcup_{\alpha=0}^{1} \alpha\, p(A_\alpha) \qquad (6)$$

which evidently results from the application of the extension principle.
Since $p(A_\alpha) \in [0, 1]$, then it may be represented as a fuzzy singleton $\{1/p(A_\alpha)\}$, and

$$P(A) = \bigcup_{\alpha=0}^{1} \alpha \{1/p(A_\alpha)\} \qquad (7)$$

Evidently, $P(A) \subseteq [0, 1]$. And since A_α's of a fuzzy event are Borel sets (Zadeh, 1968), then P(A) is well defined.

Example 2. For the same data as in Example 1, we obtain: $A_{0.1}=\{x_1,x_2,x_3\}$, $A_{0.7}=\{x_2,x_3\}$, $A_1=\{x_3\}$, and $p(A_{0.1})=1$, $p(A_{0.7})=0.4$, $p(A_1)=0.1$. Thus

$$P(A)= \bigcup_{\alpha \in \{0.1,\ 0.7,\ 1\}} \alpha \{1/p(A_\alpha)\} = = 1/0.1 + 0.7/0.4 + 0.1/1$$

For other approaches to the fuzzy probabilities of fuzzy events, see Dubois & Prade(1980). They will not be employed in the sequel.

The comparison(ordering) of fuzzy probabilities is crucial.(Evidently, this problem does not exist in the case of Zadeh's probabilities being real numbers.) A procedure proposed by Yager(1981) is apparently best suited.

The procedure consists virtually of associating with a particular fuzzy set in $[0, 1]$ some "subsuming" real number from $[0, 1]$. Then, the comparison between fuzzy sets boils down to a natural comparison of their subsuming (real) numbers.

Let $A \subseteq [0, 1]$ and $A_\alpha = \{x \in [0, 1] : \mu_A(x) \geqslant \alpha\}$, $A_\alpha \subseteq [0, 1]$. If $V=\{v_1,\dots,v_k\} \subseteq [0, 1]$, the mean value of the elements of V is

$$M(V) = (1/k) \sum_{i=1}^{k} v_i \qquad (8)$$

We introduce now the ordering function

$$F(A) = \int_0^{\alpha_{max}} M(A_\alpha)\, d\alpha \qquad (9)$$

where α_{max} is the maximum value of $\mu_A(x)$. Evidently, $F(A) \in [0, 1]$.

The ordering between $A, B \subseteq [0, 1]$ is now defined as

$$A \geqslant B \iff F(A) \geqslant F(B) \qquad (10)$$

This is well justified as demonstrated by the following relevant properties of F:

1. If A is a real number from $[0, 1]$, i.e. $A=\{a\}=\{1/a\}$, then $\alpha_{max}=1$, $A_\alpha=\{a\}$ for each $\alpha \in [0, 1]$, and
$$F(A) = \int_0^1 a\, d\alpha = a \qquad (11)$$
i.e. F(A) preserves the natural ordering between real numbers.
2. Since $M(A_\alpha) \leqslant 1$ for each $\alpha \in [0, 1]$, then
$$\int_0^1 M(A_\alpha)\, d\alpha \leqslant \int_0^1 d\alpha = 1 \qquad (12)$$
3. The maximum value of F(A) is for $A=\{1/1\}$, $F(\{1/1\}) = 1$.
4. If A is a nonfuzzy set, $A \subseteq [0, 1]$, then $A_\alpha = A$ for each $\alpha \in [0, 1]$, and
$$F(A) = \int_0^1 M(A)\, d\alpha = M(A) \qquad (13)$$
5. If $A=\{b/a\}$, then $\alpha_{max} = b$, $A_\alpha=\{a\}$ and $M(A_\alpha)=a$ for $a \geqslant b$, and
$$F(A) = \int_0^b a\, d\alpha = ab \qquad (14)$$

Thus, F makes it possible to meaningfully compare two fuzzy sets or a fuzzy set with a real number. Moreover, F may be used for fuzzy sets defined in both finite and infinite spaces.

Example 3. Let $A= 0.1/0.1+0.4/0.3+0.6/0.5+ +0.8/0.7+1/1$. Then

$A_{0.1}=\{0.1,\ 0.3,\ 0.5,\ 0.7,\ 1\}$	$M(A_{0.1})=0.52$
$A_{0.4}=\{0.3,\ 0.5,\ 0.7,\ 1\}$	$M(A_{0.4})=0.625$
$A_{0.6}=\{0.5,\ 0.7,\ 1\}$	$M(A_{0.6})=0.733$
$A_{0.3}=\{0.7,\ 1\}$	$M(A_{0.8})=0.85$
$A_1 = \{1\}$	$M(A_1) = 1$

Thus

$$F(A)= \int_0^{0.1} 0.52\, d\alpha + \int_{0.4}^{0.6} 0.733\, d\alpha + \int_{0.6}^{0.8} 0.85\, d\alpha + + \int_{0.8}^{1} d\alpha = 0.52\cdot 0.1+0.3\cdot 0.525+0.2\cdot 0.733+ +0.2\cdot 1 = 0.7561$$

Stochastic control under fuzziness with Zadeh's probability of a fuzzy event

To introduce the basic notation, let us sketch the essence of stochastic control under fuzziness with Zadeh's probability of a fuzzy event as it was used so far.
The stochastic system under control, a Markov chain, is characterized by a conditional probability $p(x_{t+1} | x_t, u_t)$, where $x_t, x_{t+1} \in X=\{s_1, \dots, s_n\}$ are the states(outputs) at time(control stage)t and t+1, respectively, and $u_t \in U=\{c_1, \dots, c_m\}$ is the control(input) at time t.
At each t, u_t is subjected to a fuzzy constraint $\mu_{C^t}(u_t)$, and on the final state x_N a fuzzy goal $\mu_{G^N}(x_N)$ is imposed. N is some termination time(horizon).
The fuzzy decision is

$$\mu_D(u_0, \dots, u_{N-1} | x_0) = \mu_{C^0}(u_0) \wedge \dots \wedge \mu_{C^{N-1}}(u_{N-1}) \wedge \mu_{G^N}(x_N) \quad (15)$$

The control problem may be formulated in two ways: find an optimal sequence of controls $u_0^*, \dots, u_{N-1}^*$, such that
- due to Bellman & Zadeh(1970):

$$\mu_D(u_0^*, \dots, u_{N-1}^* | x_0) = \max_{u_0, \dots, u_{N-1}} (\mu_{C^0}(u_0) \wedge \dots \wedge \mu_{C^{N-1}}(u_{N-1}) \wedge E\mu_{G^N}(x_N)) \quad (16)$$

i.e. such that it maximizes the probability of attainment of the fuzzy goal subject to the fuzzy constraints;
- due to Kacprzyk & Staniewski(1980):

$$\mu_D(u_0^*, \dots, u_{N-1}^* | x_0) = \max_{u_0, \dots, u_{N-1}} E(\mu_{C^0}(u_0) \wedge \dots \wedge \mu_{C^{N-1}}(u_{N-1}) \wedge \mu_{G^N}(x_N)) \quad (17)$$

i.e. such that it maximizes the expected value of $\mu_D(\cdot | \cdot)$.
Evidently, both G^N and D are considered as fuzzy events, and $E\mu_{G^N}(x_N)$ in Eq.(16) and $E\mu_D(\cdot | \cdot)$ in Eq.(17) are in the sense of Zadeh, i.e. due to Eq.(2). The solution of Eq.(16) and Eq.(17) may be obtained by dynamic-programming-based techniques.

The above basic (for the fixed and specified termination time) formulations can be extended to the cases of fuzzy and infinite termination time(see Kacprzyk, 1982a, 1983a).

STOCHASTIC CONTROL UNDER FUZZINESS WITH YAGER'S PROBABILITY OF A FUZZY EVENT

Fixed and specified termination time

As previously, the stochastic system is described by $p(x_{t+1} | x_t, u_t)$, $x_t, x_{t+1} \in X=\{s_1, \dots, s_n\}$, $u_t \in U=\{c_1, \dots, c_m\}$. At each t, u_t is subjected to a fuzzy constraint $\mu_{C^t}(u_t)$, and on x_N a fuzzy goal $\mu_{G^N}(x_N)$ is imposed. N is a fixed and specified termination time.
The problem is to find an optimal sequence of controls $u_0^*, \dots, u_{N-1}^*$, such that

$$\mu_D(u_0^*, \dots, u_{N-1}^* | x_0) = \max_{u_0, \dots, u_{N-1}} (\mu_{C^0}(u_0) \mathbin{\hat{\wedge}} \dots \mathbin{\hat{\wedge}} \mu_{C^{N-1}}(u_{N-1}) \mathbin{\hat{\wedge}} E_f\mu_{G^N}(x_N)) \quad (18)$$

where, regarding G^N as a fuzzy event in X,

$$E_f\mu_{G^N}(x_N) = \bigcup_{\alpha=0}^{1} \alpha\{1/p(G^N_\alpha)\} \quad (19)$$

and $N < \infty$ is a fixed and specified termination time.
Let us notice that for any realization of $u_0, \dots, u_{N-1}$, $\mu_{C^t}(u_t) \in [0, 1]$, while $E_f\mu_{G^N}(x_N) \subsetneqq [0, 1]$, t=0,1,...,N-1; they are incompatible and thus "$\hat{\wedge}$" instead of "$\wedge$" is written in Eq.(18).

However, if we use the function F given by Eq.(9), then for each fixed x_{N-1} and u_{N-1} we have $F(E_f u_{G^N}(x_N)) \in [0 \ \ 1]$ which is compatible with $\mu_{C^t}(u_t) \in [0, 1]$. Therefore, Eq.(18) may be written as

$$\mu_D(u_0^*, \dots, u_{N-1}^* | x_0) = \max_{u_0, \dots, u_{N-1}} (\mu_{C^0}(u_0) \wedge \dots \wedge \mu_{C^{N-1}}(u_{N-1}) \wedge F(E_f\mu_{G^N}(x_N))) \quad (20)$$

It is easy to see that the last two terms of Eq.(20) depend only on u_{N-1}, the next last - on u_{N-2}, etc. The solution may therefore be obtained via dynamic programming. The set of recurrence equations is

$$\begin{cases} \mu_{G^{N-i}}(x_{N-i}) = \max\limits_{u_{N-1}} (\mu_{C^{N-i}}(u_{N-1}) \wedge F(E_f\mu_{G^{N-i+1}}(x_{N-i+1}))) & (21a) \\ E_f\mu_{G^{N-i+1}}(x_{N-i+1}) = \bigcup\limits_{\alpha=0}^{1} \alpha\{1/p(G^{N-i+1}_\alpha)\} & (21b) \\ i=1, \dots, N & \end{cases} \quad (21)$$

where G^{N-i} may be viewed as a fuzzy goal at t=N-i induced by G^{N-i+1} at t=N-i+1.
Thus, solving Eq.(21) we obtain u_{N-i}^* or in fact the optimal policies a_{N-i}^*, such that $u_{N-i}^* = a_{N-i}^*(x_{N-i})$, i=1,...,N.

Example 4. Let $X=\{s_1, s_2, s_3\}$, $U=\{c_1, c_2\}$, N=2,

$C^0=0.6/c_1+1/c_2 \quad C^1=1/c_1+0.6/c_2$

$G^2=0.3/s_1+1/s_2+0.8/s_3$

and $p(x_{t+1} | x_t, u_t) =$

$$= x_t \overset{u_t=c_1,\ x_{t+1}}{\begin{array}{c|ccc} & s_1 & s_2 & s_3 \\ \hline s_1 & 0.8 & 0.1 & 0.1 \\ s_2 & 0 & 0.1 & 0.9 \\ s_3 & 0.8 & 0.1 & 0.1 \end{array}} \quad x_t \overset{u_t=c_2,\ x_{t+1}}{\begin{array}{c|ccc} & s_1 & s_2 & s_3 \\ \hline s_1 & 0.1 & 0.9 & 0 \\ s_2 & 0.8 & 0.1 & 0.1 \\ s_3 & 0.1 & 0 & 0.9 \end{array}}$$

From Eq.(21b) we obtain $E_f\mu_{G^2}(x_2)=$

$= x_1$	$u_1=c_1$	$u_1=c_2$
s_1	1/0.1+0.8/0.2+ +0.3/1	1/0.9+0.3/1
s_2	1/0.1+0.8/1	1/0.1+0.8/0.2+ +0.3/1
s_3	1/0.1+0.8/0.2+ +03/1	1/0+0.8/0.9+0.3/1

and from Eq.(9)

$$F(E_f\mu_{G^2}(x_2)) = u_1 \begin{array}{c} \\ c_1 \\ c_2 \end{array}\overset{x_1}{\begin{bmatrix} 0.225 & 0.46 & 0.225 \\ 0.795 & 0.225 & 0.415 \end{bmatrix}}$$

Then, using Eq.(21a)

$$\mu_{C^1}(u_1) \wedge F(E_f\mu_{G^2}(x_2)) = u_1 \begin{array}{c} \\ c_1 \\ c_2 \end{array}\begin{array}{c} s_1 \quad\; s_2 \quad\; s_3 \\ \begin{bmatrix} 0.225 & 0.46 & 0.225 \\ 0.6 & 0.225 & 0.415 \end{bmatrix} \end{array}$$

and $G^1 = 0.6/s_1 + 0.46/s_2 + 0.415/s_3$ with $a_1^*(s_1) = c_2$, $a_1^*(s_2) = c_1$, $a_1^*(s_3) = c_2$.

Next, from Eq.(21b) $E_f\mu_{G^1}(x_1) =$

$$= x_1 \begin{array}{c} s_1 \\ \\ s_2 \\ \\ s_3 \\ \\ \end{array} \overset{\qquad u_0 = c_1 \qquad\qquad\qquad u_0 = c_2}{\begin{bmatrix} 0.46/0.2+0.6/0.8+ & 0.6/0.1+0.46/1 \\ +0.415/1 & \\ 0.6/0+0.46/0.1+ & 0.46/0.2+0.6/0.8+ \\ +0.415/1 & +0.415/1 \\ 0.46/0.2+0.6/0.8+ & 0.6/0.1+0.415/1 \\ +0.415/1 & \end{bmatrix}}$$

and Eq.(9) yields

$$F(E_f\mu_{G^1}(x_1)) = u_0 \begin{array}{c} \\ c_1 \\ c_2 \end{array}\begin{array}{c} x_0 \\ s_1 \quad\; s_2 \quad\; s_3 \\ \begin{bmatrix} 0.449 & 0.161 & 0.449 \\ 0.254 & 0.449 & 0.408 \end{bmatrix} \end{array}$$

Then, from Eq.(21a)

$$\mu_{C^0}(u_0) \wedge F(E_f\mu_{G^1}(x_1)) = u_0 \begin{array}{c} \\ c_1 \\ c_2 \end{array}\begin{array}{c} x_0 \\ s_1 \quad\; s_2 \quad\; s_3 \\ \begin{bmatrix} 0.449 & 0.161 & 0.449 \\ 0.254 & 0.449 & 0.408 \end{bmatrix} \end{array}$$

Thus, $a_0^*(s_1) = c_1$, $a_0^*(s_2) = c_2$, $a_0^*(s_3) = c_1$.

The solution is therefore similar, though not identical, as in the case of Zadeh's probability (Bellman & Zadeh, 1970).

Fuzzy termination time

In practice, a precise termination time, say 5 years, may sometimes be inadequate and a vague estimation, say about 5 years, may be preferable. This leads to the concept of fuzzy termination time introduced by Fung & Fu(1977) and Kacprzyk(1977, 1978a, 1978b).

The fuzzy termination time is defined as a fuzzy set in the space of control stages, $\underset{\sim}{T} \subseteq \{1, \ldots, K-1, K, K+1, \ldots, N\}$, and $\mu_T(r)$ is viewed as a measure of how preferable a particular $r \in \{1, \ldots, N\}$ is as the termination time - the higher $\mu_T(r)$, the more preferable r; $\mu_T(r)=0$ means that the process cannot terminate at time r. We assume that $r \in \{t \in \{1, \ldots, N\} : \mu_T(t) \geqslant 0\} = \{K, K+1, \ldots, N\}$ which does not limit the generality of discussion.

The fuzzy termination time may be involved in the control problem formulation in the following two ways:

- due to Fung & Fu(1977)

$$\mu_D(u_0, \ldots, u_{r-1} \mid x_0) = \mu_{C^0}(u_0) \wedge \ldots \wedge \mu_{C^{r-1}}(u_{r-1}) \wedge \mu_G(x_r) \wedge \mu_T(r) \qquad (22)$$

- due to Kacprzyk(1977, 1978a, 1978b)

$$\mu_D(u_0, \ldots, u_{r-1} \mid x_0) = \mu_{C^0}(u_0) \wedge \ldots \wedge \mu_{C^{r-1}}(u_{r-1}) \wedge \mu_T(r) \cdot \mu_G(x_r) \qquad (23)$$

Eqs.(22) and (23) are similar; in the sequel we will use Eq.(23).

The problem is now to find an optimal termination time r^* and an optimal sequence of controls $u_0^*, \ldots, u_{r^*-1}^*$, such that

$$\mu_D(u_0^*, \ldots, u_{r^*-1}^* \mid x_0) = \max_{u_0, \ldots, u_{r-1}, r} (\mu_{C^0}(u_0) \wedge \ldots \wedge \mu_{C^{r-1}}(u_{r-1}) \wedge \mu_T(r) \cdot E\mu_G(x_r)) \qquad (24)$$

Two approaches, that due to Kacprzyk(1978a, 1978b) and due to Stein(1980), may be used to solve this problem.

And analogously, using Yager's probability of a fuzzy event, Eq.(24) becomes (employing the same reasoning as for Eqs.(18)-(20)): find r^* and $u_0^*, \ldots, u_{r^*-1}^*$, such that

$$\mu_D(u_0^*, \ldots, u_{r^*-1}^* \mid x_0) = \max_{u_0, \ldots, u_{r-1}, r} (\mu_{C^0}(u_0) \wedge \ldots \wedge \mu_{C^{r-1}}(u_{r-1}) \wedge \mu_T(r) \cdot F(E_f\mu_G(x_r))) \qquad (25)$$

where F is the ordering function defined by Eq.(9).

The solution of Eq.(25) may be obtained by employing the two approaches mentioned.

Approach of Kacprzyk(1978a, 1978b). In $u_0^*, \ldots, u_{r^*-1}^*$ its part $u_{K-1}^*, \ldots, u_{r^*-1}^*$ must itself be optimal. Therefore, the solution of Eq.(25) boils down to the solution of the following sets of recurrence equations:

- for t=K-1 to t=r-1

$$\begin{cases} \mu_{G^{r-i}}(x_{r-i}, r) = \max\limits_{u_{r-i}} (\mu_{C^{r-i}}(u_{r-i}) \wedge F(E_f\mu_{G^{r-i+1}}(x_{r-i+1}, r))) \\ E_f\mu_{G^{r-i+1}}(x_{r-i+1}, r) = \bigcup\limits_{\alpha=0}^{1} \alpha\{1/p(G_\alpha^{N-i+1})\} \\ r=K, \ldots, N;\ i=1, \ldots, r-K+1 \end{cases} \qquad (26)$$

where $\mu_{G^r}(x_r, r) = \mu_T(r) \cdot \mu_G(x_r)$;

Then, r^* is given by

$$\mu_{G^{K-1}}(x_{K-1}) = \max_r \mu_{G^{K-1}}(x_{K-1}, r) \qquad (27)$$

Thus, we obtain r^* and $u_{K-1}^*, \ldots, u_{r^*-1}^*$;

- for t=0 to t=K-2

$$\begin{cases} \mu_{G^{K-1-i}}(x_{K-1-i}) = \max\limits_{u_{K-1-i}} (\mu_{C^{K-1-i}}(u_{K-1-i}) \wedge F(E_f\mu_{G^{K-i}}(x_{K-i}))) \\ E_f\mu_{G^{K-i}}(x_{K-i}) = \bigcup\limits_{\alpha=0}^{1} \alpha\{1/p(G_\alpha^{K-i})\} \\ i=1, \ldots, K-1 \end{cases} \qquad (28)$$

which yields $u_0^*, \ldots, u_{K-1}^*$.

Evidently, we determine in fact the optimal policies a_t^*, such that $u_t^* = a_t^*(x_t)$, $t=0,1,\ldots, r^*-1$.

Approach of Stein(1980). At $t=N-i$, $i=1,\dots,N-1$, we may either terminate and obtain $\mu_T(N-i)\cdot\mu_G(x_{N-i})$, or apply u_{N-i} and obtain $\mu_{C^{N-i}}(u_{N-i}) \wedge F(E_f\mu_{G^{N-i+1}}(x_{N-i+1}))$. The better alternative should evidently be taken, hence the solution is given by the following set of equations

$$\begin{cases} \mu_{G^{N-i}}(x_{N-i}) = \mu_T(N-i)\cdot\mu_G(N-i) \vee \\ \vee \max\limits_{u_{N-i}} (\mu_{C^{N-i}}(u_{N-i}) \wedge F(E_f\mu_{G^{N-i+1}}(x_{N-i+1}))) \\ E_f\mu_{G^{N-i+1}}(x_{N-i+1}) = \bigcup\limits_{\alpha=0}^{1} \alpha\{1/p(G_\alpha^{N-i+1})\} \\ i=1,\dots,N \end{cases} \quad (29)$$

where $\mu_{G^N}(x_N) = \mu_T(N)\cdot\mu_G(x_N)$, and r^* is given by such $N-i$ for which

$$\mu_T(N-i)\cdot\mu_G(x_{N-i}) > \max_{u_{N-i}} (\mu_{C^{N-i}}(u_{N-i}) \wedge F(E_f\mu_{G^{N-i+1}}(x_{N-i+1}))) \quad (30)$$

The case of infinite termination time is very specific and requires different tools, hence it will not be dealt with in this paper. For some details , see Kacprzyk(1982a, 1983a).

CONCLUDING REMARKS

The basic models of stochastic control under fuzziness in which Zadeh's probability of a fuzzy event had been used so far were shown to be applicable in principle in the case of using Yager's probability of a fuzzy event. The solutions of the resulting control problems were however more difficult, though more computationally than conceptually.

Finite termination time models, i.e. with the fixed and specified and fuzzy termination time, were discussed. The case of infinite termination time poses some difficulty and requires different means. It will be discussed in another paper.

REFERENCES

Bellman, R.E., and L.A. Zadeh(1970). Decision-making in a fuzzy environment. Manag. Sci. 17, 151-169.

Dubois, D., and H. Prade(1980). Fuzzy Sets and Systems: Theory and Applications. Academic Press, New York.

Fung, L.W., and K.S. Fu(1977). Characterization of a class of fuzzy optimal control problems. In M.M. Gupta, G.N. Saridis, and B.R. Gaines(Eds.). Fuzzy Automata and Decision Processes. North Holland, Amsterdam.

Jacobson, D.H.(1976). On fuzzy goals and maximizing decisions in stochastic optimal control. J. Math. Anal. & Appl. 55, 434-440.

Kacprzyk, J.(1977). Control of a nonfuzzy system in a fuzzy environment with fuzzy termination time. Syst. Sci. 3, 320-331.

Kacprzyk, J.(1978a). Control of stochastic system in a fuzzy environment with fuzzy termination time. Syst. Sci. 4, 291-300.

Kacprzyk, J.(1978b). Decision-making in a fuzzy environment with fuzzy termination time. Fuzzy Sets & Syst. 1, 169-179.

Kacprzyk, J.(1982a). Multistage decision processes in a fuzzy environment: a survey. In M.M. Gupta, and E. Sanchez (Eds.). Fuzzy Information and Decision Processes. North-Holland, Amsterdam.

Kacprzyk, J.(1982b). Control of a stochastic system in a fuzzy environment with Yager's probability of fuzzy event. Techn. Rep. #MII-242, Machine Intelligence Institute, Iona College, New Rochelle, N.Y., U.S.A.

Kacprzyk, J.(1983a). Multistage Decision-Making under Fuzziness: Theory and Applications. ISR Series. Verlag TUV Rheinland, Cologne.

Kacprzyk, J.(1983b). Control of stochastic systems in fuzzy environment. In M.G. Singh(Ed.). Encyclopedia of Systems and Control. Pergamon Press, Oxford.

Kacprzyk, J., K. Safteruk, and P. Staniewski (forthcoming). On the control of stochastic systems in a fuzzy environment. Syst. Sci.

Kacprzyk, J., and P. Staniewski(1980). A new approach to the control of stochastic systems in a fuzzy environment. Arch. Autom. & Telemech. XXV, 434-444.

Kacprzyk, J., and P. Staniewski(forthcoming). Control of a deterministic system in a fuzzy environment over infinite horizon. Fuzzy Sets & Syst.

Sladky, K., J. Kacprzyk, and P. Staniewski (1982). On optimal control of stochastic system in a fuzzy environment. Proc.4-th FORMATOR Symp. Prague.

Smets, Ph.(1982). Probability of fuzzy event: an axiomatic approach. Fuzzy Sets & Syst. 7, 153-164.

Stein, W.E.(1980). Optimal stopping in fuzzy environment. Fuzzy Sets & Syst. 3, 253-259.

Yager, R.R.(1979). A note on probabilities of fuzzy events. Infor. Sci. 18, 113-129.

Yager, R.R.(1981). A procedure for ordering fuzzy subsets of the unit interval. Infor. Sci. 34, 143-161.

Zadeh, L.A.(1965). Fuzzy sets. Infor. & Contr. 8, 338-353.

Zadeh, L.A.(1968). Probability measures of fuzzy events. J. Math. Anal. & Appl. 23, 421-427.

Zadeh, L.A.(1978). Fuzzy sets as a basis for a theory of possibility. Fuzzy Sets & Syst. 1, 3-28.

PLANNING HORIZONS FOR PRODUCTION PLANNING MODELS IN THE CASE OF CONCAVE COSTS

A. Bensoussan* and J-M. Proth**

Inria, BP 105, Rocquencourt, 78153 Le Chesnay Cedex (France) and Université Paris IX Dauphine
***Inria, BP 105, Rocquencourt, 78153 Le Chesnay Cedex (France)*

RESUME. Ce papier est consacré à la recherche d'horizons de planification dans une production faisant intervenir des coûts concaves.
L'approche choisie est la programmation dynamique de type rétrograde. Cette approche, associée à un résultat que nous établissons, permet :
- de simplifier un algorithme de recherche de la solution optimale du problème à horizon fini proposé précédemment par les auteurs.
- de dégager un algorithme qui conduit à un horizon de planification dans un nombre important de cas.

Nous présentons des résultats numériques.

ABSTRACT. This paper is devoted to the search of planning horizons for production planning models in the case of concave costs.
The backward dynamic programming formulation is used. This approach, associated with a new result, allows :
- to simplify an algorithm proposed elsewhere in order to obtain an optimal solution for the finite horizon problems.
- to obtain an algorithm which leads to a planning horizon for a wide range of cases.

We also give examples.

INTRODUCTION

Many papers have been devoted to the search of planning horizons for production planning models in the case of concave inventory and production costs (see, for instance,[2], [3] and [5]).

As far as we know, these works use particular costs (constant or linear on $\mathbb{R}^{*+}$).

Our aim is to give a new approach to this problem, considering concave and non decreasing costs.

The organisation of the paper is as follows :

1. We first give the notation and definitions used.
2. We then recall the main result obtained in [1] and present an important simplification of the algorithm proposed starting from this main result.
3. We show how to use this simplification in order to obtain a planning horizon for a wide range of cases.
4. Finally, we consider the non stationary problem where the costs are linear over $\mathbb{R}^{*+}$.

DEFINITIONS, NOTATION AND ASSUMPTIONS

d_i, i = 1, 2, 3, ... is the demand at time i and $d_i \geq 0$.

v_i, i = 1, 2, 3, ... is the quantity launched in production at time i, which is available one period later at time i+1 and :

$$v_i \geq 0 \quad (1)$$

The inventory balance equations are then :

$$y_{i+1} = y_i + v_i - d_{i+1}, \; i = 0, 1, 2, \ldots \quad (2)$$

where y_i is the stock level on $[i, i+1[$.

The initial stock level y_0 is known and $y_0 \geq 0$
Moreover, backlogging is not allowed, such that :

$$y_i \geq 0, \; i = 0, 1, 2, \ldots \quad (3)$$

Note that (1) and (3) can be rewritten as :

$$v_i \geq (d_{i+1} - y_i)^+ \quad (4)$$

where :

$$(d_{i+1} - y_i)^+ = \text{Max}\,(0, d_{i+1} - y_i)$$

We also use the following notation for $n \le m$

$D_{n,m} = \{d_{n+1}, \ldots, d_m\}$, $D_{n,n} = \emptyset$

$V_{n,m} = \{v_n, \ldots, v_{m-1}\}$, $V_{n,n} = \emptyset$

the later also being called a control on $[n,m]$.

Let us consider $0 \le n < m$, and suppose that $y_n \ge 0$ (stock level on $[n, n+1[$), $y_m \ge 0$ (stock level on $[m, n+1[$) and $D_{n,m}$ are known.

If $V_{n,m}$ satisfies (4) giving (2), it is said to be an admissible control for the problem $P_{n,m}(y_n, y_m, D_{n,m})$ (or $P_{n,m}(y_n, y_m)$ if any confusion can be made).

The set of admissible controls for this problem is denoted $E_{n,m}(y_n,y_m,D_{n,m})$ (or $E_{n,m}(y_n, y_m)$).

We also use the following cost functions :

$c_i(v)$, which denotes the cost of producing, or purchasing, the lot $v \ge 0$ at time i, v being available at time $i+1$.

$f_i(y)$, which denote the cost of holding in stock a quantity y at time i for one period until $i+1$.

These functions are defined on $\mathbb{R}^+$, non decreasing, concave and take their values on $\mathbb{R}^+$.

The cost corresponding to the admissible control $V_{n,m}$ for the problem $P_{n,m}(y_n,y_m,D_{n,m})$ is denoted $K_{n,m}(y_n,y_m,D_{n,m},V_{n,m})$ or $K_{n,m}(y_n,y_m,V_{n,m})$ if any confusion can be made and :

$$K_{n,m}(y_n,y_m,D_{n,m},V_{n,m}) = \sum_{i=n}^{m-1} [c_i(v_i) + f_i(y_i)] \tag{5}$$

where y_i $(i = n+1, \ldots, m-1)$ are the stock levels corresponding to $V_{n,m}$ using (2) and knowing y_n and y_m.

$V^*_{n,m} = \{v^*_n, \ldots, v^*_{m-1}\}$ is optimal if :

$$K_{n,m}(y_n,y_m,V^*_{n,m}) = K^*_{n,m}(y_n,y_m) = \min_{V_{n,m} \in E_{n,m}(y_n,y_m)} K_{n,m}(y_n,y_m,V_{n,m}) \tag{6}$$

Furthermore, we already proved in [1] that :

$$K^*_{n,m}(y_n,0) = \min_{y_m \ge 0} K^*_{n,m}(y_n,y_m) \tag{7}$$

When there are no constraints imposed on the final stock level y_m, (7) shows that y_m is equal to zero for at least one optimal control.

In that case, the notation is $P_{n,m}(y_n,D_{n,m})$ for the problem, $E_{n,m}(y_n,D_{n,m})$ for the set of admissible controls and $K^*_{n,m}(y_n,D_{n,m})$ for the optimal cost (or $P_{n,m}(y_n)$, $E_{n,m}(y_n)$ and $K^*_{n,m}(y_n)$ if any confusion can be made).

Let us now consider $P_{0,M}(y_0,D_{0,M})$ and $0<N\le M$. If o denotes the concatenation :

$D_{0,M} = D_{0,N} \circ D_{N,M}$

The components of $D_{0,N}$ are known and, in the case of $M > N$, every component of $D_{N,M}$ varies on $[0,+\infty)$.

For every $K \in [1,N]$ which verifies $y_0 - \sum_{i=1}^{K} d_i \le 0$, we denote $V^*_{0,K}$ the optimal control of $P_{0,K}(y_0, D_{0,K})$ which leads to $y_K = 0$ (y_K is the stock level on $[K, K+1[$ corresponding to $V^*_{0,K}$).

If there exists $K \in [1,N]$ so that at least one optimal control of each problem $P_{0,M}(y_0, D_{0,M})$ is obtained by extending $V^*_{0,K}$ on $K+1, \ldots, N, \ldots, M$ in an adequate manner, then K is said to be a planning horizon for the forecast horizon $N \ge K$.

We also introduce the following definition :

$i \in \{0,1, \ldots, M\}$ is a regeneration point for the problem $P_{0,M}$ if $v^*_i > 0$, where $V^*_{0,M} = \{v^*_0, v^*_1, \ldots, v^*_{M-1}\}$.

THE FINITE HORIZON PROBLEM

Let us condider $P_{0,N}(y_0,D_{0,N})$ also called a N-horizon problem. Using the concavity of the cost functions, we proved in [1] that the backward dynamic programming equations become,

with $\sigma_{i,j} = \begin{cases} \sum_{k=i}^{j} d_k & \text{if } j \ge i \\ 0 & \text{else} \end{cases}$:

$K^*_{N,N} = 0$

For $i = N-1, N-2, \ldots, 0$

$$\left.\begin{array}{l} \text{if } x \in [0, d_{i+1}[\\ K^*_{i,N}(x) = f_i(x) + \inf_{r=i+1,\ldots,N} [c_i(\sigma_{i+1,r} - x) + K^*_{i+1,N}(\sigma_{i+2,r})] \end{array}\right\} \tag{8}$$

if $x \in [\sigma_{i+1,s}, \sigma_{i+1,s+1}[$, $(s=i+1,\ldots,N-1)$

$$\left.\begin{aligned} K^*_{i,N}(x) &= f_i(x) + \\ &+ \inf\{c_i(0)+K^*_{i+1,N}(x-d_{i+1}), \\ &\inf_{r=s+1,\ldots,N} [c_i(\sigma_{i+1,r}-x) \\ &+ K^*_{i+1,N}(\sigma_{i+2,r})]\} \end{aligned}\right\} \quad (9)$$

If $i = N-1$, this statement is void.

if $x \geq \sigma_{i+1,N}$

$$\left.\begin{aligned} K^*_{i,N}(x) &= f_i(x) + c_i(0) + \\ &+ K^*_{i+1,N}(x-d_{i+1}) \end{aligned}\right\} \quad (10)$$

Considering (8) to (10), we see that we have only to consider the following stock levels (i.e. the following values for x) in equations (8), (9) and (10) :

$$\left.\begin{aligned} x^p_i &= (\sigma_{i+1,p}, y_0 - \sigma_{1,i})^+ \\ &\text{for } p = i,i+1, \ldots, N \end{aligned}\right\} \quad (11)$$

We proved in [1] the following results :

1. if both the costs are stationary ($c_i = c$ and $f_i = f$, $\forall i$), then (10) replace (9). (12)

2. if at least one of the costs is not stationary, $K^*_{0,N}(y_0)$ remains the optimal cost of the problem $P_{0,N}(y_0,D_{0,N})$ if we replace : (9) by (10) if $x > y_0 - \sigma^i_1$ (13)

But, if we use (13), it may be that $K^*_{i,N}(x)$ is not the optimal cost for the problem $P_{i,N}(x,D_{i,N})$ for $i = 1,2,\ldots,N-1$.

Remark

(12) and (13) have been obtained using the fact that the cost functions are concave and non decreasing.

Equations (8), (9) and (10), taking into account (11), (12) and (13), lead to an algorithm which is easy to implement and run very fast. As we showed in [1], the amount of computations to make in order to obtain the optimal control of $P_{0,N}(y_0,D_{0,N})$ is proportional to $\frac{N(N+1)}{2}$.

We now give a new result which allows to reduce the amount of computations.

For $i = 0,1,\ldots,N-1$, we denote r^*_i the smallest integer so that :

$$K^*_{i,N}(x^{r^*_i}_i) = \min_{p=i,i+1,\ldots,N} K^*_{i,N}(x^p_i) \quad (14)$$

The following theorem holds :

Theorem I

For $i = 2,3,\ldots,N$:

$$\left.\begin{aligned} &\min_{p=j,j+1,\ldots,N} K^*_{j,N}(x^p_j) \\ &= \min_{p=j,\ldots,r^*_i} K^*_{j,N}(x^p_j), \\ &\forall j = 1, 2, \ldots, N-1 \end{aligned}\right\} \quad (15)$$

where r^*_i is given by (14).

Proof : We let the proof to the reader for lack of place.

Corollary 1

Equation (8) can be replaced by :

$$\left.\begin{aligned} K^*_{i,N}(x) &= f_i(x) \\ &+ \inf_{r=i+1,\ldots,r^*_i} [c_i(\sigma_{i+1,r}-x) \\ &+ K^*_{i+1,N}(\sigma_{i+2,r})] \end{aligned}\right\} \quad (16)$$

Equation (9) can be replaced by :

$$\left.\begin{aligned} K^*_{i,N}(x) &= f_i(x) \\ &+ \inf\{c_i(0)+K^*_{i+1,N}(x-d_{i+1}), \\ &\inf_{r=s+1,\ldots,r^*_i} [c_i(\sigma_{i+1,r}-x) \\ &+ K^*_{i+1,N}(\sigma_{i+2,r})]\} \end{aligned}\right\} \quad (17)$$

If $s \geq r^*_i$, (17) is replaced by (10).

Proof

Obvious if we consider Theorem I and the fact that functions c_i are non decreasing.

Remark

The more the inventory costs are "high" in comparison with the production costs, the more r^*_i is small, whatever be $i \in \{0,1,\ldots,N-1\}$ and, consequently, the less is the amount of computations. We now illustrate this result with an example.

An Example

We choose $N=150$, $d_i=4$ for $i = 1,2,\ldots,150$, $y_0=0$ and the following cost functions :

$$c_i(v) = \begin{cases} 0 & \text{if } v = 0 \\ 1+v & \text{if } v > 0 \end{cases} \quad \text{for } i=0,1,\ldots,149$$

and

$$f_i(y) = \begin{cases} 0 & \text{if } y = 0 \\ L(1+y)/10^4 & \text{if } y>0 \end{cases} \quad \text{for } i=0,1,\ldots,149$$

The problem is then stationary.

The next table gives for some values of L :

- the number of regeneration points
- the number of $K^*_{i,N}(x^p_i)$ computed ($K^*_{N,N}(0)$ = 0 included). We call this number : "Number of computations".
- the number of computations divided by the maximum number of computations which is equal to N(N+1)/2+1 = 11326 (=comparison).

K	number of regeneration points	number of computations	comparison
20	10	11326	1
100	21	10411	0.92
150	25	8347	0.74
260	37	5630	0.49
300	37	5055	0.45
450	50	3658	0.32
500	50	3287	0.29

THE PLANNING HORIZON PROBLEM

We know that, in order to determine whether a given K is a planning horizon for the forecast horizon (K ≤ N), we have to take into account the set of problems $P_{0,M}(y_0, D_{0,M})$ for each M ≥ N. $D_{0,N}$ is given and, if M > N, the components of $D_{N,M}$ vary on $[0,+\infty[$. We proved in [4] the following theorem.

Theorem II

We suppose that $y_0 = 0$ (initial stock level)

$K \in [1,N]$ is a planning horizon for the forecast horizon N if and only if $y_K = 0$ for at least one optimal control of each problem $P_{0,N+1}(y_0, D_{0,N+1})$, where $D_{0,N+1} = D_{0,N} o \{z\}$ and $z \in [0,+\infty)$ (y_K is the stock level on $[K,K+1[$ corresponding to the optimal control).

Note that, in the case of $y_0=0$, the statement "$y_K=0$" is equivalent to "K is a regeneration point".

A New Result in the General Case

We consider the $P_{0,N+1}(0, D_{0,N} o \{z\})$ problems ($z \ge 0$).

We consider (8), (9) and (10) for the N+1-horizon problem and replace

$$K^*_{i,N+1}(x) \text{ by } K^*_{i,N+1}(x,z) \qquad (18)$$

Because $y_0 = 0$, we have to consider, at time i ($i \in \{0,1,\dots,N\}$), the following values for x (see (11)) :

$$\{\sigma_{i+1,\ell}\}_{\ell=i,i+1,\dots,N} \text{ and } \sigma_{i+1,N}+z \qquad (19)$$

We now prove the main result of this paper.

Theorem III

Let us consider the condition :

$H \in \{1,2,\dots,N\}$ so that :

$$\left.\begin{aligned} &\inf_{\ell=H,\dots,N} K^*_{H,N+1}(\sigma_{H+1,\ell}, z) \\ &\le K^*_{H,N+1}(\sigma_{H+1,N}+z, z),\ \forall z \ge 0 \end{aligned}\right\} \qquad (20)$$

Let $r_H(z)$ the smallest value which verifies :

$$K^*_{H,N+1}(\sigma_{H+1,r_H(z)}, z) = \inf_{\ell=H,\dots,N} K^*_{H,N+1}(\sigma_{H+1,\ell}, z)$$

We denote

$$r^*_H = \operatorname{Max}_{z\ge 0} r_H(z) \qquad (21)$$

Suppose that (20) holds.

Then, for every problem $P_{0,N+1}(0, D_{0,N} o \{z\})$, there exists at least one optimal control with a regeneration point belonging to $\{H, \dots, r^*_H\}$.

Proof : We let the proof to the reader for lack of place

Corollary 2

Suppose that condition (20) holds.

For $L \in \{H,\dots,r^*_H\}$ we consider the problem $P_{0,L}(0, D_{0,L})$. We denote $V^*_{0,L}$ the optimal control which leads to a stock level equal to 0 on [L, L+1).

If $K \in \{1,2,\dots,H-1\}$ is a regeneration point for every control $V^*_{0,L}$, then K is a planning horizon for the forecast horizon N.

Proof

From Theorem III, we see that K is a regeneration point for $P_{0,N+1}(0, D_{0,N} o \{z\})$ whatever $z \ge 0$.

Then, from Theorem II, K is a planning horizon for the forecast horizon N. □

Remarks

1. Theorem II has been proven using the concavity of the costs. The proof of Theorem III needs only the fact that the cost functions are non decreasing.

2. Note that corollary 2 gives a sufficient, but not necessary, condition in order that K be a planning horizon for the forecast horizon N.

3. The forecast horizon N have to be a chosen by the user. If it is too small it may be that

the planning horizon cannot be found. If N increases, the amount of computations increase too.

A Particular Case : The Cost Functions are Linear on $\mathbb{R}^{**}$

Suppose that, for i=0,1,2,...,N-1,N,N+1,... :

$$c_i(v) = a_i v + \chi_{v>0}\, b_i \quad (22)$$

$$f_i(y) = w_i y + \chi_{y>0}\, s_i \quad (23)$$

where a_i, b_i, w_i and s_i are non negative $\forall i$ and

$$\chi_{u>0} = \begin{cases} 1 \text{ if } u > 0 \\ 0 \text{ if } u \leq 0 \end{cases}$$

In that case (see (9) and (13)) :

$$K^*_{H,N+1}(\sigma_{H+1,\ell}, z) = \sum_{k=H}^{\ell-1} (w_k \sigma_{k+1,\ell} + \chi_{\sigma_{k+1,\ell}>0}\, s_k)$$

$$+ \inf_{j=\ell,\dots,N} \{K^*_{\ell,j}(0,0) + a_j \sigma_{j+1,N} + \chi_{\sigma_{j+1,N}+z>0}\, b_j$$

$$+ \sum_{k=j+1}^{N} [w_k \sigma_{k+1,N} + \chi_{\sigma_{k+1,N}+z>0}\, s_k]$$

$$+ [a_j + \sum_{k=j+1}^{N} w_k] z\}$$

where $K^*_{\ell,j}(0,0)$ is the optimal cost corresponding to $P_{\ell,j}(0,0,D_{\ell,j})$.

The preceding equation leads to :

$$\left.\begin{aligned} &\inf_{\ell=H,\dots,N} K^*_{H,N+1}(\sigma_{H+1,\ell}, z) \\ &= \inf_{\substack{(\ell,j)\\ H\leq \ell \leq j \leq N}} [B_{H,\ell,j} + A_j z] \end{aligned}\right\} \quad (24)$$

where

$$A_j = a_j + \sum_{k=j+1}^{N} w_k$$

and

$$B_{H,\ell,j} = \sum_{k=H}^{\ell-1} (w_k \sigma_{k+1,\ell} + \chi_{\sigma_{k+1,\ell}>0}\, s_k)$$

$$+ K^*_{\ell,j}(0,0) + a_j \sigma_{j+1,N} + \chi_{\sigma_{j+1,N}+z>0}\, b_j$$

$$+ \sum_{k=j+1}^{N} [w_k \sigma_{k+1,N} + \chi_{\sigma_{k+1,N}+z>0}\, s_k]$$

Let be, for every $j \in \{H, \dots, N\}$:

$$B^*_{H,j} = \min_{H\leq \ell \leq j} B_{H,\ell,j} \quad (25)$$

and

$\ell^*(H,j)$ the smallest integer belonging to $[H,j]$ which verifies :

$$B_{H,\ell^*(H,j),j} = B^*_{H,j} \quad (26)$$

Starting from (25), equation (24) can be rewritten :

$$\left.\begin{aligned} &\inf_{\ell=H,\dots,N} K^*_{H,N+1}(\sigma_{H+1,\ell}, z) \\ &= \inf_{H\leq j\leq N} [B^*_{H,j} + A_j . z] \end{aligned}\right\} \quad (27)$$

On the other hand :

$$K^*_{H,N+1}(\sigma_{H+1,N}+z, z) = C_H + D_H . z \quad (28)$$

where :

$$C_H = \sum_{k=H}^{N} [w_k . \sigma_{k+1,N} + \chi_{\sigma_{k+1,N}+z>0} . s_k]$$

and

$$D_H = \sum_{k=H}^{N} w_k$$

Let $U^1_{H,H}, U^1_{H,H+1}, \dots, U^1_{H,N}$ be the set $B^*_{H,H}, B^*_{H,H+1}, \dots, B^*_{H,N}$ ordered by increasing values and $V^1_H, V^1_{H+1}, \dots, V^1_N$ the corresponding values for $A_H, A_{H+1}, \dots, A_N$.

For $k \in \{H, H+1, \dots, N-1\}$, if $U^1_{H,k} = U^1_{H,k+1}$ then :

- if $V_k \geq V_{k+1}$ we cancel $(U^1_{H,k}, V^1_k)$
- if $V_k < V_{k+1}$ we cancel $(U^1_{H,k+1}, V^1_{k+1})$

Let $\{U^2_{H,k}, V^2_H\}_{k=1,\dots,L}$, with $L \leq N-H+1$, be the remaining set of couples.

In the case of $L \geq 3$, for every $K \in \{1, 2, \dots, L-2\}$ so that :

$$\frac{U^2_{H,k+2} - U^2_{H,k}}{V^2_k - V^2_{k+2}} < \frac{U^2_{H,k+1} - U^2_{H,k}}{V^2_k - V^2_{k+1}},$$

we cancel $(U^2_{H,k+1}, V^2_{k+1})$.

We denote :

$\{U^3_{H,k}, V^3_k\}_{k=1,\dots,M}$ $(M \leq L)$

the remaining set of couples.

We now introduce Theorem IV.

Theorem IV

Let us consider the condition :

$H \in \{1, 2, \dots, N\}$ so that :

$$\left.\begin{array}{l}\inf\limits_{k=1,\ldots,M} \{U^3_{H,k} + V^3_k.z\} \\ \leq C_H + D_H.z,\ \forall z \geq 0\end{array}\right\} \quad (29)$$

We denote (see (26)) :

$$r^*_H = \underset{k \in E}{\text{Max}}\ \ell^*(H,k) \quad (30)$$

where

$E = \{k/k\in(H,\ldots,N)$ and

$\exists\, j\in(1,\ldots,M)/(A_k=V^3_j \text{ and } B_{H,k}=U^3_{H,j})\}$

If (29) holds, then for every problem $P_{0,N+1}(0,D_{0,N}o\{z\})$ there exists at least one optimal control with a regeneration point belonging to $\{H, \ldots, r^*_H\}$.

Proof

Taking into account the process which leads to the sequence $\{U^3_{H,k},V^3_k\}_{k=1,\ldots,M}$ and the relations (25) and (26), it is easy to verify that Theorem IV is equivalent to Theorem III in the case of c_i and f_i are linear on $\mathbb{R}^{*+}$.□

Note that corollary 2 holds, replacing (20) by (29). r^*_H is given by (30) instead of (21).

We now give an example obtained starting from tha above results

An example

$N = 40$ (forecast horizon), $y_0 = 0$.

$d_i = 1$ for $i = 1, 2, \ldots, 40$

$$c_i(v) = \begin{cases} v+\chi_{v>0}*0.8 \text{ for } i=0,\ldots,24 \text{ and} \\ i = 30, \ldots, 39 \\ 0.8v+\chi_{v>0}*0.5 \text{ for } i=25,\ldots,29 \end{cases}$$

$$f_i(y) = \begin{cases} 0.05y+\chi_{y\ 0}*0.01 \text{ for } i=0,\ldots,24 \text{ and} \\ i = 30, \ldots, 39 \\ 0.07y+\chi_{y>0}*0.02 \text{ for } i=25,\ldots,29 \end{cases}$$

For H=34, we obtained the following common regeneration points : 4, 9, 14, 19, 24, 28. The smallest planning horizon is then 4.

CONCLUSION

This paper starts from results obtained elsewhere using the backward dynamic programming equations and the fact that the costs are concave.

We first showed that it is often possible to reduce extensively the amount of computations using an algorithm based on a backward dynamic programming formulation. In fact, if the inventory costs are "high" in comparison with the production costs, the number of computations may be proportional to 2N instead of N^2 with the initial algorithm.

We also showed that it is often possible to find a planning horizon starting from a forecast horizon. Nevertheless, it is sometimes difficult to decide, in the most general cases, if condition (20) is true or not. But, if the costs are linear on R^{*+}, the algorithm is ea easy to obtain and run very fast.

REFERENCES

[1] Bensoussan A. - Proth J.M. (1981) "Gestion des stocks avec coûts concaves" R.A.I.R.O. Automatique/Systems Analysis and Control (vol. 15, n° 3, pp. 201-220)

[2] Eppen G.D. - Gould F.J. - PASHIGIAN B.P (1969) "Extension of the planning horizon theorem in the dynamic lot size model", Management Science, vol. 15, n° 5, January.

[3] Lundin R.A. - MORTON T.E. (1975) "Planning horizons for the dynamic lot si size model : Zabel vs. Protective procedures and computational results", Operations Research, vol. 23, n° 4, July-August.

[4] Proth J.M. "Gestion de stocks avec coûts concaves : notion d'horizon de planification", Sciences de Gestion, n° 2.

[5] Zabel E. (1964) "Some generalizations of an inventory planning horizon theorem", Man. Sc. 10, pp. 465-471.

INVERSE OPERATIONS FOR FUZZY NUMBERS

D. Dubois* and H. Prade**

**Cert/Dera, B.P. 4025, 31055 Toulouse Cedex*
***LSI, Université P. Sabatier, 31062 Toulouse Cedex*

Abstract. A fuzzy number is defined as a convex normalized fuzzy set of the real line with an upper semi-continuous membership function. The addition of fuzzy numbers denoted ⊕ , and defined via a sup- t-norm convolution is now well-understood from both theoretical and computational points of view, at least for the t-norm 'min'. Moreover, there exists an "inverse", -A, of a fuzzy number A, i.e. -A restricts the possible values of the variable -u iff A restricts the possible values of u. However, the identify A ⊕ [-A] = 0 does not hold but when A is a genuine real number. Thus, the solution X of the equation of fuzzy numbers : X ⊕ A = B, when it exists, is NOT given by X = B ⊕ [-A].

An operation)+(is introduced, which enables to express the solution, when it exists, under the form X = B)+([-A] ; for instance A)+([-A] = 0. This operation)+(is shown to be related to Gödel-Brouwer implication, when a "sup-min" convolution is used in the definition of ⊕. On crisp intervals,)+(reduces to an operation sometimes known as "MINKOWSKI subtraction" : if A and B are crisp intervals and the length of A is greater than that of B, then A)+(B is an interval whose mean value (in the sense of arithmetic mean) is the sum of the mean values of A and B, and whose length is equal to the difference of the respective lengths of A and B. In terms of error analysis,)+(corresponds to the maximal compensation of errors (i.e. the optimistic case) while ⊕ corresponds to the cumulation of errors (i.e. the pessimistic case).

)+(is also defined when a t-norm other than "min" is used in the convolution defining ⊕ . For instance, if ⊕ employs the product, then)+(is based on an implication earlier considered by Goguen and Gaines.

The same approach can be applied to the multiplication.

INTRODUCTION

Imprecise numerical quantities can be conveniently represented by fuzzy numbers. The membership function of a fuzzy number fuzzily restricts the set of values which are to some extent possible for the imprecisely known real-valued quantity represented by this fuzzy number. The degree of membership of a value to the fuzzy number can be viewed as the measure of the possibility that this value is the value of the quantity. Possibility theory, as introduced by Zadeh [19], is indeed the right framework for dealing with fuzzy numbers. Arithmetic operations have been extended to fuzzy numbers (Zadeh [18]) and their properties studied (Mizumoto, Tanaka [10], Dubois, Prade [1]). The problems of performing arithmetic operations on fuzzy numbers or of ranking them is now well-understood and completely mastered from both theoretical and computational points of view (See Dubois Prade [2]-[5]). It has been shown that arithmetic operations on fuzzy numbers generalize usual operations on real numbers as well as interval analysis, as developped by R.C.H. Young [17] and Ramon Moore [11] : real numbers and crisp intervals are particular cases of fuzzy numbers when possibility degrees are {0,1} - valued, and the arithmetic operations used in interval analysis are particular cases of the ones used on fuzzy numbers.

However, from the beginning, it has been noticed that the opposite -A of a fuzzy number A (i.e. the membership function of -A restricts the possible values of the quantity -u iff the membership function of A restricts the possible values of u) is not a genuine inverse in the sense that A ⊕ (-A) = 0 does not hold but when A is a genuine real number (here ⊕ denotes the extended addition on fuzzy numbers). A similar situation exists for the extended multiplication. As a consequence, it has been more or less believed that there does not exist a simple explicit expression of the solution X of the equation A ⊕ X = B where A and B are fuzzy numbers (see for instance

Yager [16]).

In the following, an operation denoted)+(is introduced, which enables to express the solution, when it exists, under the form X = B)+((-A). Properties of)+(is shown to be related to a multivalent implication which depends on the triangular norm which is used in the definition of $\oplus$. Then, the operation)+(is shown to be the extension to fuzzy numbers of an operation on crisp intervals called 'Minkowski subtraction' and is interpreted in terms of sensitivity analysis. Finally, the concluding remarks point out that the approach can be successfully applied to the multiplication, propose alternative definitions for the possibility and the necessity of fuzzy events.

THE INVERSE OPERATION OF THE ADDITION :

1 - Background on the addition of fuzzy numbers :

By a fuzzy number A, here we mean a fuzzy set of the real line $\mathbb{R}$ whose membership function μ_A is

i) normalized, i.e. there exists at least one (but possibly more) point of $\mathbb{R}$ whose degree of membership to A is equal to 1 ;

ii) unimodal, i.e.
$\forall\lambda\in[0,1]$, $\forall$ $x\in\mathbb{R}$, $\forall$ $y\in\mathbb{R}$,

$$\mu_A(\lambda.x+(1-\lambda).y)\geqslant\min(\mu_A(x),\mu_B(y)) \quad (1)$$

in other words, the α-cuts A_α of A, defined by
$\forall\alpha\in]0,1]$, $A_\alpha = \{x\in\mathbb{R}, \mu_A(x)\geqslant\alpha\}$

are convex in the usual sense and A is then said to be convex.

iii) upper semi-continuous, which entails, taking into account ii), that the A_α's are closed intervals ;

iv) moreover, we suppose that the support of A,
$\text{supp } A = \{x\in\mathbb{R}, \mu_A(x) > 0\}$ is bounded.

μ_A is viewed as the possibility distribution which restricts the possible values of the imprecise numerical quantity, supposed to be finite, represented by A.

Given two fuzzy sets of the real line, A and B, their sum $A \oplus B$ is defined by (Zadeh [18], Dubois, Prade [4])

$$\forall z\in\mathbb{R}, \mu_{A\oplus B}(z) = \sup_{\substack{x,y\\x+y=z}} T(\mu_A(x),\mu_B(y)) \quad (2)$$

where the use of 'sup' is coherent with the axioms of possibility theory (see Dubois, Prade [4]) and where T is a triangular norm, "t-norm" for short, i.e. T is a two-place function from [0,1]x[0,1] to [0,1] such that T is i) symmetric ii) associative, iii) non-decreasing (i.e. if $a \leqslant b$ and $c \leqslant d$, then $T(a,c) \leqslant T(b,d)$ and iv) such that T(1,a) = a (see Schweizer, Sklar [15]. The greatest triangular norm is "min" (i.e. for every t-norm T, $\forall$ a, $\forall$ b, $T(a,b) \leqslant \min(a,b)$. When the quantity u, whose imprecise value is represented by A, does not interact with the quantity v, whose imprecise value is represented by B (i.e. the value of u does not depend on the value of v and conversely), we use T = min in (2). Other t-norms than min such that T(a,b) = a.b or T(a,b) = max (0, a+b-1), may be used instead of min, if we consider that the possibility of u=x and v=y, estimated by $T(\mu_A(x),\mu_B(y))$, decreases rapidly when the possibility of u=x and the possibility of v=y, estimated respectively by $\mu_A(x)$ and $\mu_B(y)$, are both less than 1. Because of the axiom iv), the behaviors of the different t-norms coincide when the possibility of u=x or the possibility of v=y is equal to 1 ; note that $\forall(a,b)\in[0,1]^2$, $\max(0, a+b-1) \leqslant a.b$). When there is a dependence relation between u and v, it must be taken into account in (2) as an extra constraint on each pair (x,y) (see Dubois, Prade [4]).

Whatever T is, $\oplus$ is clearly commutative, associative and has $\{0\}$ as a neutral element ($\mu_{\{0\}}(x)= 0$, $x \neq 0$ and $\mu_{\{0\}}(0) = 1$).

Moreover, the sum of two fuzzy numbers is a fuzzy number provided that T is continuous. Indeed, it is clear that $\oplus$ preserves the normalization, the boundedness of the support and the upper semicontinuity of the membership function (if T is continuous). Let us show that $\oplus$ also preserves the convexity. The result is already known for T = min. Let us study the general case.

2 - Convexity and t-norm-based addition :

The membership function μ_A of a normalized, support-bounded, convex fuzzy set A of the real line can be always decomposed into two separate parts, μ^+_A and μ^-_A, such that $\mu_A^+(x)$ ranges non-decreasingly from 0 to 1 when x increases from $-\infty$ to a and $\mu_A^-(x)$ ranges non-increasingly from 1 to 0 when x increases from a to $+\infty$, where a is such that $\mu_A(a) = 1$. We have :

$\mu_A = \max(\mu_A^+, \mu_A^-)$. Let us denote by A^+ (A^- resp.) the fuzzy set whose membership function is defined by μ_A^+ (μ_A^- resp.) on $(-\infty, a]$ ($[a, +\infty)$ resp.) and is zero elsewhere.

Besides, any t-norm T is distributive on max (or min) since T is non-decreasing, i.e. $T(\max(a,b),c) = \max(T(a,c), T(b,c))$. Thus, we have for two fuzzy numbers decomposed in the above manner,

$$A \oplus B = (A^+ \cup A^-) \oplus (B^+ \cup B^-) =$$

$$(A^+ \oplus B^+) \cup (A^+ \oplus B^-) \cup (A^- \oplus B^+) \cup (A^- \oplus B^-).$$

It can be easily shown that $\mu_{A^+\oplus B^+}$ is non-decreasing on its support $(-\infty, a+b]$ and that $\mu_{A^-\oplus B^-}$ is non-increasing on its support $[a+b, +\infty)$; moreover, it can be checked that $\mu_{A^+ \oplus B^-}(x)$ ($\mu_{A^- \oplus B^+}(x)$ resp.) is equal to

$\mu_{A^+}(x-b)$ on $(-\infty, a+b]$

and to

$\mu_{B^-}(x-a)$ on $[a+b, +\infty)$ ($\mu_{A^-}(x-b)$ on $[a+b, +\infty)$ and to

$\mu_{B^+}(x-a)$ on $(-\infty, a+b]$ resp.).

Thus, $\mu_{A \oplus B}$ is non-decreasing on $(-\infty, a+b]$ and non-increasing on $[a+b, +\infty)$, i.e. $\mu_{A \oplus B}$ is unimodal and then we have the result :

Theorem 1 : When A and B are convex, $A \oplus B$ is convex whatever the t-norm T, used in the definition of $\oplus$, is. However, $A \oplus B$ may be convex even if A or B is not convex. for instance, we have $[2,4] \oplus (\{1\} \cup \{2\}) = [3,6]$ or $(\{1\} \cup [2,3]) \oplus ([1,2] \cup \{3\}) = [2,6]$.

3 - Discussing the equation $A \oplus X = B$:

The convex hull $\hat{A}$ of a fuzzy set A of the real line is defined by (see Lowen [8])

$$\forall t, \mu_{\hat{A}}(t) = \sup_{\substack{x,y \\ x \leqslant t \leqslant y}} \min(\mu_A(x), \mu_A(y)) \qquad (3)$$

$\hat{A}$ is a convex fuzzy set, it is the smallest convex fuzzy set which contains A in the sense of fuzzy set inclusion ($A \subseteq B$ iff $\mu_A \leqslant \mu_B$) ; if A is convex, then $\hat{A} = A$.

Let us consider the equation $A \oplus X = B$, i.e.

$$\forall z \in \mathbb{R}, \sup_t T(\mu_A(z-t), \mu_X(t)) = \mu_B(z) \qquad (4)$$

where A, X and B are only supposed to be fuzzy sets of the real line and where X is unknown. Let X_0 be a solution, when it exists, and $\hat{X}_0$ be its convex hull.

It can be easily checked that

$$A \oplus \hat{X}_0 \supseteq B \qquad (5)$$

Let us show that

$$\hat{B} \supseteq A \oplus \hat{X}_0 \qquad (6)$$

where $\hat{B}$ is the convex hull of B.

Proof : $\mu_{A \oplus \hat{X}_0}(z)$

$$= \sup_t T(\mu_A(z-t), \sup_{\substack{x,y \\ x \leqslant t \leqslant y}} \min(\mu_{X_0}(x), \mu_{X_0}(y)))$$

$$= \sup_{\substack{x,t,y \\ x \leqslant t \leqslant y}} \min[T(\mu_A(z-t), \mu_{X_0}(x)), T(\mu_A(z-t), \mu_{X_0}(y))]$$

$$= \sup_t \min[\sup_{\substack{x \\ x \leqslant t}} T(\mu_A(z-t), \mu_{X_0}(x)), \sup_{\substack{y \\ t \leqslant y}} T(\mu_A(z-t), \mu_{X_0}(y))]$$

$$\leqslant \sup_{\substack{u,v \\ u \leqslant z \leqslant v}} \min[\sup_x T(\mu_A(u-x), \mu_{X_0}(x)), \sup_y T(\mu_A(v-y), \mu_{X_0}(y))]$$

since, if the extremum is reached for x^* and t^* in the last but one expression, we have $z-t^* \leqslant z-x^*$ and then $\exists u \leqslant z$, $u-x^* = z-t^*$.

In the last expression, we recognize $\mu_{\hat{B}}(z)$. Q.E.D.

When B is convex, $\hat{B} = B$ and (5) and (6) yield $A \oplus \hat{X}_0 = B$. Thus, we have the result

Theorem 2 : When B is convex, if X_0 is a solution of $A \oplus X = B$, then its convex hull $\hat{X}_0$ is also a solution.

Let us show the following result.

Theorem 3 : If X_0 and X_1 are two solutions of the equation $A \oplus X = B$, then their union $X_0 \cup X_1$ is also a solution.

Proof : $\mu_B(z) = \sup_s T(\mu_A(z-s), \mu_{X_0}(s))$

$$\leqslant \sup_s T(\mu_A(z-s), \max(\mu_{X_0}(s), \mu_{X_1}(s)))$$

The right part of the inequality can be rewritten

$$\mu_{A \oplus (X_0 \cup X_1)}(s)$$

$$= \sup_s \max(T(\mu_A(z-s), \mu_{X_0}(s)), T(\mu_A(z-s), \mu_{X_1}(s)))$$

$$= \max[\sup_s T(\mu_A(z-s), \mu_{X_0}(s)), \sup_t T(\mu_A(z-t), \mu_{X_1}(t))]$$

$$= \max(\mu_B(z), \mu_B(z)) = \mu_B(z)$$

Q.E.D.

Using theorem 3, we deduce that there exists a greatest solution in the sense of the inclusion of fuzzy sets ($F \subseteq G$ iff $\mu_F \leqslant \mu_G$) if the considered t-norm T is continuous and if the equation $A \oplus X = B$ has a solution, since then a non-decreasing, upper bounded sequence of solutions can be built and the continuity of T is a sufficient condition to ensure that the limit is still a solution.

Then, from theorem 2, we deduce that, if B is convex, then the greatest solution (when there exists a solution) is convex. Moreover, when the t-norm min is used in the definition of $\oplus$, the greatest solution can be obtained as the convex hull of any existing solution, if B is convex ; this result can be proved using $\forall \alpha \in]0,1]$,

$(A \oplus_{min} B)_\alpha = A_\alpha \oplus_{min} B_\alpha$ provided that μ_A and μ_B are upper semi-continuous. However, the convex hull of a solution does not always yields the greatest solution for any t-norm ; a counter-example can be easily built using the least t-norm $T_W(a,b) =$ a if b = 1, b if a = 1, 0 otherwise, and results given in Dubois Prade [4].

Lastly, it is clear that the equation $A \oplus X = B$ has not always a solution. For instance, if μ_B is the characteristic function of a real number and if μ_A is the membership function of a genuine fuzzy set (i.e. $\exists x_0,\ 1 > \mu_A(x_0) > 0$), the equation has no solution, because A ⊕ X cannot be a real number whatever X is. Even if A and B crisp sets, X may not exist, e.g. $[0,1] \oplus X = [1,2] \cup \{3\}$.

4 - Solving the equation A ⊕ X = B :

The greatest solution $\mu_{\bar{X}}$ (when there exists a solution) of the functional equation

$$\forall_z, \mu_B(z) = \sup_x T(\mu_A(z-x), \mu_X(x)) \text{ where T is supposed}$$

to be continuous is given by

$$\mu_{\bar{X}}(x) = \inf_z \sup(\{a \in [0,1], T(\mu_A(z-x), a) \leqslant \mu_B(z)\}) \quad (7)$$

since we are looking for the greatest possible quantity a such that $T(\mu_A(z-x),a)$ is less or equal to $\mu_B(z)$ for every z. Note that the supremum in (7) does belong to the set $\{a \in [0,1],\ T(\mu_A(z-x),a) \leqslant \mu_B(z)\}$ since T is continuous. The following table yields the values of the expression $\sup(\{a \in [0,1], T(s,a) \leqslant t\})$ for the main continuous t-norms :

min(s,t)	$\begin{cases} 1 \text{ if } s \leqslant t \\ t \text{ if } s > t \end{cases}$
s.t	$\begin{cases} 1 \text{ if } s = 0 \\ \min(1,(1/s) \times t) \text{ if } s \neq 0 \end{cases}$
max(0,s+t-1)	min(1,1-s+t)

Table 1

Some well-known multivalent implication functions can be recognized in the right part of the table, namely, for t=min, the so-called Gödel-Brouwer implication ; for T = product, an implication function already considered by Goguen [7] and Gaines [8] ; for T(s,t) = max (0, s+t-1), the Łukasiewicz implication. See Dubois Prade [3] and Prade [13] for a general presentation of implication functions.

Any solution (when there exists a solution) of the equation A ⊕ X = B, where A and B are fuzzy numbers and ⊕ is based on a continuous t-norm, is normalized and support-bounded ; moreover $\bar{X}$, given by (7) is convex.

N.B.1 : For the non-continuous t-norm T_W, $\sup(\{a \in [0,1], T_W(s,a) \leqslant t\}) = 1$ if $s \neq 1$ and is equal to t if s = 1. Using T_W in (2), the sum of two non-normalized sets is the empty set. Then, it can be easily checked that $\bar{X}$ given by (7) is not a solution of A ⊕ X = Ø where A is non-normalized and ⊕ defined using T_W.

N.B.2 : The equation A ⊕ X = B is a particular case of fuzzy relation equations (see for instance Dubois Prade [2]). As such its greatest solution (7) can be obtained using the general result established by Sanchez [14] for T = min and recently extended to any t-norm by Pedrycz [12].

5 - The extended Minkowski subtraction :

As pointed out in Prade [13], the expression (7) (when A and B are crisp sets) is related to a set operation which is sometimes known as "Minkowski subtraction" (see Matheron [9] for instance). Let us recall how Minkowski subtraction is introduced.

The addition of fuzzy numbers, defined by (2), can be viewed as a generalization of the addition of crisp sets (also called "Minkowski addition") defined by

$$A \oplus B = \{x + y,\ x \in A,\ y \in B\} \quad (8)$$

which can be shown to be equal to

$$A \oplus B = \{z,\ A \cap (-B)_z \neq \emptyset\} \quad \text{where} \quad (9)$$

$(-B)_z = (-B) \oplus \{z\}$ with $-B = \{-x,\ x \in B\}$

since $z - x \in B \Leftrightarrow x \in (-B_z)$.

By duality, Minkowski subtraction is defined as

$$A\ )+(\ B = \overline{(\bar{A} \oplus B)} \quad (10)$$
$$\text{i.e. } A\ )+(\ B = \{z,\ (-B)_z \subseteq A\} \quad (11)$$

Note that $A \oplus B = \overline{(\bar{A}\ )+(\ B)}$.

Thus, $A \oplus B = \{z, \exists x \in A, \exists y \in B, x+y=z\}$ (12)
While $A\ )+(\ B = \{z, \forall y \in B, \exists x \in A, x+y=z\}$ (13)

Viewing B as the set of the possible values of an unknown but precise quantity b, A)+(B is the set of values necessarily (certainly) covered by A ⊕ {b} and A ⊕ B is the set of all the values possibly covered by A ⊕ {b}.

For instance, for A = [a,a'] and B = [b,b'], we get

A ⊕ B = [a + b, a' + b']
A)+(B = [a + b', a' + b] if a'+b ⩾ a+b'
Ø otherwise

Note that here the length of A ⊕ B, (a'+b')-(a+b) is the length of A, a'-a, plus the length of B, b'-b, while the length of A)+(B, (a'+b)-(a+b') is the length of A minus the length of B.Thus, A ⊕ B is a dilating addition while A)+(B is an eroding addition (A being eroded by B) when A)+(B ≠ Ø.

We have A)+(B ⊆ A ⊕ B

The two operations)+(and ⊕ coincide when B is a real number and they reduce to the ordinary addition + when A and B are real numbers.

It can be checked that

$$(A \oplus B)\ )+(\ (-B) = A \quad (14)$$

when A and B are closed intervals. Note that the right part of (14) is always included in the left part ; however the inclusion may be

strict :
e.g. if $A =]a,a'[$, $B =]b,b'[$,
then $(A \oplus B)$)+($(-B) = [a,a']$.
If A)+($B \neq \emptyset$,

$$(A \,)+(\, B) \oplus (-B) = A \qquad (15)$$

when A and B are closed intervals. Note that the left part of (15) is always included in the right part. For $A = \{0\}$, (14) yields

$$B \,)+(\, (-B) = \{0\} \qquad (16)$$

It can be checked that (16) holds for any bounded subset B of the real line $\mathbb{R}$.

When A and B are crisp sets, the formula (7) clearly reduces to

$$\overline{X} = \{x,\ A_X \subseteq B\} \qquad (17)$$

since $\mu_{A \oplus \{x\}}(z) = \mu_A(z-x)$

and $\sup(\{a \in [0,1], T(s,a) \leqslant t\})$ is equal to the ordinary implication function for $(s,t) \in \{0,1\}^2$.

Then, (11) and (17) give

$$\overline{X} = B \,)+(\, (-A) \qquad (18)$$

Thus, Minkowski subtraction can be extended to fuzzy sets by the following definition

$$\mu_{B\,)+(\,A}(z) = \inf_t(\mu_A(z-t) \,\alpha_T\, \mu_B(t)) \qquad (19)$$

where T is a t-norm and $\forall\ (s,t) \in [0,1]^2$,
$s \,\alpha_T\, t = \sup(\{a \in [0,1], T(s,a) \leqslant t\})$,
since $\mu_{-A}(x) = \mu_A(-x)$.

N.B. B)+($A = \emptyset$ iff $\forall\ z, \exists\ t$,

$$\mu_{(-A)_z}(t) \,\alpha_T\, \mu_B(t) = 0.$$

i.e., from table 1, if T is min or the product

$\forall\ z, \exists t, \mu_{(-A)_z}(t) > 0$ and $\mu_B(t) = 0$

$$\Longleftrightarrow \forall\ z,\ \text{supp}\ A_z \not\subseteq \text{supp}\ B \qquad (20)$$

and if $T(s,t) = \max(0, s+t-1)$ or $T(s,t) = T_W(s,t)$
$\forall\ z, \exists t, \mu_{(-A)_z}(t) = 1$ and $\mu_B(t) = 0$

$$\Longleftrightarrow \forall\ z,\ \text{core}\ (A_z) \not\subseteq \text{supp}\ B \qquad (21)$$

where core $(A) = \{x, \mu_A(x) = 1\}$.

If we consider the equation $X \oplus B = A \oplus B$ where A, B and X are fuzzy sets and $\oplus$ is based on a continuous t-norm, we know that its greatest solution is $\overline{X} = (A \oplus B)$)+($(-B)$ and then

$$(A \oplus B) \,)+(\, (-B) \supseteq A \qquad (22)$$

which generalizes (14). For T = min, if A is convex and if μ_A is upper semi-continuous, the equality holds.

Then, we have for any fuzzy set B
B)+($(-B) = \{0\}$.

Note that we may have B)+($A \neq \emptyset$ while the equation $(-A) \oplus X = B$ has no solution. This fact can be easily seen for T = min using results about the addition of fuzzy numbers. However, si B)+($A = \emptyset$, the equation has clearly no solution.

Let us now look at some properties of the operation)+(.

6 - Some other properties of the operation)+(:

The following properties hold

. $A \,)+(\, (B \cup C) = (A \,)+(\, B) \cap (A \,)+(\, C)$ (23)
. $(B \cap C) \,)+(\, A = (B \,)+(\, A) \cap (C \,)+(\, A)$ (24)
. $A \,)+(\, (B \cap C) \supseteq (A \,)+(\, B) \cup (A \,)+(\, C)$ (25)
. $(B \cup C) \,)+(\, A \supseteq (B \,)+(\, A) \cup (C \,)+(\, A)$ (26)

where $\cup$ and $\cap$ are defined using max and min respectively.

In order to prove (23) we have to check that $\forall\ (r,s,t) \in [0,1]^3$,
$\max(r,s) \,\alpha_T\, t = \min(r \,\alpha_T\, t,\ s \,\alpha_T\, t)$

which holds since if $r \geqslant s$, $r \alpha_T t \leqslant s \alpha_T t$ due to the non-decreasingness of T, α_T being defined by
$s \alpha_T t = \sup(\{a \in [0,1], T(s,a) \leqslant t\})$. (24) can be proved noticing that $\forall \in (r,s,t) \in [0,1]^3$,
$r \,\alpha_T \min(s,t) = \min(r \,\alpha_T\, s, r \,\alpha_T\, t)$. (25) and (26) can be proved in a similar way.

Moreover if $B \subseteq C$ (i.e. $\mu_B \leqslant \mu_C$), then A)+($B \supseteq A$)+(C and B)+($A \subseteq C$)+(A.

Lastly, when the equation $X \oplus (-B) \oplus (-C) = A$ has a solution, it can be easily proved that

$$A \,)+(\, (B \oplus C) = (A \,)+(\, B) \,)+(\, C. \qquad (27)$$

7 - Practical computation and interpretation :

For the easy computing of the sum of fuzzy numbers, a so-called L-R representation has been introduced (see Dubois Prade [1]-[4]). The membership function of a L-R-represented fuzzy number A can be written under the form

$$\mu_A(x) = \begin{cases} L(\frac{a-x}{\alpha}) & \text{if } x \leqslant a \\ R(\frac{x-a}{\alpha'}) & \text{if } x \geqslant a \end{cases} \qquad (28)$$

where $\alpha > 0$, $\alpha' > 0$ and L and R are non-increasing functions from [0,1] to [0,1] such that $L(0) = R(0) = 1$, $L(1) = R(1) = 0$.

Symbolically, A is written $A = (a, \alpha, \alpha')_{LR}$.
With $A = (a, \alpha, \alpha')_{LR}$ and $B = (b, \beta, \beta')_{LR}$,
we have $(-A) = (-a, \alpha', \alpha)_{RL}$
$A \oplus B = (a+b, \alpha + \beta, \alpha' + \beta')_{LR}$

Then, obviously we have for R = L
A)+($B = (a+b, \alpha - \beta', \alpha' - \beta)_{LL}$
provided that $\alpha \geqslant \beta'$ and $\alpha' \geqslant \beta$,
i.e. a L-L fuzzy number.

Particulary, with crisp intervals or L-R fuzzy numbers, the difference of nature between the two additions $\oplus$ and)+(appears to be clear. In terms of sensitivity analysis, $\oplus$ corresponds to the cumulation of errors ($\mu_{A \oplus B}(z)$ is the <u>possibility</u> that the sum of the imprecisely known quantities, represented by A and B respectively, is equal to z) while)+(corresponds to the maximal compensation of the error on the quantity represented by B, (when the equation $X \oplus (-B) = A$ has a solution, $\mu_{A\,)+(\,B}(z)$ estimates to what extent z "necessarily" belongs to the support of $A \oplus \{b\}$ where b is the imprecisely known quantity represented by B). In that respect, $\oplus$ corresponds to the pessimistic case and)+(to the optimistic one.

CONCLUDING REMARKS

The same approach can be applied to other group operations such as the multiplication of real numbers. Thus, an inverse operation)×(can be associated with the extended multiplication $\otimes$ (see Dubois Prade [1]-[3]).

(2) and (19) being respectively replaced by,

$$\mu_{A \otimes X}(z) = \sup_t T(\mu_A(z/t), \mu_B(t))$$

and $\mu_{A\ )\times(\ B}(z) = \inf_t (\mu_B(z/t)\ \alpha_T\ \mu_A(t))$.

Besides, it is worth noticing that the equality $A\ )+(\ B = \overline{(\overline{A} \oplus B)}$ no longer holds with non-crisp sets in the general case. However there exists at least one t-norm for which the equality holds for fuzzy sets, namely $T(s,t) = \max(0,s+t-1)$ since $1 - \max(0,1-s+t-1) = \min(1,1-t+s) = t\ \alpha_T s$ if we use $\mu_{\overline{A}} = 1 - \mu_A$ for defining the complementation.

But with T = min, the characteristic function of $\overline{(\overline{A} \oplus B)}$ is

$$\mu_{\overline{(\overline{A} \oplus B)}}(z) = \inf_x \max(1-\mu_B(z-x), \mu_A(x))$$

which clearly differs from (19). Thus, by duality, for T = min, we can introduce a new "addition" $\boxplus$ defined by $\overline{A}\ )+(\ B$, namely

$$\mu_{A \boxplus B}(z) = \sup_x (\mu_A(x)\ \varsigma\ \mu_B(z-x))$$

$s\ \varsigma\ t = \begin{cases} 0 & \text{if } s+t \leqslant 1 \\ s & \text{if } s+t > 1 \end{cases}$. Note that $A \boxplus B \neq B \boxplus A$

when A or B are non-crisp.

These remarks induce us to make the following suggestion.

More generally, we may introduce (in connection with the t-norm min) an alternative measure of "possibility" for a fuzzy event E :

$\Pi(E) = \sup_{x \in X} \pi(x)\ \varsigma\ \mu_E(x)$ (the usual definition being $\Pi(E) = \sup_{x \in X} \min(\mu_E(x), \pi(x))$)

and an alternative measure of "necessity"

$N(E) = \inf_{x \in X} (\pi(x)\ \alpha_{\min}\ \mu_E(x))$ (the usual definition being

$$N(E) = \inf_{x \in X} \max(\mu_E(x), 1 - \pi(x))).$$

See, for instance, Prade [13] for a discussion of the usual definition of the possibility and of the necessity of a fuzzy event.

As a final remark the close connection that there exists between several results of this paper and general results in minimax algebra (see [21]) must be pointed out. The extensive use of this connection will be made in forthcoming papers.

ACKNOWLEDGEMENTS

Special thanks are due to Elie Sanchez for his useful comments on an earlier version of this paper during its refering process. Sanchez has himself contributed several important results concerning inverse operations in a recent paper[20] which was not available at the time when this paper was written.

REFERENCES

1 Dubois,D.,Prade, H.(1978) Operations on fuzzy numbers. Int.J.of Sys.Sci., 9, 613-626.
2 Dubois,D.,Prade,H.(1979) Fuzzy real algebra : some results.Fuzzy Sets & Systems, 2,327-348.
3 Dubois,D.,Prade,H.(1980) Fuzzy Sets & Systems : Theory and Applications,Academic Press,New-York.
4 Dubois,D., Prade, H.(1981) Additions of interactive fuzzy numbers.IEEE Transactions on Automatic Control,26, 926-936.
5 Dubois,D.,Prade,H.(1982) Ranking fuzzy numbers in the setting of possibility theory. Submitted to Information Sciences.
6 Gaines,B.R.(1976) Foundations of fuzzy reasoning.Int.J.Man-Machine Stu.,8, 623-668.
7 Goguen,J.A.(1969) The logic of inexact concepts. Synthese, 19, 325-373.
8 Lowen,R.(1980) Convex fuzzy sets. Fuzzy Sets & Systems, 3, p. 291-310.
9 Matheron,G.(1975) Random Sets and Integral Geometry, Wiley.
10 Mizumoto,M.,Tanaka-K.(1979) Some properties of fuzzy numbers. in Advances in Fuzzy Set Theory and Applications(M.M.Gupta,R.K.Ragade, R.R.Yager,eds). North-Holland,153-164.
11 Moore,R.(1966) Interval analysis.Prentice-Hall
12 Pedrycz,W.(1982)Fuzzy relational equations with triangular norms and their resolutions. BUSEFAL n° 11, Toulouse, 24-32.
13 Prade,H.(1982) Modèles mathématiques de l'Imprécis et de l'Incertain en Vue d'Applications au Raisonnement Naturel.(358p.) Thèse d'Etat, Université Paul Sabatier, Toulouse.
14 Sanchez,E.(1977) Solutions in composite fuzzy relation equations : Application to medical diagnosis in Brouwerian logic. in "Fuzzy Automata and Decision Processes".(M.M. Gupta, G.N. Saridis,B.R.Gaines,eds.)North-Holland.
15 Schweizer, B. Sklar,A.(1963)Associative functions and abstract semigroups.Publ. Math. Debrecen. 10, 69-81.
16 Yager,R.R.(1980)On the lack of inverses in fuzzy arithmetic.Fuzzy Sets & Systems. 4,73-82.
17 Young, R.C.H.(1931)The algebra of many-valued quantities. Math.Ann., 104, 260-290.
18 Zadeh,L.A.(1975) The concept of a linguistic variable and its application to approximate reasoning. Info. Sci.,1 :8, 199-249, 2 : 8, 301-357, 3 : 9, 43-80.
19 Zadeh,L.A.(1978) Fuzzy sets as a basis for a theory of possibility. Fuzzy Sets & Systems,1.
20 Sanchez,E.(1982)Solution of fuzzy equations with extended operations.Memo N° UCB/ERL M82/63, Berkeley.
21 Cuninghame-Green,R.A.(1976) Projections in minimax algebra. Math. Programming, Vol.10, 111-123

ARITHMETIC OPERATIONS ON LEVEL SETS OF CONVEX FUZZY NUMBERS

W. K. Chang*, Louis R. Chow and S. K. Chang*****

**Graduate Institute of Management Sciences, Tamkang University, Taiwan, Republic of China*
***Graduate School of Information Science, Tamkang University, Taiwan, Republic of China*
Department of Electrical Engineering, Illinois Institute of Technology, Chicago, Illinois 60616 U.S.A.

Abstract. Four arithmetic operations on level sets of convex fuzzy numbers, such as addition, subtraction, multiplication and division are introduced. These operations generalize interval analysis and are computationally attractive, requiring no more computations than those cases of dealing with error intervals in conventional tolerance analysis. For those positive convex fuzzy numbers, calculations performed by the proposed operations are extremely easy to proceed.

Keywords. Arithmetic operations; fuzzy sets; convex fuzzy numbers; level sets; extension principle.

INTRODUCTION

Four arithmetic operations on level sets of convex fuzzy numbers are developed, based on the extension principle (Zadeh, 1975). These operations take advantages of the usual nonfuzzy operations but are concerned with fuzzy quantities.

An example to the resistance circuits is illustrated. It will be shown that the tolerance level of an equivalent resistance is easily obtained by the proposed operations.

NOTATIONS

A fuzzy number A in a real line X is characterized by a membership function (Kaufmann, 1975):

$$f_A : X \to [0,1] \tag{1}$$

and is denoted as:

$$A = \{ f_A(x) / x \mid x \in X \}. \tag{2}$$

The support S(A) of a fuzzy number A is an ordinary nonfuzzy set on X:

$$S(A) = \{ x \mid f_A(x) > 0 \}. \tag{3}$$

A fuzzy number A is said to be positive if its support is in a positive real line. Similarly, A is called negative if in a negative real line. If the support of A includes 0, then A is a zero fuzzy number.

Further, a fuzzy number A is convex if and only if

$$f_A(t \cdot x_1 + (1-t)\, x_2) \geq \min \{ f_A(x_1), f_A(x_2) \} \tag{4}$$

for all x_1, x_2 in X and t in [0,1]. Note that this definition does not imply that $f_A(x)$ must be a convex function (Zadeh, 1965).

The t level set of a fuzzy number A is a nonfuzzy set A_t defined by

$$A_t = \{ x \mid f_A(x) \geq t,\ t \in (0,1] \}. \tag{5}$$

Some authors (e.g., Dubois and Prade, 1980) used the name t-cut instead of t level set.

The global point G(A) of a fuzzy number A is defined as:

$$G(A) = \max \{ x \mid f_A(x) = 1 \}. \tag{6}$$

Besides, the right spread $R_t(A)$ of a fuzzy number A at t level is the maximum of its t level set, i.e.,

$$R_t(A) = \max \{ x \mid f_A(x) \geq t \}. \tag{7}$$

Similarly, the left spread $L_t(A)$ of a fuzzy number A at t level is the minimum of its t level set,

$$L_t(A) = \min \{ x \mid f_A(x) \geq t \}. \tag{8}$$

The extension principle introduced by Zadeh may extend the algebraic operations for fuzzy numbers. For instance, the extended addition, A(+)B, and multiplication, A(.)B, between two fuzzy numbers A and B are defined as

$$f_{A(+)B}(x) = \sup_{x=a+b} \min \{ f_A(a),\ f_B(b) \} \quad (9)$$

$$f_{A(.)B}(x) = \sup_{x=a.b} \min \{ f_A(a),\ f_B(b) \} \quad (10)$$

ARITHMETIC OPERATIONS

On the basis of the extension principle, we may perform various extended arithmetic operations which have the following two properties (Mizumota and Tanaka, 1976):

Theorem 1. If A and B are two normal convex fuzzy numbers, then the extended fuzzy numbers A(+)B, A(−)B and A(.)B are also normal and convex.

Theorem 2. If A and B are normal and convex, and B is not a zero fuzzy number, then the extended fuzzy number A(÷)B is a normal convex number.

Furthermore, Mizumota and Tanaka (1976) stated without proof that the following property is valid under the addition operation.

Theorem 3. If both A and B are two convex fuzzy numbers, then

$$(A\ (+)\ B)_t = A_t + B_t \quad (11)$$

Proof. Since A and B are convex, we may consider two different parts of them, i.e., nondecreasing and nonincreasing parts at both sides of G(.) (see Fig. 1).

Firstly, let us consider the nondecreasing parts. Suppose $a \in [a_1, a_2] \subset S(A)$ and $b \in [b_1, b_2] \subset S(B)$ such that $f_A(a) = f_B(b) = t$ (possibly $a_1 = a_2$ or $b_1 = b_2$). Let the sum $c = a + b$. To evaluate $f_C(c)$ at t level, we must consider every possible pair (x,y) such that $c = x + y$, for all $x \in A$, $y \in B$ and $f_C(c) = t$. But we have

(a) If $x \in [a_1, a_2]$ and $y \in [b_1, b_2]$, then

$$\min \{ f_A(x),\ f_B(y) \} = t.$$

(b) If $x < a_1$, then $f_A(x) < f_A(a_1) = t$, since $f_A(.)$ is nondecreasing on (x, G(A)). So

$$\min \{ f_A(x),\ f_B(y) \} < t.$$

(c) If $x > a_2$, then $y < b_1$, because the addition is a continuous increasing operation. Similarly since $f_B(.)$ is nondecreasing on (y, G(B)), $f_B(y) < f_B(b_1) = t$. So

$$\min \{ f_A(x),\ f_B(y) \} < t.$$

Therefore, the maximum of $\min \{f_A(x), f_B(y)\}$ is reached at $x = a$ and $y = b$. In other words, there exist $a \in [a_1, a_2] \subset S(A)$ and $b \in [b_1, b_2] \subset S(B)$ such that

$$f_C(c) = \sup_{c=x+y} \min \{ f_A(x),\ f_B(y) \}.$$
$$= t$$

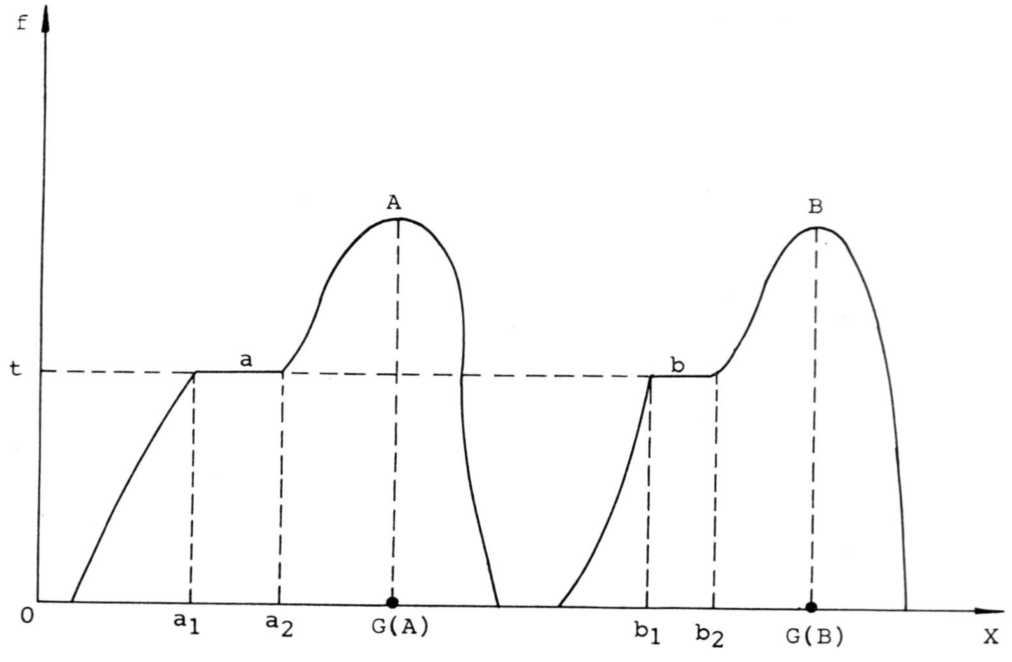

Fig. 1 Two convex fuzzy numbers

Secondly, a similar result will hold when we consider the nonincreasing parts of A and B. Thus, the desired fact is followed by Nguyen (1978). Q.E.D.

It is obvious that the extended subtraction and multiplication also hold the same property.

<u>Corollary 4</u>. If A and B are two convex fuzzy numbers, then

$$(A (-) B)_t = A_t - B_t \tag{12}$$

Proof. Since by the last theorem,

$$(A (-) B)_t = (A (+) (-B))_t$$
$$= (A (+) C)_t$$
$$= A_t + C_t$$

where $C = -B$, that is, $f_C(x) = f_B(-x)$. Then

$$C_t = \{ x \mid f_C(x) \geq t \}$$
$$= \{ x = -y \mid f_B(y) \geq t \}$$
$$= - B_t \qquad \text{Q.E.D.}$$

<u>Corollary 5</u>. If A and B are convex, then

$$(A (.) B)_t = A_t \cdot B_t \tag{13}$$

Proof. The reason is obvious. No matter what signs of A and B are, the maximal and minimal values of $(A (.) B)_t$ are determined by those of the product $A_t \cdot B_t$. Q.E.D.

<u>Theorem 6</u>. If A and B are convex, and B is not a zero fuzzy number, then

$$(A (\div) B)_t = A_t \div B_t \tag{14}$$

Proof. Since

$$(A (\div) B)_t = (A (.) (1 \div B))_t$$
$$= (A (.) C)_t$$
$$= A_t \cdot C_t, \text{ by Corollary 5,}$$

where $C = 1 \div B$. But

$$C_t = \{ x \mid f_C(x) \geq t \}$$
$$= \{ x = 1 \div y \mid f_B(y) \geq t \}$$
$$= 1 \div B_t \qquad \text{Q.E.D.}$$

In terms of the level sets, the above properties may be denoted in another way. Assume that A and B are two convex fuzzy numbers in the following.

<u>Theorem 7</u>.

$$R_t(A (+) B) = R_t(A) + R_t(B) \tag{15}$$
$$L_t(A (+) B) = L_t(A) + L_t(B) \tag{16}$$

Proof. The proof is straightforward. For t in (0,1], we have, by Theorem 3,

$$(A (+) B)_t = A_t + B_t$$

Take the maximum value of both sides:

$$\max (A (+) B)_t = R_t(A (+) B)$$

and since + is an increasing operation,

$$\max (A_t + B_t) = \max A_t + \max B_t$$
$$= R_t(A) + R_t(B).$$

The same for the minimum case. Q.E.D.

<u>Theorem 8</u>.

$$R_t(A (-) B) = R_t(A) - L_t(B) \tag{17}$$
$$L_t(A (-) B) = L_t(A) - R_t(B) \tag{18}$$

Proof. Let $A_t= [a_1, a_2]$, $B_t= [b_1, b_2]$, $a_1 \leq a_2$, $b_1 \leq b_2$. Since by Corollary 4, we have

$$(A (-) B)_t = A_t - B_t$$

$$= [a_1 - b_2, a_2 - b_1].$$

Or using the spread notation of t level set,

$$R_t(A (-) B) = \max (A (-) B)_t$$
$$= \max [a_1 - b_2, a_2 - b_1]$$
$$= a_2 - b_1$$
$$= R_t(A) - L_t(B)$$

and

$$L_t(A (-) B) = a_1 - b_2 = L_t(A) - R_t(B)$$

Note that $a_2 - b_1 \geq a_1 - b_2$ always holds. Q.E.D.

<u>Theorem 9</u>. If A and B are positive fuzzy numbers, then

$$R_t(A (.) B) = R_t(A) \cdot R_t(B) \tag{19}$$
$$L_t(A (.) B) = L_t(A) \cdot L_t(B) \tag{20}$$

Proof. The proof is analogous to that for Theorem 7, since the multiplication is an increasing operation as the addition, provided that its operands are all positive.

<u>Theorem 10</u>. If both A and B are positive, then

$$R_t(A (\div) B) = R_t(A) \div L_t(B) \tag{21}$$
$$L_t(A (\div) B) = L_t(A) \div R_t(B) \tag{22}$$

Proof. From Theorem 6, we have

$$(A (\div) B)_t = A_t \div B_t$$

Take the maximum value of both sides:

$$\max (A (\div) B)_t = R_t(A (\div) B)$$

and since $\div$ is a decreasing operation,

$$\max (A_t \div B_t) = R_t(A) \div L_t(B)$$

Similarly, the minimum case can be proved. Q.E.D.

A NUMERICAL EXAMPLE

Since the extended operations of fuzzy numbers generalize those of nonfuzzy tolerance analysis (Dubois and Prade, 1979), the extreme points of t-level set of a fuzzy number may be considered to constitute a tolernace interval at t degree of confidence, where the degree of confidence is meant to reflect the reliability of a measure.

As a simple illustration, suppose that there are three resistances having 10% of tolerance and being uniformly distributed with mean equal to 100 ohms. Alternatively, each resistance is considered as a level set with t = 1.0 :

$$R_1 = R_2 = R_3 = [90, 110].$$

Then by the proposed operations on level sets, the equivalent resistance R_A of the series combination of R_1 and R_2 is

$$R_A = R_1 + R_2 = [180, 220],$$

while the equivalent resistance R_B for the circuit in which R_1 is in series with a parallel combination of R_2 and R_3 is given as

$$R_B = R_1 + R_2 \cdot R_3 \div (R_2 + R_3)$$
$$= [127, 177].$$

It can be seen that our results are consistent with those nonfuzzy quantities, i.e., 200 ohms for R_A and 150 ohms for R_B.

Although Jain (1976) made use of the algebraic sum and the min product for addition and multiplication of fuzzy numbers, respectively, and obtained the equivalent fuzzy resistances, the number of elements in the resultant fuzzy number will grow very large and need enormous computational effort (Dubois and Prade, 1978).

CONCLUSION

So far algebraic operations on fuzzy numbers have been extensively developed by means of the extension principle. But most operations are considered through all of the support elements of a fuzzy number. In practical applications, however, it is still informative by taking level sets of fuzzy numbers, i.e., by studying the most significant part only of the support elements in order to save operation times as well as storage. The main advantage of this approach lies on the fact that level sets of a fuzzy number are no longer fuzzy. Hence usual operations may be applied immediately without much modifications.

It has been shown that we are able to obtain, without much effort, the tolerance level of an equivalent resistance in a circuit by the proposed operations on level sets. In particular, these operations are helpful in comparing two convex fuzzy numbers of the same family (Chang, 1982).

REFERENCES

Chang, W. K. A study on the ranking of fuzzy alternatives and its application to decision making, Ph.D. dissertation, Chap. 2, Tamkang Univ., May, 1982.

Dubois, D. and Prade, H. (1978). Comment on "Tolerance analysis using fuzzy sets" and "A procedure for multiple aspect decision making," Int. J. Syst. Sci., 9, 357-360.

Dubois, D. and Prade, H. (1979). Fuzzy real algebra: some results, Fuzzy Sets and Systems, 2, 327-348.

Dubois, D. and Prade, H., Fuzzy Sets and Systems: Theory and Applications, Academic Press, New York, 1980.

Jain, R. (1976). Tolerance analysis using fuzzy sets, Int. J. Syst. Sci., 7, 1393-1401.

Kaufmann, A., Introduction to the Theory of Fuzzy Subsets, Vol I, Academic Press, New York, 1975.

Mizumoto, M. and Tanaka, K. (1976). The four operations of arithmetic on fuzzy numbers, Systems, Computers and Controls, 7, 73-81.

Nguyen, H. T. (1978). A note on the extension principle for fuzzy sets, J. Math. Anal. and Appl., 64, 369-380.

Zadeh, L. A. (1965). Fuzzy sets, Inf. and Control, 8, 338-353.

Zadeh, L. A. (1975). The concepts of a linguistic variable and its application to approximate reasoning, (Part I), Inf. Sci., 8, 199-249.

RANKING OF FUZZY ALTERNATIVES IN ELECTROCARDIOGRAPHY

G. Bortolan and R. T. Degani

*Institute for Research on System Dynamics and Bioengineering (LADSEB) — CNR
Corso Stati Uniti 4 — 35100 Padova, Italy*

Abstract. A system for computerized electrocardiography, where the final classification of the ECG is obtained through a fuzzy decision, is briefly described. The arising problem of ranking fuzzy utilities is explored. Different methods for ordering fuzzy quantities are applied to the simulation of a diagnosis of inferior myocardial infarction, in order to test the validity of the results. It is concluded that in general all methods are comparable, possibly with a slight preference in this practical context for the "center of gravity" suggested by Yager.

Keywords. Computer applications; decision-making; electrocardiography; fuzzy logic; medical information processing.

INTRODUCTION

The automatic analysis of the electrocardiogram by computers is a well established procedure, which has received attention for more than 20 years. For a recent review on this subject see for example Jenkins (1981). Yet further work is being developed in this field in order to improve the quality of the results. The reason for this continuous effort is easily explained both by the ongoing medical research in the cardiac domain and by the lack of 100% correct quantitative procedures.

The main approaches to ECG classification are the deterministic simulation of the diagnostic process followed by the cardiologist and the evaluation of the a posteriori probability of a diagnosis, given the actual ECG. The first method is easily accepted by the medical community because it is transparent, i.e. the logical path from the inspection of the signal to the final decision is clearly understandable by the medical user. On the other hand the diagnostic criteria, usually described by logical trees, are not shared by the whole medical community: "the" cardiologist to be simulated is far from being a reality. The second approach learns from an annotated data-base, i.e. from a group of signals with a well defined diagnosis, derived from ECG independent criteria, the probability that a certain pathology is associated to a particular ECG appearance. If the a priori probability of the disease is known, the Bayes theorem gives the a posteriori probability necessary to reach a decision. In this way the subjectivity due to the cardiologist can be overcome. However the results are not easily accepted and the problem of obtaining sufficient and sufficiently well defined data bases for any possible combination of pathologies makes the method very difficult to be applied in practice.

We have accepted the point of view that for the moment the best data-bases are those in the minds of the cardiologists. Hence we have chosen to follow the first approach. However we have verified very soon that a number of discrepancies arise between the results of the algorithms and those obtained by the same cardiologist whose criteria are implemented by the algorithms. This is partly due to gross errors in the automatic identification and measurement of the diagnostic parameters (essentially waveform durations and amplitudes) or to some inconsistencies in the behaviour of the physician. However, once eliminated these factors, we still can observe differences which are only explained by an intrinsic difference in the behaviour of the human mind when compared to the programmed computer. The machine can easily reproduce exact procedures, but whenever a quantitative algorithm is used to represent a qualitative process, then some approximations are unavoidable and can lead to big mistakes.

This situation applies to computerized electrocardiography because the commonly used ECG features are sometimes measured too precisely by the computer, whereas the physician bases his/her decision upon measurements which, by their own nature, are ill-defined, hence fuzzily observable. In addition to this vagueness, linked to the description of the ECG, another source of fuzziness is connected to the ill-definition of the diagnostic criteria, which in turn depends from the ill-definition of the pathological classes to which any ECG can be assigned. The boundaries of these classes are not sharp and there is clearly some overlapping, possibly due to incomplete knowledge of the cardiac process. However the cardiologist takes his/her decisions in such an ill-defined situation: the reproduction of this behaviour in an overprecise environment must be avoided. With these remarks in mind we decided to use some concepts derived from the theory of fuzzy sets in order to face vagueness with a typical non vague apparatus as the digital computer.

THE FUZZY DECISION-MAKING MODEL

Our fuzzy model for ECG classification has been firstly described in Degani and Pacini (1978). The main parts are:

i) the description of the ECG in fuzzy linguistic terms;

ii) the construction of relational tables assigning to each possible description s_i the "utility" u_{ij} of choosing a related pathology d_j;

iii) the evaluation of the global utility U_j connected to the choice d_j;

iv) the ranking of the U_j's in order to choose the best alternative.

The ECG is generally represented by the values of its parameters. However we assume that the numerical representation is too strict if compared to the way the cardiologist interprets the signal. His/her description of a parameter can be at most "Q duration is about 40ms", surely not "38ms" or "42ms". The important ranges he/she has in mind are the typical normal and abnormal intervals, with possibly a transition or borderline region. Hence the important aspect for them is that the number they extract from the ECG for a certain parameter can be interpreted as normal, or abnormal, or borderline. These three words indicate regions to some extent overlapping, hence they can easily be interpreted as the labels of three fuzzy sets. For each parameter we have used the standardized terms LOW, MEDIUM, HIGH. The ECG in our model is then linguistically represented through these three labels. This means sampling the continuum where each parameter can take its values with a finite family $\{F_i,\ i=1,3\}$ of fuzzy sets: F_1=LOW, F_2=MEDIUM, F_3=HIGH. Any value x of the parameter is then viewed as a fuzzy set $\tilde{x}$ on $\{F_i\}$:

$$\tilde{x}=\mu_L(x)/L+\mu_M(x)/M+\mu_H(x)/H=\sum_i \mu_{F_i}(x)/F_i$$

This kind of presentation preserves all the important characteristics of the parameter and allows measuring simultaneously how much the given numerical value is normal, borderline and abnormal.

In general the values $x_1,\ldots,x_m,\ldots x_n$ of n parameters are relevant for a decision on the presence, possible presence or absence of a disease d. Then the representation of the ECG is of the form:

$$\tilde{s}=\sum_i \mu_{\tilde{s}}(s_i)/s_i$$

where

$$s_i=F_{i_1}\ \ldots\ F_{i_m}\ \ldots\ F_{i_n}$$

and

$$\mu_{\tilde{s}}(s_i)=\min_m \mu_{F_{i_m}}(x_m)$$

The medical knowledge is represented by means of relational tables indicating for each possible linguistic description s_i, $i=1,\ 3^n$, the utility u_{ij}, or willingness of the cardiologist, to accept that the correct decision is d_j, $j=0,2$. Usually d_0 denotes normality, d_1 possible abnormality, d_2 abnormality.

To ask a cardiologist whether his/her acceptance of a decision is 0.7 or 0.6 is clearly unfeasible. We have allowed the experts to use words to express their feelings. The vocabulary has been purposedly restricted to 5 different labels: Very Low, Low, Medium, High, Very High, whose meaning is expressed by fuzzy sets defined in the interval $[0,1]$. Hence the tables relating s_i and d_j contain in the generic ij-th cell a linguistic utility

$$\tilde{u}_{ij}\in\{VL,\ L,\ M,\ H,\ VH\}\equiv\{\tilde{w}_p\}\quad p=1,5$$

Should the state of the ECG be represented by a simple description s_i, then the choice would be for d_{jo}, such that

$$\tilde{u}_{ijo}=\max_j \tilde{u}_{ij}$$

where max indicates the operation of choosing the "best" utility, "best" being defined in some sense. However the ECG is represented by

$$\tilde{s}=\sum_i \mu_{\tilde{s}}(s_i)/s_i$$

what is then the global utility U_j of choosing d_j, when the state of the ECG is a fuzzy state $\tilde{s}$? By the extension principle

$$U_j=\sum_i \mu_{\tilde{s}}(s_i)/\tilde{u}_{ij}=\tilde{\tilde{u}}_j\ .$$

We then have a level 2 fuzzy set representing the final utility of choosing d_j, given $\tilde{s}$. In order to rank the $\tilde{\tilde{u}}_j$'s we have followed the suggestion of Jain (1976), reducing $\tilde{\tilde{u}}_j$ to an ordinary fuzzy set

$$\tilde{u}_j=\sum_k \mu_k/z_k$$

where

$$\mu_k=\bigoplus_p\left\{\min\left[\left(\bigoplus_{\substack{i\\ \tilde{u}_{ij}=\tilde{w}_p}}\mu_{\tilde{s}}(s_i)\right),\mu_{\tilde{w}_p}(z_k)\right]\right\}$$

$\oplus$ indicates the probabilistic sum: $a\oplus b = a + b - ab$. We have chosen to represent all the fuzzy sets by a finite number of points: in the case of the utilities we have sampled the unit interval by steps of 0.01. For this reason we use the representation in terms of $\sum$ instead than in terms of $\int$.

The $\tilde{u}_j$'s must be ranked in order to choose the best alternative. Again we use the method suggested by Jain (1976, 1977), which is briefly summarized here. First the support S of $\tilde{u}=\cup\tilde{u}_j$ is determined. Let z_{max} be supS. Then the maximizing set

$$\tilde{u}_{max}=\sum \mu_{max}/z$$

of S is evaluated, where

$$\mu_{max} = \left[\frac{z}{z_{max}} \right]^k \qquad k > 0$$

Finally

$$hgt\ (\tilde{u}_j \cap \tilde{u}_{max}) = \alpha_j$$

is used to compare the $\tilde{u}_j$'s. Then the choice is for d_{jo} such that

$$\alpha_{jo} = \max_j \alpha_j$$

This approach has been variously criticized by Dubois and Prade (1978, 1982), Baldwin and Guild (1979), Chang (1981) and Kerre (1982), both for the extension principle where ⊕ is used instead of the max operator ∨, and for the strong dependence of the final ranking from the choice of k for μ_{max}, and on the basis of some specific examples. As regards the use of ⊕ instead of ∨ we agree that it gives incorrect results in the continuous case: however we have only few values to which the ⊕ operator is applied, hence, as already noted by Dubois and Prade (1978) and by Gaines (1975,1976) we do not see a significant variation in the results obtained with the different operators. We will see further in the following section how the two operators act on the over-all decision policy for myocardial infarction. It is also obvious that the classification depends on the definition of the maximizing set, hence on k, but we do not feel that this is a criticism. If the results are good for a fixed choice of k (for instance we have chosen k=1), then no further problem arises on this point.

The true big problem is that we do not always know what is a "good" result. Generally intuition is used to decide whether a result is correct or not, but in some instances intuition does not seem to be so clear. On this point let us point out two citations, both referring to an example repeatedly used in the literature on comparisons of fuzzy numbers. The case, shown in Fig.1, was firstly presented by Baldwin and Guild (1979) to illustrate their point that already existing methods of comparison (Baas and Kwakernaak, 1977; Jain, 1977) were not able to discriminate otherwise discriminable alternatives.

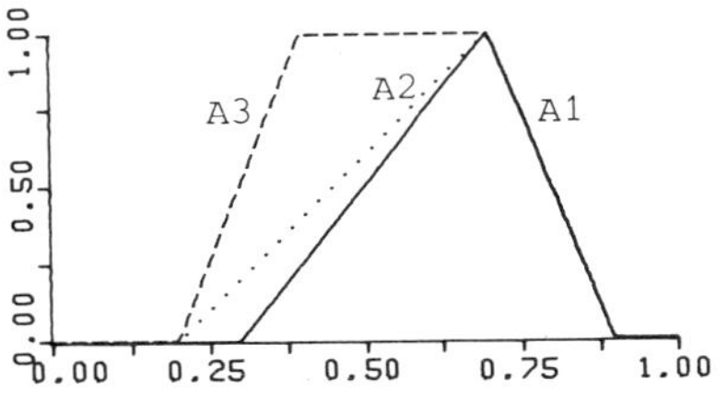

Fig. 1. Comparison of three fuzzy sets.

In this example their method prefers A1 to A2 and A3, the last two resulting at the same level. To quote them:

"... our procedure gives an equal membership to alternatives A2 and A3, while A2 is clearly preferable to A3: but should a decision between A2 and A3 be required (e.g. because A1 is rejected on some other grounds) the procedure should be repeated for this pair only, when a clear difference will result".

The same situation is discussed by Chang (1981), whose method prefers both A2 and A3, which again are not distinguishable, to A1:

"... Baldwin and Guild prefer A1 to A2 and A3, which is clearly not a satisfied result. Our procedure gives the same ranking to alternatives A2 and A3, though A3 is intuitively preferable to A2".

Obviously two opposite intuitions are applied to this example perhaps suggesting that Baas Kwakernaak's and Jain's methods, which represent a completely confused situation, are more realistic that those of Baldwin - Guild and Chang.

A review of the main methods proposed in the literature for ordering fuzzy quantities (Bortolan and Degani, 1983) has shown that problem cases always exist. We then suggest to verify practically that a method is able to emulate a decision-maker in the context of an evolving process, where the decisions can be supposed smoothly changing, then can be more or less easily predicted. In the next section we will see how the following indexes behave in the diagnosis of inferior myocardial infarction (IMI):

a) the center of gravity F_1 (Yager, 1978);
b) the index F_3 also proposed by Yager (1981);
c) the index suggested by Chang (1981);
d) the α-preference index obtained by Adamo (1980);
e) the index studied by Baas and Kwakernaak (1977);
f) the index modified by Baldwin and Guild (1979);
g) the Hamming distance from the fuzzified max used by Kerre (1982);
h) the index suggested by Jain (1976,1977);
i-to-l) the four indices PD, PSD, ND, NSD studied by Dubois and Prade (1982).

IMI CLASSIFICATION

A decision for IMI is based on two parameters: Q duration in D2, D3, aVF and Q/R amplitude ratio in the same leads. From the analysis of the relational table for IMI prepared by the cardiologist we can see a very high utility for the choice $d_2 \hat{=}$IMI is surely present, if both Q duration and Q/R ratio are HIGH; a possible diagnosis of IMI (d_1) when both parameters are MEDIUM; a very high utility for $d_0 \hat{=}$IMI is surely absent, if the parameters have LOW values.

Our aim is to study the evolution of the ranking indices for d_0, d_1, d_2 in the space of the parameters, where we think that no discontinuity should be present. For this reason we have simulated all the possible numerical values for Q duration and Q/R amplitude ratio and have evaluated the related orderings.

We remember here that the global utility $\tilde{\tilde{u}}_j$ of choosing d_j can be reduced to an ordinary fuzzy set

$$\tilde{u}_j = \sum_k \mu_k / z_k$$

both using ⊕ or V operator. We have examined the influence of these operators on Jain index, in order to estimate the possible differences. Figure 2 shows the results. The computer output represents the decision surfaces, i.e. the distributions of the indexes α_j for any combination of the parameter values.

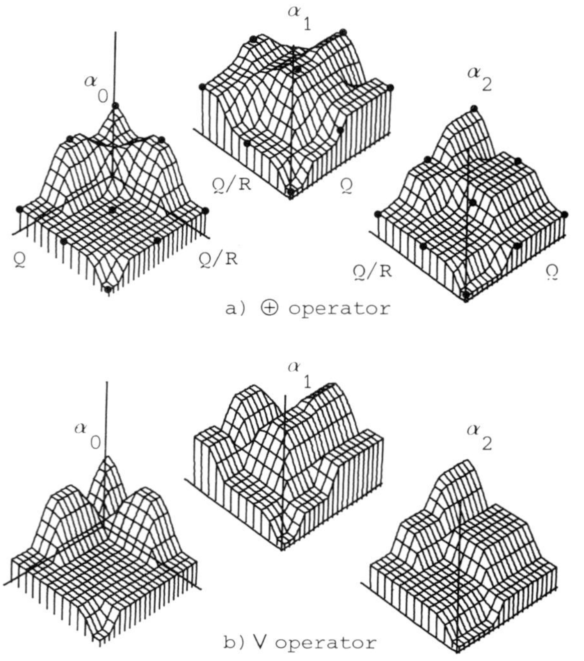

Fig. 2. Jain decision surfaces.

The impression we receive from the examination of the results is that the ⊕ operator gives smoother surfaces. This can be seen especially for d_0. However some kinds of inconsistencies can be discerned, which consist in a decreasing utility for d_j when, for a fixed value of one parameter, the other one moves in the direction of an increasing utility. We could think that a similar behaviour is present in the utility table prepared by the cardiologist. This is not true, because no incongruence can be seen in the dots of Fig.2 a), which represent the 3^2=9 linguistic descriptions directly weighted by the physician in the relational table.

For comparison with Jain ranking method using ⊕ we have utilized the same operator for all the procedures. Similar results could be however observed using V instead of ⊕ , the main difference being less smooth surfaces. The decision surfaces for d_2 are represented in Fig.3 a) to l), where the different indices have been normalized in some sense, keeping them, when this was not the case, in the interval [0,1] (Bortolan and Degani,1983).

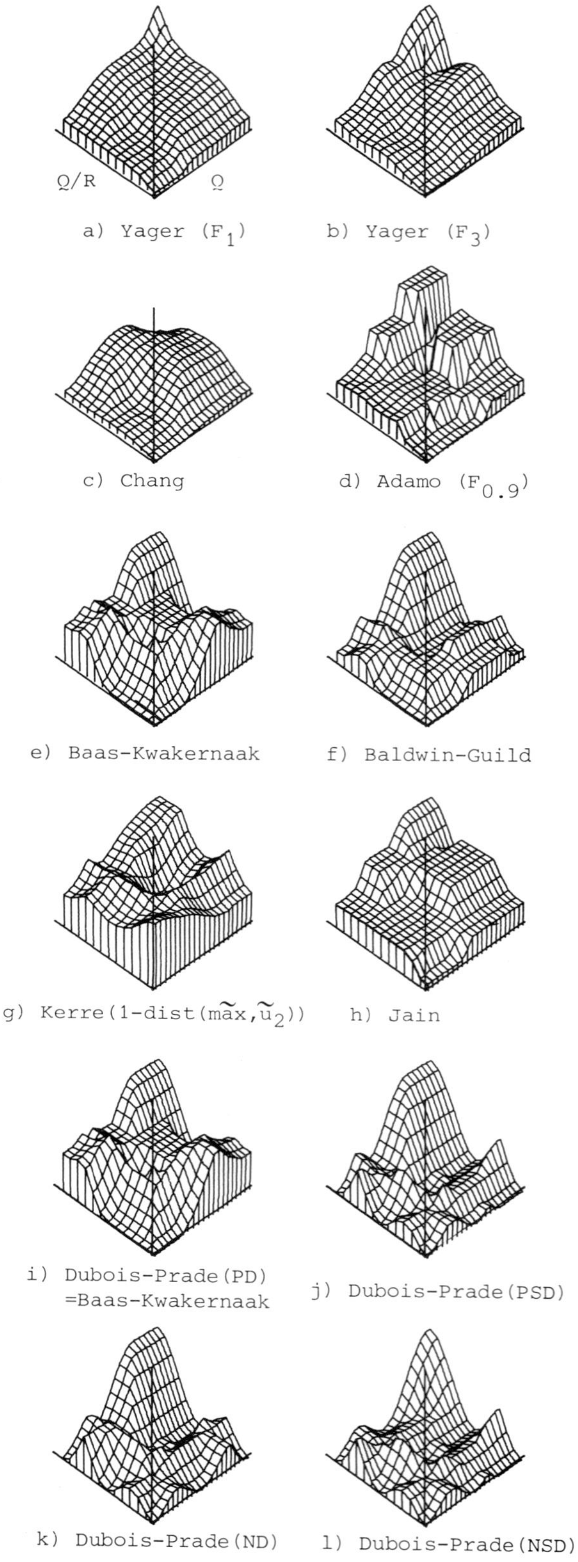

Fig. 3. Decision surfaces for d_2.

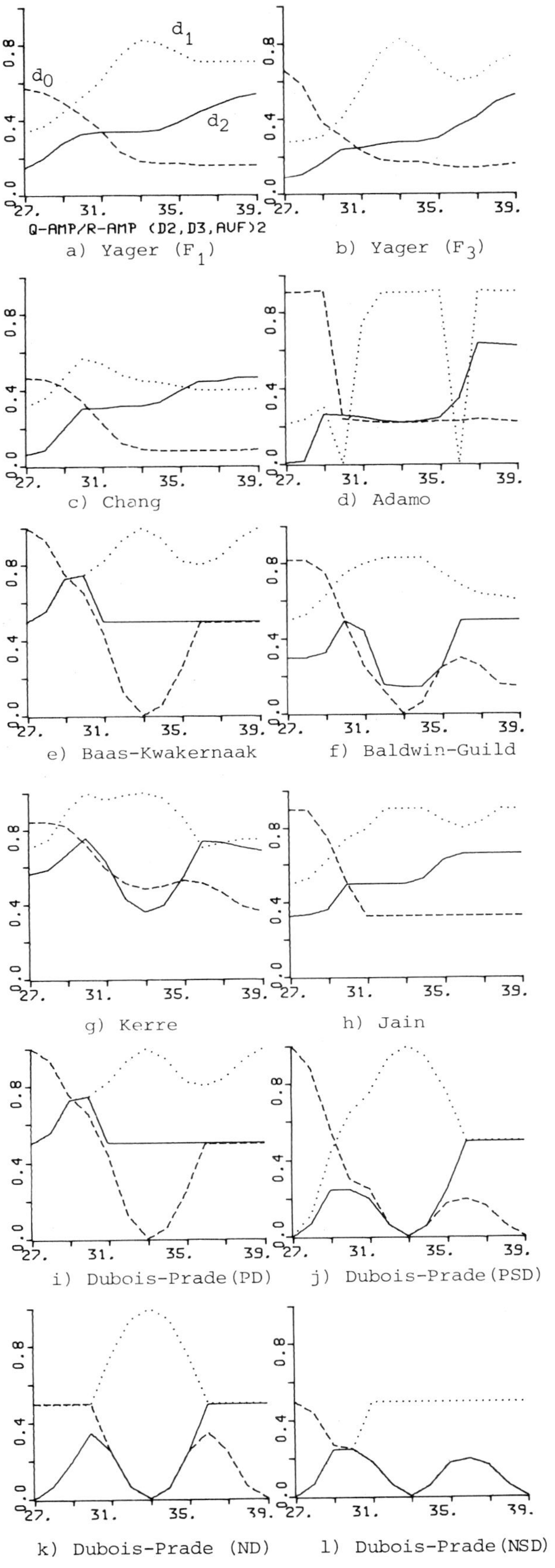

Fig. 4. Comparison of d_0, d_1, d_2 (Q=35ms).

In order to see how the different methods confront the three alternatives d_0, d_1 and d_2 we must consider together their decision surfaces. This kind of representation is however very difficult to be graphically illustrated, hence we report here only one section, obtained considering a fixed value of Q-duration=35ms, which corresponds to the description Q=1./MEDIUM+0.5/HIGH. The results of the different methods are reported in Fig.4, where we can see that the critical region where the decision between different alternatives is difficult is uniformly centered around the Q/R value of 30%.

CONCLUSIONS

From the results illustrated in the previous section we can summarize two comments. First, in this application there is no big difference between the use of the ⊕ and the V operators for the extension principle. Smoother surfaces can be obtained with ⊕, probably because ⊕ acts mainly in the transition regions, where the single memberships tend to decrease, hence reducing the fluctuations in the decision surfaces due to this fuzzy discretization of the continuum. Second, we don't see any really big difference between the results obtained with the various methods. Probably the smoothest surfaces are those obtained with Yager's center of gravity. However the critical regions tend to be centered around the same values for each procedure. This indicates that a difficult choice is always correctly identified and can be pointed out to the user.

The final conclusion is that in our model for ECG analysis we can indifferently use any of the methods reported in this paper, with the possible exception of Adamo's method, whose results clearly show that for non-normalized fuzzy sets wrong results can occur (see Fig. 4 d)). This is a general conclusion, which does not take account of some particular situations, for example the one depicted in Fig.5. Here the ranking obtained by Kerre is clearly incorrect for Q/R=33%=1./MEDIUM, because the utility table proposed by the

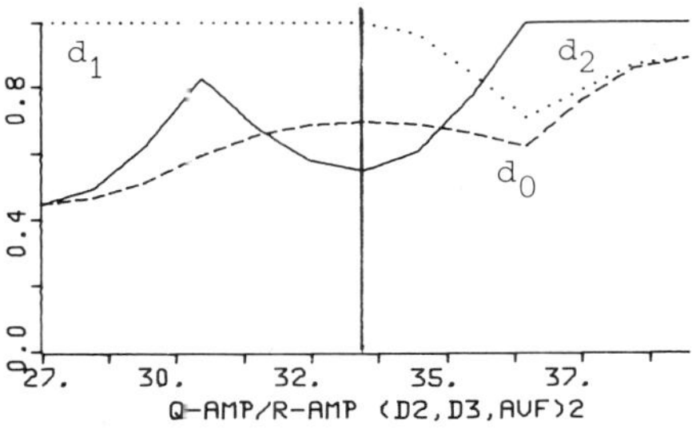

Fig. 5. Comparison of d_0,d_1,d_2 with Kerre. (Q=45=1./HIGH)

physician for

Q/R=1./MEDIUM Q=1./HIGH

gives

$\tilde{u}_0$=LOW $\tilde{u}_1$=VERY HIGH $\tilde{u}_2$=MEDIUM hence

$$\tilde{u}_1 > \tilde{u}_2 > \tilde{u}_0$$

whereas the ranking obtained with Kerre gives

$$\tilde{u}_1 > \tilde{u}_0 > \tilde{u}_2$$

This incongruence is clearly due to the difficulty already noted for this method in Bortolan and Degani (1983). However, except some special cases of this sort, our opinion that no big differences exist seems to be strongly supported by the results.

REFERENCES

Adamo, J.M. (1980). Fuzzy decision trees. Fuzzy Sets & Syst.,4, 207-219.

Baas, S.M., and H. Kwakernaak (1977). Rating and ranking of multiple-aspect alternatives using fuzzy sets. Automatica, 13, 47-58.

Baldwin, J.F., and N.C.F. Guild (1979). Comparison of fuzzy sets on the same decision space. Fuzzy Sets & Syst., 2, 213-233.

Bortolan, G., and R. Degani (1983). A review of some methods for ranking fuzzy subsets. Int. Rep. LADSEB 83/01.

Chang, W. (1981). Ranking of fuzzy utilities with triangular membership functions. Proc. Int. Conf. on Policy Analysis and Information Systems, Taipei, Taiwan, 263-272.

Degani, R., and G. Pacini (1978). Linguistic pattern recognition algorithms for computer analysis of ECG. Proc. BIOSIGMA 78, Paris, France, 1, 18-26.

Dubois, D., and H. Prade (1978). Comment on "Tolerance analysis using fuzzy sets" and "A procedure for multiple-aspect decision making". Int. J. Syst. Sci., 9, 357-360.

Dubois, D., and H. Prade (1982). Ranking of fuzzy numbers in the setting of possibility theory. Inf. & Control (in press).

Gaines, B.R. (1975). Stochastic and fuzzy logics. Electron. Lett., 11, 444-445.

Gaines, B.R. (1976). Foundations of fuzzy reasoning. Int. J. Man-Mach. Stud., 8, 623-668.

Jain, R. (1976). Decision-making in the presence of fuzzy variables. IEEE Trans. Syst., Man & Cybern., SMC-6, 698-703.

Jain, R. (1977). A procedure for multiple-aspect decision making using fuzzy sets. Int. J. Syst. Sci., 8, 1-7.

Jenkins, J.M. (1981). Computerized electrocardiography. CRC Crit. Rev. Bioeng.,6, 307-350.

Kerre, E.E. (1982). The use of fuzzy set theory in electrocardiological diagnostics. In M.M. Gupta and E. Sanchez (Eds.), Approximate Reasoning in Decision Analysis. North-Holland, Amsterdam. pp. 277-282.

Yager, R.R. (1978). Ranking fuzzy subsets over the unit interval. Proc. IEEE Conf. on Decision & Control, San Diego, Ca., 1435-1437.

Yager, R.R. (1981). A procedure for ordering fuzzy subsets of the unit interval. Inf. Sci., 24, 143-161.

LONGITUDINAL FUZZY NUMBER AND ITS APPLICATIONS

Liu Yun-feng*, Wang Shou-dao and Wei Gong-yi*****

**Institute of Mathematics, Academia Sinica, Beijing, China*
***Institute of Chemistry, Academia Sinica, Beijing, China*
****Computing Centre, Academia Sinica, Beijing, China*

Abstract. The more generalized fuzzy number -- LFN (longitudinal fuzzy number) is discussed and its E-operation (extension operation) ⊛ , P-operation (point by point operation) ∗ , C-operation (coincide position operation) ⊞ , S-operation ∪ , ∩ and ¯ are defined . Some of its applications to fuzzy control are also presented.

Keywords. Artificial intelligence; intelligent mathematics; longitudinal fuzzy number; feedback; controllers; identification; models.

THE MATHEMATICAL BACKGROUND OF LFN

HFN (horizontal fuzzy number) is ordinary fuzzy number which in longitude is confied to [0,1] , while LFN is more generalized fuzzy number which in longitude may not confined to [0,1] .

Human brains may not only do quantitatively computing of CM (classical math.), but also do qualitatively analysing of FM (fuzzy math.). The generalized math. which can fit for both may be called IM (intelligent math.). IM should be such a math., it is abstracted from HI (human intelligence) and applied to AI (artificial intelligence, Fig. 1):

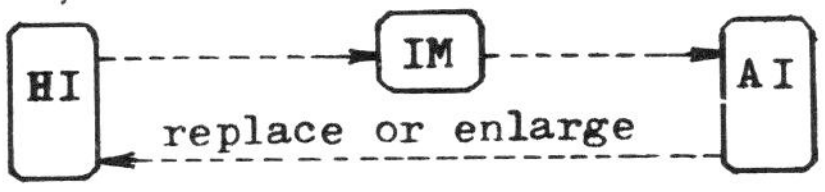

Fig. 1.

IM must contain two contents:IML (IM logic) and IMC (IM computing). Boolean type fuzzy logic [7] may be regared a IML; It is shown from some examples that the operations of LFN may be regared a IMC.

In IM, FM and CM are consistent, so that the change of qualitative description of FM into quantitative representation of CM is possible. For example,a method converting LCR (linguistic control rules) of fuzzy controller into a quantitative controller design can be given in this paper. Because of the introducing LFN, some conclusion of [4] can be advanced to some extent.

BASIC CONCEPTS

In this paper, fuzziness means degree of membership which written $\mu_{\underset{\sim}{A}}(x)$ or $\rho_{\underset{\sim}{L}}(x)$, and fuzzy number is a number set which contains fuzziness.

Definition 1. Let $X, K, M \subseteq \mathbb{R}$, $[0,1]\cdot K \subseteq M$. a longitudinal fuzzy number (LFN) $\underset{\sim}{L}$ on X is said to be given, if functions $\rho_{\underset{\sim}{L}}$, ϕ and $f_{\underset{\sim}{L}}$ are given as following:

$$\left.\begin{aligned} \rho_{\underset{\sim}{L}}&: X \longrightarrow [0,1], \\ \phi&: X \longrightarrow K, \\ f_{\underset{\sim}{L}}&: X \longrightarrow M, \\ & x \longmapsto f_{\underset{\sim}{L}}(x) = \rho_{\underset{\sim}{L}}(x)\cdot\phi(x), \end{aligned}\right\} \qquad (1)$$

where $\rho_{\underset{\sim}{L}}$, ϕ and $f_{\underset{\sim}{L}}$ are called membership function , longitudinal-stretsh function and definition function,resp. $\underset{\sim}{L}$ is denoted by

$$\underset{\sim}{L}=\{(x,f_{\underset{\sim}{L}}(x))|x\in X\}\triangleq\int_X f_{\underset{\sim}{L}}(x)/x.$$

The collection of all LFN obtained is denoted by $\underset{\sim}{\mathcal{M}}$. If $M=\mathbb{R}$,then $\underset{\sim}{L}$ is called real LFN and denoted by $\underset{\sim}{L}\in\underset{\sim}{\mathcal{R}}$, similarly, $\underset{\sim}{L}\in\underset{\sim}{\mathcal{N}}$, $\underset{\sim}{L}\in\underset{\sim}{\mathcal{Q}}$, etc. can be defined.

Definition 2. $\underset{\sim}{A}=\int_{\mathbb{R}}\mu_{\underset{\sim}{A}}(x)/x$ is called D type HFN (brief written D-HFN) , if it is a ordinary fuzzy number defined as Dubois and Prade in [1] . $\underset{\sim}{A}$ is called I type HFN (I-HFN), only if condition (a) of D-HFN is changed to

(a') For $\forall\bar{x}\in[x_1,x_2]\subseteq X,\ \mu_{\underset{\sim}{A}}(\bar{x})=1.$

Example 1. Let $\underset{\sim}{A}$ is a D-HFN or I-HFN. If set

$$f_{\underset{\sim}{L}}(x)=\mu_{\underset{\sim}{A}}(x),\quad \Phi(x)=1\ (\text{for }\forall x\in X),$$

then $f_{\underset{\sim}{L}}(x)=\mu_{\underset{\sim}{A}}(x)$, $\underset{\sim}{L}=\underset{\sim}{A}$, thus D-HFN or I-HFN is LFN.

Example 2. Let G is a graph of function $g_G(x)$ domained on X . Set

$$f_{\underset{\sim}{L}}(x)=1\ (\text{for }\forall x\in X\ ),\ \Phi(x)=g_G(x),$$

then $f_{\underset{\sim}{L}}(x)=g_G(x)$, $\underset{\sim}{L}=G$, so every graph of function G is LFN.

Definition 3. A LFN $\underset{\sim}{L}$ is called L-LFN, if $X=M=\mathbb{R}$,and $f_{\underset{\sim}{L}}$ is piecewise continous. A LFN $\underset{\sim}{L}$ is called D-LFN (I-LFN), if $f_{\underset{\sim}{L}}(x)=x\mu_{\underset{\sim}{A}}(x)$ and if $\underset{\sim}{A}$ is a D-HFN(I-HFN). is a exact number, if

$$\underset{\sim}{L}=\left\{(x,y)\,\middle|\,\begin{matrix}y=b \text{ whenever } x=a\\ y=0 \text{ whenever } x\neq a\end{matrix}\right\}.$$

A LFN $\underset{\sim}{L}$ is called positive flat-top , if for $\forall\bar{x}\in[x',x'']\subseteq X$

$$b=f_{\underset{\sim}{L}}(\bar{x})=\sup_{x\in X} f_{\underset{\sim}{L}}(x), \tag{2}$$

and if $f_{\underset{\sim}{L}}$ is increasing from 0 to b on $(-\infty,x']\cap X$, decreasing from b to 0 on $[x'',+\infty)\cap X$; A LFN $\underset{\sim}{L}$ is called negative flat-top, if

$$-\underset{\sim}{L}=\{(x,-f_{\underset{\sim}{L}}(x))\,|\,x\in X\}$$

is a positive flat-top. Flat-top LFN is denoted by $\underset{\sim}{L}\langle[x',x''],b\rangle$. A flat-top LFN is called fuzzy constant and denoted $\underset{\sim}{L}\langle X,b\rangle$ or more brief $\underset{\sim}{b}\langle X\rangle$, if $[x',x'']=X$. A flat-top LFN $\underset{\sim}{L}$ is called single-peak and denoted by $\underset{\sim}{L}\langle a,b\rangle$, if $x'=x''=a$. a and $b=f_{\underset{\sim}{L}}(a)$ are called peak position and peak height of $\underset{\sim}{L}$, resp. A single-peak LFN $\underset{\sim}{L}$ is called E-LFN, if $a=b$; in this case, it can be denoted as $\underset{\sim}{L}\langle a,a\rangle=\underset{\sim}{a}$.

Example 3. In Fig. 1, universe X is represented annual rainfall, $\underset{\sim}{D}$, $\underset{\sim}{N}$ and $\underset{\sim}{W}$ are three HFN on X , are called "drought", "normal" and "waterlogging" , resp. $\underset{\sim}{D}'$, $\underset{\sim}{N}'$ and $\underset{\sim}{W}'$ are corresponding LFN , where $\underset{\sim}{D}'$ is D-LFN, $\underset{\sim}{N}'$ and $\underset{\sim}{W}'$ are I-LFN.

Definition 4. Let $\underset{\sim}{L}_1,\cdots,\underset{\sim}{L}_n$ are D-LFN or I-LFN on X , $x_0\in X$, X may be partitioned into some interval $X_1,\cdots,X_n$ by following rule: if

$$f_{\underset{\sim}{L}_k}(x_0)=max\left(f_{\underset{\sim}{L}_1}(x_0),\cdots,f_{\underset{\sim}{L}_n}(x_0)\right),$$

then is regarded $x_0\in X_k$ ($k\in\{1,\cdots,n\}$); and if

$$\bigcup_{k=1,n}X_k=X\ ,\quad X_i\cap X_j=\Phi\ (i\neq j),$$

where Φ is empty set or single set , then X_k is called dominion of $\underset{\sim}{L}_k$ and denoted by $dom(\underset{\sim}{L}_k)=X_k$.

In Fig. 2,

$$dom(\underset{\sim}{D}')=[0,x_1],\quad dom(\underset{\sim}{N}')=[x_1,x_4],$$
$$dom(\underset{\sim}{W}')=[x_4,+\infty).$$

If $\underset{\sim}{L}_1,\cdots,\underset{\sim}{L}_n$ partition X into dominions $X_1,\cdots,X_n$, then can be defined

$$"x\in\underset{\sim}{L}_k"\longrightarrow"x\in X_k" .$$

For example, in Fig. 2,

"year of drought" ⟶

"annual rainfall in 0,1 ".

In fuzzy control, by using concept of dominion of LFN, the linguistic variable (LFN) can be changed to corresponding dominion (interval number):

$$"x\in\underset{\sim}{L}_k"\longrightarrow"x\in X_k".$$

E-OPERATION OF LFN

Definition 4. Let $\underset{\sim}{L}=\int_R f_{\underset{\sim}{L}}(x)/x$ is a D-LFN or I-LFN, then

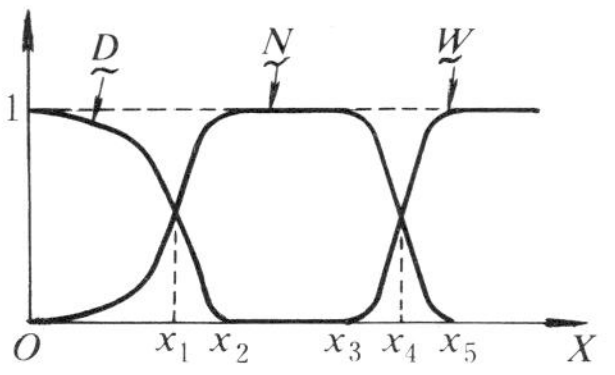

(a) HFN

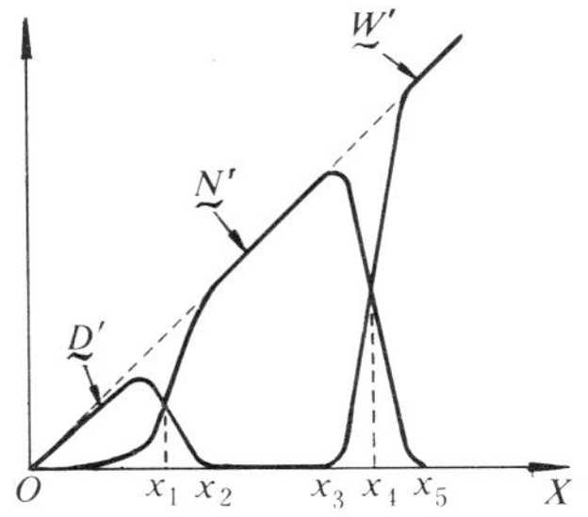

(b) Corresponding LFN

Fig. 2.

$$\rho_{\underset{\sim}{L}}(x) = \begin{cases} \frac{1}{x} f_{\underset{\sim}{L}}(x), & (x \neq 0) \\ 1, & (x = 0) \end{cases}$$

is called degree of membership of $\underset{\sim}{L}$ on X , and

$$(\underset{\sim}{L})_\alpha = \left\{ x \,\middle|\, f_{\underset{\sim}{L}}(x) \geq \alpha x,\ \alpha \in [0,1] \right\}$$

is called α-cut set of $\underset{\sim}{L}$ (Fig. 3).

D-LFN or I-LFN may contain the advantages of FM and CM:

(1) They have degree of membership, thus they can represent "drought","waterlogging" etc. qualitative fuzzy concepts as FM,

(2) They have definition function, thus for two cases which are "waterlogging" , they can make more finely quantitative distinguish as CM.

Definition 5. Let $\underset{\sim}{\mathfrak{M}}$ is collection of D-LFN or I-LFN, for $\forall\, \underset{\sim}{L}, \underset{\sim}{I} \in \underset{\sim}{\mathfrak{M}}$, E-operation on $\underset{\sim}{\mathfrak{M}}$ is defined as following

$$f_{\underset{\sim}{L} \circledast \underset{\sim}{I}}(z) \triangleq z \left[\sup_{x * y = z} \left(\rho_{\underset{\sim}{L}}(x) \wedge \rho_{\underset{\sim}{I}}(y) \right) \right], \qquad (3)$$

where $* \in \{+,-,\cdot,\div\}$.

The E-operation on $\underset{\sim}{\mathfrak{M}}$ is similar to that on D-HFN set(Fig. 4). It is convenient and feasible that HFN is replaced by D-LFN or I-LFN. This kind of E-operation also has some advantges of CM.

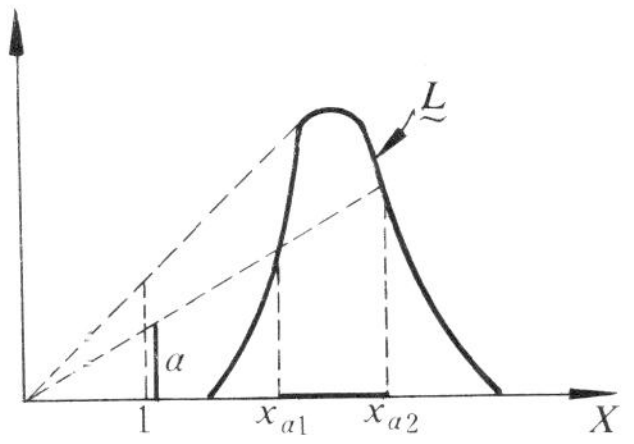

$$(\underset{\sim}{L})_\alpha = [x_{\alpha 1}, x_{\alpha 2}]$$

Fig. 3.

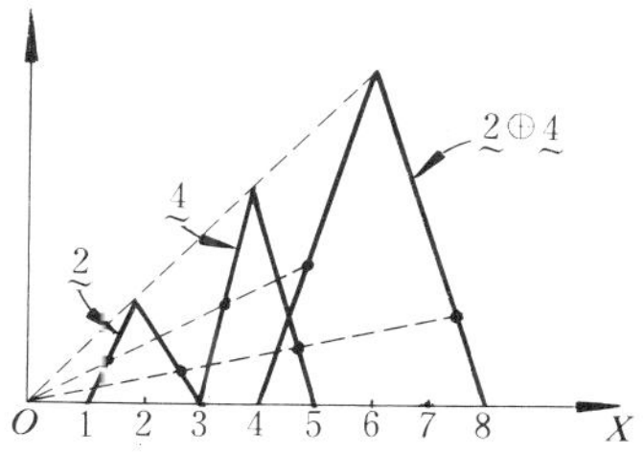

Fig. 4.

P-OPERATION OF LFN

Definition 6. Let $\underset{\sim}{\mathfrak{M}}$ is certain type LFN set, for $\forall\, \underset{\sim}{L}, \underset{\sim}{I} \in \underset{\sim}{\mathfrak{M}}$, P-operation on $\underset{\sim}{\mathfrak{M}}$ is defined as following:

$$f_{\underset{\sim}{L} * \underset{\sim}{I}}(x) = f_{\underset{\sim}{L}}(x) * f_{\underset{\sim}{I}}(x), \qquad (4)$$

$$f_{\underset{\sim}{\alpha}\langle x \rangle \cdot \underset{\sim}{I}}(x) = \alpha f_{\underset{\sim}{I}}(x), \quad (\alpha \in M), \qquad (5)$$

where $* \in \{+,-,\cdot,\div\}$,when $*$ is $\div$, $f_{\underset{\sim}{I}}(x) \neq 0$ (for $\forall x \in X$).

Definition 7. If $\underset{\sim}{\mathfrak{M}}_1$ and $\underset{\sim}{\mathfrak{M}}_2$ are two LFN sets, then function

$$\sigma: \underset{\sim}{\mathfrak{M}}_1 \longrightarrow \underset{\sim}{\mathfrak{M}}_2$$

is called fuzzy function.

P-operation of LFN can constitute some fuzzy function which represent some fuzzy restrictions.

Example 5. "Big truck must go slowly" can be represented by LFN as

$$f_{\underset{\sim}{I}}\left(f_{\underset{\sim}{L}}(x)\right) = \frac{240}{f_{\underset{\sim}{L}}(x)}, \quad \left(f_{\underset{\sim}{L}}(x) \neq 0\right), \qquad (6)$$

where $f_{\underset{\sim}{L}}(x) = x \rho_{\underset{\sim}{L}}(x)$ is definition function of "big truck of x tons", $f_{\underset{\sim}{I}}(y) = 240/y$ is definition function of "go slowly".

Obviously, when $\rho_{\underset{\sim}{L}}(x) \to 1$, $f_{\underset{\sim}{L}}(x) \to x$, so

$f_I(f_L(x)) \longrightarrow f_I(x)$. Thus fuzzy function of LFN is natural extension of classical function when $\rho_L(x) \longrightarrow 1$. Therefore, by introducing P-operation of LFN , closer relations between FM and CM can be built.

S-OPERATION OF LFN

Some control objects are complex and changeable, its state can only be roughly per ceived and approximately measured, but can be artificially controled. Thus one can regard them as black or grey boxes, and if input-output data are enough by using S-operation of LFN,then their mathematical model of IM can be identified.

Definition 8. If $L_1,\cdots,L_n$ are LFN which can be partitioned X into dominions $X_1,\cdots,X_n$, then $L_1,\cdots,L_n$ are called complementary. Set

$$\sum_{k=1,n} f_{L_k}(x) = f_U(x), \tag{7}$$

then U is called longitudinal sum (LS) of $L_1,\cdots,L_n$. If $L_1,\cdots,L_n$ ($L_k = L_k\langle a_k, b_k\rangle$, $k=1,\cdots,n$) are complementary LFN, and their LS is U ,then their S-operation is defined as following:

$$f_{L_i \cup L_j}(x) = f_{L_i}(x) + f_{L_j}(x), \tag{8}$$

$$f_{L_i \cap L_j}(x) = \begin{cases} \frac{1}{2}[f_{L_i}(x) + f_{L_j}(x), & (a_i = a_j), \\ 0, & (\text{otherwise}), \end{cases} \tag{9}$$

$$f_{\bar{L}_i}(x) = f_U(x) - f_{L_i}(x). \tag{10}$$

In fuzzy set theory, two contraditory peak may appear [4] :

$$M^+ \vee S^+ = \{(0,0.3),(1,0.8),(2,1),(3,0.7),(4,1),(5,0.7),(6,0.2)\},$$

in IM, using (8), that is using the $M'^+ + S'^+$ to replace the $M^+ \vee S^+$, above case may not appear . Therefore, to find mathematical model of fuzzy logic for feedback controller, using S-operation of LFN is more rational.

First of all, a fuzzy controller of single input x (error) and single output y (control quantity) is discussed.

The relation between LCR (linguistic control rules) which is from human experience and S-operation of LFN is

LCR	S-operation
if a_k then b_k	$L_k = L_k\langle a_k, b_k\rangle$
if a_k then b_k or if a_{k+1} then b_{k+1}	$L_k + L_{k+1}$
if a_k then b_k and if a_k then b'_k	$\frac{1}{2}\langle x\rangle\cdot(L_k - L'_k)$

where

$$a_k = \text{"}x \in a_k\text{"}, \quad b_k = \text{"}y \in b_k\text{"}, \quad (k=1,\cdots,n)$$

are called longitudinal fuzzy proposition which correspond to E-LFN a_k and b_k ,resp. $X=[x_0,x_n]$ is divided by $a_1,\cdots,a_{n-1}$ into equal intervals $X_1,\cdots,X_n$; $a_0=x_0$, $a_n=x_n$; $b_k \in \{b_1,\cdots,b_m\}$. $Y=[y_0,\cdots,y_m]$ is divided by $b_1,\cdots,b_{m-1}$ into equal intervals $Y_1,\cdots,Y_m$; $b_0=y_0$, $b_m=y_m$. Because of linguistic variable of man "$x \in a_k$" can be replaced by "$x \in X_k$" , if L_k is set to be such single-peak LFN:

its bottom width $w(L_k) = 2(x_n - x_0)/2$,

the peak position is in a_k ,

the peak height is b_k ,

the definition function is already given,

then the LCR, i.e.

if a_0 then b_0
or
....
or
if a_n then b_n

corresponds to (Fig. 5)

$$\sum_{k=0,n} L_k = L = \int_X f_L(x)/x, \tag{11}$$

where $\sum$ is "point by point addition", and

$$y = f_L(x) \tag{12}$$

is the mathematical model (approximate) obtained.

In fuzzy set theory,by increasing the grade of quantization n and m the precision of model can be increased. But n and m can not be much high,because it will result in so much computing work to do. In IM, this restrict is not necessary.

For fuzzy controller of two input x_1 and x_2 (error e and change of error $\dot{e}$) and one output y (control quantity u), similarly, its mathematical model can be give as following

$$y = f_{\underset{\sim}{L}}(x_1, x_2). \tag{13}$$

In (12), when $f_{\underset{\sim}{L}}(x) = \alpha x$ (α is constant), a proportional control is obtained

$$y = \alpha x . \tag{12'}$$

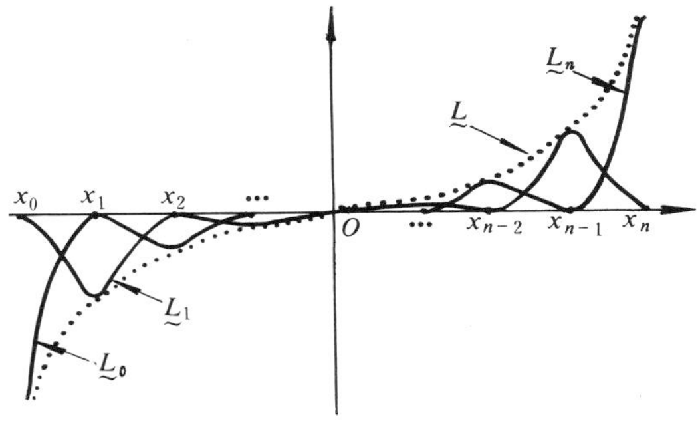

Fig. 5.

In (13), if $x_1 = x$, $x_2 = \frac{dx}{dt}$, then when

$$f_{\underset{\sim}{L}}(x_1, x_2) = \alpha x + \beta \frac{dx}{dt} ,$$

where α and β are constants, a proportional-differentional control is obtained as following

$$y = \alpha x + \beta \frac{dx}{dt} . \tag{13'}$$

Using these analysis representions one can proceed to next step of quantitative discussion as CM. For example, set $\beta = 1 - \alpha$, then the parameter regulation of adaptive fuzzy controller can be discussed [5] .

C-OPERATION OF LFN

Definition 9. Let $\underset{\sim}{\mathfrak{M}}$ is single-peak LFN set, for $\forall \underset{\sim}{L}_1\langle a_1, b_1\rangle, \underset{\sim}{L}_2\langle a_2, b_2\rangle \in \underset{\sim}{\mathfrak{M}}$, C-operation on $\underset{\sim}{\mathfrak{M}}$ is defined as following:

$$f_{\underset{\sim}{L}_1 \boxast \underset{\sim}{L}_2}(x) = f_{\underset{\sim}{L}_1}(x-(a_1 * a_2)+a_1) * f_{\underset{\sim}{L}_2}(x-(a_1 * a_2)+a_2), \tag{14}$$

$$f_{\underset{\sim}{\alpha}\langle x\rangle \boxdot \underset{\sim}{L}}(x) = \alpha f_{\underset{\sim}{L}}(x - \alpha a + a), \tag{15}$$

where $* \in \{+, -, \cdot, \div\}$, when $\boxast$ is $\boxed{\div}$, $f_{\underset{\sim}{L}}(x) \neq 0$ (for $\forall x \in X$).

One can prove following theorem.

Theorem. Let $\underset{\sim}{\mathfrak{M}}$ is positive single-peak LFN set, then $\langle \underset{\sim}{\mathfrak{M}}, \boxplus, \boxdot \rangle$ form a Abelean semiring with units $\underset{\sim}{0}\langle X\rangle$ and $\underset{\sim}{1}\langle X\rangle$.

Applications of C-operation of LFN will be discussed in other paper.

REFERENCES

[1] Dubois,D. and Prade,H., Fuzzy sets and systems, theory and applications, New York, 1980.

[2] Mizumoto,M. and Tanaka,K., Some properties of fuzzy number, Advances in fuzzy set theory and applications , 1979.

[3] Kaufmann,A., Hybrid convolution, a way to combine fuzzy numbers and random numbers, Fuzzy mathematics, 2 (1981).

[4] King,P.J. and Mamdani,E.H., The application of fuzzy control systems to industrial processes, Fuzzy automata and decision processes, 1977.

[5] Long Shengzhao, Wang Peizhuang, On self-regulation of fuzzy control rule, Fuzzy mathematics, 3 (1982).

[6] J.Y.Zhu, Control of machine tools using the fuzzy control technique, Annals of the CIRP, vol. 31/1/1982.

[7] Liu Yun-feng, An axiomatic structure of fuzzy set theory and fuzzy logic, Fuzzy mathematics, 3 (1982).

EXTENSION OF THE FUZZY DATABASE WITH FUZZY ARITHMETIC

B. P. Buckles* and F. E. Petry**

Department of Computer Science and Engineering, University of Texas at Arlington, Arlington, Texas

***Computer Science Department, Tulane University, New Orleans, Louisiana*

Abstract. The fuzzy relational database model originated by the authors permits fuzzy domain values from a discrete, finite universe. The model is extended here by demonstrating that fuzzy numbers may be employed as domain values without loss of consistency with respect to representation or the relational algebra. Where equivalence is required in an ordinary relational database, similarity is employed in a fuzzy relational database. For discrete, finite universes, similarity between atomic elements is described via a fuzzy similarity relation with max-min transitivity. Two or more fuzzy numbers are defined to be α-similar if their union forms a continuous α-level set over the real line. This convention effects the partitioning of fuzzy number domains that is necessary to assure the well-definedness of the fuzzy relational algebra.

Keywords. Computer applications; information retrieval; set theory; fuzzy sets; databases

INTRODUCTION

Uncertainty or imprecision may be purposely represented in a database in three basic ways--the data (e.g., null values, range values, incomplete values), the assignment of meaning to specific data values (e.g., linguistic terms used as values), and query resolution (e.g., statistical databases). The fuzzy relational database, originated by the authors, incorporates aspects of all three approaches. The work described here extends the capabilities of the basic model to represent uncertainty in the data itself.

In the authors' previous descriptions of the fuzzy relational database (1982a, 1982b, 1982c, 1983), it is assumed that domains are members of a powerset over a discrete, finite universe. Frequently, the atomic values of this universe are linguistic values whose meaning is partially interpreted via an explicit similarity relationship (Zadeh 1971) that relates each to other values in the universe. This universe is called a domain set. The properties of the database include well-definedness of the fuzzy relational algebra commands and the inclusion of the ordinary relational algebra as a special case.

Here, the integration of domain values consisting of fuzzy numbers is described together with the conditions that lead to the same desirable consistency characteristics. The following section establishes the framework as it exists. Section 3 describes the method of assimilating fuzzy numbers and the impact of the fuzzy relational algebra. Section 4 examines the implications of this extension on the previously defined mechanisms for query resolution.

FUZZY DATABASES

A fuzzy database is organized in a manner corresponding to the relational database model. The user view is a collection of two-dimensional tables called relations. The rows of a relation are called tuples and the columns are called domains. Individual components of a domain for a specific instance are called domain values and are selected from a universe called the domain set. Unlike ordinary relational databases, each domain set is associated with a fuzzy similarity relationship. A similarity relationship is a generalization of an equivalence relationship and the collection of similarity relationships associated with a fuzzy database has a major role in the definition of the commands used to retrieve information. The set of commands, called the fuzzy relational algebra, is based upon the ordinary relational algebra for databases. In the instances that the ordinary relational algebra requires equivalence or identity between two or more domain values, the fuzzy relational algebra uses similarity.

The construction of a relation, r, with m domains requires domain sets, D_j, $j = 1,2,\ldots,m$. For each D_j there is a similarity relationship

$$s_j : D_j \times D_j \rightarrow [0,1]$$

such that for $x,y,z \in D_j$

$$s_j(x, x) = 1 \text{ (reflexive)}$$

$$s_j(x, y) = s_j(y, x) \text{ (Symmetric)}$$

$$s_j(x, z) \geq \max_y\{\min[s_j(x, y), s_j(y, z)]\} \text{ (transitive)}$$

If $\underline{2}^{D_j}$ denotes any nonnull member of the powerset of D_j (i.e., $\underline{2}^{D_j} = 2^{D_j} - \Phi$) then r is a subset of the set cross product $\underline{2}^{D_1} \times \underline{2}^{D_2} \times \ldots \times \underline{2}^{D_m}$ and a fuzzy tuple, t, is any member of r. A domain value, d_{ij}, of tuple, t_i, is a nonnull subset of D_j. An interpretation, $(a_1, a_2, \ldots, a_m)$, of a tuple $t_i = (d_{i1}, \ldots, d_{im})$ is any assignment of value such that $a_j \in d_{ij}$ for all j.

Given a domain set, D_j, of a relation, the similarity threshold is defined as the minimun similarity over all domain values:

$$\text{THRES}(D_j) = \min\{\min_{x,y \in d_{ij}}[s_j(x, y)]\}; \ i = 1, 2, \ldots$$

A fixed, a priori threshold is called a level value and denoted LEVEL(D_j). Two tuples, $t_i = (d_{i1}, d_{i2}, \ldots, d_{im})$ and $t_k = (d_{k1}, d_{k2}, \ldots, d_{km})$, are redundant if

$$\text{LEVEL}(D_j) \leq \min_{x,y \in d_{ij} \cup d_{kj}} [s_j(x, y)]; \ j = 1, 2, \ldots, m$$

That is, if two tuples could be merged via the set union of the respective domain values while permitting THRES(D_j) $\geq$ LEVEL(D_j) for all j, then the tuples are redundant. Operationally, the fuzzy relational algebra commands are used to specify level values and the implementation of the commands requires the construction of a new relation, the resultant, having no redundant tuples.

The (ordinary) relational algebra is a procedural language for extracting information from a relational database. A command consists of an operation name, one or more relation names, one or more domain names, and an optional conditional expression that eliminates tuples from the relations prior to application of the command. For instance

(Project((FAMILY over FATHER, SON) where MOTHER = "beth") giving REL1

causes the domains FATHER and SON to be copied from a relation called FAMILY and placed in a resultant relation called REL1. Only those tuples in FAMILY for which the domain value for MOTHER is "beth" are copied to REL1.

A fuzzy relational algebra command consists of the same four parts plus a clause defining the level values. Consider the command

(Project((EMPLOYEE over APTITUDE,AGE) where SKILL = "electronics") with LEVEL(APTITUDE) = 0.75, LEVEL(AGE) = 0.80) giving REL2.

LEVEL(APTITUDE) is given as 0.75 and LEVEL (AGE) is given as 0.80. The final form of REL2 is obtained by merging as many tuples as possible without violating the constraint THRES(D_j) $\geq$ LEVEL(D_j). The process of removing redundancy by merging tuples is the simplest case in which similarity is used rather than equivalence as required in the ordinary relational algebra for which redundancy implies the presence of identical tuples. Similarity replaces equivalence in more complex ways in such commands as Intersection, Join, Select, Union, Difference, and Divide.

A generalized form of Boolean queries may also be used to retrieve information. A query, $Q(a_i, a_h, \ldots, a_k)$, is an expression of one or more factors combined by disjunctive or conjunctive Boolean operators: V_i op V_h op ... op V_k. In order to be well formed with respect to a relation, r, having domain sets $D_1, D_2, \ldots, D_m$, each factor, V_j, must be

(1) A domain element a_j, $a_j \in D_j$, where D_j is a domain set for r, or
(2) A domain element modified by one or more linguistic modifiers, e.g. NOT, VERY, MORE-OR-LESS.

Relation r may be one of the original database relations or one obtained as a result of a series of fuzzy relational algebra operations. Fuzzy semantics apply to both operators and modifiers. An example query is

MORE-OR-LESS big and NOT VERY VERY heavy

where "big" is an abbreviation of the term (SIZE = big) in a relation having a domain called SIZE. The value "heavy" is likewise an abbreviation.

A membership value of a tuple in a response relation, r, is assigned according to the possibility of its matching the query specifications. Let $a \in D_j$ be an arbitrary element. The membership value, $\mu_a(b)$, where a, $b \in D_j$ is defined as $s_j(a,b)$. The query $Q(\cdot)$ induces a membership value, $\mu_Q(t)$, for a tuple, t, in the response r as follows:

(1) Each interpretation $I = (a_1', a_2', \ldots, a_m')$ of t determines a value $\mu_{aj}(a_j')$ for each domain element a_j of $Q(a_i, a_h, \ldots, a_k)$.
(2) Evaluation of the modifiers and operators in $Q(\cdot)$ over the membership values $\mu_{aj}(a_j')$, yields $\mu_Q(I)$, the membership value of the interpretation with respect to the query.
(3) Finally, $\mu_Q(t) = \max_{I \text{ of } t} \{Q(I)\}$

In short, the membership value of a tuple represents the best matching interpretation. The response relation is then the set of tuples having nonzero membership values, although, in practice, it may be more realistic to consider only the tuple with the highest value.

PROPERTIES OF FUZZY NUMBER DOMAINS

A fuzzy number, A, defined over the real line

consists of the set

$$\{(x, \mu_A(x)) \mid x \in R, \mu_A(x) \in [0, 1]\}$$

An α-level set of a fuzzy set, A, written A_α, is the ordinary set $\{x \mid \mu_A(x) > \alpha\}$. If a fuzzy relation, r, has k domain sets consisting of fuzzy numbers it will be expressed as $r \subseteq Q_1 \times Q_2 \times \ldots \times Q_k \times D_{k+1} \times \ldots \times D_m$ where Q_j denotes a fuzzy number domain and D_j denotes a finite, discrete domain as discussed in Section 2.

Below, the symbols $\cup$ and $\cap$ will have the usual meaning of union and intersection of ordinary sets. The symbols $\sqcup$ and $\sqcap$ will be used to denote the union and intersection of fuzzy numbers:

$$q_i \sqcup q_j = \{(x, \mu_{q_i \sqcup q_j}(x)) \mid \mu_{q_i \sqcup q_j}(x) = \mu_{q_i}(x) \vee \mu_{q_j}(x)\}$$

$$q_i \sqcap q_j = \{(x, \mu_{q_i \sqcap q_j}(x)) \mid \mu_{q_i \sqcap q_j}(x) = \mu_{q_i}(x) \wedge \mu_{q_j}(x)\}$$

where v and $\wedge$ mean max and min, respectively.

LEMMA 1. $(q_i \sqcup q_j)_\alpha = (q_i)_\alpha \cup (q_j)_\alpha$

PROOF. $x \in (q_i \sqcup q_j)_\alpha$ implies $\mu_{q_i}(x) > \alpha$ or

$$\mu_{q_j}(x) > \alpha$$

Thus, either $x \in (q_i)_\alpha$ or $x \in (q_j)_\alpha$. Q.E.D.

The keystone concept of the fuzzy database is the replacement of equivalence or identity relationships among domain values with a measure of "nearness". The similarity relationship employed for discrete, finite domain sets cannot be directly extended to continuous sets because there is not a transitivity property that effects partitioning of the domain set in a manner that guarantees uniqueness of relation representation. New "nearness" relationship called α-similar and α-proximate are defined here.

Definition. q_i and q_j are α-similar, written $q_i\ S_\alpha\ q_j$, if given $\beta \in [0,1]$, $x \in (q_i \sqcup q_j)_\alpha$, $y \in (q_i \sqcup q_j)_\alpha$, and $z = \beta x + (1 - \beta)y$ then $z \in (q_i \sqcup q_j)_\alpha$.

Definition. q_i and q_j are α-proximate (with respect to set P) written $q_i\ S_\alpha^+\ q_j$, if there exists zero or more fuzzy numbers, q_h, q_k, $\ldots$, $q_p \in P$, such that $q_i\ S_\alpha\ q_h\ S_\alpha\ q_k\ S_\alpha \cdots S_\alpha\ q_p\ S_\alpha\ q_j$.

LEMMA 2. S_α^+ partitions any set of fuzzy numbers $\{q_1, q_2, \ldots, q_h\}$.

PROOF. Since $q_i\ S_\alpha^+\ q_i$ every value is in some block, H_i. Assume $q_k \in H_i \cap H_j$ for some $i \neq j$. Let q_i, q_j be any values of H_i, H_j, respectively. $q_k \in H_i$ implies $q_i\ S_\alpha \ldots S_\alpha\ q_k$ and $q_k \in H_j$ implies $q_k\ S_\alpha \ldots S_\alpha\ q_j$. Thus, $q_i\ S_\alpha^+\ q_j$ and $H_i \equiv H_j$. Thus, every fuzzy number is in some block and every pair of blocks is disjoint Q.E.D.

To extend the definition of redundant tuples given in Section 2, two tuples, t_i and t_k, are redundant if simultaneously.

(1) $d_{ij} \cup d_{kj}$ does not violate the threshold constraint for any discrete domain.
(2) q_{ij} and q_{kj} are α_j- proximate for all fuzzy number domains.

Redundant tuples are eliminated (as described earlier) by merging (via set union) the respective domains of the original tuples.

THEOREM 1. Let r be a relation

$$r \subseteq Q_1 \times Q_2 \times \ldots \times Q_k \times D_{k+1} \times \ldots \times D_m$$

where $k > 0$ and $m \geq k$. Let r' be derived by merging redundant tuples according to the level values $\alpha_1, \alpha_2, \ldots, \alpha_m$. r' is uniquely defined.

PROOF. It is sufficient to show that the tuples with which $t_i = (q_{i1}, q_{i2}, \ldots, q_{ik}, d_{ik+1}, \ldots, d_{im})$ will merge are uniquely defined. In [1] it was shown that for domains $D_{k+1}, D_{k+2}, \ldots, D_m$, the block of tuples with which t_i can merge is unique. Call this block H_{k+1}. (If $m = k$ let $H_{k+1} = r$.) Similarly, q_{i1} uniquely defines a block, H_1, by Lemma 2. q_{i2} defines H_2, $\ldots$, q_{ik} defines H_k. By definition of the merge operation t_i merges with $t_h \in H_1 \cap H_2 \cap \ldots \cap H_{k+1}$. The intersection of the blocks is uniquely defined. Q.E.D.

The major concern is the consistency and completeness of the fuzzy relational algebra. By consistent. it is meant that a command admits the ordinary relational algebra as a special case and be well-defined in the fuzzy case. The problem of well-definedness is addressed here. By complete, it is meant all ordinary relational commands have consistent fuzzy counterparts.

Of the several relational algebra commands, it is sufficient to show that Select, Project, Join, and Difference are well-defined. Union, Intersection, and Divide can be defined in terms of these.

THEOREM 2. The fuzzy relational algebra commands

(Select ((r) where < boolean clause >)
with LEVEL $(Q_1) = \alpha_1, \ldots,$ LEVEL$(D_m) = \alpha_m$)
and
(Project ((r over $Q_1, \ldots, Q_p, D_{p+1}, \ldots D_q$))
with LEVEL$(Q_1) = \alpha_1, \ldots,$ LEVEL$(D_q) = \alpha_q$)
well-defined.

PROOF. SELECT creates a new relation consisting of a subset of the tuples of an original relation. PROJECT creates a new relation consisting of a subset of the domains of an original relation. By Theorem 1, the new relation is unique upon elimination of redundant tuples. Q.E.D.

THEOREM 3. The fuzzy relational algebra command

(Join $((r_1, r_2$ over $Q_1, \ldots, Q_p, D_{p+1}, \ldots, D_p))$with LEVEL$(Q_1) = \alpha_1$, lll, LEVEL$(D_q) = \alpha_p)$ is well-defined

PROOF. Let $t_i \in r_1$. By definition of Join, t_i is concatenated with $t_j \in r_2$ if

(1) $q_{ik}\ S_{\alpha_k}\ q_{jk}$; $k = 1, 2, \ldots, p$

(2) $s_i(a_{ik}, b_{jk}) = s_j(a_{ik}, b_{jk}) \geqslant \alpha_k$;

$a_{ik} \in d_{ik}, b_{jk} \in d_{jk}, k = p+1, p+2, \ldots$

$k = p + 1, p + 2, \ldots, q$

This is a symmetric relation (i.e., Join (r_1, r_2) = Join (r_2, r_1)). Therefore, each tuple in r_1 (or r_2) is concatenated individually with each member of a uniquely determined set of tuples in r_2 (or r_1). Since the merge operation over the result is well-defined (Theorem 1), the result is well-defined. Q.E.D.

Two relations are said to be union compatible if they have the same domains. Union compatibility is a prerequisite for the Union, Intersection, or Difference operation. The complement of a fuzzy relation, r, over a closed universe, P, is denoted $r \downarrow P$ and defined as all tuples in P not α-similar to tuples in r. The well-definedness of Difference will be shown by showing that Intersection is well-defined then formulating Difference in terms of Intersection.

Given union compatible relations r_1 and r_2 with domains $Q_1, \ldots, Q_k, D_{k+1}, \ldots, D_m$, let tuples $t_i \in r_1$ and $t_h \in r_2$ be denoted as

$t_i = (q_{i1}, \ldots, q_{ik}, d_{i,k+1}, \ldots, d_{im})$

$t_h = (q_{h1}, \ldots, q_{hk}, d_{h,k+1}, \ldots, d_{hm})$

Each tuple in the result of the fuzzy relational algebra operation

(Intersection $((r_1, r_2))$ with LEVEL$(Q_1) = \alpha_1, \ldots$, LEVEL$(D_m) = \alpha_m)$

is formed from tuples t_i and t_h by appropriate domain unions

$(q_{i1} \sqcup q_{h1}, \ldots, q_{ik} \sqcup q_{hk}, d_{i,k+1} \cup d_{h,k+1}, \ldots, d_{im} \cup d_{hm})$

where the criteria

$q_{ij}\ S_{\alpha_j}\ q_{hj}$; $j = 1, 2, \ldots, k$

$\min_{x, y \in d_{ij} \cup d_{hj}}[s_j(x, y)] \geqslant \alpha_j$; $j = k+1, \ldots, m$

are simultaneously met.

THEOREM 4. The fuzzy relational algebra command, Intersection, is well-defined.

PROOF. Intersection is equivalent to a Join command over all domains. Therefore, by Theorem 3, it is well-defined. Q.E.D.

Given union compatible relations r_1 and r_2 the fuzzy relational algebra command

(Difference $((r_1, r_2))$
with LEVEL$(Q_1) = \alpha_1, \ldots$, LEVEL$(D_m) = \alpha_m)$

is defined to be the tuples $t_i \in r_1$ such that either

$\exists(j \in 1, 2, \ldots, k)(q_{ij} \not S_{\alpha_j} q_{hj})$

or

$\exists(j \in k+1, \ldots, m)(\min_{x,y \in d_{ij} \cup d_{hj}}[s_j(x,y)] < \alpha_j)$

for all $t_h \in r_2$. Difference (r_1, r_2) can obviously be reformulated as

Intersection $(r_1, r_2) \downarrow r_1$

Thus, a proof of well-definedness will be omitted.

A final property dealing with consistency of representation of a single relation is the assurance that any possible interpretation of the tuples resides in, at most, one tuple. This property is necessary to eliminate query ambiguity and assure the absence of update anomalies. In the fuzzy database as previously defined, any assignment of value

$I = (a_1, a_2, \ldots, a_m)$; $a_j \in d_{ij}$,
$j = 1, 2, \ldots, m$

was unique. When fuzzy number domains are incorporated, a membership restriction must be placed on interpretation components drawn from fuzzy number domains.

LEMMA 3. If $(q_i)_\alpha$ and $(q_j)_\alpha$ are each continuous sets over the real line and $x \in (q_i)_\alpha \cap (q_j)_\alpha$ then $q_i\ S\alpha\ q_j$.

The proof of the above lemma is obvious since they union of two continuous segments in continuous if they intersect.

THEOREM 5. Given $r \subseteq Q_1 \times Q_2 \times \ldots \times Q_k \times D_{k+1} \times \ldots \times D_m$ having no redundant tuples with respect to LEVEL$(\cdot) = (\alpha_1, \alpha_2, \ldots, \alpha_m)$ then any interpretation $I = (x_1, x_2, \ldots, x_k, a_{k+1}, \ldots, a_m)$ occurs at most once in r with $\mu(x_j) \geqslant \alpha_j$, $j = 1, 2, \ldots, k$.

PROOF. Assume $t_i, t_h \in r$, $i \neq h$, and I is an interpretation of both t_i and t_h. If $a_j \in d_{ij}$ and $a_j \in d_{hj}$ then $d_{ij} \cup d_{hj}$ does not violate the threshold constraint, LEVEL(α_j), by transitivity of the similarity relationship. If $x_j \in q_{ij}$ and $x_j \in q_{hj}$ then $q_{ij}\ S_{\alpha j}\ q_{hj}$ by Lemma 3. Therefore, t_i and t_h are redundant contrary to the assumption. Q.E.D

Thus fuzzy numbers can be incorporated in fuzzy relational databases while maintaining the consistency of representation of relations and the well-definedness of the relational algebra. No assumptions are required regarding the normality or convexity of the fuzzy number domain values. However, domain values must be defined over (have support in) a continuous segment of the real line.

FUZZY BOOLEAN QUERIES

A form of Boolean query was defined by Buckles and Petry (1983) and reviewed in Section 2 that can be translated to the procedural notation of the relational algebra. To extend the definition, query $Q(x_i, \ldots, x_h, a_j, \ldots, a_k)$ is an expression of one or more factors combined by disjunctive or conjunctive Boolean operators:

$$V_i \text{ op } \ldots \text{ op } V_h \text{ op } V_j \text{ op } \ldots \text{ op } V_k$$

Each V_i can be $x_i \in R$ or $a_j \in D_j$ modified by one more linquistic terms. An example is

cost = 27 AND weight = VERY VERY heavy

A membership value is induced for each tuple in a relation in terms of its possibilistic measure with respect to the query (see reference cited above).

An alternate approach is the direct translation of an informal query to fuzzy relational algebra operations. Consider a university course scheduling query, given the base relations SCHOLAR and FACULTY illustrated in Figure 1. The relations express the class time preferences of students and professors in terms of L - R fuzzy numbers of the form (m, α, β). The scalar m is the most preferred time of day while $L(m - x)$ and $R(x - m)$ are functions that stipulate the degree of acceptability of earlier and later times, respectively. For this example, assume that L and R are linear and $L(\alpha) = R(\beta) = 0.5$. Thus, the fuzzy number (14,2,1) specifies that 12:00 n or 3:00 P.M. are acceptable to degree 0.5.

RELATION SCHOLAR

Student	Course	Time
Weiss	Computability	(14,2,1)
Baum	Logic	(10,1,2)
Celli	Databases	(11,3,2)
Gerard	Fuzzy Sets	(8,0,3)
Morgan	Fuzzy Sets	(13,2,1)
Goldstein	Databases	(10,1,1)
Hayashi	Logic	(15,3,1)
Moritani	Computability	(12,2,2)
Herbst	Logic	(12,2,2)
Monet	Fuzzy Sets	(11,1,3)

RELATION FACULTY

Professor	Course	Time
Hebert	Databases	(15,2,2)
Buckles	Logic	(10,1,1)
Hebert	Computability	(11,2,1)
Petry	Fuzzy Sets	(14,1,2)

Fig. 1. Base Relations

The objective is to determine which students and professor have compatible time requirements for specific courses. The informal query might be: Which students and professors have roughly equal time preferences for the available courses?

Assume members of the Student, Professor, and Course domains have e-similarity (the identity relation). That is, s(x, y) = 1 if x and y are the same member and that s(x, y)=0 otherwise. Allow "roughly equal" to be interpreted as C-similar to degree 0.5. The formal fuzzy relational algebraic translation of the query is

```
(Join((SCHOLAR, FACULTY
 over COURSE, TIME))
 with LEVEL(COURSE) = 1,
      LEVEL(PROFESSOR) = 1,
      LEVEL(STUDENT) = 0,
      LEVEL(TIME) = 0.5)
 giving SCHEDULE .
```

Figure 2 shows the result. Members of the Student domain were permitted to merge without similarity protection, i.e., LEVEL (STUDENT) = 0. On the other hand, members of the Professor domain were completely protected against merging, i.e., LEVEL(PROFESSOR) = 1. Time values, on the other hand, are partially protected in accordance with the interpretation of "roughly equal." Thus, Student Goldstein was not included in Professor Hebert's class on databases due to incompatibility of time preferences.

RELATION SCHEDULE

Professor	Student	Course	Time
Hebert	{Celli}	Databases	{(11,3,2) (15,2,2)}
Buckles	{Baum Herbst}	Logic	{(10,1,1) (10,1,2) (12,2,2)}
Hebert	{Moritani Weiss}	Computability	{(11,2,1) (12,2,2) (14,2,1)}
Petry	{Monet Morgan}	Fuzzy Sets	{(11,1,3) (13,2,1) (10,1,1)}

Fig. 2. Query Result

The presence of fuzzy numbers permits fuzzy query capabilities other than simple information retrieval. Particularly, functional queries are permitted such as determining the average of a specific domain or performing Boolean tests based on the sum or difference of pairs of domains. It is well known that fuzzy numbers, even when restricted to convex, normal, and positive, form at most a commutative semiring (Dubois 1978, Dubois 1979, Mizumoto 1979) without additive or multiplicative inverses (Yager 1980). Yet many functional queries are effective despite this restriction. For example, an average of a set of domain values requires the addition of fuzzy numbers followed by division by the set power (an ordinary real number). . This does not necessitate convexity or normality of the domain values.

CONCLUSION

The representation of fuzzy numbers considerably expands the utility of fuzzy databases. Until now, the fuzzy database framework has not allowed operational consistency with respect to fuzzy numbers when using formal methods such as the relational algebra. The method described here has all the desirable properties previously proven to exist for finite, discrete domain values. Future work in the area of Boolean and functional queries and in the area of the relational calculus is needed to complete the cycle begun here.

REFERENCES

Buckles, B. P. and F. E. Petry (1982a). A fuzzy representation of data for relational databases. Fuzzy Sets & Syst., 7, 213-226.

Buckles, B. P. and F. E. Petry (1982b). Security and fuzzy databases. Proc. IEEE Int. Conf. on Cybernetics & Society, IEEE Press, 622-625.

Buckles, B. P. and F. E. Petry (1982c). Fuzzy databases and their applications. In M. Gupta and E. Sanchez (Eds.), Fuzzy Information and Decision Processes. North-Holland Pub. Co., NY, pp. 361-371.

Buckles, B. P. and F. E. Petry (1983). Information-theoretic characterization of fuzzy relational databases. IEEE Trans on Syst., Man & Cybern., SMC-13, 74-77.

Dubois,D. and H. Prade (1978). Operations on fuzzy numbers. Int. J. Syst. Sci., 9, 613-626.

Dubois, D. and H. Prade (1979). Fuzzy real algebra: some results. Fuzzy Sets & Syst., 2, 327-348.

Mizumoto, M. and K. Tanaka (1979). Some properties of fuzzy numbers. In M. Gupta, R. Ragade, and R. Yager (Eds.), Advances in Fuzzy Set Theory and Applications, North-Holland Pub. Co., NY, pp. 153-164.

Yager, R. R. (1980). On the lack of inverses in fuzzy arithmetic. Fuzzy Sets & Syst., 4, 73-82.

Zadeh, L. A. (1971). Similarity relations and fuzzy orderings. Inf. Sci., 3, 177-200.

ON SOME PROPERTIES OF FINITE AND COUNTABLE FUZZY RANDOM VARIABLES

M. Miyakoshi and M. Shimbo

Division of Information Engineering, Graduate School of Engineering, Hokkaido University, Sapporo, Japan

Abstract. Some binary operations of fuzzy numbers are reviewed and a limit of a sequence of fuzzy numbers is defined. Based on these concepts and properties, finite and countable fuzzy random variables are considered and a convergence property of a sequence of fuzzy random variables is shown.

Keywords. Fuzzy numbers; limit of a sequence of fuzzy numbers; fuzzy random variables; convergence property.

INTRODUCTION

The theory of fuzzy random variables(frv's) was first introduced by Kwakernaak(1978,1979) and he investigated their fundamental properties. In this paper we intend to show some additional properties of frv's based on the concepts of the set-representation for a fuzzy set and of fuzzy numbers. Firstly, we define a supremum and an infimum, a superior and an inferior limits, and a limit of a sequence of fuzzy numbers, and investigate their properties. These lead the definitions of a superior and an inferior limits and a limit of a sequence of frv's, and finally we discuss their measurability and a convergence property for a sequence of frv's.

FUZZY NUMBERS

A Class of Fuzzy Numbers

Fuzzy numbers are defined as fuzzy sets on $\mathbb{R}$, and their algebraic and other properties have been investigated (Mizumoto and Tanaka,1979; Nguyen,1978). We discuss on the sum(+), the subtract(-) and the product(·) of fuzzy numbers in a restrictive situation. Their definitions should be referred to Mizumoto and Tanaka (1979) and Nguyen(1978).

We denote fuzzy numbers by $\tilde{a}$, $\tilde{b}$ etc., and let $\tilde{P}_N(\mathbb{R})$ be a class of fuzzy numbers which satisfy the following four conditions:

(i) $\tilde{a}$ is a fuzzy number,

(ii) $h_{\tilde{a}}$ is upper semicontinuous,

(iii) $\tilde{a}$ is strictly normal, that is, there exists at least one element x of $\mathbb{R}$ such that $h_{\tilde{a}}(x)=1$,

(iv) supp $\tilde{a}$ is bounded, that is, $\{x|h_{\tilde{a}}(x)>0\}$ is a bounded subset in $\mathbb{R}$.

Putting $(\tilde{a})_{\bar{\alpha}} \triangleq \{x|h_{\tilde{a}}(x)\geq\alpha\}$, $\bar{a}_{\bar{\alpha}} \triangleq \sup(\tilde{a})_{\bar{\alpha}}$, and $\underline{a}_{\bar{\alpha}} \triangleq \inf(\tilde{a})_{\bar{\alpha}}$, note that the conditions (ii) and (iii) guarantee $(\tilde{a})_{\bar{\alpha}}\neq\phi$ for any α and so $\bar{a}_{\bar{\alpha}}$ and $\underline{a}_{\bar{\alpha}}$ are contained in $(\tilde{a})_{\bar{\alpha}}$.
From the definitions of the sum, the subtract and the product of fuzzy numbers, it follows immediately that for $\forall\tilde{a}$ and $\forall\tilde{b}\ \varepsilon\ \tilde{P}_N(\mathbb{R})$ and $\forall\alpha\varepsilon(0,1]$

$$(\tilde{a}+\tilde{b})_{\bar{\alpha}} = \tilde{a}_{\bar{\alpha}} + \tilde{b}_{\bar{\alpha}} ,$$

$$(\tilde{a}-\tilde{b})_{\bar{\alpha}} = \tilde{a}_{\bar{\alpha}} - \tilde{b}_{\bar{\alpha}} ,$$

and $$(\tilde{a}\cdot\tilde{b})_{\bar{\alpha}} = \tilde{a}_{\bar{\alpha}}\cdot \tilde{b}_{\bar{\alpha}} .$$

[Proposition 1]

$\tilde{P}_N(\mathbb{R})$ is closed under the binary operations +, - and · .

Limit of a Sequence of Fuzzy Numbers

Let $\{\tilde{a}_n, n\geq1\}$ be a sequence whose elements are contained in $\tilde{P}_N(\mathbb{R})$. Now we define a subset of $\mathbb{R}$, denoted by $\sup_{k\geq n}^{\Pi}(\tilde{a}_k)_{\bar{\alpha}}$,

$$\sup_{k\geq n}^{\Pi}(\tilde{a}_k)_{\bar{\alpha}} \triangleq \{\sup_{k\geq n} u_k \,|\, u_k\varepsilon(\tilde{a}_k)_{\bar{\alpha}} \text{ for } k\geq n\} ,$$

where k and n are positive integers.
From the Extension Principle (Zadeh,1975; Nguyen,1978) we get

$$h_{\sup_{k\geq n}\tilde{a}_k}(x) = \sup_{x=\sup_{k\geq n} u_k} (\inf_{k\geq n} h_{\tilde{a}_k}(u_k)) \quad \text{for } \forall x\varepsilon\mathbb{R}.$$

as the membership-functional representation of the supremum of a subsequence $\{\tilde{a}_k, k\geq n\}$.
In the set-representation , however, it is easy to show that for $\forall\alpha\varepsilon(0,1)$

$$(\sup_{k\geq n}\tilde{a}_k)_{\bar{\alpha}} \supseteq \sup_{k\geq n}^{\Pi}(\tilde{a}_k)_{\bar{\alpha}} \supseteq (\sup_{k\geq n}\tilde{a}_k)_{\alpha} ,$$

where $(\sup_{k\geq n} \tilde{a}_k)_\alpha \triangleq \{x \mid h_{\sup_{k\geq n} \tilde{a}_k}(x) > \alpha\}$.

From the above equation we obtain

$$\sup_{\alpha\varepsilon(0,1)}(\alpha \wedge \chi_{(\sup_{k\geq n} \tilde{a}_k)_{\bar{\alpha}}}(x)) \geq \sup_{\alpha\varepsilon(0,1)}(\alpha\wedge\chi_{\sup_{k\geq n}^{\Pi}(\tilde{a}_k)_{\bar{\alpha}}}(x))$$

$$\geq \sup_{\alpha\varepsilon(0,1)}(\alpha\wedge\chi_{(\sup_{k\geq n} \tilde{a}_k)_\alpha}(x)) \text{ for } \forall n\geq 1 \text{ and } \forall x\varepsilon\mathbb{R},$$

and finally we have

$$h_{\sup_{k\geq n} \tilde{a}_k}(x) = \sup_{\alpha\varepsilon[0,1]}(\alpha\wedge\chi_{\sup_{k\geq n}^{\Pi}(\tilde{a}_k)_{\bar{\alpha}}}(x)).$$

Therefore, we can use the family of subsets of $\mathbb{R}$, $\{\sup_{k\geq n}^{\Pi}(\tilde{a}_k)_{\bar{\alpha}} \mid \alpha\varepsilon(0,1]\}$, as the set-representation of $\sup_{k\geq n} \tilde{a}_k$. In a dual manner we get the following functional and set-representations of the infimum of the subsequence:

$$h_{\inf_{k\geq n} \tilde{a}_k}(x) = \sup_{\alpha\varepsilon[0,1]}(\alpha\wedge\chi_{\inf_{k\geq n}^{\Pi}(\tilde{a}_k)_{\bar{\alpha}}}(x)) \text{ for } \forall x\varepsilon\mathbb{R},$$

where $\inf_{k\geq n}^{\Pi}(\tilde{a}_k)_{\bar{\alpha}} \triangleq \{\inf_{k\geq n} u_k \mid u_k\varepsilon(\tilde{a}_k)_{\bar{\alpha}} \text{ for } k\geq n\}$.

Henceforth, for brevity, we only list the definitions and the propositions, which are used in the latter section, without proofs.

[Proposition 2]

$$\sup(\sup_{k\geq n} \tilde{a}_k)_{\bar{\alpha}} = \sup_{k\geq n} \bar{a}_{\bar{\alpha}}^{(k)}, \quad \inf(\sup_{k\geq n} \tilde{a}_k)_{\bar{\alpha}} = \sup_{k\geq n} \underline{a}_{\bar{\alpha}}^{(k)},$$

$$\sup(\inf_{k\geq n} \tilde{a}_k)_{\bar{\alpha}} = \inf_{k\geq n} \bar{a}_{\bar{\alpha}}^{(k)}, \quad \inf(\inf_{k\geq n} \tilde{a}_k)_{\bar{\alpha}} = \inf_{k\geq n} \underline{a}_{\bar{\alpha}}^{(k)},$$

where $\bar{a}_{\bar{\alpha}}^{(k)} \triangleq \sup(\tilde{a}_k)_{\bar{\alpha}}$ and $\underline{a}_{\bar{\alpha}}^{(k)} \triangleq \inf(\tilde{a}_k)_{\bar{\alpha}}$.

[Proposition 3]

Let $\{\tilde{a}_k, k\geq n\}$ be a sequence, whose elements are contained in $\tilde{P}_N(\mathbb{R})$ and convex. Then the supremum and the infimum of the sequence are also convex.

[Definition 0]

A sequence of fuzzy numbers $\{\tilde{a}_n, n\geq 1\}$ is said to be bounded if there exists a positive number c such that, for $\forall n\geq 1$ and $\forall x \varepsilon \operatorname{supp} \tilde{a}_n$, $|x| \leq c$.

[Definition 1]

Let $\{\tilde{a}_n, n\geq 1\}$ be a bounded sequence of fuzzy numbers and $\tilde{a}_n \varepsilon \tilde{P}_N(\mathbb{R})$ for $\forall n\geq 1$.

A superior and an inferior limits of the sequence are defined as follows:

$$\text{Superior limit: } \overline{\lim}\, \tilde{a}_n \triangleq \inf_{n\geq 1}(\sup_{k\geq n} \tilde{a}_k),$$

$$\text{Inferior limit: } \underline{\lim}\, \tilde{a}_n \triangleq \sup_{n\geq 1}(\inf_{k\geq n} \tilde{a}_k).$$

[Proposition 4]

$$\sup(\overline{\lim}\, \tilde{a}_n)_{\bar{\alpha}} = \overline{\lim}\, \bar{a}_{\bar{\alpha}}^{(n)}, \quad \inf(\overline{\lim}\, \tilde{a}_n)_{\bar{\alpha}} = \overline{\lim}\, \underline{a}_{\bar{\alpha}}^{(n)},$$

$$\sup(\underline{\lim}\, \tilde{a}_n)_{\bar{\alpha}} = \underline{\lim}\, \bar{a}_{\bar{\alpha}}^{(n)}, \quad \inf(\underline{\lim}\, \tilde{a}_n)_{\bar{\alpha}} = \underline{\lim}\, \underline{a}_{\bar{\alpha}}^{(n)},$$

where the symbols, $\overline{\lim}$ and $\underline{\lim}$, in the right hand sides of the equations stand for a superior and an inferior limits of a sequence of numbers, respectively.

[Definition 2]

A bounded sequence of fuzzy numbers, $\{\tilde{a}_n, n\geq 1\}$, is said to have a limit or to be convergent if $\overline{\lim}\, \tilde{a}_n = \underline{\lim}\, \tilde{a}_n$, and we call the common fuzzy number the limit of the bounded sequence and write $\lim \tilde{a}_n$ simply.

[Definition 3]

Let $\{\tilde{a}_n, n\geq 1\}$ be a sequence of fuzzy numbers such that $\tilde{a}_n \varepsilon \tilde{P}_N(\mathbb{R})$ for $\forall n\geq 1$. We define a partial sum of the first n-term fuzzy numbers, denoted by $\tilde{S}_n$, as follows:

$$\tilde{S}_n \triangleq \sum_{k=1}^{n} \tilde{a}_k \triangleq \tilde{a}_1 + \tilde{a}_2 + \ldots + \tilde{a}_n.$$

Hence $\{\tilde{S}_n, n\geq 1\}$ is a sequence such that $\tilde{S}_n \varepsilon \tilde{P}_N(\mathbb{R})$ for each n.

We define an infinite sum of a sequence of fuzzy numbers as follows: if the sequence $\{\tilde{S}_n, n\geq 1\}$ is bounded and convergent, the sequence $\{\tilde{a}_n, n\geq 1\}$ is said to have an infinite sum, denoted by $\sum_{n=1}^{\infty} \tilde{a}_n$.

FUZZY RANDOM VARIABLES

Definitions

Here we review the definitions of an frv and of the expectation of frv. The details should be referred to Kwakernaak(1978).

Let (Ω, A, P) be a probability space and the ω stands for the element of Ω.

[Definition 4]

An frv ξ is defined as a function from Ω to $\tilde{P}_N(\mathbb{R})$ satisfying the following two conditions:

(i) $\underline{X}_{\bar{\alpha}}(\omega)$, $\overline{X}_{\bar{\alpha}}(\omega) \varepsilon \xi(\omega)_{\bar{\alpha}}$ for $\forall \alpha\varepsilon(0,1]$ and $\forall \omega\varepsilon\Omega$,

(ii) $\underline{X}_{\bar{\alpha}}$ and $\overline{X}_{\bar{\alpha}}$ are A-measurable ω-functions,

where $\underline{X}_{\bar{\alpha}}(\omega) \triangleq \inf \xi(\omega)_{\bar{\alpha}}$ and $\overline{X}_{\bar{\alpha}}(\omega) \triangleq \sup \xi(\omega)_{\bar{\alpha}}$.

The $\xi(\omega)$ is called a sample value of ω with respect to ξ or briefly a sample value of ω, and it is a fuzzy number in $\tilde{P}_N(\mathbb{R})$. Since $\xi(\omega)_{\bar{\alpha}}$ is a nonempty subset of $\mathbb{R}$ for $\forall \alpha\varepsilon(0,1]$ and $\forall \omega\varepsilon\Omega$, $\underline{X}_{\bar{\alpha}}$ and $\overline{X}_{\bar{\alpha}}$ defined above are well-defined.

Now we define two ω-functions as follows:

$$\overline{X}_0(\omega) \triangleq \sup(\operatorname{supp} \xi(\omega)),$$

$$\underline{X}_0(\omega) \triangleq \inf(\operatorname{supp} \xi(\omega)).$$

It is easy to show that

$$\overline{X}_0(\omega) = \sup_{0<r} \overline{X}_{\bar{r}}(\omega) \text{ and } \underline{X}_0(\omega) = \inf_{0<r} \underline{X}_{\bar{r}}(\omega),$$

where r stands for the generic element of rational numbers in (0,1], and that $\underline{X}_0$ and $\overline{X}_0$ are A-measurable and $\underline{X}_0 \le \underline{X}_{\bar{\alpha}} \le \overline{X}_{\bar{\alpha}} \le \overline{X}_0$ for $\forall\alpha\varepsilon(0,1]$.

Henceforth we add a condition to those of the definition of frv as follows:

(iii) $\underline{X}_0$ and $\overline{X}_0$ are integrable with respect to P-measure,

and so we have $\underline{X}_{\bar{\alpha}}$ and $\overline{X}_{\bar{\alpha}}$ are integrable for $\forall\alpha\varepsilon(0,1]$.

Next we review the definition of expectation of frv (Kwakernaak, 1978).

[Definition 5]
An expectation of an frv ξ, written $E\xi$, is defined as a fuzzy number, whose set-representation is given as $(E\xi)_{\bar{\alpha}} = [E\underline{X}_{\bar{\alpha}}, E\overline{X}_{\bar{\alpha}}]$ for $\forall\alpha\varepsilon$ (0,1].

From this defintion and the condition (iii) we have $E\xi\varepsilon\tilde{P}_N(\mathbb{R})$.

Propoerties of Frv's

The fundamental properties of frv's have been investigated by Kwakernaak(1978). In this section some additional properties are shown.

We define the sum, the subtract and the product of two frv's ξ and η pointwise as follows:

[Definition 6]

$$(\xi + \eta)(\omega) = \xi(\omega) + \eta(\omega),$$
$$(\xi - \eta)(\omega) = \xi(\omega) - \eta(\omega), \qquad (1)$$
$$(\xi \cdot \eta)(\omega) = \xi(\omega)\cdot\eta(\omega).$$

It should be noted that from Proposition 1 the value of the left hand sides in Eq.(1) are contained in $\tilde{P}_N(\mathbb{R})$.

[Property 1]
If ξ and η are frv's, so are $\xi+\eta$, $\xi-\eta$ and $\xi\cdot\eta$.

Hence we get the following property.

[Property 2]

$$E(\xi + \eta) = E\xi + E\eta \ ,$$
$$E(\xi - \eta) = E\xi - E\eta \ .$$

[Defintion 7]
Let ξ and η be two frv's. Then ξ and η are said to be mutually independent (Kwakernaak,1978) if the two σ-fields, $\sigma(\underline{X}_{\bar{\alpha}}, \overline{X}_{\bar{\alpha}}; \alpha\varepsilon(0,1])$ and $\sigma(\underline{Y}_{\bar{\alpha}}, \overline{Y}_{\bar{\alpha}}; \alpha\varepsilon(0,1])$, are mutually independent, where $\underline{Y}_{\bar{\alpha}}(\omega) \triangleq \inf\ \eta(\omega)_{\bar{\alpha}}$ and $\overline{Y}_{\bar{\alpha}}(\omega) \triangleq \sup\ \eta(\omega)_{\bar{\alpha}}$.

From this definition we get the following property.

[Property 3]

If frv's ξ and η are mutually independent then

$$E(\xi\cdot\eta) \supseteqq E\xi\cdot E\eta,$$

and moreover, if ξ and η are positive or negative,

$$E(\xi\cdot\eta) = E\xi\cdot E\eta.$$

Let $\tilde{a}\varepsilon\tilde{P}_N(\mathbb{R})$ and $\xi(\omega) \triangleq \tilde{a}$ for $\forall\omega\varepsilon\Omega$. In general, however, it does not hold that $E\tilde{a}=\tilde{a}$ but that $E\tilde{a} \supseteqq \tilde{a}$. The equality holds if and only if $\tilde{a}$ is convex. Let $\tilde{a},\tilde{b}\varepsilon\tilde{P}_N(\mathbb{R})$ and ξ and η be definitely signatured, then $\tilde{a}\cdot\xi$ and $\tilde{b}\cdot\eta$ are frv's, and so is $\tilde{a}\cdot\xi + \tilde{b}\cdot\eta$. Hence it is easy to show the following property.

[Property 4]
If $\tilde{a}$, $\tilde{b}$, and ξ and η are definitely signatured then $\tilde{a}\cdot\xi + \tilde{b}\cdot\eta$ is an frv.

$$E(\tilde{a}\cdot\xi + \tilde{b}\cdot\eta) \supseteqq \tilde{a}\cdot E\xi + \tilde{b}\cdot E\eta,$$

and moreover, if $\tilde{a}$ and $\tilde{b}$ are convex, then

$$E(\tilde{a}\cdot\xi + \tilde{b}\cdot\eta) = \tilde{a}\cdot E\xi + \tilde{b}\cdot E\eta.$$

Properties of the Sequence of Frv's

Here we investigate the properties of a sequence of frv's.

Let $\{\xi_n, n\ge 1\}$ be a sequence of frv's. Now we define a superior and an inferior limits of the sequence of frv's.

[Definition 8]
The supremum and the infimum of a subsequence $\{\xi_k, k\ge n\}$ are defined pointwise as follows:

$$\text{Supremum: } (\sup_{k\ge n} \xi_k)(\omega) \triangleq \sup_{k\ge n} \xi_k(\omega),$$

$$\text{Infimum : } (\inf_{k\ge n} \xi_k)(\omega) \triangleq \inf_{k\ge n} \xi_k(\omega).$$

The superior and the inferior limits of the sequence $\{\xi_n, n\ge 1\}$ are defined pointwise as follows:

$$\text{Superior limit: } (\overline{\lim}\ \xi_n)(\omega) \triangleq \overline{\lim}\ \xi_n(\omega) = \inf_{n\ge 1}(\sup_{k\ge n} \xi_k(\omega)),$$

$$\text{Inferior limit: } (\underline{\lim}\ \xi_n)(\omega) \triangleq \underline{\lim}\ \xi_n(\omega) = \sup_{n\ge 1}(\inf_{k\ge n} \xi_k(\omega)).$$

[Definition 9]

Let $\{\xi_n, n\ge 1\}$ be a sequence of frv's. The sequence $\{\xi_n, n\ge 1\}$ is said to be integrably bounded if there exists an integrable ω-function X such that, for $\forall n\ge 1$ and $\forall\omega\varepsilon\Omega$,

$$|\underline{X}_0^{(n)}(\omega)| < X(\omega) \text{ and } |\overline{X}_0^{(n)}(\omega)| < X(\omega),$$

where $\underline{X}_0^{(n)}(\omega) \triangleq \inf(\text{ supp } \xi_n(\omega))$ and $\overline{X}_0^{(n)}(\omega) \triangleq \sup(\text{ supp } \xi_n(\omega))$.

[Property 5]

Let $\{\xi_n, n\ge 1\}$ be a sequence of frv's and be integrably bounded. Then $\sup_{k\ge n} \xi_k$ and $\inf_{k\ge n} \xi_k$ are frv's for $\forall n\ge 1$, and so are $\overline{\lim}\ \xi_n$ and $\underline{\lim}\ \xi_n$.

From this property we can calculate the expectations of $\overline{\lim}\ \xi_n$ and $\underline{\lim}\ \xi_n$. Before we show a convergence property for frv's, we have to define a limit of a sequence of frv's.

[Definition 10]
The sequence $\{\xi_n, n\geq 1\}$ is said to be convergent if the sequence of sample values of ω with respect to ξ, $\{\xi_n, n\geq 1\}$, is convergent pointwise. The limit of the sequence $\{\xi_n, n\geq 1\}$ is defined pointwise as follows:

Limit: $(\lim \xi_n)(\omega) = \lim \xi_n(\omega)$.

Note that $\lim \xi_n(\omega) = \underline{\lim}\, \xi_n(\omega) = \overline{\lim}\, \xi_n(\omega)$,

and that, if exists, the limit of the sequence is an frv.

Now we show the main result of this paper.

[Property 6]
Let $\{\xi_n, n\geq 1\}$ be a sequence of frv's and be integrably bounded.
Then it holds that

$$E\overline{\lim}\, \xi_n \geq \overline{\lim}\, E\xi_n \geq \underline{\lim}\, E\xi_n \geq E\underline{\lim}\, \xi_n, \qquad (2)$$

and if $\lim \xi_n$ exists,

$$E\lim \xi_n = \lim E\xi_n.$$

(Proof) At first we show that $E\overline{\lim}\, \xi_n \geq \overline{\lim}\, E\xi_n$. According to Zadeh(1975), if $\tilde{a}$ and $\tilde{b}$ are normal-convex fuzzy numbers whose α-level sets are closed intervals, that is, $(\tilde{a})_{\bar{\alpha}} = [\underline{a}_{\bar{\alpha}}, \overline{a}_{\bar{\alpha}}]$ and $(\tilde{b})_{\bar{\alpha}} = [\underline{b}_{\bar{\alpha}}, \overline{b}_{\bar{\alpha}}]$, the following four statements are equivalent each other:

$$\tilde{a} \geq \tilde{b},$$

$$\tilde{a} \vee \tilde{b} = \tilde{a},$$

$$\tilde{a} \wedge \tilde{b} = \tilde{b},$$

and $[\underline{a}_{\bar{\alpha}}, \overline{a}_{\bar{\alpha}}] \vee [\underline{b}_{\bar{\alpha}}, \overline{b}_{\bar{\alpha}}] = [\underline{a}_{\bar{\alpha}}, \overline{a}_{\bar{\alpha}}]$ for $\forall \alpha \in (0,1]$.
Since $(E\xi_n)_{\bar{\alpha}} = [E\underline{X}^{(n)}_{\bar{\alpha}}, E\overline{X}^{(n)}_{\bar{\alpha}}]$, from Propositions 2 and 3 and Definition 1, we get that

$$(\overline{\lim}\, E\xi_n)_{\bar{\alpha}} = [\overline{\lim}\, E\underline{X}^{(n)}_{\bar{\alpha}}, \overline{\lim}\, E\overline{X}^{(n)}_{\bar{\alpha}}].$$

On the other hand, from Property 5 and Definition 5, we have

$$(E\overline{\lim}\, \xi_n)_{\bar{\alpha}} = [E\overline{\lim}\, \underline{X}^{(n)}_{\bar{\alpha}}, E\overline{\lim}\, \overline{X}^{(n)}_{\bar{\alpha}}].$$

By the integrable boundedness of ξ_n's, we can apply Lebesgue's dominated convergence theorem to $\{\underline{X}^{(n)}_{\bar{\alpha}}, n\geq 1\}$ and $\{\overline{X}^{(n)}_{\bar{\alpha}}, n\geq 1\}$, and get that $E\overline{\lim}\, \underline{X}^{(n)}_{\bar{\alpha}} \geq \overline{\lim}\, E\underline{X}^{(n)}_{\bar{\alpha}}$ and $E\overline{\lim}\, \overline{X}^{(n)}_{\bar{\alpha}} \geq \overline{\lim}\, E\overline{X}^{(n)}_{\bar{\alpha}}$.
Hence we have

$[E\overline{\lim}\, \underline{X}^{(n)}_{\bar{\alpha}}, E\overline{\lim}\, \overline{X}^{(n)}_{\bar{\alpha}}] \vee [\overline{\lim}\, E\underline{X}^{(n)}_{\bar{\alpha}}, \overline{\lim}\, E\overline{X}^{(n)}_{\bar{\alpha}}] = [E\overline{\lim}\, \underline{X}^{(n)}_{\bar{\alpha}}, E\overline{\lim}\, \overline{X}^{(n)}_{\bar{\alpha}}]$, and this means that

$$E\overline{\lim}\, \xi_n \geq \overline{\lim}\, E\xi_n.$$

In a dual manner we get

$$\underline{\lim}\, E\xi_n \geq E\underline{\lim}\, \xi_n.$$

Next we show $\overline{\lim}\, E\xi_n \geq \underline{\lim}\, E\xi_n$.

From Proposition 4, it is clear that $(\overline{\lim}\, E\xi_n)_{\bar{\alpha}} = [\overline{\lim}\, E\underline{X}^{(n)}_{\bar{\alpha}}, \overline{\lim}\, E\overline{X}^{(n)}_{\bar{\alpha}}]$ and $(\underline{\lim}\, E\xi_n)_{\bar{\alpha}} = [\underline{\lim}\, E\underline{X}^{(n)}_{\bar{\alpha}}, \underline{\lim}\, E\overline{X}^{(n)}_{\bar{\alpha}}]$.
Since the superior limit is greater than or equal to the inferior one, it follows that $[\overline{\lim}\, E\underline{X}^{(n)}_{\bar{\alpha}}, \overline{\lim}\, E\overline{X}^{(n)}_{\bar{\alpha}}] \vee [\underline{\lim}\, E\underline{X}^{(n)}_{\bar{\alpha}}, \underline{\lim}\, E\overline{X}^{(n)}_{\bar{\alpha}}] = [\overline{\lim}\, E\underline{X}^{(n)}_{\bar{\alpha}}, \overline{\lim}\, E\overline{X}^{(n)}_{\bar{\alpha}}]$, and this means that $\overline{\lim}\, E\xi_n \geq \underline{\lim}\, E\xi_n$.

To prove the second part it is sufficient to note that, if the limit exists, both sides in Eq.(2) are equal to each other, that is, $E\overline{\lim}\, \xi_n = E\underline{\lim}\, \xi_n$, and this means that the sequence $\{\xi_n, n\geq 1\}$ is convergent and bounded and that $E\lim \xi_n = \lim E\xi_n$. Q.E.D.

The following property is obtained by applying Property 6 to a sequence $\{\eta_n, n\geq 1\}$, where η_n is a partial sum of the first n-term frv's, that is, $\eta_n = \sum_{k=1}^{n} \xi_k$.

[Property 7]
Let $\{\xi_n, n\geq 1\}$ be a sequence of frv's satisfying the conditions:

$$\sum_{n=1}^{\infty} E|\underline{X}_0^{(n)}| < \infty \text{ and } \sum_{n=1}^{\infty} E|\overline{X}_0^{(n)}| < \infty .$$

If the sequence $\{\xi_n, n\geq 1\}$ has an infinite sum $\sum_{n=1}^{\infty} \xi_n$, then the infinite sum is an frv, and it holds that

$$E\sum_{n=1}^{\infty} \xi_n = \sum_{n=1}^{\infty} E\xi_n .$$

CONCLUSION

We have investigated some properties of finite and countable frv's, and proved a dominated convergence property for frv's. Some other properties, for example, a strong law of large numbers for frv's and Fatou's lemma for conditional expectations of frv's, have also been shown, though their proofs are not shown in this paper. Properties of time-dependent frv's will be investigated in the near future.

REFERENCES

Kwakernaak, H. (1978). Fuzzy random variables I. Information Sciences, 15, 1-29.

Kwakernaak, H. (1979). Fuzzy random variables II. ibid., 17, 253-278.

Mizumoto, M., and Tanaka, K. (1979). Some properties of fuzzy numbers. In M.M. Gupta, R.K. Ragade and R.R. Yager (Ed.), Advances in fuzzy set theory and applications. North-Holland, Amsterdam. pp.153-164.

Nguyen, H.T. (1978). A note on the extension principle for fuzzy sets. J. of Mathematical Analysis and Applications, 64, 369-380.

Zadeh, L.A. (1975). The concept of a linguistic variable and its application to approximate reasoning-I. Information Sciences, 8, 199-249.

ON THE SHADOW OF A FUZZY SET

Liwen Pei and Mian Ouyang

Department of Mathematics, Wuhan University, Wuhan, Hubei, People's Republic of China

Abstract. This paper is mainley concerned with certain properties of the shadow of a fuzzy set proposed in [2], [3]. At the sane time, several corrections are made for some results in [2] and formulae for the decomposition of a fuzzy set into the cylinder of the shadow are discussed in detail. The necessary and sufficient condition for the equality of the fuzzy sets are also discussed.

Keywords. Set theory; fuzzy set; convex fuzzy set; convex hull; shadow of fuzzy set; cylinder of shadow; optimization; fuzzy restraint.

The shadow concept of a fuzzy set is very important in the optimization problem with fuzzy restraint.

Definition 1. Let A be a fuzzy set in R^n. Let H be a hyperplane in R^n, $P_0 \in R^n$. A fuzzy set S(A) in R^n is referred to be point-shadow of a fuzzy set A from P_0 to H, if the membership function of S(A) is as follows:

$$\mu_{S(A)}(h) = \begin{cases} \sup_{x \in L} \mu_A(x) & h \in H \\ 0 & h \bar{\in} H \end{cases} \quad (1)$$

Here L shows linking line of P_0 and h. From now on, S(A) is used to express a point-shadow.

Definition 2. Let C(A) be a fuzzy set. The membership function of C(A) is as follows:

$$\mu_{C(A)}(h) = \begin{cases} \inf_{x \in L} \mu_A(x) & h \in H \\ 0 & h \bar{\in} H \end{cases} \quad (2)$$

Here L expresses linking line of h and P_0. So that C(A) is called the complementary point-shadow of a fuzzy set A from P_0 to H. Therefore, on H the complementary point-shadow of A to H equals a complementary fuzzy set of point-shadow of the complementary fuzzy set of A to H. When $P_0 = \infty$, $L \perp H$, S(A) is called orthogonal shadow and C(A) is called orthogonal complementary point-shadow. Complementary point-shadow and point-shadow have similar property as follows:

If A, B are fuzzy sets, then we have

(1) Homogeneity.

$$K \in [0, 1], \quad C(KA) = KC(A), \quad (3)$$

Here KA is a fuzzy set: $\mu_{KA}(x) = K\mu_A(x)$.

(2) Monotonicity.

$$A \subseteq B \Longrightarrow C(A) \subseteq C(B), \quad (4)$$

(3) Distributivity and subdistributivity.

$$C(A \cup B) \supseteq C(A) \sqcup C(B) \qquad C(A \cap B) = C(A) \cap C(B). \quad (5)$$

(4) Nilpotentiality.

$$C^2(A) = C(\ C(A)\ ) = \Phi \quad (6)$$

(5) Keep concavity. If A is a concave fuzzy set in R^n, then orthogonal complementary shadow of A is a concave fuzzy set in H.

Definition 3. Let A be a fuzzy set in R^n. Let H be the hyperplane in R^n. S_H Shows a orthogonal shadow of A to H. C_H expresses a orthogonal complementary shadow of A to H. If $\forall x \in R^n$

$$\mu_{S^*}(x) = \mu_{S(A)}(x^*) \qquad \mu_{C^*}(x) = \mu_{C(A)}(x^*). \quad (7)$$

Here x^* is a orthogonal projection point of

x to H. The fuzzy sets S^*, C^* are called the cylined fuzzy sets generated from S_H, C_H. Clearly

$$\mu_{S^*}(x) = \sup_{h\in L_x} \mu_A(h)$$

$$\mu_{C^*}(x) = \inf_{x\in L_x} \mu_A(h) \qquad (8)$$

Here L_x is line which is perpendicular to H through x. So that

$$\bigcup_\lambda C_\lambda^* \subseteq A \subseteq \bigcap_\lambda S_\lambda^* \qquad (9)$$

<u>Theorem 1.</u> If A is a strictly convex fuzzy set, then

$$\bigcup_\lambda C_\lambda^* \subseteq A = \bigcap_\lambda S_\lambda^* \qquad (10)$$

If A is a strictly concave fuzzy set, then

$$\bigcup_\lambda C_\lambda^* = A \supseteq \bigcap_\lambda S_\lambda^* \qquad (11)$$

<u>Proof:</u> According to the known conclusion, in order to prove (10) it is sufficient to prove $A \supseteq \bigcap_\lambda S_\lambda^*$, that is

$$\mu_A(x) \geqslant \mu_{\bigcap_\lambda S_\lambda^*}(x) = \inf_\lambda(\mu_{S_\lambda^*}(x)) \quad (12)$$

So that we want only to prove that:
$\forall x_0 \in R^n$, $\exists H_{\lambda_0}$, such that for corresponding $S_{\lambda_0}^*$.

$$\mu_A(x_0) = \mu_{S_{\lambda_0}^*}(x_0).$$

Let $\mu_A(x_0) = \alpha < 1$, since A is a strictly convex fuzzy set, hence $\Gamma_\alpha = \{x \mid \mu_A(x) \geqslant \alpha\}$ is a strictly convex set. When x_0 is at the boundary of Γ_α, through x_0 we make a hyperplane π of supporting Γ_α, such that $\mu_A(x) < \alpha$ on one side of π, at π $\mu_A(x) < \alpha$, except for $\mu_A(x_0) = \alpha$.
We make a hyperplane H_{λ_0} in R^n, and H_{λ_0} with π is perpendicalar, we make orthogonal projection, from A to H_{λ_0} so that we can obtain a fuzzy cylinder $S_{\lambda_0}^*$. Because line L_{x_0} (which is a perpendicalar line with H_{λ_0}) lid in the hyperplane π, so we have

$$\mu_A(x_0) = \sup \mu_A(h) = \mu_{S_{\lambda_0}^*}(x\ \ )$$

Thus $A = \bigcap_\lambda S_\lambda^*$.

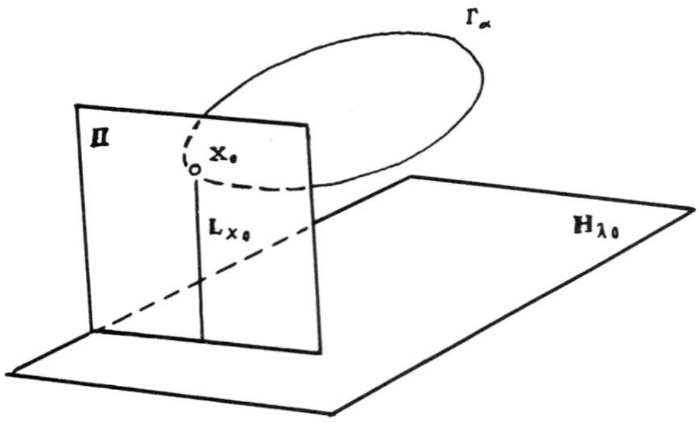

When x_0 is inside of Γ_α, take a sufficiently small $\varepsilon > 0$, then $x_0 \bar{\in} \Gamma_{\alpha+\varepsilon}$, here $\Gamma_{\alpha+\varepsilon} = \{x \mid \mu_A(x) \geqslant \alpha+\varepsilon\}$ is a strictly convex set.
When $R^n = R^2$, it is shown by the following figure.

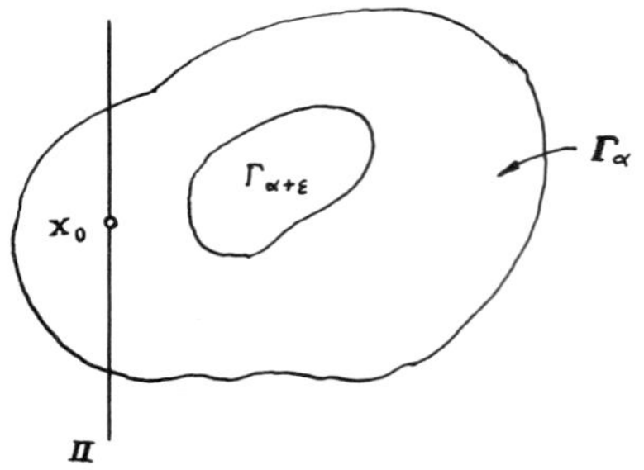

Through x_0 we can make a hyperplane π, such that $\mu_A(x_0) \leqslant \alpha+\varepsilon$ on the herperplane and one side of the herperplane.
Hence $\mu_{\bigcap_\lambda A_\lambda}(x_0) \leqslant \alpha+\varepsilon$.
Owing to $\varepsilon > 0$ is arbitrary, we obtain

$$\mu_{\bigcap_\lambda A_\lambda}(x_0) \leqslant \alpha = \mu_A(x_0)$$

When $\alpha = 1$, (12) is clear.

Generally, relationship "$\subseteq$" on the left of (10) can't be changed into "=" relationship (in the same way, at (11) "$\supseteq$" on the right can't be changed into "=").
Using the similar method we can prove (11), and also we can use (10) to prove (11).
Because $A = \int_{R^n} \mu_A(x)/x$ is a concave fuzzy set, so that $\bar{A} = \int_{R^n} (1-\mu_A(x))/x$ is a convex fuzzy set.

Owing to keep convexity of S_λ, $S_\lambda(\bar{A})$ also convex, and $\bar{A}=\bigcap_\lambda S^*_\lambda(\bar{A})$, but

$$S_\lambda(\bar{A}) = \int_{H_\lambda} \sup_{h\in L_x} (1-\mu_A(h))/x$$
$$= \int_{H_\lambda} (1-\inf_{h\in L_x} \mu_A(h))/x$$
$$= \int_{H_\lambda} (1-\mu_{C(A)}(x))/x$$
$$= \overline{C_\lambda(A)} \qquad (13)$$

$$\bar{A} = \bigcap_\lambda S_\lambda{}^* = \bigcap_\lambda \overline{C_\lambda{}^*} = \overline{\bigcup_\lambda C_\lambda{}^*} \qquad (14)$$

therefore

$$A = \bigcup_\lambda C_\lambda{}^* \qquad (15)$$

<u>Inference 1.</u> If A,B are two arbitrary strictly convex fuzzy sets in R^n, then

$$A = B \iff \forall\lambda \quad S_\lambda{}^*(A) = S_\lambda{}^*(B) \qquad (16)$$

Similarly, if A,B are two arbitrary strictly concave fuzzy sets in R^n, then

$$A = B \iff \forall\lambda \quad C_\lambda{}^*(A) = C_\lambda{}^*(B) \qquad (17)$$

Here $S_\lambda{}^*$ is a cylinder fuzzy set of a orthogonal shadow; C^*_λ is a cylinder fuzzy set of a orthogonal complementary shadow. In this case there exists decomposition as follows:

$A = B = \bigcap_\lambda S^*_\lambda$ A,B are the strictly convex fuzzy sets (18)

$A = B = \bigcup_\lambda C^*_\lambda$ A,B are the strictly concave fuzzy sets (19)

<u>Inference 2.</u> If A,B are two arbitrary strictly convex fuzzy sets in R^n, and H_1, H_2, ..., H_n are n coordinate planes, then $A = B \iff$ for an arbitrary point-shadow in R^n we have

$$S(A) = S(B)$$

If A,B are two arbitrary strictly concave fuzzy sets in R^n, then

$A = B \iff$ for an arbitrary point-shadow we have

$$C(A) = C(B)$$

Here a point-shadow is made from corresponding coordinate plane.

<u>Remark 1.</u> In [2] the conditions of the theorem 2 of the fourth chapter § 2 are not for n coordinate planes, it is only for one fixed hyperplane, this conclusion is not correct.

<u>Remark 2.</u> The fuzzy set of theorem 1 must be strictly convex (or concave). Without "strictly" condition, the conclusion may be not correct.

<u>Remark 3.</u> If the fuzzy set A is not convex, then $\bigcap_\lambda S^*_\lambda$ may not be convex.

<u>Remark 4.</u> The convex hull of a fuzzy set A (that is a mininum convex fuzzy set including A) may be not strictly convex.

<u>Remark 5.</u> Whether A is a ordinary set or a fuzzy set, the convex hull of A is denoted by convA. Suppose that A,B are two fuzzy sets, from convA = convB, we may not obtain S(A) = S(B). Here S(A), S(B) are two point-shadows.

<u>Remark 6.</u> From what the arbitrary point-shadow of fuzzy set A,B is equal, we may not obtain convA = convB.

Due to the space forbids, those examples have been omitted, useing them explains these remarks above mentioned.

<u>Definition 4.</u> Let A be a fuzzy set in R^n, and suppose the corresponding cutting set $\Gamma_\alpha = \{x \mid \mu_A(x) \geqslant \alpha\}$ of orbitrary α $(0<\alpha\leqslant 1)$ is a convex closed set, then A is called a convex closed fuzzy set. If Γ_α is a concave closed set, then A is called a concave closed fuzzy set.

<u>Theorem 2.</u> If A is a convex closed fuzzy set in R^n, then we have

$$A = \bigcap_\lambda S^*_\lambda \qquad (20)$$

If A is a concave closed fuzzy set in R^n, then we have

$$A = \bigcup_\lambda C^*_\lambda \qquad (21)$$

<u>Proof:</u> Because A is a convex closed fuzzy set, $\forall\, \alpha\in(0,1]$, Γ_α is a convex closed set in R^n. Let $\mu_A(x_0)=\alpha_0$, $0\leqslant\alpha_0<1$, take $\varepsilon>0$ such that $\alpha_0+\varepsilon<1$, and $\Gamma_{\alpha_0+\varepsilon}$ is a convex closed set, and $x_0 \bar{\in} \Gamma_{\alpha_0+\varepsilon}$. Since $\Gamma_{\alpha_0+\varepsilon}$ is closed, the distance from x_0 to $\Gamma_{\alpha_0+\varepsilon}$ is longer than zero. So that through x_0 we can make

a hyperplane π, such that not be intersected with $\Gamma_{\alpha_0+\varepsilon}$. On one side of the hyperplane, and on the hyperplane we have $\mu_A(x)<\alpha_0+\varepsilon$, so that

$$\mu_{\bigcap_\lambda S_\lambda^*}(x_0)<\alpha_0+\varepsilon$$

Owing to $\varepsilon>0$ is arbitrary, we can obtain

$$\mu_{\bigcap_\lambda S_\lambda^*}(x_0)\leqslant\alpha_0=\mu_A(x_0)$$

From (9), we have

$$\mu_A(x_0)=\mu_{\bigcap_\lambda S_\lambda^*}(x_0)$$

As for $\mu_A(x_0)=1$, according to (9) it is clear. Therefore, we all have

$$A=\bigcap_\lambda S_\lambda^*.$$

To repeat inference from (13) to (15), we can know that if A is a concave closed fuzzy set, (21) will be also correct.

Inference 1. If A,B are two arbitrary convex closed fuzzy sets in R^n, then

$$A=B \iff \forall\lambda \quad S_\lambda^*(A)=S_\lambda^*(B)$$

and there exists decomposition

$$A=B=\bigcap_\lambda S_\lambda^*$$

If A,B are two concave closed fuzzy set, then

$$A=B \iff \forall\lambda \quad C_\lambda^*(A)=C_\lambda^*(B),$$

and there exists decomposition

$$A=B=\bigcup_\lambda C_\lambda^*.$$

Inference 2. If A,B are arbitrary two convex closed fuzzy sets in R^n, then

$A=B \iff$ for any point-shadow in relation to n coordinate planes always have

$$S(A)=S(B)$$

If A,B are two concave closed fuzzy sets, then

$A=B \iff$ for any complementary point-shadow in relation to n coordinate plane always have

$$C(A)=C(B)$$

Theorem 3. If A is a convex (or concave fuzzy set in R^n, and membership function $\mu_A(x)$ is continuous in R^n, then we have decomposition

$$A=\bigcap_\lambda S_\lambda^* \qquad (\text{or } A=\bigcup_\lambda C_\lambda^*) \tag{22}$$

Proof: Because $\mu_A(x)$ is continuous, arbitrary Γ_α $(0<\alpha\leqslant 1)$ is a closed set in R^n. From convexity (or concavity), according to theorem 2 we can obtain this decomposition.

Theorem 4. Let A,B be two arbitrary fuzzy sets, and suppose $\mu_A(x)$, $\mu_B(x)$ are continuous, $\lim_{x\to\infty}\mu_A(x)=\lim_{x\to\infty}\mu_B(x)=0$, then

All point-shadows $S(A)=S(B) \Rightarrow$ convex hull $\mathrm{conv}A=\mathrm{conv}B$;

All complementary point-shadows $C(A)=C(B) \Rightarrow$ concave kernel (maxinum concave fuzzy set including A) $\mathrm{conc}A=\mathrm{conc}B$.

Proof:

$$\begin{aligned}\mathrm{conv}A&=\int_0^1\alpha\,\mathrm{conv}\,\Gamma_\alpha^A\\ \mathrm{conv}B&=\int_0^1\alpha\,\mathrm{conv}\,\Gamma_\alpha^B\end{aligned} \tag{23}$$

here

$$\Gamma_\alpha^A=\{x\mid \mu_A(x)\geqslant\alpha\}$$

$$\Gamma_\alpha^B=\{x\mid \mu_B(x)\geqslant\alpha\}$$

Because $\mu_A(x)$, $\mu_B(x)$ are continuous, Γ_α^A, Γ_α^B are two closed sets. $\mathrm{conv}\,\Gamma_\alpha^A$, $\mathrm{conv}\,\Gamma_\alpha^B$ are their convex hull.

$$\mathrm{conv}\,\Gamma_\alpha^A=\{x\mid x=\sum_{k=1}^{l}\lambda_k x_k,\ \sum_{k=1}^{l}\lambda_k=1,\ \lambda_k\geqslant 0,\ x_k\in\Gamma_\alpha^A,\ k=1,\dots,l\}$$

$$\mathrm{conv}\,\Gamma_\alpha^B=\{x\mid x=\sum_{n=1}^{m}\mu_n x_n,\ \sum_{n=1}^{m}\mu_n=1,\ \mu_n\geqslant 0,\ x_n\in\Gamma_\alpha^B,\ n=1,\dots,m\}$$

so that we obtain $\mathrm{conv}\,\Gamma_\alpha^A=\mathrm{conv}\,\Gamma_\alpha^B$.

Because otherwise, at least exist a point $x^0\in\mathrm{conv}\,\Gamma_\alpha^B$, but $x^0\bar{\in}\,\mathrm{conv}\,\Gamma_\alpha^A$.

Since $x^0\in\mathrm{conv}\,\Gamma_\alpha^B$,

$\therefore \exists\ \mu_n^0,\ x_n^0$ (n=1, ... m)

$\sum_{n=1}^{m}\mu_n^0=1$, $x_n^0\in\Gamma_\alpha^B$ (n=1, ..., m) such that

$$x^0=\sum_{n=1}^{m}\mu_n^0 x_n^0 .$$

since $x^0\bar{\in}\,\mathrm{conv}\,\Gamma_\alpha^A$, there exists at least one of $x_1^0, x_2^0, \dots, x_m^0$ (denoted by x_g^0)

such that $x_g^o \bar{\in} \text{conv}\,\Gamma_\alpha^A$ ($\because$ otherwise $x^o \in \text{conv}\,\Gamma_\alpha^A$). So that we can make a hyperplane through x_g^o such that will not be intersected with convex hull, and have $\mu_A(x) < \alpha$ on the hyperplane and on one side of the hyperplane. Since $\lim_{x\to\infty} \mu_A(x) = 0$, we have $S_\lambda(A) < \alpha$ for the hyperplane's point-shadows. But $x_g^o \in \Gamma_\alpha^B$, hence $S_\lambda(B) \geqslant \alpha$. This shows that there exists at least one point-shadow such that $S(A) = S(B)$, that is contradictory to known conditions. Hence $\text{conv}A = \text{conv}B$.

After proving the first conclusion, we can obtain the second conclusion.
Since $C(A) = C(B)$, so that

$$S(\bar{A}) = S(\bar{B})$$
$$\text{conv}(\bar{A}) = \text{conv}(\bar{B})$$
$$\overline{\text{conv}(\bar{A})} = \overline{\text{conv}(\bar{B})}$$
$$\text{conc}A = \text{conc}B$$

References

1. Frederick, A. Valentine (1964). Convex Sets. New York.

2. Mizumoto, M. (1970-1973). Fuzzy algebra and its application. Mathematical Sciences, 8-11.

3. Zadeh, L.A. (1965). Fuzzy Sets. Inform. and Control, 8, 338-353.

EXTRACTION METHOD OF THE DIFFERENCE BETWEEN FUZZY GRAPHS

M. Morioka*, H. Yamashita* and T. Takizawa**

**Waseda University, Tokyo, Japan*
***Tamagawa University, Tokyo, Japan*

Abstract. We discuss the extraction method of the structural difference between fuzzy graphs, which is useful for instructional design, sociometric analysis and so on. The method is called FRS (Fuzzy Relational Structure) Analysis based on the fuzzy decision. As a reasonable application, we describe the educational analysis of the logical structure using the test data, which provides us with effective information to evaluate the instructional program, to design CAI (Computer Aided Instruction) courseware and so on. In this paper, we present the FRS Analysis and explain its application to the instructional design.

Keywords. Computer-aided instruction; Decision theory; Fuzzy set; FRS analysis; Graph theory.

INTRODUCTION

Fuzzy theory is an effective tool to analyze instructional, sociometric and many other phenomena which are not deterministic. Here we present the extraction method of the difference between fuzzy graphs and discuss its application to the instructional design.

Instructional programs are designed by sequencing a lot of logical nodes which are complicatedly related each other. In order to investigate and modify the logical relations, we have been studying the analysis method which provides the instructors with effective information to evaluate and refine the instructional programs.

It is convenient to use graphical representation when we analyze the logical relation of the subject matters. Therefore, we use fuzzy graphs (Negoita, 1975; Nishida, 1978; Zadeh, 1965) and verify the adequacy of instructional structure, i.e.
(1) We analyze the logical structure of the subject matters and construct the logical structure graph,
(2) We construct the test structure graph by the test items attached to the logical nodes of the previous graph,
(3) We execute the test and construct the test fuzzy graph by analyzing test data (Takeya, 1980),
(4) We compare the test fuzzy graph with the logical structure graph (the test structure graph) and extract the difference between them,
(5) We analyze the causes of the difference from educational point of view and modify and improve the instructional programs.

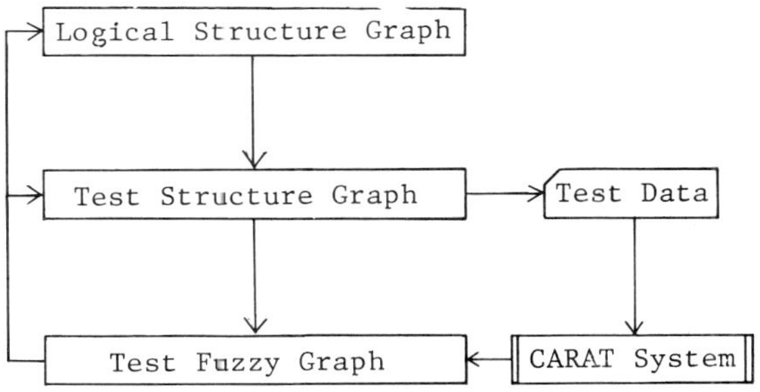

Fig.1. FRS Analysis and CARAT System

For this analysis, the supporting system CARAT (Computer-Assisted Retrieval and Analysis for Test-items) has been cooperatively developed by Waseda University, Tamagawa University and Nippon Electric Co. Ltd., and practically utilized among us (Sunouchi, 1978, 1981).

In this paper, we explain the analysis method called FRS Analysis (Fuzzy Relational Structure Analysis) and discuss the representative application to higher Mathematics.

PRELIMINARIES

Concerning connectivity, the oriented graphs are classified into the three sets which are weakly connected graphs (G_1), unilaterally connected graphs (G_2) and strongly connected graphs (G_3).

Definition 1 (Connectivity of graphs G_1, G_2 and G_3)

Graph	Definition	Examples
Strongly connected G_3	There exists a return path between any two nodes.	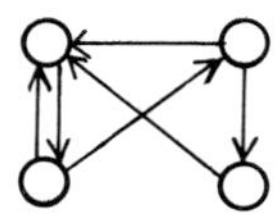
Unilaterally connected G_2	There exists at least one path between any two nodes.	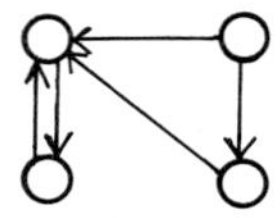
Weakly connected G_1	There exists a chain between any two nodes.	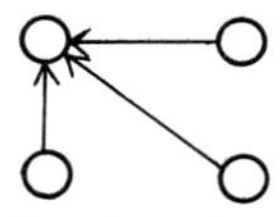

Let $R=[r_{ij}]$ $(1 \leq i,j \leq n)$ be a fuzzy graph. The operations $\vee$, $\wedge$ denote max, min, respectively. Let $\hat{R}=[\hat{r}_{ij}]$ be the max-min transitive fuzzy graph of R (Negoita, 1975), that is

$$\hat{r}_{ij}=r_{ij}^{(n-1)}$$
$$\bigvee\Big[\bigvee_{\substack{k_1,\ldots,k_{n-1}=1\\ k_r\neq k_s, i,j}}^{n}(r_{ik_1}\wedge r_{k_1k_2}\wedge\cdots\wedge r_{k_{n-1}j})\Big] \tag{1}$$

R' denotes transposed matrix of R, that is

$$R'=[r_{ji}] \quad (1 \leq i,j \leq n) \tag{2}$$

and we define $\Delta=[\Delta_{ij}]$ $(1 \leq i,j \leq n)$ in the following.

$$\Delta_{ij}=r_{ij}\vee r_{ji} \tag{3}$$

Definition 2 (Connective degree of fuzzy graph R) (Nishida, 1978)
We define the connective degree r_3, r_2, r_1 concerning G_3, G_2, G_1, respectively.

$$r_3 \triangleq \bigwedge_{i,j}\hat{r}_{ij} \tag{4}$$

i.e. the smallest value of elements in $\hat{R}$

$$r_2 \triangleq \bigwedge_{i,j}(\hat{r}_{ij}\vee\hat{r}_{ji}) \tag{5}$$

i.e. the smallest value of elements in $\hat{R}\vee\hat{R}'$

$$r_1 \triangleq \bigwedge_{i,j}\widehat{(r_{ij}\vee r_{ji})} \tag{6}$$

i.e. the smallest value of elements in $\hat{\Delta}$

Let $T=[t_{ij}]$ $(1 \leq i,j \leq n)$ be the logical structure (the test structure) graph and $Z=[z_{ij}]$ $(1 \leq i, j \leq n)$ be the test fuzzy graph. The graph T is a non-fuzzy (and/or) graph, we have

$$T=[t_{ij}], \begin{cases} t_{ij}=0,1 & (i\neq j)\\ t_{ij}=1 & (i=j)\end{cases} \tag{7}$$

and the graph Z is a fuzzy graph, we have

$$Z=[z_{ij}], \begin{cases} 0\leq z_{ij}\leq 1 & (i\neq j)\\ z_{ij}=1 & (i=j)\end{cases} \tag{8}$$

We will never need to consider loops, that is, we can assume, if we wish, that our fuzzy graph is reflexive. $\hat{T}=[\hat{t}_{ij}]$ and $\hat{Z}=[\hat{z}_{ij}]$ mean max-min transitive fuzzy graphs.

For the graph T and Z, let $\hat{T}^{(l)}$ and $\hat{Z}^{(l)}$ be the max-min transitive subgraph of the graph T and R from which one node N_l $(1 \leq l \leq n)$ and arcs connected with the node are omitted. For simplicity, the node N_l is denoted by l.

Definition 3 (Connective degree matrices concerning T, Z)
For the subgraph $T^{(l)}$, we can calculate the connective degrees τ_{11}, τ_{12}, τ_{13} concerning G_3, G_2, G_1, respectively. These connective degrees form $3\times n$ matrix $\tau=[\tau_{ij}]$ $(1 \leq i \leq n, 1 \leq j \leq 3)$ and this matrix is called the connective degree matrix concerning T. Similarly, we define the connective degree matrix concerning Z, denoted by $\Psi=[\psi_{ij}]$ $(1 \leq i \leq n,\ 1 \leq j \leq 3)$.

$$\tau \triangleq \begin{array}{c} \\ 1\\ 2\\ \vdots\\ l\\ \vdots\\ n\end{array}\begin{array}{c} \begin{array}{ccc} G_1 & G_2 & G_3\end{array}\\ \begin{pmatrix} \tau_{11} & \tau_{12} & \tau_{13}\\ \tau_{21} & \tau_{22} & \tau_{23}\\ \vdots & \vdots & \vdots\\ \tau_{l1} & \tau_{l2} & \tau_{l3}\\ \vdots & \vdots & \vdots\\ \tau_{n1} & \tau_{n2} & \tau_{n3}\end{pmatrix}\end{array} \qquad \Psi \triangleq \begin{array}{c} \\ 1\\ 2\\ \vdots\\ l\\ \vdots\\ n\end{array}\begin{array}{c} \begin{array}{ccc} G_1 & G_2 & G_3\end{array}\\ \begin{pmatrix} \psi_{11} & \psi_{12} & \psi_{13}\\ \psi_{21} & \psi_{22} & \psi_{23}\\ \vdots & \vdots & \vdots\\ \psi_{l1} & \psi_{l2} & \psi_{l3}\\ \vdots & \vdots & \vdots\\ \psi_{n1} & \psi_{n2} & \psi_{n3}\end{pmatrix}\end{array} \tag{9}$$

We have the difference between the graph T and the graph Z by their connective degrees. Similarly, we have the difference between the subgraphs $T^{(l)}$ and $Z^{(l)}$ by their matrices τ and Ψ.

Definition 4 (Difference between T and Z)
For the graphs T and Z, let the connective degrees concerning G_1, G_2, G_3 be T_1, T_2, T_3 and Z_1, Z_2, Z_3. For each connectivity G_m $(m=1,2,3)$, we define the difference measure E_m in the following.

$$E_m \triangleq T_m - Z_m \quad (m=1,2,3) \tag{10}$$

For the subgraphs $T^{(l)}$ and $Z^{(l)}$, we also define the difference measure e_{lm} in the following.

	G_1	G_2	G_3
T	T_1	T_2	T_3
Z	Z_1	Z_2	Z_3
E	E_1	E_2	E_3

$$e_{l_m} \triangleq \tau_{l_m} - \psi_{l_m} \quad (m=1,2,3) \tag{11}$$

Generally speaking, in graph theory, the type of nodes in the non-fuzzy graph is classified into Strengthening or Neutral or Weakening point (S-node or N-node or W-node). This concept is extended to the fuzzy graph.

Definition 5 (S-node, N-node, W-node) (Nishida, 1978)

$$p_{l_m} \triangleq T_m - \tau_{l_m} \tag{12}$$

If $p_{l_m}>0$, then the node l is called Strengthening node in the non-fuzzy graph T. If $p_{l_m}=0$, then the node l is called Neutral node in the graph T. If $p_{l_m}<0$, then the node l is called Weakening node in the graph T.

$$q_{l_m} \triangleq Z_m - \psi_{l_m} \tag{13}$$

Similarly, if $q_{l_m}>0$, then the node l is called S-node in the fuzzy graph Z. If $q_{l_m}=0$, then the node l is called N-node in the graph Z. If $q_{l_m}<0$, then the node l is called W-node in the graph Z.

Remark 1 Actually, $q_{l_m}>0$ means that the connective degree of the graph Z concerning G_m is larger than that of the subgraph $Z^{(l)}$, that is, the node l strengthens the connective degree of the graph Z. Therefore, such node l is called strengthening node.

And, $q_{l_m}=0$ means that the node l neither strengthens nor weakens the connective degree of the graph Z, therefore such node l is called Neutral node.

FUZZY DECISION OF STATE

We assume here that one initial state x (x=0 or 1) becomes another state y ($0 \leq y \leq 1$) by certain cause. In order to measure the difference between the state x and y, we define the fuzzy set S and F in the following.

Definition 6 (Fuzzy set $\underset{\sim}{S}$, $\underset{\sim}{F}$)

If x=0, we define (i) the fuzzy set $\underset{\sim}{S}$: the state y becomes inverse to x

$$m_{\underset{\sim}{S}}(y) \triangleq y \tag{14}$$

(ii) the fuzzy set $\underset{\sim}{F}$: the state y becomes fuzzy, that is, we determine that y becomes neither similar nor inverse to x

$$m_{\underset{\sim}{F}}(y) \triangleq 1-|1-2y| \tag{15}$$

and if x=1, instead of (14), we define $\underset{\sim}{S}$ nextly.

$$m_{\underset{\sim}{S}}(y) \triangleq 1-y \tag{16}$$

For the fuzzy set $\underset{\sim}{S}$, the fuzzy set not $\underset{\sim}{S}$ and very $\underset{\sim}{S}$ are denoted by $\bar{\underset{\sim}{S}}$ and $\underset{\sim}{S}^2$, respectively. If x=0, we have the following fuzzy decisions, $\bar{\underset{\sim}{S}}^2 \wedge \underset{\sim}{F}$ and $\underset{\sim}{S}^2 \wedge \underset{\sim}{F}$.

The fuzzy decision $\bar{\underset{\sim}{S}}^2 \wedge \underset{\sim}{F}$: the state y is very similar to x

$$m_{\bar{\underset{\sim}{S}}^2 \wedge \underset{\sim}{F}}(y) = (1-y)^2 \wedge (1-|1-2y|) \tag{17}$$

The fuzzy decision $\underset{\sim}{S}^2 \wedge \underset{\sim}{F}$: the state y is very inverse to x

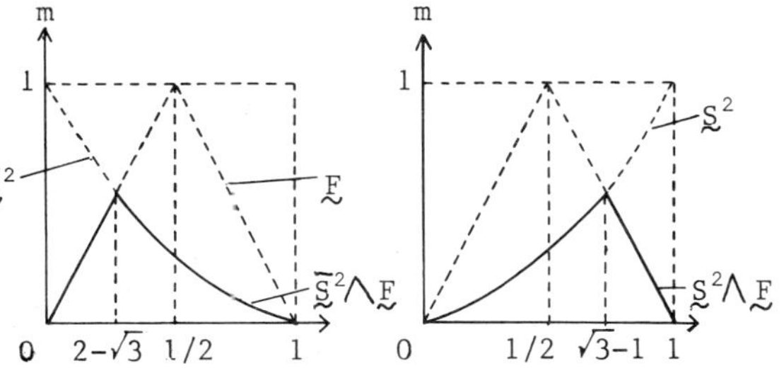

Fig.2. Fuzzy decision of state (x=0)

$$m_{\underset{\sim}{S}^2 \wedge \underset{\sim}{F}}(y) = y^2 \wedge (1-|1-2y|) \tag{18}$$

If x=1, instead of (17), (18) we have the following fuzzy decisions.

$$m_{\bar{\underset{\sim}{S}}^2 \wedge \underset{\sim}{F}}(y) = y^2 \wedge (1-|1-2y|) \tag{19}$$

$$m_{\underset{\sim}{S}^2 \wedge \underset{\sim}{F}}(y) = (1-y)^2 \wedge (1-|1-2y|) \tag{20}$$

The maximal decisions of $\bar{\underset{\sim}{S}}^2 \wedge \underset{\sim}{F}$ and $\underset{\sim}{S}^2 \wedge \underset{\sim}{F}$ are denoted by $\bar{\underset{\sim}{S}}^2 \wedge \underset{\sim}{F}|_y$ and $\underset{\sim}{S}^2 \wedge \underset{\sim}{F}|_y$. We have

$$\bar{\underset{\sim}{S}}^2 \wedge \underset{\sim}{F}|_y = 2-\sqrt{3},\ \underset{\sim}{S}^2 \wedge \underset{\sim}{F}|_y = \sqrt{3}-1 \quad (\text{if } x=0) \tag{21}$$

$$\bar{\underset{\sim}{S}}^2 \wedge \underset{\sim}{F}|_y = \sqrt{3}-1,\ \underset{\sim}{S}^2 \wedge \underset{\sim}{F}|_y = 2-\sqrt{3} \quad (\text{if } x=1) \tag{22}$$

Therefore, if x=0 and $0 \leq y \leq 2-\sqrt{3}$, then the state y is very similar to x, and if x=0 and $\sqrt{3}-1 \leq y \leq 1$, then the state y is very inverse to x.

Theorem 1 (Determinative condition of state, D-condition)

The inverse state of x is denoted by $\check{x}$ i.e. $\check{x}=1-x$. We have the ranges where the state y is similar to the initial state x (x~y), y is inverse to x ($\check{x}$~y) as follows.

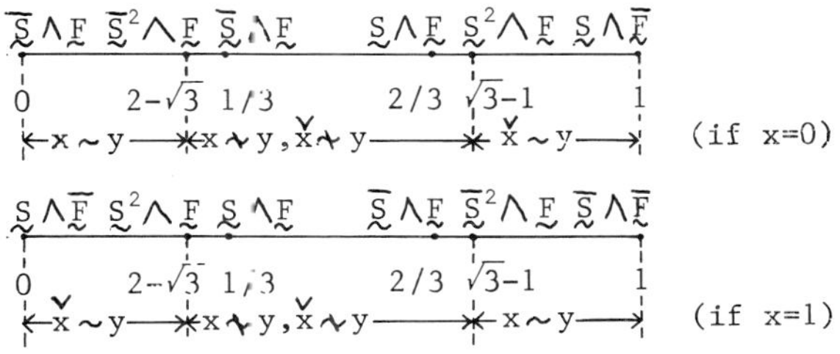

Fig.3. Determinative condition of state

FUZZY RELATIONAL STRUCTURE ANALYSIS

In this section, we show the extraction method of the difference between fuzzy graphs for structure analysis. We call it the Fuzzy Relational Structure Analysis (FRS Analysis). There exists the following relation among (10), (11), (12) and (13).

$$E_m = T_m - Z_m = (p_{l_m}+\tau_{l_m}) - (q_{l_m}+\psi_{l_m})$$
$$= (p_{l_m}-q_{l_m}) + e_{l_m} \tag{23}$$

Therefore, this equation (23) shows that the difference between the graph T and Z depends on both the difference of the type of the node l ($p_{l_m}-q_{l_m}$) and that of the connective degree between the sybgraph $T^{(l)}$ and $Z^{(l)}$ (e_{lm}). So, we can extract the difference between the graph T and Z applying the D-condition.

The extraction analysis of the difference between the graph T and Z is classified into the following four cases.

Case A : $T_m=1$ and $\tau_{lm}=1$ i.e. $p_{lm}=0$
Case B : $T_m=0$ and $\tau_{lm}=0$ i.e. $p_{lm}=0$
Case C : $T_m=1$ and $\tau_{lm}=0$ i.e. $p_{lm}>0$
Case D : $T_m=0$ and $\tau_{lm}=1$ i.e. $p_{lm}<0$

In the case A, the fuzzy graph Z becomes one of the following nine cases.

(1) (a) $|e_{lm}|<2-\sqrt{3}$ and (b) $q_{lm}>0$
(2) (a) $2-\sqrt{3}\leq|e_{lm}|\leq\sqrt{3}-1$ and (b) $q_{lm}>0$
(3) (a) $|e_{lm}|>\sqrt{3}-1$ and (b) $q_{lm}>0$
(4) (a) $|e_{lm}|<2-\sqrt{3}$ and (b) $q_{lm}=0$
(5) (a) $2-\sqrt{3}\leq|e_{lm}|\leq\sqrt{3}-1$ and (b) $q_{lm}=0$
(6) (a) $|e_{lm}|>\sqrt{3}-1$ and (b) $q_{lm}=0$
(7) (a) $|e_{lm}|<2-\sqrt{3}$ and (b) $q_{lm}<0$
(8) (a) $2-\sqrt{3}\leq|e_{lm}|\leq\sqrt{3}-1$ and $q_{lm}<0$
(9) (a) $|e_{lm}|>\sqrt{3}-1$ and $q_{lm}<0$

(1) (a) $|e_{lm}|<2-\sqrt{3}$ means that the connective degree of the subgraph $Z^{(l)}$ concerning G_m is $\psi_{lm}>\sqrt{3}-1$.
(b) $q_{lm}>0$ means that the node l is S-node in the graph Z.
In this case, we could evaluate the difference between T and Z as follows.

(a) The connective degree of the subgraph $T^{(l)}$ concerning G_m, $\tau_{lm}=1$ alters into that of the subgraph $Z^{(l)}$, $\psi_{lm}>\sqrt{3}-1$. According to the D-condition, we can determine that there scarcely exists difference of the connective degree between $T^{(l)}$ and $Z^{(l)}$.

(b) The connective degree of the graph T concerning G_m, $T_m=1$ alters into that of the graph Z, Z_m. Since $\sqrt{3}-1<\psi_{lm}<Z_m$, according to the D-condition, we could determine that there scarcely exists the difference of the connective degree between T and Z, that is, scarcely that of the arcs connected with the node l concerning T and Z.

Finally, we could determine that there scarcely exists the difference between T and Z in this case.

(2) (a) $2-\sqrt{3}\leq|e_{lm}|\leq\sqrt{3}-1$ means that $\tau_{lm}=1$ alters into $2-\sqrt{3}\leq\psi_{lm}\leq\sqrt{3}-1$. According to the D-condition, we could not determine that there scarcely exists the difference of the connective degree concerning G_m between $T^{(l)}$ and $Z^{(l)}$. Therefore, we could determine that there *might* exist some differences of the connective degree between $T^{(l)}$ and $Z^{(l)}$.

By the definition 2, for the non-transitive fuzzy graph $Z=[z_{ij}]$ $(1\leq i,j\leq n)$, we have $\widehat{Z}$, $\widehat{Z}\vee\widehat{Z}'$ and $\widehat{\Delta}_Z=[\widehat{z_{ij}\vee z_{ji}}]$. For simplicity, $\widehat{Z}$, $\widehat{Z}\vee\widehat{Z}'$ and $\widehat{\Delta}_Z$ are newly denoted by $\underset{3}{Z}=[\underset{3}{z}_{ij}]$, $\underset{2}{Z}=[\underset{2}{z}_{ij}]$ and $\underset{1}{Z}=[\underset{1}{z}_{ij}]$, respectively. Similarly, the subgraph $\widehat{Z^{(l)}}$, $\widehat{Z^{(l)}}\vee\widehat{Z'^{(l)}}$ and $\widehat{\Delta}_{Z^{(l)}}=[\widehat{z^{(l)}_{ij}\vee z^{(l)}_{ji}}]$ are newly denoted by $\underset{3}{Z}^{(l)}=[\underset{3}{z}^{(l)}_{ij}]$, $\underset{2}{Z}^{(l)}=[\underset{2}{z}^{(l)}_{ij}]$ and $\underset{1}{Z}^{(l)}=[\underset{1}{z}^{(l)}_{ij}]$, respectively.

Now, these some differences which cause $2-\sqrt{3}\leq|e_{lm}|\leq\sqrt{3}-1$ are the arcs $\underset{m}{z}^{(l)}_{ij}$ in the graph $\underset{m}{Z}^{(l)}$ s.t. $\psi_{lm}\leq\underset{m}{z}^{(l)}_{ij}\leq\sqrt{3}-1$ (m=1,2,3). Then, we can extract these arcs in $\underset{m}{Z}^{(l)}$ which *might* be different from (inverse to) the arcs $\underset{m}{t}^{(l)}_{ij}$ in the graph $\underset{m}{T}^{(l)}$. Furthermore, according to the D-condition, the arcs $\underset{m}{z}^{(l)}_{ij}$ s.t. $\underset{m}{z}^{(l)}_{ij}>\sqrt{3}-1$, if exist, are similar to the arcs $\underset{m}{t}^{(l)}_{ij}$ in $\underset{m}{T}^{(l)}$, and the arcs $\underset{m}{z}^{(l)}_{ij}$ s.t. $\underset{m}{z}^{(l)}_{ij}<2-\sqrt{3}$, if exist, are inverse to the arcs $\underset{m}{t}^{(l)}_{ij}$ in $\underset{m}{T}^{(l)}$.

(b) The connective degree $T_m=1$ alters into E_m. Since $\psi_{lm}<Z_m$, if $Z_m>\sqrt{3}-1$, then we could determine that there scarcely exists the difference of the arcs connected with the node l concerning T and Z according to the D-condition. If $\psi_{lm}<Z_m\leq\sqrt{3}-1$, we could determine that there *might* exist some differences of the arcs connected with the node l according to the D-condition. These some differences which cause $2-\sqrt{3}\leq|E_m|<\sqrt{3}-1$ are the arcs $\underset{m}{z}_{lk}$ and $\underset{m}{z}_{kl}$ in $\underset{m}{Z}$ s.t. $Z_m\leq\underset{m}{z}_{lk}$, $\underset{m}{z}_{kl}\leq\sqrt{3}-1$ $(1\leq k\leq n,\ k\neq l)$. Then, we can extract these arcs in $\underset{m}{Z}$ which *might* be different from (inverse to) the arcs $\underset{m}{t}_{lk}$ and $\underset{m}{t}_{kl}$ in the graph $\underset{m}{T}$.

(3) (a) $|e_{lm}|>\sqrt{3}-1$ means that $\tau_{lm}=1$ alters into $\psi_{lm}<2-\sqrt{3}$. According to the D-condition, we could determine that there exist some differences of the connective degree between $T^{(l)}$ and $Z^{(l)}$. These some differences which cause $|e_{lm}|>\sqrt{3}-1$ are the arcs $\underset{m}{z}^{(l)}_{ij}$ in $\underset{m}{Z}^{(l)}$ s.t. $\psi_{lm}\leq\underset{m}{z}^{(l)}_{ij}<2-\sqrt{3}$. Then, we can extract these arcs in $\underset{m}{Z}^{(l)}$ which are different from (inverse to) the arcs $\underset{m}{t}^{(l)}_{ij}$ in $\underset{m}{T}^{(l)}$.

Furthermore, according to the D-condition, the arcs $\underset{m}{z}^{(l)}_{ij}$ s.t. $2-\sqrt{3}\leq z^{(l)}_{ij}\leq\sqrt{3}-1$, if exist, *might* be different from (inverse to) the arcs $\underset{m}{t}^{(l)}_{ij}$ in $\underset{m}{T}^{(l)}$, and the arcs $\underset{m}{z}^{(l)}_{ij}$ s.t. $\underset{m}{z}^{(l)}_{ij}>\sqrt{3}-1$, if exist, are similar to the arcs $t^{(l)}_{mij}$ in $T^{(l)}_{m}$.

(b) The connective degree $T_m=1$ alters into E_m. Since $\psi_{lm}<Z_m$, if $Z_m>\sqrt{3}-1$, then we could determine that there scarcely exists the difference of the arcs connected with the node l concerning T and Z according to the D-condition. If $2-\sqrt{3}\leq Z_m\leq\sqrt{3}-1$, then we can extract the arcs z_{mlk} and z_{mkl} in Z_m s.t. $Z_m\leq z_{mlk}, z_{mkl}\leq\sqrt{3}-1$ $(1\leq k\leq n, k\neq l)$. We see that these arcs *might* be different from (inverse to) the arcs t_{mlk} and t_{mkl} in T_m. If $\psi_{lm}<Z_m<2-\sqrt{3}$, then we can extract the arcs z_{mlk} and z_{mkl} in Z_m s.t. $Z_m\leq z_{mkl}, z_{mlk}<2-\sqrt{3}$ $(1\leq k\leq n, k\neq l)$. We see that these arcs are inverse to the arcs t_{mlk} and t_{mkl} in T_m.

In this way, we can extract the difference between the graph T and Z in case of (4)~(9) and have the following result.

Theorem 2 (Extraction method of case A)
Case A : $T_m=1$ and $\tau_{lm}=1$ i.e. $p_{lm}=0$

	Graph Z	Extracted arcs in Z
(1)	(a) $\lvert e_{lm}\rvert<2-\sqrt{3}$ (b) $q_{lm}>0$ (S-node in Z)	(a) S-E-D$^{(l)}$. (b) S-E-D.
(2)	(a) $2-\sqrt{3}\leq\lvert e_{lm}\rvert\leq\sqrt{3}-1$ (b) $q_{lm}>0$	(a) $z^{(l)}_{mij}>\sqrt{3}-1$: Sim$^{(l)}$. $\psi_{lm}\leq z^{(l)}_{mij}\leq\sqrt{3}-1$: M-Inv$^{(l)}$. $z^{(l)}_{mij}<2-\sqrt{3}$: Inv$^{(l)}$. (b) $Z_m>\sqrt{3}-1$: S-E-D. $Z_m\leq z_{mlk}, z_{mkl}\leq\sqrt{3}-1$: M-Inv.
(3)	(a) $\lvert e_{lm}\rvert>\sqrt{3}-1$ (b) $q_{lm}>0$	(a) $z^{(l)}_{mij}>\sqrt{3}-1$: Sim$^{(l)}$. $2-\sqrt{3}\leq z^{(l)}_{mij}\leq\sqrt{3}-1$: M-Inv$^{(l)}$. $\psi_{lm}\leq z^{(l)}_{mij}<2-\sqrt{3}$: Inv$^{(l)}$. (b) $Z_m>\sqrt{3}-1$: S-E-D. $2-\sqrt{3}\leq Z_m\leq z_{mlk}, z_{mkl}\leq\sqrt{3}-1$: M-Inv. $Z_m\leq z_{mlk}, z_{mkl}<2-\sqrt{3}$: Inv.
(4)	(a) $\lvert e_{lm}\rvert<2-\sqrt{3}$ (b) $q_{lm}=0$ (N-node in Z)	(a) same as (1) (a). (b) same as (1) (b).
(5)	(a) $2-\sqrt{3}\leq\lvert e_{lm}\rvert\leq\sqrt{3}-1$ (b) $q_{lm}=0$	(a) same as (2) (a). (b) $z_{mlk}, z_{mkl}>\sqrt{3}-1$: Sim. $2-\sqrt{3}\leq Z_m\leq z_{mlk}, z_{mkl}\leq\sqrt{3}-1$: M-Inv. $z_{mlk}, z_{mkl}<2-\sqrt{3}$: Inv.
(6)	(a) $\lvert e_{lm}\rvert>\sqrt{3}-1$ (b) $q_{lm}=0$	(a) same as (3) (a). (b) same as (5) (b).
(7)	(a) $\lvert e_{lm}\rvert<2-\sqrt{3}$ (b) $q_{lm}<0$	(a) same as (1) (a). (b) same as (3) (b).
(8)	(a) $2-\sqrt{3}\leq\lvert e_{lm}\rvert\leq\sqrt{3}-1$ (b) $q_{lm}<0$	(a) same as (2) (a). (b) same as (5) (b).
(9)	(a) $\lvert e_{lm}\rvert>\sqrt{3}-1$ (b) $q_{lm}<0$	(a) same as (3) (a). (b) same as (5) (b).

S-E-D$^{(l)}$: There scarcely exists the difference of the arcs between $T^{(l)}_m$ and $Z^{(l)}_m$.
S-E-D : There scarcely exists the difference of the arcs connected with the node l between T_m and Z_m. Sim$^{(l)}$: similar to the arcs $t^{(l)}_{mij}$ in $T^{(l)}_m$. M-Inv$^{(l)}$: *might* be inverse to the arcs $t^{(l)}_{mij}$ in $T^{(l)}_m$. Inv$^{(l)}$: inverse to the arcs $t^{(l)}_{mij}$ in $T^{(l)}_m$. Sim : similar to the arcs t_{mlk} and t_{mkl} in T_m. M-Inv : *might* be inverse to the arcs t_{mlk} and t_{mkl} in T_m. Inv : inverse to the arcs t_{mlk} and t_{mkl} in T_m.

Similarly, we can extract the difference between the graph T and Z in case B, C and D. Since this extraction method FRS analysis is supported by the CARAT system, we do not need the waste of time for this analysis.

CASE STUDY OF FRS ANALYSIS

We analyze the logical structure of instructional matters by investigating the prerequisite relations among logical nodes. A logical node N_i is prerequisite to the node N_j ($N_i\rightarrow N_j$) if N_i is a necessary node to understand N_j. We would verify whether the logical structure is adequate or not according to student performance scores. We have already discussed the logical structure of Mathematics and FORTRAN (Sunouchi, 1980; Takeya, 1980; Yamashita, 1981). According to the logical structure of the subject matters, we construct a logical structure graph. Nextly, in order to verify the adequacy of the logical structure by student performance scores, the test structure graph T is formed by the test items attached to logical nodes of the logical structure graph.

We discuss a case study of the logical structure analysis concerning differential equation in higher Mathematics.

$$\sum_i a_i \frac{d^{n-i}y}{dx^{n-i}} = R(x) \tag{24}$$

This equation has the following properties. (a) Homogeneous, (b) Inhomogeneous and $R(x)=a^x$, (c) Inhomogeneous and $R(x)=x^p$, (d) Characteristic equation with real solutions, (e) Characteristic equation with complex solutions. Considering these five properties, we could construct the logical structure graph as shown in Fig.4. According to this graph, we could construct the test structure graph T of six nodes as shown in Fig.4.

We discuss the case study of the FRS analysis based on the test data obtained by 83 undergraduate students of the school of political science and economics in Waseda University. We first construct the test fuzzy graph Z by test data and secondly compare the test fuzzy graph with the test structure graph T by the FRS analysis. We obtain the relational value

z_{ij} $(0 \leq z_{ij} \leq 1)$ from one node to the other by analyzing student performance scores and the test fuzzy graph by these values as shown in Fig.5 (Takeya, 1980). As for the relational value z_{ij} from N_i to N_j, it is assumed that there exists the prerequisite relation $N_i \to N_j$ if it is close to 1 and there does not exist the prerequisite relation if it is close to 0.

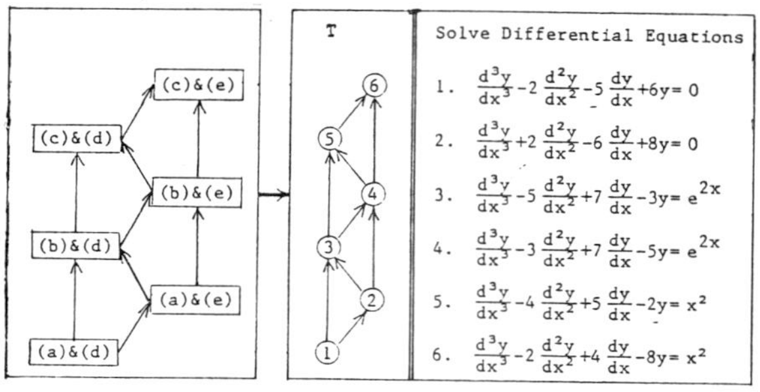

Fig.4. Logical structure graph and test structure graph T

We also obtain the connective degrees T_i, Z_i (i=1,2,3) as shown in Fig.5, and the connective degree matrices τ, ψ of the graphs T, Z as shown in Fig.6.

Z =

	1	2	3	4	5	6
1	1.00	1.00	0.83	1.00	1.00	1.00
2	0.31	1.00	0.55	0.55	0.55	0.55
3	0.31	0.36	1.00	0.91	0.76	0.73
4	0.31	0.36	0.61	1.00	0.70	0.66
5	0.31	0.36	0.59	0.59	1.00	0.63
6	0.31	0.36	0.48	0.48	0.48	1.00

	G_1	G_2	G_3
T	1.00	1.00	0.00
Z	0.91	0.55	0.31
E	0.09	0.45	-0.31

Fig.5. Test fuzzy graph Z and connective degrees of the graphs T, Z

$$\tau = \begin{array}{c} \\ 1 \\ 2 \\ 3 \\ 4 \\ 5 \\ 6 \end{array} \begin{array}{c} G_1 \quad G_2 \quad G_3 \\ \begin{pmatrix} 1 & 1 & 0 \\ 1 & 1 & 0 \\ 1 & 1 & 0 \\ 1 & 1 & 0 \\ 1 & 1 & 0 \\ 1 & 1 & 0 \end{pmatrix} \end{array} \qquad \psi = \begin{pmatrix} 0.55 & 0.55 & 0.36 \\ 0.91 & 0.63 & 0.25 \\ 1.00 & 0.55 & 0.31 \\ 0.83 & 0.42 & 0.28 \\ 0.91 & 0.55 & 0.31 \\ 0.91 & 0.55 & 0.31 \end{pmatrix}$$

Fig.6. Connective degree matrices of the graphs T,Z

According to the FRS analysis, we could obtain the following informations concerning the instructional structure of differential equation.

(1) The arcs $1 \to 3 \to 5$, $1 \to 2$ and $3 \to 4$ of the test fuzzy graph Z are similar to the arcs $1 \to 3 \to 5$, $1 \to 2$, and $3 \to 4$ of the test (logical) structure graph T. By these informations, we could recognize that there exist prerequisite relations among the logical nodes (a)&(d), (b)&(d) and (c)&(d), and that the logical node (a)&(d) and (b)&(d) are surely prerequisite to (a)&(e) and (b)&(e), respectively.

(2) The arcs $4 \to 6$ and $5 \to 6$ of the graph Z *might* be different from, that is, *might* be inverse to the arcs $4 \to 6$ and $5 \to 6$ of the graph T. By these informations, it should be necessary to investigate in detail whether the prerequisite relations among the logical nodes (b)&(e), (c)&(d) and (c)&(e) are adequate or not.

(3) The arc $2 \to 3$ of the graph Z is inverse to the arc $2 \to 3$ of the graph T, that is, the arc $2 \to 3$ disappears in the graph Z. By these informations, we could recognize the lack of fundamental knowledge of students' concerning complex numbers, especially the property of $e^{i\theta}$. In this way, we could modify and improve the instructional processes in the classroom.

CONCLUSION

We have presented the extraction method of the difference between a non-fuzzy graph and a fuzzy graph, and its application to the logical structure analysis of the instructional programs. This extraction method FRS analysis has been extended to the comparison between two fuzzy graphs. Although our case study is experimented in higher Mathematics, the FRS analysis supported by the CARAT system is available for arbitrary subjects.

ACKNOWLEDGEMENT

The authors are grateful to Prof. Haruo Sunouchi of Waseda University and Dr. Makoto Takeya of Nippon Electric Co.Ltd., for several helpful advices.

REFERENCES

Negoita, C. V., and D. A. Ralescu (1975) . *Application of fuzzy sets to system analysis.* Birkhäuser Verlag, Basel.

Nishida, T., and E.Takeda (1978). *Fuzzy sets and its applications, in Japanese.* Morikita Publishing Co., Tokyo.

Sunouchi, H., H. Yamashita, and M. Morioka (1982). *Instructional structure analysis applying fuzzy theory.* 8th International Conference on Improving University Teaching.

Sunouchi,H., H.Yamashita, and M.Takeya(1980). *Logical structure analysis of instructional programs.* 4th International Conference on Mathematical Education.

Sunouchi,H., H.Yamashita, and others (1978). *Computer-assisted retrieval and analysis for educational test-items.* Proceedings of 3rd USA-JAPAN Computer Conference.

Sunouchi,H., H.Yamashita, and others (1981). *Analysis method and supporting systems for CAI courseware development.* 3rd World Conference on Computers in Education.

Takeya, M. (1980). *Relational structure analysis among test items on performance scores.* Journal of Science Education in Japan.

Yamashita,H.,M.Yokoi, and T.Takizawa (1981). *Instructional structure analysis of FORTRAN language.* 7th International Conference on Improving University Teaching.

Zadeh,L.A. (1965). *Fuzzy sets.* Information and Control, Vol.8.

THE FRACTAL DATA OR IS FUZZY STRUCTURATION MEANINGFUL?

J. Legrand

L.I.S.T. Université Pierre et Marie Curie, Paris, France

Abstract. After briefly recalling the definitions of our mathematical model of structuration : the R-Structures and the principles of another model of structuration the Hypergraph Based Data Structures, which allows the manipulation of fuzzy information at the level of data capture, we explain how they can be drawn closer to each other. Our algorithms of correspondence which allow to turn any H.B. Data Structure into a R-Structure are associated with the H.B. Data Stucturation of the R-Structures to show that the R-Structures are a powerful tool for the H.B.D.S. manipulation. We use the Fuzzy Set Theory to consider fuzzy R-Structures. Taking as a basis the fuzzy Abstract Data Types of the H.B.D.S. we show our results include them and lead to suggest new types of fuzzy data of materialization. Then a quite new concept, fuzzy skeleton of structure, is discussed. It is shown their manipulation is made possible by the previous results. The most important point is the presentation of a type of fuzziness between materialization level and skeleton level, we named the Fractal Data. To conclude we go further the scientific and theoritical meaning we have given to the fuzzy structuration to reach an epistemological meaning.

Keywords. Canonical forms ; Data structuration ; Fuzziness ; Models ; Modelling ; Philosophical aspects.

FUNDAMENTALS OF THE R-STRUCTURES

E is a set of elements noted e_i, $E = \{e_i\}$ $i \in I$, R is a relation on E. This relation defines a matrix $M = \{\alpha_{ij}\}_{i,j \in I}$. From this one can define for each e_i <u>its trace on</u> E <u>by</u> R R-trace (e_i) noted : $z_i = \{e_j \ / \ \alpha_{ij} = 1 \ j \in I\}$
We shall denote $\{z_i\}_{i \in I}$ by $P_R(E)$;
$\{\bar{z}_i\}_{i \in I}$ by $P_{\bar{R}}(E)$.

The notion of R-Structure is introduced from three fundamental properties built with the intersection of the R-traces : (noted $\cap$)

<u>XOR</u> $(e_i, e_j) \in z_i \cap z_j \quad i,j \in I$
noted $\quad {}^{i/j}_{i,j}$ ①[R]

<u>AND</u> $(e_i, e_j) \in z_i \cap z_j \quad i,j \in I$
noted $\quad {}_{i,j}$ ②[R]

$\exists k$, $k \neq i,j$ such that $e_k \in z_i \cap z_j$
$i,j,k \in I$ noted $\quad {}^{k}_{i,j}$ ③[R]

with rigorous definitions of <u>XOR</u> and <u>AND</u> :

<u>XOR</u> $(e_i, e_j) \in z_i \cap z_j \Leftrightarrow \{e_i, e_j\} \not\subset z_i \cap z_j$
and $(e_i \in z_i \cap z_j$ or $e_j \in z_i \cap z_j)$

<u>AND</u> $(e_i\ e_j) \in z_i \cap z_j \Leftrightarrow \{e_i, e_j\} \subset z_i \cap z_j$
or $(e_i \in z_i \cap z_j$ and $e_j \in z_i \cap z_j)$

The generalization of these properties has two faces :
/A/ The intersection concerns finite families of subsets of $P_E(E)$.
/B/ The traces on E concerned in an intersection may come from different relations (e.g. we shall use here $P_{R1}(E)$, $P_{R2}(E)$).

A semantic interpretation of this model has been done by considering E as the set of the properties we use to describe the object O of a system. R 1 is a relation between the properties ($e_i \in E$) : e_i R1 e_j $\Leftrightarrow$ the property e_j perfects the description of the object O we observe with e_i.

Let $V(e_i)$ be the set of valuations of the property e_i, R2 is the induced relation on E by R'2, R'2 is the inclusion in $P(V(E))$, we define R2 by e_i R2 e_j $\Leftrightarrow$ $V(e_j) \subset V(e_i)$.

If we consider R'1 the induced relation on $P(V(E))$ by R1 such that $V(e_i)$ R'1 $V(e_j)$ $\Leftrightarrow$ e_i R1 e_j , for each $(V(e_i), V(e_j)) \subset$ R'1 we build an application

$$W_{ij} : V(e_i) \times V(e_j) \rightarrow \{0,1\};$$

By that way the relation between two properties (conceptualizations) may be transfered to the valuations of the properties (materializations).

PRINCIPLES OF THE HYPERGRAPH BASED DATA STRUCTURES

The H.B.D.S. model (Bouillé,1977), based on the concepts of set, element, property, relation leading to Class, Object, Attribute, Link owns the advantages of an associated graphical representation, data bases developments and fuzzy data processing capability.
An important notion of the model is the skeleton-structure, it sums up all the broad lines of a particular structure, it is set forth as "the part we firstly must recognize when we have to build a structure" (Bouillé, 1978). We oppose it to the notion of data of materialization, received in the skeleton during a data capture.
A definition of fuzzy data have been given by the author of the model :

<fuzzy data>:= in (Bouillé,1979)	<fuzzy attribute of object> / <fuzzy link between object> / <fuzzy object>

Definitions of the right side may be found in (Bouillé,1979).

If we consider the six basic Abstract Data Types of the H.B.D.S. (Class, Object, Attribute of Object, Link between Classes, Attribute of Class, Link between Objects), these definitions allow different types of fuzziness for data of materialization (the elements their properties and relations). On the other hand, skeleton members (the sets, their properties and relations) are concerned by the fuzziness (they accept or not fuzzy materializations) but <u>they cannot be fuzzy.</u>

A CORRESPONDENCE BETWEEN THE H.B.D.S. AND THE R-STRUCTURES

We have built two algorithms representing an application F, F : H.B.D.S. → R-Structures and its reverse F^{-1}. If we consider H.B.D.S. as a domain and F(H.B.D.S.) as a range F is a bijection but F(H.B.D.S.) ⊂ R-Structures and the equality is not proven ; we discuss this point in (Legrand,1982).
These algorithms are based on the manipulation of the Abstract Data Types of the H.B.D.S. model and the relations between the A.D.T. we gave a rigorous definition.
By that way any H.B.D. Structure may be considered a R-Structure. On the opposite the above mentionned correspondence (set, element property, relation) → (Class, Object, Attribute, Link) allows us to consider any R-Structure as data of a H.B.D. Structure we define now.
Indeed E is a set therefore a Class with the e_i as Objects ; the $V(e_i)$ are the valuations of an Attribute V of the Class E ; each Object e_i gets its Attribute $V(e_i)$; the correspondences between the valuations of two R1-related e_i (W_{ij}) are the valuations of an Attribute of Link W carried by R1. R1 is a Loop on the Class E and likewise R2 because they are Links carrying the relations between elements of E.
The figure (see next page) shows how powerful is the tool as a way of simplification of H.B.D.S. within itself. As it says that R-Structures manipulations may be done in a H.B.D.S. frame, it says that the H.B.D.S. model implementation and H.B.D. Structures manipulations may be conceived from this basic new tool.

FUZZY R-STRUCTURES

Fuzzy Set Theory (Kaufman,1973) allows to consider fuzzy R-Structures. Let $\tilde{R}$ be a fuzzy relation.
The definition of a R-trace is modified, it may have two faces :
- R-trace (e_i) of level ν :

$$z_i(\nu) \overset{i}{=} \left\{ e_j \,/\, a_{ij} \geqslant \nu \quad j \in I \right\}$$

This approach gives us a R-Structure with non fuzzy properties. This situation may be seen as a situation of statistic and probabilistic type where one works with binary logic and non binary results.

- $\tilde{R}$-trace (e_i) noted :

$$\tilde{z}_i = \left\{ e_j^i \,/\, \mu_{\tilde{R}}(e_i,e_j) \quad j \in I \right\}$$

In this case a $\tilde{R}$-trace is a fuzzy subset of and we continue the study from this point of view.
The intersection of the $\tilde{R}$-traces is now defined : $\mu_{\tilde{z}_i \cap \tilde{z}_j}(e_k) = \mathrm{MIN}(\mu_{\tilde{z}_i}(e_k),\ \mu_{\tilde{z}_j}(e_k))$
Two faces are still possible for the definition of the properties :

- The property is said of level ν when $\mu_{\tilde{R}} \geqslant \nu$ and the situation is equivalent to the R-traces of level ν .
- The property is fuzzy and given a degree :

$$^{i/j}_{i,j}\ \underline{①}[\tilde{R}] \quad \mu_{\widetilde{XOR}}(e_i,e_j) = \mathrm{MIN}(\mathrm{MAX}(\mu_{\tilde{z}_i\cap\tilde{z}_j}(e_i),\mu_{\tilde{z}_i\cap\tilde{z}_j}(e_j)),\ 1-\mathrm{MIN}(\mu_{\tilde{z}_i\cap\tilde{z}_j}(e_i),\mu_{\tilde{z}_i\cap\tilde{z}_j}(e_j)))$$

$$_{i,j}\ \underline{②}[\tilde{R}] \quad \mu_{\widetilde{AND}}(e_i,e_j) = \mathrm{MAX}(\mathrm{MIN}(\mu_{\tilde{z}_i\cap\tilde{z}_j}(e_i),\mu_{\tilde{z}_i\cap\tilde{z}_j}(e_j)),\ 1-\mathrm{MAX}(\mu_{\tilde{z}_i\cap\tilde{z}_j}(e_i),\mu_{\tilde{z}_i\cap\tilde{z}_j}(e_j)))$$

$$^{k}_{i,j}\ \underline{③}[\tilde{R}] \quad \mu_{\tilde{z}_i\cap\tilde{z}_j}(e_k) = \mathrm{MIN}(\mu_{\tilde{z}_i}(e_k),\mu_{\tilde{z}_j}(e_k))$$

Since this definition, the general definitions used in Fuzzy Set Theory may be used, we should like to stress on :

- the Hamming distance which gives the distance between $\tilde{z}_i$ and $\tilde{z}_j$
- the fuzziness indice which gives the distance between $\tilde{z}_i$ and $z_i(0{,}5)$
- the fuzziness entropy which gives an appreciation of the gaps between the $z_i(\nu)$ (for one given i)
- the decomposition theorem which gives $\tilde{z}_i$ from the $z_i(\nu)$

$$\tilde{z}_i = \mathrm{MAX}\ (\mu_{\tilde{R}}(e_i,e_j)\ z_i(\mu_{\tilde{R}}(e_i,e_j)))$$

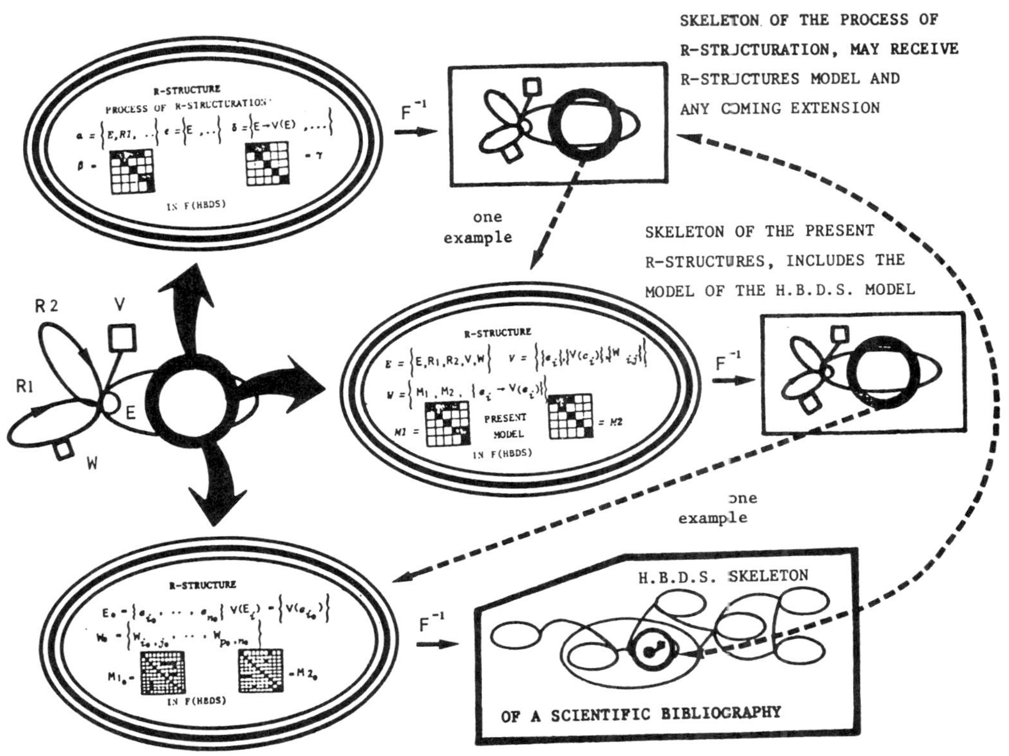

H.B.D. Structuration of the R-Structures
and R-Structuration of H.B.D.S. model

FUZZY DATA

We may consider W_{ij} (which may be seen as a relation $R''1 \subset V(e_i) \times V(e_j)$) fuzzy.

The fuzziness of W_{ij} says that, in the R-Structure corresponding to a H.B.D. Structure, the correspondences between the materializations may be fuzzy.
This case leads to the notions of fuzzy Link and Attribute of Object as defined in (Bouillé, 1979).
On the other hand the study of the fuzzy parts of a set may be done on $V(E)$, in that case the $V(e_i)$ are fuzzy subsets $\underset{\sim}{V}(e_i)$. That means the correspondences between the materializations and the realizations of A.D.T. are fuzzy.
We meet the notion of fuzzy Object as defined in (Bouillé, 1979) and new types of fuzziness too :
- non fuzzy Link between fuzzy Objects
- non fuzzy Attribute of Object with fuzzy attribution

These types of fuzziness may be seen as information which is not an answer to a question the skeleton has generated. For example observations which are not in concordence with the skeleton and which may lead to a skeleton modification must not be rejected as errors.
Till now we only study the R1 aspect of the fuzziness. An important question still remains what the inclusion relation between fuzzy sets becomes ?
If we consider $\underset{\sim}{\subseteq}$ (inclusion of Fuzzy Set Theory) with domination rule, $R'2$ is not fuzzy and $M2$ is still a binary matrix.
However to say $R'2$ is not fuzzy is not to say $R2$ induced from $R'2$ inclusion in $P(V(E))$ is the same as $R2$ induced from $R'2$ inclusion in $\underset{\sim}{P}(V(E))$.

In the first case :

$$V(e_i) \subset V(e_j) \Leftrightarrow \forall v_x \in V(e_i) \; \exists v_y \in V(e_j) \text{ such that } v_x = v_y$$

In the second case :

$$\underset{\sim}{V}(e_i) \underset{\sim}{\subseteq} \underset{\sim}{V}(e_j) \Leftrightarrow$$
$$\forall v_x / \mu_{\underset{\sim}{V}(e_i)}(v_x) > 0 \; \exists v_y / \mu_{\underset{\sim}{V}(e_j)}(v_y) > 0$$
such that
$$v_x = v_y \text{ and } \mu_{\underset{\sim}{V}(e_i)}(v_x) \leqslant \mu_{\underset{\sim}{V}(e_j)}(v_y)$$

We may conclude that the fuzziness of W_{ij} and $V(e_i)$ concerns only the data of

materialization but we have just seen above R2 says the A.D.T. of the skeleton may carry fuzziness as a potentiality. The achievement of a non fuzzy skeleton built with $\underset{\sim}{\subseteq}$ predicts some fuzzy materialization data.
This conclusion is quite close to the presentation of skeleton A.D.T. as fuzzy data control tools (Bouillé,1979).

FUZZY STRUCTURATION

If a R-Structure is achieved with $\underset{\sim}{R}1$ and $\underset{\sim}{R}2$ F^{-1} may be adapted to get a fuzzy $\underset{\sim}{\tilde{H}.B.D.S.}$ skeleton. As we have mentionned above such a concept has not been authorized by the author of the H.B.D.S. model.
At first to avoid the debate about the question " does fuzzy structure exist ? " we have to say :
It is a problem of vocabulary and a priori choices.
If we set forth :

- a modelization, we shall name a structure frame of reference, is not fuzzy by definition.
- a phenomenon is structured since one has got a structure frame of reference.
- any observation which is not perfectly suitable with the frame of reference is fuzzy with the meaning "rather incorrect".

We can conclude :

- fuzzy structuration does not exist, fuzzy data does.

The H.B.D.S. skeleton has a structure frame of reference role. But one must not forget the skeleton is not "given" it is the result of a continuous "way there and back" between structuration and insertion. This rigid frame is achieved in part with fuzzy data. More, if one admits a data structure is not "frozen" there are two categories of fuzzy data in data of materialization :

- the data with a role in the skeleton evolution
- the data <u>without any</u> role in the skeleton evolution

These categories are implicitly mentionned in the H.B.D.S. model when it is said the A.D.T. of the skeleton carry information about the accepted fuzziness. So any rejected fuzzy insertion belongs to the first category.
This division is arbitrary because the boundary between a needed modification and a "sleight of hand" is subjective.
The modelization of the phenomena consist of placing a frame of reference on the real world, any "reality" being either discussed or fuzzy (the fuzziness being the distance between the frame of reference and the "reality", zero value allowed).
Our solution is simple and far from any controversy.
The fuzzy skeletons exist but they exist only since the R-Structures have been introduced. It was difficult to cope with fuzziness in a H.B.D.S. skeleton because fuzziness is the distance to this structure frame of reference ; however one of the results exposed in our third paragraph is to turn a fuzzy skeleton into fuzzy data of a non fuzzy skeleton. The R-Structures give an expression to the notion of fuzzy structuration. An example (Bouillé,1981) has recently confirmed the theory we have developped.
After this long introduction we focuse on the consequences of $\underset{\sim}{R}1$ and $\underset{\sim}{R}2$ in a H.B.D. Structure.
In a fuzzy H.B.D. Structure the A.D.T. are not fuzzy their mutual relations are fuzzy.
We have Classes, Attribute of Classes ...etc but the set of the Attributes of a Class, the connection of a Class to a Hyperclass ...etc may be fuzzy. These aspects are the translation at the skeleton level of things we have already mentionned for the materialization level.
There is a more important aspect corresponding to a quite new concept : this one of Fractal Data. The Fractal Data are fuzzy data the fuzziness of which is neither at the level of the skeleton nor at the level of the data of materialization but belongs to both levels.
Let a very simple structure be a Class PERSON with two attributes DEFAULT and QUALITY, the value "frankness" for a member "Paul" of the Class is a fuzzy datum.
It is obvious that "frankness" is a fuzzy datum which asks for a modification in the skeleton. However the only modification is to create a compound Attribute (a generic property) with DEFAULT and QUALITY. It gives a limit to the fuzziness but seems to be only the "sleight of hand" we mentionned above.
We gave this phenomenon the name of Fractal Data because we consider it is intuitively close to the generalization of Fractal Object (Mandelbrot,1975).
The fractal dimension is a particular case, due to a geometrical approach, of the irregularity of real objects. In matter of fractal dimension B. MANDELBROT gives the formal generalization of an elementary property of the dimension : the role it plays to compute length, surface, volume.
Our step is an analogical one : the physical object with a dimension which is not an integer $1 < D < 2$ gets a property between a length and a surface. But does one know the property with meter $\sqrt{3}$ as valuation ? No and (Mandelbrot,1975) gives the proof with examples (broken line, plan with holes) where, in spite of mathematical notations, the intuitive description of the fractal objects rests on the classical dimensions.
The broken line is a point of view where there is a certain way to possess a length which gives a surface, the plan with holes is a point of view where there is a certain way to possess a surface which gives a length.
The reader may choose whether the frankness is a default with holes or a broken quality, it is obvious one has no geometrical representation of Fractal Data.
The analogy may be done with many current properties and it may be used to explain long controversies during modelization processes (see for example our work in Human Sciences (Legrand,1983).
We think the Fractal Data may be represented in a fuzzy R-Structure.
$\underset{\sim}{R}1$ gives the degree of dependence of properties, "perfects the description" we use in

our semantical interpretation has not always a strict meaning of predication.
$\underset{\sim}{R}2$ gives the degree of dependence due to the relations between sets of valuations (a simple intersection for example).
It is not sure our study take the whole of Fractal Objects into account ; we have to say the Fractal Data are not the development of a mathematical theory but an intuitive idea which may be a basis to formalize a kind of fuzziness still unknown in the data structures.

A LOOK FORWARD

To conclude this paper we try to go further the scientific and theoritical meaning we have given to the fuzzy structuration to reach an epistemological meaning. Indeed, in spite of our demonstration concerning the fuzzy skeleton and the enthousiasm such works may rise if they meet structuration user requirements, a question still remains " is not structuration the antonym of fuzziness ? ". We are going to give our point of view on the couple (structuration,fuzziness) and some explanations will show the question is not silly and the answer not far from yes.
When we say that the R-Structuration of an observable static system S is an application from S with its observations into a R-Structure (Legrand,1981) ; we mean our observation presuppose an interpretation which is in part carried by the achieved representation of the phenomenon. Following the principle, we suggest in (Legrand,1983), saying that the representation of meaning for a machine is done by the introduction of a non interpretable representation which reduces the interpretation, we consider the introduction of new signs which, by a rigorous definition, try to force the interpretation of another representation introduces an aspect said semantical.
The remaining part of interpretation measures the degree of uncertainty about the mastership (seen as a syntactical approach) over the phenomenon and not the degree of freedom of the model user.
The fuzziness in the data structures has a fundamental role because its modelization is the modelization of a certain freedom of the model user. The fuzziness does not measure the uncertainty of the phenomenon mastership but increases it in the sense it takes in part the links between the phenomenon and its environment into account.
In fact it seems that, from the taking into account of an uncertain physical measure to the notion of Fractal Data, the fuzziness is a motion parallel or indebted to a phenomenological proceeding issued from the Gestalt Philosophie.
When F. BOUILLE write :
"If one has an objective attitude in front of phenomena, one sees that most of data are fuzzy, that fuzziness is the most current form of information and that a non fuzzy datum is an exceptional case" (Bouillé,1979 page 2 free translation).
We "translate" by :
The measure device, the experimenter, the experiment achievement conditions and its further utilisations are parts of the phenomenon which cannot be isolated, separated from its environment. To turn the phenomenon into a pre-established number of binary digits is often unbearable, even in the case of scientific activities. The intolerance the user may feel in front of a computer has causes the computer science cannot suppress but it can decrease the effects by giving the user the capability of improving the model within previously established boundaries.
Are we to speak of structural phenomenology ? We should not like to run away with this idea. We think that to point out a phenomenology looking like proceeding does not allow to go, modularity and grammar in hands, for a philosophical motion which doubts of the universal capability of these very tools.
The modelization of fuzziness may receive the same criticism of rigidity. That would lead to further developments without any limit ; except if one considers their mastership may increase complexity till such a level it would make the use of computer processing useless.
We are lead to conclude the study of representations done for widespread use and large limits of tolerance would produce more for cognitive science or human knowledge than for technological improvement. Indeed the suggested complexity would lead to consider it is more efficient anybody freedom to be used to act individually than to create collective an binary brains the aim of which being this freedom preservation.
However recalling the famous joke where a serious combinatorial computation established the ratio of american phone users would lead any american woman to become ... operator, we want to be cautious.
We stop here our reflexion, which appears to be clear, on the limits of our work to avoid any unconvienient fiction.

BIBLIOGRAPHY AND REFERENCES

Bouillé, F. (1977). Un modèle universel de banque de données simultanément partageable, portable et répartie. Doctoral thesis. Université P. & M. Curie, Paris.
Bouillé, F. (1978). Fuzzy data processing with the Hypergraph-Based Data Structure. In International Conference on Cybernetics and Society, Tokyo-Kyoto, november.
Bouillé, F. (1979). Le concept de structure de données topologique floue. In First International HBDS Seminar, 18-22 june, Lisbon, Portugal.
Bouillé, F. (1981). Do hierarchical structures exist in the real world. In Second International HBDS Seminar, 9-13 march, Richmond, U.S.A.
Courtieux, G. (1979). Informatique et Idéologie. In Représentation des connaissances et du raisonnement dans les sciences de l'homme, Colloque IRIA-LISH, 17-19 sept., St Maximin, France.
Hayes, J. R. (1978). Cognitive psychology. The Dorsey press.
Kaufman, A. (1973) Introduction à la théorie

des sous ensembles flous, Tome 1, Masson, Paris.

Kitagawa, T. (1979). Generalized artificial grammars and their implications to knowledge engineering approches. Research report n°6, International Institute for advanced study of social information science, Tokyo, Japan.

Legrand, J. (1981). A proposal for a way to automatic structuration. In Second International HBDS Seminar, 9-13 march, Richmond, U.S.A.

Legrand, J. (1982). The R-Structures : a new theoritical tool for topological modelling. In Applied modelling and simulation, AMS'82, IASTED symposium, 29/06-02/07, Paris.

Legrand, J. (1983). Les R-Structures : contribution théorique d'une approche informatique des sciences juridiques à la structuration des données. Doctoral thesis, waiting to be defended.

Mandelbrot, B. (1975). Les objets fractals, forme, hasard, dimension. Flammarion, Paris.

Merleau-ponty, M. (1945) Phénoménologie de la perception. Bibliothèque des idées, Gallimard, Paris.

MULTIVARIABLE FUZZY WEIGHTED DIGRAPH-ELEMENT NETWORK DIGRAPH AND SYSTEM IDENTIFICATION

Zhao Hong*, Li Taihang and Shen Zuliang*****

**Process Automation and Instrumentation Tianjin Institute*
***Calculating Technology Shanghai Institute*
****Shanghai Industrial University, Shou Du Steel Works*

Abstract. In this paper the simplest expression of multivariable complex system fuzzy network is obtained by the network simplified method. So that a kind of the arithmetic method of the general graph theory is provided as to work at the fuzzy system identification and control of the multi-input and multi-output system.

The application of the above mentioned to build the [si] forecasting model of the blast furnace smelting process through the statistical analysis of the practical data process the theoretical frame to be basically available.

Keywords. Artificial interlligence; system theory; identification; multivariable systems; graph theory.

INTRODUCTION

For investigating the model of the large scale system F.S. Boberts [1] provided the model of graph theory. Based on man involved in system, using the theory of fuzzy sets, the semantic model built up from fuzzy conditional statements has been considered by Mamdani, Tong R.M. [2] Yoger [3], Li Bao-shou, Lu Zhr-zun [4]etc. Compositing the two ideas mentioned above, this paper provides the method of semantic model of multivariable fuzzy weighted digraph — element network graph. The relations of input—output variables are described by the simplified form of multivariable fuzzy element network graph in order to obtain output by knowing input.

I. Fuzzy Conditional Statements

If the fuzzy system is analyzed in time sequence by fuzzy conditional statements to forecast the quantity in the system, generally we have: if $U_{t-l_1}=\underset{\sim}{A}_{t-l_1}$ and $Y_{t-l_1'}=\underset{\sim}{B}_{t-l_1'}$ and $U_{t-l_2}=\underset{\sim}{A}_{t-l_2}$ and $Y_{t-l_2'}=\underset{\sim}{B}_{t-l_2'}$ ...... then $Y_t=\underset{\sim}{C}$.

Like [4], it can be simplified this way: If $\underset{\sim}{A}_{t-l_m}$ and $\underset{\sim}{B}_{t-l_n}$ then $\underset{\sim}{B}_t$, this is the general formula to identify the fuzzy system. But a problem exists in listing the fuzzy conditional statements. i.e. in the listing rule, some are sure to be compatible, but others may be not. From [3], we therefore introduce the concept 'degree of compatibility' to allow a compromise in the contradictions cases.

II. Multivariable Fuzzy Weighted Digraph — Element Network Digraph and the Correspondence Between the Digraph and the Fuzzy Conditional Statements

Che Yu-hu once advanced the definition :

Ordered three dimensional group (V, A, $\emptyset$) is called fuzzily weighted digraph, or fuzzily weighted graph in brief. We write it as $\underset{\sim}{G}$. V is the vertex sets; A is the circular sets; $\emptyset \subseteq \mathcal{F}(R)$; $\emptyset$ is the class of the fuzzy subset on R, the real number range, also the only full and simple map.

$$f: \underset{\sim}{A} \longrightarrow \underset{\sim}{\emptyset}$$

The elements in the $\underset{\sim}{\emptyset}$ are called the fuzzy weights on correspondence circular. $\underset{\sim}{\emptyset}$ is called of $\underset{\sim}{G}$.

We take the basic form of fuzzy conditional statements to show the correspondence between the basic form and the fuzzy weighted digraph. If there is "if $\underset{\sim}{A}$ then $\underset{\sim}{B}$", let the two fuzzy variables $\underset{\sim}{A}$ and $\underset{\sim}{B}$ respectively correspond to the two vertex ordered couple in fuzzy weighted digraph (Vi Vj) i, j=1, 2 ...... N. Between the two vertex fuzzy weights $\underset{\sim}{\emptyset}$ i j $\in \underset{\sim}{\emptyset}$, also $\underset{\sim}{\emptyset}$ i j=$\underset{\sim}{R}$ ij $\underset{\sim}{R}$ i j: U$\longrightarrow$Y as Fig. 2-1.

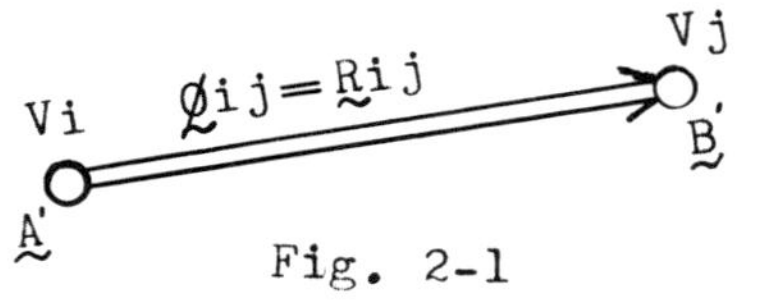

Fig. 2-1

According to the correspondence between the fuzzy weighted graph and conditional statements given by Fig. 2-1, we can obtain explanative frame graph corresponding to the formula if $\underset{\sim}{A}_{t-Li}$ and $\underset{\sim}{B}_{t-ki}$ then $\underset{\sim}{B}_t$, (i=1, 2, ... N) as following Fig. 2-2.

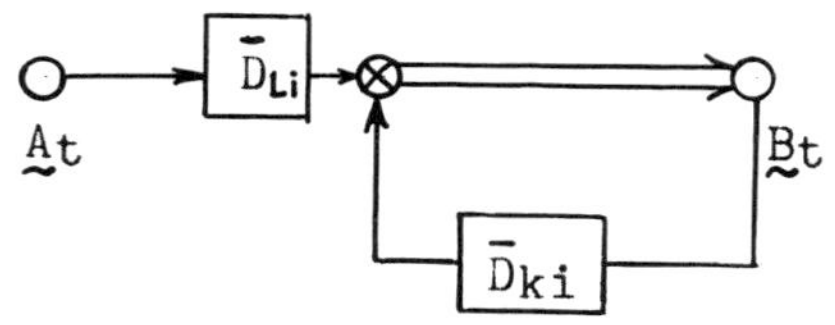

Fig. 2-2

Both $\bar{D}_{Li}$ and $\bar{D}_{Ki}$ are the lead-time operators of fuzzy variables. Just as $\underset{\sim}{A}_t$ through D_{Li} acting becomes $\underset{\sim}{A}_{t-Li}$, etc.

From the point of view of the physical significance, we give "weightes", the time rule of the fuzzy relation:

For the conditional statements as Fig. 2-1, if Li$\geqslant$Ki, then the action time which composes the systematic links of the single path in the 'weightes' is represented by the time parametes of fuzzy relation, we write it as $R_{AB(Li)}$. Conversely, if Li$\leqslant$Ki, then $R_{AB(Ki)}$. After having set up the rules, Fig. 2-2 can be simplified to represent by the following equivalent graph:

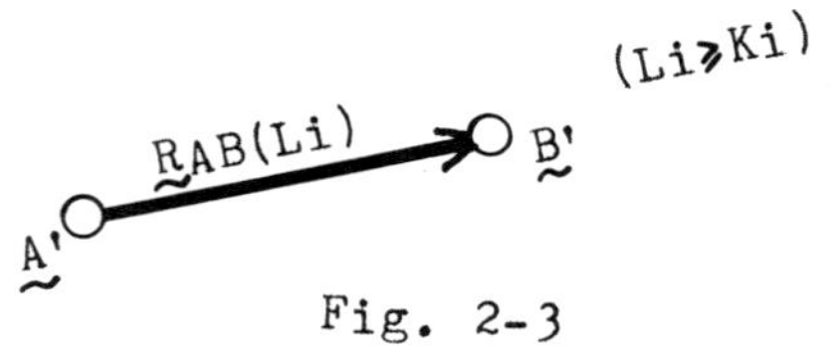

Fig. 2-3

At the same time, we set up the directed fuzzy element network graph corresponding to it:

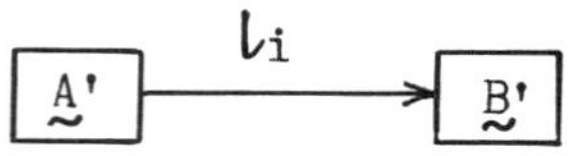

Fig. 2-4

We take the maximum bisection method as the superimposed algonithm of fuzzy ended sets of each directive arc when these arcss intersect at the same end.

We once attempted to use the method from [6] for finding the simplist gain of graph to obtain the direct connection between the graph of input — output variables. If the gain (to be equal as transfer function in [6] which joints the two nodes in the variable fuzzy weighted digraph is expressed as a fuzzy relation matrix, since matrix is the semi-group algebraic structure without convers elements, we have no way to quote the topological formula of the gain of matrix signal flow in [6]; also there are many algorithms for fuzzy relation R. It troubles some to select by man, so instead of emphasising on the simplification of R, we try to avoid these difficulties by introducing the method of simplifying on fuzzy element network in order to solve the problems.

III. The Simplification of the Fuzzy Element Network

The network and simplification of the fuzzy element are introduced from [7].

First of all, the fuzzy subsets representing the different physical quantity should be uniformize quantization since the simplifying at the same universe is the premise of the fuzzy element network simplification. That is to have the universe unified.

Unifying on the same axis of the universe by relative present. Having had the treatment, the same semantic of the different physical quantities such as 'rather high temperature' 'rather high pressure' can be differentiated through its uniformizing. That presents the different fuzzy subsets. Since the axis is unified at the universe of real number interval (0, 1), the grade of membership of fuzzy subsets formed this way can be expressed by fuzzy number.

As far as every superposition component O^{m}_{In}, the one of the output in multi-input (output) system is concerned, (m: the sign of output component n: the sign of input component) the compositive fuzzy relation $\underset{\sim}{R}^{O^m}_{I_n(Li)}$ between the input component. In and the output component O^m is obtained by the simplified method from fuzzy weighted digraph — element network graph given above. As to the algorithm of time rule will be showed later.

The accuracy result O^m of an output in system can be obtained by superimposing according to bisection method of maximum provided by II after using

$$\frac{R^{O^m}_{I_n}(Li),\ \underset{\sim}{I}_n}{\underset{\sim}{O}^m} \qquad (3\text{-}3),$$

the algorithm of fuzzy reasoning and through discriminating to obtain accuracy value $O^m_{I_n}$.

The time algorithm of the network simplification in system:

Proceeding from physical significance and concerning with rule at the time algorithm of single path given from II, we provided

(1) The total time of the paralled connected path in the network is the arithmetic superiposation of every single path (each arc which both ends are connected) time.

(2) The total time of series connected path in the network is the arithnetic mean of every branch path time.

(3) When each component of the link of output is superposing the total time effect is the maximum of the component time.

Now we mention an example of two inputs and one output as the sum-up for the discussion given above.

Let there be multivariable fuzzy weighted digraph formed by mapping of the fuzzy conditional statements as the following Fig. 3-1.

Its correspondent element network digraph is the F as Fig.3-2 denoted.

Since it is the fuzzy element network with two inputs, it should be divided into single input. We divide it into the form of minor graph $\underset{\sim}{I}_1 \rightarrow \underset{\sim}{O}_{I_1}$, and $\underset{\sim}{I}_2 \rightarrow \underset{\sim}{O}_{I_2}$, that is to search for all the sets of channels (tracks) from $\underset{\sim}{I}_1 \rightarrow \underset{\sim}{O}$ and $\underset{\sim}{I}_2 \rightarrow \underset{\sim}{O}$ as the Fig. 3-3 and Fig.3-4 denoted.

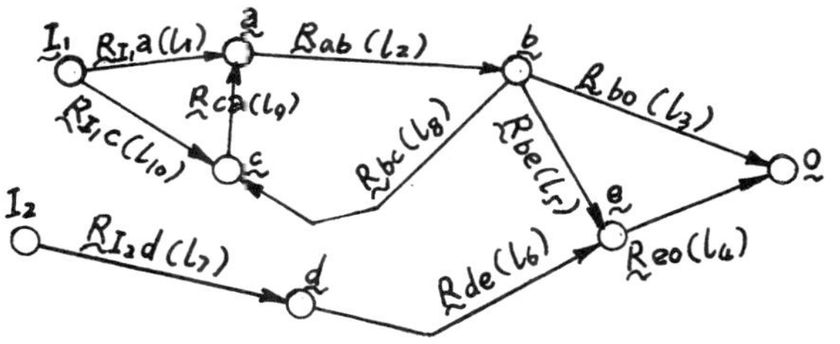

Fig. 3-1

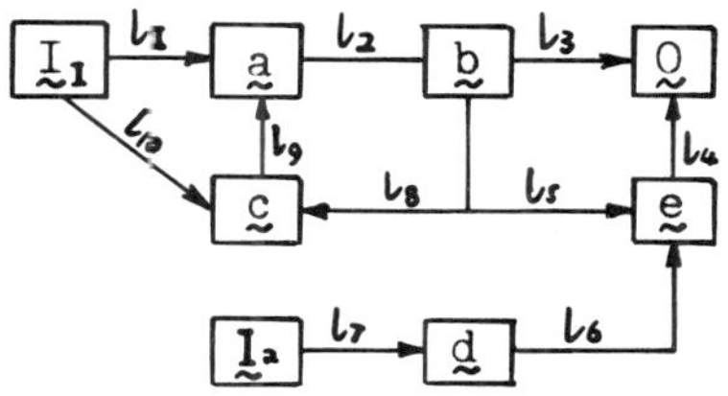

Fig. 3-2

The sets of the $\underset{\sim}{I}_1 \rightarrow \underset{\sim}{O}_{I_1}$ channel are known from Fig. 3-3.

$\{(\underset{\sim}{a}.\underset{\sim}{b}),\ (\underset{\sim}{a}.\underset{\sim}{b}.\underset{\sim}{c}.\underset{\sim}{a}.\underset{\sim}{b}),\ (\underset{\sim}{a}.\underset{\sim}{b}.\underset{\sim}{c}.\underset{\sim}{a}.\underset{\sim}{b}.\underset{\sim}{e}),\ (\underset{\sim}{a}.\underset{\sim}{b}.\underset{\sim}{e}),\ (\underset{\sim}{c}.\underset{\sim}{a}.\underset{\sim}{b}),\ (\underset{\sim}{c}.\underset{\sim}{a}.\underset{\sim}{b}.\underset{\sim}{e}),\ (\underset{\sim}{c}.\underset{\sim}{a}.\underset{\sim}{b}.\underset{\sim}{c}.\underset{\sim}{a}.\underset{\sim}{b}),\ (\underset{\sim}{c}.\underset{\sim}{a}.\underset{\sim}{b}.\underset{\sim}{c}.\underset{\sim}{a}.\underset{\sim}{b}.\underset{\sim}{e})\}$

The sets of simple channel are obtained by using idempotence to simplify:

$\{(\underset{\sim}{a}.\underset{\sim}{b}),\ (\underset{\sim}{a}.\underset{\sim}{b}.\underset{\sim}{c}),\ (\underset{\sim}{a}.\underset{\sim}{b}.\underset{\sim}{c}.\underset{\sim}{e}),\ (\underset{\sim}{a}.\underset{\sim}{b}.\underset{\sim}{e}),\ (\underset{\sim}{c}.\underset{\sim}{a}.\underset{\sim}{b}),\ (\underset{\sim}{c}.\underset{\sim}{a}.\underset{\sim}{b}.\underset{\sim}{e}),\ (\underset{\sim}{c}.\underset{\sim}{a}.\underset{\sim}{b}),\ (\underset{\sim}{c}.\underset{\sim}{a}.\underset{\sim}{b}.\underset{\sim}{e})\}$

The largest sets of simple channel are obtained by using absorption to simplify:

$$\left\{ (\underset{\sim}{a}.\underset{\sim}{b}) \big|\, l_1+l_2+l_3 \quad (\underset{\sim}{c}.\underset{\sim}{a}.\underset{\sim}{b}) \big|\, l_{10}+l_9+l_2+l_3 \right\}$$

It is showed that it consists of two series path from the input $\underset{\sim}{I}_1$ to component of the output $\underset{\sim}{O}_{I_1}$ as Fig. 3-5 denoted.

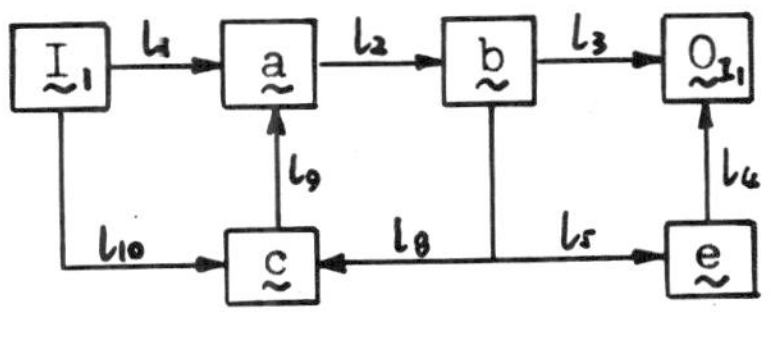

Fig. 3-3

The accuracy values of the outputs $O_{I_1(S_1)}$, $O_{I_1(S_2)}$, of the series branches I, II can be obtained respectively from the form (3-3).

As it is known by Fig.3-4 that $\underset{\sim}{I}_2 \rightarrow \underset{\sim}{O}_{I_2}$ minor graph had not simplified, $\underset{\sim}{O}_{I_2}$ can be obtained from the form (3-3) and the accuracy value $\underset{\sim}{O}_{I_2}$ is discriminated by this.

After having susperiposed $\underset{\sim}{Q}_{I_1}(s_1)$, $\underset{\sim}{Q}_{I_1}(s_2)$ and $\underset{\sim}{Q}_{I_2}$ by bisection method of maximum, the accuracy value of output O can be obtained.

Time computing:

$$t=\max\left[\left(\frac{l_1+l_2+l_3+l_{10}+l_9+l_2+l_3}{2}\right),\left(l_7+l_6+l_4\right)\right]$$

IV. Conclution

From the above example we can realize the advantage of applying the method.

1. We can chose the input (output) variable group agilely with choice of the research object in the system: Also it is allowed to have coupling connection inside input (output) variable group, thereby the couples of the different variable (link) group is obtained to be the model of the system when they are input or output.

2. After having simplified, the ring path disappeared, and a few of the simplist paths at most formed in the way of parallel — series connection joining at top of output. In this case, the algorithm can be used in recurrence, then it has finished treating complex multivariable fuzzy weighted digraph → fuzzy element network digraph with coupling feedback.

3. From the view of artificial intellingence, in some cases, could we say that the simplist path has nearly avoided the man's intuition to break out?

The melting of the blast furnace is an involved interaction process. Up to now, blast furnace is the largest single equipment of production in industry. Owing to the interval uneven distribution and same information difficult to obtain a given quantity and some other stochastic disturbance factors, we are not able to get a complete applicable analytical model. So there is a barrier in using computers to control the close in the furnace. In the past, the unsuccessful ways of building blast furnace enlightened us to consider the problem — whether we could break away from the isolated ideal but unpractical research method of theories or lab. — oratory — mid experiment — technological product in order to find a general method of study in the technological process environment directly. Of course, we do not mean to discard the analysis of metallurgy process technology mechanism, the mathematical statistic method and the testing study method. We intend to make them scientific basis to form conditional statements and to set definit acting time so as to reduce the unpractical or even unnecessary mental burden under the influence of the traditional model idea for these brachches of learning. (Such as: Could we put forward this kink of questions. Is it enough just about given first place to set up ope operating game models for similar blast furnace?)

Under the guidance of this thinking and taking for experiment object, the Blast Furnace No.3 in Shou Du Steel Works, we have got membership function of technological quantities by using the relative frequency method [9] and have collected continous melting data on the spot. We have summed up the implication list of the fuzzy conditional statements as lists 1-1; 1-2; 1-3; denoted. The correspondent fuzzy element network digraph such as Fig.4-1 ([si]n is the intended quantity of silicon which pig-iron contains in each furnace, [si]n-1 is the correspondent value of the last furnace). We use ALGOL language in programmer by the data of the practical measure the [si]n was predicted away from the product line on the CROMEMCO microcomputer.

$\underset{\sim}{A}$ / [si] n / [si] n-1	3	5	1	4	2
3 (low)	1	5	3	3	3
5 (lower)	4	1	5	3	3
1 (normal)	2	4	1	5	3
4 (higher)	2	2	4	1	5
2 (high)	2	2	2	4	1

List 1-1

$\underset{\sim}{C}$ / [si] n / [si] n-1	3	5	1	4	2
3 (low)	3	3	3	5	1
5 (lower)	3	3	5	1	4
1 (normal)	3	5	1	4	2
4 (higher)	5	1	4	2	2
2 (high)	1	4	2	2	2

List 1-2

$\underset{\sim}{G}$ [si] n / [si] n-1	3	5	1	4	2
3 (low)	1	5	3	3	3
5 (lower)	4	1	5	3	3
1 (normal)	2	4	1	5	3
4 (higher)	2	2	4	1	5
2 (high)	2	2	2	4	1

List 1-3

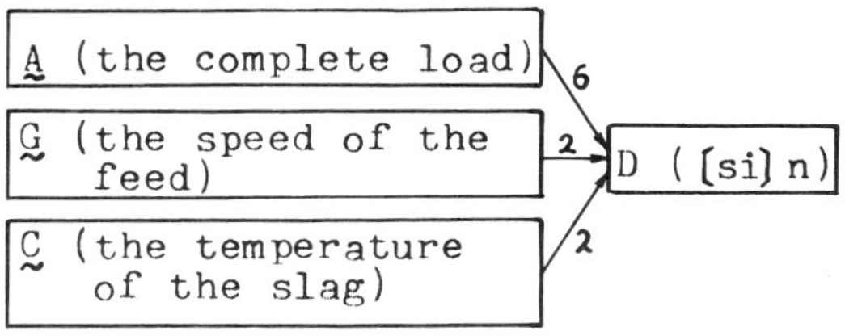

Fig. 4-1

We use hypothetic check to analize the result of the experiment with statistic, and then we can realize that when we have not added the program of self-learning correction (the purpose is for appraising fuzzy algorithm itself abviously), the rate of success of the algorithm in prediction is not lower than what man did. It is showed that the basic algorithm is available. The complex coupling network can be simplified by III, and then recurred by the basic algorithm, the following work, adding the algorithm of self-learning correction, is expected to raise the rate of success continously in prediction. (The [si] self-learning predicting result showed: more than 80 furnace practical anastomosis rate can arrive at 90%; the error allowance: ±0.05%. The detail will give later.)

REFERENCE

[1] Fred S. Roberts, Discret Mathematical Models 1976 by Prentice-Hall, Inc. Englewood Cliffs, New Jerssy.

[2] Tong. R.M. Synthesis of Fuzzy Models for Industrial Processes. Int. J. Gen., Systems 4, 1978, p.143-162

[3] R.R. Yager Building Fuzzy Systems Models Jona College New Rochelle, New York, U.S.A.

[4] Li Bao-shou, Lu Zhi-jun, Using Models of Fuzzy Theory Identification Systems (Information and Control), 1980, 3, p.32-38

[5] Signal Flow Diagram and Application, by Zhao Yong-chang, People Posts and Telecommunication Publishing Department.

[6] D.E. Riegle, P.M. Lin, Matrix Signal Flow Graphs and an Optimum Topological Method for Evaluting Their Gains, IEEE. Trans., on Circuit Theory, 1972, Sept. Vol. CT-19, No.5

[7] A Ranfmarn Instroducrion to the theory of Fuzzy Subsets, Vol.1 p.235-242

[8] 浅居, 田中, 奥田, C.V. Neyoita, D.A. Raleseu, あいまいシステムの理論入門, オーム社, 1978, 東京.

[9] Random-Phenomenon, Slave Characteristics and Probability Characteristics. Wuhan Building Material College Journal, 1981, 1, 2.

SIMULATION OF FUZZY ELECTRICAL NETWORKS

G. Elst* and B. Straube**

**Centre for Research and Technic (ZFT), VEB Robotron, German Democratic Republic*
***Central Institute for Cybernetics and Information Processes of the Academy of Sciences of the German Democratic Republic*

Abstract. Electrical networks are an important class of systems. The modelling of electrical circuts with circuit elements having tolerances, influences of technology, ageing or drift phenomena, etc. leads to networks with uncertin or fuzzy parameters. A fuzzy terminology basing on notions of general network theory is presented. Solving methods for computing fuzzy currents and voltages and for interconnected fuzzy subnetworks are given. Simple examples illustrate the approach.

Keywords. Computer-aided circuit analysis; fuzzy network; analysis of fuzzy network; tolerance analysis; large-scale systems.

INTRODUCTION

The modelling of electrical circuits having circuit elements with tolerances, uncertain influences of technology, ageing or drift phenomena, etc. leads to parameter dependent netwoks. The investigations of such networks demands special analysis methods. In the case of small variations of the parameters the sensitivity analysis gives often good results. But for the analysis of large variations of the parameters the calculation of transfer functions or stochastic techniques must be taken. In the stochastic case the distribution functions of the parameters should be known. However, in practical applications they are often unknown.
Experts have a lot of empiric knowledge of the properties of the networks and of their relationship to the parameter values from technological, physical, chemical etc. insight. This knowledge ought to be described by fuzzy sets.

In this paper we give the fuzzy terminology based on notions of a general network theory. The description of networks leads to systems of equations with fuzzy parameters. We present a solving method for computing the fuzzy voltages and fuzzy currents.
For large-scale networks a hierarchical simulation is necessary. We used cluster analysis to tear the given network into subnetworks. In this respect terminal behaviour is an important notion. Terminal behaviour is that part of the solution set which has to be taken for the analysis of the interconnection of the subnetworks. The fuzzy terminal behaviour of a subnetwork is the projection of the fuzzy voltages and currents onto the subset of the terminal variables. In this way fuzzy subnetworks are described as n-ary fuzzy relations. The analysis of a interconnected network is realized in the same way as the analysis of each of its subnetworks.

BASIC NOTIONS OF NETWORK THEORY

In short a network is characterized by

a network graph describing its topological structure

and

a voltage-current-relation describing the relationship between the branch-voltages u and the branch-currents i.

The ordered pairs (u,i) are elements of a set of time functions suitable chosen. More details about a general network theory are found in (Reibiger and Straube, 1981, 1982).
In applications the voltage-current-relation ϑ is mostly given as the set of solutions of a system of equations:

$$\mathcal{V} = \{(u,i) \mid V(u,i) = 0\} \quad (1)$$

For concentrate networks $\mathcal{V}$ is generally represented by systems of algebraic equations or by mixed systems of algebraic-differential equations.
The voltages and currents satisfying Kirchhoff's laws are represented by the set of solutions of systems of linear equations:

$$\mathcal{K} = \{(u,i) \mid B\,u = 0\ ,\ A\,i = 0\} \quad (2)$$

where B is the loop matrix and A the incidence matrix. The problem of the network analysis consists in finding all the elements of the set

$$\mathcal{L} = \mathcal{V} \cap \mathcal{K} \quad (3)$$

and making available systematic methods for it. Roughly speaking the elements of the set $\mathcal{L}$ are called the solutions of the network.
If networks are to be linked the nodes of a network which are connected with nodes of other networks are called terminals. The set of all terminal voltages and terminal currents which are consistent with the network is denoted with terminal behaviour (see Reibiger, 1981). Networks having the same terminal behaviour belongs to an equivalent class. The term n-pole denotes such a class of equivalent networks with n terminals and is a represenptive of this class. The terminal behaviour of the networks of a class is identical with the voltage-current-relation of the n-pole representing this class. This n-pole also is identical with the term 'macromodel' of an extensive network using in technical literature (e.g. Rabbat and others, 1979).

The terms and relations introduced above can be extended to $\mathcal{V}$ and $\mathcal{K}$ being fuzzy. Then the systems of equations representing $\mathcal{V}$ and $\mathcal{K}$ contain some fuzzy parameters with physical meaning. These fuzzy parameters can describe

- tolerances of cicuit elements,
- uncertainties in design,
- uncertain connections in the network model.

Then the membership values of the elements of the set $\mathcal{L}$ have to be found out in network analysis moreover. The same must be done for computing and representation of the terminal behavior of a network and of the voltage-current-relation of the pertinent n-pole.

ANALYSIS OF FUZZY NETWORKS

The analysis of networks is carried out by solving (nonlinear) algebraic or algebro-differential equations. Therefore, let us consider the following system of equations:

$$\begin{aligned} f_1(x_1,x_2,\ \ldots\ ,x_n,a_1) &= 0 \\ f_2(x_1,x_2,\ \ldots\ ,x_n,a_2) &= 0 \\ &\vdots \\ f_m(x_1,x_2,\ \ldots\ ,x_n,a_m) &= 0 \end{aligned} \quad (4)$$

where $(x_1,x_2,\ldots,x_n) \in \prod_{i\in I} X_i$, $A = \prod_{p\in M} A_p$ and the index sets $I = \{1,2,\ldots,n\}$ and $M = \{1,2\ \ldots\ ,m\}$. The mathematical properties of $\prod X_i$ depends upon the considered application. Let $a_p \in A_p$, $p \in M$ be the parameters to be described by fuzzy sets. Without loss of generality, each of the equations is assumed to have only a single a_p . If there are more than one parameter in an equation by means of a suitable substitution and extension of the system of equations including these substitution correspondences the former description will be achieved.

Given an easy single branch network with the voltage-current-relation $u - R\cdot i = 0$. The solution set of this equation is shown in Fig. 1. Let R be described by a fuzzy set in Fig. 2 the pertinent fuzzy voltage-current-relation has the (discretizied) solution set shown in Fig. 3.

The set of solutions of the equation f_p with fixed a_p is denoted by $S_p(a_p)$,

$$S_p(a_p) = \{(x_1,x_2,\ldots,x_n) \in \prod_{i\in I} X_i \mid f_p(x_1,x_2,\ldots,x_n,a_p) = 0,\ a_p \in A_p\}$$

Given $a = (a_1,\ldots,a_m)$ $\mathcal{L}(a)$ denotes the solution set of the system of equations (4). It results of the intersection of the sets $S_p(a_p)$

$$\mathcal{L}(a) = \bigcap_{p\in M} S_p(a_p)$$

The calculation of this solution set is done by a suitable numerical method (Gaussian elimination for linear algebraic systems, Newton method for nonlinear ones, integration techniques for algebro-differential systems).

Let A_p be the set of values a_p . On this set a fuzzy set $B_p : A_p \longrightarrow [0,1]$ is defined. The membership value $B_p(a_p)$ points out in which scale the parameter $a_p \in A_p$ has the modelled property B_p. For some value $a_p \in A_p$ the solution manifold $S_p(a_p)$ is paired by

the membership value $B_p(a_p)$, i.e., $(S_p(a_p), B_p(a_p))$. This fuzzy set can be written as follows:
$S_p(a_p)$ is described by its characteristic function χ_{p,a_p} with

$$\chi_{p,a_p}(x_1,\dots,x_n) = \begin{cases} 1 \text{ if } (x_1,\dots,x_n) \in S_p(a_p) \\ 0 \text{ else} \end{cases}$$

Thus, the fuzzy solution set F_p on $\prod X_i$ in regard with all possible values a_p is given by

$$F_p(x_1,\dots,x_n) = \bigvee_{a_p \in A_p} (B_p(a_p) \wedge \chi_{p,a_p}(x_1,\dots,x_n))$$

F_p is the fuzzy extension of S_p.
The fuzzy solution set L on $\prod X_i$ of the whole system is given by the fuzzy intersection

$$L = \bigwedge_{p \in M} F_p$$

If [0,1] is the set of membership values the set of all these fuzzy sets is a complete distributive lattice. Hence, the following correspondence holds for all $(x_1,\dots,x_n) \in \prod_{i \in I} X_i$

$$\begin{aligned} L(x_1,\dots,x_n) &= \bigwedge_{p \in M} F_p(x_1,\dots,x_n) \quad (5) \\ &= \bigwedge_{p \in M} \left[\bigvee_{a_p \in A_p} (B_p(a_p) \wedge \chi_{p,a_p}(x_1,\dots,x_n)) \right] \\ &= \bigvee_{a \in A} \left[\bigwedge_{p \in M} (B_p(a_p) \wedge \chi_{p,a_p}(x_1,\dots,x_n)) \right] \end{aligned}$$

In the last expression it is seen that for solving systems of fuzzy equations the deterministic technique denoted by χ_{p,a_p} is superposed by the evaluation of the membership values.
In available programs it is not difficult to extent them to the fuzzy case.

FUZZY TERMINAL BEHAVIOUR

A network having terminals is refered to as a subnetwork. Subnetworks might be created by tearing the given network. This can be done for large scale networks by clustering, see (Elst and Straube, 1981).

In (Reibiger, 1981) it is shown that the evaluation of the terminal behaviour is equivalent to the analysis of the network connected with norators in a suitable manner.
Then, the terminal behaviour is the projection of the solution set onto the subspace of all the terminal voltages and currents. In the following considerations the terminal voltages and currents are composed in x^{kl} which are referred to as terminal variables.

All the other voltages and currents in the system are called internal variables x^{in}.

The projection of a fuzzy relation is defined as follows:
Let R be a fuzzy set on $\prod_{i \in I} Y_i$. Thus R is a fuzzy relation too. Let $J \subset I$
Then $P_J R$ denotes the projection of R onto the set $\prod_{j \in J} Y_j$. $P_J R$ is a fuzzy set on $\prod_{j \in J} Y_j$, with $K = I \setminus J$

$$P_J R(pr_J y) = \bigvee_{pr_K y \in \prod_{l \in K} Y_l} R(y) \quad , \quad y \in \prod_{i \in I} Y_i$$

Example. $R : Y_1 \times Y_2 \times Y_3 \times Y_4 \longrightarrow [0,1]$

$J = \{2,3\} \qquad pr_J y = (y_2, y_3)$

$$P_J R(y_2,y_3) = \bigvee_{\{y_1,y_4\} \in Y_1 \times Y_4} R(y_1,y_2,y_3,y_4)$$

$$\forall (y_2,y_3) \in Y_2 \times Y_3$$

The fuzzy terminal behaviour results from the projection $P_J L$ of the fuzzy solution set onto the subset of all the terminal variables, i.e.,

$$P_J L(pr_J x) = \bigvee_{pr_K x \in \prod_{l \in K} X_l} L(x)$$

where $K = I \setminus J$ and $x \in \prod_{i \in I} X_i$.
Using (5) then

$$P_J L(x^{kl}) = \bigvee_{x^{in} \in \prod_{l \in K} X_l} \bigvee_{a \in A} \left[\bigwedge_{p \in M} (B_p(a_p) \wedge \chi_{p,a_p}(x)) \right] \quad (6)$$

The evaluation of the fuzzy terminal behaviour needs besides calculating the fuzzy solution set of the given network a maximum operation yet. This can be interpreted as transforming of the fuzziness of the internal variables onto the terminals.

ANALYSIS OF THE INTERCONNECTION OF FUZZY NETWORKS

The analysis of a large-scale network that exceeds the bounds of the present available analysis programs (about 2000 variables) is possible by decomposition of the network into subnetworks and a following subnetwork by subnetwork computation of the solutions in a hierarchical sequence, see e.g., (Rabbat and others, 1979; Elst and Straube, 1981).
Over this such an analysis method is more advantageous, if the terminal behaviours of (some) subnetworks were computed and are available for the analysis of the complete network. In interconnecting every subnetwork is replaced by its representative, the n-pole. In the following the known terminal behaviour of every subnetwork is assumed.

The voltage-current-relation of every pertinent n-pole s, s=1,2,...,q, is

$$\vartheta_s^{kl} = P_{J_s} \, L_s(x_s^{kl})$$

The set ϑ_s^{kl} can be represented implicitly by means of the set of solutions of a system of equations analogous (1) or explicitly by a set of 2(n+1)-tupels. This set results from the discretization and the point by point computing of the solution set. Such sets are also called table-models. In the first case the analysis of the interconnection leads again to the solution of a system of equations, where the fuzzy solution set can be computed again with (5). In the second case the analysis of the interconnection can also be reduced to the solution of equation systems with help of suitable interpolation methods.
Hence, the same method is used for both the analysis of the interconnection and the analysis of fuzzy networks. In the case of explicitly given fuzzy sets ϑ_s^{kl} it is possible to determine directly the intersection and to get the fuzzy solution with consideration of Kirchhoff's laws. Because the iteration methods for solving nonlinear systems again allows only a point by point designation of the solution set of these systems, such methods get importance, which use table-models. If the representative of ϑ_s^{kl} with a finite set of points is sufficient exactly for the application, then the following method can be used for the analysis of the interconnection.
Let

$$\vartheta_s^{kl} = \left\{ (x_{s,j}^{kl} , z_{s,j}^{kl}) , j=1,\dots,r_s \right\} \quad (7)$$

be the representative of the voltage-current-relation as a set of r_s discretization points $x_{s,j}^{kl}$ with the membership values $z_{s,j}^{kl}$. From the Kirchhoff's laws the condition

$$C \cdot \begin{pmatrix} x_1^{kl} \\ \vdots \\ x_q^{kl} \end{pmatrix} = C \cdot x^z = 0 \quad (8)$$

follows for the interconnection, with the matrix C consisting of N columns and N/2 rows. The terminal variables of all subnetworks are collected in x^z.

<u>Simplified algorithm.</u>

step 1: By means of a diagonalization method the equation (8) is removed in the form

$$\left[E , C_R \right] \cdot \begin{bmatrix} x_a^z \\ x_u^z \end{bmatrix} = 0 , \quad x^z = Q \cdot \begin{bmatrix} x_a^z \\ x_u^z \end{bmatrix}$$

E denotes the unit matrix. The necessary changes of the columns are noted in the permutation matrix Q. It is

$$x_a^z = -C_R \, x_u^z \quad (9)$$

step 2: For selected values of x_u^z from the sets ϑ_s^{kl} ,s=1,2,...,q are computed the pertinent values of the independent variables x_α^z by means of (9).

step 3: For the values of x_s^{kl} , s=1,...,q the neighbour points are searched in ϑ_s^{kl} and their membership values are taken over, for instance by averaging, others operations are possible too. Because of the necessary intersection the minimum operation has to applied to all membership values just got.

step 4: If all the values have been chosen go to step 5 else go to step 2.

step 5: The end of algorithm.

For implementation those of the points are not stored having membership values equal to 0.

In the case of special interconnections such as series-, parallel- or chain-connections of subnetworks (how to use in classical two-port theory) it is possible to calculate the intersection of the sets ϑ_s^{kl}, s=1,...,q in an easier manner, see example 1.

APPLICATIONS

In the following two examples are given. In the figures the range of membership values is the closed intervall [0,9] for a better presentation. The value 0 is omitted

The fuzzy representation of the terminal behaviour leads to some remarkable results:

It is possible to deduce whether the circuit might work according to the modelled fuzzy parameters.

If the cicuit must not exceed some thresholds of some voltages or currents the corresponding membership values of these thresholds give hints to the range of the allowable parameter values.

Example 1 : Chain connection of two inverters

Given the equations of a static MOS-inverter. Fuzzy parameters are an internal threshold voltage, the source-voltage, and two geometrical ratios. The resulting fuzzy terminal behaviour is shown in Fig. 4 in a discretizied form.
In Fig. 5 the fuzzy terminal behaviour of such two inverters in a chain connection is given.
Now let us consider the evaluation of the terminal behaviour which results from the connection of the known behaviour each of the single inverters in the subnetwork sense. The result is nearly the same as in Fig. 5 . The very small changes influenced by discretization can not be seen in the figure.
For this special example it should be noticed that the connection is equal to the composition of both the fuzzy terminal relations. This fact points to the importance of table-models for fuzzy networks.

Example 2 : Analysis of a fuzzy dynamic inverter

The dynamic properties of a bipolar inverter are modelled by voltage-dependent capacities. These capacity functions are very difficult to achieve. Hence, they are imprecise. Here, these capacities are replaced by current sources whose values are characterized by fuzzy sets.
Thus, the given dynamical network is transformed into a static one. It is always easier to analyse a static network rather than a dynamic one. Answers of time behaviour are no longer possible, of course. However, answers can be given for resulting voltages and currents. The network with fuzzy current sources instead of the capacities is shown in Fig. 6 . For comparison purposes a transient solution has been calculated where the used values of the capacities correspond to the values of the current sources. To compare with the fuzzy solution the time variable was substituted by $u_e(t)=t$. The results are shown in Fig. 7 (the heavy line is the transient).

REFERENCES

Reibiger, A., and B. Straube (1981). Zur Definition und Berechnung des Klemmenverhaltens von Netzwerken. 26.Int. Wiss. Kolloquium TH Ilmenau, GDR
Reibiger, A., and B. Straube (1982). Allgemeine Netzwerke (Teil 1). Zschr. für elektr. Inf. and Energietechnik, 12, 99-126.
Rabbat, N.B.G., A.L. Sangiovanni-Vincentelli, and h.Y. Hsieh (1979). A multilevel Newton algorithm with macromodelling and latency for the analysis of large-scale nonlinear circuits in the time domain. IEEE Trans. Circuits and Syst. CAS-26, 733-741.
Elst, G., and B. Straube (1981). Über Zerlegung und Simulation großer Systeme. Zschr. elektr. Inf. und Energietechnik, 11, 129-140.

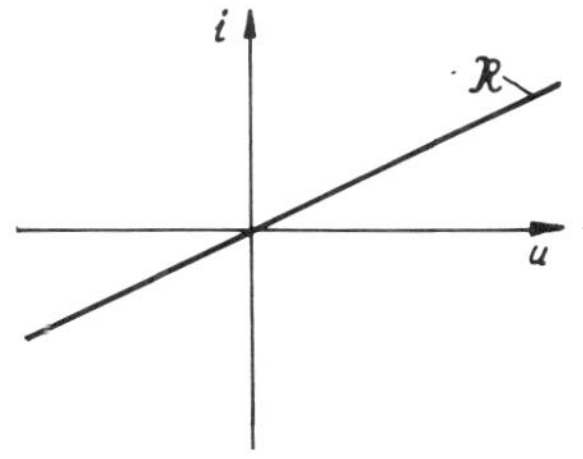

Fig. 1. Solution set $\mathcal{R}$ of the equation $u - R\, i = 0$

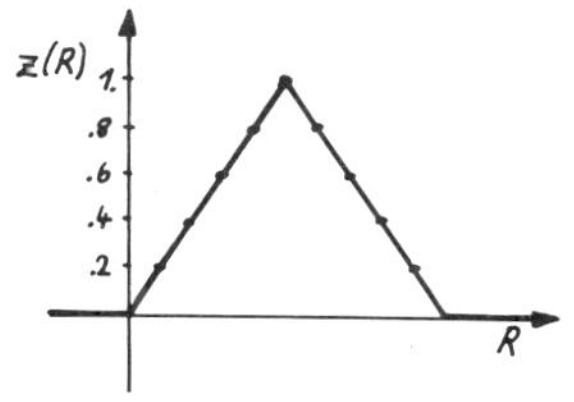

Fig. 2. Fuzzy set R

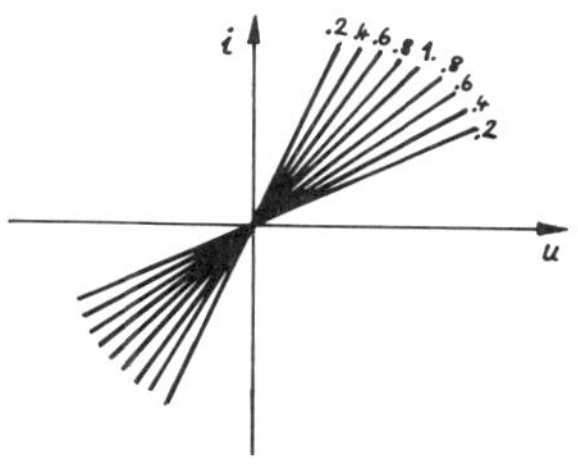

Fig. 3. Fuzzy solution set $\tilde{\mathcal{R}}$

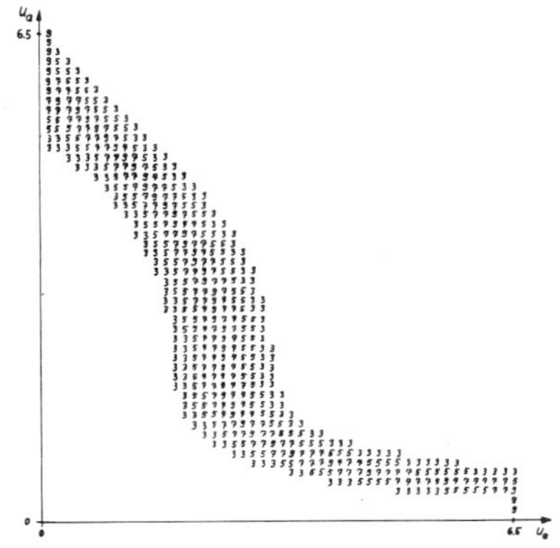

Fig. 4. Fuzzy terminal behaviour of a MOS-inverter

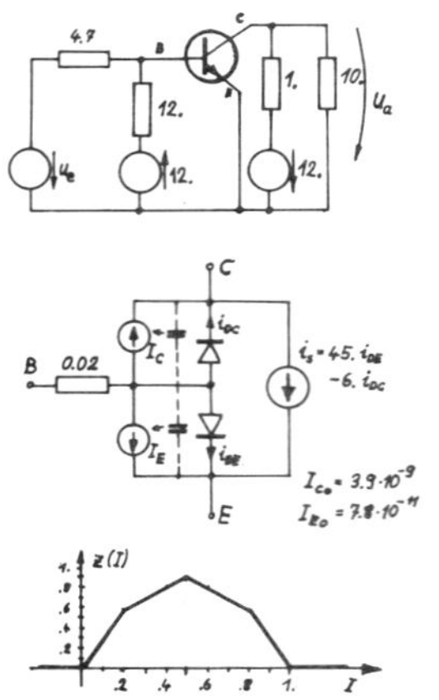

Fig. 6. Fuzzy network of an inverter

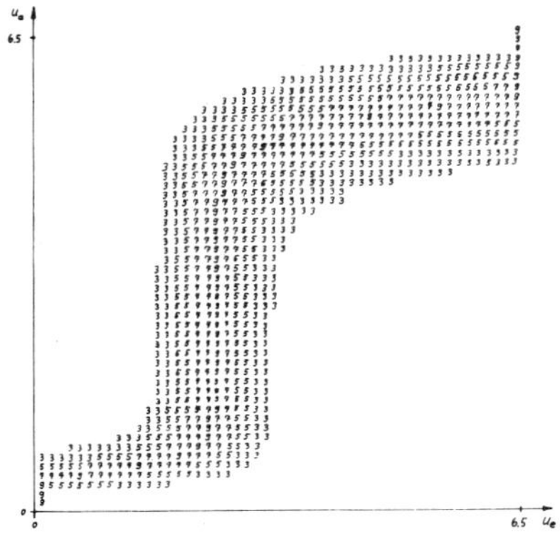

Fig. 5. Fuzzy terminal behaviour of a chain connection of two inverters

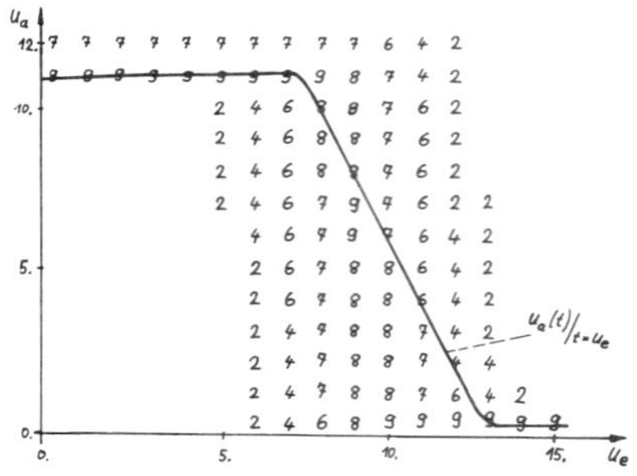

Fig. 7. The fuzzy terminal behaviour (switching on)

ON SOME DECISION PROBLEMS IN THE NETWORK WITH FUZZY PARAMETERS

S. Chanas and W. Kolodziejczyk

Institute of Management, Technical University, Wroclaw, Poland

Abstract. Presented are results of the analysis of the maximal (minimal) flow problem for the network with fuzzy arc capacities. The theorems formulated are equivalent (generalization) to classic Ford-Fulkerson theorems and relate the flow value with cut capacities in the network. The theorems make possible effective algorithms for seeking optimal - in the sense of Bellman-Zadeh's criterion - flows for integer as well as non-integer flows. It also appears that the classic bottleneck assignment problem can be treated as a particular case of the maximal flow problem in some networks with fuzzy arc capacities. Two new solution algorithms for the bottleneck assignment problem are result of this notice.

Keywords. Operations research; fuzzy optimisation; network flow; fuzzy arc capacity; maximal flow; minimal flow; bottleneck assignment problem.

INTRODUCTION

In the paper we present the main results obtained through the analysis of a flow problem in a network with fuzzy arc capacities (Chanas and Kołodziejczyk, 1982a, 1982b, 1982c). We admit in the problem, in opposition to the classical flow problem (Ford and Fulkerson, 1962), unprecise arc capacities as well as unprecise desired flow value.

The obtained results one may use for the solution of other homologous problems. Applying the results to the classical bottleneck assignment problem we can receive a very interesting and suprising outcome and take a different view of this problem. As it appears, the bottleneck assignment problem can be treated as a flow problem in a some special network with fuzzy arc capacities. As a consequence of such an approach we obtain a little bit different algorithms for solving the problem than presented for instance by Gross (1959), Garfinkel (1971) and Słomiński (1977).

FLOWS IN NETWORKS WITH FUZZY CAPACITY CONSTRAINTS

Problem Formulation

We recall shortly the formulation of the classical problem of the maximal (minimal) flow in the network (Ford and Fulkerson, 1962) and next we give its generalization.

Let $S = \langle N, A \rangle$, where N is a set of vertices and $A \subset N \times N$ is a set of arcs, denote a directed network in which each arc $(i,j) \in A$ is characterized by numbers $a_{ij}, b_{ij} \in R$ $(0 \leqslant a_{ij} \leqslant b_{ij})$ determining the lower and upper capacities of the arc.

The v-value flow from the source $s \in N$ in the destination $t \in N$ is a set of numbers related to the arcs, $x_v = \{x_{ij} / (i,j) \in A,\ x_{ij} \geqslant 0\}$, and satisfying the following balance conditions:

$$\sum_{j \in \Gamma_i^-} x_{ji} - \sum_{k \in \Gamma_i} x_{ik} = \begin{cases} -v & \text{for } i = s, \\ 0 & \text{for } i \neq s,t, \\ v & \text{for } i = t, \end{cases} \qquad (1)$$

where Γ_i^- and Γ_i denote the sets of vertices preceding and following the vertex $i \in N$, respectively.

The classical problem of the maximal (minimal) flow in the network consists in the determination of such a flow which satisfies the arc capacity constraints

$$a_{ij} \leqslant x_{ij} \leqslant b_{ij}, \quad (i,j) \in A \qquad (2)$$

as well as is characterized by the maximal (minimal) value

$$v \longrightarrow \max\ (\min). \qquad (3)$$

Now, we are interested in some generalization of the above stated problem, admitting a fuzzy determination of constraints (2) and goal (3). It means that in the new situation the decision-maker admits some violation of capacity constraints with a decreasing satisfaction degree in a certain range of tolerance; as well he admits a deviation of the flow value from an afore-said and desired level. The formal representation of the problem is as follows. To each arc $(i,j) \in A$ we relate a fuzzy number C_{ij} (Dubois and Prade, 1978) characterized by the membership function μ_{ij} as presented in Fig. 1a and 1b in the case of one-sided and two-sided arc capacity constraints, respectively. The fuzzy goal is represented by a fuzzy number, G, with the membership function μ_G as given in Fig. 1c for the maximizing of flow value and in Fig. 1d for the minimization problem. $\mu_{ij}(x_v) \triangleq \mu_{ij}(x_{ij})$ denotes in the most general case the degree of satisfying the (i,j) arc capacity constraint by the flow x_v. Similarly $\mu_G(x_v) \triangleq \mu_G(v)$ denotes the degree of realization of goal G by the flow x_v. According to the Bellman-Zadeh concept of decision-making in fuzzy conditions (Bellman and Zadeh, 1970) the problem reduces now to the determination of the flow x_v maximizing the membership function, μ_D, of the fuzzy decision, D, (D - fuzzy set in the flow space):

$$\max \mu_D(x_v) = \mu_C(x_v) \wedge \mu_G(x_v),$$

$$\text{where } D = \bigcap_{(i,j)\in A} C_{ij} \cap G, \qquad (4)$$

$$\mu_C(x_v) = \bigwedge_{(i,j)\in A} \mu_{ij}(x_v).$$

$\mu_C(x_v)$ denotes, according to the definition, the degree of simultaneous satisfaction of all arc capacity constraints by the flow x_v. On the other hand $\mu_D(x_v)$ is a "joint" (aggregated) degree of satisfying the constraints and goal by x_v.

Definition 1. The flow maximizing the membership function μ_D is called the maximal flow in the case of maximization of the flow value and the minimal flow in the case of minimization of the flow value (the form of G as in Fig. 1c and 1d, respectively).

Definition 2. The flow with the greatest (lowest) value in the set of maximal (minimal) flows is called optimal.

We present now some properties of the optimal solutions considering separately the case of integer flows and the case when also non-integer flows are admissible. But first we define a concept of cut in the network and introduce some denotations.

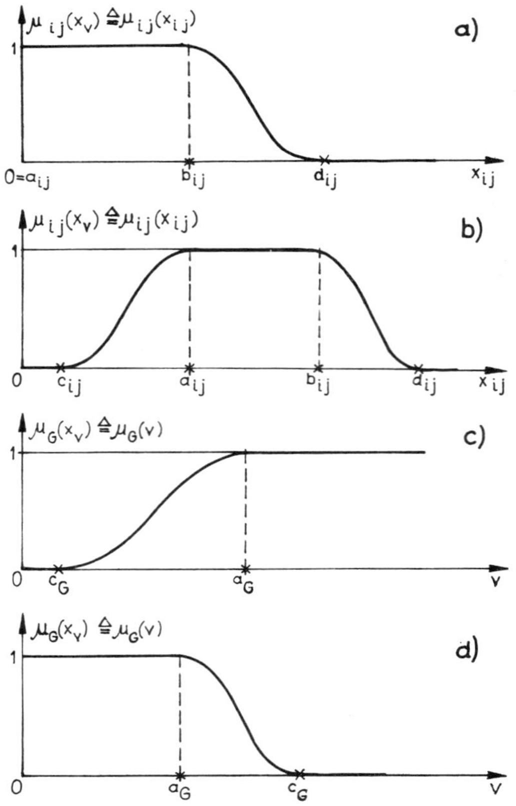

Fig. 1. The membership functions characterizing the imprecise constraints and goals in network flow problems

Definition 3. A cut in the network $S = \langle N,A \rangle$ separating the source, s, from the destination, t, is a set of all arcs, $(X,\bar{X}) \subset A$, connecting vertices $x \in X$ with vertices $y \in \bar{X}$, where $X,\bar{X} \subset N$, $X \cup \bar{X} = N$, $s \in X$, $t \in \bar{X}$.

For the sake of convenience we introduce the notations: $(\overrightarrow{X,\bar{X}})$ - the set of those arcs of cut $(X,\bar{X})$ which are directed from X to $\bar{X}$, $(\overleftarrow{X,\bar{X}})$ - the set of arcs directed from $\bar{X}$ to X.

Integer Optimal Flow

The below presented theorems may be treated as corresponding to known theorems introduced by Ford and Fulkerson (1962) which relate the optimal flow value with cut capacities in the network.

Theorem 1. If $x_v = \{x_{ij} \,/\, (i,j) \in A\}$ is the optimal integer flow in the network S (with two-sided arc capacity constraints) then there exists a cut $(X,\bar{X})$ in that network satisfying the following conditions:

1) for the maximization problem (G as in Fig. 1c)

$$(i,j) \in \underrightarrow{(X,\bar{X})} \Rightarrow \mu_{ij}(x_{ij}+1) < \mu_D(x_v),$$

$$(i,j) \in \underleftarrow{(X,\bar{X})} \Rightarrow \mu_{ij}(x_{ij}-1) < \mu_D(x_v),$$

2) for the minimization problem (G as in Fig. 1d)

$$(i,j) \in \underrightarrow{(X,\bar{X})} \Rightarrow \mu_{ij}(x_{ij}-1) < \mu_D(x_v),$$

$$(i,j) \in \underleftarrow{(X,\bar{X})} \Rightarrow \mu_{ij}(x_{ij}+1) < \mu_D(x_v).$$

<u>Theorem 2.</u> If x_v is the optimal flow in the network S with one-sided capacity constraints (C_{ij} as in Fig. 1a) then there exists in the network a cut $(X,\bar{X})$ satisfying the following conditions:

$$(i,j) \in \underrightarrow{(X,\bar{X})} \Rightarrow \mu_{ij}(x_{ij}+1) < \mu_D(x_v),$$

$$(i,j) \in \underleftarrow{(X,\bar{X})} \Longrightarrow x_{ij} = 0.$$

We recall here the algorithm (Chanas and Kołodziejczyk, 1982a) for solving of the problem in which one-sided (upper) arc capacity constraints are assumed. Chanas and Kołodziejczyk (1982b) have also described an algorithm, based on the same idea, for looking for the maximal flow in the network with two-sided fuzzy capacity constraints. In the algorithm presented below subsequently constructed flows $\bar{x}_v, \bar{x}_{v+1}, \bar{x}_{v+2}, \ldots$ are compared and each in "the best way" satisfies the capacity constraints in its class of flows with the same value i.e.

$$\mu_C(\bar{x}_k) = \max_{x_k} \mu_C(x_k) \quad \text{for } k=v,v+1,\ldots$$

<u>Algorithm 1.</u>

Step 1. Find the maximum flow $\bar{x}_v^1$ (in classical sense) in the network assuming arc capacities equal to b_{ij}. Evidently $\mu_C(\bar{x}_v^1) = 1$. If $\mu_G(v) = 1$ then stop, $\bar{x}_v^1$ is optimal. If not, go to Step 2.

Step 2. Let $\bar{x}_v^1$ be an actual flow. Generate a flow $\bar{x}_{v+1}^2$ satisfying in a maximum degree the capacity constraints. Go to Step 3.

Step 3. Check if the following condition holds: $\mu_D(\bar{x}_v^1) > \mu_D(\bar{x}_{v+1}^2)$. If so, then stop and the optimal flow remains $\bar{x}_v^1$. If not, $\bar{x}_v^1 := \bar{x}_{v+1}^2$ and go to Step 2.

The flow $\bar{x}_{v+1}^2$ in Step 2 of the algorithm is obtained by increasing the flow $\bar{x}_v^1$ by a unit on the chain r linking the source s with the destination t and maximizing the value of the expression

$$\mu^r(\bar{x}_v^1) = \bigwedge_{(i,j)\in \vec{r}} \mu_{ij}(x_{ij}^1 + 1) \wedge \bigwedge_{(i,j)\in \overleftarrow{r}} \mu_{ij}(x_{ij}^1 - 1) \longrightarrow \max,$$

where $\vec{r}$ and $\overleftarrow{r}$ denote the sets of the conforming and reverse chain's arcs, respectively.

We recall another important theorem that ensures the correctness of the above algorithm and provides us with a way of "improving" the flow admissibility in the class of flows with fixed value.

<u>Theorem 3.</u> Let be given a v-value flow, $\bar{x}_v$. If is not an "improving" cycle including arc (k,l) such that $\mu_{kl}(\bar{x}_v) = \mu_C(\bar{x}_v)$, then $\bar{x}_v$ satisfies in maximum degree the capacity constraints among all v-value flows i.e. $\mu_C(\bar{x}_v) = \max_{x_v} \mu_C(x_v)$.

Let us now explain the concept of an "improving" cycle. A cycle involving the arc (k,l) is said to be an "improving" cycle if after adding the unit to the flow values on the reverse cycle's arcs (reverse to the arc (k,l)) and after subtracting the unit from the flow values on the conforming cycle's arcs, the capacity constraints on the cycle are satisfied in a greater degree than $\mu_{kl}(\bar{x}_v)$.

The above theorem is also valid in case of two-sided arc capacity constraints.

<u>Non-integer Optimal Flows</u>

The formulated in this section theorems (4 and 5) correspond to the theorems 1 and 2. They are valid in the case of non-integer flows in networks.

<u>Theorem 4.</u> If $x_v = \{x_{ij} \,/\, (i,j) \in A\}$ is the network with two-sided capacity constraints then there exists a cut $(X,\bar{X})$ in the network satisfying the implication

$$(i,j) \in (X,\bar{X}) \Longrightarrow \mu_{ij}(x_{ij}) = \mu_D(x_v),$$

where for an arbitrary $\varepsilon > 0$ we have the following conditions:

1) for the maximization problem (G as in Fig. 1c)

$$(i,j) \in \underrightarrow{(X,\bar{X})} \Rightarrow \mu_{ij}(x_{ij}+\varepsilon) < \mu_D(x_v),$$

$$(i,j) \in \underleftarrow{(X,\bar{X})} \Rightarrow \mu_{ij}(x_{ij}-\varepsilon) < \mu_D(x_v),$$

2) for the minimization problem (G as in Fig. 1d)

$$(i,j) \in \underrightarrow{(X,\bar{X})} \Rightarrow \mu_{ij}(x_{ij}-\varepsilon) < \mu_D(x_v),$$

$$(i,j) \in \underleftarrow{(X,\bar{X})} \Rightarrow \mu_{ij}(x_{ij}+\varepsilon) < \mu_D(x_v).$$

<u>Theorem 5.</u> If x_v is the optimal flow in the network S with one-sided capacity constraints (C_{ij} as in Fig. 1a) then there exists a cut $(X,\bar{X})$ satisfying the following conditions:

$$(i,j) \in \underrightarrow{(X,\bar{X})} \Rightarrow \mu_{ij}(x_{ij}) = \mu_D(x_v),$$

$(i,j) \in (\underleftarrow{X,\bar{X}}) \Longrightarrow x_{ij} = 0.$

Formulated below theorem depicts more natural and close generalization of the Ford-Fulkerson theorems which relate a flow value with cut capacities. First we introduce two very important notions.

Definition 4. A fuzzy set, V, in R characterized by membership function $\mu_V(v) = \max_{x_v} \mu_C(x_v)$ is called a fuzzy capacity of the network S.

The value $\mu_V(v)$ can be interpreted, according to its definition, as the possibility of reaching the v-value flow in the network S, taking into account the existence of fuzzy capacity constraints.

Definition 5. A fuzzy capacity of the cut $(X,\bar{X})$ is the fuzzy number, $C(X,\bar{X})$, defined in the following way:
1) for a network with two-sided capacity constraints

$$C(X,\bar{X}) = \sum_{(i,j)\in(\underrightarrow{X,\bar{X}})} C_{ij} - \sum_{(i,j)\in(\underleftarrow{X,\bar{X}})} C_{ij} \qquad (5)$$

2) for a network with one-sided capacity constraints

$$C(X,\bar{X}) = \sum_{(i,j)\in(\underrightarrow{X,\bar{X}})} C_{ij}. \qquad (6)$$

Evidently, the signs of summation and subtraction appearing in (5) and (6) denote respective operations on fuzzy numbers (Dubois and Prade, 1978).

Theorem 6. Let W denotes the set of all cuts in the network S. The following relation is valid:

$$V = \bigcap_{(X,\bar{X})\in W} C(X,\bar{X}) \cap I, \qquad (7)$$

where I is the fuzzy set such that $\mu_I(x) = \max_v \mu_V(v)$ for all $x \in R$. It means that the fuzzy capacity of the network is equal to the intersection of cut capacities reduced to the level of $\max_v \mu_V(v)$.

The analogy of the above theorem to the Ford-Fulkerson theorems is obvious in the case of maximizing the flow under one-sided capacity constraints. In that case the fuzzy set I appearing in (7) can be left since $\max_v \mu_V(v) = 1$. Under the adopted assumption on the form of numbers C_{ij} (Fig. 1a) the intersection operation in (7) is the same in this case as the fuzzy minimum operation and can be replaced by the latter:

$$V = \min_{(X,\bar{X})} C(X,\bar{X}). \qquad (8)$$

In the case of two-sided capacity constraints both the minimization and maximization problems are treated jointly in the theorem. The left increasing part of μ_V provides information on the possibility of achieving "small" flow values in the network (determines the "minimal" fuzzy network capacity) whereas the right decreasing part of μ_V characterizes the "maximal" network capacity.

In the algorithm (Chanas and Kołodziejczyk, 1982c) for seeking the optimal flow in the network with one-sided arc capacity constraints it is assumed that C_{ij} are fuzzy numbers of the same type, that is $C_{ij} = (0,0,b_{ij},\alpha_{ij})_{LR}$ using the notation (Dubois and Prade, 1978, 1980) of the L-R type fuzzy number. Only in that case it is possible to calculate efficiently and exactly according to (6) the fuzzy cut capacity. The algorithm seeks the intersection point of functions μ_V and μ_G which determines the value and estimate of the optimal flow.

In the algorithm there are found some values denoted by b^r_{ij}, $r \in (0,1]$, meaning of which has to be explained. b^r_{ij} should be namely understood as a right end of the interval $[0,b^r_{ij}]$, where $[0,b^r_{ij}]$ is the r-set of C_{ij} i.e. $[0,b^r_{ij}] = C^r_{ij} = \{x_{ij} / \mu_{ij}(x_{ij}) \geq r\}$, $r \in (0,1]$.

Algorithm 2.

Step 1. Determine the maximal real flow x_w and the respective minimal cut $(X,\bar{X})$ (in the classical sense) in the network S with arc capacities b^1_{ij}. If $\mu_G(w) = 1$ then x_w is optimal. Otherwise go to Step 2.

Step 2. Determine the fuzzy number $C(X,\bar{X})$ and coordinates (v,r) of the intersection point of functions $\mu_{C(X,\bar{X})}$ and μ_G. Go to Step 3.

Step 3. Determine the maximal flow x_w and the respective minimal cut $(X,\bar{X})$ in S with the arc capacities b^r_{ij}. If $\mu_D(x_w) = r$ and $w = v$ then the flow x_w is optimal. Otherwise go to Step 2.

It is worth to make one important note. Namely, in the case of linear fuzzy constraints and goal that is when C_{ij} and G are, in the Dubois-Prade notation, numbers of the form (using symbols as in Fig. 1)

$$C_{ij} = (0,b_{ij},0,d_{ij}-b_{ij})_{LR},$$

$$G = (a_G,\infty,a_G-c_G,0)_{LR},$$

where $L(x) = R(x) = \max\{0,1-|x|\}$, the right parts of $\mu_{C(X,\bar{X})}$, $(X,\bar{X}) \in W$, are linear functions and the right side of

μ_V is a piecewise linear function. Hence, seeking the intersection point of $\mu_C(X,\bar{X})$ and μ_G a system of two linear equations with two unknowns is solved in Step 2 of the algorithm. Thus, if we assume that b_{ij}, d_{ij}, a_G and c_G are rational then coordinates of (v,r) generated in Step 2 are also rational and therefore the flow generated in Step 3 is rational.

In the next section we revert to the results presented in the preceding part of the paper. We describe some suprising application of them to the known classical bottleneck assignment problem.

ANOTHER LOOK AT BOTTLENECK ASSIGNMENT PROBLEM

Let us recall the formulation of the bottleneck assignment problem. Given n persons, n positions and efficiency estimates c_{ij} $(i,j=1,2,\ldots,n)$, where c_{ij} denotes the efficiency of the i-th person to the j-th position. The problem consists in the determination of an assignment (each person to one and only one position and vice versa) which maximizes the minimal efficiency. Formally the problem can be stated as follows:

$$\min\{c_{ij}\cdot x_{ij} / x_{ij} = 1\} \rightarrow \max$$

$$\sum_{i=1}^{n} x_{ij} = 1, \quad j=1,2,\ldots,n, \tag{9}$$

$$\sum_{j=1}^{n} x_{ij} = 1, \quad i=1,2,\ldots,n,$$

$$x_{ij} = 0 \text{ or } 1.$$

There are known a few methods for solving the problem (Gross, 1959; Garfinkel, 1971; Słomiński, 1977). The common property of all the algorithms consists in multiple determining, with the help of special methods (different in different algorithms), of admissible cells in the matrix of c_{ij} coefficients and in looking for a maximal set of independent admissible cells. It is equivalent to the looking for a maximal flow from sources (rows) to sinks (columns) through admissible arcs (admissible cells) what is made by the application of the Ford-Fulkerson labeling algorithm to the network of the structure presented in Fig. 2.

Let us have now another look at the bottleneck assignment problem. We join with the arcs (with arcs from the network in Fig. 2) fuzzy capacities C_{ij} characterized by membership functions:

$$\mu_{ij}(x_{ij}) = \begin{cases} 1 & \text{for } x_{ij} = 0, \\ \frac{c_{ij}}{M} & \text{for } x_{ij} = 1, \quad (i,j=1,2,\ldots,n) \\ 0 & \text{for } x_{ij} = 2,3,\ldots, \end{cases}$$

$$\mu_{si}(x_{si}) = \begin{cases} 1 & \text{for } x_{si} = 0,1, \\ & \qquad (i=1,2,\ldots,n) \\ 0 & \text{for } x_{si} = 2,3,\ldots, \end{cases}$$

$$\mu_{jt}(x_{jt}) = \begin{cases} 1 & \text{for } x_{jt} = 0,1, \\ & \qquad (j=1,2,\ldots,n) \\ 0 & \text{for } x_{jt} = 2,3,\ldots, \end{cases}$$

where $M = \max_{(i,j)} c_{ij}$.

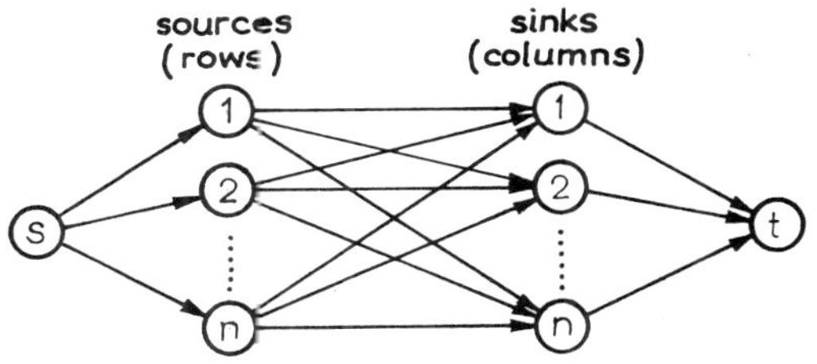

Fig. 2. Structure of the network for the bottleneck assignment problem

It is easy to observe that problem (9) is equivalent to the determination problem of the n-value flow, which in maximal degree satisfies the capacity constraints i.e. maximizes the value of the expression $\bigwedge_{(i,j)} \mu_{ij}(x_{ij})$, in such a network as presented in Fig. 2 (fuzzy arc capacities are defined by the above equations). This natural remark permits the utilization of the results previously presented for the analysis of this special problem.

We present below two new algorithms for solving of the bottleneck assignment problem resulting from the adoption of the algorithm 1 and theorem 3 to the special form of the network and from the fact that only zero-one flows are taken into consideration.

<u>Algorithm 3</u> (based on the theorem 3).

Step 1. Determine any assignment x_n of rows to columns in the matrix of coefficients c_{ij}.

Step 2. Find the element $c_{kl} = \min\{c_{ij} / x_{ij} = 1\}$.

Step 3. Applying the labeling procedure find an "improving" cycle which includes the arc (the cell) (k,l). If no "improving" cycle is found then stop. Otherwise go to Step 4.

Step 4. Change the actual assignment

erasing the unit in the cells which corresponds to the conforming cycle's arcs (conforming to arc (k,l)) and inscribing the unit in the cells which corresponds to it's reverse arcs where the cycle is determined in Step 3. Go to Step 2.

The labeling procedure may be realized in Step 3 in the following way. We join first with the column l the label "–". Then starting with the j-th labelled column we have to join the label "j" with all the not-labelled rows, for which $x_{ij} = 0$ and $c_{ij} > c_{kl}$. Starting with the labelled row i we have to label with the label "i" the column j for which $x_{ij} = 1$. The labeling procedure has to be stoped when the k-th row is labelled. If the labeling procedure has to be stoped and the k-th row is not labelled then there exist for (k,l) no "improving" cycle.

The cycle on which the "displacement" of the unit in Step 4 is done may be identified by labels. First we erase the unit in the cell (k,l). The label of the k-th row shows the number of the column where the unit has to be inscribed. In succession, the label of this column shows a number of a row in which the unit has to be erased and so on.

Algorithm 4 (the adoption of the algorithm 1).

Step 1. With all the rows having no unit join the label "–". With all the columns join the "temporary" labels of the form $(k(j),a(j))$, where $a(j) = a_{k(j)j} = \max_i a_{ij}$ and i ranges through the number of the labelled rows.

Step 2. Among the "temporary" labelled columns choose the column, say l, for which the second part of the label is maximal i.e. $a(l) = \max_j a(j)$. The label of this column becomes a "steady" label.

Step 3. If the "steady" labelled column, l, possesses no unit the labeling procedure has to be stoped – go to Step 4. Otherwise join the label $(l,a(l))$ with the k-th row for which $x_{kl} = 1$. This label becomes automaticaly a "steady" label. Starting with the k-th row improve the labels of the "temporary" labelled columns in accordance with the following rule. If in the "temporary" label of the j-th column $a(j) < \min\{a_{kj}, a(l)\}$ then this label should be replaced by the label $(k, \min\{a_{kj}, a(l)\})$, otherwise leave the label unchanged. Go to Step 2.

Step 4. Increase the flow to the sink (column) l by the unit on the chain pointed out by labels inscribing the unit in the cells which corresponds to the conforming chain's arcs and erasing the unit in the cells which corresponds to the reverse chain's arcs. If the value of the actual flow (assignment) is n then stop. Otherwise erase all the labels and go to Step 1.

Algorithm 4 seems to be more labour-consuming then the algorithm 3 but it possesses an interesting property which may be useful from the practical point of view. It enables namely the obtainment of intermediate suboptimal solutions, i.e. each partial solution x_k (k=1,2,...,n) is an optimal solution under the additional condition that only k positions are appointed from among n positions.

FINAL REMARKS

In the paper the results of the analysis of the maximal (minimal) flow in a network with fuzzy arc capacities are presented. The results one may easy utilize in other homologous problems by the application of some methods similar to that in the classical cases (Ford and Fulkerson, 1962). We mean here the problems of flows in networks with fuzzy node's capacities, the transport problems in which besides tolerances in arc's capacities also tolerances in a value of a supply (of nodes-sources) and in a value of a demand (of nodes-receivers) might be taken into account.

The presented here results which are connected with the classical bottleneck assignment problem point out that the fuzzy mathematical programming do have not only to utilize the classical solving techniques but also can be a source of new ideas for classical problems.

REFERENCES

Bellman, R., and L. A. Zadeh (1970). Decision making in a fuzzy environment. Management Sci., 17B, 141-164.

Chanas, S., and W. Kołodziejczyk (1982a). Maximum flow in a network with fuzzy arc capacities. Fuzzy Sets and Systems, 8, 165-173.

Chanas, S., and W. Kołodziejczyk (1982b). Integer flows in networks with fuzzy capacity constraints. Report 368, Inst. of Management Techn. Univ. of Wrocław.

Chanas, S., and W. Kołodziejczyk (1982c). Non-integer flows in the network with fuzzy arc capacities. Report 383, Inst. of Management Techn. Univ. of Wrocław.

Dubois, D., and H. Prade (1978). Operations on fuzzy numbers. Int. J. Syst. Sci., 9, 613-626.

Dubois, D., and H. Prade (1980). Fuzzy Sets and Systems - Theory and

applications. Academic Press, New York.
Ford, L. R., and D. R. Fulkerson (1962). Flows in Networks. A RAND Corporation Research Study, 3rd ed. Princetown University Press, Princetown, New Jersey.
Garfinkel, R.S. (1971). An improved algorithm for the bottleneck assignment problem. Oper. Res., 19, 1747-1751.
Gross, O. (1959). The bottleneck assignment problem. The RAND Corporation Paper P-1630, March 6.
Słomiński, L. (1977). An efficient approach to the bottleneck assignment problem. Bull. Acad. Polon. Sci. Ser. sci. tech., 25, 17-23.

THE APPLICATION OF A FUZZY PETRI NET FOR CONTROLLING COMPLEX INDUSTRIAL PROCESSES

H.-P. Lipp

Department of automation technique
Technical University Karl-Marx-Stadt

Abstract. At the operative control of production the regime of technology is coordinated in the sub-systems by the human operator, to reached the production goal as efficiently as possible. In this multi-decision-making process the operator applies an incomplete model, in which disturbances and control actions are recognized quickly. Such a model is the fuzzy Petri net, which is derived by fuzzification of transitions and places from the classical Petri net. The membership functions of the fuzzy sets result from intuitions and experiences of specialists.

INTRODUCTION

Complex industrial processes such as chemical reactors, pulp plants, cement kilns or glassmaking cannot be satisfactorily controlled using traditional methods of the control theory, mainly because their precise model is unknown. These processes consist of many sub-systems, which are connected by flows of workpieces, materials, energy or information. In changing production rate or in case of disturbances the dispatcher coordinates the flows between the subsystems, that the production goal is reached in the shortest time possible. In this decision process the human operator applies incomplete plant models, in which disturbances or efficiences of his control action have to be recognized quickly.
An alternative approach to the traditional control of complex processes is to investigate the control strategies employed by the human operator. In many cases the process operator can control a complex process more effectively than an automatic system. The human operator's control strategy is based on intuition and expieriences, and can be considered as a set of heuristic rules. In order to include the operator's process experiences in an automation scheme, a fuzzy Petri net is introduced. In this paper we briefly present at first modelling of complex processes by a classical Petri net, and then we describe the main ideas of the fuzzy Petri net, pointing out some specific features. This fuzzy Petri net is supported by the operator's subjective experiences and is applied as a formal model to complex processes. At last we apply it to a control system in a pulp plant.

MODELLING OF COMPLEX PROCESSES BY CLASSICAL PETRI NET

There are many ways to represent the behaviour of a complex system by a formal model. In particular such formal descriptions of dynamic systems are necessary, if the system contains interacting components that are working in parallel. Then by the unknown relative speed of these concurrent actions unexpected effects may appear. By the complexity of systems such undesired effects will increase. The Petri net model solves these problems by the following important characteristics:

- Independent actions are represented independently in the model by such components, on which they depend and where change will occur.
- Only the flow of control is modelled.
- All important properties of the model are given by a graph.

These properties are more or less common to all kinds of models. The classical Petri net consists of places and transitions, are connected by directed edges.

A Petri net is defined by

$$NP = [P, T, F, C, V, K, M, m_o]$$

$P = \{p\}$ is a finite set of places

$T = \{t\}$ is a finite set of transitions

F is a transition function $F: (PxT) \cup (TxP) \longrightarrow \{0,1\}$

$C = \{c\}$ is a finite set of colours

V is the multiplicity of tokens defined on an arc $V: (FxC) \longrightarrow \mathbb{N}$

K is the token capacity in the places $K: (PxC) \longrightarrow \mathbb{N}$

M is the set of allowable markings $M: (PxC) \longrightarrow \mathbb{N}$

$m_o \in M$ is the initial marking of the net.

The analogy between the Petri net and the components of a complex system can be exemplified by the following table:

Petri net		plant
transitions	→	actions, sub-systems, machines
places	→	conditions for ability of work in the separate sub-systems (for instance buffers, tanks, store or tolerances)
token flows	→	mass flows, flow of workpieces, energy or materials
arcs	→	structures of the plant

The different colours serve for the simplification of the plant model, when in different situations must be switched between several models. A place may have one or many tokens or is may be empty. When there is one token or many tokens in each input place of a transition according to the capacity, then this transition is ready to fire. The act of firing involves removing tokens from each input place and putting in each output place according to the multiplicity of the arcs. The places indicate the holding or not holding of a condition in the process model and transitions that describe the event of some conditions represent the control actions. Usually a Petri net is represented by a bipartibe graph, graphically we use circles for places and bars for transitions (see fig. 1).

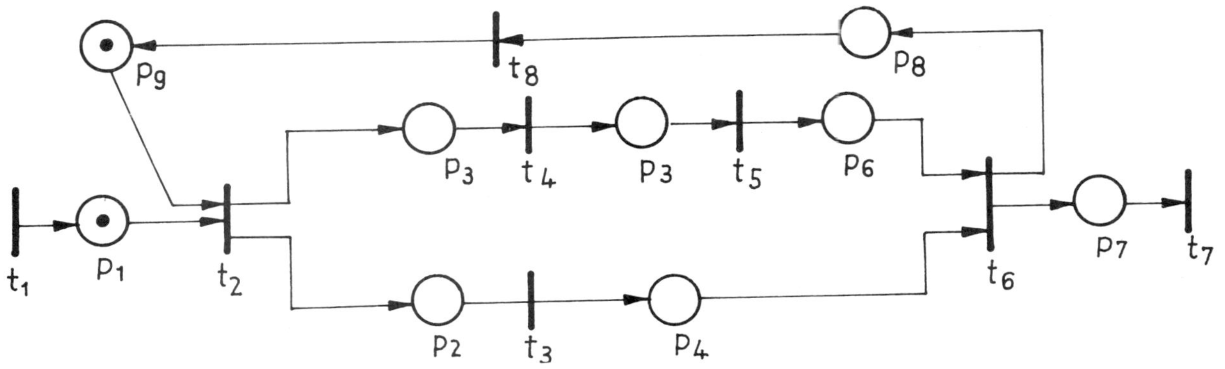

Fig. 1. A classical Petri net model of a plant

The disadvantage for this classical Petri net consists in the I/O-behaviour of places and transitions. Also coloured Petri nets can switch between idealized model components only. In many real processes a transition is not an action only, but a set of actions, for instance, actions for normal, overload, and underload situations. The conditions for ability of work in a sub-system of a real plant cannot be represented by a precise marking in some places, but by a set of markings in those components. In many real processes the different control goals and the constraints on the choice between different control alternatives are fuzzy. They cannot be represented by the classical Petri net. In order to include this fuzziness of real process in a plant model, a fuzzy Petri net is introduced.

REPRESENTATION OF THE FUZZY PETRI NET

The fuzzy Petri net is derived from the classical Petri net by fuzzification of places and transitions. A fuzzy Petri net $\widetilde{NP}$ is defined by

$$\widetilde{NP} = (\widetilde{P}, \widetilde{T}, \pi, \sigma, \eta, M, m_0)$$

$\widetilde{P} = \{\tilde{p}\}$ is a finite set of fuzzy places
$\tilde{p} = \{(p\ ; \mu_{\tilde{p}}(p))\}$

$\widetilde{T} = \{\tilde{t}\}$ is a finite set of fuzzy transitions
$\tilde{t} = \{(t\ ; \mu_{\tilde{t}}(t))\}$

σ is a transition function
$\sigma : (\widetilde{P}x\widetilde{T}) \cup (\widetilde{T}x\widetilde{P}) \longrightarrow [0,1]$

π is a start function
$\pi : (\widetilde{P}x\widetilde{T}) \longrightarrow [0,1]$

η is a final function
$\eta : (\widetilde{T}x\widetilde{P}) \longrightarrow [0,1]$

M is the set of markings in $\widetilde{PN}$
$M : P \longrightarrow \mathbb{N}$

$m_0 \in M$ is the initial marking of $\widetilde{PN}$

This fuzzy Petri net $\widetilde{PN}$ is a concentration of many classical Petri nets, which result from different models of a real process in different situations. The places and transitions are fuzzy sets, characterizing the variation of constraints or control actions in the real process. Their membership functions and the fuzzy valuation of the start, final and transition functions are based on intuitions and experiences of human operators. These membership functions can also be used for modelling the behaviour of the humans in the plant.

A fuzzy place contains many tokens and stands for all conditions of a control action in the different process situations. Their membership function describes the goodness for firing transitions (see fig. 2). Many or a few tokens in a place mark disturbances, which restrict the safety of the network. The membership function $\mu_{\tilde{p}}$ is a fuzzy measure for the capacity of places, and the number of tokens in a place is an expression for the ability to work in the adjoining transitions.

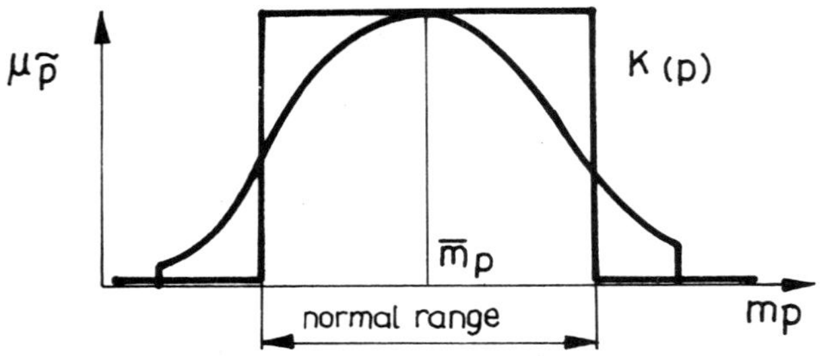

Fig. 2. Membership function $\mu_{\tilde{p}}$ of a fuzzy place

The fuzzy transitions $\tilde{T}$ of the network are a concentration of many "sharp" control actions to fuzzy sets. Their membership functions result from experiences of a human operator and are a measure of the qualification of these "sharp" actions in a special control task. Different actions of a transition $\tilde{t}$ are described by different token flows between the input and output places. The ratio of token flows of a transition $\tilde{t}$ is called the recipe of transition. By the continuous adaption of token flows between the transition the safety in the network is increased for a long time.

A high value for liveliness is described by good membership of applied control actions.

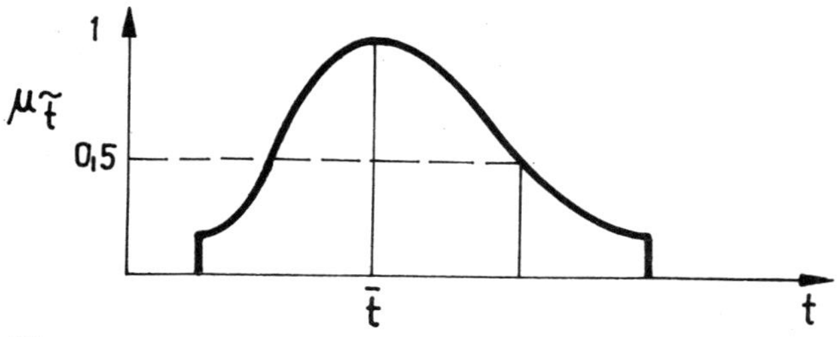

Fig. 3. Membership function of a transition $\tilde{t}$

The functions π, η and σ are used for the selection from fuzzy transitions for the improvement of safety and liveliness in the network. The transition function describes the dynamics of the transition $\tilde{t} \in \tilde{T}$, which results from the change of its token flows.

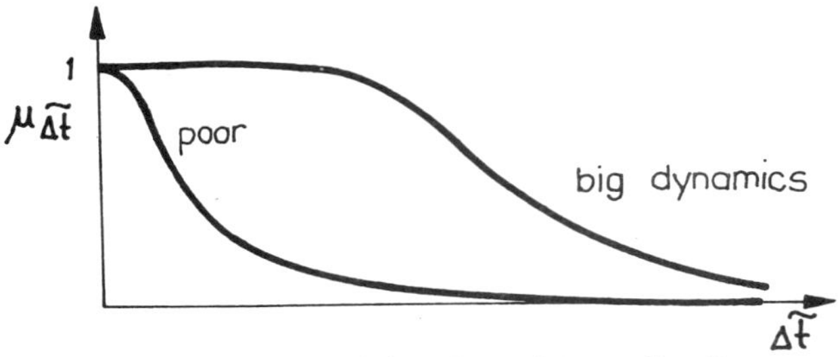

Fig. 4. Membership function $\sigma_{\tilde{t}}$ for the change of token flows

The start function $\pi_{\tilde{t}}$ describes the fuzzy ability to fire of the transition $\tilde{t} \in \tilde{T}$ in state place, and the final function represents a fuzzy measure for compromise between the goal of token flows in different transitions and desirable constraints in the places. All these membership functions result from experiences of operators consulting different specialists.

This fuzzy Petri net can be transformed to a classical Petri net, changing membership functions to sharp memberships. Hence, the modelling of continuous and discontinuous control actions is possible in the same network. The fuzzy Petri net can be deduced from many different classical Petri networks, each for special process situation. As long as the fuzzy sets are finite, the fuzzy model is equivalent to a (much larger) coloured Petri net.

DETERMINATION OF CONTROL POLICIES IN THE FUZZY PETRI NET

In this fuzzy Petri net model disturbances are identified by irregular token levels in the places.

The elimination of disturbances is possible by coordination of token flows, so that acceptable levels result in the places. The problem of a control system in a fuzzy Petri net consists in selecting the transitions, whose token flows improve the level in the places. A sequence of "sharp" control actions t_o, t_1, t_2, ..., t_m is called a control policy, when the initial marking m_o is transformed to the final marking and the liveliness of the network is guaranteed. At first this policy is determined by selecting a sequence from fuzzy transitions $\tilde{t}_o$, $\tilde{t}_1$, $\tilde{t}_2$, ..., $\tilde{t}_k$, which then is transformed to a sequence of "sharp" transitions t_o, t_1, t_2, ..., t_m.

Consequently the "sharp" transitions t_t represent the realizations of the fuzzy transitions $\tilde{t}_t$ at the time t. Their determination is performed by the adaption of fuzzy transitions to each other, so that the liveliness and the safety of the network is as much as possible.

The selection of a fuzzy transition $\tilde{t}_t$ results from the following aspects:

- selection according to the most effective place ($\mu_{\tilde{p}\tilde{t}}(\tilde{t}) = \underset{\tilde{p}}{\mathrm{MIN}}\ \mu_{\tilde{p}}$ for all $\tilde{p} \in \tilde{P}$ adjoining to $\tilde{t}$)
- selection according to the most effective control direction for compensating disturbances ($\mu_{\tilde{c}\tilde{t}}(\tilde{t})$ see fig. 5)
- selection according to the best liveliness of the transition ($\mu_{\tilde{t}}$).

Fuzzy transitions $\tilde{t}_t$ are selected if

$$\Omega(\tilde{t}_t) = \mathrm{MIN}\ \underset{\tilde{t}\in\tilde{T}}{\mathrm{MIN}}\ (\mu_{\tilde{p}\tilde{t}}, \mu_{\tilde{c}\tilde{t}}, \mu_{\tilde{t}})$$

Then the "sharp" transition t_t is an event of the fuzzy set $\tilde{t}_t = \{t\}$

$$\omega(t_t) = \mathrm{MAX}\ \underset{t\in\tilde{t}_t}{\mathrm{MIN}}\ (\pi_{\tilde{t}_t}, \sigma_{\tilde{t}_t}, \eta_{\tilde{t}_t})$$

Then all elements of the policy are determined in an iterative process in this way.

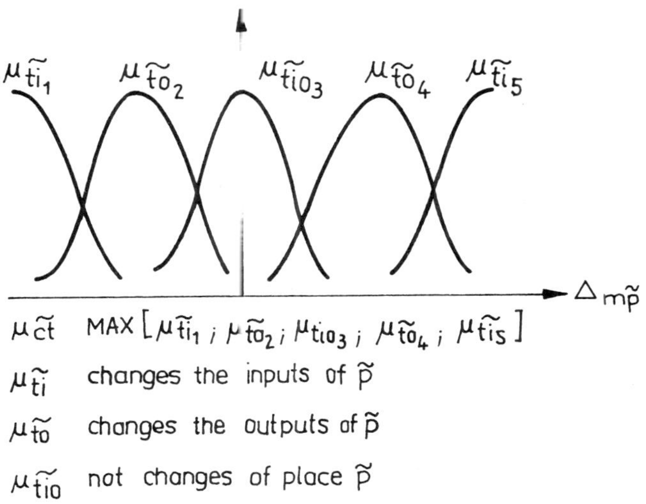

Fig. 5. Membership functions for selecting fuzzy transitions $\tilde{t}$ in place $\tilde{p}$

APPLICATION OF THE FUZZY PETRI NET FOR A CONTROL SYSTEM IN A PULP PLANT

The control of a pulp plant is very complicated, because this plant is a very complex system with large time constants. A real-time computer is used for process control and production planning, which will essentially coordinate the different sub-processes in the plant. The technological line in the pulp plant is a very complicated dynamic system, which must be approximated by a simple model. The main elements of the line are tanks and sub-systems, which are connected by mass flows, such as pulp, chemicals, liquor, steam or energy flows. The problem of control is formulated by calculation of production schedules, involving

- mass flows coordination for the elimination of disturbances
- mass flows coordination in repairing situations
- changing in the production rates of the process

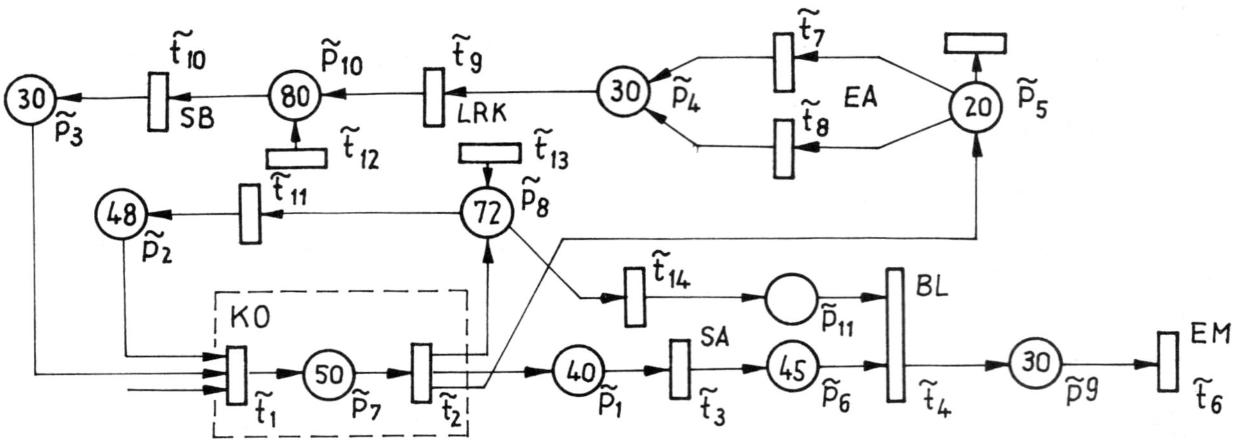

Fig. 6. A fuzzy Petri net model of a pulp plant

- possibility of indirect storage of steam
- acceptable tank levels for a stable production process at the end of the planning period.

The fuzzy Petri net model of the pulp plant consists of 15 transitions and 20 places, which are connected by 46 directed arcs (see fig. 6). In this model the places stand for tanks or buffer, and the transitions represent sub-processes, such as digester, bleach plant, paper plant and so on. The token flows of the arcs describe the structural dependences between the sub-systems, such as pulp, liquor, chemical and steam flows. The ratio between token flows around a transition is built in an updating process and is dependent on the production rate or the quality of product.

If a temporary limitation of the desired production rate is caused by disturbances, then the level in the tanks represents a measure of process stability. When the limitation is removed the dispatcher system must coordinate the process units, that the system works in the previous optimal steady state for a long time. In dialouge with the dispatcher the computer calculates production schedules, involving for tank levels and mass flows in time steps of 30 minutes for a process time in 24 hours. This strategy protocol is calculated in real-time processing by computer in a short cycle. It describes the efficiency of control actions and disturbances and helps the dispatcher in his decision process.

CONCLUSION

In this paper a fuzzy Petri net for modelling a complex process was presented. This fuzzy Petri net was derived from the classical Petri net by fuzzification of places and transitions. The fuzzy membership functions result from subjective experiences of the operator and describes the behaviour of the humans who work in the plant. This Petri net model permits the representation of control actions in their normal, overload or underload application. With these features it can be used for modelling real industrial processes.

REFERENCES

Genrich, H. E., K. Lautenbach (1979). Net theory and applications. Lectures Notes in Computer Sci., No 84, Springer, Berlin

Lipp, H.-P. (1982). Application of fuzzy Petri nets for description of control policies in complex production systems, Wiss. Z. d. TH Karl-Marx-Stadt p 633-639

Benlahcen, D. and M. Lamotte (1981). A fuzzy automation synthesis Method. Automatica (GB), Vol 17, No 2

Gluss, B. (1973). Fuzzy multi-stage decision-making, fuzzy state and and terminal regulators and their relationship to non-fuzzy quadratic state and terminal regulators, Int. J. Control, Vol 17, No 1, 177-192

Braae, M. and Rutherford, D. A. (1979) Theoretical and linguistic aspects of the fuzzy-logic-controller. Automatica (GB), Vol 15, pp. 523-527

Mamdani, E. H. and Procyk, T. J.(1979) A linguistic self-organizing process controller. Automatica (GB), Vol 15, pp. 15-30

AUTHOR INDEX

AKASHI: Control Science and Technology for the Progress of Society, 7 Volumes

ALONSO-CONCHEIRO: Real Time Digital Control Applications

ATHERTON: Multivariable Technological Systems

BABARY & LE LETTY: Control of Distributed Parameter Systems (1982)

BANKS & PRITCHARD: Control of Distributed Parameter Systems (1977)

BAYLIS: Safety of Computer Control Systems (1983)

BEKEY & SARIDIS: Identification and System Parameter Estimation (1982)

BINDER: Components and Instruments for Distributed Computer Control Systems

BULL: Real Time Programming (1983)

CAMPBELL: Control Aspects of Prosthetics and Orthotics

Van CAUWENBERGHE: Instrumentation and Automation in the Paper, Rubber, Plastics and Polymerisation Industries (1980) (1983)

CICHOCKI & STRASZAK: Systems Analysis Applications to Complex Programs

CRONHJORT: Real Time Programming (1978)

CUENOD: Computer Aided Design of Control Systems

De GIORGIO & ROVEDA: Criteria for Selecting Appropriate Technologies under Different Cultural, Technical and Social Conditions

DUBUISSON: Information and Systems

ELLIS: Control Problems and Devices in Manufacturing Technology (1980)

FERRATE & PUENTE: Software for Computer Control (1982)

FLEISSNER: Systems Approach to Appropriate Technology Transfer

GELLIE & TAVAST: Distributed Computer Control Systems (1982)

GHONAIMY: Systems Approach for Development (1977)

HAASE: Real Time Programming (1980)

HAIMES & KINDLER: Water and Related Land Resource Systems

HALME: Modelling and Control of Biotechnical Processes

HARDT: Information Control Problems in Manufacturing Technology (1982)

HARRISON: Distributed Computer Control Systems (1979)

HASEGAWA: Real Time Programming (1981)

HASEGAWA & INOUE: Urban, Regional and National Planning — Environmental Aspects

HERBST: Automatic Control in Power Generation Distribution and Protection

ISERMANN: Identification and System Parameter Estimation (1979)

ISERMANN & KALTENECKER: Digital Computer Applications to Process Control

JANSSEN, PAU & STRASZAK: Dynamic Modelling and Control of National Economies (1980)

JOHANNSEN & RIJNSDORP: Analysis, Design, and Evaluation of Man-Machine Systems

KLAMT & LAUBER: Control in Transportation Systems

LANDAU: Adaptive Systems in Control and Signal Processing

LAUBER: Safety of Computer Control Systems (1979)

LEININGER: Computer Aided Design of Multivariable Technological Systems

LEONHARD: Control in Power Electronics and Electrical Drives (1977)

LESKIEWICZ & ZAREMBA: Pneumatic and Hydraulic Components and Instruments in Automatic Control

MAHALANABIS: Theory and Application of Digital Control

MARTIN: Design of Work in Automated Manufacturing Systems

MILLER: Distributed Computer Control Systems (1981)

MUNDAY: Automatic Control in Space (1979)

NAJIM & ABDEL-FATTAH: Systems Approach for Development (1980)

NIEMI: A Link Between Science and Applications of Automatic Control

NOVAK: Software for Computer Control (1979)

O'SHEA & POLIS: Automation in Mining, Mineral and Metal Processing (1980)

OSHIMA: Information Control Problems in Manufacturing Technology (1977)

PAU & BASAR: Dynamic Modelling and Control of National Economies (1983)

PONOMARYOV: Artificial Intelligence

RAUCH: Applications of Nonlinear Programming to Optimization and Control

RAUCH: Control Applications of Nonlinear Programming

REMBOLD: Information Control Problems in Manufacturing Technology (1979)

RIJNSDORP: Case Studies in Automation related to Humanization of Work

RIJNSDORP & PLOMP: Training for Tomorrow — Educational Aspects of Computerised Automation

RODD: Distributed Computer Control Systems (1983)

SANCHEZ: Fuzzy Information, Knowledge Representation and Decision Analysis

SAWARAGI & AKASHI: Environmental Systems Planning, Design and Control

SINGH & TITLI: Control and Management of Integrated Industrial Complexes

SMEDEMA: Real Time Programming (1977)

STRASZAK: Large Scale Systems: Theory and Applications (1983)

SUBRAMANYAM: Computer Applications in Large Scale Power Systems

TITLI & SINGH: Large Scale Systems: Theory and Applications (1980)

WESTERLUND: Automation in Mining, Mineral and Metal Processing (1983)

Van WOERKOM: Automatic Control in Space (1982)

ZWICKY: Control in Power Electronics and Electrical Drives (1983)